Particle Physics Reference Library

Herwig Schopper

Editor

Particle Physics Reference Library

Volume 1: Theory and Experiments

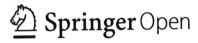 Springer Open

Editor
Herwig Schopper
CERN
Geneva, Switzerland

ISBN 978-3-030-38209-4
https://doi.org/10.1007/978-3-030-38207-0

ISBN 978-3-030-38207-0   (eBook)

This book is an open access publication.

This Springer imprint is published by the registered company Springer Nature Switzerland AG.
The registered company address is: Gewerbestrasse 11, 6330 Cham, Switzerland

Preface

For many years the *Landolt-Börnstein—Group I Elementary Particles, Nuclei and Atoms*: Vol. 21A (*Physics and Methods Theory and Experiments*, 2008), Vol. 21B1 (*Elementary Particles Detectors for Particles and Radiation. Part 1: Principles and Methods*, 2011), Vol. 21B2 (*Elementary Particles Detectors for Particles and Radiation. Part 2: Systems and Applications*), and Vol. 21C (*Elementary Particles Accelerators and Colliders*, 2013) has served as a major reference work in the field of high-energy physics.

When, not long after the publication of the last volume, open access (OA) became a reality for HEP journals in 2014, discussions between Springer and CERN intensified to find a solution for the "Labö" which would make the content available in the same spirit to readers worldwide. This was helped by the fact that many researchers in the field expressed similar views and their readiness to contribute.

Eventually, in 2016, on the initiative of Springer, CERN and the original Labö volume editors agreed in tackling the issue by proposing to the contributing authors a new OA edition of their work. From these discussions, a compromise emerged along the following lines: transfer as much as possible of the original material into open access; add some new material reflecting new developments and important discoveries, such as the Higgs boson; and adapt to the conditions due to the change from copyright to a CC BY 4.0 license.

Some authors were no longer available for making such changes, having either retired or, in some cases, deceased. In most such cases, it was possible to find colleagues willing to take care of the necessary revisions. A few manuscripts could not be updated and are therefore not included in this edition.

We consider that this new edition essentially fulfills the main goal that motivated us in the first place—there are some gaps compared to the original edition, as explained, as there are some entirely new contributions. Many contributions have been only minimally revised in order to make the original status of the field available as historical testimony. Others are in the form of the original contribution being supplemented with a detailed appendix relating to recent developments in the field. However, a substantial fraction of contributions has been thoroughly revisited by their authors resulting in true new editions of their original material.

   We would like to express our appreciation and gratitude to the contributing authors, to the colleagues at CERN involved in the project, and to the publisher, who has helped making this very special endeavor possible.

Vienna, Austria
Geneva, Switzerland
Geneva, Switzerland
July 2020

Christian Fabjan
Stephen Myers
Herwig Schopper

Contents

1 Introduction 1
   Herwig Schopper

2 Gauge Theories and the Standard Model 7
   Guido Altarelli and Stefano Forte

3 The Standard Model of Electroweak Interactions 35
   Guido Altarelli and Stefano Forte

4 QCD: The Theory of Strong Interactions 83
   Guido Altarelli and Stefano Forte

5 QCD on the Lattice 137
   Hartmut Wittig

6 The Discovery of the Higgs Boson at the LHC 263
   Peter Jenni and Tejinder S. Virdee

7 Relativistic Nucleus-Nucleus Collisions and the QCD Matter
   Phase Diagram 311
   Reinhard Stock

8 Beyond the Standard Model 455
   Eliezer Rabinovici

9 Symmetry Violations and Quark Flavour Physics 519
   Konrad Kleinknecht and Ulrich Uwer

10 The Future of Particle Physics: The LHC and Beyond 625
   Ken Peach

Correction to: Particle Physics Reference Library C1
Herwig Schopper

About the Editor

Herwig Schopper joined as a research associate at CERN since 1966 and returned in 1970 as leader of the Nuclear Physics Division and went on to become a member of the directorate responsible for the coordination of CERN's experimental program. He was chairman of the ISR Committee at CERN from 1973 to 1976 and was elected as member of the Scientific Policy Committee in 1979. Following Léon Van Hove's and John Adams' years as Director-General for research and executive Director-General, Schopper became the sole Director-General of CERN in 1981.

Schopper's years as CERN's Director-General saw the construction and installation of the Large Electron-Positron Collider (LEP) and the first tests of four detectors for the LEP experiments. Several facilities (including ISR, BEBC, and EHS) had to be closed to free up resources for LEP.

Chapter 1
Introduction

Herwig Schopper

Since old ages it has been one of the noble aspirations of humankind to understand the world in which we are living. In addition to our immediate environment, planet earth, two more remote frontiers have attracted interest: the infinitely small and the infinitely large. A flood of new experimental and theoretical results obtained during the past decades has provided a completely new picture of the micro- and macrocosm and surprisingly intimate relations have been discovered between the two. It turned out that the understanding of elementary particles and the forces acting between them is extremely relevant for our perception of the cosmological development. Quite often scientific research is supported because it is the basis for technical progress and for the material well-being of humans. The exploration of the microcosm and the universe contributes to this goal only indirectly by the development of better instruments and new techniques. However, it tries to answer some fundamental questions which are essential to understand the origins, the environment and the conditions for the existence of humankind and thus is an essential part of the cultural heritage.

One of the fundamental questions concerns the nature of matter, the substance of which the stars, the planets and living creatures are made, or to put it in another way—can the many phenomena which we observe in nature be explained on the basis of a few elementary building blocks and forces which act between them. The first attempts go back 2000 years when the Greek philosophers speculated about

The original version of this chapter was revised. A correction to this chapter can be found at https://doi.org/10.1007/978-3-030-38207-0_11

H. Schopper (✉)
CERN, Geneva, Switzerland
e-mail: Herwig.Schopper@cern.ch

indestructible atoms, like Democritus, or the four elements and the regular bodies of Plato.

Since Newton who introduced infinitely hard smooth balls as constituents of matter[1] and who described gravitation as the first force acting between them, the concept of understanding nature in terms of 'eternal' building blocks hold together by forces has not changed during the past 200 years. What has changed was the nature of the elementary building blocks and new forces were discovered. The chemists discovered the atoms of the 92 elements which, however, contrary to their name, were found to be divisible consisting of a nucleus surrounded by an electron cloud. Then it was found that the atomic nuclei contain protons and neutrons. Around 1930 the world appeared simple with everything consisting of these three particles: protons, neutrons and electrons.

Then came the 'annus mirabilis' 1931 with the discovery of the positron as the first representative of antimatter and the mysterious neutrino in nuclear beta-decay indicating a new force, the weak interaction. In the following decades the 'particle zoo' with all its newly discovered mesons, pions and 'strange' particles was leading to great confusion. Simplicity was restored when all these hundreds of 'elementary 'particles could be understood in terms of a new kind of elementary particles, the quarks and their antiquarks. The systematics of these particles is mainly determined by the strong nuclear force, well described today by the quantum chromodynamics QCD. Whether quarks and gluons (the binding particles of the strong interaction) exist only inside the atomic nuclei or whether a phase transition into a quark-gluon plasma is possible, is one the intriguing questions which still needs an answer.

Impressive progress was made in another domain, in the understanding of the weak nuclear force responsible for radioactive beta-decay and the energy production in the sun. Three kinds of neutrinos (with their associated antiparticles) were found and recently it could be shown that the neutrinos are not massless as had been originally assumed. The mechanism of the weak interaction could be clarified to a large extent by the discovery of its carriers, the W- and Z-particles. All the experimental results obtained so far will be summarized in this volume and the beautiful theoretical developments will be presented. The climax is the establishment of the 'Standard Model of Particle Physics' SM which has been shown to be a renormalizable gauge theory mainly by the LEP precision experiments. The LEP experiments have also shown that there are only three families of quarks and leptons (electron, muon, tau-particle and associated neutrinos), a fact not yet understood.

All the attempts to find experimental deviations from the SM have failed so far. However, the SM cannot be the final theory for the understanding of the microcosm. Its main fault is that it has too many arbitrary parameters (e.g. masses of the particles, values of the coupling constants of the forces, number of quark and lepton families) which have to be determined empirically by experiment. An underlying theory based on first principles is still missing and possible ways into the future will be discussed below.

---

[1] Isaac Newton, Optics, Query 31, London 1718.

Returning to the 'naïve' point of ultimate building blocks one might ask whether the quarks and leptons are fundamental indivisible particles or whether they have a substructure. Here we are running into a dilemma which was recognised already by the philosopher Immanuel Kant.[2] Either ultimate building blocks are mathematical points and cannot be divided, but then it is difficult to understand how they can have a mass and carry charges and spin. Alternatively, the building blocks might have spatial extension, but then it is hard to understand why they could not be divided into smaller parts. Whenever one meets such a paradox in science it is usually resolved by recognising that a wrong question was asked.

Indeed the recent developments of particle physics indicate that the naïve concept of ultimate building blocks of matter has to be abandoned. The smaller the 'building blocks' are, the higher energies are necessary to break them up. This is simply a consequence of the Heisenberg uncertainty principle of quantum mechanics. In the case of quarks their binding energies become so strong that any energy applied to break them apart is used to produce new quark-antiquark pairs.[3] The existence of antimatter implies also that matter does not have an 'eternal' existence. When matter meets antimatter the two annihilate by being converted into 'pure' energy and in the reverse mode matter can be produced[4] from energy in the form of particle-antiparticle pairs.

One of the most exciting development of physics or in science in general is a change of paradigms. Instead of using building blocks and forces acting between them, it was progressively recognised that symmetry principles are at the basis of our understanding of nature. It seems obvious that laws of nature should be invariant against certain transformations since 'nature does not know' how we observe it. When we make experiments we have to choose the origin of the coordinate system, its orientation in space and the moment in time when we start the observation. These choices are arbitrary and the laws deduced from the observations should not depend on them. It is known since a long time that the invariance of laws of nature against the continuous transformations, i.e. translations and rotations in space and time, give rise to the conservation of momentum, angular momentum and energy, the most basic laws of classical physics.[5] The mirror transitions (i.e. spatial reflection, particle-antiparticle exchange and time reversal) lead to the conservation of parity P, charge parity C and detailed balance rules in reactions, all of which are essential ingredients of quantum mechanics.

The detection of complete parity violation in weak interactions in 1957 was one of the most surprising observations. Many eminent physicists, including Wolfgang

---

[2]Immanuel Kant, Kritik der reinen Vernunft, 1781, see, e.g., Meiner Verlag, Hamburg 1998, or translation by N.K. Smith, London, MacMillan 1929.

[3]The binding energies are comparable to mc^2, where m is the rest mass of a quark and c is the velocity of light.

[4]When Pope Paul John II visited CERN and I explained to him that we can 'create' matter his response was: you can 'produce' matter, but 'creation' is my business.

[5]Emmy Noether, Nachr. d. königl. Gesellschaft d. Wissenschaften zu Göttingen, 1918, page 235.

Pauli, thought that this symmetry could not be violated. Such a believe indeed goes back to Emanuel Kant[2] who claimed that certain 'a priori' concepts have to be valid so that we would be able to explore nature. Since it seemed obvious that nature does not know whether we observe it directly or through a mirror a violation of mirror symmetry seemed unacceptable. This phenomenon is still not understood, although the fact that also C conservation is completely violated and the combined symmetry PC seemed to hold has reduced somewhat the original surprise. The whole situation has become more complicated by the detection that PC is also violated, although very little. A deep understanding of the violation of these 'classical' symmetries is still missing. So far experiments show that the combined symmetry PCT still holds as is required by a very general theorem.

In field theories another class of more abstract symmetries has become important—the gauge symmetries. As is well known from Maxwell's equations the electrodynamic fields are fully determined by the condition that gauge symmetry holds, which means that the electric and magnetic fields are independent against gauge transformations of their potentials. It was discovered that analogous gauge symmetries determine the fields of the strong and weak interactions in which case the (spontaneous) breaking of the symmetries plays a crucial role.

In summary, we have abandoned the description of nature in terms of hard indestructible spheres in favour of abstract ideas—the symmetries and there breaking. From a philosophical point of view one might, in an over-simplistic way, characterize the development as moving away from Democritus to Plato.

Finally, it should be mentioned that in particle physics progress was only possible by an intimate cooperation between theory and experiments. The field has become so complex that by chance discoveries are extremely rare. The guidance by theory is necessary to be able to put reasonable questions to nature. This does not exclude great surprises since many theoretical predictions turned out to be wrong. Indeed most progress could be made by verifying or disproving theories.

Although the Standard Model of Particle Physics SM (with some extensions, e.g. allowing for masses of neutrinos) has achieved a certain maturity by being able to reproduce all experimental results obtained so far, it leaves open many fundamental questions. One particular problem one has gotten accustomed to, concerns P and C violations which are put into the SM 'by hand'. And as has been mentioned above the SM leaves open many other questions which indicate that it cannot be a final theory.

In 2008 I wrote the concluding paragraph of this introduction as "Many arguments indicate that a breakthrough in the understanding of the microcosm will happen when the results of LHC at CERN will become available. LHC will start operation in 2008, but it will probably take several years before the experiments will have sufficient data and one will be able to analyse the complicated events before a major change of our picture will occur, although surprises are not excluded. Hence it seems to be an appropriate time to review the present situation of our understanding of the microcosm". Meanwhile, more than 10 years later, and with the Higgs boson discovered in 2014 at the LHC, the extended SM has been confirmed with unprecedented precision yet the outstanding questions, in particular which path to follow beyond the SM, have remained with us.

Chapter 2
Gauge Theories and the Standard Model

Guido Altarelli and Stefano Forte

2.1 Introduction to Chaps. 2, 3 and 4

Stefano Forte

The presentation of the Standard Model in Chap. 2, Chaps. 3 and 4 was originally written by Guido Altarelli in 2007. In this introduction we provide a brief update (with references), and a discussion of the main developments which have taken place since the time of the writing.

Chapter 2 presents the architecture of the Standard Model, the way symmetries are realized and the way this can is described at the quantum level. The structure of the Standard Model is now well-established since half a century or so. The presentation in this chapter highlights the experimental (and thus, to a certain extent, historical Chap. 2) origin of the main structural aspects of the theory. The only aspects of the presentation which require (minimal) updating are the numerical values given for parameters, such as the Fermi coupling constant G_F, see Eq. (2.3). All of these parameters have been known quite accurately since the early 2000s (with the exception of neutrino masses, see Sect. 3.7 of Chap. 3), and thus their values are quite stable. The numbers given below are taken from the then-current edition of the Particle Data Book (PDG) [7]. At any given time, in order to have the

The original version of this chapter was revised. A correction to this chapter can be found at https://doi.org/10.1007/978-3-030-38207-0_11
The author "G. Altarelli" is deceased at the time of publication.

G. Altarelli
University of Rome 3, Rome, Italy

S. Forte (✉)
Dipartimento di Fisica, Università di Milano, Milano, Italy

most recent and accurate values, the reader should consult the most recent edition of the PDG [25], preferably using the web-based version [26], which is constantly updated.

Chapter 3 presents the Electroweak sector of the Standard Model, which was established as a successful theory by extensive experimentation at the LEP electron-positron collider of CERN in the last decade of the past century, including some aspects of the theory, such as the CKM mechanism for mass mixing (see Sect. 3.6) which were originally often considered to be only approximate. The discovery, at the turn of the century, of neutrino mixing, and thus non-vanishing neutrino masses (see Sect. 3.7) has been the only significant addition to the minimal version of the electroweak theory as formulated in the sixties and seventies of the past century. The general understanding of electroweak interactions was thus essentially settled at the time of the writing of this chapter.

From the experimental point of view, the main development since then is the successful completion of the first two runs of the LHC, which have provided further confirmation of the standard Electroweak theory (see Ref. [27] for a recent review). From a theoretical point of view, the main surprise (from the LHC, but also a number of other experiments) is that there have been no surprises.

First and foremost, the Higgs sector of the Standard Model: after discovery of the Higgs boson in 2012 [28, 29] the Higgs sector has turned out so far to be in agreement with the minimal one-doublet structure presented in Sect. 3.5. The discussion presented there, as well as the phenomenology of the Standard Model Higgs of Sect. 3.13, remain thus essentially unchanged by the Higgs discovery. A theoretical introduction with more specific reference to the LHC can be found in Ref. [30], while the current experimental status of Higgs properties can be found in the continually updated pages of the CERN Higgs cross-section working group [31]. Perhaps, the only real surprise in the Higgs sector of the Standard Model is the extreme closeness of the measured Higgs mass to the critical value required for vacuum stability (see Sect. 3.13.1 below)—a fact with interesting cosmological implications [32]. The discovery of the Higgs has changed somewhat the nature of global fits of Standard Model parameters discussed in Sect. 3.12: with the value of the Higgs mass known, the fit is over-constrained—though the conclusion of global consistency remains unchanged. An updated discussion is given in Ref. [27], as well as in the review on the Electroweak Model by Erler and Freitas in the PDG [26].

Besides Higgs discovery, the general trend of the last several years has been that of the gradual disappearance of all anomalies—instances of discrepancy between Standard Model predictions and the data—either due to more accurate theory calculations (or even the correction of errors: see Sect. 3.9), or to more precise measurements. A case in point is that of the measurements of the electroweak mixing angle, discussed in Sect. 3.12: the tensions or signals of disagreement which are discussed there have all but disappeared, mostly thanks to more accurate theoretical calculations. Another case in which the agreement between Standard Model and experiment is improving (albeit perhaps more slowly) is that of lepton anomalous magnetic moments, discussed in Sect. 3.9. In both cases, updates on the current situation can again been found in Ref. [27], and in the aforementioned PDG review by Erler and Freitas.

Finally, there is a number of cases in which data from LHC experiments (as well as other experiments, specifically in the fields of flavor physics and neutrino physics) have brought more accuracy and more stringent tests, without changing the overall picture. These include gauge boson couplings, discussed in Sects. 3.3–3.4, for which we refer to Ref. [27]; the CKM matrix and flavor physics, discussed in Sect. 3.6, for which we refer to the review by Ceccucci, Ligeti and Sakai in the PDG [26]; neutrino masses and mixings, discussed in Sect. 3.7, for which we refer to the PDG review by Gonzalez-Garcia and Yokohama [26].

This perhaps unexpected success of the Standard Model, and the failure to find any evidence so far of new physics (and in particular supersymmetry) at the LHC has somewhat modified the perspective on the limitations of the Standard Model discussed in Sect. 3.14. Specifically, the significance of the hierarchy problem—the so-called "naturalness" issue—must be questioned, given that it entails new physics which has not be found: a suggestive discussion of this shift in perspective is in Ref. [33]. Yet, the classification of possible new physics scenarios of Sect. 3.14 remains essentially valid: recent updates are in Ref. [34] for supersymmetric models, and in Ref. [35] for non-supersymmetric ones. Consequently, looking for new physics has now become a precision exercise, and this has provided a formidable stimulus to the study of Electroweak radiative corrections, which has been the subject of very intense activity beyond the classic results discussed in Sect. 3.10: a recent detailed review is in Ref. [36].

Chapter 4 is devoted to the theory of strong interactions, Quantum Chromodynamics (QCD). This theory has not changed since its original formulation in the second half of the past century. Specifically, its application to hard processes, which allows for the use of perturbative methods, is firmly rooted in the set of classic results and techniques discussed in Sect. 4.5 below. What did slowly change over the years is the experimental status of QCD. What used to be, in the past century, a theory established qualitatively, has gradually turned into a theory firmly established experimentally—though, at the time this chapter was written, not quite tested to the same precision as the electroweak theory (see Sect. 4.7). Now, after the first two runs of the LHC, it can be stated that the whole of the Standard Model, QCD and the Electroweak theory, are tested to the same very high level of accuracy and precision, typically at the percent or sub-percent level.

Turning QCD into a precision theory has been a pre-requisite for successful physics at the LHC, a hadron collider in which every physical process necessarily involves the strong interaction, since the colliding objects are protons (or nuclei). This has grown into a pressing need as the lack of discovery of new particles or major deviations from Standard Model predictions has turned the search for new physics signals into a precision exercise: it has turned the LHC from an "energy frontier" to a "rarity/accuracy frontier" machine—something that was deemed inconceivable just before the start of its operation [37].

This rapid progress has happened thanks to an ever-increasing set of computational techniques, which, building upon the classic results presented in this chapter, has allowed for an enormous expansion of the set of perturbative computations of processes at colliders which are introduced in Sect. 4.5.4, and discussed in more detail in the context of LHC (and specifically Higgs) physics in Ref. [30].

To begin with, basic quantities such as the running of the coupling, discussed in Sect. 4.4, and $R_{e^+e^-}$, discussed in Sect. 4.5.1 are now know to one extra perturbative order (see the QCD review of the PDG [26] for the current state of the art and full references). These are five-loop perturbative calculations, now made possible thanks to the availability of powerful computing resources. Furthermore, the set of processes discussed in Sect. 4.5.4 has now been extended to include essentially all relevant hadron collider processes, which have been routinely computed to third perturbative order, while the first fourth-order calculations have just started appearing. Again, the QCD review of the PDG [26] provides a useful status update, including comparison between computation and experiment, which refer to cross-sections which span about ten orders of magnitude in size.

This progress has been happening thanks to the development of a vast new set of computational techniques, which, rooted in perturbative QCD, have now spawned a dedicated research field: that of amplitudes [38], which relates phenomenology, quantum field theory, and mathematics. The classic set of methods for "resummation"—the sum of infinite classes of perturbative contributions, discussed specifically in Sect. 4.5.3.1 for deep-inelastic scattering, has been extended well beyond the processes and accuracy discussed in Sect. 4.5.4—an up-to-date list is in the QCD review of the PDG [26]. Moreover, an entirely new set of resummation techniques has been developed, using the methodology of effective field theories: the so-called soft-collinear effective theory (SCET) which provides an extra tool in the resummation box [39]. One remarkable consequence of all these developments is that it is now possible to understand in detail the structure of pure strong interaction events, in which jets of hadrons are produced in the final state, by looking inside these events and tracing their structure in terms of the fundamental fields of QCD—quarks and gluons [40].

One topic in which things have changed rather less is the determination of the strong coupling, discussed in Sect. 4.7. Whereas the agreement between predicted and observed scaling violations discussed in Sect. 4.6.3 is ever more impressive (see the review on structure functions of the PDG [26]) the accuracy on the determination of the strong coupling itself has not improved much. Updated discussions can be found in the QCD review of the PDG, as well as in Ref. [41]. Progress is likely to come from future, more accurate LHC data, as well as from non-perturbative calculations [42] (not discussed here) soon expected to become competitive.

All in all, the dozen or so years since the original writing of these chapter have seen a full vindication of the Standard Model as a correct and accurate theory, and have stimulated a vast number of highly sophisticated experimental and theoretical results which build upon the treatment presented below.

2.2 Introduction

The ultimate goal of fundamental physics is to reduce all natural phenomena to a set of basic laws and theories that, at least in principle, can quantitatively reproduce and predict the experimental observations. At microscopic level all the phenomenology of matter and radiation, including molecular, atomic, nuclear and subnuclear physics, can be understood in terms of three classes of fundamental interactions: strong, electromagnetic and weak interactions. In atoms the electrons are bound to nuclei by electromagnetic forces and the properties of electron clouds explain the complex phenomenology of atoms and molecules. Light is a particular vibration of electric and magnetic fields (an electromagnetic wave). Strong interactions bind the protons and neutrons together in nuclei, being so intensively attractive at short distances that they prevail over the electric repulsion due to the equal sign charges of protons. Protons and neutrons, in turn, are composites of three quarks held together by strong interactions to which quarks and gluons are subject (hence these particles are called "hadrons" from the Greek word for "strong"). To the weak interactions are due the beta radioactivity that makes some nuclei unstable as well as the nuclear reactions that produce the enormous energy radiated by the stars and by our Sun in particular. The weak interactions also cause the disintegration of the neutron, the charged pions, the lightest hadronic particles with strangeness, charm, and beauty (which are "flavour" quantum numbers) as well as the decay of the quark top and of the heavy charged leptons (the muon μ^- and the tau τ^-). In addition all observed neutrino interactions are due to weak forces.

All these interactions are described within the framework of quantum mechanics and relativity, more precisely by a local relativistic quantum field theory. To each particle, described as pointlike, is associated a field with suitable (depending on the particle spin) transformation properties under the Lorentz group (the relativistic space-time coordinate transformations). It is remarkable that the description of all these particle interactions is based on a common principle: "gauge" invariance. A "gauge" symmetry is invariance under transformations that rotate the basic internal degrees of freedom but with rotation angles that depend on the space-time point. At the classical level gauge invariance is a property of the Maxwell equations of electrodynamics and it is in this context that the notion and the name of gauge invariance were introduced. The prototype of all quantum gauge field theories, with a single gauged charge, is QED, Quantum Electro-Dynamics, developed in the years from 1926 until about 1950, which indeed is the quantum version of Maxwell theory. Theories with gauge symmetry, at the renormalizable level, are completely determined given the symmetry group and the representations of the interacting fields. The whole set of strong, electromagnetic and weak interactions is described by a gauge theory, with 12 gauged non-commuting charges, which is called "the Standard Model" of particle interactions (SM). Actually only a subgroup of the SM symmetry is directly reflected in the spectrum of physical states. A part of the electroweak symmetry is hidden by the Higgs mechanism for the spontaneous symmetry breaking of a gauge symmetry.

For all material bodies on the Earth and in all geological, astrophysical and cosmological phenomena a fourth interaction, the gravitational force, plays a dominant role, while it is instead negligible in atomic and nuclear physics. The theory of general relativity is a classic (in the sense of non quantum mechanical) description of gravitation that goes beyond the static approximation described by Newton law and includes dynamical phenomena like, for example, gravitational waves. The problem of the formulation of a quantum theory of gravitational interactions is one of the central problems of contemporary theoretical physics. But quantum effects in gravity become only important for energy concentrations in space-time which are not in practice accessible to experimentation in the laboratory. Thus the search for the correct theory can only be done by a purely speculative approach. All attempts at a description of quantum gravity in terms of a well defined and computable local field theory along similar lines as for the SM have so far failed to lead to a satisfactory framework. Rather, at present the most complete and plausible description of quantum gravity is a theory formulated in terms of non pointlike basic objects, the so called "strings", extended over distances much shorter than those experimentally accessible, that live in a space-time with 10 or 11 dimensions. The additional dimensions beyond the familiar 4 are, typically, compactified which means that they are curled up with a curvature radius of the order of the string dimensions. Present string theory is an all-comprehensive framework that suggests a unified description of all interactions together with gravity of which the SM would be only a low energy or large distance approximation.

A fundamental principle of quantum mechanics, the Heisenberg indetermination principle, implies that, for studying particles with spatial dimensions of order Δx or interactions taking place at distances of order Δx, one needs as a probe a beam of particles (typically produced by an accelerator) with impulse $p \gtrsim \hbar/\Delta x$, where $\hbar$ is the reduced Planck constant ($\hbar = h/2\pi$). Accelerators presently in operation or available in the near future, like the Large Hadron Collider at CERN near Geneva, allow to study collisions between two particles with total center of mass energy up to $2E \sim 2pc \lesssim 14$ TeV. These machines, in principle, can allow to study physics down to distances $\Delta x \gtrsim 10^{-18}$ cm. Thus, on the basis of results from experiments at existing accelerators, we can confirm that, down to distances of that order of magnitude, indeed electrons, quarks and all the fundamental SM particles do not show an appreciable internal structure and look elementary and pointlike. We expect that quantum effects in gravity will certainly become important at distances $\Delta x \gtrsim 10^{-33}$ cm corresponding to energies up to $E \sim M_{Pl}c^2 \sim 10^{19}$ GeV, where M_{Pl} is the Planck mass, related to Newton constant by $G_N = \hbar c/M_{Pl}^2$. At such short distances the particles that so far appeared as pointlike could well reveal an extended structure, like for strings, and be described by a more detailed theoretical framework of which the local quantum field theory description of the SM would be just a low energy/large distance limit.

From the first few moments of the Universe, after the Big Bang, the temperature of the cosmic background went down gradually, starting from $kT \sim M_{Pl}c^2$, where $k = 8.617\ldots 10^{-5}$ eV K^{-1} is the Boltzmann constant, down to the present situation where $T \sim 2.725$ K. Then all stages of high energy physics from string theory, which is a purely speculative framework, down to the SM phenomenology, which is directly accessible to experiment and well tested, are essential for the reconstruction of the evolution of the Universe starting from the Big Bang. This is the basis for the ever increasing relation between high energy physics and cosmology.

2.3 Overview of the Standard Model

The SM is a gauge field theory based on the symmetry group $SU(3) \otimes SU(2) \otimes U(1)$. The transformations of the group act on the basic fields. This group has 8+3+1= 12 generators with a non trivial commutator algebra (if all generators commute the gauge theory is said to be "abelian", while the SM is a "non abelian" gauge theory). $SU(3)$ is the "colour" group of the theory of strong interactions (QCD: Quantum Chromo-Dynamics [1–3]). $SU(2) \otimes U(1)$ describes the electroweak (EW) interactions [4–6] and the electric charge Q, the generator of the QED gauge group $U(1)_Q$, is the sum of T_3, one of the $SU(2)$ generators and of $Y/2$, where Y is the $U(1)$ generator: $Q = T_3 + Y/2$.

In a gauge theory to each generator T is associated a vector boson (also said gauge boson) with the same quantum numbers as T, and, if the gauge symmetry is unbroken, this boson is of vanishing mass. These vector (i.e. of spin 1) bosons act as mediators of the corresponding interactions. For example, in QED the vector boson associated to the generator Q is the photon γ. The interaction between two charged particles in QED, for example two electrons, is mediated by the exchange of one (or seldom more than one) photon emitted by one electron and reabsorbed by the other one. Similarly in the SM there are 8 massless gluons associated to the $SU(3)$ colour generators, while for $SU(2) \otimes U(1)$ there are 4 gauge bosons W^+, W^-, Z^0 and γ. Of these, only the photon γ is massless because the symmetry induced by the other 3 generators is actually spontaneously broken. The masses of W^+, W^- and Z^0 are quite large indeed on the scale of elementary particles: $m_W \sim 80.4$ GeV, $m_Z \sim 91.2$ GeV are as heavy as atoms of intermediate size like rubidium and molibdenum, respectively. In the electroweak theory the breaking of the symmetry is of a particular type, denoted as spontaneous symmetry breaking. In this case charges and currents are as dictated by the symmetry but the fundamental state of minimum energy, the vacuum, is not unique and there is a continuum of degenerate states that all together respect the symmetry (in the sense that the whole vacuum orbit is spanned by applying the symmetry transformations). The symmetry breaking is due to the fact that the system (with infinite volume and infinite number of degrees of freedom) is found in one particular vacuum state, and this choice, which for the SM occurred in the first instants of the Universe life, makes the symmetry violated in the spectrum of states. In a gauge theory like the SM the spontaneous symmetry breaking is realized by the Higgs mechanism (described in detail in Sect. (2.7)): there are a number of scalar (i.e. of zero spin) Higgs bosons with a potential that produces an orbit of degenerate vacuum states. One or more of these scalar Higgs particles must necessarily be present in the spectrum of physical states with masses very close to the range so far explored. It is expected that the Higgs particle(s) will be found at the LHC thus completing the experimental verification of the SM.

The fermionic (all of spin 1/2) matter fields of the SM are quarks and leptons. Each type of quark is a colour triplet (i.e. each quark flavour comes in three colours) and also carries electroweak charges, in particular electric charges +2/3 for up-type quarks and $-1/3$ for down-type quarks. So quarks are subject to all SM interactions.

Leptons are colourless and thus do not interact strongly (they are not hadrons) but have electroweak charges, in particular electric charges -1 for charged leptons (e^-, μ^- and τ^-) while it is 0 for neutrinos (ν_e, ν_μ and ν_τ). Quarks and leptons are grouped in 3 "families" or "generations" with equal quantum numbers but different masses. At present we do not have an explanation for this triple repetition of fermion families:

$$\begin{bmatrix} u & u & u & \nu_e \\ d & d & d & e \end{bmatrix}, \quad \begin{bmatrix} c & c & c & \nu_\mu \\ s & s & s & \mu \end{bmatrix}, \quad \begin{bmatrix} t & t & t & \nu_\tau \\ b & b & b & \tau \end{bmatrix}. \tag{2.1}$$

The QCD sector of the SM has a simple structure but a very rich dynamical content, including the observed complex spectroscopy with a large number of hadrons. The most prominent properties of QCD are asymptotic freedom and confinement. In field theory the effective coupling of a given interaction vertex is modified by the interaction. As a result, the measured intensity of the force depends on the transferred (four)momentum squared, Q^2, among the participants. In QCD the relevant coupling parameter that appears in physical processes is $\alpha_s = e_s^2/4\pi$ where e_s is the coupling constant of the basic interaction vertices of quark and gluons: qqg or ggg (see Eq. (2.30)). Asymptotic freedom means that the effective coupling becomes a function of Q^2: $\alpha_s(Q^2)$ decreases for increasing Q^2 and vanishes asymptotically. Thus, the QCD interaction becomes very weak in processes with large Q^2, called hard processes or deep inelastic processes (i.e. with a final state distribution of momenta and a particle content very different than those in the initial state). One can prove that in 4 space-time dimensions all pure-gauge theories based on a non commuting group of symmetry are asymptotically free and conversely. The effective coupling decreases very slowly at large momenta with the inverse logarithm of Q^2: $\alpha_s(Q^2) = 1/b \log Q^2/\Lambda^2$ where b is a known constant and Λ is an energy of order a few hundred MeV. Since in quantum mechanics large momenta imply short wavelengths, the result is that at short distances the potential between two colour charges is similar to the Coulomb potential, i.e. proportional to $\alpha_s(r)/r$, with an effective colour charge which is small at short distances. On the contrary the interaction strength becomes large at large distances or small transferred momenta, of order $Q \lesssim \Lambda$. In fact all observed hadrons are tightly bound composite states of quarks (baryons are made of qqq and mesons of $q\bar{q}$), with compensating colour charges so that they are overall neutral in colour. In fact, the property of confinement is the impossibility of separating colour charges, like individual quarks and gluons or any other coloured state. This is because in QCD the interaction potential between colour charges increases at long distances linearly in r. When we try to separate the quark and the antiquark that form a colour neutral meson the interaction energy grows until pairs of quarks and antiquarks are created from the vacuum and new neutral mesons are coalesced and observed in the final state instead of free quarks. For example, consider the process $e^+e^- \rightarrow q\bar{q}$ at large center of mass energies. The final state quark and antiquark have large energies, so they separate in opposite directions very fast. But the colour confinement forces create new pairs in between them. What is observed is two back-to-back jets of colourless hadrons with a number of slow pions that make the exact separation of the two jets impossible. In some cases a third well separated jet of hadrons is also observed: these events correspond to the radiation of an energetic gluon from the parent quark-antiquark pair.

In the EW sector the SM inherits the phenomenological successes of the old $(V - A) \otimes (V - A)$ four-fermion low-energy description of weak interactions, and provides a well-defined and consistent theoretical framework including weak interactions and quantum electrodynamics in a unified picture. The weak interactions derive their name from their intensity. At low energy the strength of the effective four-fermion interaction of charged currents is determined by the Fermi coupling constant G_F. For example, the effective interaction for muon decay is given by

$$\mathcal{L}_{\text{eff}} = (G_F/\sqrt{2}) \left[\bar{\nu}_\mu \gamma_\alpha (1 - \gamma_5)\mu\right] \left[\bar{e}\gamma^\alpha (1 - \gamma_5)\nu_e\right] , \tag{2.2}$$

with [7]

$$G_F = 1.16639(1) \times 10^{-5} \, \text{GeV}^{-2} . \tag{2.3}$$

In natural units $\hbar = c = 1$, G_F has dimensions of $(\text{mass})^{-2}$. As a result, the intensity of weak interactions at low energy is characterized by $G_F E^2$, where E is the energy scale for a given process ($E \approx m_\mu$ for muon decay). Since

$$G_F E^2 = G_F m_p^2 (E/m_p)^2 \simeq 10^{-5}(E/m_p)^2 , \tag{2.4}$$

where m_p is the proton mass, the weak interactions are indeed weak at low energies (up to energies of order a few ten's of GeV). Effective four fermion couplings for neutral current interactions have comparable intensity and energy behaviour. The quadratic increase with energy cannot continue for ever, because it would lead to a violation of unitarity. In fact, at large energies the propagator effects can no longer be neglected, and the current–current interaction is resolved into current–W gauge boson vertices connected by a W propagator. The strength of the weak interactions at high energies is then measured by g_W, the $W - -\mu$-ν_μ coupling, or, even better, by $\alpha_W = g_W^2/4\pi$ analogous to the fine-structure constant α of QED (in Chap. 3, g_W is simply denoted by g or g_2). In the standard EW theory, we have

$$\alpha_W = \sqrt{2} \, G_F \, m_W^2/\pi \cong 1/30 . \tag{2.5}$$

That is, at high energies the weak interactions are no longer so weak.

The range r_W of weak interactions is very short: it is only with the experimental discovery of the W and Z gauge bosons that it could be demonstrated that r_W is non-vanishing. Now we know that

$$r_W = \frac{\hbar}{m_W c} \simeq 2.5 \times 10^{-16} \, \text{cm}, \tag{2.6}$$

corresponding to $m_W \simeq 80.4$ GeV. This very large value for the W (or the Z) mass makes a drastic difference, compared with the massless photon and the infinite range of the QED force. The direct experimental limit on the photon mass is [7] $m_\gamma < 6 \ 10^{-17}$ eV. Thus, on the one hand, there is very good evidence that the photon is massless. On the other hand, the weak bosons are very heavy. A unified theory of EW interactions has to face this striking difference.

Another apparent obstacle in the way of EW unification is the chiral structure of weak interactions: in the massless limit for fermions, only left-handed quarks and leptons (and right-handed antiquarks and antileptons) are coupled to W's. This clearly implies parity and charge-conjugation violation in weak interactions.

The universality of weak interactions and the algebraic properties of the electromagnetic and weak currents [the conservation of vector currents (CVC), the partial conservation of axial currents (PCAC), the algebra of currents, etc.] have been crucial in pointing to a symmetric role of electromagnetism and weak interactions at a more fundamental level. The old Cabibbo universality [8] for the weak charged current:

$$J_\alpha^{\text{weak}} = \bar{\nu}_\mu \gamma_\alpha (1 - \gamma_5)\mu + \bar{\nu}_e \gamma_\alpha (1 - \gamma_5)e + \cos\theta_c \ \bar{u}\gamma_\alpha(1 - \gamma_5)d +$$
$$+ \sin\theta_c \ \bar{u}\gamma_\alpha(1 - \gamma_5)s + \dots , \tag{2.7}$$

suitably extended, is naturally implied by the standard EW theory. In this theory the weak gauge bosons couple to all particles with couplings that are proportional to their weak charges, in the same way as the photon couples to all particles in proportion to their electric charges [in Eq. (2.7), $d' = \cos\theta_c \ d + \sin\theta_c \ s$ is the weak-isospin partner of u in a doublet. The (u, d') doublet has the same couplings as the (ν_e, ℓ) and (ν_μ, μ) doublets].

Another crucial feature is that the charged weak interactions are the only known interactions that can change flavour: charged leptons into neutrinos or up-type quarks into down-type quarks. On the contrary, there are no flavour-changing neutral currents at tree level. This is a remarkable property of the weak neutral current, which is explained by the introduction of the Glashow-Iliopoulos-Maiani (GIM) mechanism [9] and has led to the successful prediction of charm.

The natural suppression of flavour-changing neutral currents, the separate conservation of e, μ and τ leptonic flavours that is only broken by the small neutrino masses, the mechanism of CP violation through the phase in the quark-mixing matrix [10], are all crucial features of the SM. Many examples of new physics tend to break the selection rules of the standard theory. Thus the experimental study of rare flavour-changing transitions is an important window on possible new physics.

The SM is a renormalizable field theory which means that the ultra-violet divergences that appear in loop diagrams can be eliminated by a suitable redefinition of the parameters already appearing in the bare lagrangian: masses, couplings and field normalizations. As it will be discussed later, a necessary condition for a theory to be renormalizable is that only operator vertices of dimension not larger than 4 (that is m^4 where m is some mass scale) appear in the lagrangian density $\mathcal{L}$ (itself of dimension 4, because the action S is given by the integral of $\mathcal{L}$ over d^4x and is dimensionless in natural units: $\hbar = c = 1$). Once this condition is added to the specification of a gauge group and of the matter field content the gauge theory lagrangian density is completely specified. We shall see the precise rules to write down the lagrangian of a gauge theory in the next Section.

2.4 The Formalism of Gauge Theories

In this Section we summarize the definition and the structure of a gauge Yang–Mills theory [11]. We will list here the general rules for constructing such a theory. Then these results will be applied to the SM.

Consider a lagrangian density $\mathcal{L}[\phi, \partial_\mu \phi]$ which is invariant under a D dimensional continuous group of transformations:

$$\phi'(x) = U(\theta^A)\phi(x) \qquad (A = 1, 2, \ldots, D) . \tag{2.8}$$

with:

$$U(\theta^A) = \exp[ig \sum_A \theta^A T^A] \sim 1 + ig \sum_A \theta^A T^A + \ldots, \tag{2.9}$$

The quantities θ^A are numerical parameters, like angles in the particular case of a rotation group in some internal space. The approximate expression on the right is valid for θ^A infinitesimal. Then, g is the coupling constant and T^A are the generators of the group Γ of transformations (2.8) in the (in general reducible) representation of the fields ϕ. Here we restrict ourselves to the case of internal symmetries, so that T^A are matrices that are independent of the space-time coordinates and the arguments of the fields ϕ and ϕ' in Eq. (2.8) is the same. If U is unitary, then the generators T^A are Hermitian, but this need not be the case in general (though it is true for the SM). Similarly if U is a group of matrices with unit determinant, then the traces of T^A vanish: $\text{tr}(T^A) = 0$. The generators T^A are normalized in such a way that for the lowest dimensional non-trivial representation of the group Γ (we use t^A to denote the generators in this particular representation) we have

$$\text{tr}(t^A t^B) = \frac{1}{2}\delta^{AB} . \tag{2.10}$$

The generators satisfy the commutation relations

$$[T^A, T^B] = iC_{ABC}T^C . \tag{2.11}$$

For A, B, C .... up or down indices make no difference: $T^A = T_A$ etc. The structure constants C_{ABC} are completely antisymmetric in their indices, as can be easily seen. In the following, for each quantity f^A we define

$$\mathbf{f} = \sum_A T^A f^A .$$ (2.12)

For example, we can rewrite Eq. (2.9) in the form:

$$U(\theta^A) = \exp[ig\boldsymbol{\theta}] \sim 1 + ig\boldsymbol{\theta} + \dots,$$ (2.13)

If we now make the parameters θ^A depend on the space–time coordinates $\theta^A = \theta^A(x_\mu)$, $\mathcal{L}[\phi, \partial_\mu \phi]$ is in general no longer invariant under the gauge transformations $U[\theta^A(x_\mu)]$, because of the derivative terms: indeed $\partial_\mu \phi' = \partial_\mu (U\phi) \neq U\partial_\mu \phi$. Gauge invariance is recovered if the ordinary derivative is replaced by the covariant derivative:

$$D_\mu = \partial_\mu + ig\mathbf{V}_\mu ,$$ (2.14)

where V_μ^A are a set of D gauge vector fields (in one-to-one correspondence with the group generators) with the transformation law

$$\mathbf{V}'_\mu = U\mathbf{V}_\mu U^{-1} - (1/ig)(\partial_\mu U)U^{-1} .$$ (2.15)

For constant θ^A, $\mathbf{V}$ reduces to a tensor of the adjoint (or regular) representation of the group:

$$\mathbf{V}'_\mu = U\mathbf{V}_\mu U^{-1} \simeq \mathbf{V}_\mu + ig[\boldsymbol{\theta}, \mathbf{V}_\mu] \dots,$$ (2.16)

which implies that

$$V_\mu^{\prime C} = V_\mu^C - gC_{ABC}\theta^A V_\mu^B \dots,$$ (2.17)

where repeated indices are summed up.

As a consequence of Eqs. (2.14) and (2.15), $D_\mu \phi$ has the same transformation properties as ϕ:

$$(D_\mu \phi)' = U(D_\mu \phi) .$$ (2.18)

In fact

$$(D_\mu \phi)' = (\partial_\mu + ig\mathbf{V}'_\mu)\phi' = (\partial_\mu U)\phi + U\partial_\mu \phi + igU\mathbf{V}_\mu \phi - (\partial_\mu U)$$
$$\phi = U(D_\mu \phi) .$$ (2.19)

Thus $\mathcal{L}[\phi, D_\mu \phi]$ is indeed invariant under gauge transformations. But, at this stage, the gauge fields V_μ^A appear as external fields that do not propagate. In order to construct a gauge-invariant kinetic energy term for the gauge fields V_μ^A, we consider

$$[D_\mu, D_\nu]\phi = ig\{\partial_\mu \mathbf{V}_\nu - \partial_\nu \mathbf{V}_\mu + ig[\mathbf{V}_\mu, \mathbf{V}_\nu]\}\phi \equiv ig\mathbf{F}_{\mu\nu}\phi \,, \tag{2.20}$$

which is equivalent to

$$F_{\mu\nu}^A = \partial_\mu V_\nu^A - \partial_\nu V_\mu^A - g C_{ABC} V_\mu^B V_\nu^C \,. \tag{2.21}$$

From Eqs. (2.8), (2.18) and (2.20) it follows that the transformation properties of $F_{\mu\nu}^A$ are those of a tensor of the adjoint representation

$$\mathbf{F}_{\mu\nu}' = U\mathbf{F}_{\mu\nu}U^{-1} \,. \tag{2.22}$$

The complete Yang–Mills lagrangian, which is invariant under gauge transformations, can be written in the form

$$\mathcal{L}_{\mathrm{YM}} = -\frac{1}{2} Tr\mathbf{F}_{\mu\nu}\mathbf{F}^{\mu\nu} + \mathcal{L}[\phi, D_\mu \phi] = -\frac{1}{4}\sum_A F_{\mu\nu}^A F^{A\mu\nu} + \mathcal{L}[\phi, D_\mu \phi] \,. \tag{2.23}$$

Note that the kinetic energy term is an operator of dimension 4. Thus if $\mathcal{L}$ is renormalizable, also $\mathcal{L}_{\mathrm{YM}}$ is renormalizable. In fact it is the most general gauge invariant and renormalizable lagrangian density. If we give up renormalizability then more gauge invariant higher dimension terms could be added. It is already clear at this stage that no mass term for gauge bosons of the form $m^2 V_\mu V^\mu$ is allowed by gauge invariance.

For an abelian theory, as for example QED, the gauge transformation reduces to $U[\theta(x)] = \exp[ieQ\theta(x)]$, where Q is the charge generator. The associated gauge field (the photon), according to Eq. (2.15), transforms as

$$V_\mu' = V_\mu - \partial_\mu \theta(x) \,. \tag{2.24}$$

and the familiar gauge transformation by addition of a 4-gradient of a scalar function is recovered. The QED lagrangian density is given by:

$$\mathcal{L} = -\frac{1}{4} F^{\mu\nu} F_{\mu\nu} + \sum_\psi \bar{\psi}(i\not{D} - m_\psi)\psi \,. \tag{2.25}$$

Here $\not{D} = D_\mu \gamma^\mu$, where γ^μ are the Dirac matrices and the covariant derivative is given in terms of the photon field A_μ and the charge operator Q by:

$$D_\mu = \partial_\mu + ieA_\mu Q \tag{2.26}$$

and

$$F_{\mu\nu} = \partial_\mu A_\nu - \partial_\nu A_\mu \tag{2.27}$$

Note that in QED one usually takes the e^- to be the particle, so that $Q = -1$ and the covariant derivative is $D_\mu = \partial_\mu - ieA_\mu$ when acting on the electron field. In this case, the $F_{\mu\nu}$ tensor is linear in the gauge field V_μ so that in the absence of matter fields the theory is free. On the other hand, in the non abelian case the $F_{\mu\nu}^A$ tensor contains both linear and quadratic terms in V_μ^A, so that the theory is non-trivial even in the absence of matter fields.

2.5 Application to QCD

According to the formalism of the previous section, the statement that QCD is a renormalizable gauge theory based on the group $SU(3)$ with colour triplet quark matter fields fixes the QCD lagrangian density to be

$$\mathcal{L} = -\frac{1}{4} \sum_{A=1}^{8} F^{A\mu\nu} F_{\mu\nu}^A + \sum_{j=1}^{n_f} \bar{q}_j (i \not{D} - m_j) q_j \tag{2.28}$$

Here q_j are the quark fields (of n_f different flavours) with mass m_j and D_μ is the covariant derivative:

$$D_\mu = \partial_\mu + ie_s \mathbf{g}_\mu; \tag{2.29}$$

e_s is the gauge coupling and later we will mostly use, in analogy with QED

$$\alpha_s = \frac{e_s^2}{4\pi}. \tag{2.30}$$

Also, $\mathbf{g}_\mu = \sum_A t^A g_\mu^A$ where g_μ^A, $A = 1, 8$, are the gluon fields and t^A are the $SU(3)$ group generators in the triplet representation of quarks (i.e. t_A are 3×3 matrices acting on q); the generators obey the commutation relations $[t^A, t^B] = iC_{ABC}t^C$ where C_{ABC} are the complete antisymmetric structure constants of $SU(3)$ (the normalisation of C_{ABC} and of e_s is specified by $Tr[t^A t^B] = \delta^{AB}/2$);

$$F_{\mu\nu}^A = \partial_\mu g_\nu^A - \partial_\nu g_\mu^A - e_s C_{ABC} g_\mu^B g_\nu^C \tag{2.31}$$

Chapter 4 is devoted to a detailed description of the QCD as the theory of strong interactions. The physical vertices in QCD include the gluon-quark-antiquark vertex, analogous to the QED photon-fermion-antifermion coupling, but also the 3-gluon and 4-gluon vertices, of order e_s and e_s^2 respectively, which have no analogue in an abelian theory like QED. In QED the photon is coupled to all electrically charged particles but itself is neutral. In QCD the gluons are coloured hence self-coupled. This is reflected in the fact that in QED $F_{\mu\nu}$ is linear in the gauge field, so that the term $F_{\mu\nu}^2$ in the lagrangian is a pure kinetic term, while in QCD $F_{\mu\nu}^A$ is quadratic in the gauge field so that in $F_{\mu\nu}^{A2}$ we find cubic and quartic vertices beyond the kinetic term. Also instructive is to consider the case of scalar QED:

$$\mathcal{L} = -\frac{1}{4}F^{\mu\nu}F_{\mu\nu} + (D_\mu\phi)^\dagger(D^\mu\phi) - m^2(\phi^\dagger\phi) \tag{2.32}$$

For $Q = 1$ we have:

$$(D_\mu\phi)^\dagger(D^\mu\phi) = (\partial_\mu\phi)^\dagger(\partial^\mu\phi) + ieA_\mu[(\partial^\mu\phi)^\dagger\phi - \phi^\dagger(\partial^\mu\phi)] + e^2A_\mu A^\mu\phi^\dagger\phi \tag{2.33}$$

We see that for a charged boson in QED, given that the kinetic term for bosons is quadratic in the derivative, there is a two-gauge vertex of order e^2. Thus in QCD the 3-gluon vertex is there because the gluon is coloured and the 4-gluon vertex because the gluon is a boson.

2.6 Quantization of a Gauge Theory

The lagrangian density $\mathcal{L}_{YM}$ in Eq. (2.23) fully describes the theory at the classical level. The formulation of the theory at the quantum level requires that a procedure of quantization, of regularization and, finally, of renormalization is also specified. To start with, the formulation of Feynman rules is not straightforward. A first problem, common to all gauge theories, including the abelian case of QED, can be realized by observing that the free equation of motion for V_μ^A, as obtained from Eqs. ((2.21), (2.23)), is given by

$$[\partial^2 g_{\mu\nu} - \partial_\mu\partial_\nu]V^{A\nu} = 0 \tag{2.34}$$

Normally the propagator of the gauge field should be determined by the inverse of the operator $[\partial^2 g_{\mu\nu} - \partial_\mu\partial_\nu]$ which, however, has no inverse, being a projector over the transverse gauge vector states. This difficulty is removed by fixing a particular gauge. If one chooses a covariant gauge condition $\partial^\mu V_\mu^A = 0$ then a gauge fixing term of the form

$$\Delta\mathcal{L}_{GF} = -\frac{1}{2\lambda}\sum_A |\partial^\mu V_\mu^A|^2 \tag{2.35}$$

has to be added to the lagrangian ($1/\lambda$ acts as a lagrangian multiplier). The free equations of motion are now modified as follows:

$$[\partial^2 g_{\mu\nu} - (1 - 1/\lambda)\partial_\mu \partial_\nu] V^{A\nu} = 0. \tag{2.36}$$

This operator now has an inverse whose Fourier transform is given by:

$$D_{\mu\nu}^{AB}(q) = \frac{i}{q^2 + i\epsilon} [-g_{\mu\nu} + (1 - \lambda)\frac{q_\mu q_\nu}{q^2 + i\epsilon}] \delta^{AB} \tag{2.37}$$

which is the propagator in this class of gauges. The parameter λ can take any value and it disappears from the final expression of any gauge invariant, physical quantity. Commonly used particular cases are $\lambda = 1$ (Feynman gauge) and $\lambda = 0$ (Landau gauge).

While in an abelian theory the gauge fixing term is all that is needed for a correct quantization, in a non abelian theory the formulation of complete Feynman rules involves a further subtlety. This is formally taken into account by introducing a set of D fictitious ghost fields that must be included as internal lines in closed loops (Faddeev-Popov ghosts [12]). Given that gauge fields connected by a gauge transformation describe the same physics, clearly there are less physical degrees of freedom than gauge field components. Ghosts appear, in the form of a transformation Jacobian in the functional integral, in the process of elimination of the redundant variables associated with fields on the same gauge orbit [13]. The correct ghost contributions can be obtained from an additional term in the lagrangian density. For each choice of the gauge fixing term the ghost langrangian is obtained by considering the effect of an infinitesimal gauge transformation $V_\mu^{'C} = V_\mu^C - gC_{ABC}\theta^A V_\mu^B - \partial_\mu\theta^C$ on the gauge fixing condition. For $\partial^\mu V_\mu^C = 0$ one obtains:

$$\partial^\mu V_\mu^{'C} = \partial^\mu V_\mu^C - gC_{ABC}\partial^\mu(\theta^A V_\mu^B) - \partial^2\theta^C = -[\partial^2\delta_{AC} + gC_{ABC}V_\mu^B\partial^\mu]\theta^A \tag{2.38}$$

where the gauge condition $\partial^\mu V_\mu^C = 0$ has been taken into account in the last step. The ghost lagrangian is then given by:

$$\Delta\mathcal{L}_{Ghost} = \bar{\eta}^C[\partial^2\delta_{AC} + gC_{ABC}V_\mu^B\partial^\mu]\eta^A \tag{2.39}$$

where η^A is the ghost field (one for each index A) which has to be treated as a scalar field except that a factor (-1) for each closed loop has to be included as for fermion fields.

Starting from non covariant gauges one can construct ghost-free gauges. An example, also important in other respects, is provided by the set of "axial" gauges: $n^\mu V_\mu^A = 0$ where n_μ is a fixed reference 4-vector (actually for n_μ spacelike one has axial gauge proper, for $n^2 = 0$ one speaks of a light-like gauge and for n_μ

timelike one has a Coulomb or temporal gauge). The gauge fixing term is of the form:

$$\Delta \mathcal{L}_{GF} = -\frac{1}{2\lambda} \sum_A |n^\mu V_\mu^A|^2 \tag{2.40}$$

With a procedure that can be found in QED textbooks [14] the corresponding propagator, in Fourier space, is found to be:

$$D_{\mu\nu}^{AB}(q) = \frac{i}{q^2 + i\epsilon} \left[-g_{\mu\nu} + \frac{n_\mu q_\nu + n_\nu q_\mu}{(nq)} - \frac{n^2 q_\mu q_\nu}{(nq)^2} \right] \delta^{AB} \tag{2.41}$$

In this case there are no ghost interactions because $n^\mu V_\mu^{'A}$, obtained by a gauge transformation from $n^\mu V_\mu^A$, contains no gauge fields, once the gauge condition $n^\mu V_\mu^A = 0$ has been taken into account. Thus the ghosts are decoupled and can be ignored.

The introduction of a suitable regularization method that preserves gauge invariance is essential for the definition and the calculation of loop diagrams and for the renormalization programme of the theory. The method that is by now currently adopted is dimensional regularization [15] which consists in the formulation of the theory in n dimensions. All loop integrals have an analytic expression that is actually valid also for non integer values of n. Writing the results for $n = 4 - \epsilon$ the loops are ultraviolet finite for $\epsilon > 0$ and the divergences reappear in the form of poles at $\epsilon = 0$.

2.7 Spontaneous Symmetry Breaking in Gauge Theories

The gauge symmetry of the SM was difficult to discover because it is well hidden in nature. The only observed gauge boson that is massless is the photon. The gluons are presumed massless but cannot be directly observed because of confinement, and the W and Z weak bosons carry a heavy mass. Indeed a major difficulty in unifying the weak and electromagnetic interactions was the fact that e.m. interactions have infinite range ($m_\gamma = 0$), whilst the weak forces have a very short range, owing to $m_{W,Z} \neq 0$.

The solution of this problem is in the concept of spontaneous symmetry breaking, which was borrowed from statistical mechanics.

Consider a ferromagnet at zero magnetic field in the Landau–Ginzburg approximation. The free energy in terms of the temperature T and the magnetization $\mathbf{M}$ can be written as

$$F(\mathbf{M}, T) \simeq F_0(T) + 1/2\, \mu^2(T)\mathbf{M}^2 + 1/4\, \lambda(T)(\mathbf{M}^2)^2 + \dots . \tag{2.42}$$

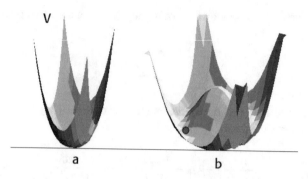

Fig. 2.1 The potential $V = 1/2\,\mu^2\mathbf{M}^2 + 1/4\,\lambda(\mathbf{M}^2)^2$ for positive (**a**) or negative μ^2 (**b**) (for simplicity, $\mathbf{M}$ is a 2-dimensional vector). The small sphere indicates a possible choice for the direction of $\mathbf{M}$

This is an expansion which is valid at small magnetization. The neglect of terms of higher order in $\vec{M}^2$ is the analogue in this context of the renormalizability criterion. Also, $\lambda(T) > 0$ is assumed for stability; F is invariant under rotations, i.e. all directions of $\mathbf{M}$ in space are equivalent. The minimum condition for F reads

$$\partial F/\partial M_i = 0, \quad [\mu^2(T) + \lambda(T)\mathbf{M}^2]\mathbf{M} = 0 . \tag{2.43}$$

There are two cases, shown in Fig. 2.1. If $\mu^2 \gtrsim 0$, then the only solution is $\mathbf{M} = 0$, there is no magnetization, and the rotation symmetry is respected. In this case the lowest energy state (in a quantum theory the vacuum) is unique and invariant under rotations. If $\mu^2 < 0$, then another solution appears, which is

$$|\mathbf{M}_0|^2 = -\mu^2/\lambda . \tag{2.44}$$

In this case there is a continuous orbit of lowest energy states, all with the same value of $|\mathbf{M}|$ but different orientations. A particular direction chosen by the vector $\mathbf{M}_0$ leads to a breaking of the rotation symmetry.

For a piece of iron we can imagine to bring it to high temperature and let it melt in an external magnetic field $\mathbf{B}$. The presence of $\mathbf{B}$ is an explicit breaking of the rotational symmetry and it induces a non zero magnetization $\mathbf{M}$ along its direction. Now we lower the temperature while keeping $\mathbf{B}$ fixed. The critical temperature T_{crit} (Curie temperature) is where $\mu^2(T)$ changes sign: $\mu^2(T_{\text{crit}}) = 0$. For pure iron T_{crit} is below the melting temperature. So at $T = T_{\text{crit}}$ iron is a solid. Below T_{crit} we remove the magnetic field. In a solid the mobility of the magnetic domains is limited and a non vanishing M_0 remains. The form of the free energy becomes rotationally invariant as in Eq. (2.43). But now the system allows a minimum energy state with non vanishing $\mathbf{M}$ in the direction where $\mathbf{B}$ was. As a consequence the symmetry is broken by this choice of one particular vacuum state out of a continuum of them.

We now prove the Goldstone theorem [16]. It states that when spontaneous symmetry breaking takes place, there is always a zero-mass mode in the spectrum.

In a classical context this can be proven as follows. Consider a lagrangian

$$\mathcal{L} = \frac{1}{2}|\partial_\mu \phi|^2 - V(\phi).$$ (2.45)

The potential $V(\phi)$ can be kept generic at this stage but, in the following, we will be mostly interested in a renormalizable potential of the form (with no more than quartic terms):

$$V(\phi) = -\frac{1}{2}\mu^2 \phi^2 + \frac{1}{4}\lambda \phi^4.$$ (2.46)

Here by ϕ we mean a column vector with real components ϕ_i (1=1,2...N) (complex fields can always be decomposed into a pair of real fields), so that, for example, $\phi^2 = \sum_i \phi_i^2$. This particular potential is symmetric under a NxN orthogonal matrix rotation $\phi' = O\phi$, where O is a SO(N) transformation. For simplicity, we have omitted odd powers of ϕ, which means that we assumed an extra discrete symmetry under $\phi \leftrightarrow -\phi$. Note that, for positive μ^2, the mass term in the potential has the "wrong" sign: according to the previous discussion this is the condition for the existence of a non unique lowest energy state. More in general, we only assume here that the potential is symmetric under the infinitesimal transformations

$$\phi \rightarrow \phi' = \phi + \delta\phi, \quad \delta\phi_i = i\delta\theta^A t_{ij}^A \phi_j .$$ (2.47)

where $\delta\theta^A$ are infinitesimal parameters and t_{ij}^A are the matrices that represent the symmetry group on the representation of the fields ϕ_i (a sum over A is understood). The minimum condition on V that identifies the equilibrium position (or the vacuum state in quantum field theory language) is

$$(\partial V/\partial\phi_i)(\phi_i = \phi_i^0) = 0 .$$ (2.48)

The symmetry of V implies that

$$\delta V = (\partial V/\partial\phi_i)\delta\phi_i = i\delta\theta^A(\partial V/\partial\phi_i)t_{ij}^A \phi_j = 0 .$$ (2.49)

By taking a second derivative at the minimum $\phi_i = \phi_i^0$, given by the previous equation, we obtain that, for each A:

$$\frac{\partial^2 V}{\partial\phi_k\partial\phi_i}(\phi_i = \phi_i^0)t_{ij}^A\phi_j^0 + \frac{\partial V}{\partial\phi_i}(\phi_i = \phi_i^0)t_{ik}^A = 0 .$$ (2.50)

The second term vanishes owing to the minimum condition, Eq. (2.48). We then find

$$\frac{\partial^2 V}{\partial\phi_k\partial\phi_i}(\phi_i = \phi_i^0)t_{ij}^A\phi_j^0 = 0 .$$ (2.51)

The second derivatives $M_{ki}^2 = (\partial^2 V / \partial \phi_k \partial \phi_i)(\phi_i = \phi_i^0)$ define the squared mass matrix. Thus the above equation in matrix notation can be written as

$$M^2 t^A \phi^0 = 0 . \tag{2.52}$$

In the case of no spontaneous symmetry breaking the ground state is unique, all symmetry transformations leave it invariant, so that, for all A, $t^A \phi^0 = 0$. On the contrary, if, for some values of A, the vectors $(t^A \phi^0)$ are non-vanishing, i.e. there is some generator that shifts the ground state into some other state with the same energy (hence the vacuum is not unique), then each $t^A \phi^0 \neq 0$ is an eigenstate of the squared mass matrix with zero eigenvalue. Therefore, a massless mode is associated with each broken generator. The charges of the massless modes (their quantum numbers in quantum language) differ from those of the vacuum (usually taken as all zero) by the values of the t^A charges: one says that the massless modes have the same quantum numbers of the broken generators, i.e. those that do not annihilate the vacuum.

The previous proof of the Goldstone theorem has been given in the classical case. In the quantum case the classical potential corresponds to tree level approximation of the quantum potential. Higher order diagrams with loops introduce quantum corrections. The functional integral formulation of quantum field theory [13, 17] is the most appropriate framework to define and compute, in a loop expansion, the quantum potential which specifies, exactly as described above, the vacuum properties of the quantum theory. If the theory is weakly coupled, e.g. if λ is small, the tree level expression for the potential is not too far from the truth, and the classical situation is a good approximation. We shall see that this is the situation that occurs in the electroweak theory if the Higgs is moderately light (see Chap. 3, Sect. 3.13.1).

We note that for a quantum system with a finite number of degrees of freedom, for example one described by the Schrödinger equation, there are no degenerate vacua: the vacuum is always unique. For example, in the one dimensional Schrödinger problem with a potential:

$$V(x) = -\mu^2/2 \, x^2 + \lambda \, x^4/4 , \tag{2.53}$$

there are two degenerate minima at $x = \pm x_0 = \sqrt{(\mu^2/\lambda)}$ which we denote by $|+\rangle$ and $|-\rangle$. But the potential is not diagonal in this basis: the off diagonal matrix elements:

$$\langle +|V|-\rangle = \langle -|V|+\rangle \sim \exp(-khd) = \delta \tag{2.54}$$

are different from zero due to the non vanishing amplitude for a tunnel effect between the two vacua, proportional to the exponential of the product of the distance d between the two vacua and the height h of the barrier with k a constant (see Fig. 2.2). After diagonalization the eigenvectors are $(|+\rangle + |-\rangle)/\sqrt{2}$ and $(|+\rangle - |-\rangle)/\sqrt{2}$,

Fig. 2.2 A Schrödinger
potential $V(x)$ analogous to
the Higgs potential

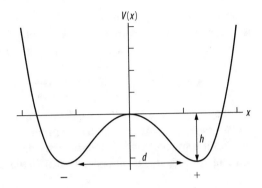

with different energies (the difference being proportional to δ). Suppose now that
you have a sum of n equal terms in the potential, $V = \sum_i V(x_i)$. Then the
transition amplitude would be proportional to δ^n and would vanish for infinite n: the
probability that all degrees of freedom together jump over the barrier vanishes. In
this example there is a discrete number of minimum points. The case of a continuum
of minima is obtained, always in the Schrödinger context, if we take

$$V = 1/2 \, \mu^2 \mathbf{r}^2 + 1/4 \, \lambda(\mathbf{r}^2)^2 \,, \tag{2.55}$$

with $\mathbf{r} = (x, y, z)$. Also in this case the ground state is unique: it is given by
a state with total orbital angular momentum zero, an s-wave state, whose wave
function only depends on $|\mathbf{r}|$, independent of all angles. This is a superposition of
all directions with the same weight, analogous to what happened in the discrete
case. But again, if we replace a single vector $\mathbf{r}$, with a vector field $\mathbf{M}(x)$, that is a
different vector at each point in space, the amplitude to go from a minimum state in
one direction to another in a different direction goes to zero in the limit of infinite
volume. In simple words, the vectors at all points in space have a vanishing small
amplitude to make a common rotation, all together at the same time. In the infinite
volume limit all vacua along each direction have the same energy and spontaneous
symmetry breaking can occur.

The massless Goldstone bosons correspond to a long range force. Unless the
massless particles are confined, as for the gluons in QCD, these long range forces
would be easily detectable. Thus, in the construction of the EW theory we cannot
accept massless physical scalar bosons. Fortunately, when spontaneous symmetry
breaking takes place in a gauge theory, the massless Goldstone modes exist, but they
are unphysical and disappear from the spectrum. Each of them becomes, in fact, the
third helicity state of a gauge boson that takes mass. This is the Higgs mechanism
(it should be called Englert-Brout-Higgs mechanism [18], because an equal merit
should be credited to the simultaneous paper by Englert and Brout). Consider, for
example, the simplest Higgs model described by the lagrangian

$$\mathcal{L} = -\frac{1}{4} F_{\mu\nu}^2 + |(\partial_\mu + ieA_\mu Q)\phi|^2 + \mu^2 \phi^* \phi - \frac{\lambda}{2}(\phi^* \phi)^2 \,. \tag{2.56}$$

Note the 'wrong' sign in front of the mass term for the scalar field ϕ, which is necessary for the spontaneous symmetry breaking to take place. The above lagrangian is invariant under the $U(1)$ gauge symmetry

$$A_\mu \to A'_\mu = A_\mu - \partial_\mu \theta(x), \quad \phi \to \phi' = \exp[ieQ\theta(x)] \, \phi. \tag{2.57}$$

For the U(1) charge Q we take $Q\phi = -\phi$, like in QED, where the particle is e^-. Let $\phi^0 = v \neq 0$, with v real, be the ground state that minimizes the potential and induces the spontaneous symmetry breaking. In our case v is given by $v^2 = \mu^2/\lambda$. Making use of gauge invariance, we can do the change of variables

$$\phi(x) \to [v + h(x)/\sqrt{2}] \exp[-i\zeta(x)/v\sqrt{2}],$$

$$A_\mu(x) \to A_\mu - \partial_\mu \zeta(x)/ev\sqrt{2}. \tag{2.58}$$

Then $h = 0$ is the position of the minimum, and the lagrangian becomes

$$\mathcal{L} = -\frac{1}{4}F_{\mu\nu}^2 + e^2v^2A_\mu^2 + \frac{1}{2}e^2h^2A_\mu^2 + \sqrt{2}e^2hvA_\mu^2 + \mathcal{L}(h). \tag{2.59}$$

The field $\zeta(x)$ is the would-be Goldstone boson, as can be seen by considering only the ϕ terms in the lagrangian, i.e. setting $A_\mu = 0$ in Eq. (2.56). In fact in this limit the kinetic term $\partial_\mu \zeta \partial^\mu \zeta$ remains but with no ζ^2 mass term. Instead, in the gauge case of Eq. (2.56), after changing variables in the lagrangian, the field $\zeta(x)$ completely disappears (not even the kinetic term remains), whilst the mass term $e^2v^2A_\mu^2$ for A_μ is now present: the gauge boson mass is $M = \sqrt{2}ev$. The field h describes the massive Higgs particle. Leaving a constant term aside, the last term in Eq. (2.59) is given by:

$$\mathcal{L}(h) = \frac{1}{2}\partial_\mu h \partial^\mu h - h^2\mu^2 + \ldots. \tag{2.60}$$

where the dots stand for cubic and quartic terms in h. We see that the h mass term has the "right" sign, due to the combination of the quadratic terms in h that, after the shift, arise from the quadratic and quartic terms in ϕ. The h mass is given by $m_h^2 = 2\mu^2$.

The Higgs mechanism is realized in well-known physical situations. It was actually discovered in condensed matter physics by Anderson [19]. For a superconductor in the Landau–Ginzburg approximation the free energy can be written as

$$F = F_0 + \frac{1}{2}\mathbf{B}^2 + |(\nabla - 2ie\mathbf{A})\phi|^2/4m - \alpha|\phi|^2 + \beta|\phi|^4. \tag{2.61}$$

Here $\mathbf{B}$ is the magnetic field, $|\phi|^2$ is the Cooper pair (e^-e^-) density, $2e$ and $2m$ are the charge and mass of the Cooper pair. The 'wrong' sign of α leads to $\phi \neq 0$ at the minimum. This is precisely the non-relativistic analogue of the Higgs model of the previous example. The Higgs mechanism implies the absence of propagation of massless phonons (states with dispersion relation $\omega = kv$ with constant v). Also the mass term for **A** is manifested by the exponential decrease of **B** inside the superconductor (Meissner effect).

2.8 Quantization of Spontaneously Broken Gauge Theories: R_ξ Gauges

We have discussed in Sect. 2.6 the problems arising in the quantization of a gauge theory and in the formulation of the correct Feynman rules (gauge fixing terms, ghosts etc.). Here we give a concise account of the corresponding results for spontaneously broken gauge theories. In particular we describe the R_ξ gauge formalism [13, 17, 20]: in this formalism the interplay of transverse and longitudinal gauge boson degrees of freedom is made explicit and their combination leads to the cancellation from physical quantities of the gauge parameter ξ. We work out in detail an abelian example that later will be easy to generalize to the non abelian case.

We restart from the abelian model of Eq. (2.56) (with $Q = -1$). In the treatment presented there the would be Goldstone boson $\zeta(x)$ was completely eliminated from the lagrangian by a non linear field transformation formally identical to a gauge transformation corresponding to the $U(1)$ symmetry of the lagrangian. In that description, in the new variables we eventually obtain a theory with only physical fields: a massive gauge boson A_μ with mass $M = \sqrt{2}ev$ and a Higgs particle h with mass $m_h = \sqrt{2}\mu$. This is called a "unitary" gauge, because only physical fields appear. But for a massive gauge boson the propagator:

$$i D_{\mu\nu}(k) = -i\frac{g_{\mu\nu} - k_\mu k_\nu/M^2}{k^2 - M^2 + i\epsilon} , \tag{2.62}$$

has a bad ultraviolet behaviour due to the second term in the numerator. This choice does not prove to be the most convenient for a discussion of the ultraviolet behaviour of the theory. Alternatively one can go to an alternative formulation where the would be Goldstone boson remains in the lagrangian but the complication of keeping spurious degrees of freedom is compensated by having all propagators with good ultraviolet behaviour ("renormalizable" gauges). To this end we replace the non linear transformation for ϕ in Eq. (2.58) with its linear equivalent (after all perturbation theory deals with the small oscillations around the minimum):

$$\phi(x) \rightarrow [v + h(x)/\sqrt{2}] \exp[-i\zeta(x)/v\sqrt{2}] \sim [v + h(x)/\sqrt{2} - i\zeta(x)/\sqrt{2}] . \tag{2.63}$$

Here we have only applied a shift by the amount v and separated the real and imaginary components of the resulting field with vanishing vacuum expectation value. If we leave A_μ as it is and simply replace the linearized expression for ϕ, we obtain the following quadratic terms (those important for propagators):

$$\mathcal{L}_{\text{quad}} = -\frac{1}{4} \sum_A F^A_{\mu\nu} F^{A\mu\nu} + \frac{1}{2} M^2 A_\mu A^\mu +$$

$$+ \frac{1}{2}(\partial_\mu \zeta)^2 + M A_\mu \partial^\mu \zeta + \frac{1}{2}(\partial_\mu h)^2 - h^2 \mu^2 \tag{2.64}$$

The mixing term between A_μ and $\partial_\mu \zeta$ does not allow to directly write diagonal mass matrices. But this mixing term can be eliminated by an appropriate modification of the covariant gauge fixing term given in Eq. (2.35) for the unbroken theory. We now take:

$$\Delta \mathcal{L}_{GF} = -\frac{1}{2\xi}(\partial^\mu A_\mu - \xi M \zeta)^2 . \tag{2.65}$$

By adding $\Delta \mathcal{L}_{GF}$ to the quadratic terms in Eq. (2.64) the mixing term cancels (apart from a total derivative that can be omitted) and we have:

$$\mathcal{L}_{\text{quad}} = -\frac{1}{4} \sum_A F^A_{\mu\nu} F^{A\mu\nu} + \frac{1}{2} M^2 A_\mu A^\mu - \frac{1}{2\xi}(\partial^\mu A_\mu)^2 +$$

$$+ \frac{1}{2}(\partial_\mu \zeta)^2 - \frac{\xi}{2} M^2 \zeta^2 + \frac{1}{2}(\partial_\mu h)^2 - h^2 \mu^2 \tag{2.66}$$

We see that the ζ field appears with a mass $\sqrt{\xi} M$ and its propagator is:

$$i D_\zeta = \frac{i}{k^2 - \xi M^2 + i\epsilon}. \tag{2.67}$$

The propagators of the Higgs field h and of gauge field A_μ are:

$$i D_h = \frac{i}{k^2 - 2\mu^2 + i\epsilon}, \tag{2.68}$$

$$i D_{\mu\nu}(k) = \frac{-i}{k^2 - M^2 + i\epsilon}\left(g_{\mu\nu} - (1-\xi)\frac{k_\mu k_\nu}{k^2 - \xi M^2}\right). \tag{2.69}$$

As anticipated, all propagators have a good behaviour at large k^2. This class of gauges are called "R_ξ gauges" [20]. Note that for $\xi = 1$ we have a sort of generalization of the Feynman gauge with a Goldstone of mass M and a gauge propagator:

$$i D_{\mu\nu}(k) = \frac{-i g_{\mu\nu}}{k^2 - M^2 + i\epsilon}. \tag{2.70}$$

Also for $\xi \to \infty$ the unitary gauge description is recovered in that the Goldstone propagator vanishes and the gauge propagator reproduces that of the unitary gauge in Eq. (2.62). All ξ dependence, including the unphysical singularities of the ζ and A_μ propagators at $k^2 = \xi M^2$, present in individual Feynman diagrams, must cancel in the sum of all contributions to any physical quantity.

An additional complication is that a Faddeev-Popov ghost is also present in R_ξ gauges (while it is absent in an unbroken abelian gauge theory). In fact under an infinitesimal gauge transformation with parameter $\theta(x)$:

$$A_\mu \to A_\mu - \partial_\mu \theta$$
$$\phi \to (1 - ie\theta)[v + h(x)/\sqrt{2} - i\zeta(x)/\sqrt{2}] , \qquad (2.71)$$

so that:

$$\delta A_\mu = -\partial_\mu \theta, \quad \delta h = -e\zeta\theta, \quad \delta\zeta = e\theta\sqrt{2}(v + h/\sqrt{2}) . \qquad (2.72)$$

The gauge fixing condition $\partial_\mu A^\mu - \xi M \zeta = 0$ undergoes the variation:

$$\partial_\mu A^\mu - \xi M \zeta \to \partial_\mu A^\mu - \xi M \zeta - [\partial^2 + \xi M^2 (1 + h/v\sqrt{2})]\theta , \qquad (2.73)$$

where we used $M = \sqrt{2}ev$. From this, recalling the discussion in Sect. 2.6, we see that the ghost is not coupled to the gauge boson (as usual for an abelian gauge theory) but has a coupling to the Higgs field h. The ghost lagrangian is:

$$\Delta\mathcal{L}_{Ghost} = \bar{\eta}[\partial^2 + \xi M^2 (1 + h/v\sqrt{2})]\eta . \qquad (2.74)$$

The ghost mass is seen to be $m_{gh} = \sqrt{\xi}M$ and its propagator is:

$$iD_{gh} = \frac{i}{k^2 - \xi M^2 + i\epsilon}. \qquad (2.75)$$

The detailed Feynman rules follow from all the basic vertices involving the gauge boson, the Higgs, the would be Goldstone boson and the ghost and can be easily derived, with some algebra, from the total lagrangian including the gauge fixing and ghost additions. The generalization to the non abelian case is in principle straightforward, with some formal complications involving the projectors over the space of the would be Goldstone bosons and over the orthogonal space of the Higgs particles. But for each gauge boson that takes mass M_a we still have a corresponding would be Goldstone boson and a ghost with mass $\sqrt{\xi}M_a$. The Feynman diagrams, both for the abelian and the non abelian case, are listed explicitly, for example, in the Cheng and Li textbook in ref.[17].

We conclude that the renormalizability of non abelian gauge theories, also in presence of spontaneous symmetry breaking, was proven in the fundamental works of t'Hooft and Veltman [21] and discussed in detail in [22].

References

1. H. Fritzsch, M. Gell-Mann and H. Leutwyler, Phys. Lett. B **47**, 365 (1973)
2. D. Gross and F. Wilczek, Phys. Rev. Lett. **30**, 1343 (1973), Phys. Rev. D **8**, 3633 (1973); H.D. Politzer, Phys. Rev. Lett. **30**, 1346 (1973)
3. S. Weinberg, Phys. Rev. Lett. **31**, 494 (1973)
4. S.L. Glashow, Nucl. Phys. **22**, 579 (1961)
5. S. Weinberg, Phys. Rev. Lett. **19**, 1264 (1967)
6. A. Salam, in *Elementary Particle Theory*, ed. N. Svartholm (Almquist and Wiksells, Stockholm, 1969), p. 367
7. Particle Data Group, J. Phys. G **33**, 1 (2006)
8. N. Cabibbo, Phys. Rev. Lett. **10**, 531 (1963)
9. S.L. Glashow, J. Iliopoulos and L. Maiani, Phys. Rev. **96**, 1285 (1970)
10. M. Kobayashi and T. Maskawa, Prog. Theor. Phys. **49**, 652 (1973)
11. C.N. Yang and R. Mills, Phys. Rev. **96**, 191 (1954)
12. R. Feynman, Acta Phys. Pol. **24**, 697 (1963); B. De Witt, Phys. Rev. **162**, 1195, 1239 (1967); L.D. Faddeev and V.N. Popov, Phys. Lett. B **25**, 29 (1967)
13. E.S. Abers and B.W. Lee, Phys. Rep. **9**, 1 (1973)
14. J.D. Bjorken and S. Drell, *Relativistic Quantum Mechanics/Fields*, Vols. I, II, McGraw-Hill, New York, (1965)
15. G.'t Hooft and M. Veltman, Nucl. Phys. B **44**, 189 (1972); C.G. Bollini and J.J. Giambiagi, Phys. Lett. B **40**, 566 (1972); J.F. Ashmore, Nuovo Cim. Lett. **4**, 289 (1972); G.M. Cicuta and E. Montaldi, Nuovo Cim. Lett. **4**, 329 (1972)
16. J. Goldstone, Nuovo Cim. **19**, 154 (1961)
17. C. Itzykson and J. Zuber, *Introduction to Quantum Field Theory*, McGraw-Hill, New York, (1980); T.P. Cheng and L.F. Li, *Gauge Theory of Elementary Particle Physics*, Oxford Univ. Press, New York (1984); M.E. Peskin and D.V. Schroeder, *An Introduction to Quantum Field Theory*, Perseus Books, Cambridge, Mass. (1995); S. Weinberg, *The Quantum Theory of Fields*, Vols. I, II, Cambridge Univ. Press, Cambridge, Mass. (1996); A. Zee, *Quantum Field Theory in a Nutshell*, Princeton Univ. Press, Princeton, N.J. (2003); C.M. Becchi and G. Ridolfi, *An Introduction to Relativistic Processes and the Standard Model of Electroweak Interactions*, Springer, (2006)
18. F. Englert, R. Brout, Phys. Rev. Lett. **13**, 321 (1964); P.W. Higgs, Phys. Lett. **12**, 132 (1964)
19. P.W. Anderson, Phys. Rev. **112**, 1900 (1958); Phys. Rev. **130**, 439 (1963)
20. K. Fujikawa, B.W. Lee and A. Sanda, Phys. Rev. D **6**, 2923 (1972); Y.P. Yao, Phys. Rev. D **7**, 1647 (1973)
21. M. Veltman, Nucl. Phys. B **21**, 288 (1970); G.'t Hooft, Nucl. Phys. B **33**, 173 (1971); **35**, 167 (1971)
22. B.W. Lee and J. Zinn-Justin, Phys. Rev. D **5**, 3121; 3137 (1972); **7**, 1049 (1973)
23. I. J. R. Aitchison and A. J. G. Hey, "Gauge theories in particle physics: A practical introduction. Vol. 1: From relativistic quantum mechanics to QED,"
24. M. Maggiore, "A Modern introduction to quantum field theory," Oxford University Press, 2005. (Oxford Series in Physics, 12. ISBN 0 19 852073 5)
25. M. Tanabashi *et al.* [Particle Data Group], Phys. Rev. D **98** (2018) no.3, 030001.
26. http://pdg.web.cern.ch/pdg/
27. G. Hamel de Monchenault, "Electroweak measurements at the LHC," Ann. Rev. Nucl. Part. Sci. **67** (2017) 19.
28. G. Aad *et al.* [ATLAS Collaboration], Phys. Lett. B **716** (2012) 1 [arXiv:1207.7214 [hep-ex]].
29. S. Chatrchyan *et al.* [CMS Collaboration], Phys. Lett. B **716** (2012) 30 [arXiv:1207.7235 [hep-ex]].
30. T. Plehn, "Lectures on LHC Physics," Lect. Notes Phys. **886** (2015).
31. https://twiki.cern.ch/twiki/bin/view/LHCPhysics/LHCHXSWG

32. J. R. Espinosa, "Cosmological implications of Higgs near-criticality," Phil. Trans. Roy. Soc. Lond. A **376** (2018) no.2114, 20170118.
33. G. F. Giudice, "The Dawn of the Post-Naturalness Era," arXiv:1710.07663 [physics.hist-ph].
34. H. E. Haber and L. Stephenson Haskins, "Supersymmetric Theory and Models," arXiv:1712.05926 [hep-ph].
35. Csáki, Csaba, S. Lombardo and O. Telem, "TASI Lectures on Non-supersymmetric BSM Models," arXiv:1811.04279 [hep-ph].
36. Ansgar Denner and Stefan Dittmaier, "Electroweak Radiative Corrections for Collider Physics" arXiv:1912.06823[hep-ph].
37. Freeman Dyson, "Leaping into the Grand Unknown", New York Rev. Books **56** (2009)
38. J. M. Henn and J. C. Plefka, "Scattering Amplitudes in Gauge Theories," Lect. Notes Phys. **883** (2014)
39. T. Becher, A. Broggio and A. Ferroglia, "Introduction to Soft-Collinear Effective Theory," Lect. Notes Phys. **896** (2015) [arXiv:1410.1892 [hep-ph]].
40. S. Marzani, G. Soyez and M. Spannowsky, "Looking inside jets: an introduction to jet substructure and boosted-object phenomenology," Lect. Notes Phys. **958** (2019) [arXiv:1901.10342 [hep-ph]].
41. G. P. Salam, "The strong coupling: a theoretical perspective," arXiv:1712.05165 [hep-ph].
42. S. Aoki *et al.* [Flavour Lattice Averaging Group], "FLAG Review 2019," arXiv:1902.08191 [hep-lat].

Chapter 3
The Standard Model of Electroweak Interactions

Guido Altarelli and Stefano Forte

3.1 Introduction

In this chapter,[1] we summarize the structure of the standard EW theory [1] and specify the couplings of the intermediate vector bosons $W^\pm$, Z and of the Higgs particle with the fermions and among themselves, as dictated by the gauge symmetry plus the observed matter content and the requirement of renormalizability. We discuss the realization of spontaneous symmetry breaking and of the Higgs mechanism [2]. We then review the phenomenological implications of the EW theory for collider physics (that is we leave aside the classic low energy processes that are well described by the "old" weak interaction theory (see, for example, [3])). Moreover, a detailed description of experiments for precision tests of the EW theory is presented in Chap. 6.

For this discussion we split the lagrangian into two parts by separating the terms with the Higgs field:

$$\mathcal{L} = \mathcal{L}_{\text{gauge}} + \mathcal{L}_{\text{Higgs}} . \tag{3.1}$$

Both terms are written down as prescribed by the $SU(2) \otimes U(1)$ gauge symmetry and renormalizability, but the Higgs vacuum expectation value (VEV) induces the

---

The author "G. Altarelli" is deceased at the time of publication.

[1] See Chap. 2 for a general introduction to Chap. 2–4 with updated references.

G. Altarelli
University of Rome 3, Rome, Italy

S. Forte (✉)
Dipartimento di Fisica, Università di Milano, Milano, Italy

© The Author(s) 2020
H. Schopper (ed.), *Particle Physics Reference Library*,
https://doi.org/10.1007/978-3-030-38207-0_3

spontaneous symmetry breaking responsible for the non vanishing vector boson and fermion masses.

3.2 The Gauge Sector

We start by specifying $\mathcal{L}_{\text{gauge}}$, which involves only gauge bosons and fermions, according to the general formalism of gauge theories discussed in Chap. 2:

$$\mathcal{L}_{\text{gauge}} = -\frac{1}{4}\sum_{A=1}^{3} F_{\mu\nu}^A F^{A\mu\nu} - \frac{1}{4}B_{\mu\nu}B^{\mu\nu} + \bar{\psi}_L i\gamma^\mu D_\mu \psi_L + \bar{\psi}_R i\gamma^\mu D_\mu \psi_R \ . \tag{3.2}$$

This is the Yang–Mills lagrangian for the gauge group $SU(2) \otimes U(1)$ with fermion matter fields. Here

$$B_{\mu\nu} = \partial_\mu B_\nu - \partial_\nu B_\mu \quad \text{and} \quad F_{\mu\nu}^A = \partial_\mu W_\nu^A - \partial_\nu W_\mu^A - g\epsilon_{ABC}\, W_\mu^B W_\nu^C \tag{3.3}$$

are the gauge antisymmetric tensors constructed out of the gauge field B_μ associated with $U(1)$, and W_μ^A corresponding to the three $SU(2)$ generators; ϵ_{ABC} are the group structure constants (see Eqs. (3.8, 3.9)) which, for $SU(2)$, coincide with the totally antisymmetric Levi-Civita tensor (recall the familiar angular momentum commutators). The normalization of the $SU(2)$ gauge coupling g is therefore specified by Eq. (3.3).

The fermion fields are described through their left-hand and right-hand components:

$$\psi_{L,R} = [(1 \mp \gamma_5)/2]\psi, \quad \bar{\psi}_{L,R} = \bar{\psi}[(1 \pm \gamma_5)/2]\,, \tag{3.4}$$

with γ_5 and other Dirac matrices defined as in the book by Bjorken–Drell [4]. In particular, $\gamma_5^2 = 1$, $\gamma_5^\dagger = \gamma_5$. Note that, as given in Eq. (3.4),

$$\bar{\psi}_L = \psi_L^\dagger \gamma_0 = \psi^\dagger[(1-\gamma_5)/2]\gamma_0 = \bar{\psi}\gamma_0[(1-\gamma_5)/2]\gamma_0 = \bar{\psi}[(1+\gamma_5)/2]\,.$$

The matrices $P_\pm = (1 \pm \gamma_5)/2$ are projectors. They satisfy the relations $P_\pm P_\pm = P_\pm$, $P_\pm P_\mp = 0$, $P_+ + P_- = 1$.

The sixteen linearly independent Dirac matrices can be divided into γ_5-even and γ_5-odd according to whether they commute or anticommute with γ_5. For the γ_5-even, we have

$$\bar{\psi}\Gamma_E\psi = \bar{\psi}_L\Gamma_E\psi_R + \bar{\psi}_R\Gamma_E\psi_L \qquad (\Gamma_E \equiv 1, i\gamma_5, \sigma_{\mu\nu})\,, \tag{3.5}$$

whilst for the γ_5-odd,

$$\bar{\psi}\Gamma_O\psi = \bar{\psi}_L\Gamma_O\psi_L + \bar{\psi}_R\Gamma_O\psi_R \qquad (\Gamma_O \equiv \gamma_\mu, \gamma_\mu\gamma_5) . \tag{3.6}$$

The standard EW theory is a chiral theory, in the sense that ψ_L and ψ_R behave differently under the gauge group (so that parity and charge conjugation non conservation are made possible in principle). Thus, mass terms for fermions (of the form $\bar{\psi}_L\psi_R$ + h.c.) are forbidden in the symmetric limit. In particular, in the Minimal Standard Model (MSM: i.e. the model that only includes all observed particles plus a single Higgs doublet), all ψ_L are $SU(2)$ doublets while all ψ_R are singlets. But for the moment, by $\psi_{L,R}$ we mean column vectors, including all fermion types in the theory that span generic reducible representations of $SU(2) \otimes U(1)$.

In the absence of mass terms, there are only vector and axial vector interactions in the lagrangian and those have the property of not mixing ψ_L and ψ_R. Fermion masses will be introduced, together with $W^\pm$ and Z masses, by the mechanism of symmetry breaking. The covariant derivatives $D_\mu\psi_{L,R}$ are explicitly given by

$$D_\mu\psi_{L,R} = \left[\partial_\mu + ig\sum_{A=1}^{3} t^A_{L,R}W^A_\mu + ig'\frac{1}{2}Y_{L,R}B_\mu\right]\psi_{L,R} , \tag{3.7}$$

where $t^A_{L,R}$ and $1/2Y_{L,R}$ are the $SU(2)$ and $U(1)$ generators, respectively, in the reducible representations $\psi_{L,R}$. The commutation relations of the $SU(2)$ generators are given by

$$[t^A_L, t^B_L] = i\,\epsilon_{ABC}t^C_L \quad\text{and}\quad [t^A_R, t^B_R] = i\epsilon_{ABC}t^C_R . \tag{3.8}$$

We use the normalization (3.8) [in the fundamental representation of $SU(2)$]. The electric charge generator Q (in units of e, the positron charge) is given by

$$Q = t^3_L + 1/2\,Y_L = t^3_R + 1/2\,Y_R . \tag{3.9}$$

Note that the normalization of the $U(1)$ gauge coupling g' in (3.7) is now specified as a consequence of (3.9). Note that $t^i_R\psi_R = 0$, given that, for all known quark and leptons, ψ_R is a singlet. But in the following, we keep $t^i_R\psi_R$ for generality, in case 1 day a non singlet right-handed fermion is discovered.

3.3 Couplings of Gauge Bosons to Fermions

All fermion couplings of the gauge bosons can be derived directly from Eqs. (3.2) and (3.7). The charged W_μ fields are described by $W^{1,2}_\mu$, while the photon A_μ and weak neutral gauge boson Z_μ are obtained from combinations of W^3_μ and B_μ. The charged-current (CC) couplings are the simplest. One starts from the $W_\mu^{1,2}$ terms in Eqs. (3.2) and (3.7) which can be written as:

$$g(t^1 W_\mu^1 + t^2 W_\mu^2) = g \left\{ [(t^1 + it^2)/\sqrt{2}](W_\mu^1 - iW_\mu^2)/\sqrt{2}] + \text{h.c.} \right\}$$

$$= g \left\{ [(t^+ W_\mu^-)/\sqrt{2}] + \text{h.c.} \right\}, \tag{3.10}$$

where $t^\pm = t^1 \pm it^2$ and $W^\pm = (W^1 \pm iW^2)/\sqrt{2}$. By applying this generic relation to L and R fermions separately, we obtain the vertex

$$V_{\bar\psi\psi W} = g\bar\psi\gamma_\mu \left[(t_L^+/\sqrt{2})(1 - \gamma_5)/2 + (t_R^+/\sqrt{2})(1 + \gamma_5)/2 \right] \psi W_\mu^- + \text{h.c.} \tag{3.11}$$

Given that $t_R = 0$ for all fermions in the SM, the charged current is pure $V - A$. In the neutral-current (NC) sector, the photon A_μ and the mediator Z_μ of the weak NC are orthogonal and normalized linear combinations of B_μ and W_μ^3:

$$A_\mu = \cos\theta_W B_\mu + \sin\theta_W W_\mu^3,$$

$$Z_\mu = -\sin\theta_W B_\mu + \cos\theta_W W_\mu^3. \tag{3.12}$$

and conversely:

$$W_\mu^3 = \sin\theta_W A_\mu + \cos\theta_W Z_\mu,$$

$$B_\mu = \cos\theta_W A_\mu - \sin\theta_W Z_\mu. \tag{3.13}$$

Equations (3.12) define the weak mixing angle θ_W. We can rewrite the W_μ^3 and B_μ terms in Eqs. (3.2) and (3.7) as follows:

$$gt^3 W_\mu^3 + g'Y/2 B_\mu = [gt^3 \sin\theta_W + g'(Q - t^3)\cos\theta_W]A_\mu +$$

$$+ [gt^3 \cos\theta_W - g'(Q - t^3)\sin\theta_W]Z_\mu, \tag{3.14}$$

where Eq. (3.9) for the charge matrix Q was also used. The photon is characterized by equal couplings to left and right fermions with a strength equal to the electric charge. Thus we immediately obtain

$$g \sin\theta_W = g' \cos\theta_W = e, \tag{3.15}$$

or equivalently,

$$\text{tg } \theta_W = g'/g \tag{3.16}$$

Once θ_W has been fixed by the photon couplings, it is a simple matter of algebra to derive the Z couplings, with the result

$$V_{\bar\psi\psi Z} = \frac{g}{2\cos\theta_W}\bar\psi\gamma_\mu[t_L^3(1-\gamma_5)+t_R^3(1+\gamma_5)-2Q\sin^2\theta_W]\psi Z^\mu\,, \qquad (3.17)$$

where $V_{\bar\psi\psi Z}$ is a notation for the vertex. Once again, recall that in the MSM, $t_R^3 = 0$ and $t_L^3 = \pm1/2$.

In order to derive the effective four-fermion interactions that are equivalent, at low energies, to the CC and NC couplings given in Eqs. (3.11) and (3.17), we anticipate that large masses, as experimentally observed, are provided for $W^\pm$ and Z by $\mathcal{L}_{\text{Higgs}}$. For left–left CC couplings, when the momentum transfer squared can be neglected, with respect to m_W^2, in the propagator of Born diagrams with single W exchange (see, for example, the diagram for μ decay in Fig. 3.1, from Eq. (3.11) we can write

$$\mathcal{L}_{\text{eff}}^{\text{CC}} \simeq \frac{g^2}{8m_W^2}[\bar\psi\gamma_\mu(1-\gamma_5)t_L^+\psi][\bar\psi\gamma^\mu(1-\gamma_5)t_L^-\psi]\,. \qquad (3.18)$$

By specializing further in the case of doublet fields such as $\nu_e - e^-$ or $\nu_\mu - \mu^-$, we obtain the tree-level relation of g with the Fermi coupling constant G_F precisely measured from μ decay (see Chap. 2, Eqs. (2), (3)):

$$G_F/\sqrt{2} = g^2/8m_W^2\,. \qquad (3.19)$$

By recalling that $g\sin\theta_W = e$, we can also cast this relation in the form

$$m_W = \mu_{\text{Born}}/\sin\theta_W\,, \qquad (3.20)$$

with

$$\mu_{\text{Born}} = (\pi\alpha/\sqrt{2}G_F)^{1/2} \simeq 37.2802\,\text{GeV}\,, \qquad (3.21)$$

where α is the fine-structure constant of QED ($\alpha \equiv e^2/4\pi = 1/137.036$).

In the same way, for neutral currents we obtain in Born approximation from Eq. (3.17) the effective four-fermion interaction given by

$$\mathcal{L}_{\text{eff}}^{\text{NC}} \simeq \sqrt{2}\,G_F\rho_0\bar\psi\gamma_\mu[\ldots]\psi\bar\psi\gamma^\mu[\ldots]\psi\,, \qquad (3.22)$$

Fig. 3.1 The Born diagram for μ decay

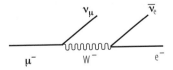

where

$$[\ldots] \equiv t_L^3(1 - \gamma_5) + t_R^3(1 + \gamma_5) - 2Q \sin^2 \theta_W \tag{3.23}$$

and

$$\rho_0 = \frac{m_W^2}{m_Z^2 \cos^2 \theta_W}. \tag{3.24}$$

All couplings given in this section are obtained at tree level and are modified in higher orders of perturbation theory. In particular, the relations between m_W and $\sin \theta_W$ (Eqs. (3.20) and (3.21)) and the observed values of ρ ($\rho = \rho_0$ at tree level) in different NC processes, are altered by computable EW radiative corrections, as discussed in Sect. (3.11).

The partial width $\Gamma(W \to \bar{f} f')$ is given in Born approximation by the simplest diagram in Fig. 3.2 and one readily obtains from Eq. (3.11) with $t_R = 0$, in the limit of neglecting the fermion masses and summing over all possible f' for a given f:

$$\Gamma(W \to \bar{f} f') = N_C \frac{G_F m_W^3}{6\pi \sqrt{2}} = N_C \frac{\alpha m_W}{12 \sin^2 \theta_W}, \tag{3.25}$$

where $N_C = 3$ or 1 is the number of colours for quarks or leptons, respectively, and the relations Eqs. (3.15, 3.19) have been used. Here and in the following expressions for the Z widths the one loop QCD corrections for the quark channels can be absorbed in a redefinition of N_C: $N_C \to 3[1 + \alpha_s(m_Z)/\pi + \ldots]$. Note that the widths are particularly large because the rate already occurs at order g^2 or G_F. The experimental values of the W total width and the leptonic branching ratio (the average of e, μ and τ modes) are [5, 8] (see Chap. 6):

$$\Gamma_W = 2.147 \pm 0.060 \, \text{GeV}, \qquad B(W \to l\nu_l) = 10.80 \pm 0.09. \tag{3.26}$$

The branching ratio B is in very good agreement with the simple approximate formula, derived from Eq. (3.25):

$$B(W \to l\nu_l) \sim \frac{1}{2 \cdot 3 \cdot (1 + \alpha_s(m_Z^2)/\pi) + 3} \sim 10.8\%. \tag{3.27}$$

Fig. 3.2 Diagrams for (**a**) the W and (**b**) the Z widths in Born approximation

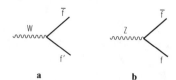

The denominator corresponds to the sum of the final states $d'\bar{u}$, $s'\bar{c}$, $e^-\bar{\nu}_e$, $\mu^-\bar{\nu}_\mu$, $\tau^-\bar{\nu}_\tau$ (for the definition of d' and s' see Eq. (3.66)).

For $t_R = 0$ the Z coupling to fermions in Eq. (3.17) can be cast into the form:

$$V_{\bar{\psi}_f \psi_f Z} = \frac{g}{2 \cos\theta_W} \bar{\psi}_f \gamma_\mu [g_V^f - g_A^f \gamma_5] \psi_f Z^\mu , \tag{3.28}$$

with:

$$g_A^f = t_L^{3f} , \quad g_V^f / g_A^f = 1 - 4|Q_f| \sin^2\theta_W . \tag{3.29}$$

and $t_L^{3f} = \pm 1/2$ for up-type or down-type fermions. In terms of $g_{A,V}$ given in Eqs. (3.29) (the widths are proportional to $(g_V^2 + g_A^2)$), the partial width $\Gamma(Z \to \bar{f}f)$ in Born approximation (see the diagram in Fig. 3.2), for negligible fermion masses, is given by:

$$\Gamma(Z \to \bar{f}f) = N_C \frac{\alpha m_Z}{12 \sin^2 2\theta_W}[1 + (1 - 4|Q_f|\sin^2\theta_W)^2]$$

$$= N_C \rho_0 \frac{G_F m_Z^3}{24\pi\sqrt{2}}[1 + (1 - 4|Q_f|\sin^2\theta_W)^2]. \tag{3.30}$$

where $\rho_0 = m_W^2/m_Z^2 \cos^2\theta_W$ is given in Eq. (3.55). The experimental values of the Z total width and of the partial rates into charged leptons (average of e, μ and τ), into hadrons and into invisible channels are [5, 8] (see Chap. 6):

$$\Gamma_Z = 2.4952 \pm 0.0023 \, \text{GeV},$$

$$\Gamma_{l^+l^-} = 83.985 \pm 0.086 \, \text{MeV},$$

$$\Gamma_h = 1744.4 \pm 2.0 \, \text{MeV},$$

$$\Gamma_{inv} = 499.0 \pm 1.5 \, \text{MeV}. \tag{3.31}$$

The measured value of the Z invisible width, taking radiative corrections into account, leads to the determination of the number of light active neutrinos (see Chap. 6):

$$N_\nu = 2.9841 \pm 0.0083, \tag{3.32}$$

well compatible with the three known neutrinos ν_e, ν_μ and ν_τ; hence there exist only the three known sequential generations of fermions (with light neutrinos), a result with important consequences also in astrophysics and cosmology.

At the Z peak, besides total cross sections, various types of asymmetries have been measured. The results of all asymmetry measurements are quoted in terms of the asymmetry parameter A_f, defined in terms of the effective coupling constants,

g_V^f and g_A^f, as:

$$A_f = 2\frac{g_V^f g_A^f}{g_V^{f2} + g_A^{f2}} = 2\frac{g_V^f/g_A^f}{1 + (g_V^f/g_A^f)^2}, \qquad A_{FB}^f = \frac{3}{4}A_e A_f. \quad (3.33)$$

The measurements are: the forward-backward asymmetry ($A_{FB}^f = (3/4)A_e A_f$), the tau polarization (A_τ) and its forward backward asymmetry (A_e) measured at LEP, as well as the left-right and left-right forward-backward asymmetry measured at SLC (A_e and A_f, respectively). Hence the set of partial width and asymmetry results allows the extraction of the effective coupling constants: widths measure ($g_V^2 + g_A^2$) and asymmetries measure g_V/g_A.

The top quark is heavy enough that it can decay into a real bW pair, which is by far its dominant decay channel. The next mode, $t \to sW$, is suppressed in rate by a factor $|V_{ts}|^2 \sim 1.7 \cdot 10^{-3}$, see Eqs. (3.71–3.73). The associated width, neglecting m_b effects but including 1-loop QCD corrections in the limit $m_W = 0$, is given by (we have omitted a factor $|V_{tb}|^2$ that we set equal to 1):

$$\Gamma(t \to bW^+) = \frac{G_F m_t^3}{8\pi\sqrt{2}}(1 - \frac{m_W^2}{m_t^2})^2(1 + 2\frac{m_W^2}{m_t^2})[1 - \frac{\alpha_s(m_Z)}{3\pi}(\frac{2\pi^2}{3} - \frac{5}{2}) + \ldots].$$
$$(3.34)$$

The top quark lifetime is so short, about $0.5 \cdot 10^{-24}$ s, that it decays before hadronizing or forming toponium bound states.

3.4 Gauge Boson Self-interactions

The gauge boson self-interactions can be derived from the $F_{\mu\nu}$ term in $\mathcal{L}_{\text{gauge}}$, by using Eq. (3.12) and $W^\pm = (W^1 \pm iW^2)/\sqrt{2}$.

Defining the three-gauge-boson vertex as in Fig. 3.3 (with all incoming lines), we obtain ($V \equiv \gamma, Z$)

$$V_{W^-W^+V} = ig_{W^-W^+V}[g_{\mu\nu}(p - q)_\lambda + g_{\mu\lambda}(r - p)_\nu + g_{\nu\lambda}(q - r)_\mu], \quad (3.35)$$

with

$$g_{W^-W^+\gamma} = g\,\sin\theta_W = e \quad \text{and} \quad g_{W^-W^+Z} = g\,\cos\theta_W. \quad (3.36)$$

Note that the photon coupling to the W is fixed by the electric charge, as imposed by QED gauge invariance. The ZWW coupling is larger by a $\tan\theta_W$ factor. This form of the triple gauge vertex is very special: in general, there could be departures from the above SM expression, even restricting us to Lorentz invariant, em gauge

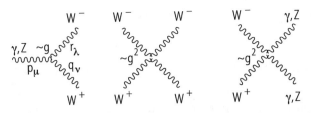

Fig. 3.3 The three- and four-gauge boson vertices. The cubic coupling is of order g, while the quartic one is of order g^2

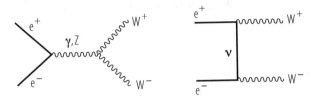

Fig. 3.4 The three- and four-gauge boson vertices. The cubic coupling is of order g, while the quartic one is of order g^2

symmetric and C and P conserving couplings. In fact some small corrections are already induced by the radiative corrections. But, in principle, more important could be the modifications induced by some new physics effect. The experimental testing of the triple gauge vertices has been done mainly at LEP2 and at the Tevatron. At LEP2 the crosssection and angular distributions for the process $e^+e^- \rightarrow W^+W^-$ have been studied (see Chap. 6).

In Born approximation the Feynman diagrams for the LEP2 process are shown in Fig. 3.4 [6]. Besides neutrino exchange which only involves the well established charged current vertex, the triple weak gauge vertices $V_{W^-W^+V}$ appear in the γ and Z exchange diagrams. The Higgs exchange is negligible because the electron mass is very small. The analytic cross section formula in Born approximation can be found, for example, in Ref. [5]. The experimental data are compared with the SM prediction in Chap. 6 [7]. The agreement is very good. Note that the sum of all three exchange amplitudes has a better high energy behaviour. This is due to cancellations among the amplitudes implied by gauge invariance, connected to the fact that the theory is renormalizable (the crosssection can be seen as a contribution to the imaginary part of the $e^+e^- \rightarrow e^+e^-$ amplitude).

The quartic gauge coupling is proportional to $g^2 \epsilon_{ABC} W^B W^C \epsilon_{ADE} W^D W^E$. Thus in the term with $A = 3$ we have four charged W's. For $A = 1$ or two we have two charged W's and 2 W_3's, each W_3 being a combination of γ and Z according to Eq. (3.13). With a little algebra the quartic vertex can be cast in the form:

$$V_{WWVV} = igwwvv[2g_{\mu\nu}g_{\lambda\rho} - g_{\mu\lambda}g_{\nu\rho} - g_{\mu\rho}g_{\nu\lambda}], \qquad (3.37)$$

where, μ and ν refer to W^+W^+ in the $4W$ vertex and to VV in the $WWVV$ case and:

$$g_{WWWW} = g^2, \quad g_{WW\gamma\gamma} = -e^2, \quad g_{WW\gamma Z} = -eg\cos\theta_W, \quad g_{WWZZ} = -g^2\cos^2\theta_W .$$
(3.38)

In order to obtain these result for the vertex the reader must duly take into account the factor of $-1/4$ in front of $F_{\mu\nu}^2$ in the lagrangian and the statistical factors which are equal to two for each pair of identical particles (like W^+W^+ or $\gamma\gamma$, for example). The quartic coupling, being quadratic in g, hence small, could not be directly tested so far.

3.5 The Higgs Sector

We now turn to the Higgs sector of the EW lagrangian. The Higgs lagrangian is specified by the gauge principle and the requirement of renormalizability to be

$$\mathcal{L}_{\text{Higgs}} = (D_\mu\phi)^\dagger(D^\mu\phi) - V(\phi^\dagger\phi) - \bar{\psi}_L\Gamma\psi_R\phi - \bar{\psi}_R\Gamma^\dagger\psi_L\phi^\dagger ,$$
(3.39)

where ϕ is a column vector including all Higgs fields; it transforms as a reducible representation of the gauge group. The quantities Γ (which include all coupling constants) are matrices that make the Yukawa couplings invariant under the Lorentz and gauge groups. Without loss of generality, here and in the following, we take Γ to be γ_5-free. The potential $V(\phi^\dagger\phi)$, symmetric under $SU(2)\otimes U(1)$, contains, at most, quartic terms in ϕ so that the theory is renormalizable:

$$V(\phi^\dagger\phi) = -\mu^2\phi^\dagger\phi + \frac{1}{2}\lambda(\phi^\dagger\phi)^2$$
(3.40)

As discussed in Chap. 2, spontaneous symmetry breaking is induced if the minimum of V, which is the classical analogue of the quantum mechanical vacuum state (both are the states of minimum energy), is obtained for non-vanishing ϕ values. Precisely, we denote the vacuum expectation value (VEV) of ϕ, i.e. the position of the minimum, by v (which is a doublet):

$$\langle 0|\phi(x)|0\rangle = v = \begin{pmatrix} 0 \\ v \end{pmatrix} \neq 0 .$$
(3.41)

The reader should be careful that the same symbol is used for the doublet and the only non zero component of the same doublet. The fermion mass matrix is obtained from the Yukawa couplings by replacing $\phi(x)$ by v:

$$M = \bar{\psi}_L \mathcal{M}\psi_R + \bar{\psi}_R\mathcal{M}^\dagger\psi_L ,$$
(3.42)

with

$$\mathcal{M} = \Gamma \cdot v \,. \tag{3.43}$$

In the MSM, where all left fermions ψ_L are doublets and all right fermions ψ_R are singlets, only Higgs doublets can contribute to fermion masses. There are enough free couplings in Γ, so that one single complex Higgs doublet is indeed sufficient to generate the most general fermion mass matrix. It is important to observe that by a suitable change of basis we can always make the matrix $\mathcal{M}$ Hermitian and diagonal. In fact, we can make separate unitary transformations on ψ_L and ψ_R according to

$$\psi'_L = U\psi_L, \quad \psi'_R = W\psi_R \tag{3.44}$$

and consequently

$$\mathcal{M} \to \mathcal{M}' = U^\dagger \mathcal{M} W \,. \tag{3.45}$$

This transformation does not alter the structure of the fermion couplings in $\mathcal{L}_{\text{symm}}$ (because both the kinetic terms and the couplings to gauge bosons do not mix L and R spinors) except that it leads to the phenomenon of mixing, as we shall see in Sect. (3.6).

If only one Higgs doublet is present, the change of basis that makes $\mathcal{M}$ diagonal will at the same time diagonalize the fermion–Higgs Yukawa couplings. Thus, in this case, no flavour-changing neutral Higgs vertices are present. This is not true, in general, when there are several Higgs doublets. But one Higgs doublet for each electric charge sector i.e. one doublet coupled only to u-type quarks, one doublet to d-type quarks, one doublet to charged leptons (and possibly one for neutrino Dirac masses) would also be all right, because the mass matrices of fermions with different charges are diagonalized separately. For several Higgs doublets in a given charge sector it is also possible to generate CP violation by complex phases in the Higgs couplings. In the presence of six quark flavours, this CP-violation mechanism is not necessary. In fact, at the moment, the simplest model with only one Higgs doublet seems adequate for describing all observed phenomena.

We now consider the gauge-boson masses and their couplings to the Higgs. These effects are induced by the $(D_\mu \phi)^\dagger (D^\mu \phi)$ term in $\mathcal{L}_{\text{Higgs}}$ (Eq. (3.39)), where

$$D_\mu \phi = \left[\partial_\mu + ig \sum_{A=1}^{3} t^A W_\mu^A + ig'(Y/2) B_\mu \right] \phi \,. \tag{3.46}$$

Here t^A and $Y/2$ are the $SU(2) \otimes U(1)$ generators in the reducible representation spanned by ϕ. Not only doublets but all non-singlet Higgs representations can contribute to gauge-boson masses. The condition that the photon remains massless is equivalent to the condition that the vacuum is electrically neutral:

$$Q|v\rangle = (t^3 + \frac{1}{2}Y)|v\rangle = 0 .$$ (3.47)

We now explicitlly consider the case of a single Higgs doublet:

$$\phi = \begin{pmatrix} \phi^+ \\ \phi^0 \end{pmatrix}, \quad v = \begin{pmatrix} 0 \\ v \end{pmatrix} ,$$ (3.48)

The charged W mass is given by the quadratic terms in the W field arising from $\mathcal{L}_{\text{Higgs}}$, when $\phi(x)$ is replaced by v in Eq. (3.41). By recalling Eq. (3.10), we obtain

$$m_W^2 W_\mu^+ W^{-\mu} = g^2 |(t^+ v/\sqrt{2})|^2 W_\mu^+ W^{-\mu} ,$$ (3.49)

whilst for the Z mass we get [recalling Eqs. (3.12–3.14)]

$$\frac{1}{2} m_Z^2 Z_\mu Z^\mu = |[g \cos\theta_W t^3 - g' \sin\theta_W (Y/2)]v|^2 Z_\mu Z^\mu ,$$ (3.50)

where the factor of 1/2 on the left-hand side is the correct normalization for the definition of the mass of a neutral field. By using Eq. (3.47), relating the action of t^3 and $Y/2$ on the vacuum v, and Eqs. (3.16), we obtain

$$\frac{1}{2} m_Z^2 = (g \cos\theta_W + g' \sin\theta_W)^2 |t^3 v|^2 = (g^2/\cos^2\theta_W)|t^3 v|^2 .$$ (3.51)

For a Higgs doublet, as in Eq. (3.48), we have

$$|t^+ v|^2 = v^2, \quad |t^3 v|^2 = 1/4 v^2 ,$$ (3.52)

so that

$$m_W^2 = 1/2 g^2 v^2, \quad m_Z^2 = 1/2 g^2 v^2/\cos^2\theta_W .$$ (3.53)

Note that by using Eq. (3.19) we obtain

$$v = 2^{-3/4} G_F^{-1/2} = 174.1 \text{ GeV} .$$ (3.54)

It is also evident that for Higgs doublets

$$\rho_0 = \frac{m_W^2}{m_Z^2 \cos^2 \theta_W} = 1 \, . \tag{3.55}$$

This relation is typical of one or more Higgs doublets and would be spoiled by the existence of Higgs triplets etc. In general,

$$\rho_0 = \frac{\sum_i ((t_i)^2 - (t_i^3)^2 + t_i) v_i^2}{\sum_i 2(t_i^3)^2 v_i^2} \tag{3.56}$$

for several Higgs bosons with VEVs v_i, weak isospin t_i, and z-component t_i^3. These results are valid at the tree level and are modified by calculable EW radiative corrections, as discussed in Sect. (3.7).

The measured values of the W and Z masses are [5, 8] (see Chap. 6):

$$m_W = 80.398 \pm 0.025 \, \text{GeV}, \qquad m_Z = 91.1875 \pm 0.0021 \, \text{GeV}. \tag{3.57}$$

In the minimal version of the SM only one Higgs doublet is present. Then the fermion–Higgs couplings are in proportion to the fermion masses. In fact, from the Yukawa couplings $g_{\phi \bar{f} f}(\bar{f}_L \phi f_R + h.c.)$, the mass m_f is obtained by replacing ϕ by v, so that $m_f = g_{\phi \bar{f} f} v$. In the minimal SM three out of the four Hermitian fields are removed from the physical spectrum by the Higgs mechanism and become the longitudinal modes of W^+, W^-, and Z. The fourth neutral Higgs is physical and should be found. If more doublets are present, two more charged and two more neutral Higgs scalars should be around for each additional doublet.

The couplings of the physical Higgs H can be simply obtained from $\mathcal{L}_{\text{Higgs}}$, by the replacement (the remaining three hermitian fields correspond to the would be Goldstone bosons that become the longitudinal modes of $W^\pm$ and Z):

$$\phi(x) = \begin{pmatrix} \phi^+(x) \\ \phi^0(x) \end{pmatrix} \to \begin{pmatrix} 0 \\ v + (H/\sqrt{2}) \end{pmatrix} , \tag{3.58}$$

[so that $(D_\mu \phi)^\dagger (D^\mu \phi) = 1/2 (\partial_\mu H)^2 + \ldots$], with the results

$$\mathcal{L}[H, W, Z] = g^2 \frac{v}{\sqrt{2}} W_\mu^+ W^{-\mu} H + \frac{g^2}{4} W_\mu^+ W^{-\mu} H^2 +$$

$$+ g^2 \frac{v}{2\sqrt{2} \cos^2 \theta_W} Z_\mu Z^\mu H + \frac{g^2}{8 \cos^2 \theta_W} Z_\mu Z^\mu H^2 \, . \tag{3.59}$$

Note that the trilinear couplings are nominally of order g^2, but the adimensional coupling constant is actually of order g if we express the couplings in terms of the masses according to Eqs. (3.53):

$$\mathcal{L}[H, W, Z] = g m_W W_\mu^+ W^{-\mu} H + \frac{g^2}{4} W_\mu^+ W^{-\mu} H^2 +$$

$$+ \frac{g m_Z}{2 \cos^2 \theta_W} Z_\mu Z^\mu H + \frac{g^2}{8 \cos^2 \theta_W} Z_\mu Z^\mu H^2 . \qquad (3.60)$$

Thus the trilinear couplings of the Higgs to the gauge bosons are also proportional to the masses. The quadrilinear couplings are genuinely of order g^2. Recall that to go from the lagrangian to the Feynman rules for the vertices the statistical factors must be taken into account: for example, the Feynman rule for the $ZZHH$ vertex is $i g_{\mu\nu} g^2 / 2 \cos^2 \theta_W$.

The generic coupling of H to a fermion of type f is given by (after diagonalization):

$$\mathcal{L}[H, \bar{\psi}, \psi] = \frac{g_f}{\sqrt{2}} \bar{\psi} \psi H, \qquad (3.61)$$

with

$$\frac{g_f}{\sqrt{2}} = \frac{m_f}{\sqrt{2} v} = 2^{1/4} G_F^{1/2} m_f . \qquad (3.62)$$

The Higgs self couplings are obtained from the potential in Eq. (3.40) by the replacement in Eq. (3.58). Given that, from the minimum condition:

$$v = \sqrt{\frac{\mu^2}{\lambda}} \qquad (3.63)$$

one obtains:

$$V = -\mu^2 (v + \frac{H}{\sqrt{2}})^2 + \frac{\mu^2}{2 v^2} (v + \frac{H}{\sqrt{2}})^4 = -\frac{\mu^2 v^2}{2} + \mu^2 H^2 + \frac{\mu^2}{\sqrt{2} v} H^3 + \frac{\mu^2}{8 v^2} H^4 \qquad (3.64)$$

The constant term can be omitted in our context. We see that the Higgs mass is positive (compare with Eq. (3.40)) and is given by:

$$m_H^2 = 2\mu^2 = 2\lambda v^2 \qquad (3.65)$$

We see that for $\sqrt{\lambda} \sim o(1)$ the Higgs mass should be of the order of the weak scale.

The difficulty of the Higgs search is due to the fact that it is heavy and coupled in proportion to mass: it is a heavy particle that must be radiated by another heavy particle. So a lot of phase space and luminosity is needed. At LEP2 the main process for Higgs production was the Higgs-strahlung process $e^+ e^- \rightarrow ZH$ shown in

Fig. 3.5 Higgs production diagrams in Born approximation: (**a**) The Higgs-strahlung process $e^+e^- \rightarrow ZH$, (**b**) the WW fusion process $e^+e^- \rightarrow Hv\bar{v}$

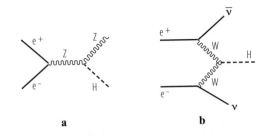

Fig. 3.5 [9]. The alternative process $e^+e^- \rightarrow Hv\bar{v}$, via WW fusion, also shown in Fig. 3.5 [10], has a smaller crosssection at LEP2 energies but would become important, even dominant at higher energy e^+e^- colliders, like the ILC or CLIC (the corresponding ZZ fusion process has a much smaller crosssection). The analytic formulae for the crosssections of both processes can be found, for example, in [11]. The direct experimental limit on m_H from LEP2 is $m_H \gtrsim 114\,\text{GeV}$ at 95% c.l. (see Chap. 6).

3.6 The CKM Matrix

Weak charged currents are the only tree level interactions in the SM that change flavour: for example, by emission of a W an up-type quark is turned into a down-type quark, or a v_l neutrino is turned into a l^- charged lepton (all fermions are letf-handed). If we start from an up quark that is a mass eigenstate, emission of a W turns it into a down-type quark state d' (the weak isospin partner of u) that in general is not a mass eigenstate. The mass eigenstates and the weak eigenstates do not coincide and a unitary transformation connects the two sets:

$$D' = \begin{pmatrix} d' \\ s' \\ b' \end{pmatrix} = V \begin{pmatrix} d \\ s \\ b \end{pmatrix} = VD \tag{3.66}$$

V is the Cabibbo-Kobayashi-Maskawa (CKM) matrix [12] (and similarly we can denote by U the column vector of the three up quark mass eigenstates). Thus in terms of mass eigenstates the charged weak current of quarks is of the form:

$$J_\mu^+ \propto \bar{U}\gamma_\mu(1 - \gamma_5)t^+ VD \tag{3.67}$$

where

$$V = U_u^\dagger U_d \tag{3.68}$$

Here U_u and U_d are the unitary matrices that operate on left-handed doublets in the diagonalization of the u and d quarks, respectively (see Eq. (3.44)). Since V is unitary (i.e. $VV^\dagger = V^\dagger V = 1$) and commutes with T^2, T_3 and Q (because all d-type quarks have the same isospin and charge), the neutral current couplings are diagonal both in the primed and unprimed basis (if the down-type quark terms in the Z current are written in terms of weak isospin eigenvectors as $\bar{D}'\Gamma D'$, then by changing basis we get $\bar{D}V^\dagger\Gamma V D$ and V and Γ commute because, as seen from Eq. (3.23), Γ is made of Dirac matrices and of T_3 and Q generator matrices). It follows that $\bar{D}'\Gamma D' = \bar{D}\Gamma D$. This is the GIM mechanism [13] that ensures natural flavour conservation of the neutral current couplings at the tree level.

For N generations of quarks, V is a N×N unitary matrix that depends on N^2 real numbers (N^2 complex entries with N^2 unitarity constraints). However, the $2N$ phases of up- and down-type quarks are not observable. Note that an overall phase drops away from the expression of the current in Eq. (3.67), so that only $2N - 1$ phases can affect V. In total, V depends on $N^2 - 2N + 1 = (N - 1)^2$ real physical parameters. A similar counting gives $N(N - 1)/2$ as the number of independent parameters in an orthogonal N×N matrix. This implies that in V we have $N(N - 1)/2$ mixing angles and $(N - 1)^2 - N(N - 1)/2 = (N - 1)(N - 2)/2$ phases: for $N = 2$ one mixing angle (the Cabibbo angle θ_C) and no phases, for $N = 3$ three angles (θ_{12}, θ_{13} and θ_{23}) and one phase φ etc.

Given the experimental near diagonal structure of V a convenient parametrisation is the one proposed by Maiani [14]. It can be cast in the form of a product of three independent 2×2 block matrices (s_{ij} and c_{ij} are shorthands for $\sin\theta_{ij}$ and $\cos\theta_{ij}$):

$$V = \begin{pmatrix} 1 & 0 & 0 \\ 0 & c_{23} & s_{23} \\ 0 & -s_{23} & c_{23} \end{pmatrix} \begin{pmatrix} c_{13} & 0 & s_{13}e^{i\varphi} \\ 0 & 1 & 0 \\ -s_{13}e^{-i\varphi} & 0 & c_{13} \end{pmatrix} \begin{pmatrix} c_{12} & s_{12} & 0 \\ -s_{12} & c_{12} & 0 \\ 0 & 0 & 1 \end{pmatrix}. \quad (3.69)$$

The advantage of this parametrization is that the three mixing angles are of different orders of magnitude. In fact, from experiment we know that $s_{12} \equiv \lambda$, $s_{23} \sim o(\lambda^2)$ and $s_{13} \sim o(\lambda^3)$, where $\lambda = \sin\theta_C$ is the sine of the Cabibbo angle, and, as order of magnitude, s_{ij} can be expressed in terms of small powers of λ. More precisely, following Wolfenstein [15] one can set:

$$s_{12} \equiv \lambda, \qquad s_{23} = A\lambda^2, \qquad s_{13}e^{-i\phi} = A\lambda^3(\rho - i\eta) \quad (3.70)$$

As a result, by neglecting terms of higher order in λ one can write down:

$$V = \begin{bmatrix} V_{ud} & V_{us} & V_{ub} \\ V_{cd} & V_{cs} & V_{cb} \\ V_{td} & V_{ts} & V_{tb} \end{bmatrix} \sim \begin{bmatrix} 1 - \frac{\lambda^2}{2} & \lambda & A\lambda^3(\rho - i\eta) \\ -\lambda & 1 - \frac{\lambda^2}{2} & A\lambda^2 \\ A\lambda^3(1 - \rho - i\eta) & -A\lambda^2 & 1 \end{bmatrix} + o(\lambda^4). \quad (3.71)$$

It has become customary to make the replacement $\rho, \eta \rightarrow \bar{\rho}, \bar{\eta}$ with:

$$\rho - i\eta = \frac{\bar{\rho} - i\bar{\eta}}{\sqrt{1 - \lambda^2}} \sim (\bar{\rho} - i\bar{\eta})(1 + \lambda^2/2 + \ldots). \tag{3.72}$$

Present values of the CKM parameters as obtained from experiment are [16] [17] (a survey of the current status of the CKM parameters can also be found in Ref. [5]):

$$\lambda = 0.2258 \pm 0.0014$$

$$A = 0.818 \pm 0.016$$

$$\bar{\rho} = 0.164 \pm 0.029; \quad \bar{\eta} = 0.340 \pm 0.017 \tag{3.73}$$

A more detailed discussion of the experimental data is given in Chap. 10.

In the SM the non vanishing of the η parameter (related to the phase φ in Eqs. 3.69 and 3.70) is the only source of CP violation. Unitarity of the CKM matrix V implies relations of the form $\sum_a V_{ba} V_{ca}^* = \delta_{bc}$. In most cases these relations do not imply particularly instructive constraints on the Wolfenstein parameters. But when the three terms in the sum are of comparable magnitude we get interesting information. The three numbers which must add to zero form a closed triangle in the complex plane, with sides of comparable length. This is the case for the t-u triangle (unitarity triangle) shown in Fig. 3.6 (or, what is equivalent in first approximation, for the d-b triangle):

$$V_{td} V_{ud}^* + V_{ts} V_{us}^* + V_{tb} V_{ub}^* = 0 \tag{3.74}$$

All terms are of order λ^3. For $\eta = 0$ the triangle would flatten down to vanishing area. In fact the area of the triangle, J of order $J \sim \eta A^2 \lambda^6$, is the Jarlskog invariant [18] (its value is independent of the parametrization). In the SM all CP violating observables must be proportional to J, hence to the area of the triangle or to η. A direct and by now very solid evidence for J non vanishing is obtained from the measurements of ϵ and ϵ' in K decay. Additional direct evidence is being obtained from the experiments on B decays at beauty factories and at the TeVatron where the angles β (the most precisely measured), α and γ have been determined. Together with the available information on the magnitude of the sides all the measurements

Fig. 3.6 The unitarity triangle corresponding to Eq. (3.74)

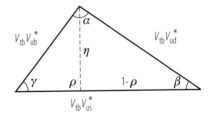

Fig. 3.7 Box diagrams describing $K^0 - \bar{K}^0$ mixing at the quark level at 1-loop

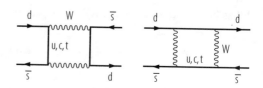

are in good agreement with the predictions from the SM unitary triangle [16, 17] (see Chap. 10).

As we have discussed, due to the GIM mechanism, there are no flavour changing neutral current (FCNC) transitions at tree level in the SM. Transitions with $|\Delta F| = 1, 2$ are induced at one loop level. In particular, meson mixing, i.e. $M \rightarrow \bar{M}$ off diagonal $|\Delta F| = 2$ mass matrix elements (with $M = K$, D or B neutral mesons), are obtained from box diagrams. For example, in the case of $K^0 - \bar{K}^0$ mixing the relevant transition is $\bar{s}d \rightarrow s\bar{d}$ (see Fig. 3.7). In the internal quark lines all up-type quarks are exchanged. In the amplitude, two vertices and the connecting propagator (with virtual four momentum p_μ) at one side contribute a factor ($u_i = u, c, t$):

$$F_{GIM} = \sum_i V_{u_i s}^* \frac{1}{\not{p} - m_{ui}} V_{u_i d} , \qquad (3.75)$$

which, in the limit of equal m_{ui}, is clearly vanishing due to the unitarity of the CKM matrix V. Thus the result is proportional to mass differences. For $K^0 - \bar{K}^0$ mixing the contribution of virtual u quarks is negligible due to the small value of m_u and the contribution of the t quark is also small due to the mixing factors $V_{ts}^* V_{td} \sim o(A^2\lambda^5)$. The dominant c quark contribution to the real part of the box diagram quark-level amplitude is approximately of the form (see, for example, [19]):

$$Re H_{box} = \frac{G_F^2}{16\pi^2} m_c^2 Re(V_{cs}^* V_{cd})^2 \eta_1 O^{\Delta s=2} , \qquad (3.76)$$

where $\eta_1 \sim 0.85$ is a QCD correction factor and $O^{\Delta s=2} = \bar{d}_L \gamma_\mu s_L \, \bar{s}_L \gamma_\mu d_L$ is the 4-quark dimension six relevant operator. To obtain the $K^0 - \bar{K}^0$ mixing its matrix element between meson states must be taken which is parametrized in terms of a "B_K parameter" which is defined in such a way that $B_K = 1$ for vacuum state insertion between the two currents:

$$\langle K^0 | O^{\Delta s=2} | \bar{K}^0 \rangle = \frac{16}{3} f_K m_K^2 B_K , \qquad (3.77)$$

where $f_K \sim 113 MeV$ is the kaon pseudoscalar constant. Clearly to the charm contribution in Eq. (3.76) non perturbative additional contributions must be added, some of them of $o(m_K^2/m_c^2)$, because the smallness of m_c makes a completely partonic dominance inadequate. In particular, B_K is best evaluated by QCD lattice simulations. In Eq. (3.76) the factor $o(m_c^2/m_W^2)$ is the "GIM suppression" factor

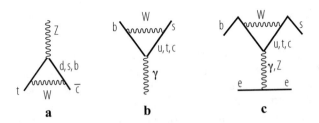

Fig. 3.8 Examples of $|\Delta F| = 1$ transitions at the quark level at 1-loop: (**a**) Diagram for a $Z \rightarrow t\,\bar{c}$ vertex, (**b**) $b \rightarrow s\,\gamma$, (**c**) a "penguin" diagram for $b \rightarrow s\,e^{+}e^{-}$

($1/m_W^2$ is hidden in G_F according to Eq. (3.19)). For B mixing the dominant contribution is from the t quark. In this case, the partonic dominance is more realistic and the GIM factor $o(m_t^2/m_W^2)$ is actually larger than one.

All sorts of transitions with $|\Delta F| = 1$ are also induced at loop level. For example, an effective vertex $Z \rightarrow t\bar{c}$, which does not exist at tree level, is generated at 1-loop (see Fig. 3.8). Similarly, transitions involving photons or gluons are also possible, like $t \rightarrow c\,g$ or $b \rightarrow s\,\gamma$ (Fig. 3.8) or $b \rightarrow s\,g$. For light fermion exchange in the loop the GIM suppression is also effective in $|\Delta F| = 1$ amplitudes. For example, analogous leptonic transitions like $\mu \rightarrow e\,\gamma$ or $\tau \rightarrow \mu\,\gamma$ also exist but are extremely small in the SM because the tiny neutrino masses enter in the GIM suppression factor. But new physics effects could well make these rare processes accessible to experiments in the near future. The external Z, photon or gluon can be attached to a pair of light fermions, giving rise to an effective four fermion operator, as in "penguin diagrams" like the one shown in Fig. 3.8 for $b \rightarrow s\,l^{+}l^{-}$. The inclusive rate $B \rightarrow X_s\,\gamma$ with X_s a hadronic state containing a unit of strangeness corresponding to an s-quark, has been precisely measured. The world average result for the branching ratio with $E_\gamma > 1.6\,\text{GeV}$ is [5]:

$$B(B \rightarrow X_s\,\gamma)_{exp} = (3.55 \pm 0.26) \cdot 10^{-4}\,. \tag{3.78}$$

The theoretical prediction for this inclusive process is to a large extent free of uncertainties from hadronisation effects and is accessible to perturbation theory as the b-quark is heavy enough. The most complete result at order α_s^2 is at present [20] (and refs. therein):

$$B(B \rightarrow X_s\,\gamma)_{th} = (2.98 \pm 0.26) \cdot 10^{-4}\,. \tag{3.79}$$

Note that the theoretical value has recently become smaller than the experimental value. The fair agreement between theory and experiment imposes stringent constraints on possible new physics effects.

3.7 Neutrino Masses

In the minimal version of the SM the right handed neutrinos ν_{iR}, which have no gauge interactions, are not present at all. With no ν_R no Dirac mass is possible for neutrinos. If lepton number conservation is also imposed, then no Majorana mass is allowed either and, as a consequence, all neutrinos are massless. But, at present, from neutrino oscillation experiments (see Chapter 11 of the present work), we know that at least 2 out of the 3 known neutrinos have non vanishing masses: the two mass squared differences measured from solar (Δm_{12}^2) and atmospheric oscillations (Δm_{23}^2) are given by $\Delta m_{12}^2 \sim 8 \; 10^{-5} \; eV^2$ and $\Delta m_{23}^2 \sim 2.5 \; 10^{-3}$ [21]. The absolute values of the masses are very small, with an upper limit of a fraction of eV, obtained from laboratory experiments (tritium β decay near the end point: $m_\nu \lesssim 2$ eV [5], absence of visible neutrinoless double β decay : $|m_{ee}| \lesssim 0.3 - 0.7$ eV (m_{ee} is a combination of neutrino masses; for a review, see, for example [22]) and from cosmological observations: $m_\nu \lesssim 0.1 - 0.7$ eV (depending on the cosmological model assumptions) [23]. If ν_{iR} are added to the minimal model and lepton number is imposed by hand, then neutrino masses would in general appear as Dirac masses, generated by the Higgs mechanism, like for any other fermion. But, for Dirac neutrinos, to explain the extreme smallness of neutrino masses, one should allow for very small Yukawa couplings. However, we stress that, in the SM, baryon B and lepton L number conservation, which are not guaranteed by gauge symmetries (as is the case for the electric charge Q), are understood as "accidental" symmetries, due to the fact that, out of the SM fields, it is not possible to construct gauge invariant operators which are renormalizable (i.e. of operator dimension $d \leq 4$) and violate B and/or L. In fact the SM lagrangian should contain all terms allowed by gauge symmetry and renormalizability. The most general renormalizable lagrangian, built from the SM fields, compatible with the SM gauge symmetry, in absence of ν_{iR}, is automatically B and L conserving. But in presence of ν_{iR}, this is no more true and the right handed Majorana mass term is allowed:

$$M_{RR} = \bar{\nu}_{iR}^c M_{ij} \nu_{jR} = \nu_{iR}^T C M_{ij} \nu_{jR} \, , \tag{3.80}$$

where $\nu_{iR}^c = C \bar{\nu}_{iR}^T$ is the charge conjugated neutrino field and C is the charge conjugation matrix in Dirac spinor space. The Majorana mass term is an operator of dimension $d = 3$ with $\Delta L = 2$. Since the ν_{iR} are gauge singlets the Majorana mass M_{RR} is fully allowed by the gauge symmetry and a coupling with the Higgs is not needed to generate this type of mass. As a consequence, the entries of the mass matrix M_{ij} do not need to be of the order of the EW symmetry breaking scale v and could be much larger. If one starts from the Dirac and RR Majorana mass terms for neutrinos, the resulting mass matrix, in the L, R space, has the form:

$$m_\nu = \begin{bmatrix} 0 & m_D \\ m_D & M \end{bmatrix} \tag{3.81}$$

where m_D and M are the Dirac and Majorana mass matrices (M is the matrix M_{ij} in Eq. (3.80)). The corresponding eigenvalues are three very heavy neutrinos with masses of order M and three light neutrinos with masses

$$m_\nu = -m_D^T M^{-1} m_D , \tag{3.82}$$

which are possibly very small if M is large enough. This is the see-saw mechanism for neutrino masses [24]. Note that if no ν_{iR} exist a Majorana mass term could still be built out of ν_{jL}. But ν_{jL} have weak isospin 1/2, being part of the left handed lepton doublet l. Thus, the left handed Majorana mass term has total weak isospin equal to one and needs two Higgs fields to make a gauge invariant term. The resulting mass term:

$$O_5 = \lambda l_i^T \lambda_{ij} l_j H H / M , \tag{3.83}$$

with M a large scale (apriori comparable to the scale of M_{RR}) and λ a dimensionless coupling generically of o(1), is a non renormalizable operator of dimension 5. The corresponding mass terms are of the order $m_\nu \sim \lambda v^2 / M$, hence of the same generic order of the light neutrino masses from Eq. (3.82).

In conclusion, neutrino masses are believed to be small because neutrinos are Majorana particles with masses inversely proportional to the large scale M of energy where L non conservation is induced. It is interesting that the observed magnitudes of the mass squared splittings of neutrinos are well compatible with a scale M remarkably close to the Grand Unification scale, where in fact L non conservation is naturally expected.

In the previous Section we have discussed flavour mixing for quarks. But, clearly, given that non vanishing neutrino masses have been established, a similar mixing matrix is also introduced in the leptonic sector, but will not be discussed here (see Chapter 11).

3.8   Renormalization of the Electroweak Theory

The Higgs mechanism gives masses to the Z, the $W^\pm$ and to fermions while the lagrangian density is still symmetric. In particular the gauge Ward identities and the symmetric form of the gauge currents are preserved. The validity of these relations is an essential ingredient for renormalizability. In the previous Sections we have specified the Feynman vertices in the "unitary" gauge where only physical particles appear. However, as discussed in Chap. 2, in this gauge the massive gauge boson propagator would have a bad ultraviolet behaviour:

$$W_{\mu\nu} = \frac{-g_{\mu\nu} + \frac{q_\mu q_\nu}{m_W^2}}{q^2 - m_W^2} . \tag{3.84}$$

A formulation of the standard EW theory with good apparent ultraviolet behaviour can be obtained by introducing the renormalizable or R_ξ gauges, in analogy with the abelian case discussed in detail in Chap. 2. One parametrizes the Higgs doublet as:

$$\phi = \begin{pmatrix} \phi^+ \\ \phi^0 \end{pmatrix} = \begin{pmatrix} \phi_1 + i\phi_2 \\ \phi_3 + i\phi_4 \end{pmatrix} = \begin{pmatrix} -iw^+ \\ v + \frac{H+iz}{\sqrt{2}} \end{pmatrix}, \tag{3.85}$$

and similarly for $\phi^\dagger$, where w^- appears. The scalar fields $w^\pm$ and z are the pseudo Goldstone bosons associated with the longitudinal modes of the physical vector bosons $W^\pm$ and Z. The R_ξ gauge fixing lagrangian has the form:

$$\Delta\mathcal{L}_{GF} = -\frac{1}{\xi}|\partial^\mu W_\mu - \xi m_W w|^2 - \frac{1}{2\eta}(\partial^\mu Z_\mu - \eta m_Z z)^2 - \frac{1}{2\alpha}(\partial^\mu A_\mu)^2. \tag{3.86}$$

The $W^\pm$ and Z propagators, as well as those of the scalars $w^\pm$ and z, have exactly the same general forms as for the abelian case in Eqs. (67)–(69) of Chap. 2, with parameters ξ and η, respectively (and the pseudo Goldstone bosons $w^\pm$ and z have masses ξm_W and ηm_Z). In general, a set of associated ghost fields must be added, again in direct analogy with the treatment of R_ξ gauges in the abelian case of Chap. 2. The complete Feynman rules for the standard EW theory can be found in a number of textbooks (see, for example, [25]).

The pseudo Goldstone bosons $w^\pm$ and z are directly related to the longitudinal helicity states of the corresponding massive vector bosons $W^\pm$ and Z. This correspondence materializes in a very interesting "equivalence theorem": at high energies of order E the amplitude for the emission of one or more longitudinal gauge bosons V_L (with $V = W, Z$) becomes equal (apart from terms down by powers of m_V/E) to the amplitude where each longitudinal gauge boson is replaced by the corresponding Goldstone field $w^\pm$ or z [26]. For example, consider top decay with a longitudinal W in the final state: $t \to bW_L^+$. The equivalence theorem asserts that we can compute the dominant contribution to this rate from the simpler $t \to bw^+$ matrix element:

$$\Gamma(t \to bW_L^+) = \Gamma(t \to bw^+)[1 + o(m_W^2/m_t^2)]. \tag{3.87}$$

In fact one finds:

$$\Gamma(t \to bw^+) = \frac{h_t^2}{32\pi}m_t = \frac{G_F m_t^3}{8\pi\sqrt{2}}, \tag{3.88}$$

where $h_t = m_t/v$ is the Yukawa coupling of the top quark (numerically very close to 1), and we used $1/v^2 = 2\sqrt{2}G_F$ (see Eq. (3.54)). If we compare with Eq. (3.34), we see that this expression coincides with the total top width (i.e. including all polarizations for the W in the final state), computed at tree level, apart from terms down by powers of $o(m_W^2/m_t^2)$. In fact, the longitudinal W is dominant in the final state because $h_t \gg g^2$. Similarly the equivalence theorem can be applied to find the dominant terms at large $\sqrt{s}$ for the crosssection $e^+e^- \rightarrow W_L^+ W_L^-$, or the leading contribution in the limit $m_H \gg m_V$ to the width for the decay $\Gamma(H \rightarrow VV)$.

The formalism of the R_ξ gauges is also very useful in proving that spontaneously broken gauge theories are renormalizable. In fact, the non singular behaviour of propagators at large momenta is very suggestive of the result. Nevertheless to prove it is by far not a simple matter. The fundamental theorem that in general a gauge theory with spontaneous symmetry breaking and the Higgs mechanism is renormalizable was proven by 't Hooft and Veltman [27, 28].

For a chiral theory like the SM an additional complication arises from the existence of chiral anomalies. But this problem is avoided in the SM because the quantum numbers of the quarks and leptons in each generation imply a remarkable (and, from the point of view of the SM, mysterious) cancellation of the anomaly, as originally observed in Ref. [29]. In quantum field theory one encounters an anomaly when a symmetry of the classical lagrangian is broken by the process of quantization, regularization and renormalization of the theory. Of direct relevance for the EW theory is the Adler-Bell-Jackiw (ABJ) chiral anomaly [30]. The classical lagrangian of a theory with massless fermions is invariant under a U(1) chiral transformations $\psi\prime = e^{i\gamma_5\theta}\psi$. The associated axial Noether current is conserved at the classical level. But, at the quantum level, chiral symmetry is broken due to the ABJ anomaly and the current is not conserved. The chiral breaking is produced by a clash between chiral symmetry, gauge invariance and the regularization procedure.

The anomaly is generated by triangular fermion loops with one axial and two vector vertices (Fig. 3.9). For example, for the Z the axial coupling is proportional to the third component of weak isospin t_3, while the vector coupling is proportional to a linear combination of t_3 and the electric charge Q. Thus in order for the chiral anomaly to vanish all traces of the form $tr\{t_3 QQ\}$, $tr\{t_3 t_3 Q\}$, $tr\{t_3 t_3 t_3\}$ (and also $tr\{t_+ t_- t_3\}$ when charged currents are also included) must vanish, where the trace is extended over all fermions in the theory that can circulate in the loop. Now all these traces happen to vanish for each fermion family separately. For example take $tr\{t_3 QQ\}$. In one family there are, with $t_3 = +1/2$, three colours of up quarks with

Fig. 3.9 Triangle diagram that generates the ABJ anomaly

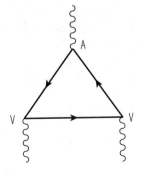

charge $Q = +2/3$ and one neutrino with $Q = 0$ and, with $t_3 = -1/2$, three colours of down quarks with charge $Q = -1/3$ and one l^- with $Q = -1$. Thus we obtain $tr\{t_3 Q Q\} = 1/2 \cdot 3 \cdot 4/9 - 1/2 \cdot 3 \cdot 1/9 - 1/2 \cdot 1 = 0$. This impressive cancellation suggests an interplay among weak isospin, charge and colour quantum numbers which appears as a miracle from the point of view of the low energy theory but is in fact understandable from the point of view of the high energy theory. For example, in Grand Unified Theories (GUTs) (for reviews, see, for example, [31]) there are similar relations where charge quantization and colour are related: in the five of SU(5) we have the content $(d, d, d, e^+, \bar{\nu})$ and the charge generator has a vanishing trace in each SU(5) representation (the condition of unit determinant, represented by the letter S in the SU(5) group name, translates into zero trace for the generators). Thus the charge of d quarks is $-1/3$ of the positron charge because there are three colours. A whole family fits perfectly in one 16 of SO(10) which is anomaly free. So GUTs can naturally explain the cancellation of the chiral anomaly.

An important implication of chiral anomalies together with the topological properties of the vacuum in non abelian gauge theories is that the conservation of the charges associated to baryon (B) and lepton (L) numbers is broken by the anomaly [32], so that B and L conservation is actually violated in the standard electroweak theory (but B-L remains conserved). B and L are conserved to all orders in the perturbative expansion but the violation occurs via non perturbative instanton effects [33] (the amplitude is proportional to the typical non perturbative factor $\exp -c/g^2$, with c a constant and g the $SU(2)$ gauge coupling). The corresponding effect is totally negligible at zero temperature T, but becomes relevant at temperatures close to the electroweak symmetry breaking scale, precisely at $T \sim o(TeV)$. The non conservation of B+L and the conservation of B−L near the weak scale plays a role in the theory of baryogenesis that quantitatively aims at explaining the observed matter antimatter asymmetry in the Universe (for a recent review, see, for example, [34]; see also Chap. 9).

3.9   QED Tests: Lepton Anomalous Magnetic Moments

The most precise tests of the electroweak theory apply to the QED sector. Here we discuss some recent developments. The anomalous magnetic moments of the electron and of the muon are among the most precise measurements in the whole of physics. The magnetic moment $\vec{\mu}$ and the spin $\vec{S}$ are related by $\vec{\mu} = -g e \vec{S}/2m$, where g is the gyromagnetic ratio ($g = 2$ for a pointlike Dirac particle). The quantity $a = (g-2)/2$ measures the anomalous magnetic moment of the particle. Recently there have been new precise measurements of a_e and a_μ for the electron [35] and the muon [36]:

$$a_e^{exp} = 11596521808.5(7.6) \cdot 10^{-13}, \qquad a_\mu^{exp} = 11659208.0(6.3) \cdot 10^{-10}.$$

$$(3.89)$$

Fig. 3.10 The hadronic
contributions to the
anomalous magnetic moment:
vacuum polarization (left)
and light by light scattering
(right)

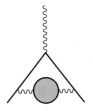
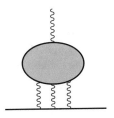

The theoretical calculations in general contain a pure QED part plus the sum of hadronic and weak contribution terms:

$$a = a^{QED} + a^{hadronic} + a^{weak} = \sum_i C_i (\frac{\alpha}{\pi})^i + a^{hadronic} + a^{weak}. \qquad (3.90)$$

The QED part has been computed analytically for $i = 1, 2, 3$, while for $i = 4$ there is a numerical calculation with an error (see, for example, [38] and refs therein). Some terms for $i = 5$ have also been estimated for the muon case. The hadronic contribution is from vacuum polarization insertions and from light by light scattering diagrams (see Fig. 3.10). The weak contribution is from W or Z exchange.

For the electron case the weak contribution is essentially negligible and the hadronic term ($a_e^{hadronic} \sim (16.71 \pm 0.19) \cdot 10^{-13}$) does not introduce an important uncertainty. As a result this measurement can be used to obtain the most precise determination of the fine structure constant [37]:

$$\alpha^{-1} \sim 137.035999710(96) , \qquad (3.91)$$

with an uncertainty about 10 times smaller than the previous determination. However, very recently a theoretical error in the α^4 terms was corrected [39]. As a result the value of α^{-1} in Eq. (3.91) is shifted by $-6.41180(73) \, 10^{-7}$ (about 7 σ's). This change has a minor impact in the following discussion of the muon ($g - 2$).

In the muon case the experimental precision is less by about three orders of magnitude, but the sensitivity to new physics effects is typically increased by a factor $(m_\mu/m_e)^2 \sim 4 \cdot 10^4$ (one mass factor arises because the effective operator needs a chirality flip and the second one is because, by definition, one must factor out the Bohr magneton $e/2m$). From the theory side, the QED term (using the value of α from a_e in Eq. (3.91)), and the weak contribution are affected by small errors and are given by (all theory number are taken here from the review [40])

$$a_\mu^{QED} = (116584718.09 \pm 1.6) \cdot 10^{-11}, \qquad a_\mu^{weak} = (154 \pm 2.2) \cdot 10^{-11} \qquad (3.92)$$

The dominant ambiguities arise from the hadronic term. The lowest order (LO) vacuum polarization contribution can be evaluated from the measured cross sections in $e^+e^- \to$ hadrons at low energy via dispersion relations (the largest contribution is from the $\pi\pi$ final state), with the result $a_\mu^{LO} \cdot 10^{-11} = 6909 \pm 44$. The higher order (HO) vacuum polarization contribution (from 2-loop diagrams containing an hadronic insertion) is given by: $a_\mu^{HO} \cdot 10^{-11} = -98 \pm 1$. The contribution of the light by light (LbL) scattering diagrams is estimated to be: $a_\mu^{LbL} \cdot 10^{-11} = 120 \pm 35$. Adding the above contributions up the total hadronic result is reported as:

$$a_\mu^{hadronic} = (6931 \pm 56) \cdot 10^{-11}. \tag{3.93}$$

At face value this would lead to a 3.3σ deviation from the experimental value a_μ^{exp} in Eq. (3.89):

$$a_\mu^{exp} - a_\mu^{th(e^+e^-)} = (275 \pm 84) \cdot 10^{-11}. \tag{3.94}$$

However, the error estimate on the LbL term, mainly a theoretical uncertainty, is not compelling, and it could well be somewhat larger (although probably not by as much as to make the discrepancy to completely disappear). Another puzzle is the fact that, using the conservation of the vector current (CVC) and isospin invariance, which are well established tools at low energy, a_μ^{LO} can also be evaluated from τ decays. But the results on the hadronic contribution from e^+e^- and from τ decay, nominally of comparable accuracy, do not match well, and the discrepancy would be much attenuated if one takes the τ result [41]. Since it is difficult to find a theoretical reason for the e^+e^- vs τ difference, one must conclude that there is something which is not understood either in the data or in the assessment of theoretical errors. The prevailing view is to take the e^+e^- determination as the most directly reliable, which leads to Eq. (3.94), but doubts certainly remain. Finally, we note that, given the great accuracy of the a_μ measurement and the relative importance of the non QED contributions, it is not unreasonable that a first signal of new physics can appear in this quantity.

3.10 Large Radiative Corrections to Electroweak Processes

Since the SM theory is renormalizable higher order perturbative corrections can be reliably computed. Radiative corrections are very important for precision EW tests. The SM inherits all successes of the old V-A theory of charged currents and of QED. Modern tests have focussed on neutral current processes, the W mass and the measurement of triple gauge vertices. For Z physics and the W mass the state of the art computation of radiative corrections include the complete one loop diagrams and selected dominant two loop corrections. In addition some resummation techniques are also implemented, like Dyson resummation of vacuum polarization functions and important renormalization group improvements for large QED and QCD logarithms. We now discuss in more detail sets of large radiative corrections which are particularly significant (for reviews of radiative corrections for LEP1 physics, see, for example: [42]).

Even leaving aside QCD corrections, a set of important quantitative contributions to the radiative corrections arise from large logarithms [e.g. terms of the form $(\alpha/\pi \ln (m_Z/m_{fll}))^n$ where f_{ll} is a light fermion]. The sequences of leading and close-to-leading logarithms are fixed by well-known and consolidated techniques (β functions, anomalous dimensions, penguin-like diagrams, etc.). For example, large logarithms from pure QED effects dominate the running of α from m_e, the electron mass, up to m_Z. Similarly large logarithms of the form $[\alpha/\pi \ln (m_Z/\mu)]^n$ also enter, for example, in the relation between $\sin^2 \theta_W$ at the scales m_Z (LEP, SLC) and μ (e.g. the scale of low-energy neutral-current experiments). Also, large logs from initial state radiation dramatically distort the line shape of the Z resonance as observed at LEP1 and SLC and this effect was accurately taken into account for the measurement of the Z mass and total width. The experimental accuracy on m_Z obtained at LEP1 is $\delta m_Z = \pm 2.1\,\text{MeV}$ (see Chap. 6). Similarly, a measurement of the total width to an accuracy $\delta \Gamma = \pm 2.3\,\text{MeV}$ has been achieved. The prediction of the Z line-shape in the SM to such an accuracy has posed a formidable challenge to theory, which has been successfully met. For the inclusive process $e^+ e^- \to f \bar{f} X$, with $f \neq e$ (for a concise discussion, we leave Bhabha scattering aside) and X including γ's and gluons, the physical cross-section can be written in the form of a convolution [42]:

$$\sigma(s) = \int_{z_0}^{1} dz\, \hat{\sigma}(zs) G(z, s) , \qquad (3.95)$$

where $\hat{\sigma}$ is the reduced cross-section, and $G(z, s)$ is the radiator function that describes the effect of initial-state radiation; $\hat{\sigma}$ includes the purely weak corrections, the effect of final-state radiation (of both γ's and gluons), and also non-factorizable terms (initial- and final-state radiation interferences, boxes, etc.) which, being small, can be treated in lowest order and effectively absorbed in a modified $\hat{\sigma}$. The radiator $G(z, s)$ has an expansion of the form

$$G(z, s) = \delta(1 - z) + \alpha/\pi (a_{11} L + a_{10}) + (\alpha/\pi)^2 (a_{22} L^2 + a_{11} L + a_{20}) + \cdots +$$

$$+ (\alpha/\pi)^n \sum_{i=0}^{n} a_{ni} L^i , \qquad (3.96)$$

where $L = \ln s/m_e^2 \simeq 24.2$ for $\sqrt{s} \simeq m_Z$. All first- and second-order terms are known exactly. The sequence of leading and next-to-leading logs can be exponentiated (closely following the formalism of structure functions in QCD). For $m_Z \approx 91\,\text{GeV}$, the convolution displaces the peak by $+110\,\text{MeV}$, and reduces it by a factor of about 0.74. The exponentiation is important in that it amounts to an additional shift of about 14 MeV in the peak position with respect to the one loop radiative correction.

Among the one loop EW radiative corrections, a very remarkable class of contributions are those terms that increase quadratically with the top mass. The sensitivity of radiative corrections to m_t arises from the existence of these terms. The quadratic dependence on m_t (and on other possible widely broken isospin multiplets from new physics) arises because, in spontaneously broken gauge theories, heavy virtual particles do not decouple. On the contrary, in QED or QCD, the running of α and α_s at a scale Q is not affected by heavy quarks with mass $M \gg Q$. According to an intuitive decoupling theorem [43], diagrams with heavy virtual particles of mass M can be ignored at $Q \ll M$ provided that the couplings do not grow with M and that the theory with no heavy particles is still renormalizable. In the spontaneously broken EW gauge theories both requirements are violated. First, one important difference with respect to unbroken gauge theories is in the longitudinal modes of weak gauge bosons. These modes are generated by the Higgs mechanism, and their couplings grow with masses (as is also the case for the physical Higgs couplings). Second the theory without the top quark is no more renormalizable because the gauge symmetry is broken as the (t,b) doublet would not be complete (also the chiral anomaly would not be completely cancelled). With the observed value of m_t the quantitative importance of the terms of order $G_F m_t^2 / 4\pi^2 \sqrt{2}$ is substancial but not dominant (they are enhanced by a factor $m_t^2 / m_W^2 \sim 5$ with respect to ordinary terms). Both the large logarithms and the $G_F m_t^2$ terms have a simple structure and are to a large extent universal, i.e. common to a wide class of processes. In particular the $G_F m_t^2$ terms appear in vacuum polarization diagrams which are universal (virtual loops inserted in gauge boson internal lines are independent of the nature of the vertices on each side of the propagator) and in the $Z \to b\bar{b}$ vertex which is not. This vertex is specifically sensitive to the top quark which, being the partner of the b quark in a doublet, runs in the loop. Instead all types of heavy particles could in principle contribute to vacuum polarization diagrams. The study of universal vacuum polarization contributions, also called "oblique" corrections, and of top enhanced terms is important for an understanding of the pattern of radiative corrections. More in general, the important consequence of non decoupling is that precision tests of the electroweak theory may apriori be sensitive to new physics even if the new particles are too heavy for their direct production, but aposteriori no signal of deviation has clearly emerged.

While radiative corrections are quite sensitive to the top mass, they are unfortunately much less dependent on the Higgs mass. If they were sufficiently sensitive by now we would precisely know the mass of the Higgs. But the dependence of one loop diagrams on m_H is only logarithmic: $\sim G_F m_W^2 \log(m_H^2 / m_W^2)$. Quadratic terms $\sim G_F^2 m_H^2$ only appear at two loops [44] and are too small to be detectable. The difference with the top case is that the splitting $m_t^2 - m_b^2$ is a direct breaking of the gauge symmetry that already affects the 1- loop corrections, while the Higgs couplings are "custodial" SU(2) symmetric in lowest order.

3.11   Electroweak Precision Tests in the SM and Beyond

For the analysis of electroweak data in the SM one starts from the input parameters: as is the case in any renormalizable theory, masses and couplings have to be specified from outside. One can trade one parameter for another and this freedom is used to select the best measured ones as input parameters. Some of them, α, G_F and m_Z, are very precisely known, as we have seen, some other ones, m_{flight}, m_t and $\alpha_s(m_Z)$ are less well determined while m_H is largely unknown. Among the light fermions, the quark masses are badly known, but fortunately, for the calculation of radiative corrections, they can be replaced by $\alpha(m_Z)$, the value of the QED running coupling at the Z mass scale. The value of the hadronic contribution to the running, embodied in the value of $\Delta\alpha_{had}^{(5)}(m_Z^2)$ (see Table 3.1, [8]) is obtained through dispersion relations from the data on $e^+e^- \to$ hadrons at moderate centre-of-mass energies. From the input parameters one computes the radiative corrections to a sufficient precision to match the experimental accuracy. Then one compares the theoretical predictions with the data for the numerous observables which have been measured [45], checks the consistency of the theory and derives constraints on m_t, $\alpha_s(m_Z)$ and m_H. A detailed discussion of all experimental aspects of precision tests of the EW theory is presented in Chap. 6.

The basic tree level relations:

$$\frac{g^2}{8m_W^2} = \frac{G_F}{\sqrt{2}}, \qquad g^2 \sin^2 \theta_W = e^2 = 4\pi\alpha \qquad (3.97)$$

can be combined into

$$\sin^2 \theta_W = \frac{\pi\alpha}{\sqrt{2}G_F m_W^2} \qquad (3.98)$$

Always at tree level, a different definition of $\sin^2 \theta_W$ is from the gauge boson masses:

$$\frac{m_W^2}{m_Z^2 \cos^2 \theta_W} = \rho_0 = 1 \implies \sin^2 \theta_W = 1 - \frac{m_W^2}{m_Z^2} \qquad (3.99)$$

where $\rho_0 = 1$ assuming that there are only Higgs doublets. The last two relations can be put into the convenient form

$$(1 - \frac{m_W^2}{m_Z^2})\frac{m_W^2}{m_Z^2} = \frac{\pi\alpha}{\sqrt{2}G_F m_Z^2} \qquad (3.100)$$

Beyond tree level, these relations are modified by radiative corrections:

$$(1 - \frac{m_W^2}{m_Z^2})\frac{m_W^2}{m_Z^2} = \frac{\pi\alpha(m_Z)}{\sqrt{2}G_F m_Z^2}\frac{1}{1 - \Delta r_W}$$

$$\frac{m_W^2}{m_Z^2 \cos^2\theta_W} = 1 + \Delta\rho_m \tag{3.101}$$

The Z and W masses are to be precisely defined in terms of the pole position in the respective propagators. Then, in the first relation the replacement of α with the running coupling at the Z mass $\alpha(m_Z)$ makes Δr_W completely determined at 1-loop by purely weak corrections (G_F is protected from logarithmic running as an indirect consequence of (V-A) current conservation in the massless theory). This relation defines Δr_W unambigously, once the meaning of $\alpha(m_Z)$ is specified (for example, $\bar{MS}$). On the contrary, in the second relation $\Delta\rho_m$ depends on the definition of $\sin^2\theta_W$ beyond the tree level. For LEP physics $\sin^2\theta_W$ is usually defined from the $Z \to \mu^+\mu^-$ effective vertex. At the tree level the vector and axial-vector couplings g_V^μ and g_A^μ are given in Eqs. (3.29). Beyond the tree level a corrected vertex can be written down in terms of modified effective couplings. Then $\sin^2\theta_W \equiv \sin^2\theta_{eff}$ is in general defined through the muon vertex:

$$g_V^\mu/g_A^\mu = 1 - 4\sin^2\theta_{eff}$$

$$\sin^2\theta_{eff} = (1 + \Delta k)s_0^2, \qquad s_0^2 c_0^2 = \frac{\pi\alpha(m_Z)}{\sqrt{2}G_F m_Z^2}$$

$$g_A^{\mu 2} = \frac{1}{4}(1 + \Delta\rho) \tag{3.102}$$

We see that s_0^2 and c_0^2 are "improved" Born approximations (by including the running of α) for $\sin^2\theta_{eff}$ and $\cos^2\theta_{eff}$. Actually, since in the SM lepton universality is only broken by masses and is in agreement with experiment within the present accuracy, in practice the muon channel can be replaced with the average over charged leptons.

We can write a symbolic equation that summarizes the status of what has been computed up to now for the radiative corrections (we list some recent work on each item from where older references can be retrieved) Δr_W [46], $\Delta\rho$ [47] and Δk [48]:

$$\Delta r_W, \Delta\rho, \Delta k = g^2\frac{m_t^2}{m_W^2}(1 + \alpha_s + \alpha_s^2) + g^2(1 + \alpha_s + \sim\alpha_s^2) + g^4\frac{m_t^4}{m_W^4} + g^4\frac{m_t^2}{m_W^2} + \dots \tag{3.103}$$

The meaning of this relation is that the one loop terms of order g^2 are completely known, together with their first order QCD corrections (the second order QCD corrections are only estimated for the g^2 terms not enhanced by m_t^2/m_W^2), and the terms of order g^4 enhanced by the ratios m_t^4/m_W^4 or m_t^2/m_W^2 are also known.

In the SM the quantities Δr_W, $\Delta\rho$, Δk, for sufficiently large m_t, are all dominated by quadratic terms in m_t of order $G_F m_t^2$. The quantity $\Delta\rho_m$ is not independent and can expressed in terms of them. As new physics can more easily be disentangled if not masked by large conventional m_t effects, it is convenient to keep $\Delta\rho$ while trading Δr_W and Δk for two quantities with no contributions of order $G_F m_t^2$. One thus introduces the following linear combinations (epsilon parameters) [49]:

$$\epsilon_1 = \Delta\rho,$$

$$\epsilon_2 = c_0^2\Delta\rho + \frac{s_0^2\Delta r_W}{c_0^2 - s_0^2} - 2s_0^2\Delta k,$$

$$\epsilon_3 = c_0^2\Delta\rho + (c_0^2 - s_0^2)\Delta k. \tag{3.104}$$

The quantities ϵ_2 and ϵ_3 no longer contain terms of order $G_F m_t^2$ but only logarithmic terms in m_t. The leading terms for large Higgs mass, which are logarithmic, are contained in ϵ_1 and ϵ_3. To complete the set of top-enhanced radiative corrections one adds ϵ_b defined from the loop corrections to the $Zb\bar{b}$ vertex. One modifies g_V^b and g_A^b as follows:

$$g_A^b = -\frac{1}{2}(1 + \frac{\Delta\rho}{2})(1 + \epsilon_b),$$

$$\frac{g_V^b}{g_A^b} = \frac{1 - 4/3\sin^2\theta_{eff} + \epsilon_b}{1 + \epsilon_b}. \tag{3.105}$$

ϵ_b can be measured from $R_b = \Gamma(Z \to b\bar{b})/\Gamma(Z \to \text{hadrons})$ (see Table 3.1). This is clearly not the most general deviation from the SM in the $Z \to b\bar{b}$ vertex but ϵ_b is the quantity where the large m_t corrections are located in the SM. Thus, summarizing, in the SM one has the following "large" asymptotic contributions:

$$\epsilon_1 = \frac{3G_F m_t^2}{8\pi^2\sqrt{2}} - \frac{3G_F m_W^2}{4\pi^2\sqrt{2}}\tan^2\theta_W \ln\frac{m_H}{m_Z} + \ldots,$$

$$\epsilon_2 = -\frac{G_F m_W^2}{2\pi^2\sqrt{2}}\ln\frac{m_t}{m_Z} + \ldots,$$

$$\epsilon_3 = \frac{G_F m_W^2}{12\pi^2\sqrt{2}}\ln\frac{m_H}{m_Z} - \frac{G_F m_W^2}{6\pi^2\sqrt{2}}\ln\frac{m_t}{m_Z}\ldots,$$

$$\epsilon_b = -\frac{G_F m_t^2}{4\pi^2\sqrt{2}} + \ldots. \tag{3.106}$$

The ϵ_i parameters vanish in the limit where only tree level SM effects are kept plus pure QED and/or QCD corrections. So they describe the effects of quantum corrections (i.e. loops) from weak interactions. A similar set of parameters are the S, T, U parameters [50]: the shifts induced by new physics on S, T and U are proportional to those induced on ϵ_3, ϵ_1 and ϵ_2, respectively. In principle, with no model dependence, one can measure the four ϵ_i from the basic observables of LEP physics $\Gamma(Z \to \mu^+\mu^-)$, A_{FB}^{μ} and R_b on the Z peak plus m_W. With increasing model dependence, one can include other measurements in the fit for the ϵ_i. For example, use lepton universality to average the μ with the e and τ final states, or include all lepton asymmetries and so on. The present experimental values of the ϵ_i, obtained from a fit of all LEP1-SLD measurements plus m_W, are given by The LEP Electroweak Working Group [8]:

$$\epsilon_1 \cdot 10^3 = 5.4 \pm 1.0, \quad \epsilon_2 \cdot 10^3 = -8.9 \pm 1.2,$$
$$\epsilon_3 \cdot 10^3 = 5.34 \pm 0.94, \quad \epsilon_b \cdot 10^3 = -5.0 \pm 1.6. \tag{3.107}$$

Note that the ϵ parameters are of order a few in 10^{-3} and are known with an accuracy in the range 15–30%. As discussed in the next Section, these values are in agreement with the SM with a light Higgs. All models of new physics must be compared with these findings and pass this difficult test.

3.12   Results of the SM Analysis of Precision Tests

The electroweak Z pole measurements, combining the results of all the experiments, are summarised in Table 3.1. The various asymmetries determine the effective electroweak mixing angle for leptons with highest sensitivity. The weighted average of these results, including small correlations, is:

$$\sin^2 \theta_{eff} = 0.23153 \pm 0.00016, \tag{3.108}$$

Note, however, that this average has a χ^2 of 11.8 for 5 degrees of freedom, corresponding to a probability of a few %. The χ^2 is pushed up by the two most precise measurements of $\sin^2 \theta_{eff}$, namely those derived from the measurements of A_l by SLD, dominated by the left-right asymmetry A_{LR}^0, and of the forward-backward asymmetry measured in $b\bar{b}$ production at LEP, $A_{FB}^{0,b}$, which differ by about 3σs.

We now discuss fitting the data in the SM. One can think of different types of fit, depending on which experimental results are included or which answers one wants to obtain. For example, in Table 3.2 we present in column 1 a fit of all Z pole data plus m_W and Γ_W (this is interesting as it shows the value of m_t obtained indirectly from radiative corrections, to be compared with the value of m_t measured in production experiments), in column 2 a fit of all Z pole data plus

Table 3.1 Summary of electroweak precision measurements at high Q^2 [8]

Observable	Measurement	SM fit
m_Z [GeV]	91.1875 ± 0.0021	91.1875
Γ_Z [GeV]	2.4952 ± 0.0023	2.4957
σ_h^0 [nb]	41.540 ± 0.037	41.477
R_l^0	20.767 ± 0.025	20.744
$AFB^{0,l}$	0.01714 ± 0.00095	0.01645
A_l (SLD)	0.1513 ± 0.0021	0.1481
A_l (P_τ)	0.1465 ± 0.0032	0.1481
R_b^0	0.21629 ± 0.00066	0.21586
R_c^0	0.1721 ± 0.0030	0.1722
$A_{FB}^{0,b}$	0.0992 ± 0.0016	0.1038
$A_{FB}^{0,c}$	0.0707 ± 0.0035	0.0742
A_b	0.923 ± 0.020	0.935
A_c	0.670 ± 0.027	0.668
$\sin^2 \theta_{eff}$ (Q_{FB}^{had})	0.2324 ± 0.0012	0.2314
m_W [GeV]	80.398 ± 0.025	80.374
Γ_W [GeV]	2.140 ± 0.060	2.091
m_t [GeV ($p\bar{p}$)]	170.9 ± 1.8	171.3
$\Delta\alpha_{had}^{(5)}(m_Z^2)$	0.02758 ± 0.00035	0.02768

The first block shows the Z-pole measurements. The second block shows additional results from other experiments: the mass and the width of the W boson measured at the Tevatron and at LEP-2, the mass of the top quark measured at the Tevatron, and the contribution to α of the hadronic vacuum polarization. The SM fit results are derived from the SM analysis of these results

m_t (here it is m_W which is indirectly determined), and, finally, in column 3 a fit of all the data listed in Table 3.1 (which is the most relevant fit for constraining m_H). From the fit in column 1 of Table 3.2 we see that the extracted value of m_t is in good agreement with the direct measurement (see Table 3.1). Similarly we see that the experimental measurement of m_W in Table 3.1 is larger by about one standard deviation with respect to the value from the fit in column 2. We have seen that quantum corrections depend only logarithmically on m_H. In spite of this small sensitivity, the measurements are precise enough that one still obtains a quantitative indication of the mass range. From the fit in column 3 we obtain: $\log_{10} m_H(\text{GeV}) = 1.88 \pm 0.16$ (or $m_H = 76^{+34}_{-24}$ GeV). This result on the Higgs mass is particularly remarkable. The value of $\log_{10} m_H(\text{GeV})$ is compatible with the small window between ~ 2 and ~ 3 which is allowed, on the one side, by the direct search limit ($m_H > 114$ GeV from LEP-2 [8]), and, on the other side, by the theoretical upper limit on the Higgs mass in the minimal SM, $m_H \lesssim 600 - 800$ GeV [51].

Thus the whole picture of a perturbative theory with a fundamental Higgs is well supported by the data on radiative corrections. It is important that there is a clear indication for a particularly light Higgs: at 95% c.l. $m_H \lesssim 182$ GeV (including

Table 3.2 Standard Model fits of electroweak data [8]

Fit	1	2	3
Measurements	m_W	m_t	m_t, m_W
m_t (GeV)	178.9^{+12}_{-9}	170.9 ± 1.8	171.3 ± 1.7
m_H (GeV)	145^{+240}_{-81}	99^{+52}_{-35}	76^{+34}_{-24}
$\log[m_H(\text{GeV})]$	$2.16 \pm +0.39$	2.00 ± 0.19	1.88 ± 0.16
$\alpha_s(m_Z)$	0.1190 ± 0.0028	0.1189 ± 0.0027	0.1185 ± 0.0026
m_W (MeV)	80385 ± 19	80360 ± 20	80374 ± 15

All fits use the Z pole results and $\Delta\alpha^{(5)}_{had}(m_Z^2)$ as listed in Table 3.1. In addition, the measurements listed on top of each column are included as well. The fitted W mass is also shown [8] (the directly measured value is $m_W = 80398 \pm 25$ MeV)

the input from the direct search result). This is quite encouraging for the ongoing search for the Higgs particle. More general, if the Higgs couplings are removed from the Lagrangian the resulting theory is non renormalizable. A cutoff Λ must be introduced. In the quantum corrections $\log m_H$ is then replaced by $\log \Lambda$ plus a constant. The precise determination of the associated finite terms would be lost (that is, the value of the mass in the denominator in the argument of the logarithm). A heavy Higgs would need some unfortunate accident: the finite terms, different in the new theory from those of the SM, should by chance compensate for the heavy Higgs in a few key parameters of the radiative corrections (mainly ϵ_1 and ϵ_3, see, for example, [49]). Alternatively, additional new physics, for example in the form of effective contact terms added to the minimal SM lagrangian, should accidentally do the compensation, which again needs some sort of conspiracy.

To the list of precision tests of the SM one should add the results on low energy tests obtained from neutrino and antineutrino deep inelastic scattering (NuTeV [52]), parity violation in Cs atoms (APV [53]) and the recent measurement of the parity-violating asymmetry in Moller scattering [54] (see Chap. 6). When these experimental results are compared with the SM predictions the agreement is good except for the NuTeV result that shows a deviation by three standard deviations. The NuTeV measurement is quoted as a measurement of $\sin^2 \theta_W = 1 - m_W^2/m_Z^2$ from the ratio of neutral to charged current deep inelastic cross-sections from ν_μ and $\bar{\nu}_\mu$ using the Fermilab beams. But it has been argued and it is now generally accepted that the NuTeV anomaly probably simply arises from an underestimation of the theoretical uncertainty in the QCD analysis needed to extract $\sin^2 \theta_W$. In fact, the lowest order QCD parton formalism on which the analysis has been based is too crude to match the experimental accuracy.

When confronted with these results, on the whole the SM performs rather well, so that it is fair to say that no clear indication for new physics emerges from the data. However, as already mentioned, one problem is that the two most precise measurements of $\sin^2 \theta_{\text{eff}}$ from A_{LR} and A^b_{FB} differ by about 3σs. In general, there appears to be a discrepancy between $\sin^2 \theta_{\text{eff}}$ measured from leptonic asymmetries $((\sin^2 \theta_{\text{eff}})_l)$ and from hadronic asymmetries $((\sin^2 \theta_{\text{eff}})_h)$. In fact, the result from

Fig. 3.11 The data for $\sin^2 \theta_{\text{eff}}^{\text{lept}}$ are plotted vs m_H. The theoretical prediction for the measured value of m_t is also shown. For presentation purposes the measured points are shown each at the m_H value that would ideally correspond to it given the central value of m_t (updated from [55])

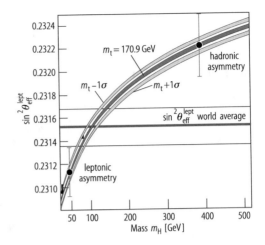

A_{LR} is in good agreement with the leptonic asymmetries measured at LEP, while all hadronic asymmetries, though their errors are large, are better compatible with the result of A_{FB}^b. These two results for $\sin^2 \theta_{\text{eff}}$ are shown in Fig. 3.11 [55]. Each of them is plotted at the m_H value that would correspond to it given the central value of m_t. Of course, the value for m_H indicated by each $\sin^2 \theta_{\text{eff}}$ has an horizontal ambiguity determined by the measurement error and the width of the $\pm 1\sigma$ band for m_t. Even taking this spread into account it is clear that the implications on m_H are sizably different. One might imagine that some new physics effect could be hidden in the $Zb\bar{b}$ vertex. Like for the top quark mass there could be other non decoupling effects from new heavy states or a mixing of the b quark with some other heavy quark. However, it is well known that this discrepancy is not easily explained in terms of some new physics effect in the $Zb\bar{b}$ vertex. A rather large change with respect to the SM of the b-quark right handed coupling to the Z is needed in order to reproduce the measured discrepancy (precisely a ∼30% change in the right-handed coupling), an effect too large to be a loop effect but which could be produced at the tree level, e.g., by mixing of the b quark with a new heavy vectorlike quark [56]), or some mixing of the Z with ad hoc heavy states [57]. But then this effect should normally also appear in the direct measurement of A_b performed at SLD using the left-right polarized b asymmetry, even within the moderate precision of this result. The measurements of neither A_b at SLD nor R_b confirm the need of a new effect. Alternatively, the observed discrepancy could be simply due to a large statistical fluctuation or an unknown experimental problem. As a consequence of this problem, the ambiguity in the measured value of $\sin^2 \theta_{\text{eff}}$ is in practice larger than the nominal error, reported in Eq. 3.108, obtained from averaging all the existing determinations, and the interpretation of precision tests is less sharp than it would otherwise be.

We have already observed that the experimental value of m_W (with good agreement between LEP and the Tevatron) is a bit high compared to the SM prediction (see Fig. 3.12). The value of m_H indicated by m_W is on the low side, just in the same interval as for $\sin^2 \theta_{\text{eff}}^{\text{lept}}$ measured from leptonic asymmetries.

Fig. 3.12 The data for m_W are plotted vs m_H. The theoretical prediction for the measured value of m_t is also shown (updated from [55])

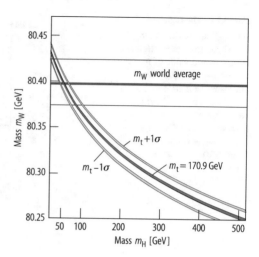

In conclusion, overall the validity of the SM has been confirmed to a level that we can say was unexpected at the beginning. In the present data there is no significant evidence for departures from the SM, no compelling evidence of new physics. The impressive success of the SM poses strong limitations on the possible forms of new physics.

3.13 Phenomenology of the SM Higgs

The Higgs problem is really central in particle physics today. On the one hand, the experimental verification of the Standard Model (SM) cannot be considered complete until the structure of the Higgs sector is not established by experiment. On the other hand, the Higgs is also related to most of the major problems of particle physics, like the flavour problem and the hierarchy problem, the latter strongly suggesting the need for new physics near the weak scale. In turn the discovery of new physics could clarify the dark matter identity. It is clear that the fact that some sort of Higgs mechanism is at work has already been established. The W or the Z with longitudinal polarization that we observe are not present in an unbroken gauge theory (massless spin-1 particles, like the photon, are transversely polarized). The longitudinal degree of freedom for the W or the Z is borrowed from the Higgs sector and is an evidence for it. Also, it has been verified that the gauge symmetry is unbroken in the vertices of the theory: all currents and charges are indeed symmetric. Yet there is obvious evidence that the symmetry is instead badly broken in the masses. Not only the W and the Z have large masses, but the large splitting of, for example, the t-b doublet shows that even a global weak SU(2) is not at all respected by the fermion spectrum. This is a clear signal of spontaneous symmetry breaking and the implementation of spontaneous symmetry breaking in a gauge theory is via the Higgs mechanism. The big remaining questions are about the nature and the properties of the Higgs particle(s). The present experimental information on the Higgs sector, is surprisingly limited and can be summarized in a few lines, as follows. First, the relation $M_W^2 = M_Z^2 \cos^2 \theta_W$, Eq. (3.55), modified by small, computable radiative corrections, has been experimentally proven. This relation means that the effective Higgs (be it fundamental or composite) is indeed a weak isospin doublet. The Higgs particle has not been found but, in the SM, its mass can well be larger than the present direct lower limit $m_H \gtrsim 114\,\text{GeV}$ (at 95% c.l.) obtained from searches at LEP-2. The radiative corrections computed in the SM when compared to the data on precision electroweak tests lead to a clear indication for a light Higgs, not too far from the present lower bound. The exact experimental upper limit for m_H in the SM depends on the value of the top quark mass m_t. The CDF and D0 combined value after Run II is at present [8] $m_t = 170.9 \pm 1.8\,\text{GeV}$ (it went down with respect to the value $m_t = 178 \pm 4.3\,\text{GeV}$ from Run I and also the experimental error is now sizably reduced). As a consequence the present limit on m_H is more stringent [8]: $m_H < 182\,\text{GeV}$ (at 95% c.l., after including the information from the $114\,\text{GeV}$ direct bound). On the Higgs the LHC will address the following questions : one doublet, more doublets, additional singlets? SM Higgs or SUSY Higgses? Fundamental or composite (of fermions, of WW...)? Pseudo-Goldstone boson of an enlarged symmetry? A manifestation of large extra dimensions (5th component of a gauge boson, an effect of orbifolding or of boundary conditions...)? Or some combination of the above or something so far unthought of? Here in the following we will summarize the main properties of the SM Higgs that provide an essential basis for the planning and the interpretation of the LHC Higgs programme. We start from the mass, then the width and the branching ratios and, finally, the most important production channels.

3.13.1 Theoretical Bounds on the SM Higgs Mass

It is well known [58–60] that in the SM with only one Higgs doublet a lower limit on m_H can be derived from the requirement of vacuum stability (or, in milder form, of a moderate instability, compatible with the lifetime of the Universe [61]). The limit is a function of m_t and of the energy scale Λ where the model breaks down and new physics appears. The Higgs mass enters because it fixes the initial value of the quartic Higgs coupling λ for its running up to the large scale Λ. Similarly an upper bound on m_H (with mild dependence on m_t) is obtained, as described in [62] and refs. therein, from the requirement that for λ no Landau pole appears up to the scale Λ, or in simpler terms, that the perturbative description of the theory remains valid up to Λ. We now briefly recall the derivation of these limits.

The possible instability of the Higgs potential $V[\phi]$ is generated by the quantum loop corrections to the classical expression of $V[\phi]$. At large ϕ the derivative

$V'[\phi]$ could become negative and the potential would become unbound from below. The one-loop corrections to $V[\phi]$ in the SM are well known and change the dominant term at large ϕ according to $\lambda\phi^4 \to (\lambda + \gamma \log \phi^2/\Lambda^2)\phi^4$. This one-loop approximation is not enough in this case, because it fails at large enough ϕ, when $\gamma \log \phi^2/\Lambda^2$ becomes of order one. The renormalization group improved version of the corrected potential leads to the replacement $\lambda\phi^4 \to \lambda(\Lambda)\phi'^4(\Lambda)$ where $\lambda(\Lambda)$ is the running coupling and $\phi'(\mu) = \phi \exp \int^t \gamma(t')dt'$, with $\gamma(t)$ being an anomalous dimension function and $t = \log\Lambda/v$ (v is the vacuum expectation value $v = (2\sqrt{2}G_F)^{-1/2}$). As a result, the positivity condition for the potential amounts to the requirement that the running coupling $\lambda(\Lambda)$ never becomes negative. A more precise calculation, which also takes into account the quadratic term in the potential, confirms that the requirements of positive $\lambda(\Lambda)$ leads to the correct bound down to scales Λ as low as ~ 1 TeV. The running of $\lambda(\Lambda)$ at one loop is given by:

$$\frac{d\lambda}{dt} = \frac{3}{4\pi^2}[\lambda^2 + 3\lambda h_t^2 - 9h_t^4 + \text{small gauge and Yukawa terms}] , \qquad (3.109)$$

with the normalization such that at $t = 0$, $\lambda = \lambda_0 = m_H^2/2v^2$ and the top Yukawa coupling $h_t^0 = m_t/v$. We see that, for m_H small and m_t fixed at its measured value, λ decreases with t and can become negative. If one requires that λ remains positive up to $\Lambda = 10^{15}$–10^{19} GeV, then the resulting bound on m_H in the SM with only one Higgs doublet is given by, (also including the effect of the two-loop beta function terms) [60] :

$$m_H(\text{GeV}) > 128.4 + 2.1 [m_t - 170.9] - 4.5 \frac{\alpha_s(m_Z) - 0.118}{0.006} . \qquad (3.110)$$

Note that this limit is evaded in models with more Higgs doublets. In this case the limit applies to some average mass but the lightest Higgs particle can well be below, as it is the case in the minimal SUSY extension of the SM (MSSM).

The upper limit on the Higgs mass in the SM is clearly important for assessing the chances of success of the LHC as an accelerator designed to solve the Higgs problem. The upper limit [62] arises from the requirement that the Landau pole associated with the non asymptotically free behaviour of the $\lambda\phi^4$ theory does not occur below the scale Λ. The initial value of λ at the weak scale increases with m_H and the derivative is positive at large λ (because of the positive λ^2 term—the $\lambda\varphi^4$ theory is not asymptotically free—which overwhelms the negative top-Yukawa term). Thus, if m_H is too large, the point where λ computed from the perturbative beta function becomes infinite (the Landau pole) occurs at too low an energy. Of course in the vicinity of the Landau pole the 2-loop evaluation of the beta function is not reliable. Indeed the limit indicates the frontier of the domain where the theory is well described by the perturbative expansion. Thus the quantitative evaluation of the limit is only indicative, although it has been to some extent supported by simulations of the Higgs sector of the EW theory on the lattice. For the upper limit on m_H one finds [62]

$$m_H \lesssim 180\,GeV \text{ for } \Lambda \sim M_{GUT} - M_{Pl}$$
$$m_H \lesssim 0.5 - 0.8\,TeV \text{ for } \Lambda \sim 1\,TeV. \tag{3.111}$$

In conclusion, for $m_t \sim 171\,$GeV, only a small range of values for m_H is allowed, $130 < m_H < \sim 200\,$GeV, if the SM holds up to $\Lambda \sim M_{GUT}$ or M_{Pl}.

An additional argument indicating that the solution of the Higgs problem cannot be too far away is the fact that, in the absence of a Higgs particle or of an alternative mechanism, violations of unitarity appear in some scattering amplitudes at energies in the few TeV range [63]. In particular, amplitudes involving longitudinal gauge bosons (those most directly related to the Higgs sector) are affected. For example, at tree level in the absence of Higgs exchange, for $s >> m_Z^2$ one obtains:

$$A(W_L^+ W_L^- \to Z_L Z_L)_{no\ Higgs} \sim i\frac{s}{v^2} \tag{3.112}$$

In the SM this unacceptable large energy behaviour is quenched by the Higgs exchange diagram contribution:

$$A(W_L^+ W_L^- \to Z_L Z_L)_{Higgs} \sim -i\frac{s^2}{v^2(s - m_H^2)} \tag{3.113}$$

Thus the total result in the SM is:

$$A(W_L^+ W_L^- \to Z_L Z_L)_{SM} \sim -i\frac{sm_H^2}{v^2(s - m_H^2)} \tag{3.114}$$

which at large energies saturates at a constant value. To be compatible with unitarity bounds one needs $m_H^2 < 4\pi\sqrt{2}/G_F$ or $m_H < 1.5\,$TeV. Both the Landau pole and the unitarity argument show that, if the Higgs is too heavy, the SM becomes a non perturbative theory at energies of o(1 TeV). In conclusion, these arguments imply that the SM Higgs cannot escape detection at the LHC.

3.13.2   SM Higgs Decays

The total width and the branching ratios for the SM Higgs as function of m_H are given in Figs. 3.13 and 3.14, respectively [64].

Since the couplings of the Higgs particle are in proportion to masses, when m_H increases the Higgs becomes strongly coupled. This is reflected in the sharp rise of the total width with m_H. For m_H near its present lower bound of 114 GeV, the width is below 5 MeV, much less than for the W or the Z which have a comparable mass.

Fig. 3.13 The total width of the SM Higgs boson [64]

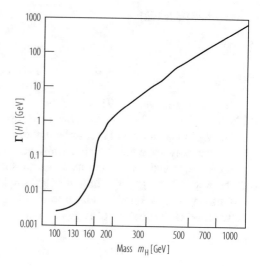

Fig. 3.14 The branching ratios of the SM Higgs boson [65]

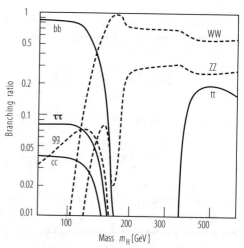

The dominant channel for such a Higgs is $H \to b\bar{b}$. In Born approximation the partial width into a fermion pair is given by Djouadi [64] and Haber [66]:

$$\Gamma(H \to f\bar{f}) = N_C \frac{G_F}{4\pi\sqrt{2}} m_H m_f^2 \beta_f^3 \qquad (3.115)$$

where $\beta_f = (1 - 4m_f^2/m_H^2)^{1/2}$. The factor of β^3 appears because the fermion pair must be in a p-state of orbital angular momentum for a Higgs with scalar coupling, because of parity (this factor would be β for a pseudoscalar coupling). We see that the width is suppressed by a factor m_f^2/m_H^2 with respect to the natural size $G_F m_H^3$

for the width of a particle of mass m_H decaying through a diagram with only one weak vertex.

A glance to the branching ratios shows that the branching ratio into τ pairs is larger by more than a factor of two with respect to the $c\bar{c}$ channel. This is at first sight surprising because the colour factor N_C favours the quark channels and the masses of τ's and of D mesons are quite similar. This is due to the fact that the QCD corrections replace the charm mass at the scale of charm with the charm mass at the scale m_H, which is lower by about a factor of 2.5. The masses run logarithmically in QCD, similar to the coupling constant. The corresponding logs are already present in the 1-loop QCD correction that amounts to the replacement $m_q^2 \to m_q^2[1 + 2\alpha_s/\pi (\log m_q^2/m_H^2 + 3/2)] \sim m_q^2(m_H^2)$.

The Higgs width sharply increases as the WW threshold is approached. For decay into a real pair of V's, with $V = W, Z$, one obtains in Born approximation [64, 66]:

$$\Gamma(H \to VV) = \frac{G_F m_H^3}{16\pi\sqrt{2}} \delta_V \beta_W (1 - 4x + 12x^2) \qquad (3.116)$$

where $\beta_W = \sqrt{1 - 4x}$ with $x = m_V^2/m_H^2$ and $\delta_W = 2$, $\delta_Z = 1$. Much above threshold the VV channels are dominant and the total width, given approximately by:

$$\Gamma_H \sim 0.5 \text{ TeV} (\frac{m_H}{1 \text{ TeV}})^3 \qquad (3.117)$$

becomes very large, signalling that the Higgs sector is becoming strongly interacting (recall the upper limit on the SM Higgs mass in Eq. (3.111)). The VV dominates over the $t\bar{t}$ because of the β threshold factors that disfavour the fermion channel and, at large m_H, by the cubic versus linear behaviour with m_H of the partial widths for VV versus $t\bar{t}$. Below the VV threshold the decays into virtual V particles is important: VV^* and V^*V^*. Note in particular the dip of the ZZ branching ratio just below the ZZ threshold: this is due to the fact that the W is lighter than the Z and the opening of its threshold depletes all other branching ratios. When the ZZ threshold is also passed then the ZZ branching fraction comes back to the ratio of approximately 1:2 with the WW channel (just the number of degrees of freedom: two hermitian fields for the W, one for the Z).

The decay channels into $\gamma\gamma$, $Z\gamma$ and gg proceed through loop diagrams, with the contributions from W (only for $\gamma\gamma$ and $Z\gamma$) and from fermion loops (for all) (Fig. 3.15).

We reproduce here the results for $\Gamma(H \to \gamma\gamma)$ and $\Gamma(H \to gg)$ [64, 66]:

$$\Gamma(H \to \gamma\gamma) = \frac{G_F \alpha^2 m_H^3}{128\pi^3\sqrt{2}} |A_W(\tau_W) + \sum_f N_C Q_f^2 A_f(\tau_f)|^2 \qquad (3.118)$$

$$\Gamma(H \to gg) = \frac{G_F \alpha_s^2 m_H^3}{64\pi^3\sqrt{2}} |\sum_{f=Q} A_f(\tau_f)|^2 \qquad (3.119)$$

Fig. 3.15 One-loop diagrams for Higgs decay into $\gamma\gamma$, $Z\gamma$ and gg

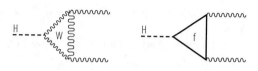

where $\tau_i = m_H^2/4m_i^2$ and:

$$A_f(\tau) = \frac{2}{\tau^2}[\tau + (\tau - 1)f(\tau)]$$

$$A_W(\tau) = -\frac{1}{\tau^2}[2\tau^2 + 3\tau + 3(2\tau - 1)f(\tau)] \tag{3.120}$$

with:

$$f(\tau) = \arcsin^2\sqrt{\tau} \quad \text{for } \tau \leq 1$$

$$f(\tau) = -\frac{1}{4}[\log\frac{1 + \sqrt{1 - \tau^{-1}}}{1 - \sqrt{1 - \tau^{-1}}} - i\pi]^2 \quad \text{for } \tau > 1 \tag{3.121}$$

For $H \rightarrow \gamma\gamma$ (as well as for $H \rightarrow Z\gamma$) the W loop is the dominant contribution at small and moderate m_H. We recall that the $\gamma\gamma$ mode can be a possible channel for Higgs discovery only for m_H near its lower bound (i.e for $114 < m_H < 150\,\text{GeV}$). In this domain of m_H we have $\Gamma(H \rightarrow \gamma\gamma) \sim 6\text{–}23\,\text{KeV}$. For example, in the limit $m_H \ll 4m_i^2$, or $\tau \rightarrow 0$, we have $A_W(0) = -7$ and $A_f(0) = 4/3$. The two contributions become comparable only for $m_H \sim 650\,\text{GeV}$ where the two amplitudes, still of opposite sign, nearly cancel. The top loop is dominant among fermions (lighter fermions are suppressed by m_f^2/m_H^2 modulo logs) and, as we have seen, it approaches a constant for large m_t. Thus the fermion loop amplitude for the Higgs would be sensitive to effects from very heavy fermions, in particular the $H \rightarrow gg$ effective vertex would be sensitive to all possible very heavy coloured quarks. As discussed in the QCD Chapter (Chap. 4) the $gg \rightarrow H$ vertex provides one of the main production channels for the Higgs at hadron colliders.

3.14 Limitations of the Standard Model

No signal of new physics has been found neither in electroweak precision tests nor in flavour physics. Given the success of the SM why are we not satisfied with this theory? Why not just find the Higgs particle, for completeness, and declare that particle physics is closed? The reason is that there are both conceptual problems and phenomenological indications for physics beyond the SM. On the conceptual side the most obvious problems are that quantum gravity is not included in the SM and the related hierarchy problem. Among the main phenomenological hints for new physics we can list coupling unification, dark matter, neutrino masses (discussed in Sect. (3.7)), baryogenesis and the cosmological vacuum energy.

The computed evolution with energy of the effective SM gauge couplings clearly points towards the unification of the electro-weak and strong forces (GUT's) at scales of energy $M_{GUT} \sim 10^{15} - 10^{16}$ GeV [31] which are close to the scale of quantum gravity, $M_{Pl} \sim 10^{19}$ GeV. One is led to imagine a unified theory of all interactions also including gravity (at present superstrings provide the best attempt at such a theory). Thus GUT's and the realm of quantum gravity set a very distant energy horizon that modern particle theory cannot ignore. Can the SM without new physics be valid up to such large energies? One can imagine that some obvious problems could be postponed to the more fundamental theory at the Planck mass. For example, the explanation of the three generations of fermions and the understanding of fermion masses and mixing angles can be postponed. But other problems must find their solution in the low energy theory. In particular, the structure of the SM could not naturally explain the relative smallness of the weak scale of mass, set by the Higgs mechanism at $\mu \sim 1/\sqrt{G_F} \sim 250$ GeV with G_F being the Fermi coupling constant. This so-called hierarchy problem is due to the instability of the SM with respect to quantum corrections. This is related to the presence of fundamental scalar fields in the theory with quadratic mass divergences and no protective extra symmetry at $\mu = 0$. For fermion masses, first, the divergences are logarithmic and, second, they are forbidden by the $SU(2) \otimes U(1)$ gauge symmetry plus the fact that at $m = 0$ an additional symmetry, i.e. chiral symmetry, is restored. Here, when talking of divergences, we are not worried of actual infinities. The theory is renormalizable and finite once the dependence on the cut off Λ is absorbed in a redefinition of masses and couplings. Rather the hierarchy problem is one of naturalness. We can look at the cut off as a parameterization of our ignorance on the new physics that will modify the theory at large energy scales. Then it is relevant to look at the dependence of physical quantities on the cut off and to demand that no unexplained enormously accurate cancellations arise.

The hierarchy problem can be put in very practical terms (the "little hierarchy problem"): loop corrections to the Higgs mass squared are quadratic in Λ. The most pressing problem is from the top loop. With $m_h^2 = m_{bare}^2 + \delta m_h^2$ the top loop gives

$$\delta m_{h|top}^2 \sim -\frac{3 G_F}{2\sqrt{2}\pi^2} m_t^2 \Lambda^2 \sim -(0.2\Lambda)^2 \tag{3.122}$$

If we demand that the correction does not exceed the light Higgs mass indicated by the precision tests, Λ must be close, $\Lambda \sim o(1\,TeV)$. Similar constraints arise from the quadratic Λ dependence of loops with gauge bosons and scalars, which, however, lead to less pressing bounds. So the hierarchy problem demands new physics to be very close (in particular the mechanism that quenches the top loop). Actually, this new physics must be rather special, because it must be very close, yet its effects are not clearly visible neither in precision electroweak tests (the "LEP Paradox" [67]) nor in flavour changing processes and CP violation. Examples of proposed classes of solutions for the hierarchy problem are: (1) Supersymmetry

[68]. In the limit of exact boson-fermion symmetry the quadratic divergences of bosons cancel so that only log divergences remain. However, exact SUSY is clearly unrealistic. For approximate SUSY (with soft breaking terms), which is the basis for all practical models, Λ is replaced by the splitting of SUSY multiplets, $\Lambda^2 \sim m_{SUSY}^2 - m_{ord}^2$. In particular, the top loop is quenched by partial cancellation with s-top exchange, so the s-top cannot be too heavy. (2) Technicolor [69]. The Higgs system is a condensate of new fermions. There is no fundamental scalar Higgs sector, hence no quadratic divergences associated to the μ^2 mass in the scalar potential. This mechanism needs a very strong binding force, $\Lambda_{TC} \sim 10^3 \Lambda_{QCD}$. It is difficult to arrange that such nearby strong force is not showing up in precision tests. Hence this class of models has been disfavoured by LEP, although some special class of models have been devised aposteriori, like walking TC, top-color assisted TC etc (for recent reviews, see, for example, [69]). (3) Extra dimensions (for a recent review, see, for example, [70]). The idea is that M_{Pl} appears very large, or equivalently that gravity appears very weak, because we are fooled by hidden extra dimensions so that either the real gravity scale is reduced down to a lower scale, even possibly down to $o(1\,TeV)$ or the intensity of gravity is red shifted away by an exponential warping factor [71]. This possibility is very exciting in itself and it is really remarkable that it is compatible with experiment. It provides a very rich framework with many different scenarios. (4) "Little Higgs" models [72]. In these models the Higgs is a pseudo-Goldstone boson and extra symmetries allow $m_h \neq 0$ only at two-loop level, so that Λ can be as large as $o(10\,\text{TeV})$ with the Higgs within present bounds (the top loop is quenched by exchange of heavy vectorlike new quarks with charge 2/3). The physics beyond the SM will be discussed in Chap. 8.

Acknowledgments I am very grateful to Giuseppe Degrassi, Paolo Gambino, Martin Grunewald, Vittorio Lubicz for their help and advise.

References

1. S.L. Glashow, Nucl. Phys. **22** (1961) 579; S. Weinberg, Phys. Rev. Lett. **19** (1967) 1264; A. Salam, in *Elementary Particle Theory*, ed. N. Svartholm (Almquist and Wiksells, Stockholm, 1969), p. 367.
2. F. Englert, R. Brout, Phys.Rev.Lett.**13**, 321 (1964); P.W. Higgs, Phys.Lett. **12**,132 (1964).
3. E.D.Commins, *Weak interactions*, Mc Graw Hill, 1973; L. V. Okun, *Leptons and Quarks*, North Holland, 1982; D. Bailin, *Weak Interactions*, 2nd e., Hilger, 1982; H. M. Georgi, *Weak interactions and modern particle theory*, Benjamin, 1984.
4. J.D. Bjorken and S. Drell, *Relativistic Quantum Mechanics/Fields*, Vols. I, II, McGraw-Hill, New York, (1965).
5. Particle Data Group, The Journal of Physics G 33(2006)1.
6. G.Altarelli, T.Sjöstrand and F.Zwirner (eds.), "Physics at LEP2", CERN Report 95-03.
7. The ALEPH, DELPHI, L3, OPAL, SLD Collaborations and the LEP Electroweak Working Group, *A Combination of Preliminary Electroweak Measurements and Constraints on the Standard Model*, hep-ex/0312023, and references therein.
8. The LEP Electroweak Working Group, http://lepewwg.web.cern.ch/LEPEWWG/.

9. J. Ellis, J. M. Gaillard and D. V. Nanopoulos, Nucl. Phys. B106 (1976) 292; J. D. Bjorken, SLAC Report 198 (1976).

10. G. Altarelli, B. Mele and F. Pitolli, Nucl. Phys. B287 (1987) 205.

11. G. Altarelli, T. Sjostrand and F. Zwirner, *Physics at LEP2*, CERN Report 96-01, (1996), pag. 361–366.

12. N. Cabibbo, Phys. Rev. Lett., 10 (1963) 531; M. Kobayashi and T. Maskawa, Progr. Th. Phys. 49 (1973) 652.

13. S.L. Glashow, J. Iliopoulos and L. Maiani, Phys. Rev. D2 (1970) 1285.

14. L. Maiani, *Proc. Int. Symp. on Lepton and Photon Interactions at High Energy*, Hamburg,1977.

15. L. Wolfenstein, Phys. Rev. Lett., 51 (1983) 1945.

16. The Unitary Triangle Fit Group, http://www.utfit.org

17. http://ckmfitter.in2p3.fr/

18. C. Jarlskog, Phys. Rev. Lett., 55 (1985) 1039.

19. J. F. Donoghue, E. Golowich and B. Holstein, *Dynamics of the standard model*, Cambridge Univ. Press, 1992.

20. T. Becher and M. Neubert, Phys.Rev.Lett.98:022003,2007, [hep-ph 0610067]; see also M. Misiak et al, Phys.Rev.Lett.98:022002,2007 [hep-ph/0609232].

21. G.L. Fogli et al, *Proceedings of the 40th Rencontres de Moriond on Electroweak Interactions and Unified Theories*, La Thuile, Italy, hep-ph/0506307; M. Maltoni et al, New J. Phys. 6 (2004) 122, [hep-ph/0405172].

22. K. Zuber, Acta Phys.Polon.B37:1905–1921,2006; [nucl-ex/0610007].

23. J. Lesgourgues and S. Pastor, Phys.Rept.429:307–379,2006; [astro-ph/0603494].

24. P. Minkowski, Phys. Lett. B67 (1977) 421; T. Yanagida, in *Proc. of the Workshop on Unified Theory and Baryon Number in the Universe*, KEK (1979); S. L. Glashow, in *Quarks and Leptons'*, Cargèse, ed. M. Lévy et al., Plenum (1980); M. Gell-Mann, P. Ramond and R. Slansky, in *Supergravity*, Stony Brook (1979); R. N. Mohapatra and G. Senjanovic, Phys. Rev. Lett., 44 (1980) 912.

25. Ta-Pei Cheng and Ling-Fong Li, *Gauge theory of elementary particle physics*, Oxford Univ. Press, (1988); D. Bailin and A. Love,*Introduction to gauge field theory (Revised edition)*, (1993); D. Y. Bardin and G. Passarino, *The standard model in the making : precision study of the electroweak interactions*; Oxford Clarendon Press, (1999).

26. J.M. Cornwall, D.N. Levin and G. Tiktopoulos, Phys. Rev. **D10**, 1145 (1974); C.E. Vayonakis, Lett. Nuovo Cim. **17**, 383 (1976); B.W. Lee, C. Quigg and H. Thacker, Phys. Rev. **D16**, 1519 (1977); M. S. Chanowitz and M. K. Gaillard, Nucl. Phys.**B261**, 379 (1985).

27. M. Veltman, Nucl. Phys.**B21**, 288 (1970); G.'t Hooft, Nucl. Phys. **B33**, 173 (1971); **B35**, 167 (1971).

28. B.W. Lee and J. Zinn-Justin, Phys. Rev. **D5**, 3121; 3137 (1972); **D7**, 1049 (1973).

29. C. Bouchiat, J. Iliopoulos and P. Meyer, Phys.Lett.B38:519,1972.

30. S. L. Adler, Phys.Rev.177:2426,1969; S. L. Adler and W. A. Bardeen, Phys.Rev.182:1517,1969; J.S. Bell and R. Jackiw, Nuovo Cim. A60:47,1969; W. A. Bardeen, Phys. Rev. **184** (1969) 1848.

31. G.G.Ross, "Grand Unified Theories", Benjamin, 1985; R.N.Mohapatra, "Unification and Supersymmetry" Springer-Verlag, 1986; S. Raby, hep-ph/0608183. [32] G. 't Hooft, Phys.Rev.Lett.37:8,1976; Phys. Rev. D **14** (1976) 3432 [Erratum-ibid. D **18** (1978) 2199].

32. G. 't Hooft, Phys.Rev.Lett.37:8,1976.

33. A. Belavin, A. Polyakov, A. Shvarts and Y. Tyupkin, Phys. Lett. B **59** (1975) 85.

34. W. Buchmuller, R.D. Peccei and T. Yanagida, Ann.Rev.Nucl.Part.Sci.55:311,2005; [hep-ph 0502169].

35. B. Odom, D. Hanneke, B. D'Urso and G. Gabrielse, Phys. Rev. Lett. 97: 030801 (2006).

36. Muon g-2 Collab., G. W. Bennett et al, Phys. Rev. D73, 072003 (2006).

37. G. Gabrielse et al, Phys.Rev.Lett. 97:030802,2006.

38. T. Kinoshita and N. Nio, Phys. Rev. D 73, 013003(2006).
39. T. Aoyama, M. Hayakawa, T. Kinoshita and M. Nio, Phys. Rev. Lett. **99** (2007) 110406 [arXiv:0706.3496 [hep-ph]]. T. Aoyama, M Hayakawa, T. Kinoshita and N. Nio, hep-ph/0706.3496.
40. M. Passera, Nucl. Phys. Proc. Suppl. **169** (2007) 213 [hep-ph/0702027 [HEP-PH]].
41. M. Davier, Nucl. Phys. Proc. Suppl. **169** (2007) 288 [hep-ph/0701163].
42. G. Altarelli, R. Kleiss and C.Verzegnassi (eds.), Z Physics at LEP1 (CERN 89- 08, Geneva, 1989), Vols. 1–3; Precision Calculations for the Z Resonance, ed. by D.Bardin, W.Hollik and G.Passarino, CERN Rep 95-03 (1995); M.I. Vysotskii, V.A. Novikov, L.B. Okun and A.N. Rozanov, hep-ph/9606253 or Phys.Usp.39:503–538,1996.
43. T. Appelquist and J. Carazzone, Phys.Rev.D11:2856,1975.
44. J. van der Bij and M.J.G. Veltman, Nucl.Phys.B231:205,1984.
45. The LEP and SLD Collaborations, Phys. Rep. **427** (2006) 257, hep-ex/0509008.
46. M. Awramik and M. Czakon, Phys. Rev. Lett. **89** (2002) 241801, [hep-ph/0208113]. M. Awramik, M. Czakon, A. Onishchenko and O. Veretin, Phys. Rev. D **68** (2003) 053004, [hep-ph/0209084]. M. Awramik and M. Czakon, Phys. Lett. B **568** (2003) 48, [hep-ph/0305248]. A. Freitas, W. Hollik, W. Walter and G. Weiglein, Phys. Lett. B **495** (2000) 338, [Erratum-ibid. B **570** (2003) 260], [hep-ph/0007091]. A. Freitas, W. Hollik, W. Walter and G. Weiglein, Nucl. Phys. B **632** (2002) 189, [Erratum-ibid. B **666** (2003) 305], [hep-ph/0202131]. A. Onishchenko and O. Veretin, Phys. Lett. B **551** (2003) 111 [hep-ph/0209010].
47. K. G. Chetyrkin, M. Faisst, J. H. Kuhn, P. Maierhofer and C. Sturm, Phys. Rev. Lett. **97** (2006) 102003, hep-ph/0605201; R. Boughezal and M. Czakon, Nucl. Phys. B **755** (2006) 221, [hep-ph/0606232].
48. M. Awramik, M. Czakon, A. Freitas and G. Weiglein, Phys. Rev. Lett. **93** (2004) 201805, [hep-ph/0407317]; M. Awramik, M. Czakon and A. Freitas, hep-ph/0608099; Phys. Lett. B **642** (2006) 563, [hep-ph/0605339]; W. Hollik, U. Meier and S. Uccirati, hep-ph/0610312.
49. G.Altarelli, R.Barbieri and F.Caravaglios, Int. J. Mod. Phys. A 13(1998)1031 and references therein.
50. M.E. Peskin and T. Takeuchi, Phys. Rev. Lett. 65 (1990) 964; Phys. Rev. D46 (1991) 381.
51. See, for example, M. Lindner, Z. *Phys.* **31**, 295 (1986); T. Hambye and K. Riesselmann, *Phys. Rev.* **D55**, 7255 (1997); hep-ph/9610272.
52. The NuTeV Collaboration, G.P. Zeller et al., Phys. Rev. Lett. **88** (2002) 091802.
53. M. Yu. Kuchiev and V. V. Flambaum, hep-ph/0305053.
54. The SLAC E158 Collaboration, P.L. Anthony et al., hep-ex/0312035, hep-ex/0403010.
55. P. Gambino, Int.J.Mod.Phys.A19:808,2004, [hep-ph/0311257].
56. D. Choudhury, T.M.P. Tait and C.E.M. Wagner, Phys. Rev. D 65 (2002) 053002, hep-ph/0109097.
57. A. Djouadi, G. Moreau and F. Richard, Nucl. Phys. B **773** (2007) 43 [hep-ph/0610173].
58. N. Cabibbo, L. Maiani, G. Parisi and R. Petronzio, *Nucl. Phys.* **B158**,295 (1979).
59. M. Sher, *Phys. Rep.* **179**, 273 (1989); *Phys. Lett.* **B317**, 159 (1993).
60. G. Altarelli and G. Isidori, *Phys. Lett.* **B337**, 141 (1994); J.A. Casas, J.R. Espinosa and M. Quirós, *Phys. Lett.* **B342**, 171 (1995); J.A. Casas et al., *Nucl. Phys.* **B436**, 3 (1995); E**B439**, 466 (1995); M. Carena and C.E.M. Wagner, *Nucl. Phys.* **B452**, 45 (1995).
61. G. Isidori, G. Ridolfi and A. Strumia; *Nucl. Phys.* **B609**,387 (2001).
62. T. Hambye and K.Riesselmann, Phys.Rev.D55:7255,1997.
63. B. W. Lee, C. Quigg and H.B. Thacker, Phys.Rev. D16:1519,1977.
64. A. Djouadi, hep-ph/0503172.
65. E. Accomando et al., Phys. Rep. 299 (1998) 1.
66. H. E. Haber, G. Kane, S. Dawson and J. F. Gunion, *The Higgs Hunter's Guide*, Westview, 1990.
67. R. Barbieri and A. Strumia, hep-ph/0007265.
68. H.P. Nilles, Phys. Rep. C110 (1984) 1; H.E. Haber and G.L. Kane, Phys. Rep. C117 (1985) 75; R. Barbieri, Riv. Nuovo Cim. 11 (1988) 1; S. P. Martin, hep-ph/9709356; M. Drees, R. Godbole and P. Roy, *Theory and Phenomenology of Sparticles*, World Sci. (2004).
69. K. Lane, hep-ph/0202255; R.S. Chivukula, hep-ph/0011264.

70. R. Rattazzi, hep-ph/ 0607055.
71. L. Randall and R. Sundrum, Phys. Rev. Lett. 83 (1999) 3370; 83 (1999) 4690. W.D. Goldberger and M. B. Wise, Phys. Rev. Letters 83 (1999) 4922.
72. M. Schmaltz, hep-ph/0210415;H. C. Cheng and I. Low, JHEP **0408** (2004) 061 [hep-ph/0405243]; J. Hubisz, P. Meade, A. Noble and M. Perelstein, JHEP **0601** (2006) 135 [hep-ph/0506042].

Chapter 4
QCD: The Theory of Strong Interactions

Guido Altarelli and Stefano Forte

4.1 Introduction

This Chapter[1] is devoted to a concise introduction to Quantum Chromo-Dynamics (QCD), the theory of strong interactions [1–3]. We start with a general introduction where a broad overview of the strong interactions is presented. The basic principles and the main applications of perturbative QCD will be discussed first (for reviews of the subject, see, for example, [4–6]). Then the methods of non perturbative QCD will be introduced, first the analytic approaches and then the simulations of the theory on a discrete space-time lattice. The main emphasis will be on ideas with a minimum of technicalities.

As discussed in Chap. 2 the QCD theory of strong interactions is an unbroken gauge theory based on the group $SU(3)$ of colour. The eight massless gauge bosons are the gluons g_μ^A and matter fields are colour triplets of quarks q_i^a (in different flavours i). Quarks and gluons are the only fundamental fields of the Standard Model (SM) with strong interactions (hadrons). As discussed in Chap. 2, the statement that

---

The author "G. Altarelli" is deceased at the time of publication.

[1] See Chap. 2 for a general introduction to Chaps. 2–4 with updated references.

G. Altarelli
University of Rome 3, Rome, Italy

S. Forte (✉)
Dipartimento di Fisica, Università di Milano, Milano, Italy

QCD is a renormalisable gauge theory based on the group $SU(3)$ with colour triplet quark matter fields [7] fixes the QCD lagrangian density to be:

$$\mathcal{L} = -\frac{1}{4} \sum_{A=1}^{8} F^{A\mu\nu} F_{\mu\nu}^{A} + \sum_{j=1}^{n_f} \bar{q}_j (i\slashed{D} - m_j) q_j \qquad (4.1)$$

Here: q_j are the quark fields (of n_f different flavours) with mass m_j; $\slashed{D} = D_\mu \gamma^\mu$, where γ^μ are the Dirac matrices and D_μ is the covariant derivative:

$$D_\mu = \partial_\mu + i e_s \mathbf{g}_\mu; \qquad (4.2)$$

e_s is the gauge coupling, later we will mostly use, in analogy with QED

$$\alpha_s = \frac{e_s^2}{4\pi}; \qquad (4.3)$$

$\mathbf{g}_\mu = \sum_A t^A g_\mu^A$ where g_μ^A, $A = 1, 8$, are the gluon fields and t^A are the $SU(3)$ group generators in the triplet representation of quarks (i.e. t_A are 3×3 matrices acting on q); the generators obey the commutation relations $[t^A, t^B] = iC_{ABC}t^C$ where C_{ABC} are the complete antisymmetric structure constants of $SU(3)$ (the normalisation of C_{ABC} and of e_s is specified by $Tr[t^A t^B] = \delta^{AB}/2$);

$$F_{\mu\nu}^A = \partial_\mu g_\nu^A - \partial_\nu g_\mu^A - e_s C_{ABC} g_\mu^B g_\nu^C \qquad (4.4)$$

For quantisation the classical Lagrangian in Eq. (4.1) must be enlarged to contain gauge fixing and ghost terms, as described in Chap. 2. The Feynman rules of QCD are listed in Fig. 4.1. The physical vertices in QCD include the gluon-quark-antiquark vertex, analogous to the QED photon-fermion-antifermion coupling, but also the 3-gluon and 4-gluon vertices, of order e_s and e_s^2 respectively, which have no analogue in an abelian theory like QED.

The QCD lagrangian in Eq. (4.1) has a simple structure but a very rich dynamical content. It gives rise to a complex spectrum of hadrons, it implies the striking properties of confinement and asymptotic freedom, is endowed with an approximate chiral symmetry which is spontaneously broken, has a highly non trivial topological vacuum structure (instantons, $U(1)_A$ symmetry breaking, strong CP violation (which is a problematic item in QCD possibly connected with new physics, like axions, ...), an intriguing phase transition diagram (colour deconfinement, quark-gluon plasma, chiral symmetry restoration, colour superconductivity, ...).

Confinement is the property that no isolated coloured charge can exist but only colour singlet particles. For example, the potential between a quark and an antiquark has been studied on the lattice. It has a Coulomb part at short distances and a linearly rising term at long distances:

$$V_{q\bar{q}} \approx C_F[\frac{\alpha_s(r)}{r} + \dots + \sigma r] \qquad (4.5)$$

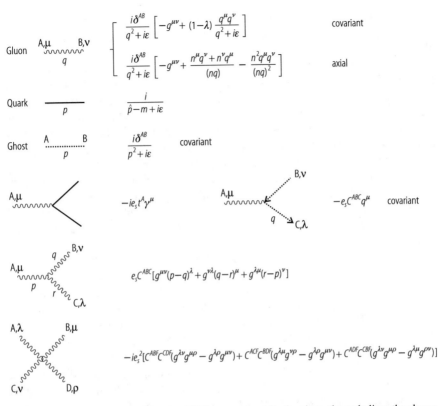

Fig. 4.1 Feynman rules for QCD. The solid lines represent the fermions, the curly lines the gluons, and the dotted lines represent the ghosts (see Chap. 2). The gauge parameter is denoted by λ. The 3-gluon vertex is written as if all gluon lines are outgoing

where

$$C_F = \frac{1}{N_C} \sum_A t^A t^A = \frac{N_C^2 - 1}{2N_C} \tag{4.6}$$

with N_C the number of colours ($N_C = 3$ in QCD). The scale dependence of α_s (the distance r is Fourier-conjugate to momentum transfer) will be explained in detail in the following. The understanding of the confinement mechanism has much improved thanks to lattice simulations of QCD at finite temperatures and densities. The slope decreases with increasing temperature until it vanishes at a critical temperature T_C. Above T_C the slope remains zero. The phase transitions of colour deconfinement and of chiral restauration appear to happen together on the lattice. A rapid transition is observed in lattice simulations where the energy density $\epsilon(T)$ is seen to sharply increase near the critical temperature for deconfinement and chiral restauration. The critical parameters and the nature of the phase transition depend on the number of quark flavours n_f and on their masses. For example, for $n_f = 2$ or $2 + 1$ (i.e. two light u and d quarks and one heavier s quark), $T_C \sim 175\,\text{MeV}$ and $\epsilon(T_C) \sim 0.5 - 1.0\,\text{GeV/fm}^3$. For realistic values of the masses m_s and $m_{u,d}$ the phase transition appears to be a second order one, while it becomes first order for very small or very large $m_{u,d,s}$. The hadronic phase and the deconfined phase are separated by a crossover line at small densities and by a critical line at high densities. Determining the exact location of the critical point in T and μ_B is an important challenge for theory which is also important for the interpretation of heavy ion collision experiments. At high densities the colour superconducting phase is also present with bosonic diquarks acting as Cooper pairs.

A large investment is being done in experiments of heavy ion collisions with the aim of finding some evidence of the quark gluon plasma phase. Many exciting results have been found at the CERN SPS in the past years and more recently at RHIC. The status of the experimental search for the quark-gluon plasma will be reviewed in Chap. 7.

The linearly rising term in the potential makes it energetically impossible to separate a $q - \bar{q}$ pair. If the pair is created at one space-time point, for example in e^+e^- annihilation, and then the quark and the antiquark start moving away from each other in the center of mass frame, it soon becomes energetically favourable to create additional pairs, smoothly distributed in rapidity between the two leading charges, which neutralise colour and allow the final state to be reorganised into two jets of colourless hadrons, that communicate in the central region by a number of "wee" hadrons with small energy. It is just like the familiar example of the broken magnet: if you try to isolate a magnetic pole by stretching a dipole, the magnet breaks down and two new poles appear at the breaking point.

Confinement is essential to explain why nuclear forces have very short range while massless gluon exchange would be long range. Nucleons are colour singlets and they cannot exchange colour octet gluons but only colourless states. The lightest colour singlet hadronic particles are pions. So the range of nuclear forces is fixed by the pion mass $r \simeq m_\pi^{-1} \simeq 10^{-13}\,cm : V \approx \exp(-m_\pi r)/r$.

Why $SU(N_C = 3)_{colour}$? The selection of $SU(3)$ as colour gauge group is unique in view of a number of constraints. (a) The group must admit complex representations because it must be able to distinguish a quark from an antiquark. In fact there are meson states made up of $q\bar{q}$ but not analogous qq bound states. Among simple groups this restricts the choice to $SU(N)$ with $N \geq 3$, $SO(4N + 2)$ with $N \geq 2$ (taking into account that $SO(6)$ has the same algebra as $SU(4)$) and $E(6)$. (b) The group must admit a completely antisymmetric colour singlet baryon made up of 3 quarks: qqq. In fact, from the study of hadron spectroscopy we know that the low lying baryons, completing an octet and a decuplet of (flavour) $SU(3)$ (the approximate symmetry that rotate the three light quarks u, d and s), are made up of three quarks and are colour singlets. The qqq wave function must be completely antisymmetric in colour in order to agree with Fermi statistics. Indeed if we consider, for example, a N^{*++} with spin z-component $+3/2$, this is made up of $(u \Uparrow u \Uparrow u \Uparrow)$ in an s-state. Thus its wave function is totally symmetric in space, spin and flavour so that complete antisymmetry in colour is required by Fermi statistics. In QCD this requirement is very simply satisfied by $\epsilon_{abc}q^a q^b q^c$ where a, b, c are $SU(3)_{colour}$ indices. (c) The choice of $SU(N_C = 3)_{colour}$ is confirmed by many processes that directly measure N_C. Some examples are listed here. The total rate for hadronic production in e^+e^- annihilation is linear in N_C. Precisely if we consider $R = \sigma(e^+e^- \to hadrons)/\sigma_{point}(e^+e^- \to \mu^+\mu^-)$ above $b\bar{b}$ threshold and below m_Z and we neglect small computable radiative corrections (that will be discussed later in Sect. 4.5) we have a sum of individual contributions (proportional to Q^2, where Q is the electric charge in units of the proton charge) from $q\bar{q}$ final states with $q = u, c, d, s, b$:

$$R \approx N_C[2 \cdot \frac{4}{9} + 3 \cdot \frac{1}{9}] \approx N_C\frac{11}{9} \tag{4.7}$$

The data neatly indicate $N_C = 3$ as seen from Fig. 4.2 [9]. The slight excess of the data with respect to the value 11/3 is due to the QCD radiative corrections (Sect. 4.5). Similarly we can consider the branching ratio $B(W^- \to e^-\bar{\nu})$, again in Born approximation. The possible fermion-antifermion ($f\bar{f}$) final states are for $f = e^-, \mu^-, \tau^-, d, s$ (there is no $f = b$ because the top quark is too heavy for $b\bar{t}$ to occur). Each channel gives the same contribution, except that for quarks we have N_C colours:

$$B(W^- \to e^-\bar{\nu}) \approx \frac{1}{3 + 2N_C} \tag{4.8}$$

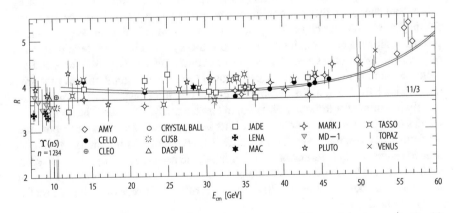

Fig. 4.2 Comparison of the data on $R = \sigma(e^+e^- \to hadrons)/\sigma_{point}(e^+e^- \to \mu^+\mu^-)$ with the QCD prediction [9]. $N_C = 3$ is indicated

For $N_C = 3$ we obtain $B = 11\%$ and the experimental number is $B = 10.7\%$. Another analogous example is the branching ratio $B(\tau^- \to e^- \bar{v}_e v_\tau)$. From the final state channels with $f = e^-$, μ^-, d we find

$$B(\tau^- \to e^- \bar{v}_e v_\tau) \approx \frac{1}{2 + N_C} \tag{4.9}$$

For $N_C = 3$ we obtain $B = 20\%$ and the experimental number is $B = 18\%$ (the less accuracy in this case is explained by the larger radiative and phase-space corrections because the mass of τ^- is much smaller than m_W). An important process that is quadratic in N_C is the rate $\Gamma(\pi^0 \to 2\gamma)$. This rate can be reliably calculated from a solid theorem in field theory which has to do with the chiral anomaly:

$$\Gamma(\pi^0 \to 2\gamma) \approx (\frac{N_C}{3})^2 \frac{\alpha^2 m_{\pi^0}^3}{32\pi^3 f_\pi^2} = (7.73 \pm 0.04)(\frac{N_C}{3})^2 \text{ eV} \tag{4.10}$$

where the prediction is obtained for $f_\pi = (130.7 \pm 0.37)\,\text{MeV}$. The experimental result is $\Gamma = (7.7\pm0.5)\,\text{eV}$ in remarkable agreement with $N_C = 3$. There are many more experimental confirmations that $N_C = 3$: for example the rate for Drell-Yan processes (see Sect. 5.4) is inversely proportional to N_C.

How do we get testable predictions from QCD? On the one hand there are non perturbative methods. The most important at present is the technique of lattice simulations: it is based on first principles, it has produced very valuable results on confinement, phase transitions, bound states, hadronic matrix elements and so on, and it is by now an established basic tool. The main limitation is from computing power and therefore there is continuous progress and a lot of good perspectives for the future. Another class of approaches is based on effective lagrangians which provide simpler approximations than the full theory, valid in some definite domain of physical conditions. Chiral lagrangians are based on soft pion theorems and are valid for suitable processes at energies below 1 GeV. Heavy quark effective theories are obtained from expanding in inverse powers of the heavy quark mass and are mainly important for the study of b and, to less accuracy, c decays. The approach of QCD sum rules has led to interesting results but appears to offer not much potential for further development. Similarly specific potential models for quarkonium have a limited range of application. On the other hand, the perturbative approach, based on asymptotic freedom, still remains the main quantitative connection to experiment, due to its wide range of applicability to all sorts of "hard" processes. To perturbative QCD will be devoted the next sections.

4.2 Massless QCD and Scale Invariance

As discussed in Chap. 2, the QCD lagrangian in Eq. (4.1) only specifies the theory at the classical level. The procedure for quantisation of gauge theories involves a number of complications that arise from the fact that not all degrees of freedom of gauge fields are physical because of the constraints from gauge invariance which can be used to eliminate the dependent variables. This is already true for abelian theories and we are familiar with the QED case. One introduces a gauge fixing term (an additional term in the lagrangian density that acts as a Lagrange multiplier in the action extremisation). One can choose to preserve manifest Lorentz invariance. In this case, one adopts a covariant gauge, like the Lorentz gauge, and in QED one proceeds according to the formalism of Gupta-Bleuler. Or one can give up explicit formal covariance and work in a non covariant gauge, like the Coulomb or the axial gauges, and only quantise the physical degrees of freedom (in QED the transverse components of the photon field). While this is all for an abelian gauge theory, in the non-abelian case some additional complications arise, in particular the necessity to introduce ghosts for the formulation of Feynman rules. As we have seen, there are in general as many ghost fields as gauge bosons and they appear in the form of a transformation Jacobian in the Feynman diagram functional integral. Ghosts only propagate in closed loops and their vertices with gluons can be included as additional terms in the lagrangian density which are fixed once the gauge fixing terms and their infinitesimal gauge transformations are specified. Finally the complete Feynman rules in a given gauge can be obtained and they appear in Fig. 4.1.

Once the Feynman rules are derived we have a formal perturbative expansion but loop diagrams generate infinities. First a regularisation must be introduced, compatible with gauge symmetry and Lorentz invariance. This is possible in QCD. In principle one can introduce a cut-off K (with dimensions of energy), for example, a' la Pauli-Villars. But at present the universally adopted regularisation procedure is dimensional regularisation that we will briefly describe later on. After regularisation the next step is renormalisation. In a renormalisable theory (like for all gauge theories in four spacetime dimensions and for QCD in particular) the dependence on the cutoff can be completely reabsorbed in a redefinition of particle masses, of gauge coupling(s) and of wave function normalisations. After renormalisation is achieved the perturbative definition of the quantum theory that corresponds to a classical lagrangian like in Eq. (4.1) is completed. In the QCD Lagrangian of Eq. (4.1) quark masses are the only parameters with physical dimensions (we work in the natural system of units $\hbar = c = 1$). Naively we would expect that massless QCD is scale invariant. This is actually true at the classical level. Scale invariance implies that dimensionless observables should not depend on the absolute scale of energy but only on ratios of energy-dimensional variables. The massless limit should be relevant for the asymptotic large energy limit of processes which are non singular for $m \to 0$.

The naive expectation that massless QCD should be scale invariant is false in the quantum theory. The scale symmetry of the classical theory is unavoidably destroyed by the regularisation and renormalisation procedure which introduce a dimensional parameter in the quantum version of the theory. When a symmetry of the classical theory is necessarily destroyed by quantisation, regularisation and renormalisation one talks of an "anomaly". So, in this sense, scale invariance in massless QCD is anomalous.

While massless QCD is finally not scale invariant, the departures from scaling are asymptotically small, logarithmic and computable. In massive QCD there are additional mass corrections suppressed by powers of m/E, where E is the energy scale (for non singular processes in the limit $m \rightarrow 0$). At the parton level (q and g) we can conceive to apply the asymptotics from massless QCD to processes and observables (we use the word "processes" for both) with the following properties ("hard processes"). (a) All relevant energy variables must be large:

$$E_i = z_i Q, \qquad Q >> m_j; \qquad z_i: \text{scaling variables o(1)} \qquad (4.11)$$

(b) There should be no infrared singularities (one talks of "infrared safe" processes). (c) The processes concerned must be finite for $m \rightarrow 0$ (no mass singularities). To possibly satisfy these criteria processes must be as "inclusive" as possible: one should include all final states with massless gluon emission and add all mass degenerate final states (given that quarks are massless also $q - \bar{q}$ pairs can be massless if "collinear", that is moving together in the same direction at the common speed of light).

In perturbative QCD one computes inclusive rates for partons (the fields in the lagrangian, that is, in QCD, quarks and gluons) and takes them as equal to rates for hadrons. Partons and hadrons are considered as two equivalent sets of complete states. This is called "global duality" and it is rather safe in the rare instance of a totally inclusive final state. It is less so for distributions, like distributions in the invariant mass M ("local duality") where it can be reliable only if smeared over a sufficiently wide bin in M.

Let us discuss more in detail infrared and collinear safety. Consider, for example, a quark virtual line that ends up into a real quark plus a real gluon (Fig. 4.3). For the propagator we have:

$$\text{propagator} = \frac{1}{(p+k)^2 - m^2} = \frac{1}{2(p \cdot k)} = \frac{1}{2E_k E_p} \cdot \frac{1}{1 - \beta_p \cos \theta} \qquad (4.12)$$

Since the gluon is massless, E_k can vanish and this corresponds to an infrared singularity. Remember that we have to take the square of the amplitude and integrate over the final state phase space, or, in this case, all together, dE_k/E_k. Indeed we get $1/E_k^2$ from the squared amplitude and $d^3k/E_k \sim E_k dE_k$ from the phase space.

Fig. 4.3 The splitting of a virtual quark into a quark and a gluon

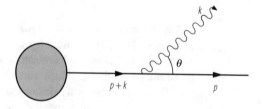

$p+k$ $\qquad\qquad$ θ $\qquad$ k $\qquad$ p

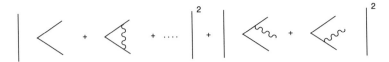

Fig. 4.4 The diagrams contributing to the total cross-section $e^+e^- \rightarrow$ hadrons at order α_s. For simplicity, only the final state quarks and (virtual or real) gluons are drawn

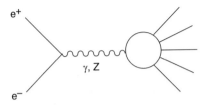

Fig. 4.5 The total cross-section $e^+e^- \rightarrow$ hadrons

Also, for $m \rightarrow 0$, $\beta_p = \sqrt{1 - m^2/E_p^2} \rightarrow 1$ and $(1 - \beta_p \cos \theta)$ vanishes at $\cos \theta = 1$. This leads to a collinear mass singularity.

There are two very important theorems on infrared and mass singularities. The first one is the Bloch-Nordsieck theorem [8]: infrared singularities cancel between real and virtual diagrams (see Fig. 4.4) when all resolution indistinguishable final states are added up. For example, for each real detector there is a minimum energy of gluon radiation that can be detected. For the cancellation of infrared divergences, one should add all possible gluon emission with a total energy below the detectable minimum. The second one is the Kinoshita-Lee, Nauenberg theorem [10]: mass singularities connected with an external particle of mass m are canceled if all degenerate states (that is with the same mass) are summed up. That is for a final state particle of mass m we should add all final states that in the limit $m \rightarrow 0$ have the same mass, also including gluons and massless pairs. If a completely inclusive final state is taken, only the mass singularities from the initial state particles remain (we shall see that they will be absorbed inside the non perturbative parton densities, which are probability densities of finding the given parton in the initial hadron).

Hard processes to which the massless QCD asymptotics can possibly apply must be infrared and collinear safe, that is they must satisfy the requirements from the Bloch-Nordsieck and the Kinoshita-Lee-Nauenberg theorems. We give now some examples of important hard processes. One of the simplest hard processes is the totally inclusive cross section for hadron production in e^+e^- annihilation, Fig. 4.5, parameterised in terms of the already mentioned dimensionless observable $R = \sigma(e^+e^- \rightarrow hadrons)/\sigma_{point}(e^+e^- \rightarrow \mu^+\mu^-)$. The pointlike cross section in the denominator is given by $\sigma_{point} = 4\pi\alpha^2/3s$, where $s = Q^2 = 4E^2$ is the squared total center of mass energy and Q is the mass of the exchanged virtual gauge boson. At parton level the final state is $(q\bar{q} + n\,g + n'\,q'\bar{q}')$ and n and n' are limited at each order of perturbation theory. It is assumed that the conversion of partons into

Fig. 4.6 Deep inelastic
lepto-production

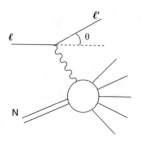

hadrons does not affect the rate (it happens with probability 1). We have already mentioned that in order for this to be true within a given accuracy an averaging over a sufficiently large bin of Q must be understood. The binning width is larger in the vicinity of thresholds: for example when one goes across the charm $c\bar{c}$ threshold the physical cross-section shows resonance bumps which are absent in the smooth partonic counterpart which however gives an average of the cross-section.

A very important class of hard processes is Deep Inelastic Scattering (DIS)

$$l + N \to l' + X \qquad l = e^{\pm}, \mu^{\pm}, \nu, \bar{\nu} \tag{4.13}$$

which has played and still plays a very important role for our understanding of QCD and of nucleon structure. For the processes in Eq. (4.13), shown in Fig. 4.6, we have, in the lab system where the nucleon of mass m is at rest:

$$Q^2 = -q^2 = -(k - k')^2 = 4EE' \sin^2 \theta/2; \qquad m\nu = (p.q); \qquad x = \frac{Q^2}{2m\nu} \tag{4.14}$$

In this case the virtual momentum q of the gauge boson is spacelike. x is the familiar Bjorken variable. The DIS processes in QCD will be extensively discussed in Sect. 4.5

4.3 The Renormalisation Group and Asymptotic Freedom

In this section we aim at providing a reasonably detailed introduction to the renormalisation group formalism and the concept of running coupling which leads to the result that QCD has the property of asymptotic freedom. We start with a summary on how renormalisation works.

In the simplest conceptual situation imagine that we implement regularisation of divergent integrals by introducing a dimensional cut-off K that respects gauge and Lorentz invariance. The dependence of renormalised quantities on K is eliminated by absorbing it into a redefinition of m (the quark mass: for simplicity we assume a single flavour here), the gauge coupling e (can be e in QED or e_s in QCD) and the wave function renormalisation factors $Z_{q,g}^{1/2}$ for q and g, using suitable renormalisation conditions (that is precise definitions of m, g and Z that can be implemented order by order in perturbation theory). For example we can define the renormalised mass m as the position of the pole in the quark propagator and, similarly, the normalisation Z_q as the residue at the pole:

$$\text{Propagator} = \frac{Z_q}{p^2 - m^2} + \text{no} - \text{pole terms} \qquad (4.15)$$

The renormalised coupling e can be defined in terms of a renormalised 3-point vertex at some specified values of the external momenta. Precisely, we consider a one particle irreducible vertex (1PI). We recall that a connected Green function is the sum of all connected diagrams, while 1PI Green functions are the sum of all diagrams that cannot be separated into two disconnected parts by cutting only one line.

We now become more specific by concentrating on the case of massless QCD. If we start from a vanishing mass at the classical (or "bare") level, $m_0 = 0$, the mass is not renormalised because it is protected by a symmetry, chiral symmetry. The conserved currents of chiral symmetry are axial currents: $\bar{q}\gamma_\mu\gamma_5 q$. The divergence of the axial current gives, by using the Dirac equation, $\partial^\mu(\bar{q}\gamma_\mu\gamma_5 q) = 2m\bar{q}\gamma_5 q$. So the axial current and the corresponding axial charge are conserved in the massless limit. Since QCD is a vector theory we have not to worry about chiral anomalies in this respect. So one can choose a regularisation that preserves chiral symmetry besides gauge and Lorentz symmetry. Then the renormalised mass remains zero. The renormalised propagator has the form in Eq. (4.15) with $m = 0$.

The renormalised coupling e_s can be defined from the renormalised 1PI 3-gluon vertex at a scale $-\mu^2$ (Fig. 4.7):

$$V_{bare}(p^2, q^2, r^2) = Z V_{ren}(p^2, q^2, r^2), \quad Z = Z_g^{-3/2}, \quad V_{ren}(-\mu^2, -\mu^2, -\mu^2) \to e_s \qquad (4.16)$$

We could as well use the quark-gluon vertex or any other vertex which coincides with e_0 in lowest order (even the ghost-gluon vertex, if we want). With a regularisation and renormalisation that preserves gauge invariance we are guaranteed that all these different definitions are equivalent.

Here V_{bare} is what is obtained from computing the Feynman diagrams including, for example, the 1-loop corrections at the lowest non trivial order (V_{bare} is defined

Fig. 4.7 Diagrams contributing to the 1PI 3-gluon vertex at the one-loop approximation level

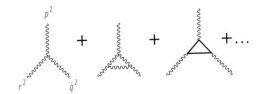

as the scalar function multiplying the vertex tensor, normalised in such a way that it coincides with e_{s0} in lowest order). V_{bare} contains the cut-off K but does not know about μ. Z is a factor that depends both on the cut-off and on μ but not on momenta. Because of infrared singularities the defining scale μ cannot vanish. The negative value $-\mu^2 < 0$ is chosen to stay away from physical cuts (a gluon with negative virtual mass cannot decay). Similarly, in the massless theory, we can define Z_g^{-1} as the inverse gluon propagator (the 1PI 2-point function) at the same scale $-\mu^2$ (the vanishing mass of the gluon is guaranteed by gauge invariance).

After computing all 1-loop diagrams indicated in Fig. 4.7 we have:

$$
\begin{aligned}
V_{bare}(p^2, p^2, p^2) &= e_{0s}[1 + c\alpha_{0s} \cdot \log \frac{K^2}{p^2} + \ldots] = \\
&= [1 + c\alpha_s \cdot \log \frac{K^2}{-\mu^2} + \ldots]e_{0s}[1 + c\alpha_s \cdot \log \frac{-\mu^2}{p^2} + \ldots] \\
&= Z_V^{-1} e_{0s}[1 + c\alpha_s \cdot \log \frac{-\mu^2}{p^2} + \ldots] \\
&= [1 + d\alpha_s \cdot \log \frac{K^2}{-\mu^2} + \ldots]e_s[1 + c\alpha_s \cdot \log \frac{-\mu^2}{p^2} + \ldots] \\
&= Z_g^{-3/2} V_{ren}
\end{aligned}
\tag{4.17}
$$

Note the replacement of e_0 with e in the second step, compensated by changing c into d in the first bracket (corresponding to $e_0 = Z_g^{-3/2} Z_V e$). The definition of e_s demands that one precisely specifies what is included in Z. For this, in a given renormalisation scheme, a prescription is fixed to specify the finite terms that go into Z (i.e. the terms of order α_s that accompany $\log K^2$). Then V_{ren} is specified and the renormalised coupling is defined from it according to Eq. (4.16). For example, in the momentum subtraction scheme we define $V_{ren}(p^2, p^2, p^2) = e_s + V_{bare}(p^2, p^2, p^2) - V_{bare}(-\mu^2, -\mu^2, -\mu^2)$, which is equivalent to say, at 1-loop, that all finite terms that do not vanish at $p^2 = -\mu^2$ are included in Z.

A crucial observation is that V_{bare} depends on K but not on μ, which is only introduced when Z, V_{ren} and hence α_s are defined. (From here on, for shorthand, we write α to indicate either the QED coupling or the QCD coupling α_s). More in general for a generic Green function G, we similarly have:

$$
G_{bare}(K^2, \alpha_0, p_i^2) = Z_G G_{ren}(\mu^2, \alpha, p_i^2)
\tag{4.18}
$$

so that we have:

$$
\frac{dG_{bare}}{d \log \mu^2} = \frac{d}{d \log \mu^2}[Z_G G_{ren}] = 0
\tag{4.19}
$$

or

$$Z_G[\frac{\partial}{\partial \log \mu^2} + \frac{\partial \alpha}{\partial \log \mu^2}\frac{\partial}{\partial \alpha} + \frac{1}{Z_G}\frac{\partial Z_G}{\partial \log \mu^2}]G_{ren} = 0 \qquad (4.20)$$

Finally the renormalisation group equation (RGE) can be written as:

$$[\frac{\partial}{\partial \log \mu^2} + \beta(\alpha)\frac{\partial}{\partial \alpha} + \gamma_G(\alpha)]G_{ren} = 0 \qquad (4.21)$$

where

$$\beta(\alpha) = \frac{\partial \alpha}{\partial \log \mu^2} \qquad (4.22)$$

and

$$\gamma_G(\alpha) = \frac{\partial \log Z_G}{\partial \log \mu^2} \qquad (4.23)$$

Note that $\beta(\alpha)$ does not depend on which Green function G we are considering, but it is a property of the theory and the renormalisation scheme adopted, while $\gamma_G(\alpha)$ also depends on G. Strictly speaking the RGE as written above is only valid in the Landau gauge ($\lambda = 0$). In other gauges an additional term that takes the variation of the gauge fixing parameter λ should also be included. We omit this term, for simplicity, as it is not relevant at the 1-loop level.

Assume that we want to apply the RGE to some hard process at a large scale Q, related to a Green function G that we can always take as dimensionless (by multiplication by a suitable power of Q). Since the interesting dependence on Q will be logarithmic we introduce the variable t as :

$$t = \log \frac{Q^2}{\mu^2} \qquad (4.24)$$

Then we can write $G_{ren} \equiv F(t, \alpha, x_i)$ where x_i are scaling variables (we often omit to write them in the following). In the naive scaling limit F should be independent of t. To find the actual dependence on t, we want to solve the RGE

$$[-\frac{\partial}{\partial t} + \beta(\alpha)\frac{\partial}{\partial \alpha} + \gamma_G(\alpha)]G_{ren} = 0 \qquad (4.25)$$

with a given boundary condition at $t = 0$ (or $Q^2 = \mu^2$): $F(0, \alpha)$.

We first solve the RGE in the simplest case that $\gamma_G(\alpha) = 0$. This is not an unphysical case: for example, it applies to $R_{e^+e^-}$ where the vanishing of γ is related to the non renormalisation of the electric charge in QCD (otherwise the proton and the electron charge would not exactly compensate: this will be better explained in Sect. 4.5). So we consider the equation:

$$[-\frac{\partial}{\partial t} + \beta(\alpha)\frac{\partial}{\partial \alpha}]G_{ren} = 0 \qquad (4.26)$$

The solution is simply

$$F(t, \alpha) = F[0, \alpha(t)] \qquad (4.27)$$

where the "running coupling" $\alpha(t)$ is defined by:

$$t = \int_{\alpha}^{\alpha(t)} \frac{1}{\beta(\alpha')}d\alpha' \qquad (4.28)$$

Note that from this definition it follows that $\alpha(0) = \alpha$, so that the boundary condition is also satisfied. To prove that $F[0, \alpha(t)]$ is indeed the solution, we first take derivatives with respect to t and α (the two independent variables) of both sides of Eq. (4.28). By taking d/dt we obtain

$$1 = \frac{1}{\beta(\alpha(t)}\frac{\partial \alpha(t)}{\partial t} \qquad (4.29)$$

We then take $d/d\alpha$ and obtain

$$0 = -\frac{1}{\beta(\alpha)} + \frac{1}{\beta(\alpha(t)}\frac{\partial \alpha(t)}{\partial \alpha} \qquad (4.30)$$

These two relations make explicit the dependence of the running coupling on t and α:

$$\frac{\partial \alpha(t)}{\partial t} = \beta(\alpha(t)) \qquad (4.31)$$

$$\frac{\partial \alpha(t)}{\partial \alpha} = \frac{\beta(\alpha(t))}{\beta(\alpha)} \qquad (4.32)$$

Using these two equations one immediately checks that $F[0, \alpha(t)]$ is indeed the solution.

Similarly, one finds that the solution of the more general equation with $\gamma \neq 0$, Eq. (4.25), is given by:

$$F(t, \alpha) = F[0, \alpha(t)]\exp\int_{\alpha}^{\alpha(t)} \frac{\gamma(\alpha')}{\beta(\alpha')}d\alpha' \qquad (4.33)$$

In fact the sum of the two derivatives acting on the factor $F[0, \alpha(t)]$ vanishes and the exponential is by itself a solution of the complete equation. Note that the boundary condition is also satisfied.

The important point is the appearance of the running coupling that determines the asymptotic departures from scaling. The next step is to study the functional form of the running coupling. From Eq. (4.31) we see that the rate of change with t of the running coupling is determined by the β function. In turn $\beta(\alpha)$ is determined by the μ dependence of the renormalised coupling through Eq. (4.22). Clearly there is no dependence on μ of the basic 3-gluon vertex in lowest order (order e). The dependence starts at 1-loop, that is at order e^3 (one extra gluon has to be emitted and reabsorbed). Thus we obtain that in perturbation theory:

$$\frac{\partial e}{\partial \log \mu^2} \propto e^3 \tag{4.34}$$

Recalling that $\alpha = e^2/4\pi$, we have:

$$\frac{\partial \alpha}{\partial \log \mu^2} \propto 2e \frac{\partial e}{\partial \log \mu^2} \propto e^4 \propto \alpha^2 \tag{4.35}$$

Thus the behaviour of $\beta(\alpha)$ in perturbation theory is as follows:

$$\beta(\alpha) = \pm b\alpha^2[1 + b'\alpha + \ldots] \tag{4.36}$$

Since the sign of the leading term is crucial in the following discussion, we stipulate that always $b > 0$ and we make the sign explicit in front.

Let us make the procedure for computing the 1-loop beta function in QCD (or, similarly, in QED) more precise. The result of the 1loop 1PI diagrams for V_{ren} can be written down as (we denote e_s and α_s by e and α, for shorthand):

$$V_{ren} = e[1 + \alpha B_{3g} \log \frac{\mu^2}{-p^2} + \ldots] \tag{4.37}$$

V_{ren} satisfies the RGE:

$$[\frac{\partial}{\partial \log \mu^2} + \beta(\alpha)\frac{\partial e}{\partial \alpha}\frac{\partial}{\partial e} - \frac{3}{2}\gamma_g(\alpha)]V_{ren} = 0 \tag{4.38}$$

With respect to Eq. (4.21) the beta function term has been rewritten taking into account that V_{ren} starts with e and the anomalous dimension term arises from a factor $Z_g^{-1/2}$ for each gluon leg. In general for a n-leg 1PI Green function $V_{n,bare} = Z_g^{-n/2}V_{n,ren}$, if all external legs are gluons. Note that in the particular case of $V = V_3$ that is used to define e other Z factors are absorbed in the replacement

$Z_V^{-1} Z_g^{3/2} e_0 = e$. At 1-loop accuracy we replace $\beta(\alpha) = -b\alpha^2$ and $\gamma_g(\alpha) = \gamma_g^{(1)}\alpha$. All together one obtains:

$$b = 2(B_{3g} - \frac{3}{2}\gamma_g^{(1)})$$
(4.39)

Similarly we can write the diagrammatic expression and the RGE for the 1PI 2-gluon Green function which is the inverse gluon propagator Π (a scalar function after removing the gauge invariant tensor):

$$\Pi_{ren} = [1 + \alpha B_{2g} \log \frac{\mu^2}{-p^2} + \ldots]$$
(4.40)

and

$$[\frac{\partial}{\partial \log \mu^2} + \beta(\alpha)\frac{\partial}{\partial \alpha} - \gamma_g(\alpha)]\Pi_{ren} = 0$$
(4.41)

Notice that the normalisation and the phase of Π are specified by the lowest order term being one. In this case the β function term is negligible being of order α^2 (because Π is a function of e only through α). and we obtain:

$$\gamma_g^{(1)} = B_{2g}$$
(4.42)

Thus, finally:

$$b = 2(B_{3g} - \frac{3}{2}B_{2g})$$
(4.43)

By direct calculation at 1-loop one finds:

$$\text{QED}: \qquad \beta(\alpha) \sim +b\alpha^2 + \ldots \qquad b = \sum_i \frac{N_C Q_i^2}{3\pi}$$
(4.44)

where $N_C = 3$ for quarks and $N_C = 1$ for leptons and the sum runs over all fermions of charge $Q_i e$ that are coupled. Also, one finds:

$$\text{QCD}: \qquad \beta(\alpha) \sim -b\alpha^2 + \ldots \qquad b = \frac{11N_C - 2n_f}{12\pi}$$
(4.45)

where, as usual, n_f is the number of coupled flavours of quarks (we assume here that $n_f \leq 16$ so that $b > 0$ in QCD). If $\alpha(t)$ is small we can compute $\beta(\alpha(t))$ in perturbation theory. The sign in front of b then decides the slope of the coupling: $\alpha(t)$ increases with t (or Q^2) if β is positive at small α (QED), or $\alpha(t)$ decreases with t (or Q^2) if β is negative at small α (QCD). A theory like QCD where the running coupling vanishes asymptotically at large Q^2 is called (ultraviolet) "asymptotically free". An important result that has been proven is that in four spacetime dimensions all and only non-abelian gauge theories are asymptotically free.

Going back to Eq. (4.28) we replace $\beta(\alpha) \sim \pm b\alpha^2$, do the integral and perform a simple algebra. We find

$$\text{QED}: \quad \alpha(t) \sim \frac{\alpha}{1 - b\alpha t} \tag{4.46}$$

and

$$\text{QCD}: \quad \alpha(t) \sim \frac{\alpha}{1 + b\alpha t} \tag{4.47}$$

A slightly different form is often used in QCD. Defining $1/\alpha = b \log \mu^2/\Lambda_{QCD}^2$ we can write:

$$\alpha(t) \sim \frac{1}{\frac{1}{\alpha} + bt} = \frac{1}{b \log \frac{\mu^2}{\Lambda_{QCD}^2} + b \log \frac{Q^2}{\mu^2}} = \frac{1}{b \log \frac{Q^2}{\Lambda_{QCD}^2}} \tag{4.48}$$

We see that $\alpha(t)$ decreases logarithmically with Q^2 and that one can introduce a dimensional parameter Λ_{QCD} that replaces μ. Often in the following we will simply write Λ for Λ_{QCD}. Note that it is clear that Λ depends on the particular definition of α, not only on the defining scale μ but also on the renormalisation scheme (see, for example, the discussion in the next session). Through the parameter b, and in general through the β function, it also depends on the number n_f of coupled flavours. It is very important to note that QED and QCD are theories with "decoupling": up to the scale Q only quarks with masses $m \ll Q$ contribute to the running of α. This is clearly very important, given that all applications of perturbative QCD so far apply to energies below the top quark mass m_t. For the validity of the decoupling theorem [11] it is necessary that the theory where all the heavy particle internal lines are eliminated is still renormalisable and that the coupling constants do not vary with the mass. These requirements are true for the mass of heavy quarks in QED and QCD, but are not true in the electroweak theory where the elimination of the top would violate $SU(2)$ symmetry (because the t and b left quarks are in a doublet) and the quark couplings to the Higgs multiplet (hence to the longitudinal gauge bosons) are proportional to the mass. In conclusion, in QED and QCD, quarks with $m \gg Q$ do not contribute to n_f in the coefficients of the relevant β function. The effects of heavy quarks are power suppressed and can be taken separately into account. For example, in e^+e^- annihilation for $2m_c < Q < 2m_b$ the relevant asymptotics is for $n_f = 4$, while for $2m_b < Q < 2m_t$ $n_f = 5$. Going accross the b threshold the β function coefficients change, so the $\alpha(t)$ slope changes. But $\alpha(t)$ is continuous, so that Λ changes so as to keep constant $\alpha(t)$ at the matching point at $Q \sim o(2m_b)$. The effect on Λ is large: approximately $\Lambda_5 \sim 0.65\Lambda_4$.

Note the presence of a pole in Eqs. (4.46, 4.47) at $\pm b\alpha t = 1$, called the Landau pole, who realised its existence in QED already in the '50's. For $\mu \sim m_e$ (in QED) the pole occurs beyond the Planck mass. In QCD the Landau pole is located for negative t or at $Q < \mu$ in the region of light hadron masses. Clearly the issue of the definition and the behaviour of the physical coupling (which is always finite, when defined in terms of some physical process) in the region around the perturbative Landau pole is a problem that lies outside the domain of perturbative QCD.

The non leading terms in the asymptotic behaviour of the running coupling can in principle be evaluated going back to Eq. (4.36) and computing b' at 2-loops and so on. But in general the perturbative coefficients of $\beta(\alpha)$ depend on the definition of the renormalised coupling α (the renormalisation scheme), so one wonders whether it is worthwhile to do a complicated calculation to get b' if then it must be repeated for a different definition or scheme. In this respect it is interesting to remark that actually both b and b' are independent of the definition of α, while higher order coefficients do depend on that. Here is the simple proof. Two different perturbative definitions of α are related by $\alpha' \sim \alpha(1 + c_1\alpha + \ldots)$. Then we have:

$$
\begin{aligned}
\beta(\alpha') = \frac{d\alpha'}{d\log\mu^2} &= \frac{d\alpha}{d\log\mu^2}(1 + 2c_1\alpha + \ldots) \\
&= \pm b\alpha^2(1 + b'\alpha + \ldots)(1 + 2c_1\alpha + \ldots) \\
&= \pm b\alpha'^2(1 + b'\alpha' + \ldots)
\end{aligned}
\tag{4.49}
$$

which shows that, up to the first subleading order, $\beta(\alpha')$ has the same form as $\beta(\alpha)$. In QCD ($N_C = 3$) one has calculated:

$$
b' = \frac{153 - 19n_f}{2\pi(33 - 2n_f)}
\tag{4.50}
$$

By taking b' into account one can write the expression of the running coupling at next to the leading order (NLO):

$$
\alpha(Q^2) = \alpha_{LO}(Q^2)[1 - b'\alpha_{LO}(Q^2)\log\log\frac{Q^2}{\Lambda^2} + \ldots]
\tag{4.51}
$$

where $\alpha_{LO}^{-1} = b\log Q^2/\Lambda^2$ is the LO result (actually at NLO the definition of Λ is modified according to $b\log\mu^2/\Lambda^2 = 1/\alpha + b'\log b\alpha$).

Summarizing, we started from massless classical QCD which is scale invariant. But we have seen that the procedure of quantisation, regularisation and renormalisation necessarily breaks scale invariance. In the quantum QCD theory there is a scale of energy, Λ, which from experiment is of the order of a few hundred MeV, its precise value depending on the definition, as we shall see in detail. Dimensionless quantities depend on the energy scale through the running coupling which is a logarithmic function of Q^2/Λ^2. In QCD the running coupling decreases logarithmically at large Q^2 (asymptotic freedom), while in QED the coupling has the opposite behaviour.

4.4 More on the Running Coupling

In the previous section we have introduced the renormalised coupling α in terms of the 3-gluon vertex at $p^2 = -\mu^2$ (momentum subtraction). The Ward identities of QCD then ensure that the coupling defined from other vertices like the $\bar{q}qg$ vertex are renormalised in the same way and the finite radiative corrections are related. But at present the universally adopted definition of α_s is in terms of dimensional regularisation because of computational simplicity which is essential given the great complexity of present day calculations. So we now briefly review the principles of dimensional regularisation and the definition of Minimal Subtraction (MS) and Modified Minimal Subtraction ($\overline{MS}$). The $\overline{MS}$ definition of α_s is the one most commonly adopted in the literature and a value quoted for it is nomally referring to this definition.

Dimensional Regularisation (DR) is a gauge and Lorentz invariant regularisation that consists in formulating the theory in $D < 4$ spacetime dimensions in order to make loop integrals ultraviolet finite. In DR one rewrites the theory in D dimensions (D is integer at the beginning, but then we will see that the expression of diagrams makes sense at all D except for isolated singularities). The metric tensor is extended into a $D \times D$ matrix $g_{\mu\nu} = diag(1, -1, -1, \ldots, -1)$ and 4-vectors are given by $k^\mu = (k^0, k^1, \ldots, k^{D-1})$. The Dirac γ^μ are $f(D) \times f(D)$ matrices and it is not important what is the precise form of the function $f(D)$. It is sufficient to extend the usual algebra in a straightforward way like $\{\gamma_\mu, \gamma_\nu\} = 2g_{\mu,\nu}I$, with I the D-dimensional identity matrix, $\gamma^\mu \gamma^\nu \gamma_\mu = -(D-2)\gamma^\nu$ or $Tr(\gamma^\mu \gamma^\nu) = f(D)g_{\mu\nu}$.

The physical dimensions of fields change in D dimensions and, as a consequence, the gauge couplings become dimensional $e_D = \mu^\epsilon e$, where e is dimensionless, $D = 4 - 2\epsilon$ and μ is a scale of mass (this is how a scale of mass is introduced in the DR of massless QCD!). The dimension of fields is determined by requiring that the action $S = \int d^D x \mathcal{L}$ is dimensionless. By inserting for $\mathcal{L}$ terms like $m\bar{\Psi}\Psi$ or $m^2 \phi^\dagger \phi$ or $e\bar{\Psi}\gamma^\mu \Psi A_\mu$ the dimensions of the fields and coupling are determined as: $m, \Psi, \phi, A_\mu, e = 1, (D-1)/2, (D-2)/2, (D-2)/2, (4-D)/2$, respectively. The formal expression of loop integrals can be written for any D. For example:

$$\int \frac{d^D k}{(2\pi)^D} \frac{1}{(k^2 - m^2)^2} = \frac{\Gamma(2 - D/2)(-m^2)^{D/2-2}}{(4\pi)^{D/2}} \tag{4.52}$$

For $D = 4 - 2\epsilon$ one can expand using:

$$\Gamma(\epsilon) = \frac{1}{\epsilon} - \gamma_E + o(\epsilon), \qquad \gamma_E = 0.5772\ldots \tag{4.53}$$

For some Green function G, normalised to one in lowest order, (like V/e with V the 3-g vertex function at the symmetric point $p^2 = q^2 = r^2$, considered in the previous section) we typically find at 1-loop:

$$G_{bare} = 1 + \alpha_0 (\frac{-\mu^2}{p^2})^\epsilon [B(\frac{1}{\epsilon} + \log 4\pi - \gamma_E) + A + o(\epsilon)] \qquad (4.54)$$

In $\overline{MS}$ one rewrites this at 1-loop accuracy (diagram by diagram: this is a virtue of the method):

$$G_{bare} = Z G_{ren}$$

$$Z = 1 + \alpha [B(\frac{1}{\epsilon} + \log 4\pi - \gamma_E)]$$

$$G_{ren} = 1 + \alpha [B \log \frac{-\mu^2}{p^2} + A] \qquad (4.55)$$

Here Z stands for the relevant product of renormalisation factors. In the original MS prescription only $1/\epsilon$ was subtracted (that clearly plays the role of a cutoff) and not also $\log 4\pi$ and γ_E. Later, since these constants always appear from the expansion of Γ functions it was decided to modify MS into $\overline{MS}$. Note that the $\overline{MS}$ definition of α is different than that in the momentum subtraction scheme because the finite terms (those beyond logs) are different. In particular here δG_{ren} does not vanish at $p^2 = -\mu^2$.

The third [12] and fourth [13] coefficients of the QCD β function are also known in the $\overline{MS}$ prescription (recall that only the first two coefficients are scheme independent). The calculation of the last term involved the evaluation of some 50,000 4-loop diagrams. Translated in numbers, for $n_f = 5$ one obtains :

$$\beta(\alpha) = -0.610\alpha^2[1 + 1.261\ldots\frac{\alpha}{\pi} + 1.475\ldots(\frac{\alpha}{\pi})^2 + 9.836\ldots(\frac{\alpha}{\pi})^3\ldots]$$
$$(4.56)$$

It is interesting to remark that the expansion coefficients are all of order 1 or (10 for the last one), so that the $\overline{MS}$ expansion looks reasonably well behaved.

It is important to keep in mind that the QED and QCD perturbative series, after renormalisation, have all their coefficients finite, but the expansion does not converge. Actually the perturbative series are not even Borel summable. After Borel resummation for a given process one is left with a result which is ambiguous by terms typically down by $\exp -n/(b\alpha)$, with n an integer and b the first β function coefficient. In QED these corrective terms are extremely small and not very important in practice. On the contrary in QCD $\alpha = \alpha_s(Q^2) \sim 1/(b \log Q^2/\Lambda^2)$ and the ambiguous terms are of order $(1/Q^2)^n$, that is are power suppressed. It is interesting that, through this mechanism, the perturbative version of the theory is able to somehow take into account the power suppressed corrections. A sequence of diagrams with factorial growth at large order n is made up by dressing gluon propagators by any number of quark bubbles together with their gauge completions (renormalons).The problem of the precise relation between the ambiguities of the perturbative expansion and the higher twist corrections has been discussed in recent years [14].

4.5 Application to Hard Processes

4.5.1 $R_{e^+e^-}$ and Related Processes

The simplest hard process is $R_{e^+e^-}$ that we have already started to discuss. R is dimensionless and in perturbation theory is given by $R = N_C \sum_i Q_i^2 F(t, \alpha_s)$, where $F = 1 + o(\alpha_s)$. We have already mentioned that for this process the "anomalous dimension" function vanishes: $\gamma(\alpha_s) = 0$ because of electric charge non renormalisation by strong interactions. Let us review how this happens in detail. The diagrams that are relevant for charge renormalisation in QED at 1-loop are shown in Fig. 4.8. The Ward identity that follows from gauge invariance in QED imposes that the vertex (Z_V) and the self-energy (Z_f) renormalisation factors cancel and the only divergence remains in Z_γ, the vacuum polarization of the photon. So the charge is only renormalised by the photon blob, hence it is universal (the same factor for all fermions, independent of their charge) and is not affected by QCD at 1-loop. It is true that at higher orders the photon vacuum polarization diagram is affected by QCD (for example, at 2-loops we can exchange a gluon between the quarks in the photon loop) but the renormalisation induced by the vacuum polarisation diagram remains independent of the nature of the fermion to which the photon line is attached. The gluon contributions to the vertex (Z_V) and to the

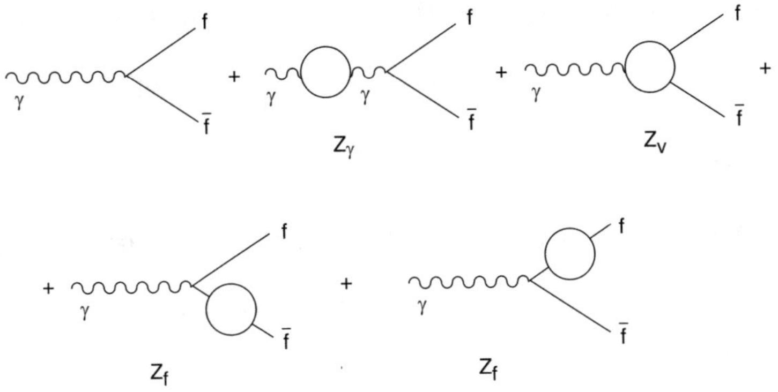

Fig. 4.8 Diagrams for charge renormalisation in QED at 1-loop

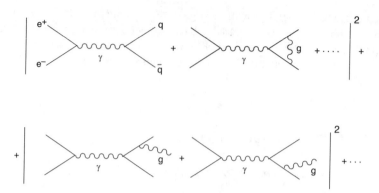

Fig. 4.9 Real and virtual diagrams relevant for the computation of R at 1-loop accuracy

self-energy (Z_f) cancel because they have exactly the same structure as in QED, so that $\gamma(\alpha_s) = 0$.

At 1-loop the diagrams relevant for the computation of R are shown in Fig. 4.9. There are virtual diagrams and real diagrams with one additional gluon in the final state. Infrared divergences cancel between the interference term of the virtual diagrams and the absolute square of the real diagrams, according to the Bloch-Nordsieck theorem. Similarly there are no mass singularities, in agreement with the Kinoshita-Lee-Nauenberg theorem, because the initial state is purely leptonic and all degenerate states that can appear at the given order are included in the final state. Given that $\gamma(\alpha_s) = 0$ the RGE prediction is simply given, as we have already seen, by $F(t, \alpha_s) = F[0, \alpha_s(t)]$. This means that if we do, for example, a 2-loop calculation, we must obtain a result of the form:

$$F(t, \alpha_s) = 1 + c_1 \alpha_s (1 - b\alpha_s t) + c_2 \alpha_s^2 + o(\alpha_s^3) \tag{4.57}$$

In fact we see that this form, taking into account that from Eq. (4.47) we have:

$$\alpha_s(t) \sim \frac{\alpha_s}{1 + b\alpha_s t} \sim \alpha_s (1 - b\alpha_s t + \ldots) \tag{4.58}$$

can be rewritten as

$$F(t, \alpha_s) = 1 + c_1 \alpha_s(t) + c_2 \alpha_s^2(t) + o(\alpha_s^3(t)) = F[0, \alpha_s(t)] \tag{4.59}$$

The content of the RGE prediction is, at this order, that there are no $\alpha_s t$ and $(\alpha_s t)^2$ terms (the leading log sequence must be absent) and the term of order $\alpha_s^2 t$ has the coefficient that allows to reabsorb it in the transformation of α_s into $\alpha_s(t)$.

At present the first three coefficients have been computed in the $\overline{MS}$ scheme [15]. Clearly $c_1 = 1/\pi$ does not depend on the definition of α_s but c_2 and c_3 do. The subleading coefficients also depend on the scale choice: if instead of expanding in $\alpha_s(Q)$ we decide to choose $\alpha_s(Q/2)$ the coefficients c_2 and c_3 change. In the $\overline{MS}$ scheme, for γ-exchange and $n_f = 5$, which are good approximations for $2m_b <<$ $Q << m_Z$, one has:

$$F[0, \alpha_s(t)] = 1 + \frac{\alpha_s(t)}{\pi} + 1.409\ldots(\frac{\alpha_s(t)}{\pi})^2 - 12.8\ldots(\frac{\alpha_s(t)}{\pi})^3 +\ldots \quad (4.60)$$

Similar perturbative results at 3-loop accuracy also exist for $R_Z = \Gamma(Z \to hadrons)/\Gamma(Z \to leptons)$, $R_\tau = \Gamma(\tau \to \nu_\tau + hadrons)/\Gamma(\tau \to \nu_\tau + leptons)$, etc. We will discuss these results later when we deal with measurements of α_s.

The perturbative expansion in powers of $\alpha_s(t)$ takes into account all contributions that are suppressed by powers of logarithms of the large scale Q^2 ("leading twist" terms). In addition there are corrections suppressed by powers of the large scale Q^2 ("higher twist" terms). The pattern of power corrections is controlled by the light-cone Operator Product Expansion (OPE) [16] which (schematically) leads to:

$$F = \text{pert.} + r_2\frac{m^2}{Q^2} + r_4\frac{<0|Tr[F_{\mu\nu}F^{\mu\nu}]|0>}{Q^4} + \ldots + r_6\frac{<0|O_6|0>}{Q^6} + \ldots$$
$$(4.61)$$

Here m^2 generically indicates mass corrections, notably from b quarks, for example (t quark mass corrections only arise from loops, vanish in the limit $m_t \to \infty$ and are included in the coefficients as those in Eq. (4.60) and the analogous ones for higher twist terms), $\mathbf{F}_{\mu\nu} = \sum_A F^A_{\mu\nu}t^A$, O_6 is typically a 4-fermion operator, etc. For each possible gauge invariant operator the corresponding power of Q^2 is fixed by dimensions.

We now consider the light-cone OPE in some more detail. $R_{e^+e^-} \sim \Pi(Q^2)$ where $\Pi(Q^2)$ is the scalar spectral function related to the hadronic contribution to the imaginary part of the photon vacuum polarization $T_{\mu\nu}$:

$$T_{\mu\nu} = (-g_{\mu\nu}Q^2 + q_\mu q_\nu)\Pi(Q^2) = \int \exp iqx <0|J^\dagger_\mu(x)J_\nu(0)|0> dx =$$
$$= \sum_n <0|J^\dagger_\mu(0)|n><n|J_\nu(0)|0> (2\pi)^4\delta^4(q - p_n) \quad (4.62)$$

For $Q^2 \to \infty$ the $x^2 \to 0$ region is dominant. To all orders in perturbation theory the OPE can be proven. Schematically, dropping Lorentz indices, for simplicity, near $x^2 \sim 0$ we have:

$$J^\dagger(x)J(0) = I(x^2) + E(x^2)\sum_{n=0}^{\infty} c_n(x^2)x^{\mu_1}\ldots x^{\mu_n} \cdot O^n_{\mu_1\ldots\mu_n}(0)$$

$$+ \text{less sing. terms} \quad (4.63)$$

Here $I(x^2)$, $E(x^2)$,..., $c_n(x^2)$ are c-number singular functions, O^n is a string of local operators. $E(x^2)$ is the singularity of free field theory, $I(x^2)$ and $c_n(x^2)$ contain powers of $\log \mu^2 x^2$ in interaction. Some O^n are already present in free field theory, other ones appear when interactions are switched on. Given that $\Pi(Q^2)$ is related to the Fourier transform of the vacuum expectation value of the product of currents, less singular terms in x^2 lead to power suppressed terms in $1/Q^2$. The perturbative terms come from $I(x^2)$ which is the leading twist term. The logarithmic scaling violations induced by the running coupling are the logs in $I(x^2)$.

4.5.2 The Final State in e^+e^- Annihilation

Experiments on e^+e^- annihilation at high energy provide a remarkable possibility of systematically testing the distinct signatures predicted by QCD for the structure of the final state averaged over a large number of events. Typical of asymptotic freedom is the hierarchy of configurations emerging as a consequence of the smallness of $\alpha_s(Q^2)$. When all corrections of order $\alpha_s(Q^2)$ are neglected one recovers the naive parton model prediction for the final state: almost collinear events with two back-to-back jets with limited transverse momentum and an angular distribution as $(1 + \cos^2 \theta)$ with respect to the beam axis (typical of spin 1/2 parton quarks: scalar quarks would lead to a $\sin^2 \theta$ distribution). At order $\alpha_s(Q^2)$ a tail of events is predicted to appear with large transverse momentum $p_T \sim Q/2$ with respect to the thrust axis (the axis that maximizes the sum of the absolute values of the longitudinal momenta of the final state particles). This small fraction of events with large p_T mostly consists of three-jet events with an almost planar topology. The skeleton of a three-jet event, at leading order in $\alpha_s(Q^2)$, is formed by three hard partons $q\bar{q}g$, the third being a gluon emitted by a quark or antiquark line. The distribution of three-jet events is given by:

$$\frac{1}{\sigma} \frac{d\sigma}{dx_1 dx_2} = \frac{2\alpha_s}{3\pi} \frac{x_1^2 + x_2^2}{(1 - x_1)(1 - x_2)} \tag{4.64}$$

here $x_{1,2}$ refer to energy fractions of massless quarks: $x_i = 2E_i/\sqrt{s}$ with $x_1 + x_2 + x_3 = 2$. At order $\alpha_s^2(Q^2)$ a hard perturbative non planar component starts to build up and a small fraction of four-jet events $q\bar{q}gg$ or $q\bar{q}q\bar{q}$ appear, and so on.

A quantitatively specified definition of jet counting must be introduced for precise QCD tests and for measuring α_s, which must be infrared safe (i.e. not altered by soft particle emission or collinear splittings of massless particles) in order to be computable at parton level and as much as possible insensitive to the transformation of partons into hadrons. One introduces a resolution parameter y_{cut} and a suitable pair variable; for example [17]:

$$y_{ij} = \frac{min(E_i^2, E_j^2)(1 - \cos \theta_{ij})}{s} \tag{4.65}$$

The particles i,j belong to different jets for $y_{ij} > y_{cut}$. Clearly the number of jets becomes a function of y_{cut}: there are more jets for smaller y_{cut}. Measurements of $\alpha_s(Q^2)$ have been performed starting from jet multiplicities, the largest error coming from the necessity of correcting for non-perturbative hadronisation effects.

4.5.3 Deep Inelastic Scattering

Deep Inelastic Scattering (DIS) processes have played and still play a very important role for our understanding of QCD and of nucleon structure. This set of processes actually provides us with a rich laboratory for theory and experiment. There are several structure functions that can be studied, $F_i(x, Q^2)$, each a function of two variables. This is true separately for different beams and targets and different polarizations. Depending on the charges of 1 and 1' (see Eq. (4.13)) we can have neutral currents (γ,Z) or charged currents in the 1-1' channel (Fig. 4.6). In the past DIS processes were crucial for establishing QCD as the theory of strong interactions and quarks and gluons as the QCD partons. At present DIS remains very important for quantitative studies and tests of QCD. The theory of scaling violations for totally inclusive DIS structure functions, based on operator expansion or diagrammatic techniques and renormalisation group methods, is crystal clear and the predicted Q^2 dependence can be tested at each value of x. The measurement of quark and gluon densities in the nucleon, as functions of x at some reference value of Q^2, which is an essential starting point for the calculation of all relevant hadronic hard processes, is performed in DIS processes. At the same time one measures $\alpha_s(Q^2)$ and the DIS values of the running coupling can be compared with those obtained from other processes. At all times new theoretical challenges arise from the study of DIS processes. Recent examples are the so-called "spin crisis" in polarized DIS and the behaviour of singlet structure functions at small x as revealed by HERA data. In the following we will review the past successes and the present open problems in the physics of DIS.

The cross-section $\sigma \sim L^{\mu\nu} W_{\mu\nu}$ is given in terms of the product of a leptonic ($L^{\mu\nu}$) and a hadronic ($W_{\mu\nu}$) tensor. While $L^{\mu\nu}$ is simple and easily obtained from the lowest order electroweak (EW) vertex plus QED radiative corrections, the complicated strong interaction dynamics is contained in $W_{\mu\nu}$. The latter is proportional to the Fourier transform of the forward matrix element between the nucleon target states of the product of two EW currents:

$$ W_{\mu\nu} = \int dx \, \exp iqx \, < p|J_\mu^\dagger(x)J_\nu(0)|p > \tag{4.66} $$

Structure functions are defined starting from the general form of $W_{\mu\nu}$ given Lorentz invariance and current conservation. For example, for EW currents between unpolarized nucleons we have:

$$W_{\mu\nu} = (-g_{\mu\nu} + \frac{q_\mu q_\nu}{q^2}) \, W_1(\nu, Q^2) + (p_\mu - \frac{m\nu}{q^2}q_\mu)(p_\nu - \frac{m\nu}{q^2}q_\nu) \, \frac{W_2(\nu, Q^2)}{m^2} -$$
$$- \frac{i}{2m^2}\epsilon_{\mu\nu\lambda\rho}p^\lambda q^\rho \, W_3(\nu, Q^2)$$

W_3 arises from VA interference and is absent for pure vector currents. In the limit $Q^2 >> m^2$, x fixed, the structure functions obey approximate Bjorken scaling which in reality is broken by logarithmic corrections that can be computed in QCD:

$$mW_1(\nu, Q^2) \rightarrow F_1(x)$$
$$\nu W_{2,3}(\nu, Q^2) \rightarrow F_{2,3}(x) \tag{4.67}$$

The $\gamma - N$ cross-section is given by ($W_i = W_i(Q^2, \nu)$):

$$\frac{d\sigma^\gamma}{dQ^2 d\nu} = \frac{4\pi\alpha^2 E'}{Q^4 E} \cdot [2\sin^2\frac{\theta}{2}W_1 + \cos^2\frac{\theta}{2}W_2] \tag{4.68}$$

while for the $\nu - N$ or $\bar{\nu} - N$ cross-section one has:

$$\frac{d\sigma^{\nu,\bar{\nu}}}{dQ^2 d\nu} = \frac{G_F^2 E'}{2\pi E}(\frac{m_W^2}{Q^2 + m_W^2})^2 \cdot [2\sin^2\frac{\theta}{2}W_1 + \cos^2\frac{\theta}{2}W_2 \pm \frac{E + E'}{m}\sin^2\frac{\theta}{2}W_3] \tag{4.69}$$

(W_i for photons, ν and $\bar{\nu}$ are all different, as we shall see in a moment).

In the scaling limit the longitudinal and transverse cross sections are given by:

$$\sigma_L \sim \frac{1}{s}[\frac{F_2(x)}{2x} - F_1(x)]$$

$$\sigma_{RH,LH} \sim \frac{1}{s}[F_1(x) \pm F_3(x)]$$

$$\sigma_T = \sigma_{RH} + \sigma_{LH} \tag{4.70}$$

where L, RH, LH refer to the helicity 0, 1, −1, respectively, of the exchanged gauge vector boson.

In the '60's the demise of hadrons from the status of fundamental particles to that of bound states of constituent quarks was the breakthrough that made possible the construction of a renormalisable field theory for strong interactions. The presence of an unlimited number of hadrons species, many of them with large spin values, presented an obvious dead-end for a manageable field theory. The evidence for constituent quarks emerged clearly from the systematics of hadron spectroscopy.

The complications of the hadron spectrum could be explained in terms of the quantum numbers of spin 1/2, fractionally charged, u, d and s quarks. The notion of colour was introduced to reconcile the observed spectrum with Fermi statistics. But confinement that forbids the observation of free quarks was a clear obstacle towards the acceptance of quarks as real constituents and not just as fictitious entities describing some mathematical pattern (a doubt expressed even by Gell-Mann at the time). The early measurements at SLAC of DIS dissipated all doubts: the observation of Bjorken scaling and the success of the "naive" (not so much after all) parton model of Feynman imposed quarks as the basic fields for describing the nucleon structure (parton quarks).

In the language of Bjorken and Feynman the virtual γ (or, in general, any gauge boson) sees the quark partons inside the nucleon target as quasi-free, because their (Lorentz dilated) QCD interaction time is much longer than $\tau_\gamma \sim 1/Q$, the duration of the virtual photon interaction. Since the virtual photon 4-momentum is spacelike, we can go to a Lorentz frame where $E_\gamma = 0$ (Breit frame). In this frame $q = (E_\gamma = 0; 0, 0, Q)$ and the nucleon momentum, neglecting the mass $m << Q$, is $p = (Q/2x; 0, 0, -Q/2x)$ (note that this correctly gives $q^2 = -Q^2$ and $x = Q^2/2(p \cdot q)$). Consider (Fig. 4.10) the interaction of the photon with a quark carrying a fraction y of the nucleon 4-momentum: $p_q = yp$ (we are neglecting the transverse components of p_q which are of order m). The incoming parton with $p_q = yp$ absorbs the photon and the final parton has 4-momentum p'_q. Since in the Breit frame the photon carries no energy but only a longitudinal momentum Q, the photon can only be absorbed by those partons with $y = x$: then the longitudinal component of $p_q = yp$ is $-yQ/2x = -Q/2$ and can be flipped into $+Q/2$ by the photon. As a result, the photon longitudinal momentum $+Q$ disappears, the parton quark momentum changes of sign from $-Q/2$ into $+Q/2$ and the energy is not changed. So the structure functions are proportional to the density of partons with fraction x of the nucleon momentum, weighted with the squared charge. Also, recall that the helicity of a massless quark is conserved in a vector (or axial vector) interaction. So when the momentum is reversed also the spin must flip. Since the process is collinear there is no orbital contribution and only a photon with helicity ± 1 (transverse photon) can be absorbed. Alternatively, if partons were spin zero only longitudinal photons would instead contribute.

Using these results, which are maintained in QCD at leading order, the quantum numbers of the quarks were confirmed by early experiments. The observation that $R = \sigma_L/\sigma_T \to 0$ implies that the charged partons have spin 1/2. The quark charges

Fig. 4.10 Schematic diagram for the interaction of the virtual photon with a parton quark in the Breit frame

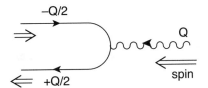

were derived from the data on the electron and neutrino structure functions:

$$F_{ep} = 4/9u(x) + 1/9d(x) + \ldots.; \qquad F_{en} = 4/9d(x) + 1/9u(x) + \ldots.$$
$$F_{vp} = F_{\bar{v}n} = 2d(x) + \ldots.; \qquad F_{vn} = F_{\bar{v}p} = 2u(x) + \ldots. \qquad (4.71)$$

where $F \sim 2F_1 \sim F_2/x$ and $u(x)$, $d(x)$ are the parton number densities in the proton (with fraction x of the proton longitudinal momentum), which, in the scaling limit, do not depend on Q^2. The normalisation of the structure functions and the parton densities are such that the charge relations hold:

$$\int_0^1 [u(x) - \bar{u}(x)]dx = 2, \quad \int_0^1 [d(x) - \bar{d}(x)]dx = 1, \quad \int_0^1 [s(x) - \bar{s}(x)]dx = 0 \tag{4.72}$$

Also it was proven by experiment that at values of Q^2 of a few GeV^2, in the scaling region, about half of the nucleon momentum, given by the momentum sum rule:

$$\int_0^1 [\sum_i (q_i(x) + \bar{q}_i(x)) + g(x)]xdx = 1 \tag{4.73}$$

is carried by neutral partons (gluons).

In QCD there are calculable log scaling violations induced by $\alpha_s(t)$. The parton rules just introduced can be summarised in the formula:

$$F(x, t) = \int_x^1 dy \frac{q_0(y)}{y} \sigma_{point}(\frac{x}{y}, \alpha_s(t)) + o(\frac{1}{Q^2}) \tag{4.74}$$

Before QCD corrections $\sigma_{point} = e^2\delta(x/y - 1)$ and $F = e^2q_0(x)$ (here we denote by e the charge of the quark in units of the positron charge, i.e. $e = 2/3$ for the u quark). QCD modifies σ_{point} at order α_s via the diagrams of Fig. 4.11. Note that the integral is from x to 1, because the energy can only be lost by radiation before interacting with the photon (which eventually wants to find a fraction x, as we have

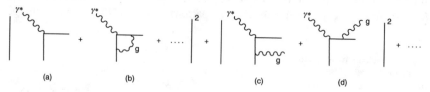

Fig. 4.11 First order QCD corrections to the virtual photon-quark cross-section: (a) leading order with (b) one-loop virtual correction; (c-d) next-to-leading order real emission

explained). From a direct computation of the diagrams one obtains a result of the following form:

$$\sigma_{point}(z, \alpha_s(t)) \simeq e^2[\delta(z-1) + \frac{\alpha_s}{2\pi}(t \cdot P(z) + f(z))] \tag{4.75}$$

For $y > x$ the correction arises from diagrams with real gluon emission. Only the sum of the two real diagrams in Fig. 4.11 is gauge invariant, so that the contribution of one given diagram is gauge dependent. There is a special form of axial gauge, called physical gauge, where, among real diagrams, the diagram of Fig. 4.11c gives the whole t-proportional term. It is obviously not essential to go to this gauge, but this diagram has a direct physical interpretation: a quark in the proton has a fraction $y > x$ of the parent 4-momentum; it then radiates a gluon and looses energy down to a fraction x before interacting with the photon. The log arises from the virtual quark propagator, according to the discussion of collinear mass singularities in Eq. (4.12). In fact in the massless limit one has:

$$\text{propagator} = \frac{1}{r^2} = \frac{1}{(k-h)^2} = \frac{-1}{2E_k E_h} \cdot \frac{1}{1-\cos\theta}$$

$$= \frac{-1}{4E_k E_h} \cdot \frac{1}{\sin^2\theta/2} \propto \frac{-1}{p_T^2} \tag{4.76}$$

where p_T is the transverse momentum of the virtual quark. So the square of the propagator goes like $1/p_T^4$. But there is a p_T^2 factor in the numerator, because in the collinear limit, when $\theta = 0$ and the initial and final quarks and the emitted gluon are all aligned, the quark helicity cannot flip (vector interaction) so that the gluon should carry helicity zero but a real gluon can only have ± 1 helicity. Thus the numerator vanishes as p_T^2 in the forward direction and the cross-section behaves as:

$$\sigma \sim \int^{Q^2} \frac{1}{p_T^2} dp_T^2 \sim \log Q^2 \tag{4.77}$$

Actually the log should be read as $\log Q^2/m^2$ because in the massless limit a genuine mass singularity appears. In fact the mass singularity connected with the initial quark line is not cancelled because we do not have the sum of all degenerate initial states, but only a single quark. But in correspondence to the initial quark we have the (bare) quark density $q_0(y)$ that appear in the convolution integral. This is a non perturbative quantity that is determined by the nucleon wave function. So we can factorize the mass singularity in a redefinition of the quark density: we replace $q_0(y) \to q(y,t) = q_0(y) + \Delta q(y,t)$ with:

$$\Delta q(x, t) = \frac{\alpha_s}{2\pi} t \int_x^1 dy \frac{q_0(y)}{y} \cdot P(\frac{x}{y}) \tag{4.78}$$

Here the factor of t is a bit symbolic: it stands for $\log Q^2/km^2$ and what we exactly put below Q^2 depends on the definition of the renormalised quark density, which also fixes the exact form of the finite term $f(z)$ in Eq. (4.75).

The effective parton density $q(y, t)$ that we have defined is now scale dependent. In terms of this scale dependent density we have the following relations, where we have also replaced the fixed coupling with the running coupling according to the prescription derived from the RGE:

$$F(x, t) = \int_x^1 dy \frac{q(y, t)}{y} e^2 [\delta(\frac{x}{y} - 1) + \frac{\alpha_s(t)}{2\pi} f(\frac{x}{y}))] = e^2 q(x, t) + o(\alpha_s(t))$$

$$\frac{d}{dt} q(x, t) = \frac{\alpha_s(t)}{2\pi} \int_x^1 dy \frac{q(y, t)}{y} \cdot P(\frac{x}{y}) + o(\alpha_s(t)^2) \tag{4.79}$$

We see that in lowest order we reproduce the naive parton model formulae for the structure functions in terms of effective parton densities that are scale dependent. The evolution equations for the parton densities are written down in terms of kernels (the "splitting functions") that can be expanded in powers of the running coupling. At leading order, we can interpret the evolution equation by saying that the variation of the quark density at x is given by the convolution of the quark density at y times the probability of emitting a gluon with fraction x/y of the quark momentum.

It is interesting that the integro-differential QCD evolution equation for densities can be transformed into an infinite set of ordinary differential equations for Mellin moments [2]. The moment f_n of a density $f(x)$ is defined as:

$$f_n = \int_0^1 dx x^{n-1} f(x) \tag{4.80}$$

By taking moments of both sides of the second of Eqs. (4.79) one finds, with a simple interchange of the integration order, the simpler equation for the n-th moment:

$$\frac{d}{dt} q_n(t) = \frac{\alpha_s(t)}{2\pi} \cdot P_n \cdot q_n(t) \tag{4.81}$$

To solve this equation we observe that:

$$\log \frac{q_n(t)}{q_n(0)} = \frac{P_n}{2\pi} \int_0^t \alpha_s(t) dt = \frac{P_n}{2\pi} \int_{\alpha_s}^{\alpha_s(t)} \frac{d\alpha'}{-b\alpha'} \tag{4.82}$$

where we used Eq. (4.31) to change the integration variable from dt to $d\alpha(t)$ (denoted as $d\alpha'$) and $\beta(\alpha) \simeq -b\alpha^2 + \ldots$. Finally the solution is:

$$q_n(t) = [\frac{\alpha_s}{\alpha_s(t)}]^{\frac{P_n}{2\pi b}} \cdot q_n(0) \tag{4.83}$$

The connection of these results with the RGE general formalism occurs via the light cone OPE (recall Eq. (4.66) for $W_{\mu\nu}$ and Eq. (4.63) for the OPE of two currents). In the case of DIS the c-number term $I(x^2)$ does not contribute, because we are interested in the connected part $< p|\ldots|p > - < 0|\ldots|0 >$. The relevant terms are:

$$J^\dagger(x)J(0) = E(x^2)\sum_{n=0}^{\infty}c_n(x^2)x^{\mu_1}\ldots x^{\mu_n}\cdot O^n_{\mu_1\ldots\mu_n}(0) + \text{less sing. terms}$$

$$(4.84)$$

A formally intricate but conceptually simple argument (Ref. [6], page 28) based on the analiticity properties of the forward virtual Compton amplitude shows that the Mellin moments M_n of structure functions are related to the individual terms in the OPE, precisely to the Fourier transform $c_n(Q^2)$ (we will write it as $c_n(t,\alpha)$) of the coefficient $c_n(x^2)$ times a reduced matrix element h_n from the operators O^n:
$< p|O^n_{\mu_1\ldots\mu_n}(0)|p >= h_n p_{\mu_1}\ldots p_{\mu_n}$:

$$c_n < p|O^n|p > \to M_n = \int_0^1 dx\, x^{n-1}F(x)$$

$$(4.85)$$

Since the matrix element of the products of currents satisfy the RGE so do the moments M_n. Hence the general form of the Q^2 dependence is given by the RGE solution (see Eq. (4.33)):

$$M_n(t,\alpha) = c_n[0,\alpha(t)]\exp\int_\alpha^{\alpha(t)}\frac{\gamma_n(\alpha')}{\beta(\alpha')}d\alpha'\cdot h_n(\alpha)$$

$$(4.86)$$

In lowest order, identifying in the simplest case M_n with q_n, we have:

$$\gamma_n(\alpha) = \frac{P_n}{2\pi}\alpha + \ldots, \qquad \beta(\alpha) = -b\alpha^2 + \ldots$$

$$(4.87)$$

and

$$q_n(t) = q_n(0)\exp\int_\alpha^{\alpha(t)}\frac{\gamma_n(\alpha')}{\beta(\alpha')}d\alpha' = [\frac{\alpha_s}{\alpha_s(t)}]^{\frac{P_n}{2\pi b}}\cdot q_n(0)$$

$$(4.88)$$

which exactly coincides with Eq. (4.83).

Up to this point we have implicitly restricted our attention to non-singlet (under the flavour group) structure functions. The Q^2 evolution equations become non diagonal as soon as we take into account the presence of gluons in the target. In fact the quark which is seen by the photon can be generated by a gluon in the target (Fig. 4.12).

Fig. 4.12 Lowest order diagram for the interaction of the virtual photon with a parton gluon

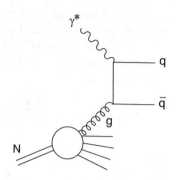

The quark evolution equation becomes:

$$\frac{d}{dt}q_i(x,t) = \frac{\alpha_s(t)}{2\pi}[q_i \otimes P_{qq}] + \frac{\alpha_s(t)}{2\pi}[g \otimes P_{qg}] \tag{4.89}$$

where we introduced the shorthand notation:

$$[q \otimes P] = [P \otimes q] = \int_x^1 dy \frac{q(y,t)}{y} \cdot P(\frac{x}{y}) \tag{4.90}$$

(it is easy to check that the convolution, like an ordinary product, is commutative). At leading order, the interpretation of Eq. (4.89) is simply that the variation of the quark density is due to the convolution of the quark density at a higher energy times the probability of finding a quark in a quark (with the right energy fraction) plus the gluon density at a higher energy times the probability of finding a quark (of the given flavour i) in a gluon. The evolution equation for the gluon density, needed to close the system, can be obtained by suitably extending the same line of reasoning to a gedanken probe sensitive to colour charges, for example a virtual gluon. The resulting equation is of the form:

$$\frac{d}{dt}g(x,t) = \frac{\alpha_s(t)}{2\pi}[\sum_i(q_i + \bar{q}_i) \otimes P_{gq}] + \frac{\alpha_s(t)}{2\pi}[g \otimes P_{gg}] \tag{4.91}$$

The explicit form of the splitting functions in lowest order [18, 19] can be directly derived from the QCD vertices [19]. They are a property of the theory and do not depend on the particular process the parton density is taking part in. The results are:

$$P_{qq} = \frac{4}{3}[\frac{1+x^2}{(1-x)_+} + \frac{3}{2}\delta(1-x)] + o(\alpha_s)$$

$$P_{gq} = \frac{4}{3}\frac{1+(1-x)^2}{x} + o(\alpha_s)$$

$$P_{qg} = \frac{1}{2}[x^2 + (1-x)^2] + o(\alpha_s)$$

$$P_{gg} = 6[\frac{x}{(1-x)_+} + \frac{1-x}{x} + x(1-x)] + \frac{33 - 2n_f}{6}\delta(1-x) + o(\alpha_s) \quad (4.92)$$

For a generic non singular weight function $f(x)$, the "+" distribution is defined as:

$$\int_0^1 \frac{f(x)}{(1-x)_+}dx = \int_0^1 \frac{f(x) - f(1)}{1 - x}dx \quad (4.93)$$

The $\delta(1-x)$ terms arise from the virtual corrections to the lowest order tree diagrams. Their coefficient can be simply obtained by imposing the validity of charge and momentum sum rules. In fact, from the request that the charge sum rules in Eq. (4.72) are not affected by the Q^2 dependence one derives that

$$\int_0^1 P_{qq}(x)dx = 0 \quad (4.94)$$

which can be used to fix the coefficient of the $\delta(1-x)$ terms of P_{qq}. Similarly, by taking the t-derivative of the momentum sum rule in Eq. (4.73) and imposing its vanishing for generic q_i and g, one obtains:

$$\int_0^1 [P_{qq}(x) + P_{gq}(x)]xdx = 0, \qquad \int_0^1 [2n_f P_{qg}(x) + P_{gg}(x)]xdx = 0. \quad (4.95)$$

At higher orders the evolution equations are easily generalised but the calculation of the splitting functions rapidly becomes very complicated. For many years the splitting functions were only completely known at NLO accuracy [20]: $\alpha_s P \sim \alpha_s P_1 + \alpha_s^2 P_2 + \ldots$. Then in recent years the NNLO results P_3 have been first derived in analytic form for the first few moments and, then the full NNLO analytic calculation, a really monumental work, was completed in 2004 by Moch, Vermaseren and Vogt [21]. Beyond leading order a precise definition of parton densities should be specified. One can take a physical definition (for example, quark densities can be defined as to keep the LO expression for the structure function F_2 valid at all orders, the so called DIS definition [22], and the gluon density could be defined starting from F_L, the longitudinal structure function, or a more abstract specification (for example, in terms of the $\overline{MS}$ prescription). Once the definition of parton densities is fixed, the coefficients that relate the different structure functions to the parton densities at each fixed order can be computed. Similarly the higher order splitting functions also depend, to some extent, from the definition of parton densities, and a consistent set of coefficients and splitting functions must be used at each order.

The scaling violations are clearly observed by experiment and their pattern is very well reproduced by QCD fits at NLO. Examples are seen in Fig. 4.13a–d [23].

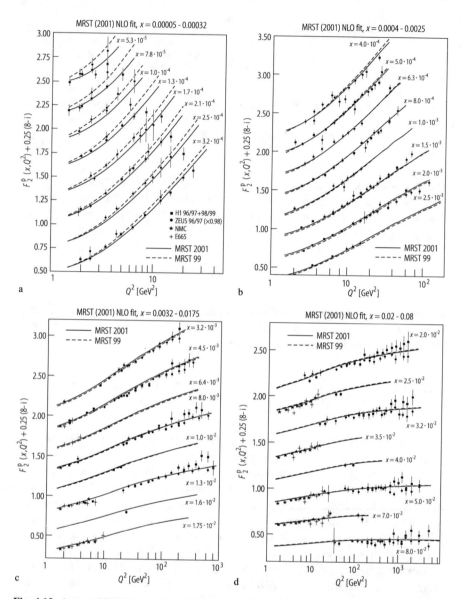

Fig. 4.13 A recent NLO fit of scaling violations from Ref. [23], for different x ranges, as functions of Q^2

These fits provide an impressive confirmation of a quantitative QCD prediction, a measurement of $q_i(x, Q_0^2)$ and $g(x, Q_0^2)$ at some reference value Q_0^2 of Q^2 and a precise measurement of $\alpha_s(m_Z^2)$.

4.5.3.1 Resummation for Deep Inelastic Structure Functions

At small or at large values of x (with Q^2 large) those terms of higher order in α_s in either the coefficients or the splitting functions which are multiplied by powers of $\log 1/x$ or $\log(1 - x)$ eventually become important and should be taken into account. Fortunately the sequences of leading and subleading logs can be evaluated at all orders by special techniques and resummed to all orders.

For large x resummation [24] I refer to the recent papers [25, 26] (the latter also involving higher twist corrections, which are important at large x) where a list of references to previous work can be found.

Here we will briefly summarise the small-x case for the singlet structure function which is the dominant channel at HERA, dominated by the sharp rise of the gluon and sea parton densities at small x. The small x data collected by HERA can be fitted reasonably well even at the smallest measured values of x by the NLO QCD evolution equations, so that there is no dramatic evidence in the data for departures. This is surprising also in view of the fact that the NNLO effects in the evolution have recently become available and are quite large. Resummation effects have been shown to resolve this apparent paradox. For the singlet splitting function the coefficients of all LO and NLO corrections of order $[\alpha_s(Q^2) \log 1/x]^n$ and $\alpha_s(Q^2)[\alpha_s(Q^2) \log 1/x]^n$, respectively, are explicitly known from the BFKL analysis of virtual gluon-virtual gluon scattering [27, 28]. But the simple addition of these higher order terms to the perturbative result (with subtraction of all double counting) does not lead to a converging expansion (the NLO logs completely overrule the LO logs in the relevant domain of x and Q^2). A sensible expansion is only obtained by a proper treatment of momentum conservation constraints, also using the underlying symmetry of the BFKL kernel under exchange of the two external gluons, and especially, of the running coupling effects (see the recent papers [29, 30] and refs. therein). In Fig. 4.14 we present the results for the dominant singlet splitting function $xP(x, \alpha_s(Q^2))$ for $\alpha_s(Q^2) \sim 0.2$. We see that while the NNLO perturbative splitting function sharply deviates from the NLO approximation at small x, the resummed result only shows a moderate dip with respect to the NLO perturbative splitting function in the region of HERA data, and the full effect of the true small x asymptotics is only felt at much smaller values of x. The related effects are not very important for processes at the LHC but could become relevant for next generation hadron colliders.

Fig. 4.14 The dominant singlet splitting function $xP(x, \alpha_s(Q^2))$ for $\alpha_s(Q^2) \sim 0.2$. The resummed result from Ref. [29] is compared with the LO, NLO and NNLO perturbative results

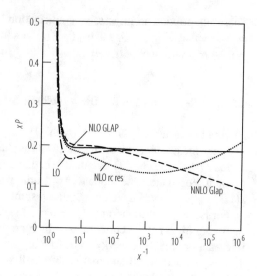

4.5.3.2 Polarized Deep Inelastic Scattering

In polarized DIS one main question is how the proton helicity is distributed among quarks, gluons and orbital angular momentum: $1/2\Delta\Sigma + \Delta g + L_z = 1/2$ (for a recent review, see, for example, [31]). For a parton density p (either a quark or a gluon) Δp indicates the first moment of the net helicity difference $p_+ - p_-$ in a polarized proton with helicity $+1/2$ or:

$$\Delta p(Q^2) = \int_0^1 dx[p_+(x, Q^2) - p_-(x, Q^2)] \tag{4.96}$$

Experiments have shown that the quark moment $\Delta\Sigma$ is small (the "spin crisis"): values from a recent fit [32] are $\Delta\Sigma_{exp} \sim 0.21 \pm 0.14$ and $\Delta g_{exp} \sim 0.50 \pm 1.27$ at $Q^2 = 1\,\text{GeV}^2$ (see also [33]). This is surprising because $\Delta\Sigma$ is conserved in perturbation theory at LO (i.e. it does not evolve in Q^2). For conserved quantities we would expect that they are the same for constituent and for parton quarks. But actually the conservation of $\Delta\Sigma$ is broken by the axial anomaly. In perturbation theory the conserved density is actually $\Delta\Sigma' = \Delta\Sigma + n_f/2\pi\alpha_s \Delta g$ [34]. Note that also $\alpha_s \Delta g$ is conserved in LO, that is $\Delta g \sim logQ^2$. This behaviour is not controversial but it will take long before the log growth of Δg will be confirmed by experiment! But to establish this behaviour would show that the extraction of Δg from the data is correct and that the QCD evolution works as expected. If Δg is large enough it could account for the difference between partons ($\Delta\Sigma$) and constituents ($\Delta\Sigma'$). From the spin sum rule it is clear that the log increase should cancel between Δg and L_z. This cancelation is automatic as a consequence of helicity conservation in the basic QCD vertices. From the spin sum rule one obtains that either $\Delta g + L_z$ is large or there are contributions to $\Delta\Sigma$ at very small x outside of the measured region.

Δg can be measured indirectly by scaling violations and directly from asymmetries, e.g. in $c\bar{c}$ production. Existing measurements by Hermes, Compass, and at RHIC are still crude but show no hint of a large Δg at accessible value of x and Q^2. Present data are consistent with Δg large enough to sizeably contribute to the spin sum rule but there is no indication that $\alpha_s \Delta g$ can explain the difference between constituents and parton quarks. The perspectives of better measurements are good at Compass and RHIC in the near future.

4.5.4 Factorisation and the QCD Improved Parton Model

The parton densities defined and measured in DIS are instrumental to compute hard processes initiated by hadronic collisions via the Factorisation Theorem (FT). Suppose you have a hadronic process of the form $h_1 + h_2 \rightarrow X + all$ where h_i are hadrons and X is some triggering particle or pair of particles which specify the large scale Q^2 relevant for the process, in general somewhat, but not much, smaller than s, the total c.o.m. squared mass. For example, in pp or $p\bar{p}$ collisions, X can be a W or a Z or a virtual photon with large Q^2, or a jet at large transverse momentum p_T, or a pair of heavy quark-antiquark of mass M. By "all" we mean a totally inclusive collection of gluons and light quark pairs. The FT states that for the total cross-section or some other sufficiently inclusive distribution we can write, apart from power suppressed corrections, the expression:

$$\sigma(s, \tau) = \sum_{AB} \int dx_1 dx_2 p_{1A}(x_1, Q^2) p_{2B}(x_2, Q^2) \sigma_{AB}(x_1 x_2 s, \tau) \qquad (4.97)$$

Here $\tau = Q^2/s$ is a scaling variable, p_{iC} are the densities for a parton of type C inside the hadron h_i, σ_{AB} is the partonic cross-section for parton-A + parton-B$\rightarrow X + all'$. This result is based on the fact that the mass singularities that are associated with the initial legs are of universal nature, so that one can reproduce the same modified parton densities, by absorbing these singularities into the bare parton densities, as in deep inelastic scattering. Once the parton densities and α_s are known from other measurements, the prediction of the rate for a given hard process is obtained with not much ambiguity (e.g from scale dependence or hadronisation effects). The NLO calculation of the reduced partonic cross-section is needed in order to correctly specify the scale and in general the definition of the parton densities and of the running coupling in the leading term. The residual scale and scheme dependence is often the most important source of theoretical error. In the following we consider a few examples.

A comparison of data and predictions on the production of jets at large $\sqrt{s}$ and p_T in pp or $p\bar{p}$ collisions is shown in Fig. 4.15 [9, 35].

This is a particularly significant test because the rates at different c.o.m. energies and, for each energy, at different values of p_T span over many orders of magnitude.

Fig. 4.15 Jet production cross-section at pp or $p\bar{p}$ colliders, as function of p_T [9]

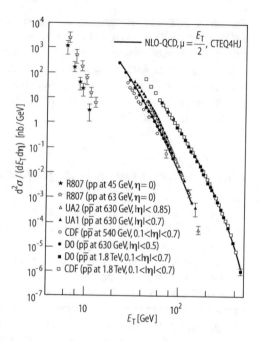

This steep behaviour is determined by the sharp falling of the parton densities at large x. Also the corresponding values of $\sqrt{s}$ and p_T are large enough to be well inside the perturbative region. The overall agreement of the data from ISR, UA1,2 and CDF and D0 is spectacular. Similar results also hold for the production of photons at large p_T. The collider data [36], shown in Fig. 4.16 [9], are in fair agreement with the theoretical predictions. For the same process less clear a situation is found with fixed target data. Here, first of all, the experimental results show some internal discrepancies. Also, the p_T accessible values being smaller, the theoretical uncertainties are larger. But it is true that the agreement is poor, so that the necessity of an "intrinsic" transverse momentum of partons inside the hadron of over 1 GeV has been claimed, which theoretically is not justified (rather, given the sharp falling down at large p_T, it could be interpreted as a correction for p_T calibration errors).

For heavy quark production at colliders [42] the agreement is very good for the top crosssection at the Tevatron (Fig. 4.17) [43, 44]. The bottom production at the Tevatron has for some time represented a problem [45]. The total rate and the p_T distribution of b quarks observed at CDF appeared in excess of the prediction, up to the largest measured values of p_T. But this is a complicated case, with different scales being present at the same time: $\sqrt{s}$, p_T, m_b. Finally the problem has been solved (Fig. 4.18) by better taking into account a number of small effects from resummation of large logarithms, the difference between b hadrons and b partons, the inclusion of better fragmentation functions etc. [46].

Fig. 4.16 Single photon
production in $p\bar{p}$ colliders as
function of p_T [9]

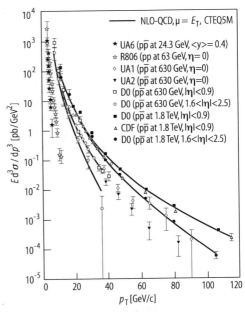

Fig. 4.17 The t production
cross-section at the Tevatron
$p\bar{p}$ collider [44]

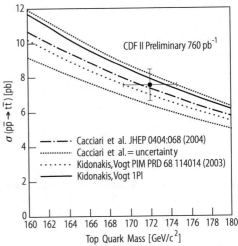

Drell-Yan processes, including lepton pair production via virtual γ, W or Z exchange, offer a particularly good opportunity to test QCD. The process is quadratic in parton densities, and the final state is totally inclusive, while the large scale is specified and measured by the invariant mass squared Q^2 of the lepton pair which itself is not strongly interacting (so there no dangerous hadronisation effects). The QCD improved parton model leads directly to a prediction for the total rate as a function of Q^2. The value of the LO cross-section is inversely proportional to the number of colours N_C because a quark of given colour can only annihilate with an

Fig. 4.18 The b production p_T distribution at the Tevatron $p\bar{p}$ collider [47]. The data from CDF also include systematics and correlations. The theoretical curve with the uncertainty range is from Ref. [46]

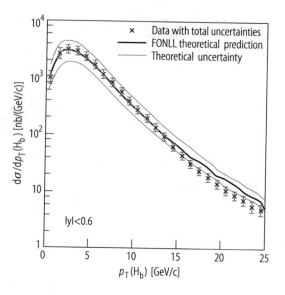

antiquark of the same colour to produce a colourless lepton pair. The order $\alpha_s(Q^2)$ corrections to the total rate were computed long ago and found to be particularly large [22, 38], when the quark densities are defined from the structure function F_2 measured in DIS at $q^2 = -Q^2$. The ratio $\sigma_{corr}/\sigma_{LO}$ of the corrected and the Born cross-sections, was called K-factor, because it is almost a constant in rapidity. In recent years also the NLO full calculation of the K-factor was completed, a very remarkable calculation [37]. The QCD predictions have been best tested for W and Z production at CERN $Sp\bar{p}S$ and Tevatron energies. $Q \sim m_{W,Z}$ is large enough to make the prediction reliable (with a not too large K-factor) and the ratio $\sqrt{\tau} = Q/\sqrt{s}$ is not too small. Recall that in lowest order $x_1 x_2 s = Q^2$ so that the parton densities are probed at x values around $\sqrt{\tau}$. We have $\sqrt{\tau} = 0.13 - 0.15$ (for W and Z production, respectively) at $\sqrt{s} = 630$ GeV (CERN $Sp\bar{p}S$ Collider) and $\sqrt{\tau} = 0.04 - 0.05$ at the Tevatron. In this respect the prediction is more delicate at the LHC, where $\sqrt{\tau} \sim 5.7 - 6.5 \cdot 10^{-3}$. One comparison of the experimental total rates at the Tevatron [48] with the QCD predictions is shown in Fig. 4.19, together with the expected rates at the LHC (based on the structure functions obtained in [23]).

The calculation of the W/Z p_T distribution has been a classic problem in QCD. For large p_T, for example $p_T \sim o(m_W)$, the p_T distribution can be reliably computed in perturbation theory, which was done up to NLO in the late '70's and early '80's. A problem arises in the intermediate range $\Lambda_{QCD} \ll p_T \ll m_W$, where the bulk of the data is concentrated, because terms of order $\alpha_s(p_T^2) \log m_W^2/p_T^2$ become of order one and should included to all orders [39]. At order α_s we have:

$$\frac{1}{\sigma_0}\frac{d\sigma_0}{dp_T^2} = (1+A)\delta(p_T^2) + \frac{B}{p_T^2}\log\frac{m_W^2}{p_T^2}\bigg|_+ + \frac{C}{(p_T^2)_+} + D(p_T^2) \qquad (4.98)$$

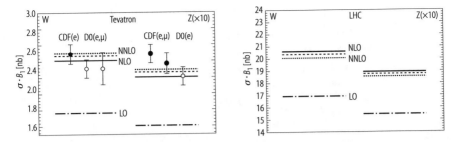

Fig. 4.19 Data vs. theory for W and Z production at the Tevatron ($\sqrt{s} = 1.8$ TeV) together with the corresponding predictions for the LHC ($\sqrt{s} = 1.4$ TeV) [48]

where A, B, C, D are coefficients of order α_s. The "+" distribution is defined in complete analogy with Eq. (4.93):

$$\int_0^{p_{TMAX}^2} g(z)f(z)_+ dz = \int_0^{p_{TMAX}^2} [g(z) - g(0)]f(z)dz \qquad (4.99)$$

The content of this, at first sight mysterious, definition is that the singular "+" terms do not contribute to the total cross-section. In fact for the cross-section the weight function $g(z) = 1$ and we obtain:

$$\sigma = \sigma_0[(1 + A) + \int_0^{p_{TMAX}^2} D(z)dz] \qquad (4.100)$$

The singular terms, of infrared origin, are present at the non completely inclusive level but disappear in the total cross-section. Arguments have been given that these singularities are expected to exponentiate. Explicit calculations in low order support the exponentiation which leads to the following expression:

$$\frac{1}{\sigma_0}\frac{d\sigma_0}{dp_T^2} = \int \frac{d^2b}{4\pi}\exp(-ib \cdot p_T)(1 + A)\exp S(b) \qquad (4.101)$$

with:

$$S(b) = \int_0^{p_{TMAX}} \frac{d^2k_T}{2\pi}[\exp ik_T \cdot b - 1][\frac{B}{k_T^2}\log\frac{m_W^2}{k_T^2} + \frac{C}{k_T^2}] \qquad (4.102)$$

At large p_T the LO perturbative expansion is recovered. At intermediate p_T the infrared p_T singularities are resummed (the Sudakov log terms, which are typical of vector gluons, are related to the fact that for a charged particle in acceleration it is impossible not to radiate, so that the amplitude for no soft gluon emission is exponentially suppressed). However this formula has problems at small p_T, for example, because of the presence of α_s under the integral for $S(b)$:

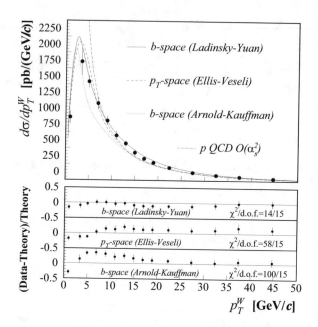

Fig. 4.20 QCD predictions for the W p_T distribution compared with recent D0 data at the Tevatron ($\sqrt{s} = 1.8$ TeV)[49] [40]

presumably the relevant scale is of order k_T^2. So it must be completed by some non perturbative ansatz or an extrapolation into the soft region. All the formalism has been extended to NLO accuracy, where one starts from the perturbative expansion at order α_s^2, and generalises the resummation to also include NLO terms of order $\alpha_s(p_T^2)^2 \log m_W^2/p_T^2$ (see, for example, [40]). The comparison with the data is very impressive. In Fig. 4.20 we see the p_T distribution as predicted in QCD (with a number of variants that mainly differ in the approach to the soft region) compared with some recent data at the Tevatron [49].

A great effort is being devoted to the preparation to the LHC. Calculations for specific processes are being completed. A very important example is Higgs production via $g + g \rightarrow H$. The amplitude is dominated by the top quark loop, as discussed in Chap. 3 [51]. The NLO corrections turn out to be particularly large [52], as seen in Fig. 4.21. Higher order corrections can be computed either in the effective lagrangian approach, where the heavy top is integrated away and the loop is shrunk down to a point [53] [the coefficient of the effective vertex is known to α_s^4 accuracy [54]], or in the full theory. At the NLO the two approaches agree very well for the rate as a function of m_H [55]. The NNLO corrections have been computed in the effective vertex approximation [56] (see Fig. 4.21). Beyond fixed order resummation of large logs were carried out [57]. Also the NLO EW contributions have been computed [58]. Rapidity (at NNLO) [59] and p_T distributions (at NLO) [60] have also been evaluated. At smaller p_T the large logarithms $[log(p_T/m_H)]^n$ have been

Fig. 4.21 The Higgs gluon fusion cross section in LO, NLO and NLLO

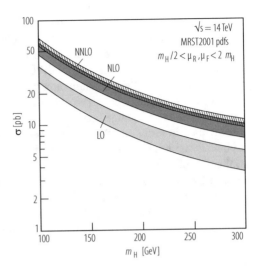

resummed in analogy with what was done long ago for W and Z production [61]. For additional recent works on Higgs physics at colliders see, for example, [62].

The activity on event simulation also received a big boost from the LHC preparation (see, for example, the review [50]). General algorithms for performing NLO calculations numerically (requiring techniques for the cancellation of singularities between real and virtual diagrams) have been developed (see, for example, [65]). The matching of matrix element calculation of rates together with the modeling of parton showers has been realised in packages, as for example in the MC@NLO [63] or POWHEG [64] based on HERWIG. The matrix element calculation, improved by resummation of large logs, provides the hard skeleton (with large p_T branchings) while the parton shower is constructed by a sequence of factorized collinear emissions fixed by the QCD splitting functions. In addition, at low scales a model of hadronisation completes the simulation. The importance of all the components, matrix element, parton shower and hadronisation can be appreciated in simulations of hard events compared with the Tevatron data.

At different places in the previous pages we have seen examples of resummation of large logs. This is a very important chapter of modern QCD. The resummation of soft gluon logs enter in different problems and the related theory is subtle. I refer the reader here to some recent papers where additional references can be found [66]. A particularly interesting related development has to do with the so called non global logs (see, for example, [67]). If in the measurement of an observable some experimental cuts are introduced, which is a very frequent case, then a number of large logs can arise from the corresponding breaking of inclusiveness. The discussion of this problem has led to rethinking the theory of final state observables. It is also important to mention the development of software for the automated implementation of resummation (see, for example, [68]).

Before closing this section I would like to mention some very interesting developments at the interface between string theory and QCD, twistor calculus. A precursor work was the Parke-Taylor result in 1986 [69] on the amplitudes for n incoming gluons with given helicities [70]. Inspired by dual models, they derived a compact formula for the maximum non vanishing helicity violating amplitude (with n−2 plus and 2 minus helicities) in terms of spinor products. Using the relation between strings and gauge theories in twistor space Witten developed in '03 [71] a formalism in terms of effective vertices and propagators that allows to compute all helicity amplitudes. The method, alternative to other modern techniques for the evaluation of Feynman diagrams [73], leads to very compact results. Since then rapid progress followed (for reviews, see [72]): for tree level processes powerful recurrence relations were established [74], the method was extended to include massless external fermions [75] and also external EW vector bosons [76] and Higgs particles [77]. The level already attained is already important for multijet events at the LHC. And the study of loop diagrams has been started. In summary, this road looks new and promising.

4.6 Measurements of α_s

Very precise and reliable measurements of $\alpha_s(m_Z^2)$ are obtained from e^+e^- colliders (in particular LEP) and from deep inelastic scattering.

4.6.1 α_s from e^+e^- Colliders

The main methods at e^+e^- colliders are: (a) Inclusive hadronic Z decay, R_l, σ_h, σ_l, Γ_Z. (b) Inclusive hadronic τ decay. (c) Event shapes and jet rates.

As we have seen, for a quantity like R_l we can write a general expression of the form:

$$R_l = \frac{\Gamma(Z, \tau \to hadrons)}{\Gamma(Z, \tau \to leptons)} \sim R^{EW}(1 + \delta_{QCD} + \delta_{NP}) + \dots \quad (4.103)$$

where R^{EW} is the electroweak-corrected Born approximation, δ_{QCD}, δ_{NP} are the perturbative (logarithmic) and non perturbative (power suppressed) QCD corrections. For a measurement of α_s at the Z (in the following we always refer to the $\overline{MS}$ definition of α_s) one can use all info from R_l, $\Gamma_Z = 3\Gamma_l + \Gamma_h$ and (f=h or l) $\sigma_f = 12\pi\Gamma_l\Gamma_f/(m_Z^2\Gamma_Z^2)$. In the past the measurement from R_l was preferred (by itself it leads to $\alpha_s(m_Z) = 0.1226 + 0.0058 - 0.0036$) but at LEP there is no reason for that. In all these quantities α_s enters through Γ_h, but the measurements of, say, Γ_Z, R_l and σ_l are really independent (they are affected by entirely different systematics: Γ_Z is extracted from the line shape, R_l and σ_l are measured at the peak but R_l does not depend on the absolute luminosity while σ_l does). The most sensitive single quantity is σ_l. The combined value from the measurements at the Z (assuming the validity of the SM and a light Higgs mass) is [78]:

$$\alpha_s(m_Z) = 0.119 \pm 0.003 \qquad (4.104)$$

For a relatively light Higgs (even if not as light as from the fit to EW observables) the final error is mainly experimental with a theoretical component from our ignorance of m_H, of higher orders in the QCD expansion [79] and also from uncertainties on the Bhabha luminometer (which affect $\sigma_{h,l}$) [80]. By adding all other electroweak precision electroweak tests (in particular m_W) one similarly finds [41]:

$$\alpha_s(m_Z) = 0.1185 \pm 0.003 \qquad (4.105)$$

We now consider the measurement of $\alpha_s(m_Z)$ from τ decay. R_τ has a number of advantages that, at least in part, tend to compensate for the smallness of $m_\tau = 1.777\,\text{GeV}$. First, R_τ is maximally inclusive, more than $R_{e^+e^-}(s)$, because one also integrates over all values of the invariant hadronic squared mass:

$$R_\tau = \frac{1}{\pi} \int_0^{m_\tau^2} \frac{ds}{m_\tau^2} \left(1 - \frac{s}{m_\tau^2}\right)^2 Im\,\Pi_\tau(s) \qquad (4.106)$$

The perturbative contribution is known at NNLO. Analyticity can be used to transform the integral into one on the circle at $|s| = m_\tau^2$:

$$R_\tau = \frac{1}{2\pi i} \oint_{|s|=m_\tau^2} \frac{ds}{m_\tau^2} \left(1 - \frac{s}{m_\tau^2}\right)^2 \Pi_\tau(s) \qquad (4.107)$$

Also, the factor $(1 - \frac{s}{m_\tau^2})^2$ is important to kill the sensitivity the region $Re[s] = m_\tau^2$ where the physical cut and the associated thresholds are located. Still the quoted result [81] looks a bit too precise:

$$\alpha_s(m_\tau) = 0.345 \pm 0.010 \qquad (4.108)$$

or

$$\alpha_s(m_Z) = 0.1215 \pm 0.0012 \qquad (4.109)$$

This precision is obtained by taking for granted that corrections suppressed by $1/m_\tau^2$ are negligible. This is because, in the massless theory, the light cone expansion is given by:

$$\delta_{NP} = \frac{ZERO}{m_\tau^2} + c_4 \cdot \frac{<O_4>}{m_\tau^4} + c_6 \cdot \frac{<O_6>}{m_\tau^6} + \cdots \qquad (4.110)$$

In fact there are no dim-2 Lorentz and gauge invariant operators. For example, $g_\mu g^\mu$ is not gauge invariant. In the massive theory, the ZERO is replaced by light quark mass-squared m^2. This is still negligible if m is taken as a lagrangian mass of a few MeV. If on the other hand the mass were taken to be the constituent mass of order Λ_{QCD}, this term would not be at all negligible and would substantially affect the result (note that $\alpha_s(m_\tau)/\pi \sim 0.1 \sim (0.6\,\text{GeV}/m_\tau)^2$ and that Λ_{QCD} for three flavours is large). For example, the PDG value and estimate of the error is [9]:

$$\alpha_s(m_Z) = 0.120 \pm 0.003. \tag{4.111}$$

Most people believe the optimistic version. I am not convinced that the gap is not filled up by ambiguities of $0(\Lambda_{QCD}^2/m_\tau^2)$ from δ_{pert} [82]. In any case, one can discuss the error, but it is true and remarkable, that the central value from τ decay, obtained at very small Q^2, is in good agreement with all other precise determinations of α_s at more typical LEP values of Q^2.

Important determinations of $\alpha_s(m_Z)$ are obtained from different infrared safe observables related to event rates and jet shapes in e^+e^- annihilation. The main problem of these measurements is the large impact of non perturbative hadronization effects on the result and therefore on the theoretical error. The perturbative part is known at NLO. One advantage is that the same measurements can be repeated at different $\sqrt{s}$ values (e.g. with the same detectors at LEP1 or LEP2) allowing for a direct observation of the energy dependence. A typical result, from jets and event shapes at LEP, quoted in Ref. [83], is given by:

$$\alpha_s(m_Z) = 0.121 \pm 0.005. \tag{4.112}$$

Recently the rate of 4-jet events (proportional to α_s^2) at LEP as function of y_{cut} has been used [84], for which a NLO theoretical calculation exists [85]. The quoted result is $\alpha_s(m_Z) = 0.1176 \pm 0.0022$ (the actual error could be somewhat larger because the ambiguity from hadronisation modeling is always debatable).

4.6.2 α_s from Deep Inelastic Scattering

QCD predicts the Q^2 dependence of $F(x, Q^2)$ at each fixed x, not the x shape. But the Q^2 dependence is related to the x shape by the QCD evolution equations. For each x-bin the data allow to extract the slope of an approximately straight line in $dlogF(x, Q^2)/dlogQ^2$: the log slope. The Q^2 span and the precision of the data are not much sensitive to the curvature, for most x values. A single value of Λ_{QCD} must be fitted to reproduce the collection of the log slopes. For the determination of α_s the scaling violations of non-singlet structure functions would be ideal, because of the minimal impact of the choice of input parton densities. We can write the non-singlet evolution equations in the form:

$$\frac{d}{dt} \log F(x,t) = \frac{\alpha_s(t)}{2\pi} \int_x^1 \frac{dy}{y} \frac{F(y,t)}{F(x,t)} P_{qq}(\frac{x}{y}, \alpha_s(t)) \tag{4.113}$$

where P_{qq} is the splitting function. At present NLO and NNLO corrections are known. It is clear from this form that, for example, the normalisation error on the input density drops away, and the dependence on the input is reduced to a minimum (indeed, only a single density appears here, while in general there are quark and gluon densities). Unfortunately the data on non-singlet structure functions are not very accurate. If we take the difference of data on protons and neutrons, $F_p - F_n$, experimental errors add up in the difference and finally are large. The $F_{3\nu N}$ data are directly non-singlet but are not very precise. A determination of α_s from the CCFR data on $F_{3\nu N}$ has led to [86]:

$$\alpha_s(m_Z) = 0.119 \pm 0.006 \tag{4.114}$$

A recent analysis of the same data leads to $\alpha_s(m_Z) = 0.119 \pm 0.002$ [87], but the theoretical error associated with the method and with the choice adopted for the scale ambiguities is not considered. A fit to non singlet structure functions in electro- or muon-production extracted from proton and deuterium data at the NNLO level was performed in Ref. [88] with the result:

$$\alpha_s(m_Z) = 0.114 \pm 0.002 \tag{4.115}$$

When one measures α_s from scaling violations on F_2 from e or μ beams, the data are abundant, the errors small but there is an increased dependence on input parton densities and especially a strong correlation between the result on α_s and the input on the gluon density. There are complete and accurate derivations of α_s from scaling violations in F_2. In a well known analysis by Santiago and Yndurain [89], the data on protons from SLAC, BCDMS, E665 and HERA are used with NLO kernels plus the NNLO first few moments. The analysis is based on an original method that uses projections on a specially selected orthogonal basis, the Bernstein polynomials. The quoted result is given by:

$$\alpha_s(m_Z) = 0.1163 \pm 0.0014 \tag{4.116}$$

(these authors also quote $\alpha_s(m_Z) = 0.115 \pm 0.006$ for F_3 data in νN scattering). A different analysis by Alekhin [90] of existing data off proton and deuterium targets with NNLO kernels and a more conventional method leads to

$$\alpha_s(m_Z) = 0.114 \pm 0.002 \tag{4.117}$$

In both analyses the dominant error is theoretical and, in my opinion, should be somewhat larger than quoted. An interesting perspective on theoretical errors can be obtained by comparing analyses with different methods. We add the following examples. From truncated moments (but with a limited set of proton data and NLO kernels) [91]: $\alpha_s(m_Z) = 0.122 \pm 0.006$, from Nachtmann moments (which take into account some higher twist terms) and proton data [92]: $\alpha_s(m_Z) = 0.1188 \pm 0.0017$. A combination of measurements at HERA by H1 and Zeus, also including final state jet observables, leads to $\alpha_s(m_Z) = 0.1186 \pm 0.0051$ [93], most of the error being theoretical. Finally, to quote a number that appears to me as a good summary of the situation of $\alpha_s(m_Z)$ from DIS one can take the result from a NNLO analysis of available data by the MRST group [94] as quoted by Particle Data Group, W.-M. Yao et al. [9]:

$$\alpha_s(m_Z) = 0.1167 \pm 0.0036 \qquad (4.118)$$

If we compare these results on α_s from DIS with the findings at the Z, given by Eq. (4.105), we see that the agreement is good, with the value of α_s from the most precise DIS measurements a bit on the low side with respect to e^+e^-.

4.6.3   Summary on α_s

There are a number of other determinations of α_s which are important because they arise from qualitatively different observables and methods. For example [9, 83], some are obtained from the Bjorken sum rule and the scaling violations in polarized DIS, from Υ decays, from the 4-jet rate in e^+e^-. A special mention deserves the "measurement" of α_s from lattice QCD [95]. A number of hadronic observables, in particular $\Upsilon' - \Upsilon$ splitting, pion and kaon decay constants, the B_s mass and the Ω baryon mass are used to fix the lattice spacing and to accurately tune the QCD simulation. The value of α_s is then obtained by computing non perturbatively a number of quantities related to Wilson loops that can also be given in perturbation theory. The result is then evolved with state of the art beta functions to obtain $\alpha_s(m_Z) = 0.1170 \pm 0.0012$. This result is interesting for its really special nature but it is not clear that the systematics due to the lattice technology is as small as claimed.

Summarising: there is very good agreement among many different measurements of α_s. In Fig. 4.22 [83], a compilation of the data is reported with each measurement plotted at the scale of the experiment, which shows the consistency of the measurements and the running of α_s. This is a very convincing, quantitative test of QCD. If I take the values of $\alpha_s(m_Z)$ from precision electroweak data, Eq. (4.105), from τ decay with the central value as in Eq. (4.109) but the larger error as in Eq. (4.111), from jets in e^+e^-, Eq. (4.112), and from DIS, Eq. (4.118), the average is :

$$\alpha_s(m_Z) = 0.119 \pm 0.002 \qquad (4.119)$$

Fig. 4.22 The running of α_s as determined from present data [83]

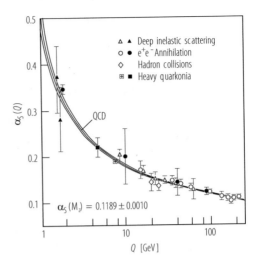

For comparison, the average value quoted by PDG 2006 is $\alpha_s(m_Z) = 0.1176 \pm 0.0020$ while Ref. [83] gives $\alpha_s(m_Z) = 0.1189 \pm 0.0010$.

The value of Λ (for $n_f = 5$) which corresponds to Eq. (4.119) is:

$$\Lambda_5 = 221 \pm 25 \text{ MeV} \tag{4.120}$$

Λ is the scale of mass that finally appears in massless QCD. It is the scale where $\alpha_s(\Lambda)$ is of order one. Hadron masses are determined by Λ. Actually the ρ mass or the nucleon mass receive little contribution from the quark masses (the case of pseudoscalar mesons is special, as they are the pseudo Goldstone bosons of broken chiral invariance). Hadron masses would be almost the same in massless QCD.

4.7   Conclusion

We have seen that perturbative QCD based on asymptotic freedom offers a rich variety of tests and we have described some examples in detail. QCD tests are not as precise as for the electroweak sector. But the number and diversity of such tests has established a very firm experimental foundation for QCD as a theory of strong interactions. The field of QCD appears as one of great maturity but also of robust vitality with many rich branches and plenty of new blossoms. The physics content of QCD is very large and our knowledge, especially in the non perturbative domain, is still very limited but progress both from experiment (Tevatron, RHIC, LHC......) and from theory is continuing at a healthy rate. And all the QCD predictions that we were able to formulate and to test appear to be in very good agreement with experiment.

Acknowledgments I am very grateful to Richard Ball, Keith Ellis, Stefano Forte, Ashutosh Kotwal, Vittorio Lubicz, Lorenzo Magnea, Michelangelo Mangano, Silvano Simula, Hartmut Wittig for their help and advise.

References for 4

Disclaimer: This list of references is by far incomplete. It is just meant to provide a guide to the vast literature on QCD. So it includes only a few old papers and, for most of the issues, some relatively recent works where to look for a more complete list of references.

References

1. M. Gell-Mann, Acta Phys. Austriaca Suppl. IX (1972) 733; H. Fritzsch and M. Gell-Mann, Proc. XVI Int. Conf. on High Energy Physics, Chicago-Batavia, 1972; H. Fritzsch, M. Gell-Mann, H. Leutwyler, Phys. Lett. B47 (1973) 365.
2. D.J. Gross, F. Wilczek, Phys. Rev. Lett. 30 (1973) 1343; Phys. Rev. D8 (1973) 3633; H.D. Politzer, Phys. Rev. Lett. 30 (1973) 1346.
3. S. Weinberg, Phys. Rev. Lett. 31 (1973) 494.
4. G. Altarelli, Phys. Rep. 81(1982)1.
5. G. Altarelli, Ann. Rev. Nucl. Part. Sci. 39(1989)357.
6. G. Altarelli, *The Development of Perturbative QCD*, World Scientific, (1994); W. Greiner, S. Schramm and E. Stein, *Quantum Chromodynamics*, Springer, 1994; *Handbuch of QCD*. ed. by M. Shifman, Vol.1-4, World Sci. (2001); R. K. Ellis, W. J. Stirling, and B. R. Webber, *QCD and Collider Physics*, Cambridge Monographs, (2003); G. Dissertori, I. Knowles and M. Schmelling, *Quantum Chromodynamics: High Energy Experiments and Theory*, Oxford Univ. Press, 2003.
7. M. Gell-Mann, Phys. Lett. 8 (1964) 214 ; G. Zweig, CERN TH 401 and 412 (1964). O.W. Greenberg,Phys. Rev. Lett. 13 (1964) 598.
8. F. Bloch and H. Nordsieck, Phys. Rev. 52(1937)54.
9. Particle Data Group, W.-M. Yao et al., Journal of Physics G 33, 1 (2006).
10. T. Kinoshita, J. Math. Phys. 3(1962)650; T.D. Lee and M. Nauenberg, Phys. Rev. 133(1964)1549.
11. T. Applequist and J. Carazzone, Phys. Rev. D11 (1975)2856.
12. O.V. Tarasov, A.A. Vladimirov and A. Yu. Zharkov, Phys. Lett. 93B(1980)429.
13. T. van Ritbergen, J.A.M. Vermaseren and S.A. Larin, Phys.Lett.B400(1997)379; see also M. Czakon, Nucl. Phys. B710 (2005) 485.
14. G. Altarelli, Proceedings of the E. Majorana Summer School, Erice, 1995, Plenum Press, ed. by A. Zichichi; M. Beneke and V. M. Braun, in *Handbuch of QCD*. ed. by M. Shifman, Vol.3, pag. 1719, World Sci. (2001).
15. S.G. Gorishny, A.L. Kataev and S.A. Larin, Phys. Lett. 259B (1991) 144; L. R. Surguladze and M. A. Samuel, Phys. Rev. Lett. 66 (1991) 560.
16. R. Brandt and G. Preparata,Nucl. Phys. 27B (1971) 541.
17. Yu. Dokshitzer, J. Phys.G17 (1991)1572; N. Brown and W. J. Stirling, Z. Phys. C53 (1992) 629.
18. V.N. Gribov and L.N. Lipatov, Sov. J. Nucl. Phys. 15 (1972) 438; see also Yu. L. Dokshitzer, Sov. Phys. JETP 46 (1977) 641.
19. G.Altarelli and G. Parisi, Nucl. Phys. B126 (1977) 298.
20. E. G. Floratos, D.A. Ross and C. T. Sachrajda, Nucl. Phys. B129 (1977) 66, E B139 (1978) 545; B152 (1979) 493; A. Gonzales-Arroyo, C. Lopez and F. J. Yndurain, Nucl. Phys. B153 (1979) 161; G. Curci, W. Furmanski and R. Petronzio, Nucl. Phys. B175 (1980) 27; W. Furmanski

and R. Petronzio, Phys. Lett. 97B (1980) 438; E.G. Floratos, R. Lacaze and C. Kounnas, Phys. Lett. 99B (1981) 89, 285; R. T. Herrod and S. Wada, Z. Phys. C9 (1981) 351.

21. S. Moch, J. A. M. Vermaseren and A. Vogt, Nucl. Phys. B **688**, 101 (2004), [arXiv:hep-ph/0403192]; A. Vogt, S. Moch and J. A. M. Vermaseren, Nucl. Phys. B **691**, 129 (2004), [arXiv:hep-ph/0404111].

22. 5) G. Altarelli, R.Keith Ellis, G. Martinelli, Nucl.Phys.B143:521,1978, E.ibid.B146:544,1978; Nucl.Phys.B157:461,1979.

23. A. D. Martin, R. G. Roberts, W. J. Stirling and R. S. Thorne, hep-ph/0110215, hep-ph/0201127.

24. G. Sterman, Nucl. Phys. B281 (1987) 310; S. Catani and L. Trentadue, Nucl. Phys. B327 (1989) 323; N. Kidonakis and G. Sterman, Nucl. Phys. B505 (1997) 321.

25. S. Forte and G. Ridolfi, Nucl.Phys. B650:229,2003; S. Moch, J. A. M. Vermaseren and A. Vogt, Nucl. Phys. B **726** (2005) 317 [arXiv:hep-ph/0506288]; T. Becher, M. Neubert and B. D. Pecjak, JHEP **0701** (2007) 076 [arXiv:hep-ph/0607228].

26. E. Gardi, G. P. Korchemsky, D. A. Ross and S. Tafat, Nucl. Phys. B **636** (2002) 385 [arXiv:hep-ph/0203161]; E. Gardi and R. G. Roberts, Nucl. Phys. B **653** (2003) 227 [arXiv:hep-ph/0210429].

27. L.N. Lipatov, Sov. Jour. Nucl. Phys. 23 (1976) 338; V.S. Fadin, E.A. Kuraev and L.N. Lipatov, Phys. Lett. 60B (1975) 50; Sov. Phys. JETP 44 (1976) 443; 45(1977)199; Y.Y. Balitski and L.N.Lipatov, Sov. Jour. Nucl. Phys. 28 (1978) 822.

28. V.S. Fadin and L.N. Lipatov, Phys. Lett. B429 (1998) 127; B429 (1998) 127; V.S. Fadin et al, Phys. Lett. B359 (1995)181; B387 (1996)593; Nucl. Phys. B406 (1993) 259; Phys. Rev. D50 (1994) 5893; Phys. Lett. B389 (1996) 737; Nucl. Phys. B477 (1996) 767; Phys. Lett. B415 (1997) 97; B422 (1998) 287; G. Camici and M. Ciafaloni, Phys. Lett. B412 (1997) 396; Phys. Lett. B430 (1998) 349. V. del Duca, Phys. Rev. D54 (1996) 989; D54 (1996) 4474; V. del Duca and C.R. Schmidt, Phys. Rev. D57 (1998) 4069; Z. Bern, V. del Duca and C.R. Schmidt, Phys. Lett. B445 (1998) 168.

29. G. Altarelli, R. Ball and S. Forte, Nucl. Phys. B742 (2006) 1.

30. M. Ciafaloni, D. Colferai, G. P. Salam and A. M. Stasto, Phys. Lett. B587 (2004) 87; see also G. P. Salam, hep-ph/0501097; C. D. White and R. S. Thorne, Phys. Rev. D **75** (2007) 034005 [hep-ph/0611204].

31. S. Forte and Y. Goto, *Proceedings of DIS 2006*, Tsukuba, Japan, World Scientific, (2007), p.908 [hep-ph/0609127].

32. M. Hirai, S. Kumano and N. Saito, Phys.Rev. D69 (2004) 05402.

33. G. Mallot, hep-ex/0612055.

34. G.Altarelli and G. G. Ross, Phys. Lett. B212 (1988) 391; A. V. Efremov and O. V. Terayev, *Proceedings of the Czech Hadron Symposium*, 1988, p. 302; R. D. Carlitz, J. C. Collins and A. H. Mueller, Phys. Lett. B214 (1988) 229; G. Altarelli and B. Lampe, Z. Phys. C47 (1990) 315.

35. CDF Run II Collaboration, Phys.Rev.D74:071103,2006; e-Print: hep-ex/0512020 the D0 Collaboration, Phys.Rev.Lett.86:2523,2001; Phys.Lett.B639:151,2006.

36. The CDF Collaboration, Phys.Rev.Lett.73:2662–2666,1994, Erratum-ibid.74:1891–1893,1995; Phys.Rev.D65:112003,2002; Phys.Rev.D70:074008,2004;

37. R. Ha the D0 Collaboration, Phys.Rev.Lett.87:251805,2001; Phys.Rev.Lett.84:2786,2000.

38. J. Kubar-Andre' and F. Paige, Phys. Rev. D19(1979) 221. mberg, W. L. van Neerven and T. Matsuura, Nucl. Phys. B359 (1991) 343; W. L. van Neerven and E. B. Zijlstra, Nucl. Phys. B382(1992) 11.

39. V.V. Sudakov, Sov. Phys. JETP, 3 (1956) 75; Yu. Dokshitzer, D. Dyakonov and S. Troyan, Phys. Lett. 76B (1978) 290, Phys. Rep. 58 (1980 269; G. Parisi and R. Petronzio, Nucl. Phys. B154 ((1979) 427; G. Curci, M. Greco and Y. Srivastava, Phys. Rev. Lett. 43 (1979), Nucl. Phys. B159 (1979) 451; J. Collins and D. Soper, Nucl. Phys. B139 (1981) 381, B194 (1982) 445, B197 (1982)446; J. Kodaira and L. Trentadue, Phys. Lett. 112B (1982) 66, 123B (1983)335; J. Collins, D. Soper and G. Sterman, Nucl. Phys. B250 (1985) 199; C. Davies, B. Webber and J. Stirling, Nucl. Phys. B256 (1985) 413.

40. P. Arnold and R. Kauffman, Nucl.Phys.B349:381,1991; G.A. Ladinsky and C.P. Yuan, Phys.Rev.D50:4239,1994; R.K. Ellis, S. Veseli, Nucl.Phys.B511:649,1998.

41. For up-to-date results see, The LEP Electroweak Group, http://lepewwg.web.cern.ch/ LEPEWWG/.
42. P. Nason, S. Dawson and R. K. Ellis, Nucl. Phys. B **303** (1988) 607; W. Beenakker, H. Kuijf, W. L. van Neerven and J. Smith, Phys. Rev. D **40** (1989) 54; P. Nason, S. Dawson and R. K. Ellis, Nucl. Phys. B **327** (1989) 49 [Erratum-ibid. B **335** (1989) 260]; M. L. Mangano, P. Nason and G. Ridolfi, Nucl. Phys. B **373** (1992) 295; R. Bonciani, S. Catani, M. L. Mangano and P. Nason, Nucl.Phys. B529:424,1998; M. Cacciari et al, JHEP 0404:068, 2004; N. Kidonakis and A. Vogt, Phys. Rev. D68 114014(2003); A. Banfi, G. P. Salam and G. Zanderighi, JHEP **0707** (2007) 026 [arXiv:0704.2999 [hep-ph]].
43. The CDF II Collaboration, Phys.Rev.D74:072005,2006; Phys.Rev.D74:072006,2006; Phys.Rev.D71:072005,2005; the CDF Collaboration, Phys.Rev.Lett.97:082004,2006; Phys.Rev.Lett.96:202002,2006; Phys.Rev.D72:052003,2005; Phys.Rev.D71:052003,2005; Phys.Rev.Lett.93:142001,2004. The D0 Collaboration, Phys.Rev.D74:112004,2006; Phys.Lett.B626:55,2005; Phys.Lett.B626:35,2005; Phys.Lett.B626:45,2005.
44. The CDF Collaboration, CDF Note 8148, v. 1.0, 2006.
45. The CDF Collaboration, Phys.Rev.D75:012010,2007; Phys.Rev.D71:032001,2005; Phys.Rev.D65:052005,2002; the D0 Collaboration, Phys.Rev.Lett.94:232001,2005; Phys.Rev.Lett.85:5068,2000; Phys.Lett.B487:264-272,2000.
46. M. Cacciari et al, JHEP 0407:033, 2004.
47. D. Bauer (for the CDF and D0 Collaborations), Nucl. Phys. B (Proc. Suppl.) 156 (2006) 226.
48. The CDF II Collaboration, Phys.Rev.Lett. 94:091803, 2005; the D0 Collaboration, Phys.Rev. D60:052003,1999.
49. The D0 collaboration, Phys. Rev. Lett. 84 (2000) 2792; Phys.Rev. D61 (2000) 032004; see also the CDF Collaboration, Phys.Rev.Lett. 84 (2000) 845.
50. J.M. Campbell, J.W. Huston and W.J. Stirling, Rept.Prog.Phys. 70 (2007) 89
51. H.Georgi, S.Glashow, M.Machacek and D.Nanopoulos,Phys.Rev.Lett. **40**,692(1978)
52. S. Dawson, Nucl. Phys. B **359**, 283 (1991); A. Djouadi, M. Spira and P. M. Zerwas, Phys. Lett. B **264**, 440 (1991); M. Spira, A. Djouadi, D. Graudenz and P. Zerwas, Nucl. Phys. B **453**, 17 (1995).
53. J. Ellis, M. Gaillard and D. Nanopoulos, Nucl. Phys. B **106**, 292 (1976)
54. K. Chetyrkin, B. Kniehl and M. Steinhauser, Phys. Rev. Lett. **79**, 353 (1997)
55. M. Kramer, E. Laenen and M. Spira, Nucl. Phys. B **511**, 523 (1998)
56. S.Catani, D.de Florian and M.Grazzini, JHEP**0105**,025(2001); JHEP**0201**,015(2002); R. V. Harlander, Phys. Lett. B **492**, 74 (2000); V. Ravindran, J. Smith and W. L. van Neerven, Nucl. Phys. B **704**, 332 (2005); R. V. Harlander and W. B. Kilgore, Phys. Rev. D **64**, 013015 (2001); R. V. Harlander and W. B. Kilgore, Phys. Rev. Lett. **88**, 201801 (2002); C. Anastasiou and K. Melnikov, Nucl. Phys. B **646**, 220 (2002); V. Ravindran, J. Smith and W. L. van Neerven, Nucl. Phys. B **665**, 325 (2003).
57. S. Catani, D. de Florian, M. Grazzini and P. Nason, JHEP **0307**, 028 (2003); S. Moch and A. Vogt, Phys. Lett. B **631**, 48 (2005); V. Ravindran, Nucl. Phys. B **752**, 173 (2006) (and references therein).
58. U. Aglietti, R. Bonciani, G. Degrassi and A. Vicini, Phys. Lett. B **595**, 432 (2004); G. Degrassi and F. Maltoni, Phys. Lett. B **600**, 255 (2004).
59. C. Anastasiou, K. Melnikov and F. Petriello, Nucl. Phys. B **724**, 197 (2005); V. Ravindran, J. Smith and W. L. van Neerven, Nucl. Phys. B **767** (2007) 100 [hep-ph/0608308].
60. D. de Florian, M. Grazzini and Z. Kunszt, Phys. Rev. Lett. **82**, 5209 (1999); V. Ravindran, J. Smith and W. L. Van Neerven, Nucl. Phys. B **634**, 247 (2002); C. J. Glosser and C. R. Schmidt, JHEP **0212**, 016 (2002).
61. G. Bozzi, S. Catani, D. de Florian and M. Grazzini, Phys. Lett. B **564**, 65 (2003); Nucl. Phys. B **737**, 73 (2006); A. Kulesza, G. Sterman and W. Vogelsang, Phys. Rev. D **69**, 014012 (2004).
62. R.K. Ellis, W.T. Giele and G. Zanderighi, Phys.Rev. D72 (2005) 054018; E. D74 (2006) 079902; J. M. Campbell, R. K. Ellis and G. Zanderighi, JHEP 0610 (2006) 028; C. Anastasiou et al, JHEP 0701 (2007) 082; G. Davatz et al, JHEP 0607 (2006) 037; S. Catani and M. Grazzini, Phys. Rev. Lett. 98 (2007) 222002; C. Anastasiou, G. Dissertori and F. Stoeckli,

JHEP **0709** (2007) 018 [arXiv:0707.2373 [hep-ph]]; P. M. Nadolsky, C. Balazs, E. L. Berger and C.-P. Yuan, Phys. Rev. D **76** (2007) 013008 [hep-ph/0702003 [HEP-PH]]; C. Balazs, E. L. Berger, P. M. Nadolsky and C.-P. Yuan, Phys. Rev. D **76** (2007) 013009 [arXiv:0704.0001 [hep-ph]]; M. Ciccolini, A. Denner and S. Dittmaier, Phys. Rev. Lett. **99** (2007) 161803 [arXiv:0707.0381 [hep-ph]].

63. S. Frixione and B. Webber, hep-ph/0612272 and references therein.
64. S. Frixione, P. Nason and G. Ridolfi, arXiv:0707.3081; S. Frixione, P. Nason and G. Ridolfi, JHEP **0709** (2007) 126 [arXiv:0707.3088 [hep-ph]]
65. K. Fabricius, I. Schmitt, G. Kramer and G. Schierholz, Z. Phys. C **11** (1981) 315; G. Kramer and B. Lampe, Fortsch. Phys. **37** (1989) 161; R. K. Ellis, D. A. Ross and A. E. Terrano, Nucl. Phys. B **178** (1981) 421; S. Frixione, Z. Kunszt and A. Signer, Nucl.Phys.B467:399,1996; D. Kosower, Phys.Rev.D71:045016,2005; S. Catani, S. Dittmaier, M. H. Seymour and Z. Trócsányi, Nucl.Phys.B627,189 (2002).
66. A. Banfi and M. Dasgupta, JHEP **0401** (2004) 027; P Bolzoni, S. Forte and G. Ridolfi, Nucl.Phys.B731:85,2005; Y. Delenda, R. Appleby, M. Dasgupta and A. Banfi, JHEP **0612** (2006) 044; P. Bolzoni, Phys.Lett.B643:325,2006; S. Mert Aybat, L. J. Dixon and G. Sterman, Phys. Rev. D **74** (2006) 074004; Yu. L. Dokshitzer and G. Marchesini, JHEP **0601** (2006) 007; E. Laenen and L. Magnea, Phys. Lett. B **632** (2006) 270; C. Lee and G. Sterman, Phys. Rev. D **75**, 014022 (2007); D. de Florian and W. Vogelsang, Phys. Rev. D **76** (2007) 074031 [arXiv:0704.1677 [hep-ph]]; R. Abbate, S. Forte and G. Ridolfi, Phys. Lett. B **657** (2007) 55 [arXiv:0707.2452 [hep-ph]].
67. M. Dasgupta and G. P. Salam, Phys. Lett. B **512** (2001) 323; M. Dasgupta and G. P. Salam, J. Phys. G **30** (2004) R143; A. Banfi, G. Corcella, M. Dasgupta, Y. Delenda, G. P. Salam and G. Zanderighi, hep-ph/0508096; J. R. Forshaw, A. Kyrieleis and M. H. Seymour, JHEP **0608** (2006) 059.
68. A. Banfi, G. P. Salam and G. Zanderighi, JHEP **0503** (2005) 073.
69. S.J. Parke and T.R. Taylor, Phys.Rev.Lett.56:2459,1986.
70. F.A. Berends and W. Giele, Nucl. Phys. B294 (1987) 700; M. L. Mangano, S.J. Parke and Z. Xu, Nucl. Phys. B298 (1988) 653.
71. E. Witten, Commun.Math.Phys.252:189,2004, [hep-th/0312171].
72. F. Cachazo and P. Svrcek, PoS RTN2005:004,2005, [hep-th/0504194]; L. J. Dixon, *Proceedings of the EPS International Europhysics Conference on High Energy Physics*, Lisbon, Portugal, 2005, PoS HEP2005:405,2006, [hep-ph/0512111] and references therein.
73. M. Dinsdale, M. Ternick and S. Weinzierl, JHEP 0603:056,2006, [hep-ph/0602204].
74. R. Roiban, M. Spradlin and A. Volovich, Phys. Rev. Lett. 94 102002 (2005); R. Britto, F. Cachazo and B. Feng, Nucl. Phys. B715 (2005) 499; R. Britto, F. Cachazo, B. Feng and E. Witten, Phys. Rev. Lett. 94 181602 (2005).
75. G. Georgiou and V. V. Khoze, JHEP 0405 (2004) 015; J. B. Wu and C.J. Zhu, JHEP 0409 (2004) 063; G. Georgiou, E. W. N. Glover and V. V. Khoze, JHEP 0407 (2004) 048.
76. Z. Bern, D. Forde, D. A. Kosower and P. Mastrolia, Phys.Rev.D72:025006,2005.
77. L. J. Dixon, E. W. N. Glover and V. V. Khoze, JHEP 0412 (2004) 015; S. D. Badger, E. W. N. Glover and V. V. Khoze, JHEP 0503 (2005) 023; S.D. Badger, E.W.N. Glover and K. Risager, Acta Phys.Polon.B38:2273–2278,2007.
78. S. Kluth, Rept. Prog. Phys. 69 (2006) 1771.
79. H. Stenzel, JHEP 0507:0132,2005.
80. W. de Boer and C. Sander, Phys.Lett.B585:276,2004.
81. M. Davier, A. Hocker and Z. Zhang, Rev.Mod.Phys.78:1043,2006, [hep-ph/0507078].
82. G Altarelli, P. Nason and G. Ridolfi, Z.Phys.C68:257,1995.
83. S. Bethke, Prog.Part.Nucl.Phys.58:351,2007, [hep-ex/0606035].
84. ALEPH Collaboration, A. Heister et al, Eur. J. Phys. C 27 (2003) 1; DELPHI Collaboration, J. Abdallah et al, Eur. J. Phys. C 38 (2005) 413; OPAL Collaboration, G. Abbiendi et al, hep-ex/0601048.
85. Z. Nagy and Z. Trocsanyi, Phys. Rev. D59 (1999) 14020, E. D62 (2000) 099902.

86. A.L. Kataev, G. Parente and A.V. Sidorov, Nucl.Phys. B573 (2000) 405; Nucl.Phys.Proc.Suppl.116:105,2003,[hep-ph/0211151].

87. P. M. Brooks and C.J. Maxwell, Nucl.Phys.B780:76, 2007, [hep-ph/0610137].

88. J. Blumlein, H. Bottcher and A. Guffanti, Nucl.Phys.B774:182,2007, [hep-ph/0607200].

89. J. Santiago and F.J. Yndurain, Nucl.Phys. B563 (1999) 45; Nucl. Phys. Proc. Suppl. 86 (2000)69; Nucl.Phys. B611 (2001) 45.

90. S.I. Alekhin, Phys.Rev. D59:114016,1999.

91. S. Forte, J. I. Latorre, L. Magnea and A. Piccione, Nucl.Phys.B643:477,2002, [hep-ph/0205286].

92. S. Simula, M. Osipenko, Nucl.Phys.B675:289,2003, [hep-ph/0306260].

93. C. Glasman, hep-ex/0506035.

94. A.D. Martin, R.G. Roberts, W.J. Stirling and R.S. Thorne, Eur.Phys.J.C35:325, 2004, [hep-ph/0308087].

95. Q. Mason et al, Phys. Rev. Lett. 95 (2005)052002, hep-lat/0503005.

Chapter 5
QCD on the Lattice

Hartmut Wittig

5.1 Introduction and Outline

Since Wilson's seminal papers of the mid-1970s, the lattice approach to Quantum Chromodynamics has become increasingly important for the study of the strong interaction at low energies, and has now turned into a mature and established technique. In spite of the fact that the lattice formulation of Quantum Field Theory has been applied to virtually all fundamental interactions, it is appropriate to discuss this topic in a chapter devoted to QCD, since by far the largest part of activity is focused on the strong interaction. Lattice QCD is, in fact, the only known method which allows ab initio investigations of hadronic properties, starting from the QCD Lagrangian formulated in terms of quarks and gluons.

5.1.1 Historical Perspective

In order to illustrate the wide range of applications of the lattice formulation, we give a brief historical account below.

First applications of the lattice approach in the late 1970s employed analytic techniques, predominantly the strong coupling expansion, in order to investigate colour confinement and also the spectrum of glueballs. While these attempts gave valuable insights, it soon became clear that in the case of non-Abelian gauge theories such expansions were not sufficient to produce quantitative results.

H. Wittig (✉)
Johannes-Gutenberg-Universität Mainz, Mainz, Germany
e-mail: hartmut.wittig@uni-mainz.de

© The Author(s) 2020
H. Schopper (ed.), *Particle Physics Reference Library*,
https://doi.org/10.1007/978-3-030-38207-0_5

First numerical investigations via Monte Carlo simulations, focusing in particular on the confinement mechanism in pure Yang–Mills theory, were carried out around 1980. The following years saw already several valiant attempts to study QCD numerically, yet it was realized that the available computer power was grossly inadequate to incorporate the effects of dynamical quarks. It was then that the so-called "quenched approximation" of QCD was proposed as a first step to solving full QCD numerically. This approximation rests on the *ad hoc* assumption that the dominant non-perturbative effects are mediated by the gluon field. Hadronic observables can then be computed on a pure gauge background with far less numerical effort compared to the real situation where quarks have a feedback on the gluon field. The main focus of activity during the 1980s was on bosonic theories: numerical simulations were used to compute the glueball spectrum in pure Yang–Mills theory. Another important result during this period concerned ϕ^4-theory and the implications of its supposed "triviality" for the Higgs-Yukawa sector of the Standard Model. Using a combination of analytic and numerical techniques, the triviality of ϕ^4 theory could be rigorously established.

Except for a brief spell of activity around the turn of the decade to simulate QCD with dynamical fermions, most projects in the 1990s were devoted to explore quenched QCD. Having recognized that the available computers and the efficiency of known algorithms were by far not sufficient to perform "realistic" simulations of QCD with controlled errors, lattice physicists resorted to exploring the quenched approximation and its limitations for a number of phenomenologically interesting quantities. Although the systematic error that arises by neglecting dynamical quarks could not be quantified reliably, many important quantities, such as quark and hadron masses, the strong coupling constant and weak hadronic matrix elements, were computed for the first time. One of the icons of that period was surely a plot of the masses of the lightest hadrons in the continuum limit of quenched QCD, produced by the CP-PACS Collaboration: their results indicated that the quenched approximation works surprisingly well (at least for these quantities), since the computed spectrum agreed with experimental determinations at the level of 10%. Simultaneously, a number of sophisticated techniques have been developed during the 1990s, thereby helping to control systematic effects, mainly pertaining to the influence of lattice artefacts, as well as the renormalization of local operators in the lattice regularized theory and their relation to continuum schemes such as $\overline{\text{MS}}$. Perhaps the most significant development at the end of the 1990s was the clarification of the issue of chiral symmetry and lattice regularization. Following this work it is now understood under which conditions the lattice formulation is compatible with chiral symmetry. The importance of this development extends far beyond QCD and implies new prospects for the non-perturbative study of chiral gauge theories.

Since 2000 the focus has decidedly shifted from the quenched approximation to serious attempts to simulate QCD with dynamical quarks, thereby tackling the biggest remaining systematic uncertainty. Progress in this area has not just been determined by the vast increase in computer power since the very first Monte Carlo simulations, but rather by the development of new algorithmic ideas, combined with the use of alternative discretizations that are numerically more efficient. At the time of writing this contribution (2007), the whole field is actually in a state of transition: although the quenched approximation is being abandoned, the latest results from simulations with dynamical quarks have not yet reached the same level of accuracy in regard to controlling systematic errors due to lattice artefacts and effects from renormalization, as compared to earlier quenched calculations. It can thus be expected that many of the results discussed later in this chapter will soon be superseded by more accurate numbers. In turn, the quenched approximation will be completely obsolete in a few years time, except perhaps to test new ideas or for exploratory studies of more complex quantities.

5.1.2 Outline

We begin with an introduction of the basic concepts of the lattice formulation of QCD. This shall include the field theoretical foundations, discretizations of the QCD Lagrangian, as well as simulation algorithms and other technical aspects related to the actual calculation of physical observables from suitable correlation functions. The following sections deal with various applications. Lattice calculations of the hadron spectrum are described in Sect. 5.3. Section 5.4 is devoted to lattice investigations of the confinement phenomenon. Determinations of the fundamental parameters of QCD, namely the strong coupling constant and quark masses are a major focus of this article, and are presented in Sect. 5.5. Another important property of QCD, namely the spontaneously broken chiral symmetry, is discussed in some detail in Sect. 5.6, which also includes a brief introduction into analytical non-perturbative approaches to the strong interaction, based on effective field theories. Lattice calculations of weak hadronic matrix elements, which serve to pin down the elements of the Cabibbo–Kobayashi–Maskawa matrix, are covered in Sect. 5.7. We end this contribution with a few concluding remarks.

In addition to the topics listed above, lattice simulations of QCD have also made important contributions to the determination of the phase structure of QCD, including results for the critical temperature of the deconfinement phase transition. Nevertheless, in this chapter we restrict the discussion to QCD at zero temperature and refer the reader to other parts of this volume.

5.2 The Lattice Approach to QCD

The essential features of the lattice formulation can be summarized by the following statement:

> Lattice QCD is the non-perturbative approach to the gauge theory of the strong interaction through regularized, Euclidean functional integrals. The regularization is based on a discretization of the QCD action which preserves gauge invariance at all stages.

This definition includes all basic ingredients: starting from the functional integral itself avoids any particular reference to perturbation theory. This is what we mean when we call lattice QCD an ab initio method. The Euclidean formulation, which is obtained by rotating to imaginary time, reveals the close relation between Quantum Field Theory and Statistical Mechanics. In particular, the Euclidean functional integral is equivalent to the partition function of the corresponding statistical system. This equivalence is particularly transparent if the field theory is formulated on a discrete space-time lattice. Via this relation, the whole toolkit of condensed matter physics, including high-temperature expansions, and, perhaps most importantly, Monte Carlo simulations, are at the disposal of the field theorist.

Many of the basic concepts introduced in this section are discussed in several common textbooks on the subject [1–4], which can be consulted for further details.

5.2.1 Euclidean Quantization

The generic steps in the Euclidean quantization procedure of a lattice field theory are the following:

1. Define the classical, Euclidean field theory in the continuum;
2. Discretize the corresponding Lagrangian;
3. Quantize the theory by defining the functional integral;
4. Determine the particle spectrum from Euclidean correlation functions.

We shall now illustrate this procedure for a simple example, namely the theory for a neutral scalar field.

Step 1 Consider a real, classical field $\phi(x)$, with $x = (x^0, x^1, x^2, x^3)$, whose time variable x^0 is obtained by analytically continuing t to $-ix^0$. The Euclidean action $S_E[\phi]$ is defined as

$$S_E[\phi] = \int d^4x \left\{ \frac{1}{2} \partial_\mu \phi(x) \partial_\mu \phi(x) + V(\phi) \right\}, \qquad \partial_\mu \equiv \frac{\partial}{\partial x_\mu}, \qquad (5.1)$$

where

$$V(\phi) = \frac{1}{2} m^2 \phi(x)^2 + \frac{\lambda}{4!} \phi(x)^4. \qquad (5.2)$$

Step 2 In order to discretize the theory, a hyper-cubic lattice, Λ_E, is introduced as the set of discrete space-time points, i.e.

$$\Lambda_E = \left\{ x \in \mathbb{R}^4 \,\middle|\, x^0/a = 1, \ldots, N_t; \; x^j/a = 1, \ldots, N_s, \; j = 1, 2, 3 \right\},$$

$$T = N_t a, \quad L = N_s a. \qquad (5.3)$$

Thus, any space-time point is an integer multiple of the lattice spacing a. The total number of lattice sites is $N_t \times N_s^3$, while the physical space-time volume is $T \times L^3$. The discretized action is then given by

$$S_E[\phi] = a^4 \sum_{x \in \Lambda_E} \left\{ \frac{1}{2} d_\mu \phi(x) d_\mu \phi(x) + \frac{1}{2} m^2 \phi(x)^2 + \frac{\lambda}{4!} \phi(x)^4 \right\}, \qquad (5.4)$$

where the lattice derivatives can be defined as

$$d_\mu \phi(x) := \frac{1}{a} \left(\phi(x + a\hat{\mu}) - \phi(x) \right) \qquad \text{``forward'' derivative,} \qquad (5.5)$$

$$d_\mu^* \phi(x) := \frac{1}{a} \left(\phi(x) - \phi(x - a\hat{\mu}) \right) \qquad \text{``backward'' derivative.} \qquad (5.6)$$

Here and below $\hat{\mu}$ denotes a unit vector in direction of μ. Via a Fourier transform, the Euclidean lattice Λ_E is related to the dual lattice, Λ_E^*, defined by

$$\Lambda_E^* = \left\{ p \in \mathbb{R}^4 \,\middle|\, p_0 = \frac{2\pi}{T} n^0, \; p_j = \frac{2\pi}{L} n^j \right\}$$

$$n^0 = -\frac{N_t}{2}, -\frac{N_t}{2} + 1, \ldots, \frac{N_t}{2} - 1, \quad n^j = -\frac{N_s}{2}, -\frac{N_s}{2} + 1, \ldots, \frac{N_s}{2} - 1. \quad (5.7)$$

This not only implies that the momenta p_0 and p_j are quantized in units of $2\pi/T$ and $2\pi/L$, respectively, but also that a momentum cutoff has been introduced, since

$$-\frac{\pi}{a} \le p_\mu \le \frac{\pi}{a}. \qquad (5.8)$$

As we shall see below, this way of introducing a momentum cutoff can be extended to gauge theories in such a way that gauge invariance is respected. An important point to realize is that the lattice action is not unique: it is only required that the discretized expression for S_E reproduces the continuum result as the lattice spacing a is taken to zero.

Step 3 The theory is quantized via the Euclidean functional integral

$$Z_E := \int D[\phi] \, e^{-S_E[\phi]}, \qquad D[\phi] = \prod_{x \in \Lambda_E} d\phi(x). \qquad (5.9)$$

Here one sees explicitly that the discretization procedure has given a mathematical meaning to the integration measure, which reduces to that of an ordinary, multiple-dimensional integration.

One can now define Euclidean correlation functions of local fields through

$$\langle \phi(x_1) \cdots \phi(x_n) \rangle = \frac{1}{Z_E} \int D[\phi] \phi(x_1) \cdots \phi(x_n) e^{-S_E[\phi]}. \tag{5.10}$$

In the continuum limit, these correlation functions approach the Schwinger functions, which encode the physical information about the spectrum within the Euclidean formulation. Osterwalder and Schrader [5] have laid down the general criteria which must be satisfied such that the information in Minkowskian space-time can be reconstructed from the Schwinger functions.

Step 4 The particle spectrum is extracted from the exponential fall-off of the Euclidean two-point correlation function. To this end, one must define the Euclidean time evolution operator. The *transfer matrix* T describes time propagation by a finite Euclidean time interval a. The functional integral can be expressed in terms of the transfer matrix as

$$Z_E = \mathrm{Tr}\, T^{N_t}, \tag{5.11}$$

where the trace is taken over the basis $|\alpha\rangle$ of the Hilbert space of physical states. In order to obtain expressions which are more reminiscent of those in Minkowski space-time, one can define a Hamiltonian H_E by

$$T =: e^{-a H_E}. \tag{5.12}$$

If $|\alpha\rangle$ denotes an eigenstate of the transfer matrix with eigenvalue λ_α, i.e.

$$T|\alpha\rangle = \lambda_\alpha |\alpha\rangle = e^{-a E_\alpha} |\alpha\rangle, \tag{5.13}$$

then one can work out the spectral decomposition of the two-point correlation function, viz.

$$\langle \phi(x) \phi(y) \rangle = \frac{1}{Z_E} \int D[\phi] \phi(x) \phi(y) e^{-S_E[\phi]} \tag{5.14}$$

$$= \sum_\alpha e^{-(E_\alpha - E_0)(x_0 - y_0)} \langle \alpha | \hat{\phi}(0, \vec{y}) | 0 \rangle \langle 0 | \hat{\phi}(0, \vec{x}) | \alpha \rangle. \tag{5.15}$$

Here, the quantity $(E_\alpha - E_0)$ is the so-called mass gap, i.e. the energy of the state $|\alpha\rangle$ above the vacuum. For large Euclidean time separations $(x_0 - y_0)$ the lowest state dominates the two-point function, i.e. all higher states die out exponentially. The spectral decomposition of the two-point function forms the basis for numerical simulations of lattice field theories, as the mass (or energy) of a given state is given by the dominant exponential fall-off at large Euclidean times (see Sect. 5.2.3).

5.2.2   Lattice Actions for QCD

Our goal now is to find a lattice transcription of the Euclidean QCD action in the continuum, i.e.

$$S_{\text{QCD}} = \int d^4x \left\{ -\frac{1}{2g_0^2} \text{Tr}\,(F_{\mu\nu} F_{\mu\nu}) + \sum_{f=u,d,s\ldots} \bar{\psi}_f \left(\gamma_\mu D_\mu + m_f \right) \psi_f \right\}, \quad (5.16)$$

where g_0 denotes the gauge coupling, and our conventions are chosen such that the covariant derivative is defined through

$$D_\mu = \partial_\mu + A_\mu, \quad (5.17)$$

while the field tensor reads

$$F_{\mu\nu} = \partial_\mu A_\nu - \partial_\nu A_\mu + [A_\mu, A_\nu], \qquad A_\mu^\dagger = -A_\mu. \quad (5.18)$$

Before attempting to write down a discretized version, we must first elucidate the notion of a lattice gauge field in a non-Abelian theory. In fact, in this case it turns out that the gauge potential A_μ must be abandoned when the theory is discretized. The reason is that the familiar non-Abelian transformation law, i.e.

$$A_\mu(x) \to g(x)A_\mu(x)g(x)^{-1} + g(x)\partial_\mu(x)g(x)^{-1}, \quad g(x) \in \text{SU}(3), \quad (5.19)$$

no longer holds exactly when ∂_μ is replaced by its discrete counterpart d_μ of Eq. (5.5). Strict gauge invariance at the level of the regularized theory cannot be maintained in this fashion.

The definition of a lattice gauge field relies on the concept of the parallel transporter. If a quark moves in the presence of a background gauge field from y to x, it picks up a non-Abelian phase factor, given by

$$U(x, y) = \text{P.O.}\exp\left\{ -\int_y^x dz_\mu\, A_\mu(z) \right\}, \quad (5.20)$$

where "P.O." denotes path ordering, as a consequence of the non-Abelian nature of the gauge field. By contrast to the gauge potential A_μ, which is an element of the Lie algebra of SU(3), the parallel transporter $U(x, y)$ is an element of the gauge group itself. On the lattice, the parallel transporter between neighbouring lattice sites x and $x + a\hat{\mu}$ is called *link variable*:

$$U(x, x + a\hat{\mu}) \equiv U_\mu(x), \qquad U(x + a\hat{\mu}, x) = U(x, x + a\hat{\mu})^{-1} = U_\mu(x)^{-1}. \quad (5.21)$$

·A consistent and manifestly gauge invariant discretization of QCD is obtained by identifying the gauge degrees of freedom with the link variables $U_\mu(x)$, which transform under the gauge group as

$$U_\mu(x) \to g(x) U_\mu(x) g(x + a\hat{\mu})^{-1}, \quad g(x), g(x + a\hat{\mu}) \in SU(3). \quad (5.22)$$

The connection with the gauge potential $A_\mu(x)$ is somewhat subtle: if $U_\mu(x)$ denotes a *given* link variable in the discretized theory, it can be used to *define* a vector field $A_\mu(x)$ as an element of the Lie algebra of SU(3) via

$$e^{aA_\mu(x)} \equiv U_\mu(x). \quad (5.23)$$

In turn, if A_μ^c is a *given* gauge potential in the continuum theory, one can always find a link variable which approximates A_μ^c up to cutoff effects.

Now we turn to the problem of defining a discretized version of the Yang–Mills action. To this end we define the *plaquette* $P_{\mu\nu}(x)$ as the product of link variables around an elementary square of the lattice:

$$P_{\mu\nu}(x) \equiv U_\mu(x)U_\nu(x + a\hat{\mu})U_\mu(x + a\hat{\nu})^{-1}U_\nu(x)^{-1}. \quad (5.24)$$

A graphical representation is shown in Fig. 5.1. Using the transformation property in Eq. (5.22), it is easy to convince oneself that this object is manifestly gauge invariant. Moreover, it serves to define the simplest discretization of the Yang–Mills action, the Wilson plaquette action [6]

$$S_G[U] = \beta \sum_{x \in \Lambda_E} \sum_{\mu < \nu} \left(1 - \frac{1}{3}\text{Re Tr } P_{\mu\nu}(x)\right). \quad (5.25)$$

It has become a standard textbook exercise to verify that for small lattice spacings

$$S_G[U] \longrightarrow -\frac{1}{2g_0^2} \int d^4x \, \text{Tr}\,(F_{\mu\nu}F_{\mu\nu}) + O(a), \quad (5.26)$$

provided that one relates the parameter β to the bare gauge coupling via $\beta = 6/g_0^2$ in Eq. (5.25). We have remarked already that the discretization of a field theory is not

Fig. 5.1 Graphical representation of the plaquette $P_{\mu\nu}(x)$ in the (μ, ν)-plane. The arrow between sites $x + a\hat{\mu}$ and x denotes the link variable $U_\mu(x)$

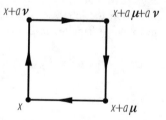

unique, and hence one is free to add further gauge invariant terms to the plaquette action which formally vanish as $a \to 0$, but which produce a discretization with an accelerated rate of convergence to the continuum limit. The most widely chosen alternatives are the Symanzik [7] and Iwasaki [8] actions.

Quark and antiquark fields, $\psi(x)$ and $\bar{\psi}(x)$, are associated with the lattice sites and transform under the gauge group as

$$\psi(x) \to g(x)\psi(x), \qquad \bar{\psi}(x) \to \bar{\psi}(x)g(x)^{-1}. \tag{5.27}$$

Using the transformation property of the link variables, it is straightforward to write down a discretized version of the covariant derivative, i.e.

$$\nabla_\mu \psi(x) := \frac{1}{a}\left(U_\mu(x)\psi(x + a\hat{\mu}) - \psi(x)\right)$$

$$\nabla_\mu^* \psi(x) := \frac{1}{a}\left(\psi(x) - U_\mu(x - a\hat{\mu})^{-1}\psi(x - a\hat{\mu})\right), \tag{5.28}$$

where ∇_μ and ∇_μ^* denote the "forward" and "backward" derivatives, respectively. Finally, we note that in Euclidean space-time, the Dirac matrices can be defined to satisfy $\{\gamma_\mu, \gamma_\nu\} = 2\delta_{\mu\nu}$.

Before we attempt to construct the fermionic part of the action of lattice QCD, it is useful to identify the basic properties that the discretized, massless Dirac operator, D, should satisfy:

(a) D is local;
(b) $\widetilde{D}(p) = i\gamma_\mu p_\mu + O(ap^2)$;
(c) $\widetilde{D}(p)$ is invertible for $p \neq 0$;
(d) $\gamma_5 D + D\gamma_5 = 0$.

Locality, i.e. the absence of long-ranged interactions, is a basic property of any quantum field theory describing elementary particles. Property (b) implies that the correct continuum behaviour of the quark-gluon interaction is reproduced. Furthermore, condition (c) ensures that the correct fermion spectrum is obtained: fermion masses are associated with poles of $\{\widetilde{D}(p)\}^{-1}$, which, in the continuum theory, only occur at vanishing four-momentum. Finally, property (d) ensures that the massless theory respects chiral symmetry.

Using the definition of the covariant derivative and the conventions for the Dirac matrices in Euclidean space-time, we can now write down the simplest discretized version of the massless lattice Dirac operator:

$$D_{\text{disc}} = \tfrac{1}{2}\gamma_\mu(\nabla_\mu + \nabla_\mu^*). \tag{5.29}$$

It turns out, however, that this "naïve" discretization violates condition (c) and therefore produces spurious fermionic degrees of freedom. This is the so-called fermion doubling problem, which is most easily explained by considering D_{disc} in momentum space for the free theory. The Fourier transform yields

$$\widetilde{D}_{\mathrm{disc}}(p) = i\gamma_\mu \frac{1}{a}\sin(ap_\mu) = i\gamma_\mu p_\mu + O(a^2). \tag{5.30}$$

The discretization procedure has thus replaced p_μ by a sine function. While the Taylor expansion guarantees that condition (b) is satisfied, the occurrence of $\sin(ap_\mu)$ implies that $\widetilde{D}_{\mathrm{disc}}(p)$ vanishes not only at $p_\mu = 0$, but also at π/a for $\mu = 0, \dots, 3$ in the permitted range of momenta, thereby violating condition (c). The massless propagator $\{\widetilde{D}_{\mathrm{disc}}(p)\}^{-1}$ therefore has $2^4 = 16$ poles, and thus there is a 16-fold degeneracy of the fermion spectrum.

As we shall see below, the fermion doubling problem is closely linked with the issue of chiral symmetry on the lattice. For now we simply list the various methods that have been devised to address fermion doubling. Historically the first was due to Wilson ("Wilson fermions") [6]. Here, the degeneracy is lifted completely, but the price to pay is the explicit breaking of chiral symmetry at the level of the regularized theory. Another method, due to Kogut and Susskind ("staggered fermions") [9], is based on the idea of spreading individual spinor components over the corners of an elementary hypercube of the lattice. Although the degeneracy is only lifted partially (from 16 to 4), this formulation has the advantage of leaving a subgroup of chiral symmetry unbroken. More recent developments include the use of so-called "domain wall" [10, 11] or "overlap" [12] fermions. These formulations leave chiral symmetry unbroken in principle, and also succeed in lifting the degeneracy completely. Finally, there are the so-called "perfect" actions [13], which are based on a renormalization group approach and which are in principle completely free of lattice artefacts. An exact realization of the perfect action which can be used in simulations is, however, difficult to obtain. In practice, one typically uses a so-called truncated fixed point action. Domain wall and overlap fermions, as well as perfect actions are particular realizations of a class of discretizations dubbed "Ginsparg-Wilson fermions". They have the remarkable feature that chiral symmetry is preserved, while the fermion doubling problem is completely avoided. We shall come back to this issue in more detail below.

For now we turn specifically to Wilson's treatment of the fermion doubling problem. It exploits the fact that the discretization is not unique. Thus, one can add a term to D_{disc}, which formally vanishes as $a \to 0$, but which pushes the masses of the unwanted doubler states to the cutoff scale at any non-zero value of the lattice spacing. Explicitly, the massless Wilson-Dirac operator D_{w} reads

$$D_{\mathrm{w}} = \tfrac{1}{2}\gamma_\mu(\nabla_\mu + \nabla_\mu^*) + ar\nabla_\mu^*\nabla_\mu, \tag{5.31}$$

where r is the so-called Wilson parameter, which is usually set to one. The Fourier transform of D_{w} for a trivial gauge field reads

$$\widetilde{D}_{\mathrm{w}}(p) = i\gamma_\mu \frac{1}{a}\sin(ap_\mu) + \frac{2r}{a}\sin^2\left(\frac{ap_\mu}{2}\right), \tag{5.32}$$

which explicitly demonstrates (for the free theory, at least) that the poles at $p_\mu = \pi/a$ receive additional contributions proportional to r/a, which is of order of the cutoff for $r = O(1)$. Although this procedure leads to a complete lifting of the degeneracy,[1] it has a number of unwanted features: first, it should be noted that the Wilson fermion action differs from the classical action in the continuum by terms of order a, as a result of adding the counterterm proportional to r. By contrast, the leading discretization effects of the Wilson plaquette action for Yang–Mills theory are only $O(a^2)$. The Wilson fermion formulation will thus have a reduced rate of convergence towards the continuum limit. Secondly, the addition of the Wilson term results in an explicit breaking of chiral symmetry, since the massless theory is no longer invariant under global axial rotations, such as

$$\psi(x) \rightarrow e^{i\alpha\gamma_5}\psi(x), \qquad \bar{\psi}(x) \rightarrow \bar{\psi}(x)e^{i\alpha\gamma_5}, \qquad (5.33)$$

which implies that property (d) is violated. While the rate of convergence to the continuum limit can be accelerated by employing what is known as "$O(a)$ improvement" (see below), the explicit breaking of chiral symmetry cannot be cured within the Wilson theory. Thus, quantities like the quark condensate, which arises from the spontaneous breaking of chiral symmetry, cannot be studied in a conceptually "clean" manner using Wilson fermions. A detailed discussion how this can be achieved with the help of a more sophisticated fermionic discretization ("Ginsparg-Wilson fermions") is presented in Sect. 5.6. However, for most applications of lattice QCD, explicit chiral symmetry breaking is merely an inconvenience, but no serious obstacle.

We have already remarked when discussing the discretized Yang–Mills part of the QCD action that the non-uniqueness of the discretization opens the possibility to construct lattice actions with an accelerated rate of convergence towards the continuum limit. A systematic way how to do this is the so-called Symanzik improvement programme [14], in which lattice artefacts can be removed order by order in the lattice spacing. In a nutshell, the improvement programme amounts to extending the renormalization procedure of a field theory to the level of irrelevant operators, i.e. operators that formally vanish as $a \rightarrow 0$. In this sense one adds suitable counterterms, which for any non-zero value of a produce a cancellation of the cutoff effects at a given order, provided that their coefficients are tuned appropriately. For QCD with Wilson fermions, Sheikholeslami and Wohlert [15] have shown that the Symanzik improvement programme to lowest order is realized by adding one $O(a)$ counterterm to the Wilson-Dirac operator D_{w}. The resulting expression in the massless case reads

$$D_{\mathrm{sw}} = D_{\mathrm{w}} + \frac{ia}{4} c_{\mathrm{sw}} \sigma_{\mu\nu} \widehat{F}_{\mu\nu}, \qquad (5.34)$$

---

[1]That the degeneracy is indeed completely lifted in the presence of a non-trivial gauge field can be verified in numerical simulations.

Fig. 5.2 Four plaquettes that must be summed over to yield the quantity $Q_{\mu\nu}(x)$ in the lattice definition of the field strength tensor. The site x is at the center of the "clover" leaf

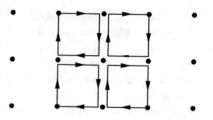

where $\sigma_{\mu\nu} = \frac{i}{2}[\gamma_\mu, \gamma_\nu]$, and $\widehat{F}_{\mu\nu}$ is a lattice transcription of the gluon field strength tensor $F_{\mu\nu}$. A suitable representation of $\widehat{F}_{\mu\nu}$ in terms of plaquette variables is given by

$$\widehat{F}_{\mu\nu}(x) = \frac{1}{8a^2}\left(Q_{\mu\nu}(x) - Q_{\nu\mu}(x)\right), \tag{5.35}$$

where $Q_{\mu\nu}(x)$ is the sum of the four plaquettes emanating from the site x, as depicted in Fig. 5.2. The object $Q_{\mu\nu}(x)$ is aptly called "clover" leaf. In order to remove all lattice artefacts of order a in hadron masses, the improvement coefficient c_{sw} must be fixed by imposing a suitable improvement condition. Without going into details here, we note that it is possible to find such a condition, which can also be evaluated at the non-perturbative level [16, 17]. The resulting, non-perturbatively $O(a)$ improved Wilson action can then be used to compute, say, hadron masses whose values differ from the continuum result by terms of only $O(a^2)$.

The Wilson-Dirac operator for a quark with bare mass m_0 is simply $(D_w + m_0)$. However, the form of the Wilson fermion action, $S_F^W[U, \bar{\psi}, \psi]$, which is found in the literature is usually expressed in terms of the "hopping parameter" κ rather than m_0. By rescaling the fermion fields according to

$$\psi(x) \to \sqrt{2\kappa}\,\psi(x), \qquad \bar{\psi}(x) \to \bar{\psi}(x)\sqrt{2\kappa}, \tag{5.36}$$

one obtains

$$S_F^W[U, \bar{\psi}, \psi] \equiv a^4 \sum_{x \in \Lambda_E} \bar{\psi}(x)(D_w + m_0)\psi(x)$$

$$= a^4 \sum_{x \in \Lambda_E} \left\{ -\kappa \sum_{\mu=0}^{3} \frac{1}{a}\Big[\bar{\psi}(x)(r - \gamma_\mu)U_\mu(x)\psi(x + a\hat{\mu})\right.$$

$$\left. +\bar{\psi}(x + a\hat{\mu})(r + \gamma_\mu)U_\mu(x)^{-1}\psi(x)\Big]$$

$$+ \bar{\psi}(x)\psi(x) \right\}. \tag{5.37}$$

The hopping parameter κ is related to the bare mass m_0 via

$$\kappa = \frac{1}{2am_0 + 8r},$$ (5.38)

while the dimensionless parameter r is usually set to one. Taken together with the plaquette action of Eq. (5.25), the Wilson action for QCD is thus conveniently parameterized in terms of the bare parameters (β, κ), with $\beta = 6/g_0^2$ and κ as above, instead of the bare gauge coupling and quark mass (g_0, m_0).

Another consequence of adding the Wilson term to the naïve lattice action is the resulting additive renormalization of the quark mass. In other words, the point where the quark mass vanishes is a priori unknown. The value that must be subtracted is called the critical quark mass, which corresponds to the critical value of the hopping parameter, κ_c. The bare subtracted quark mass is then given by

$$m = \frac{1}{2a}\left(\frac{1}{\kappa} - \frac{1}{\kappa_c}\right).$$ (5.39)

From Eq. (5.38) one easily infers that the critical value of κ in the free theory occurs at

$$\kappa_c = \frac{1}{8}, \qquad r = 1,$$ (5.40)

while for non-zero g_0 the value of κ_c must be determined, for instance, by adjusting κ to the point where the pion mass vanishes.

We now turn to discussing one alternative to using Wilson's solution to the fermion doubling problem, namely the so-called "staggered" (or Kogut-Susskind) fermions. One might think that the doubling problem arises since there are too many fermion degrees of freedom in the discretized theory, if one associates a four-component Dirac spinor with each individual lattice site. Pictorially, the main idea of Kogut and Susskind was to "thin out" the degrees of freedom by distributing single spinor components over different lattice sites. In their particular formulation, the 16 corners of a four-dimensional hypercube serve to accommodate the individual components of four Dirac spinors. Therefore, if these hypercubes are regarded as the main building blocks for the fermionic discretization, rather than the lattice sites themselves, this procedure will result in a partial lifting of the degeneracy from 16 fermion species down to four. It is clear, though, that a simple distribution of spinor components is not sufficient to define the action, since the Dirac matrices mix different spinor components. Thus, the staggered fermion action is only obtained after performing a diagonalization in spinor space, which then decouples the individual components.

Rather than describing the details of this procedure, which can be found in most textbooks, we simply state the result. Starting from the usual four-component spinor and performing a spin-diagonalization, the lattice action for staggered fermions with bare mass m_0 coupled to the gauge field is derived as

$$S_F^{\text{stagg}}[U, \bar{\chi}, \chi] = a^4 \sum_{x \in \Lambda_E} \sum_{\alpha=1}^{4} \left\{ m_0 \bar{\chi}_\alpha(x) \chi_\alpha(x) \right.$$

$$\left. + \frac{1}{2a} \sum_{\mu=0}^{3} \eta_\mu(x) \left[\bar{\chi}_\alpha(x) U_\mu(x) \chi_\alpha(x + a\hat{\mu}) - \bar{\chi}_\alpha(x + a\hat{\mu}) U_\mu(x)^{-1} \chi_\alpha(x) \right] \right\}, \quad (5.41)$$

where χ_α denotes a one-component Grassmann variables. The spin-diagonalization has thus replaced the Dirac matrices γ_μ by real, position-dependent phase factors $\eta_\mu(x)$, which are given by

$$\eta_0(x) = 1, \qquad \eta_j(x) = (-1)^{n_0 + \ldots + n_{j-1}}, \qquad n_j = x_j/a. \qquad (5.42)$$

At the level of the classical action, the spinor components are completely decoupled, and the action is decomposed into four identical pieces. In order to occupy all 16 corners of a four-dimensional hypercube with one-component Grassmann variables, one needs four Dirac spinors, each of which contributes a term like Eq. (5.41) to the overall action. This produces the fourfold degeneracy of staggered fermions, with the remnant doubler states being referred to as "tastes", in order to distinguish them from physical flavours. The formulation using the one-component fields within a hypercube can be re-expressed in terms of the spin-taste basis [18], from which one can infer directly that the taste symmetry is broken. However, one axial generator of the taste symmetry remains unbroken. The fermion mass in the staggered approach is therefore protected against any additive renormalization through the associated global axial U(1) symmetry, unlike the case of the Wilson action. While the various tastes decouple in the continuum limit, non-vanishing interactions between the tastes at $O(a^2)$ in the lattice spacing are induced, leading to large lattice artefacts. The Symanzik improvement programme can be employed to reduce these taste-changing interactions [19], and the resulting "improved staggered fermions" (the so-called "Asqtad"-action being one particular example [20]) have been widely used in a series of simulations.

For a long time lattice physicists have struggled to find a fermionic discretization which would both solve the doubling problem and be compatible with chiral symmetry. In fact, physicists grew increasingly doubtful that this could be achieved, following the proof of a "No-Go theorem" by Nielsen and Ninomiya [21], which stated that the conditions (a)–(d) mentioned above could not be satisfied simultaneously. Since one does not want to give up locality and property (b), this would imply that either (c) or (d) must be violated. Indeed, the Wilson and staggered discretizations seem to confirm this expectation: while the Wilson fermion action removes all doublers, it breaks chiral symmetry, leading to an additive renormalization of the quark mass, as well as several other consequences. By contrast, the staggered formulation preserves a U(1) subgroup of chiral symmetry at the price of only partially removing the spurious degrees of freedom.

A way to circumvent the Nielsen–Ninomiya theorem was already pointed out by Ginsparg and Wilson in 1982 [22], when they suggested to relax condition (d) in favour of

$$\gamma_5 D + D\gamma_5 = a D\gamma_5 D. \tag{5.43}$$

However, it was not before 1997 that this condition—now commonly referred to as the Ginsparg-Wilson relation—was confronted with a non-trivial solution. It was shown [23] that the so-called "perfect action" constructed from a renormalization group approach satisfied equation (5.43). It was also realized that any lattice Dirac operator, which is a solution to the Ginsparg-Wilson relation, also satisfies the Atiyah–Singer index theorem, i.e.

$$\{\gamma_5, D\} = a D\gamma_5 D \quad \Leftrightarrow \quad \text{index}(D) = a^5 \sum_{x \in \Lambda_E} \tfrac{1}{2}\text{Tr}\,(\gamma_5 D) = n_- - n_+, \tag{5.44}$$

such that the operator D exhibits $|n_- - n_+|$ exact chiral zero modes. Finally, it was shown [24] that the Ginsparg-Wilson relation implies an exact symmetry of the associated action, with infinitesimal variations proportional to

$$\delta\psi = \gamma_5(1 - aD)\psi, \qquad \delta\bar{\psi} = \bar{\psi}\gamma_5. \tag{5.45}$$

Moreover, this symmetry reproduces the correct chiral anomaly in the flavour singlet case, and therefore all the hallmarks of the correct chiral behaviour are present in the lattice theory: chiral zero modes, an exact index theorem and the chiral anomaly derived from the Ward identities associated with the exact symmetry.

Another line in the development of lattice fermion actions that preserve chiral symmetry goes back to Kaplan's domain wall fermion approach [10], which was subsequently applied to QCD by Furman and Shamir [11]. Without going into detail, we state that the basic idea is to introduce an extra, fifth dimension and to couple the fermions to a mass defect (the so-called "domain wall height") in that extra dimension. To make this more explicit, let x, y denote the coordinates in the four-dimensional bulk, and s, t the coordinates in the 5th dimension, which has finite length N_5. The gauge fields are trivial in the 5th direction, and the Dirac operator then has the general structure

$$D_{\text{dwf}}(x, s; y, t) = D^\parallel(x, y)\delta_{st} + \delta(x - y)D^\perp_{st} \tag{5.46}$$

where $D^\parallel(x, y)$ is the usual Wilson-Dirac operator with a *negative* mass term, $-M$, which represents the domain wall height. The operator $D^\perp_{st}$ couples fermions in the 5th dimension and contains the physical bare quark mass m_0. It can then be shown that for $m_0 = 0$ and in the limit $N_5 \to \infty$ there are no fermion doublers and, more importantly, chiral modes of opposite chirality are trapped in the four-dimensional domain walls at $s = 1, N_5$.

However, in a real lattice simulation of domain wall fermions, one has to work with a finite value of N_5, so that the decoupling of chiral modes is not exact. One expects, though, an exponential suppression of the remnant chiral symmetry breaking effects, and this has been confirmed in several simulations. Furthermore, the rate of suppression may be accelerated by optimizing the choice of lattice action for the gauge fields. Hence, the domain wall formulation of QCD offers a method to realize almost exact chiral symmetry at non-zero lattice spacing at the expense of simulating a five-dimensional theory.

Another operator which correctly reproduces the chiral properties of QCD at non-zero lattice spacing was constructed by Neuberger [12]. Its definition is

$$D_N = \frac{1}{a}\left(1 - \frac{A}{\sqrt{A^\dagger A}}\right), \quad A = 1 + s - a D_w, \quad \bar{a} = \frac{a}{1+s}, \tag{5.47}$$

where D_w is the massless Wilson-Dirac operator, and $|s| < 1$ is a tunable parameter. By defining $Q = -\gamma_5 A$, one can rewrite Eq. (5.47) as

$$D_N = \frac{1}{a}\left(1 + \gamma_5 \text{sign}(Q)\right). \tag{5.48}$$

The Neuberger-Dirac operator D_N removes all doublers from the spectrum, and can easily be shown to satisfy the Ginsparg-Wilson relation [12]. The occurrence of an inverse square root in D_N raises two issues. First, it is a priori not clear whether or not D_N is local. Second, the application of D_N in a computer program is potentially very costly, since the sign-function of the matrix Q must be implemented using, for instance, a polynomial approximation.

In order to qualify as a viable discretization of the quark action, "strict" locality, meaning that only fields in a local neighbourhood of a given lattice site are coupled, is not actually required. If $D(x, y)$ denotes a generic lattice Dirac operator which couples fields at sites x and y, then a sufficient condition for locality of D is the exponential suppression of non-local interactions, i.e.

$$\|D(x, y)\| \le e^{-\gamma |x-y|/a}, \tag{5.49}$$

where $|x - y|$ is the distance between sites and $\| \cdot \|$ denotes a suitably defined matrix norm. In Ref. [25] it was shown that the Neuberger-Dirac operator D_N is local in the sense of Eq. (5.49), provided that the lattice spacing in physical units[2] is not larger than about 0.13 fm. As far as the issue of numerical efficiency is concerned, we note that the most widely used approximations of $\text{sign}(Q)$ with good convergence properties include Chebysheff or Zolotarev polynomials, as well as rational fractions.

---

[2] So far we have not discussed how to assign physical units to the lattice spacing a. This is described in Sect. 5.2.4.

The last fermionic discretization we wish to mention here was originally constructed to address another problem of Wilson's discretization, namely the fact that they are not protected against the occurrence of zero modes for *any* non-zero value of the bare quark mass. These unphysical zero modes manifest themselves as "exceptional" configurations, which occur with a certain frequency in numerical simulations with Wilson quarks and which can lead to strong statistical fluctuations. The problem can be cured by introducing a so-called "chirally twisted" mass term, after which the fermionic part of the QCD action in the continuum assumes the form

$$S_F^{\text{tm; cont}} = \int d^4x \ \bar{\psi}(x)(\gamma_\mu D_\mu + m + i\mu_q\gamma_5\tau^3)\psi(x). \tag{5.50}$$

Here, μ_q is the twisted mass parameter, and τ^3 is a Pauli matrix. The standard action in the continuum can be recovered via a global chiral field rotation:

$$\psi'(x) = e^{i\alpha\gamma_5\tau^3/2}\psi(x), \qquad \bar{\psi}'(x) = \bar{\psi}(x)e^{i\alpha\gamma_5\tau^3/2}. \tag{5.51}$$

Fixing the twist angle α by requiring that $\tan\alpha = \mu_q/m$ one finds

$$S_F' = \int d^4x \ \bar{\psi}'(x)(\gamma_\mu D_\mu + M)\psi'(x), \qquad M = \sqrt{m^2 + \mu_q^2}, \tag{5.52}$$

which demonstrates the complete equivalence of the twisted formulation with "ordinary" QCD. The lattice action of twisted mass QCD for $N_f = 2$ flavours is defined as [26]

$$S_F^{\text{tm}}[U, \bar{\psi}, \psi] = a^4 \sum_{x \in \Lambda_E} \bar{\psi}(x)(D_w + m_0 + i\mu_q\gamma_5\tau^3)\psi(x). \tag{5.53}$$

Although this formulation breaks physical parity and flavour symmetries, is has a number of advantages over standard Wilson fermions. In particular, the presence of the twisted mass parameter μ_q protects the discretized theory against unphysical zero modes. Another attractive feature of twisted mass lattice QCD is the fact that the leading lattice artefacts are of order a^2 without the need to add the Sheikholeslami-Wohlert term [27], even though the Wilson-Dirac operator is used in Eq. (5.53). Although the problem of explicit chiral symmetry breaking remains, the twisted formulation is particularly useful to circumvent some of the problems that are encountered in connection with the renormalization of local operators on the lattice. Recent review of twisted mass lattice QCD can be found in [28, 29].

We wish to end this part with a few general remarks. Although we have discussed discretizations of the QCD action in some detail, including the most recent developments, many more variants of the basic types of action—including several different combinations of fermionic and pure gauge parts—can be found in the literature. This reflects the fact that the discretization is not unique. The actual choice of lattice action in a particular simulation will influence the convergence rate to the continuum limit, the algorithmic efficiency, the renormalization properties of local operators, or—in the case of domain wall fermions—the extent to which chiral symmetry is realized. Depending on the properties of a particular discretization, the choice of lattice action can be optimized for the physics one wishes to study.

5.2.3 Functional Integral and Observables

The lattice formulation provides a regularization of non-Abelian gauge theories whilst preserving the gauge invariance at all stages of the calculation. This comes at a price, since all continuous space-time symmetries are broken explicitly and must be recovered in the continuum limit. Nevertheless, the lattice regularized theory inherits all consequences of gauge invariance, including renormalizability. Moreover, the lattice regularizes the theory without any reference to perturbation theory. By contrast, in continuum schemes like the $\overline{\text{MS}}$ scheme of dimensional regularization the cutoff is only defined after fixing the order of the perturbative expansion. As we shall see below, observables in lattice QCD are directly given in terms of functional integrals, which can be evaluated stochastically using Monte Carlo integration. In this way, any use of perturbation theory is completely avoided.

For concreteness, let us assume that we have made a particular choice for the Yang–Mills part $S_G[U]$ and the fermionic part $S_F[U, \bar{\psi}, \psi]$, for instance, the Wilson plaquette action and Wilson fermions. Let Ω denote an observable, which is represented by a polynomial in the quark and antiquark fields and the link variables. The expectation value, $\langle \Omega \rangle$, is defined through the Euclidean functional integral[3]

$$\langle \Omega \rangle = \frac{1}{Z} \int D[U] D[\bar{\psi}, \psi] \, \Omega \, e^{-S_G[U] - S_F[U, \bar{\psi}, \psi]}, \qquad (5.54)$$

where Z is fixed by the condition $\langle \mathbb{1} \rangle = 1$. The functional integral involves an integration over the gauge group and over all fermionic degrees of freedom, the latter being represented by anti-commuting (Grassmann) variables. Since the fermionic action, $S_F[U, \bar{\psi}, \psi]$ is bilinear in the quark and antiquark fields, the integration over the Grassmann variables is Gaussian and can be performed analytically. This yields

$$\langle \Omega \rangle = \frac{1}{Z} \int \prod_{x \in \Lambda_E} \prod_{\mu=0}^{3} dU_\mu(x) \, \widetilde{\Omega} \, \{\det D_{\text{lat}}\}^{N_f} \, e^{-S_G[U]}. \qquad (5.55)$$

---

[3] Here and in the following we drop the subscript "E" on the partition function Z.

Equation (5.55) requires some further explanation:

- $\widetilde{\Omega}$ denotes the representation of Ω in the (effective) theory, where the quark fields have been integrated out and only the link variables remain in the functional integral measure;
- D_{lat} denotes a generic, massive lattice Dirac operator. For instance, for Wilson quarks one has $D_{\text{lat}} = D_w + m_0$. For simplicity we have displayed the expression for QCD with N_f flavours of equal mass m_0, which accounts for the power N_f. In the case of non-degenerate quarks $\{\det D_{\text{lat}}\}^{N_f}$ must be replaced by a product of determinants, in which each factor represents the contribution from a single flavour;
- The lattice formulation has given a well-defined meaning to the measure $D[U]$. The integration over the gauge degrees of freedom reduces to a finite-dimensional integration over the gauge group, based on the invariant group (Haar) measure.

The numerical evaluation of $\langle \Omega \rangle$ via Monte Carlo integration proceeds as follows. One starts by generating a set of gauge configurations using a computer program. One configuration in the set represents the collection of all link variables on a given lattice, i.e.

$$\{U_\mu(x) \,|\, x \in \Lambda_E, \ \mu = 0, \ldots, 3\}, \tag{5.56}$$

for which we shall use the shorthand $\{U_\mu(x)\}$ below. A collection of an infinite number of configurations is called an *ensemble*. The statistical weight, W, of an individual configuration is given by

$$W = \{\det D_{\text{lat}}\}^{N_f}\, e^{-S_G[U]}. \tag{5.57}$$

In other words, the composition of the ensemble is determined by a probability distribution, which is given by the negative exponentiated classical action in the integrand of the Euclidean functional integral. Owing to the weight factor, the integrand of the functional integral will be strongly peaked around those configurations for which W is large. This particular feature makes the expectation value amenable to a Monte Carlo treatment. The key idea is to replace the ensemble by a *finite* sample of N_{cfg} gauge configurations, which is dominated by those configurations for which W is large. Provided that one can construct a suitable algorithm, the sample will then consist predominantly of those configurations which give a large contribution to the Euclidean functional integral and thus $\langle \Omega \rangle$. Such a procedure is called *importance sampling*.

Technically, the sample is produced by generating a sequence of configurations via a *Markov process:*

$$\{U_\mu(x)\}_1 \longrightarrow \{U_\mu(x)\}_2 \longrightarrow \ldots \longrightarrow \{U_\mu(x)\}_{N_{\text{cfg}}}. \tag{5.58}$$

One assigns a probability for the transition from $\{U_\mu(x)\}_i$ to $\{U_\mu(x)\}_{i+1}$, which is usually a function of the statistical weights of the two configurations, W_i and W_{i+1}, respectively. For each individual configuration in the sequence one then evaluates the observable, which yields the estimates Ω_i, $i = 1, \ldots, N_{cfg}$. The expectation value $\langle\Omega\rangle$ is related to the mean value $\overline{\Omega}$ via

$$\langle\Omega\rangle = \lim_{N_{cfg}\to\infty} \overline{\Omega}, \qquad \overline{\Omega} = \frac{1}{N_{cfg}} \sum_{i=1}^{N_{cfg}} \Omega_i. \qquad (5.59)$$

In other words, in the limit of infinite statistics the mean value converges to the ensemble average which is identical to the expectation value. An important consequence of approximating the ensemble average by the sample average is a non-zero value of the *variance*. Hence, in order to specify the results from a Monte Carlo integration completely, one must also quote the statistical error which is given by the square root of the variance.

In the standard algorithms that implement Markov processes (such as the Metropolis algorithm [30]), the transition probabilities for going from one configuration to another are determined by comparing the statistical weights for *local* variations in the field variables. This guarantees computational efficiency, since the variation of individual link variables does not involve global information from the entire lattice. In Eq. (5.55) the dynamical effects of the quark fields are incorporated via the determinant of the lattice Dirac operator. The determinant, however, is a non-local object, which is expensive to compute. When the first efforts were made to compute óbservables in QCD in the 1980s, the available computer power did not allow for the inclusion of the quark determinant. Instead, lattice physicists resorted to what is known as the "quenched approximation", which is based on the assumption that the bulk of non-perturbative contributions is carried by the gauge field, so that the determinant is set to a constant:

$$\text{Quenched approximation:} \qquad \det D_{lat} = 1 \quad \Leftrightarrow \quad N_f = 0. \qquad (5.60)$$

The resulting gain in computer time amounts to several orders of magnitude. In the quenched approximation the effects of virtual quark loops are entirely suppressed. As a consequence, results for observables are afflicted with an unknown systematic error. As we shall see later, there are several quantities (for instance, the masses of the lightest hadrons) for which the quenching error amounts to just 10–15%. Although this justifies the use of the quenched approximation to some extent, it is clear that dynamical quark effects must be taken into account, in order to arrive at reliable, non-perturbative predictions with a total accuracy at the percent level.

Modern algorithms for dynamical quarks, such as the Hybrid Monte Carlo algorithm [31], do not evaluate the quark determinant directly. Rather, one exploits the property that the determinant can be rewritten as a functional integral over bosonic fields, which is then evaluated stochastically. Thereby one avoids computing a global object, but the computational effort involved in the stochastic estimation of the quark determinant is still large compared with the quenched approximation. More details can be found in Sect. 5.2.6 below.

Correlation functions, i.e. the expectation values of polynomials in the quark and gluon fields, are the most important quantities, since they determine implicitly the particle spectrum of the theory. As was discussed already in Sect. 5.2.1, the link between correlation functions and the particle spectrum is provided by the transfer matrix T. For lattice QCD with Wilson fermions, the existence of a positive transfer matrix was rigorously established [32].

As a concrete example we shall discuss the two-point correlation function of a charged kaon. A polynomial of quark fields with the quantum numbers of the kaon is given by

$$\phi_K(x) = (\bar{u}\gamma_5 s)(x), \tag{5.61}$$

where the parentheses indicate summation over spinor and colour components of the fields. Mostly one is interested in correlation functions in which all spatial points have been summed over and which therefore only depend on the Euclidean time separation. We define

$$C_K(x_0; \vec{p}) = \sum_{\vec{x}} e^{i\vec{p}\cdot\vec{x}} \left\langle \phi_K(x)\phi_K^\dagger(0) \right\rangle. \tag{5.62}$$

The inclusion of the phase factor in conjunction with the summation over $\vec{x}$ amounts to a projection onto spatial momentum $\vec{p}$. On a finite lattice with periodic boundary conditions $C_K(x_0; \vec{p})$ must be symmetric under $x_0 \leftrightarrow T - x_0$. Therefore, the spectral decomposition of $C_K(x_0; \vec{p})$ reads

$$C_K(x_0; \vec{p}) = \sum_{\alpha} \frac{|\langle 0|\phi_K(0)|\alpha\rangle|^2}{2\epsilon_\alpha(\vec{p})} \left\{ e^{-\epsilon_\alpha(\vec{p})x_0} + e^{-\epsilon_\alpha(\vec{p})(T-x_0)} \right\}, \tag{5.63}$$

where the sum runs over all states in the kaon channel with fixed momentum $\vec{p}$, and $\epsilon_\alpha(\vec{p})$ is the mass gap (see Sect. 5.2.1).[4] For large Euclidean times x_0 the ground state dominates. If we further set $\vec{p} = 0$, then the asymptotic form of the two-point function reads

$$\lim_{x_0 \to \infty} C_K(x_0; \vec{p}) = \frac{|\langle 0|\phi_K(0)|K\rangle|^2}{m_K} e^{-m_K T/2} \cosh(m_K(T/2 - x_0)), \tag{5.64}$$

where $m_K = \epsilon_0(\vec{p})|_{\vec{p}=0}$ is the mass of the kaon, and the sum of the two exponentials has been re-expressed using the cosh function. Owing to the ordering $\epsilon_0(\vec{p}) < \epsilon_1(\vec{p}) < \ldots$, the higher excited states are exponentially suppressed. The functional

---

[4]In the commonly normalization of hadron states one includes a factor $2\epsilon_\alpha(\vec{p})$ in the denominator.

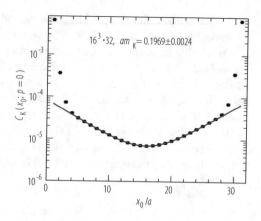

Fig. 5.3 Two-point correlation function for a pseudoscalar meson. The curve denotes a fit to Eq. (5.64) in the interval $6 \leq x_0/a \leq 26$

form of Eq. (5.64) is nicely illustrated by the plot in Fig. 5.3, where simulation data for $C_K(x_0; \vec{p} = 0)$ are compared to its asymptotic form. The data show indeed the expected cosh-behaviour. Furthermore, one observes how the contributions from higher excited states, which are clearly visible at small values of x_0/a, quickly die out as the time separation increases. From the two-point function we can extract two important quantities: the fall-off of $C_K(x_0; \vec{p} = 0)$ is characteristic of the kaon mass, i.e. the energy of the ground state. Moreover, the pre-factor of the cosh-function yields the transition amplitude between a kaon state and the vacuum, and thus contains information on the kaon's decay properties.

5.2.4 Continuum Limit, Scale Setting and Renormalization

In Sect. 5.2.2 we have discussed how to discretize the QCD action. The main principle for their construction was the condition that the corresponding expressions reproduce the continuum action in the formal limit $a \to 0$, regardless of the values of the bare parameters (such as β and the hopping parameter κ in the case of QCD with Wilson fermions). If one goes beyond the classical theory this is not possible anymore: it is a general property of quantum field theory that the parameters of the regularized theory (masses and couplings) must be adjusted as the regulator is removed. In the context of lattice QCD this implies that the continuum limit, $a \to 0$, is reached by a suitable tuning of the bare parameters.

To make this statement more precise, we shall invoke the close connection between Euclidean lattice field theory and a system in statistical mechanics. Models in statistical physics (think of the Ising model as an example) usually have a phase structure. Depending on the choice of parameters, the different phases may exhibit entirely different physical properties. The analogy with lattice field theory then implies that a particular discretization of QCD also possesses a phase structure in the space of bare parameters (β and κ, for example).[5] We shall now explain that the continuum limit of QCD is associated with a critical point in the phase diagram, which corresponds to a second-order phase transition. In the previous section we have considered hadronic two-point correlation functions, and how the mass in a given channel can be extracted from the asymptotic behaviour at large Euclidean times. Actually, this procedure yields the dimensionless combination (aM), i.e. the hadron mass in lattice units. In order to take the continuum limit, one must take $a \to 0$, while the physical mass M must remain constant. This implies

$$\frac{1}{(aM)} \equiv \xi \to 0. \tag{5.65}$$

In other words, the correlation length ξ diverges in the continuum limit. In the language of statistical physics, a divergent correlation length signals a second-order phase transition. The existence of the continuum limit in lattice QCD is therefore equivalent to the existence of a second-order transition in the space of bare parameters.

For simplicity we shall now consider Yang–Mills theory on the lattice, which we choose to describe by Wilson's plaquette action and the bare coupling parameter $\beta \equiv 6/g_0^2$. The existence of a second-order phase transition corresponds to a critical value of the bare gauge coupling, $g_{0,c}$. Furthermore, it implies that the bare coupling g_0 and the lattice spacing a (or, equivalently, the correlation length ξ) cannot be varied independently when the continuum limit is approached.[6] In this way we may regard the bare coupling as a function of the lattice spacing, $g_0(a)$, such that

$$\lim_{a \to 0} g_0(a) = g_{0,c}. \tag{5.66}$$

Let P be an observable, computed for a particular value of g_0, i.e. $P = P(g_0, a)$. Since P is a physical quantity it must stay constant as the continuum limit is taken, i.e.

$$a \frac{\mathrm{d}}{\mathrm{d}a} P(g_0, a) = 0. \tag{5.67}$$

This leads to the Callan–Symanzik equation

$$\left\{ a \frac{\partial}{\partial a} + a \frac{\partial g_0}{\partial a} \frac{\partial}{\partial g_0} \right\} P(g_0, a) = 0. \tag{5.68}$$

---

[5]This phase diagram must not be confused with the physical phase diagram of QCD in the plane defined by the temperature and the chemical potential, which is explored at heavy-ion colliders.

[6]Otherwise, an arbitrarily chosen value of g_0 would always correspond the a critical point.

We can define the renormalization group β-function β_{lat} as

$$\beta_{\text{lat}}(g_0) := -a\frac{\partial g_0}{\partial a}, \tag{5.69}$$

which describes the change in g_0 when a is varied. Note that β_{lat} depends on the choice of discretization. In perturbation theory, however, one recovers the familiar universal coefficients at one- and two-loop order. For gauge group $SU(N)$ one has

$$\beta_{\text{lat}}(g_0) = -b_0 g_0^3 - b_1 g_0^5 + O(g_0^7), \tag{5.70}$$

where

$$b_0 = \frac{1}{(4\pi)^2}\left\{\frac{11}{3}N - \frac{2}{3}N_f\right\}, \quad b_1 = \frac{1}{(4\pi)^4}\left\{\frac{34}{3}N^2 - N_f\left(\frac{13}{3}N - \frac{1}{N}\right)\right\}, \tag{5.71}$$

and $N_f = 0$ in pure Yang–Mills theory. Starting from the perturbative expansion of β_{lat} one can integrate the Callan–Symanzik equation, which gives

$$a\Lambda_{\text{lat}} = (b_0 g_0)^{-b_1/(2b_0^2)} e^{-1/(2b_0 g_0)}\left\{1 + O(g_0^2)\right\}, \tag{5.72}$$

where the integration constant Λ_{lat} represents a characteristic scale of the theory. The above expression establishes the connection between the lattice spacing and the bare coupling in perturbation theory. One reads off that

$$a \to 0 \quad \Leftrightarrow \quad g_0 \to 0, \tag{5.73}$$

and hence the critical point occurs at $g_{0,c} = 0$. These findings are a consequence of asymptotic freedom. Taking Eq. (5.72) at face value one would conclude that the relation between $P(a, g_0)$ and $P(a', g_0')$, computed for two different values of the bare coupling g_0 and g_0' near the critical point, was simply given by the ratio of Eq. (5.72) evaluated for g_0 and g_0'. However, actual simulations do not confirm this expectation. The reason for the failure to observe "asymptotic scaling", i.e. a variation of $P(a, g_0)$ with g_0 which is consistent with Eq. (5.72), is that the accessible values of g_0 in simulations are by far not near enough the critical point, in order for perturbation theory to be a good approximation.

Let P and P' be two different observables that both satisfy Eq. (5.68). Then, regardless of whether or not asymptotic scaling holds, one would expect the ratio $aP(a, g_0)/aP'(a, g_0)$ to be equal to the physical ratio P/P' for all values of g_0. However, even this weaker scaling criterion is usually not observed, the reason being that the right-hand side of Eq. (5.68) is not strictly zero. Rather one has

$$\left\{a\frac{\partial}{\partial a} - \beta_{\text{lat}}(g_0)\frac{\partial}{\partial g_0}\right\} P(g_0, a) = O(a^p), \tag{5.74}$$

Table 5.1 Most widely used discretizations of the Dirac operator and some of their properties

Action	Doublers	Leading artefacts	Chiral symmetry
Wilson	None	$O(a)$	Broken
Clover	None	$O(a^2)$	Broken
Staggered	4	$O(a^2)$	$U(1) \otimes U(1)$ subgroup unbroken
Neuberger	None	$O(a^2)$	Preserved
Domain Wall	None	$O(a^2)$	Remnant breaking exponentially suppressed
Twisted Mass Wilson	None	$O(a^2)$	Broken

where p is a positive integer. These so-called scaling violations on the right-hand side depend both on the lattice action and the observable in question. As a consequence, the ratio considered above behaves like

$$\frac{a P(a, g_0)}{a P'(a, g_0)} = O(a^p). \tag{5.75}$$

In other words, as g_0 is tuned towards zero, dimensionless ratios of observables converge to the continuum limit with a rate proportional to a^p, where the power p is characteristic of the particular discretization employed in the lattice calculation. In Table 5.1 we have already listed the leading scaling violations (lattice artefacts) for several widely used fermionic lattice actions. Discretizations of the Yang–Mills part, such as the plaquette action, have leading lattice artefacts of $O(a^2)$. The Symanzik improvement programme allows to construct lattice actions with an accelerated rate of convergence to the continuum limit.

In actual lattice calculations, the continuum limit must be taken by performing simulations at several different values of the lattice spacing and extrapolating the results to $a = 0$. The functional form of the extrapolation is chosen such that it is consistent with the leading discretization errors for a given lattice action. Such a procedure is only viable if the relation between the lattice spacing in physical units and the dimensionless coupling parameter g_0 (which is an input parameter in the simulation) is known with good accuracy. Since the perturbative formula Eq. (5.72) is not of any practical use, the relation between the scale and the coupling must be mapped out non-perturbatively. To this end one picks a value for g_0 and computes in a Monte Carlo simulation a dimensionful quantity Q, whose value is known from experiment. Common choices for Q in the pure gauge theory are the string tension or the hadronic radius r_0 [33, 34], while in full QCD one may choose the mass of the nucleon. The Monte Carlo procedure yields Q in lattice units, (aQ), and the calibration of the lattice spacing is achieved via

$$a^{-1} [\text{MeV}] = \frac{Q|_{\text{exp}} [\text{MeV}]}{(aQ)|_{g_0}}. \tag{5.76}$$

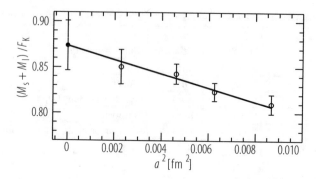

Fig. 5.4 Continuum extrapolation of the dimensionless ratio of quark masses and the kaon decay constant [35]

Knowledge of (aQ) over a range of bare couplings is a prerequisite for performing the continuum extrapolation. In Fig. 5.4 we show a particular example, namely the continuum extrapolation of the combination $M_s + \frac{1}{2}(M_u + M_d)$ of quark masses, normalized by the kaon decay constant, computed using $O(a)$ improved Wilson fermions in the quenched approximation [35]. The expected linear convergence in a^2 is clearly exhibited by the lattice data.

So far we have restricted the discussion to the pure gauge theory which contains only one bare parameter, the gauge coupling g_0 (sometimes expressed in terms of $\beta = 6/g_0^2$). When quarks are incorporated, the set of parameters must be enlarged by the values of the bare masses, one for each flavour. Lattice QCD is thus parameterized by the set of bare parameters

$$\{g_0; m_u, m_d, m_s, m_c, m_b, m_t\}.$$

In order to be predictive, the theory must be renormalized, by expressing the bare parameters in terms of renormalized ones.

A convenient and practical method for lattice QCD is based on so-called *hadronic renormalization schemes*. Here the bare coupling and quark masses are eliminated in favour of renormalized quantities such as hadron masses or decay constants. An example how this works in the pure gauge theory was already given in the preceding discussion on scale setting, where the bare coupling was eliminated by assigning a value in physical units to the lattice spacing. In the process one has to choose a quantity that sets the scale and which cannot be predicted anymore.

Replacing the values of the bare quark masses $m_u, m_d, \ldots$ in favour of hadronic quantities works as follows. Like the bare coupling, the bare quark mass is an input parameter for the simulation and thus freely adjustable. Therefore, simulations yield hadron masses (in lattice units) as a function of the input quark masses. For instance, $am_{PS}(m_1, m_2)$ denotes the mass in lattice units of a generic pseudoscalar meson composed of a quark and antiquark with bare masses m_1 and m_2, respectively. Let us assume that the lattice spacing a has been calibrated using some input quantity

Q. If we further assume exact isospin symmetry we can then determine the value of the bare isospin-symmetrized light quark mass $\hat{m} = \frac{1}{2}(m_u + m_d)$, by requiring that

$$\frac{m_{PS}(m_1, m_2)}{Q} = \frac{m_\pi}{Q}\bigg|_{exp}, \qquad m_1 = m_2, \tag{5.77}$$

i.e. the value of $\hat{m}$ is fixed by adjusting the input mass m_1 until $m_{PS}(m_1, m_2)/Q$ coincides with the experimental result. We can extend this procedure to include more massive flavours. The bare strange quark mass is found by tuning m_2 such that

$$\frac{m_{PS}(\hat{m}, m_2)}{Q} = \frac{m_K}{Q}\bigg|_{exp}. \tag{5.78}$$

Alternatively one can fix m_s via the condition $m_V(\hat{m}, m_2)/Q = m_K^*/Q|_{exp}$, where m_V denotes the mass in the vector channel. An example of a particular hadronic renormalization scheme is shown below:

Parameter	Quantity
g_0	f_π
$\frac{1}{2}(m_u + m_d)$	m_π
m_s	m_K
m_c	m_{D_s}
m_b	m_{B_s}

All quantities in a lattice calculation are genuine predictions, except for those that are listed in the right-hand column of the table, which are used to eliminate the bare parameters.

Given the multitude of hadronic states, it is obvious that there is considerable freedom in choosing hadronic renormalization schemes. Usually, masses or mass splittings of hadrons that are stable in QCD are suitable to define a scheme. Resonances, such as the ρ, should be avoided, since they do not have a sharply defined energy, owing to their large width.

5.2.5 Limitations and Systematic Effects

The lattice formulation is the basis for an exact non-perturbative treatment of QCD. The accuracy of lattice results is chiefly limited by the algorithmic performance and the available computer power. In particular, the set of bare parameters that can be simulated efficiently for a given number of lattice sites is restricted. This has the important consequence that the quark masses at the very extremes of the physical mass scale (i.e. the up/down quarks and the b-quark) cannot be simulated directly with currently available methods and machines. These technical limitations are usually translated into a systematic error, which is quoted alongside the statistical one. The most important systematic effects are due to

- lattice artefacts (cutoff effects),
- finite volume effects, and
- extrapolations in the quark mass.

In order to have sufficient control over these effects, the simulation parameters must be chosen such that the following inequalities are satisfied:

$$\frac{1}{am_{\text{had}}} \ll \frac{L}{a}, \qquad m_{\text{had}} \ll a^{-1}, \tag{5.79}$$

where m_{had} is the mass of a generic hadron in physical units computed in the simulation. The inequality on the left of (5.79) states that the hadron's correlation length must be much smaller than the linear extent of the spatial box (in lattice units), as otherwise its value will be strongly distorted by finite volume effects. The inequality on the right states that the hadron mass must be significantly smaller than the inverse lattice spacing. If this is not the case, lattice artefacts will be uncontrollably large, meaning that the presence of higher-order cutoff effects cannot be excluded, so that a reliable extrapolation to the continuum limit as a linear function of the leading power of lattice artefacts cannot be performed. With current algorithms and machines, lattice sizes of up to $L/a = 48$ and lattice spacings down to 0.05 fm are affordable, even if dynamical quarks are included. Since $a = 0.05$ fm corresponds to $a^{-1} \approx 4$ GeV, it is obvious that the b-quark mass is too large to be simulated directly. Several techniques have been devised to address this problem, and a brief account can be found in Sect. 5.7.2.

In the light quark sector, the primary limitation that forbids making direct contact with the physical values of the up and down quarks is mostly due to algorithmic performance, rather than finite size effects. A detailed discussion of the algorithmic difficulties associated with the simulation of light dynamical quarks is presented separately in the following section. Moreover, it is difficult even in the quenched approximation to reach quark masses significantly smaller than half the physical strange quark mass, in particular with Wilson fermions. This is related to the occurrence of arbitrarily small eigenvalues in the spectrum of the Wilson-Dirac operator, even for small but non-vanishing values of the bare mass. As a result, observables computed on individual, so-called "exceptional" configurations may differ from the Monte Carlo average by orders of magnitude, and thus a reliable determination of the result and its error is virtually impossible. As already mentioned in Sect. 5.2.2, the problem of exceptional configuration can be cured by employing alternative discretizations such as twisted mass QCD or the overlap operator. A related problem arising from the particular spectral properties of the Wilson-Dirac operator is the bad performance of standard algorithms for dynamical quarks, discussed in the next section.

Due to these reasons, many simulations (quenched and unquenched) were restricted to quark masses not much smaller than $m_s/2$. This value translates into a minimum mass of about 490 MeV in the pseudoscalar meson channel, so that in these simulations the pion is as heavy as the physical kaon. In this region of parameter space it is known empirically that a spatial lattice length of 2–3 fm is sufficient to satisfy the first inequality in (5.79) and to rule out significant finite volume effects. Moreover, an important analytic result derived by Lüscher [36], implies that the asymptotic convergence to the result in infinite volume is exponential.

In order to make contact with the chiral regime, lattice results must be extrapolated to the physical values of the up and down quark masses. The functional form for the dependence of observables on the quark mass is usually provided by Chiral Perturbation Theory (ChPT). For instance, at lowest order the relation between the mass of a pseudoscalar meson composed of quarks with masses m_1 and m_2 is

$$m_{PS}^2 = B_0(m_1 + m_2) + \ldots, \tag{5.80}$$

where the ellipses represent higher orders in the chiral expansion. Similar expressions are derived for vector meson and baryon masses, e.g.

$$m_V = m_V^0 + C M^2 + \ldots, \qquad m_N = m_N^0 + k M^2 + \ldots, \tag{5.81}$$

and also for other quantities such as pseudoscalar decay constants. In the above formulae, $M^2 \equiv B_0(m_1 + m_2)$, and m_V^0 and m_N^0 denote the (non-vanishing) masses in the chiral limit. A more formal introduction to the basic concepts of ChPT is presented in Sect. 5.6.1.

It remains largely unknown whether or not the expressions of ChPT considered at a given order in the expansion can be applied in the quark mass range that is accessible in current simulations. Therefore, chiral extrapolations can lead to substantial systematic uncertainties. For instance, lattice predictions for the ratio of decay constants of the B and B_s mesons, f_{B_s}/f_B, may differ by 10%, depending on whether the LO or NLO expressions are used as an *ansatz* for the extrapolation from quark masses around $m_s/2$. Currently it is estimated that pseudoscalar meson masses of 300 MeV and below must be reached in simulations, in order that ChPT at one- or even two-loop order provides an accurate prediction for the quark mass dependence of hadron masses and matrix elements.

In the quenched approximation, chiral extrapolations are particularly problematic, since the chiral limit is intrinsically pathological, due to the appearance of singularities in the quark mass dependence. This is illustrated by the NLO expression for the ratio $m_{PS}^2/(m_1 + m_2)$, i.e.

$$\frac{m_{PS}^2}{m_1 + m_2} = B_0 \left\{ 1 - \left(\delta - \tfrac{2}{3} \alpha_\Phi y \right) (\ln y + 1) + \left[(2\alpha_8 - \alpha_5) - \tfrac{1}{3} \alpha_\Phi \right] y \right\}, \tag{5.82}$$

where B_0, α_5, α_8, δ and α_Φ are low-energy constants. For notational convenience we have introduced

$$y = \frac{M^2}{(4\pi F_0)^2}, \qquad M^2 = B_0(m_1 + m_2), \tag{5.83}$$

where F_0 denotes the pion decay constant in the chiral limit. The low-energy constants δ and α_Φ, which multiply the so-called "quenched chiral logarithms", have no counterpart in the unquenched case. Since δ has a non-zero value [37], the quenched chiral logarithm in Eq. (5.82) gives rise to a singularity in the chiral limit. For many applications, the singularity can be ignored, since its effect is numerically small even at the physical pion mass. However, it signals that the quenched approximation suffers from fundamental conceptual problems.

5.2.6 Simulations with Dynamical Quarks

Although one may argue that the quenched approximation describes hadronic properties fairly well, it is clearly unsatisfactory, both from a conceptual point of view, and also because it introduces an unknown systematic error. Below we shall discuss some general issues relating to simulations with dynamical quarks. It must be stressed that several different techniques how to treat the quark determinant of Eq. (5.57) efficiently are currently being explored. A preferred or clearly superior method has not emerged so far, and it is likely that some of the approaches presented below may become obsolete in the years to come.

In order to illustrate the main difficulties, we start by introducing the Hybrid Monte Carlo (HMC) algorithm [31], which has been the standard algorithm for simulations with dynamical quarks for many years. In order to produce one step in the Markov chain, the algorithm evolves the link variables according to the equations of motion of a classical Hamiltonian system. To this end one introduces a conjugate momentum variable $\Pi_\mu(x)$ for every link $U_\mu(x)$. The Hamiltonian is defined as

$$H[U, \Pi] = \tfrac{1}{2} \sum_{x \in \Lambda_E} \sum_{\mu=0}^{3} \Pi_\mu(x)\Pi_\mu(x) + S_G[U] + S_F^{\text{eff}}[U, \phi^*, \phi], \tag{5.84}$$

where $S_G[U]$ is the lattice gauge action, and $S_F^{\text{eff}}[U, \phi^*, \phi]$ denotes an effective lattice fermion action, which is obtained by rewriting the quark determinant as a functional integral over complex *bosonic* fields $\phi(x)$ and $\phi^*(x)$. Explicitly, for $N_f = 2$ one has

$$(\det D_{\text{lat}})^2 = \int D[\phi^*, \phi] \exp\left\{ -\sum_{x \in \Lambda_E} \phi^*(x) \left[(D_{\text{lat}}^\dagger D_{\text{lat}})^{-1} \phi \right](x) \right\}. \tag{5.85}$$

For each step in the Markov chain, the conjugate momenta are drawn randomly from a Gaussian distribution ("momentum refreshment"). The Hamiltonian $H[U, \Pi]$ governs the dynamics of the variables $U_\mu(x)$ and $\Pi_\mu(x)$ with respect to "simulation time" τ, which parameterizes the evolution of $U_\mu(x)$ and $\Pi_\mu(x)$ as the simulation algorithm progresses. The evolution is described by Hamilton's equations, which read

$$\frac{d}{d\tau}U_\mu(x) = \Pi_\mu(x)U_\mu(x), \qquad \frac{d}{d\tau}\Pi_\mu(x) = -F_{G,\mu}(x) - F_{F,\mu}(x), \qquad (5.86)$$

where

$$F_{G,\mu}(x) = \frac{\partial S_G[U]}{\partial U_\mu(x)}, \qquad F_{F,\mu}(x) = \frac{\partial}{\partial U_\mu(x)} \sum_{x \in \Lambda_E} \phi^*(x) \left[(D_{\text{lat}}^\dagger D_{\text{lat}})^{-1} \phi \right](x)$$

$$(5.87)$$

are the forces associated with the gluon and quark fields, respectively. The algorithm then proceeds by integrating the equations of motion numerically. As in any numerical integration scheme, the total time interval is divided into a number of sub-intervals of finite length $\Delta\tau$, which is called the step size. Starting from an initial gauge configuration $\{U_\mu(x)\}$ and a set of conjugate momenta $\{\Pi_\mu(x)\}$, one obtains new sets $\{U'_\mu(x)\}$, $\{\Pi'_\mu(x)\}$ after the integration. In the language of classical mechanics, the variables $U_\mu(x)$ and $\Pi_\mu(x)$ evolve along a trajectory in phase space which connects the initial and final configurations. However, since numerical integration is not exact, owing to the finite step size, the energy is not conserved. In the HMC algorithm this is rectified by introducing a global accept/reject step: if ΔH denotes the energy difference between the initial and final configurations, i.e.

$$\Delta H \equiv H[U', \Pi'] - H[U, \Pi], \qquad (5.88)$$

then the new configuration $\{U'_\mu(x)\}$ is accepted with probability[7]

$$P\{U \to U'\} = \min(1, e^{-\Delta H}). \qquad (5.89)$$

In other words, a configuration $\{U'_\mu(x)\}$ associated with a large value for the energy violation ΔH is less likely to be accepted. This final step completes the Monte Carlo update. The name "Hybrid Monte Carlo" reflects the fact that one combines a deterministic classical dynamics procedure with a pseudo-random accept/reject step.

One major problem which has plagued simulations with dynamical quarks over many years is the fact that the efficiency of the conventional HMC algorithm

---

[7]In practice this is achieved by drawing a random number r, with $0 < r \leq 1$. If $r < e^{-\Delta H}$ the new configuration is rejected.

deteriorates sharply when the lattice spacing is decreased and the masses of the light (up and down) quarks are tuned to their physical values. The poor scaling behaviour is driven by the condition number of the lattice Dirac operator D_{lat}, i.e. the ratio of the largest to the smallest eigenvalue. This quantity is known to grow inversely proportional to the lattice spacing and the quark mass. In particular, the HMC algorithm scales with the second, perhaps the third power of the light quark mass. Thus, simulations based on the Wilson-Dirac operator were found to be unpractical for lattice spacings below 0.1 fm and quark masses significantly smaller than half of the strange quark mass.[8] This is related to the afore-mentioned fact that even the massive Wilson-Dirac operator is not protected against arbitrarily small eigenvalues. Its condition number may thus fluctuate strongly in the course of the simulation, leading not only to numerical instabilities, but also to large fluctuations in the quark force term $F_{\text{F},\mu}(x)$, and, in turn, ΔH. In order to keep a reasonably large acceptance rate of well over 75%, one must reduce the step size $\Delta\tau$ accordingly, and thus the numerical effort to integrate the equations of motion for an interval τ of fixed length, increases.

Two basic strategies to address this problems have been followed: the first is based on using fermionic discretizations that avoid the problem of arbitrarily small eigenvalues, while the aim of the second approach is to improve the simulation algorithms.

Staggered fermions have been advocated as a numerically more efficient alternative to the Wilson-Dirac formulation: since the staggered Dirac operator couples one-component Grassmann fields rather than four-component spinors, fewer floating point operations are required for one application of the operator. Moreover, the residual $U(1) \otimes U(1)$ symmetry protects the quark mass against additive renormalization and thus prevents the occurrence of very small eigenvalues. However, the fact that the staggered formulation describes four "tastes" per quark flavour makes a physical interpretation difficult. Technically, the degeneracy implies that the statistical weight of the quark determinant is too large compared with that of one physical flavour. An *ad hoc* method to compensate for this is to take fractional powers of the staggered quark determinant. For instance, to simulate QCD with a doublet of degenerate up and down quarks with mass $\hat{m}$, and a single heavier (strange) quark with mass m_s, the probability measure is taken as

$$P = \frac{1}{Z} \left\{ \det\left(D_{\text{stagg}} + \hat{m}\right) \right\}^{1/2} \left\{ \det\left(D_{\text{stagg}} + m_s\right) \right\}^{1/4} e^{-S_G[U]}, \qquad (5.90)$$

where D_{stagg} is the massless staggered Dirac operator. This procedure is known as the "fourth root trick". The main question, which has been hotly debated, is whether or not the rooted staggered operator corresponds to a local field theory, or whether it induces spurious interactions among the fermionic degrees of freedom, which might lead to a violation of the universality of the continuum limit. A thorough analysis

---

[8]This should be compared to the physical mass ratio of $\hat{m} \approx m_s/24$ [38].

of this problem was given in [39], but so far no firm conclusion has been reached. Nevertheless, the probability measure Eq. (5.90) and the "rooting trick" it is based on, have been employed in large-scale simulations (see, e.g. Ref. [40]).

Discretizations based on twisted mass QCD have also been proposed as a numerically more efficient quark action. Here, the twisted mass parameter μ_q protects the operator against arbitrarily small eigenvalues. The smallest mass in the pion channel that has been reached with this formulation was as low as 300 MeV [41]. This corresponds to a physical quark mass of about $m_s/5$, which may be sufficient to enter the regime where the quark mass behaviour of observables can be described analytically using Chiral Perturbation Theory.

Owing to several major algorithmic improvements, simulations based on the Wilson-Dirac operator can now be performed much more efficiently. Without going into much detail, we simply state that most of the gain is due to the use of suitably chosen factorizations of the Wilson-Dirac operator into its low- and high-frequency parts. The various factors are then "better conditioned". In particular, fluctuations in the condition number can be controlled via a separate and optimized treatment of the low-energy part. In this way the step size $\Delta\tau$ can be increased whilst keeping a reasonably high acceptance rate for fixed total trajectory length τ. Algorithmic implementations of factorization range from Hasenbusch's "mass preconditioning" [42, 43], Lüscher's domain decomposition technique based on the Schwarz Alternating Procedure (DD-HMC algorithm) [44], to factorizations based on mass preconditioning combined with rational approximations of the contributions from multiple pseudo-fermion fields [45]. Thanks to these developments, it appears that the spectral properties of the Wilson-Dirac operator are no longer an obstacle to the efficient simulation of lattice QCD with light dynamical quarks. At the same time, large-scale simulations employing the recent algorithmic improvements are only just starting.

5.3 Hadron Spectroscopy

The determination of the spectrum of hadrons, i.e. mesons, baryons, glueballs, and possibly "exotic" hadronic states, starting from the underlying gauge theory of quarks and gluons has traditionally been one of the main applications of lattice QCD. The rôle of lattice calculations in this context is twofold: first, the determination of the experimentally known values of hadron masses from first principles represents a stringent test of QCD. Second, lattice calculations can make predictions for the masses of undiscovered or poorly established states. For instance, lattice results have been instrumental in the search for glueball candidates, and have also contributed significantly to the debate on the existence of pentaquarks.

The principles of hadronic mass calculations have already been outlined at the end of Sect. 5.2.3: After defining a suitable interpolating operator with the quantum numbers of the desired hadronic channel, one computes its Euclidean two-point function. The mass (energy) of the ground state in that channel is then extracted from the exponential fall-off of the correlation function at large Euclidean times. The detailed functional form of the asymptotic behaviour depends on the choice of boundary conditions. Thus, it is not always described by a cosh function, as in the example of a pseudoscalar meson on a lattice with periodic boundary conditions in time, c.f. Eq. (5.64). In the limit of infinite temporal lattice size T, the effect of the boundary conditions is sufficiently weak, so that one may approximate the functional form of the correlation function for a generic interpolating operator $\phi_{\text{had}}(x)$ by

$$C_{\text{had}}(x_0; \vec{p}) = \sum_{\vec{x}} e^{i\vec{p}\cdot\vec{x}} \left\langle \phi_{\text{had}}(x)\phi_{\text{had}}^{\dagger}(0) \right\rangle \overset{T\to\infty}{=} \sum_{\alpha} w_{\alpha}(\vec{p}) e^{-\epsilon_{\alpha}(\vec{p})x_0}. \tag{5.91}$$

Here, the quantity $w_{\alpha}(\vec{p})$ is referred to as the spectral weight of the state $|\alpha\rangle$. A large value for the spectral weight of the ground state, $w_1(\vec{p})$, will lead to an early domination of the correlation function by the ground state energy. The choice of ϕ_{had} in a given channel can be optimized such that

$$w_1(\vec{p}) \gg w_i(\vec{p}), \qquad i = 2, 3, \ldots. \tag{5.92}$$

An optimal choice of interpolating operator is not only important to ensure a reliable determination of the ground state energy: In order to determine the energies in the excitation spectrum, the associated spectral weights must be maximized by specifying appropriate operators.

Below we provide examples for interpolating operators in several mesonic and baryonic channels:

$$
\begin{aligned}
K\text{-meson}: \quad & \phi_K = (s\gamma_5\bar{u}), \quad (s\gamma_0\gamma_5\bar{u}) \\
K^*\text{-meson}: \quad & \phi_{K^*} = (s\gamma_j\bar{u}), \qquad j = 1, 2, 3 \\
\text{nucleon}: \quad & \phi_N = \varepsilon_{abc}\{u^{aT} C\gamma_5 d^b\}u^c \\
\Delta: \quad & \phi_{\Delta} = \varepsilon_{abc}\{u^{aT} C\gamma_{\mu} d^b\}u^c
\end{aligned}
\tag{5.93}
$$

Here, parentheses indicate summation over spinor and colour indices, while curly brackets denote that only spinor indices are summed over.

5.3.1 Light Hadron Spectrum

The determination of the spectrum of light hadrons was historically one of the first attempts to compute hadronic properties on the lattice. Since the masses of the low-lying hadrons are known from experiment, such calculations serve as benchmarks to test the intrinsic accuracy of the lattice approach.

The quenched approximation has been widely used to compute a number of quantities that are of great phenomenological interest. However, these results are of limited value, since the inherent quenching error is left undetermined. A precise calculation of the masses of the lowest lying hadrons in quenched QCD will expose the typical magnitude of the systematic error incurred by neglecting dynamical quark effects. To this end, several calculations of the quenched light hadron spectrum, using different lattice actions, have been performed [46–51].

In Ref. [47], the CP-PACS Collaboration presented a comprehensive study of the masses of the lowest pseudoscalar and vector mesons, as well as octet and decuplet baryons. The Wilson fermion action without $O(a)$ improvement was used at four different values of the lattice spacing, and a continuum extrapolation linear in a has been performed for all quantities. CP-PACS adopted a hadronic renormalization scheme in which the lattice scale was fixed using the mass of the ρ-meson. The average up and down quark mass was set using m_π. In order to fix m_s, either the kaon mass ("K"-input) or the mass of the ϕ-meson ("ϕ"-input) was used. Chiral extrapolations were either based on the form expected from quenched Chiral Perturbation Theory at NLO (see Eq. (5.82)), or on the leading-order formula supplemented by a quadratic term in the quark mass. The resulting (small) differences in the extrapolated values were added as systematic errors in the final results, which are summarily displayed in Fig. 5.5. Although the lattice results are in remarkable overall agreement with the experimentally observed spectrum, one finds significant deviations. For instance, the ratio of the nucleon and the ρ-meson masses is determined as

$$\frac{m_N}{m_\rho} = 1.143 \pm 0.033 \pm 0.018, \tag{5.94}$$

where the first error is statistical, and the second is an estimate of systematic uncertainties other than quenching. The above value is 6.7% (2.5 standard deviations) below the experimental value of 1.218. Similarly, vector-pseudoscalar mass

Fig. 5.5 Quenched light hadron spectrum computed in [47], compared with experiment. The statistical error and the sum of the statistical and systematic errors are indicated

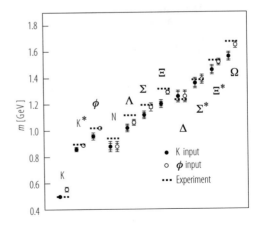

splittings, such as $m_{K^*} - m_K$, are underestimated by 10–15% (4–6σ), depending on whether m_K or m_ϕ was used to fix the strange quark mass.

The findings reported by CP-PACS, which were based on unimproved Wilson fermions, have been broadly confirmed by other collaborations employing different lattice actions [48–51]. Thereby, the universality of the continuum limit of quenched QCD has been established: although different discretizations may yield statistically inconsistent results at non-zero lattice spacing, they converge to a common continuum limit, provided that the same hadronic renormalization scheme has been employed. The latter requirement is important, as there is considerable freedom in choosing a particular scheme. This leads to ambiguities in the quenched approximation, since different quantities are affected in different way by quark loops. In Ref. [51] it was found that, by using only stable or narrow states to define the hadronic renormalization scheme, the discrepancies between the quenched and experimental spectra could be shifted to the broad resonances, ρ, Δ, N^*, while the agreement for states like K, ϕ, N, Ω could be improved. Yet this observation does not alter the conclusion that the quenched approximation is unable to reproduce the spectrum of light hadrons with an accuracy better than 10%.

The obvious question is whether sea quark effects can account for the observed deviation between the quenched and experimental spectra. Owing to the larger numerical effort required to simulate QCD with dynamical quarks, unquenched studies have not yet reached the same level of control over systematic effects— notably lattice artefacts and chiral extrapolations—compared with the quenched benchmark [47]. Thus, a "definitive" unquenched calculation of the light hadron spectrum is still lacking, and thus we refrain from presenting an overview of recent results.

Nevertheless, the observed tendency in all simulations performed to date is that dynamical quarks "do the right thing", i.e. the deviation from experiment is decreased. An example is shown in Fig. 5.6, where continuum extrapolations of meson masses in the quenched and unquenched theories are compared. The plot

Fig. 5.6 Continuum extrapolations of the masses of the K^* and ϕ mesons in full ($N_f = 2$, full symbols) and quenched QCD (open symbols), compared with experiment (diamonds) [52]

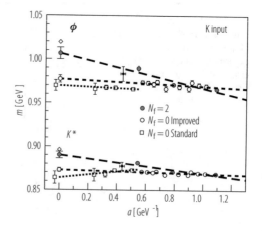

shows that the data obtained for $N_f = 2$ are closer to the experimental results in the continuum limit in comparison with their quenched counterparts. However, the figure also shows that the extrapolation of unquenched data is not well constrained, since only three data points are available. Clearly, additional simulations at smaller lattice spacings and quark masses are required for a solid determination of the total error in unquenched calculations of the light hadron spectrum.

It should also be noted that the various discretizations of the quark action have complementary advantages and shortcomings. While simulations with Wilson quarks have in the past been restricted to quark masses not much smaller than half the strange quark mass for algorithmic reasons, the use of staggered fermions in conjunction with the rooting procedure may be afflicted with conceptual problems (see the discussion in Sect. 5.2.6). Domain wall and overlap fermions are *per se* more expensive to simulate. In simulations based on tmQCD the incorporation of a third, heavier quark flavour is quite complicated. Thus, progress in this area is likely to be made through the combined information from different discretizations.

5.3.2 Glueballs

In addition to bound states composed of a quark-antiquark pair or, alternatively, three quarks, QCD is also widely believed to support the existence of glueballs, i.e. bound states consisting mainly of gluonic degrees of freedom. Although several candidates for such states have been proposed (e.g. the $f_0(1370)$, $f_0(1500)$ and $f_0(1710)$), the experimental difficulty consists in their unambiguous identification as glueballs. To this end, they need to be distinguished from "conventional" flavour-singlet meson resonances in the scalar channel. Predictions for the masses and widths of glueballs from lattice QCD provide crucial input for this task.

The basic principles of mass calculations for glueballs in lattice QCD are the same as for bound states composed of quark degrees of freedom: first one must define an interpolating operator with the appropriate quantum numbers of the glueball state in question. That is, the operator must transform correctly under spin, parity and charge conjugation. At this point a complication arises: the lattice breaks all continuous space-time symmetries, such that Lorentz-invariance or—in the language of Euclidean field theory—rotational invariance is only recovered in the continuum limit. At non-zero lattice spacing the spin assignment is therefore ambiguous. Since the gluon field is represented by link variables, any glueball operator must be constructed from particular combinations of Wilson loops, i.e. products of link variables along closed paths on a hypercubic lattice (see Fig. 5.7).

Operators constructed in this way transform under irreducible representations (IRs) of the octahedral group O_h, which are conventionally labelled A_1, A_2, E, T_1 and T_2. By computing the relations between the IRs of O_h and SU(2) one finds that each IR in the set $\{A_1, A_2, E, T_1, T_2\}$ corresponds to infinitely many spins in the continuum. For instance, A_1 transforms not only like a scalar (spin 0) state, but also contributes to spin 4 and yet higher spin states. Similarly, the lowest states to

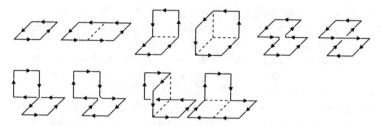

Fig. 5.7 Wilson loops used in the construction of glueball operators (from Ref. [53])

which T_1 makes a contribution are spin 1 and spin 3, while E corresponds to spins 2, 4, 5,.... In order to fully classify lattice glueball operators, the representations of O_h are supplemented by the transformation properties under parity and charge conjugation, in full analogy with the usual J^{PC}-assignment in the continuum. For example, an operator labelled A_1^{++} corresponds to the scalar channel 0^{++} in the continuum.

The above discussion implies that the two-point correlation function of an operator transforming under A_1^{++}, which is used to describe the scalar glueball, will be contaminated by contributions from a spin 4 state. However, in accordance with Regge theory one may expect that the latter dies out quickly, since higher spin states are more massive.

Another technical complication arises from the empirical observation that the spectral weight, $w_1(\vec{p})$, of the ground state in Eq. (5.91) is usually quite small. This implies that the asymptotic behaviour of the two-point correlation function is only isolated at large Euclidean times. However, the statistical accuracy deteriorates quickly as x_0 is increased, and in the asymptotic regime the correlation function is numerically comparable to the statistical noise. This precludes a precise determination of the mass of the ground state. A heuristic explanation for the small spectral weight can be given by noting that the operators constructed from the usual link variables are point-like and thus have little projection onto an extended object such as a glueball. The situation can be much improved if the links in the Wilson loops of Fig. 5.7 are replaced by so-called "smeared" or "fuzzed" links [54, 55]. For instance, the approach of [54] replaces the spatial link $U_j(x)$ by the combination

$$U_j(x) \equiv U_j^0(x) \longrightarrow \mathcal{P}\left\{ U_j^0(x) + \alpha \sum_{\pm k=1, k\neq j}^{3} U_k(x)U_j(x+a\hat{k})U_k(x+a\hat{j})^{-1}\right\}, \quad j=1,2,3,$$

$$(5.95)$$

where α is a real, tunable parameter, and the symbol $\mathcal{P}$ denotes the projection back into the group manifold of SU(3). The procedure can be iterated, so that links at smearing level s, i.e. $U_j^s(x)$, are constructed from those at level $s-1$ via Eq. (5.95). One may say that smearing reduces the UV fluctuations of the gauge field, so that the smeared, extended link variables are better suited to project onto the IR regime, i.e. the long-distance properties. It should be stressed that the links in the temporal direction do *not* undergo the fuzzing procedure: Fuzzed temporal links will alter the transfer matrix and the spectral information it contains.

In order to obtain detailed information on the glueball spectrum one also seeks to determine the masses of the excited states in a given channel. This requires another level of refinement, since one normally hopes that excited state contributions die out quickly, while they now become the very focus of interest. A widely used method to gain information on the higher excitations is to construct a whole set of interpolating operators $\{O_1, \ldots, O_r\}$ in a given channel, say, A_1^{++}. This is achieved either by considering different shapes of Wilson loops that share the same transformation properties, or by applying several different smearing levels to one particular Wilson loop. Thus, each individual member of the set $\{O_1, \ldots, O_r\}$ is a perfectly valid operator in a given channel, but the projection properties, i.e. the associated spectral weights $w_\alpha^{(i)}$ for a particular state α in the spectral sum will in general be different for each member $i = 1, \ldots, r$. One then computes the matrix

$$C_{ij}(x_0) := \sum_{\vec{x}} \left\langle O_i(x) O_j^\dagger(0) \right\rangle, \qquad i, j = 1, \ldots, r, \tag{5.96}$$

whose elements consist of the correlations of all combinations of operators in the set. The diagonalization of the matrix correlator then yields the appropriate linear combination of operators which correspond to the states $\alpha = 1, 2, \ldots$ in the spectral decomposition. Diagonalization is achieved by solving the generalized eigenvalue problem

$$C_{ij}(x_0)\phi_j = \lambda_i(x_0, x_0') C_{ik}(x_0')\phi_k, \qquad x_0' < x_0, \tag{5.97}$$

where ϕ denotes a vector, x_0' is fixed, and $C(x_0)$, $C(x_0')$ denote the matrix correlators taken at Euclidean times x_0 and x_0', respectively. As shown in [56], the set of eigenvalues $\lambda(x_0, x_0')$ converges rapidly towards

$$\lambda_\alpha(x_0, x_0') = e^{-(x_0 - x_0')\epsilon_\alpha}, \qquad \alpha = 1, \ldots, r, \tag{5.98}$$

where ϵ_α is the mass (energy) of the state α in the spectral sum.

After all these technicalities, we now report on the status of glueball calculations. Recent results obtained in the quenched approximation were published in [53, 57–60]. In Fig. 5.8 we show the results from Ref. [57]. The three lowest-lying states are the scalar (0^{++}), tensor (2^{++}) and the 0^{-+} glueballs, whose masses are determined as

$$m_{0^{++}} = 1710(50)(80) \text{ MeV}, \qquad m_{2^{++}} = 2390(30)(120) \text{ MeV},$$
$$m_{0^{-+}} = 2560(35)(120) \text{ MeV}. \tag{5.99}$$

Fig. 5.8 Glueball spectrum in quenched QCD (from Ref. [57])

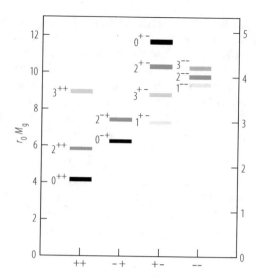

Here, the first error is statistical, while the second is an estimate of systematic uncertainties, which is dominated by the ambiguity in the scale setting in the quenched approximation.

While it is tempting to identify the experimentally established resonance $f_0(1710)$ as a scalar glueball in the light of the above results, the situation is more complicated. Since lattice predictions for the mass of the lightest glueballs fall into the mass range of conventional scalar mesons, mixing of glueballs with conventional $q\bar{q}$ states in conjunction with the observed decay patterns must be considered before drawing any definite conclusions. More details on the current phenomenological and experimental situation can be found in [61, 62]. So far, there have been only exploratory attempts to study glueball-meson mixing directly on the lattice. Any meaningful investigation must inevitably include dynamical quark effects, whose influence on the glueball spectrum have so far only been poorly understood.

5.4 Confinement and String Breaking

The empirical fact that quarks and gluons are not observed as free particles is commonly referred to as confinement. Since all experimentally observed states are singlets under $SU(3)_{colour}$, confinement is tantamount to saying that isolated colour charges are not allowed. A theoretical understanding of this phenomenon must inevitably go beyond the perturbative level, since QCD is a strongly coupled theory.

In Ref. [6], Wilson formulated a criterion for the confinement of colour charges known as the "area law". Let $U(C)$ denote the product of link variables around a

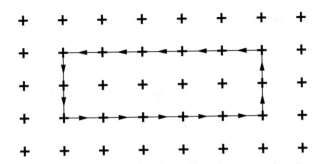

Fig. 5.9 Oriented product of link variables around a rectangle of area $r \cdot t$

closed loop $\mathcal{C}$ on a hyper-cubic lattice. The trace over colour indices is called the "Wilson loop", i.e.

$$W(\mathcal{C}) = \operatorname{tr}\{U(\mathcal{C})\}. \tag{5.100}$$

The area law then states that colour charges are confined if the expectation value of $W(\mathcal{C})$ decays exponentially with a rate proportional to the area $A(\mathcal{C})$ enclosed by the curve $\mathcal{C}$, i.e.

$$\langle W(\mathcal{C})\rangle \equiv \langle \operatorname{tr}\{U(\mathcal{C})\}\rangle \propto e^{-\sigma A(\mathcal{C})}, \tag{5.101}$$

where σ is a constant. An example for a rectangular Wilson loop is shown in Fig. 5.9.

The interpretation of the area law rests on the observation that a Wilson loop of area $r \cdot t$ is equal to the Euclidean correlator which describes the propagation of a static, i.e. infinitely heavy, quark-antiquark pair separated by a distance r over a Euclidean time interval t. If t is taken to infinity at fixed r, the correlator yields the energy of the quark-antiquark pair:

$$\langle W(\mathcal{C})\rangle \overset{t \gg 0}{\sim} e^{-V(r)t}. \tag{5.102}$$

The area law then implies $\sigma A(\mathcal{C}) = V(r)t$, and for a rectangular loop one obtains

$$V(r) \sim \sigma r. \tag{5.103}$$

Hence the energy of a static quark-antiquark pair increases linearly with the distance r. To achieve a full separation of static colour sources would therefore require an infinite amount of energy.

It has long been believed that SU(3) gauge theory is related to some kind of string theory. Heuristically, confinement may be viewed as due to the formation of a narrow tube of chromo-electric and -magnetic flux between static colour charges, the dynamics of which can be described by a string theory. The bosonic string model yields an asymptotic expansion for the static quark potential

$$V(r) = \sigma r + V_0 + \frac{c}{r} + O(1/r^2), \qquad (5.104)$$

where $V_0 = $ const, and the universal coefficient c has been computed as [63]

$$c = -\frac{\pi}{12} \qquad (5.105)$$

in the four-dimensional theory. The proportionality factor σ is called the "string tension". Instead of the potential one often considers the force, $F(r) \equiv dV(r)/dr$. The *ansatz* Eq. (5.104) yields

$$F(r) = \sigma - \frac{c}{r^2} + O(1/r^3), \qquad (5.106)$$

so that the string tension is obtained as the limiting value of the force, as $r \to \infty$,

$$\sigma = \lim_{r \to \infty} F(r). \qquad (5.107)$$

String models of hadrons have been known since the late 1960s, and a phenomenological value for σ has been determined from Regge theory, $\sqrt{\sigma} = 440\,\mathrm{MeV}$.

In QCD with light sea quarks the linear rise of the potential cannot persist for arbitrarily large distances. Instead, the creation of a light quark-antiquark pair from the vacuum will cause the hadronization of the static colour charges, leading to the formation of two static-light mesonic states. Thus, the string or flux-tube is expected to "break" when the two-meson state is energetically favoured over the linearly rising potential. The breaking of the string should set in at a characteristic value for the separation distance, r_b, causing the potential to flatten off for $r \gtrsim r_b$, since the energy of a state of two mesons is independent of their separation.

Lattice simulations have been instrumental for establishing that the area law, the string picture of confinement, as well as string breaking (i.e. hadronization) are indeed properties of SU(3) gauge theory and/or QCD. However, computations of large Wilson loops in lattice simulations suffer from the same problem encountered in glueball mass calculations: due to the strong exponential fall-off, the correlator in the asymptotic region, $r, t \to \infty$, is of the same order of magnitude than the statistical noise. Consequently, the same techniques have been applied, namely the smearing of link variables and the variational approach, which is based on the diagonalization of a matrix correlator. By combining these techniques with procedures designed to reduce statistical fluctuations [64] in the computation of large Wilson loops, one could verify the linear rise of the potential up to distances of $r \lesssim 1.5\,\mathrm{fm}$ [65, 66] (See Fig. 5.10).

Since a phenomenological value for $\sqrt{\sigma}$ could be inferred from Regge theory, the string tension used to be a popular quantity to set the lattice scale. However, as lattice calculations became increasingly precise, it was realized that the extrapolation

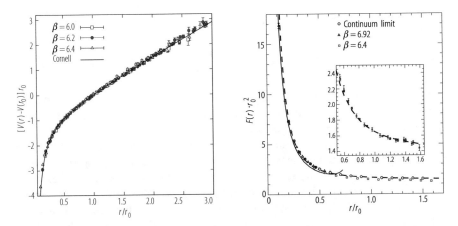

Fig. 5.10 Left panel: static quark potential in SU(3) gauge theory (from Ref. [66]). Right panel: force (from Ref. [65]) compared to the bosonic string model (dashed curve) and perturbation theory (solid curve). To compare results at different lattice spacings, all dimensionful quantities have been expressed in units of the hadronic radius $r_0 = 0.5\,\text{fm}$ (see text)

$r \to \infty$ is not easy to perform on the basis of lattice data restricted to $r \lesssim 1.5\,\text{fm}$. An alternative, conceptually much more reliable scale is obtained from the force between static colour charges [33]. The hadronic radius r_0 is defined by requiring that the force $F(r)$ evaluated at $r = r_0$ assumes a given reference value. The latter is fixed by matching $F(r)$ to phenomenological, non-relativistic potential models for heavy quarkonia. The scale r_0 is defined as the solution of

$$F(r)r^2\Big|_{r=r_0} = 1.65, \tag{5.108}$$

where the constant on the right-hand side is chosen such that r_0 has a value of $r = 0.5\,\text{fm}$ in QCD. Choosing r_0 to set the scale avoids the systematic uncertainty associated with the extrapolation of the force to infinite distance. Furthermore, r_0 remains well-defined in QCD with dynamical quarks, where string breaking must occur and the concept of a string tension as the limiting value of the force is intrinsically flawed. The quantity r_0/a has been determined numerically with good statistical accuracy over a wide range of bare couplings, corresponding to lattice spacings between $0.026 - 0.17\,\text{fm}$ [34, 65].

To test whether the bosonic string model for confinement is consistent with lattice data, one must confront the value of the Coulombic coefficient c in Eq. (5.104) with the predicted value of $c = -\pi/12$. As in the case for the string tension, such a comparison is difficult to perform reliably, since $-\pi/12$ represents the asymptotic value at infinite distance, which must be determined from data computed over a narrow range of accessible distances. Using highly accurate data for the potential $V(r)$, generated by an algorithm which allows for an exponential suppression of statistical fluctuations at large r and t, it could be shown [67] that the quantity

$$c_{\text{eff}}(r) = \frac{1}{2} r^3 \frac{d^2 V(r)}{dr^2} \qquad (5.109)$$

indeed converges towards the predicted value of $-\pi/12$. This result confirms the string picture of confinement and suggests that string-like behaviour already sets in at rather small distances of $r \gtrsim 0.5$ fm.

The incorporation of dynamical quarks should drastically change the string picture beyond a characteristic scale r_b, where due to $q\bar{q}$ pair creation string breaking occurs, since a two-meson state is energetically favoured over the flux-tube. However, the static quark potential determined from Wilson loops on dynamical configurations typically does not show any clear signs of flattening off, even at distances as large as 1 fm, where one expects hadronization to set in. This is attributed to the Wilson loop having little overlap onto the state of a broken string, such that the spectral weight associated with the broken string is extremely small. Therefore, extracting its energy reliably would require large Euclidean time separations, for which the statistical signal is usually lost.

It was thus proposed to address this problem by constructing a matrix correlator of Wilson loops supplemented by operators that directly project onto a two-meson state, and to consider their cross-correlations with the unbroken flux-tube. This strategy was first applied to Higgs models, i.e. non-Abelian gauge theory coupled to bosonic matter fields ("scalar QCD"), which are computationally much more efficient, whilst preserving the mechanism for string breaking to occur [68, 69]. The method was later extended to QCD with two flavours of dynamical quarks [70]. The plots in Fig. 5.11 clearly show that the ground state energy at short distances is linearly rising, while the first excited state (i.e. the two-meson state) is constant in r.

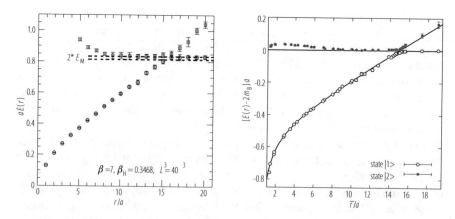

Fig. 5.11 Ground state and first excited state of the static quark potential computed using matrix correlators in the SU(2) Higgs model [68] (left panel) and QCD with $N_f = 2$ flavours of dynamical quarks [70] (right panel)

At a certain separation r_b one observes a crossing of energy levels and a continuing flat behaviour of the ground state energy. Near the crossing point one actually observes a repulsion of the energy levels, which is characteristic for the breaking phenomenon. The diagonalization of the matrix correlator also yields information on the composition of the states in the spectral decomposition. Indeed, for distances $r < r_b$ the combination of operators describing the ground state is dominated by Wilson loops, whereas for $r > r_b$, two-meson operators are the most relevant.

5.5 Fundamental Parameters of QCD

We have noted already that QCD is parameterized in terms of the gauge coupling and the masses of the quarks. In order to make predictions for cross sections, decay rates and other observables, their values must be fixed from experiment. As was discussed in detail in Sect. 4.3 , the renormalization of QCD leads to the concept of a "running" coupling constant, which depends on some momentum (energy) scale μ, and the same applies to the quark masses[9]:

$$\alpha_s(\mu) \equiv \frac{\bar{g}^2(\mu)}{4\pi}, \quad \bar{m}_u(\mu), \bar{m}_d(\mu), \bar{m}_s(\mu), \bar{m}_c(\mu), \bar{m}_b(\mu), \bar{m}_t(\mu). \qquad (5.110)$$

The property of asymptotic freedom implies that the coupling becomes weaker as the energy scale μ is increased. This explains why the perturbative expansion of cross sections in the high-energy domain allows for an accurate determination of α_s from experimental data.

The scale dependence of the coupling and the quark masses is encoded in the renormalization group (RG) equations, which are formulated in terms of the β-function and the anomalous dimension τ,

$$\mu\frac{\partial\bar{g}(\mu)}{\partial\mu} = \beta(\bar{g}), \qquad \mu\frac{\partial\bar{m}(\mu)}{\partial\mu} = \bar{m}\tau(\bar{g}). \qquad (5.111)$$

At high enough energy the RG functions β and τ admit perturbative expansions according to

$$\beta(\bar{g}) = -b_0\bar{g}^3 - b_1\bar{g}^5 + \dots, \qquad \tau(\bar{g}) = -d_0\bar{g}^2 - d_1\bar{g}^4 + \dots. \qquad (5.112)$$

Here, b_0, b_1 and $d_0 = 8/(4\pi)^2$ are universal, while the higher coefficients depend on the adopted renormalization scheme.

---

[9] As usual we denote the running parameters by a bar across the symbol.

From the asymptotic scaling behaviour at high energies one can extract the fundamental scale parameter of QCD via

$$\Lambda = \lim_{\mu \to \infty} \left\{ \mu (b_0 \bar{g}^2)^{-b_1/2b_0^2} e^{-1/2b_0 \bar{g}^2} \right\}, \qquad \bar{g} \equiv \bar{g}(\mu). \qquad (5.113)$$

Like the running coupling itself, the Λ-parameter depends on the chosen renormalization scheme.[10] A related, but less commonly used variable is the renormalization group invariant (RGI) quark mass

$$M_f = \lim_{\mu \to \infty} \left\{ \bar{m}_f (2b_0 \bar{g}^2)^{-d_0/2b_0} \right\}, \qquad f = u, d, s, \dots, \qquad \bar{m} \equiv \bar{m}(\mu). \qquad (5.114)$$

Unlike Λ, the RGI quark masses are scheme-independent quantities. Instead of using the running coupling and quark masses of Eq. (5.110), one can parameterize QCD in an entirely equivalent way through the set

$$\Lambda, \; M_u, \; M_d, \; M_s, \; M_c, \; M_b, \; M_t. \qquad (5.115)$$

At the non-perturbative level these quantities represent the most appropriate parameterization of QCD, since their values are defined without any truncation of perturbation theory.

The perturbative renormalization of QCD is accomplished by replacing the bare parameters with renormalized ones, whose values are fixed by considering the high-energy behaviour of Green's functions, usually computed in the $\overline{\text{MS}}$-scheme of dimensional regularization. However, at low energies it is convenient to adopt a hadronic renormalization scheme, in which the bare parameters are eliminated in favour of quantities such as hadron masses and decay constants (see Sect. 5.2.4). Since QCD is expected to describe both the low- and high-energy regimes of the strong interaction, one should be able to express the quantities of Eq. (5.115), which are determined from the high-energy behaviour, in terms of hadronic quantities. In other words, by matching a hadronic renormalization scheme to a perturbative scheme like $\overline{\text{MS}}$ one achieves the non-perturbative renormalization of QCD at all scales. In particular, one can express the fundamental parameters of QCD (running coupling and masses, or, equivalently, the Λ-parameter and RGI quark masses) in terms of low-energy, hadronic quantities. This amounts to predicting the values of these fundamental parameters from first principles.

---

[10]The expressions for b_0 and b_1, as well as the Λ-parameter have already been shown in Sect. 5.2.4.

5.5.1 Non-perturbative Renormalization

To illustrate the problem of matching hadronic and perturbative schemes like $\overline{\text{MS}}$, it is instructive to discuss the determination of the light quark masses. A convenient starting point is the PCAC relation, which for a charged kaon can be written as

$$f_K m_K^2 = (\bar{m}_u + \bar{m}_s) \langle 0|(\bar{u}\gamma_5 s)|K^+\rangle. \tag{5.116}$$

In order to determine the sum of quark masses $(\bar{m}_u + \bar{m}_s)$, using the experimentally determined values of f_K and m_K, it suffices to compute the matrix element $\langle 0|\bar{u}\gamma_5 s|K^+\rangle$ in a lattice simulation, as outlined in Sect. 5.2.3 (see Eq. (5.64)). The dependence on the renormalization scale and scheme cancels in Eq. (5.116), since the quantities on the left hand side are physical observables. Thus, in order to determine the combination $(\bar{m}_u + \bar{m}_s)$ in the $\overline{\text{MS}}$-scheme, one must compute the relation between the bare matrix element of the pseudoscalar density evaluated on the lattice and its counterpart in the $\overline{\text{MS}}$-scheme:

$$(\bar{u}\gamma_5 s)_{\overline{\text{MS}}} = Z_P(g_0, a\mu)(\bar{u}\gamma_5 s)_{\text{lat}}. \tag{5.117}$$

Here, μ is the subtraction point (renormalization scale) in the $\overline{\text{MS}}$-scheme. Provided that Z_P and the matrix element of $(\bar{u}\gamma_5 s)_{\text{lat}}$ are known, one can use Eq. (5.116) to compute $(\bar{m}_u + \bar{m}_s)/f_K$, which is just the ratio of a renormalized fundamental parameter expressed in terms of a hadronic quantity, up to lattice artefacts. In Fig. 5.4 we have already shown the continuum extrapolation of this ratio.[11]

The factor Z_P is obtained by imposing a suitable renormalization condition involving Green's functions of the pseudoscalar densities in the $\overline{\text{MS}}$ as well as the hadronic scheme. Since the $\overline{\text{MS}}$-scheme is intrinsically perturbative, in the sense that masses and couplings are only defined at a given order in the perturbative expansion, it is actually impossible to formulate such a condition at the non-perturbative level. In perturbation theory at one loop one finds

$$Z_P(g_0, a\mu) = 1 + \frac{g_0^2}{4\pi}\left\{\frac{2}{\pi}\ln(a\mu) + C\right\} + O(g_0^4), \tag{5.118}$$

where C is a constant that depends on the chosen discretization of the QCD action. Expressions like these are actually not very useful, since perturbation theory formulated in terms of the bare coupling g_0 converges rather slowly, so that reliable estimates of renormalization factors at one- or even two-loop order in the expansion cannot be obtained. Thus it seems that the problem of non-perturbative renormalization is severely hampered by the intrinsically perturbative

---

[11]The figure actually shows the ratio for the RGI quark masses, instead of those renormalized in the $\overline{\text{MS}}$-scheme.

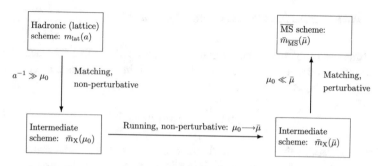

Fig. 5.12 Sketch of the matching of quark masses computed in lattice regularization and the $\overline{\text{MS}}$-scheme, via an intermediate renormalization scheme X

nature of the $\overline{\text{MS}}$ scheme in conjunction with the bad convergence properties of lattice perturbation theory.

This problem can, in fact, be resolved by introducing an intermediate renormalization scheme. Schematically, the matching procedure for the pseudoscalar density (or, equivalently, the quark mass) via such a scheme is sketched in Fig. 5.12. At low energies, corresponding to typical hadronic scales, it involves computing a non-perturbative matching relation between the hadronic and the intermediate scheme X at some scale μ_0. This matching step can be performed reliably if μ_0 is much smaller than the regularization scale a^{-1}. In the following step one computes the scale dependence within the intermediate scheme non-perturbatively from μ_0 up to a scale $\bar{\mu} \gg \mu_0$, which is large enough so that perturbation theory can be safely applied. At that point one may then determine the matching relation to the $\overline{\text{MS}}$-scheme perturbatively. Alternatively, one can continue to compute the scale dependence within the intermediate scheme to infinite energy via a numerical integration of the perturbative RG functions. According to Eq. (5.114) this yields the relation to the RGI quark mass. Since the latter is scale- and scheme-independent, one can use directly the perturbative RG functions, which in the $\overline{\text{MS}}$-scheme are known to four-loop order [71], to compute the relation to $\bar{m}_{\overline{\text{MS}}}$ at some chosen reference scale. By applying this procedure, the direct perturbative matching between between the hadronic and $\overline{\text{MS}}$-schemes (upper two boxes in Fig. 5.12), using the expression in Eq. (5.118) is thus completely avoided.

Decay constants of pseudoscalar mesons provide another example for which the renormalization of local operators is a relevant issue. For instance, the kaon decay constant is defined by the matrix element of the axial current, i.e.

$$f_K m_K = \langle 0 | (\bar{u} \gamma_0 \gamma_5 s)(0) | K^+ \rangle . \tag{5.119}$$

If the matrix element on the right hand side is evaluated in a lattice simulation, then the axial current in the discretized theory must be related to its counterpart in the continuum via a renormalization factor Z_A:

$$(\bar{u} \gamma_0 \gamma_5 s) = Z_A(g_0)(\bar{u} \gamma_0 \gamma_5 s)_{\text{lat}} . \tag{5.120}$$

Normally one would expect that the chiral Ward identities ensure that the axial current does not get renormalized. However, this no longer applies if the discretization conflicts with the symmetries of the classical action. This is clearly the case for Wilson fermions, which break chiral symmetry, such that the resulting short-distance corrections must be absorbed into a renormalization factor Z_A. Similar considerations apply to the vector current: if the discretization does not preserve chiral symmetry, current conservation is only guaranteed if the vector current is suitably renormalized by a factor Z_V, which must be considered even in the massless theory. Unlike the case of the renormalization factor of the pseudoscalar density, Z_A and Z_V are scale-independent, i.e. they only depend on the bare coupling g_0. From the above discussion it is obvious that perturbative estimates of Z_A and Z_V are inadequate in order to compute hadronic matrix elements of the axial and vectors currents with controlled errors. A non-perturbative determination of Z_A and Z_V can be achieved by imposing the chiral Ward identities as a renormalization condition.

Two widely used intermediate schemes, namely the *Schrödinger functional* (SF) and the *Regularization independent momentum subtraction* (RI/MOM) schemes are briefly reviewed in the following. We strongly recommend that the reader consult the original articles (Refs. [72–75] for the SF, and [76] for RI/MOM) for further details.

5.5.2 Finite Volume Scheme: The Schrödinger Functional

The Schrödinger functional is based on the formulation of QCD in a finite volume of size $L^3 \cdot T$—regardless of whether space-time is discretized or not—with suitable boundary conditions. Assuming that lattice regularization is employed, one imposes periodic boundary conditions on the fields in all spatial directions, while Dirichlet boundary conditions are imposed at Euclidean times $x_0 = 0$ and $x_0 = T$. In order to make this more precise, let C and C' denote classical configurations of the gauge potential. For the link variables at the temporal boundaries one then imposes

$$U_k(x)|_{x_0=0} = e^{aC}, \qquad U_k(x)|_{x_0=T} = e^{aC'}. \tag{5.121}$$

In other words, the links assume prescribed values at the temporal boundaries, but remain unconstrained in the bulk (see Fig. 5.13).

Quark fields are easily incorporated into the formalism. Since the Dirac equation is first order, only two components of a full Dirac spinor can be fixed at the boundaries. By defining the projection operator $P_\pm = \frac{1}{2}(1 \pm \gamma_0)$, one requires that the quark fields at the boundaries satisfy

$$P_+\psi(x)|_{x_0=0} = \rho(\vec{x}), \qquad P_-\psi(x)|_{x_0=T} = \rho'(\vec{x}),$$
$$\bar{\psi}(x)P_-\big|_{x_0=0} = \bar{\rho}(\vec{x}), \qquad \bar{\psi}(x)P_+\big|_{x_0=T} = \bar{\rho}'(\vec{x}), \tag{5.122}$$

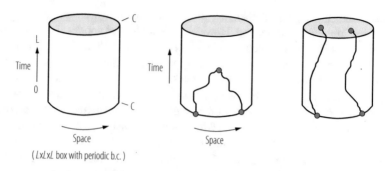

Time ↑ L, 0

Space

(LxLxL box with periodic b.c.)

Time ↑

Space

Fig. 5.13 Left panel: sketch of the SF geometry, indicating the classical gauge potentials at the temporal boundaries. Middle panel: correlation function of boundary quark fields $\zeta, \bar{\zeta}$ with a fermionic bilinear operator in the bulk. Right panel: boundary-to-boundary correlation function

where $\rho, \ldots, \bar{\rho}'$ denote prescribed values of the fields. The functional integral over all dynamical fields in a finite volume with the above boundary conditions is called the Schrödinger functional of QCD:

$$\mathcal{Z}[C', \rho', \bar{\rho}'; C, \rho, \bar{\rho}] = \int D[U]D[\bar{\psi}, \psi]\, e^{-S}. \tag{5.123}$$

The classical field configurations at the boundaries are not integrated over. Using the transfer matrix formalism, one can show that this expression is the quantum mechanical amplitude for going from the classical field configuration $\{C, \rho, \bar{\rho}\}$ at $x_0 = 0$ to $\{C', \rho', \bar{\rho}'\}$ at $x_0 = T$.

Functional derivatives with respect to $\rho, \ldots, \bar{\rho}'$ behave like quark fields located at the temporal boundaries, and hence one may identify

$$\zeta(\vec{x}) = \frac{\delta}{\delta\bar{\rho}(\vec{x})}, \quad \bar{\zeta}(\vec{x}) = -\frac{\delta}{\delta\rho(\vec{x})}, \quad \zeta'(\vec{x}) = \frac{\delta}{\delta\bar{\rho}'(\vec{x})}, \quad \bar{\zeta}'(\vec{x}) = -\frac{\delta}{\delta\rho'(\vec{x})}. \tag{5.124}$$

The boundary fields $\zeta, \bar{\zeta}, \ldots$ can be combined with local composite operators (such as the axial current or the pseudoscalar density) of fields in the bulk to define correlation functions. Particular examples are the correlation function of the pseudoscalar density, f_P and the boundary-to-boundary correlation f_1

$$f_P(x_0) = -\frac{a^6}{3} \sum_{\vec{y},\vec{z}} \left\langle \bar{\psi}(x)\gamma_5 \tfrac{1}{2}\tau^a \psi(x)\bar{\zeta}(\vec{y})\gamma_5 \tfrac{1}{2}\tau^a \zeta(\vec{z}) \right\rangle,$$

$$f_1 = -\frac{a^{12}}{3L^6} \sum_{\vec{u},\vec{v},\vec{y},\vec{z}} \left\langle \bar{\zeta}'(\vec{u})\gamma_5 \tfrac{1}{2}\tau^a \zeta'(\vec{v})\bar{\zeta}(\vec{y})\gamma_5 \tfrac{1}{2}\tau^a \zeta(\vec{z}) \right\rangle, \tag{5.125}$$

which are shown schematically in the middle and right panels of Fig. 5.13. In the above expressions, the Pauli matrices act on the first two flavour components of the fields.

The specific boundary conditions of the Schrödinger functional ensure that the Dirac operator has a minimum eigenvalue proportional to $1/T$ in the massless case [73]. As a consequence, renormalization conditions can be imposed at vanishing quark mass. If the aspect ratio T/L is set to some fixed value, the spatial length L is the only scale in the theory, and thus the masses and couplings in the SF scheme run with the box size. The recursive finite-size scaling study described below can then be used to map out the scale dependence of running quantities non-perturbatively from low to high energies. It is important to realize that in this way the relevant scale for the RG running (the box size L) is decoupled from the regularization scale (the lattice cutoff a). It is this features which ensures that the running of masses and couplings can be obtained in the continuum limit.

Let us now return to our earlier example of the renormalization of quark masses. The transition from lattice regularization and the associated hadronic scheme to the SF scheme is achieved by computing the scale-dependent renormalization factor which links the pseudoscalar density in the intermediate scheme to the bare one, i.e.

$$(\bar{s}\gamma_5 u)_{\mathrm{SF}}(\mu_0) = Z_{\mathrm{P}}(g_0, a\mu_0)\, (\bar{s}\gamma_5 u)_{\mathrm{lat}}(a). \tag{5.126}$$

A renormalization condition that defines Z_{P} can be formulated in terms of SF correlation functions:

$$Z_{\mathrm{P}}(g_0, a\mu_0) = c\, \left.\frac{\sqrt{f_1}}{f_{\mathrm{P}}(x_0)}\right|_{x_0 = T/2}, \qquad \mu_0 = 1/L_{\max}, \tag{5.127}$$

where the constant c must be chosen such that $Z_{\mathrm{P}} = 1$ in the free theory. In order to determine the RG running of the quark mass non-perturbatively one can perform a sequence of finite-size scaling steps, as illustrated in Fig. 5.14. To this end one simulates pairs of lattices with box lengths L and $2L$, at fixed lattice spacing a. The ratio of Z_{P} evaluated for each box size yields the ratio $\bar{m}_{\mathrm{SF}}(L)/\bar{m}_{\mathrm{SF}}(2L)$ (upper horizontal step in Fig. 5.14), which amounts to the change in the quark mass when the volume is scaled by a factor 2. In a subsequent step, the physical volume can be doubled once more, which gives $\bar{m}_{\mathrm{SF}}(2L)/\bar{m}_{\mathrm{SF}}(4L)$. The important point to realize is that the lattice spacing can be adjusted for a given physical box size. In this way the number of lattice sites can be kept at a manageable level, while the physical volume is gradually scaled over several orders of magnitude, as indicated by the zig-zag pattern in Fig. 5.14. Furthermore, each horizontal step can be performed for several lattice resolutions, so that the continuum limit can be taken. By contrast, if one attempted to scale the physical volume for fixed lattice spacing, one would, after only a few iterations, end up with systems so large that they would not fit into any computer's memory.

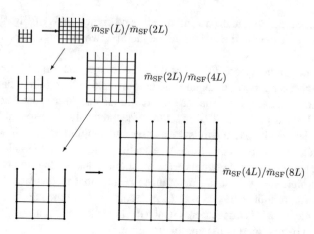

$\bar{m}_{SF}(L)/\bar{m}_{SF}(2L)$

$\bar{m}_{SF}(2L)/\bar{m}_{SF}(4L)$

$\bar{m}_{SF}(4L)/\bar{m}_{SF}(8L)$

Fig. 5.14 Illustration of the recursive finite-size scaling procedure to determine the running of $\bar{m}(L)$ for $L \to 2L \to 4L \to 8L$. In any horizontal step L is scaled by a factor 2 for fixed lattice spacing a. In every diagonal shift one keeps the physical box size L fixed and increases a by an appropriate tuning of the bare coupling g_0

In an entirely analogous fashion one can set up the finite-size scaling procedure for the running coupling constant in the SF scheme, $\bar{g}_{SF}(L)$.[12] Setting a value for the coupling actually corresponds to fixing the box size L, since the renormalization scale and the coupling in a particular scheme are in one-to-one correspondence. The sequence of scaling steps begins at the matching scale $\mu_0 = 1/L_{max}$ between the hadronic and SF schemes, and in order to express the scale evolution in physical units, the maximum box size L_{max} must be determined in terms of some hadronic quantity, such as f_π or r_0. In typical applications of the method, L_{max} corresponds to an energy scale of about 250 MeV. After n steps, the box size has decreased by a factor 2^n (typically $n = 7-9$), and at this point one is surely in the regime where the perturbative approximations to the RG functions are reliable enough to extract the Λ-parameter (in the SF scheme) and the RGI quark masses according to Eqs. (5.113) and (5.114). The transition to the $\overline{MS}$-scheme is easily performed, since the ratios $\Lambda_{SF}/\Lambda_{\overline{MS}}$, as well as $\bar{m}_{\overline{MS}}/M$ are computable in perturbation theory. At that point one has completed the steps in Fig. 5.12, and all reference to the intermediate SF scheme has dropped out in the final result.

As examples we show the running coupling and quark mass in the SF scheme from actual simulations of lattice QCD for $N_f = 2$ flavours of dynamical quarks in Fig. 5.15. The numerical data points in these plots originate from simulations with two flavours of O(a)-improved Wilson fermions and have been extrapolated to the continuum limit.

---

[12]The precise definition of $\bar{g}_{SF}$ is specified in Sect. 5.5.5 below.

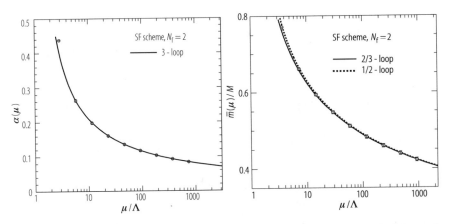

Fig. 5.15 Running of α_s (left panel) [77] and quark mass in units of the RGI mass M (right panel) [78] in the SF scheme. The results from simulations (full circles) are compared to the integration of the perturbative RG equations

5.5.3 Regularization-Independent Momentum Subtraction Scheme

An alternative choice of intermediate renormalization scheme is based on imposing renormalization conditions in terms of Green's functions of external quark states in momentum space, evaluated in a fixed gauge (e.g. Landau gauge) [76]. The external quark fields are off-shell, and their virtualities are identified with the momentum scale. Here we summarize the basic steps in this procedure by considering a quark bilinear non-singlet operator $O_\Gamma = \bar{\psi}_1 \Gamma \psi_2$, where Γ denotes a generic Dirac structure, e.g. $\Gamma = \gamma_5$ in the case of the pseudoscalar density. The corresponding renormalization factor Z_Γ is fixed by requiring that a suitably chosen renormalized vertex function $\Lambda_{\Gamma,\mathrm{R}}(p)$ be equal to its tree-level counterpart:

$$\Lambda_{\Gamma,\mathrm{R}}(p)\big|_{p^2=\mu^2} = Z_\Gamma Z_\psi^{-1} \Lambda_\Gamma(p)\big|_{p^2=\mu^2} = \Lambda_{\Gamma,0}(p). \qquad (5.128)$$

This condition defines Z_Γ up to quark field renormalization. Such a prescription can be formulated in any chosen regularization, which is why the method is said to define a regularization-independent momentum subtraction (RI/MOM) scheme. However, Z_Γ does depend on the external states and the gauge.

In order to connect to our previous example of the renormalization of quark fields, we consider the pseudoscalar density for concreteness: $\Gamma = \gamma_5 = $ "P". In this case, $\Lambda_{P,0} = \gamma_5 \otimes \mathbb{1}_{\mathrm{colour}}$, and Eq. (5.128) can be cast into the form

$$Z_P^{\mathrm{MOM}}(g_0, a\mu)\, Z_\psi^{-1}(g_0, ap)\frac{1}{12}\mathrm{Tr}\,\{\Lambda_P(p)\gamma_5\}\bigg|_{p^2=\mu^2} = 1, \qquad (5.129)$$

where the trace is taken over Dirac and colour indices.

In practice, the unrenormalized vertex function $\Lambda_P(p)$ is obtained by computing the quark propagator in a fixed gauge in momentum space and using it to amputate the external legs of the Green's function of the operator in question, evaluated between quark states, i.e.

$$\Lambda_P(p) = S(p)^{-1} G_P(p) S(p)^{-1}, \quad S(p) = \int d^4x \, e^{-ipx} \langle S(x,0) \rangle,$$

$$G_P(p) = \int d^4x \, d^4y \, e^{-ip(x-y)} \langle \psi_1(x) \left(\bar{\psi}_1(0)\gamma_5\psi_2(0) \right) \bar{\psi}_2(y) \rangle. \tag{5.130}$$

The quark field renormalization constant $Z_\psi^{1/2}$ can be fixed, e.g. via the vertex function of the vector current[13]:

$$Z_\psi = \frac{1}{48}\text{Tr}\left\{ \Lambda_{V_\mu^C}(p)\gamma_\mu \right\}\bigg|_{p^2=\mu^2}. \tag{5.131}$$

The numerical evaluation of the Green's function and quark propagators in momentum space is performed on a finite lattice with periodic boundary conditions. Unlike the situation encountered in the Schrödinger functional, there is thus no additional infrared scale, so that the renormalization conditions cannot be evaluated directly at vanishing bare quark mass. A chiral extrapolation is then required to determine mass-independent renormalization factors.

Equation (5.128) is also imposed to define the subsequent matching of the RI/MOM and $\overline{\text{MS}}$ schemes. In this case, the unrenormalized vertex function on the left-hand side is evaluated to a given oder in perturbation theory, using the $\overline{\text{MS}}$-scheme of dimensional regularization. For a generic quark bilinear this yields the factor $Z_\Gamma^{\overline{\text{MS}}}(\bar{g}_{\overline{\text{MS}}}(\mu))$. In our specific example of the pseudoscalar density operator in the PCAC relation, Eq. (5.116), the transition between the RI/MOM and $\overline{\text{MS}}$ schemes is provided by

$$(\bar{u}\gamma_5 s)_{\overline{\text{MS}}}(\mu) = R_P(\bar{\mu}/\mu)Z_P^{\text{MOM}}(g_0, a\mu)(\bar{u}\gamma_5 s)_{\text{lat}}(a). \tag{5.132}$$

The ratio R_P admits a perturbative expansion in terms of the coupling in the $\overline{\text{MS}}$-scheme, i.e.

$$R_P(\bar{\mu}/\mu) \equiv \frac{Z_P^{\overline{\text{MS}}}(\bar{g}_{\overline{\text{MS}}}(\mu))}{Z_P^{\text{MOM}}(g_0, a\mu)} = 1 + R_P^{(1)} \bar{g}_{\overline{\text{MS}}}^2 + O(\bar{g}_{\overline{\text{MS}}}^4), \tag{5.133}$$

---

[13] In this expression V_μ^C denotes the conserved lattice vector current, which involves quark fields at neighbouring lattice sites, and which is known not to undergo any finite renormalization, such that $Z_V \equiv 1$.

which is not afflicted with the bad convergence properties encountered in the direct matching of hadronic and $\overline{\text{MS}}$-schemes. Finally, for the whole method to work, one must be able to fix the virtualities μ of the external fields such that

$$\Lambda_{\text{QCD}} \ll \mu \ll 1/a. \tag{5.134}$$

In other words, the method relies on the existence of a "window" of scales in which lattice artefacts in the numerical evaluation are controlled, $\mu \ll 1/a$, and where μ is also large enough such that the perturbative matching to the $\overline{\text{MS}}$ scheme can be performed reliably. In the ideal situation one expects that the dependence of $Z_\Gamma^{\text{MOM}}(g_0, a\mu)$ on the virtuality μ inside the "window" is well described by the perturbative RG function.

The RI/MOM prescription is a flexible method to introduce an intermediate renormalization scheme and can easily be adapted to a range of operators and lattice actions. In particular, the extension to discretizations of the quark action based on the Ginsparg-Wilson relation is straightforward. This contrasts with the situation encountered in the Schrödinger functional, where extra care must be taken to ensure that imposing Schrödinger functional boundary conditions is compatible with the Ginsparg-Wilson relation [79–81]. On the other hand, the non-perturbative scale evolution, for which the Schrödinger functional is tailored, is not so easy to incorporate into the RI/MOM framework. Hence, the matching between RI/MOM and $\overline{\text{MS}}$ schemes is usually performed at fairly low scales, i.e. $\bar{\mu} = \mu_0$ in the notation of Fig. 5.12. Furthermore, the accessible momentum scales in the matching of hadronic and RI/MOM schemes are typically quite narrow, i.e. $a\mu_0 \approx 1$. Special care must also be taken when one considers operators that couple to the pion, such as the pseudoscalar density. In this case the vertex function receives a contribution from the Goldstone pole, which for $p \equiv \mu = 0$ diverges in the limit of vanishing quark mass. The fact that the chiral limit is ill-defined may spoil a reliable determination of the renormalization factor, in particular when the accessible "window" is narrow such that μ cannot be set to large values.

5.5.4 Mean-Field Improved Perturbation Theory

Another widely used strategy is to avoid the introduction of an intermediate renormalization scheme altogether and attempt the direct, perturbative matching between hadronic and $\overline{\text{MS}}$ schemes via an effective resummation of higher orders in the expansion. In this sense one regards the bare coupling and masses as parameters that run with the cutoff scale a^{-1}.

The bad convergence properties of perturbative expansions such as Eq. (5.118) has been attributed to the presence of large gluonic tadpole contributions in the relation between the link variable $U_\mu(x)$ and the continuum gauge potential $A_\mu(x)$. It was already suggested by Parisi [82] that the convergence of lattice perturbation theory could be accelerated by replacing the bare coupling g_0^2 by an "improved" coupling $\tilde{g}^2 \equiv g_0^2/u_0^4$, where u_0^4 denotes the average plaquette:

$$u_0^4 = \tfrac{1}{3}\mathrm{Re}\,\langle \mathrm{tr}\, P\rangle, \qquad P \equiv \frac{1}{6}\sum_{\mu,\nu,\nu<\mu} P_{\mu\nu}. \qquad (5.135)$$

A more systematic extension of the idea of setting up such a "tadpole" or "mean-field" improved version of lattice perturbation theory was presented in Ref. [83]. The main strategy is to factor out tadpole contributions through a redefinition of the link variable:

$$U_\mu(x) \to \tilde{U}_\mu(x) \equiv U_\mu(x)/u_0, \qquad (5.136)$$

where u_0 is the average link, defined e.g. via the average plaquette. A factor of u_0 is then absorbed into the normalization of the quark fields. According to [83], the mismatch between non-perturbative estimates for u_0 and its expression in lattice perturbation theory can be used to improve the convergence properties of lattice perturbation theory via a relative rescaling of quark fields in the continuum and lattice formulations. To make this more explicit, we consider Wilson fermions (see Sect. 5.2.2). Factoring out the average link u_0 modifies the quark field normalization of Eq. (5.36) according to

$$\psi^{\mathrm{cont}}(x) = \sqrt{2\kappa u_0}\,\psi(x), \qquad \bar{\psi}^{\mathrm{cont}}(x) = \bar{\psi}(x)\sqrt{2\kappa u_0}. \qquad (5.137)$$

The general expression for the perturbative expansion of Z_P in powers of the bare coupling reads

$$Z_P(g_0, a\mu) = 1 + g_0^2 Z_P^{(1)}(a\mu) + O(g_0^4), \qquad (5.138)$$

where $Z_P^{(1)}(a\mu)$ denotes the one-loop expansion coefficient. The convergence of Eq. (5.138) can be accelerated by dividing out u_0 in the rescaling factors of the quark and antiquark fields using its perturbative expansion and replacing it by its non-perturbative estimate computed in simulations. In other words, the rescaling of the quark fields is exploited to divide out the relative mismatch between the perturbative and non-perturbative estimates for the average link in expressions like Eq. (5.138):

$$1 = u_0(u_0)^{-1} \simeq u_0\left\{1 - u_0^{(1)}g_0^2 + O(g_0^4)\right\}, \qquad (5.139)$$

where the one-loop coefficient $u_0^{(1)} = -1/12$ for the average plaquette. In this way, i.e. by combining non-perturbatively determined values for u_0 with its perturbative expansion, and after replacing the bare coupling by $\tilde{g}^2$, one arrives at the mean-field improved version of Eq. (5.138), viz.

$$Z_P^{\mathrm{mf}} = u_0\left\{1 + \left[Z_P^{(1)}(a\mu) - u_0^{(1)}\right]\tilde{g}^2\right\}. \qquad (5.140)$$

Instead of Parisi's "boosted" coupling $\tilde{g}$ other expansion parameters have been suggested, which are expected to accelerate the convergence of the perturbative series [83]. While mean-field improvement is a general procedure, which is easily adapted to a wide range of actions and operators, it is difficult to estimate the effectiveness of the resummation and, in turn, the size of higher-order corrections. Also, a principal problem is the identification of the running scale with the cutoff, since it is difficult to separate renormalization effects from lattice artefacts.

5.5.5 The Running Coupling from the Lattice

Having discussed the non-perturbative renormalization of QCD in detail, we shall now present results for the running coupling constant, α_s, from two different approaches. This complements the discussion in Sect. 4.6, where determination of α_s from experimental data has been described in detail. Any lattice calculation of α_s proceeds along the following steps:

1. A non-perturbative definition of the coupling must be provided in terms of some quantity which can be evaluated in lattice simulations with high precision. This amounts to specifying the running coupling in a particular renormalization scheme, $\alpha_X(a\mu_0)$, which can be related to the $\overline{\text{MS}}$ scheme of dimensional regularization.
2. *Scale setting:* the matching to a hadronic scheme is performed via the calibration of the lattice spacing, which yields the scale μ_0 at which α_X is evaluated in units of some physical quantity Q:

$$\mu_0 \,[\text{MeV}] = (a\mu_0) \cdot a^{-1}\,[\text{MeV}] = (a\mu_0) \cdot \frac{Q\,[\text{MeV}]}{(aQ)}. \tag{5.141}$$

3. *Running and matching:* provided that the energy scale at which α_X has been determined is large enough, one can use perturbation theory to relate α_X to the coupling in the $\overline{\text{MS}}$ scheme, e.g.

$$\alpha_{\overline{\text{MS}}}(\bar{\mu}) = \alpha_X(\mu) + c_X^{(1)}(\bar{\mu}/\mu)\alpha_X(\mu)^2 + \dots. \tag{5.142}$$

4. The Λ-parameter can be determined from the asymptotic behaviour of α_X via Eq. (5.113).

The attentive reader has surely noticed that the above steps follow closely the general strategy for non-perturbative renormalization via an intermediate renormalization scheme outlined in Sect. 5.5.1 and Fig. 5.12.

First we discuss the determination of α_s from the Schrödinger functional. The definition of the running coupling is somewhat technical in this case. The starting point is the effective action of Eq. (5.123); the classical field configurations at the boundaries at $x_0 = 0$, T can be parameterized in terms of a real variable η:

$$C = C(\eta), \qquad C' = C'(\eta). \tag{5.143}$$

For explicit expressions we refer the reader to the original article [84]. The associated effective action is defined by

$$\Gamma(\eta) = -\ln \mathcal{Z}[C'(\eta), 0, 0; \ C(\eta), 0, 0] \tag{5.144}$$

and admits a perturbative expansion in terms of the bare coupling g_0, viz.

$$\Gamma(\eta) = \frac{1}{g_0^2}\Gamma_0 + \Gamma_1 + g_0^2\Gamma_2 + \dots. \tag{5.145}$$

A renormalized coupling can then be defined in terms of the effective action via

$$\frac{1}{\bar{g}_{\mathrm{SF}}^2(L)} = \left\{ \frac{\partial}{\partial\eta}\Gamma(\eta) \ \bigg/ \ \frac{\partial}{\partial\eta}\Gamma_0(\eta) \right\}_{\eta=0,\, m=0}. \tag{5.146}$$

This definition is imposed at vanishing quark mass, $m = 0$, and provided that the aspect ratio T/L has been fixed, the spatial dimension is the only scale in the theory, such that $\bar{g}_{\mathrm{SF}}(L)$ runs with the box size L. From the perturbative expansion of $\Gamma(\eta)$ one easily infers that $\bar{g}_{\mathrm{SF}}^2(L) = g_0^2$ at tree level. The quantity on the right-hand side is given in terms of plaquettes attached to the SF boundaries and can be computed with good statistical precision.

If $L_{\max}$ denotes the largest box size for which $\bar{g}_{\mathrm{SF}}$ is computed, then the scale is set by expressing $L_{\max}$ in terms of some known dimensionful quantity, for instance, by computing the combination $L_{\max}/r_0$ in the continuum limit and using $r_0 = 0.5\,\mathrm{fm}$.

The finite-size scaling procedure described earlier in Sect. 5.5.1 allows to compute the scale evolution of $\bar{g}_{\mathrm{SF}}$ over several orders of magnitude. In particular, each of the horizontal steps in Fig. 5.14 can be repeated for several values of the lattice spacing, so that the continuum limit is reached by taking $a/L \to 0$ for fixed physical box size L. The resulting scale evolution of $\alpha_{\mathrm{SF}} \equiv \bar{g}_{\mathrm{SF}}^2/4\pi$ is shown in Fig. 5.15 and compared to the perturbative evolution. Although the non-perturbatively determined points are described very well by perturbation theory, using the three-loop expression for the RG function, one should realize that this behaviour may be specific to the SF scheme and should not be generalized to other schemes.

Starting from $\mu_0 = 1/L_{max}$ one obtains the coupling at $\mu = 2^9/L_{max}$ after nine steps in the scaling procedure. At that point one can extract the Λ-parameter by evaluating the exact expression

$$\Lambda_{SF} = \mu \left(b_0 \bar{g}^2(\mu) \right)^{-b_1/(2b_0^2)} e^{-1/(2b_0\bar{g}^2(\mu))} \exp\left\{ -\int_0^{\bar{g}(\mu)} dx \left[\frac{1}{\beta(x)} + \frac{1}{b_0 x^3} - \frac{b_1}{b_0^2 x} \right] \right\},$$
(5.147)

where $\mu = 2^9/L_{max}$. The integral can be computed using the three-loop approximation to the RG β-function in the SF scheme. Equation (5.147) yields the combination $\Lambda_{SF}L_{max}$, and knowledge of L_{max} in physical units allows to express the Λ-parameter in MeV. Conversion to the $\overline{MS}$ scheme is easily achieved, since the ratio of Λ-parameters in two different schemes is computable via a one-loop calculation in which $\bar{g}^2_{\overline{MS}}$ is expanded in powers of $\bar{g}^2_{SF}$. This gives

$$\Lambda_{\overline{MS}} = \Lambda_{SF} \cdot c_\Lambda.$$
(5.148)

The entire procedure of determining the Λ-parameter via the Schrödinger functional has so far been carried out for the pure SU(3) gauge theory ($N_f = 0$) and for QCD with two flavours of dynamical quarks. The values of the coefficient c_Λ are 2.04872(4) for $N_f = 0$ [84] and $c_\Lambda = 2.382035(3)$ for $N_f = 2$ [85], and the resulting values for $\Lambda_{\overline{MS}}$ are [75, 77]

$$\Lambda_{\overline{MS}}^{(0)} r_0 = 0.602 \pm 0.048 \qquad \Leftrightarrow \quad \Lambda_{\overline{MS}}^{(0)} = 238 \pm 19\,\text{MeV}$$
$$\Lambda_{\overline{MS}}^{(2)} r_0 = 0.62 \pm 0.04 \pm 0.04 \qquad \Leftrightarrow \quad \Lambda_{\overline{MS}}^{(2)} = 245 \pm 16 \pm 16\,\text{MeV},$$
(5.149)

where $r_0 = 0.5\,\text{fm}$ is used to convert into physical units. There is room for improvement in several respects: for $N_f = 2$ the extrapolation to the continuum limit can be made more reliable by including simulations at smaller lattice spacings, which should reduce the first of the two quoted errors. Also, the conversion into physical units should be performed in terms of a quantity such as f_π, which is directly accessible in experiment. Finally, the calculation must be repeated with more dynamical quark flavours, in order to allow for a direct comparison with phenomenology, since all experimental determinations yield the Λ-parameter for $N_f = 4$ or 5 quark flavours.

The determination of α_s and $\Lambda_{\overline{MS}}$ via the Schrödinger functional is quite involved. However, it is the only method so far, which allows to map out the running of α_s in a completely non-perturbative manner, including the systematic elimination of lattice artefacts. In particular, perturbation theory is used only for energy scales well above 50 GeV.

The second method that we will discussed here in some detail is the determination of α_s via heavy quarkonia. Below we present an account of the calculation published in [86]. Here, the dynamical quark effects of the light (u, d, s) quarks have been accounted for in simulations with improved staggered quarks employing the fourth-root trick (see Sect. 5.2.6). In this approach, the coupling constant is defined in the so-called "V-scheme" via the heavy quark potential in momentum space:

$$V(q) = -C_F \frac{4\pi}{q^2} \alpha_V(q). \tag{5.150}$$

Small Wilson loops such as the plaquette can be expanded in powers of α_V

$$-\ln \tfrac{1}{3} \langle \mathrm{Re\,tr}\, P \rangle = c_P^{(1)} \alpha_V(s_P/a) + c_P^{(2)} [\alpha_V(s_P/a)]^2 + \dots, \tag{5.151}$$

where s_P is a real dimensionless variable which can be chosen to optimize the convergence properties of the expansion [83]. Equation (5.151) thus provides the link between the coupling and a quantity that is easily computed in lattice simulations. The above expression can be generalized to (small) rectangular Wilson loops W_{rt} with area $r \cdot t$:

$$-\ln \tfrac{1}{3} \langle W_{rt} \rangle = \sum_{k=0}^{\infty} c_{rt}^{(k)} [\alpha_V(s_{rt}/a)]^k. \tag{5.152}$$

Knowledge of the expansion coefficients in conjunction with lattice data for the quantity on the left hand side allows for the determination of α_V.

The second step, namely the calibration of the momentum scale which appears in the argument of α_V, is done by determining the lattice spacing from mass splittings in the bottomonium system. Here one typically considers the mass differences between the Υ and Υ', or alternatively, between the χ_b and Υ states. Of course, any other low-energy quantity like f_π or r_0 could be used. It can be argued, however, that mass splittings in heavy quarkonia are a natural choice for setting the scale in this particular approach, chiefly because of their relative insensitivity to the exact value of the heavy quark mass. Since the b-quark mass of $m_b \approx 4\,\mathrm{GeV}$ is greater than typical values of the inverse lattice spacing, a^{-1} one must employ special techniques to deal with heavy quarks on the lattice. In [86] this is done via an approach based on non-relativistic QCD. A detailed discussion of the specific treatment of heavy quarks in lattice simulations is deferred to Sect. 5.7.2.

After setting the scale, the Wilson loops $\langle W_{rt} \rangle$ computed on ensembles with $N_f = 3$ flavours of rooted staggered quarks are used to determine α_V via a global fit involving data at three different values of the lattice spacing. This yields

$$\alpha_V^{(3)}(7.5\,\mathrm{GeV}) = 0.2082 \pm 0.0040, \tag{5.153}$$

where the superscript on the coupling reminds us that the result is valid in the three-flavour theory. The relation to the coupling in the $\overline{\mathrm{MS}}$-scheme at the Z-pole is determined in perturbation theory, by employing the third-order expansion of $\alpha_{\overline{\mathrm{MS}}}$

in terms of α_V [87]:

$$\alpha_{\overline{MS}}^{(3)}(e^{-5/6}q) = \alpha_V^{(3)}(q) + \frac{2}{\pi}\left[\alpha_V^{(3)}(q)\right]^2 - (0.3111\ldots)\left[\alpha_V^{(3)}(q)\right]^3, \qquad (5.154)$$

which yields $\alpha_{\overline{MS}}^{(3)}(3.26\,\text{GeV})$. This coupling is then translated to $\alpha_{\overline{MS}}^{(5)}(M_Z)$ via the numerical integration of the four-loop RG β-function, including the effects from quark mass thresholds at m_c and m_b, which finally yields

$$\alpha_{\overline{MS}}^{(5)}(M_Z) = 0.1170 \pm 0.0012. \qquad (5.155)$$

This result is included in the world average of $\alpha_{\overline{MS}}^{(5)}(M_Z) = 0.1176 \pm 0.002$ in Ref. [61]. It is also in very good agreement with the non-lattice global estimate of $\alpha_{\overline{MS}}^{(5)}(M_Z) = 0.1182 \pm 0.0027$ [88].

The running and matching in this approach is done perturbatively, involving energy scales from M_Z down to m_c. In this sense the method may be regarded as similar in spirit to, say, the determination of α_s from the semi-leptonic branching ratio of τ decays, as in both cases the coupling is extracted from the perturbative expansion of a particular observable. While for τ-lepton decays an experimentally measured quantity is considered, it is the non-perturbatively computed data for the Wilson loops in the lattice approach which are expressed in terms of the running coupling. This contrasts with the Schrödinger functional approach, where also the running is computed non-perturbatively, albeit with considerable numerical effort.

The error on the result in Eq. (5.155) is rather small. It is left for future studies to confirm this level of precision, which must entail further investigations into the influence of lattice artefacts, as well as the validity of the fourth root trick.

5.5.6 Light Quark Masses

We shall now apply the general framework of non-perturbative renormalization to the determination of quark masses. Typically one distinguishes the "light" u, d, s quarks from the "heavy" c, b, t quarks. At first, this distinction may seem rather arbitrary. It is actually based on the relative magnitude of the quark masses compared with the chiral symmetry breaking scale Λ_χ, which separates "soft" from "hard" momentum scales. Masses and momenta well below Λ_χ break chiral symmetry only softly, so that spontaneous chiral symmetry breaking still dominates over the explicit breaking generated by non-zero values of the quark masses. Gasser and Leutwyler [89, 90] have demonstrated that QCD with u, d, s flavours can be studied via an "effective" theory of Goldstone boson fields. This approach, called Chiral Perturbation Theory (ChPT), has an $SU(3)_L \otimes SU(3)_R$ chiral symmetry, which is spontaneously broken to the SU(3) vector subgroup. The associated Goldstone bosons are then identified with the pions, kaons and η-mesons, whose masses are indeed small compared to typical hadronic scales, such as the mass of the nucleon, for instance. Thus, the magnitude of Λ_χ is identified with a value close to 1 GeV. In ChPT, quantities like hadron masses, decay rates or cross sections are computed through an expansion in powers of quark masses (and 4-momenta) about the chiral limit. The inclusion of the charm quark into the formalism is rather useless, since the masses if the lightest charmed pseudoscalar mesons are far greater than $\Lambda_\chi \approx 1$ GeV.

The top quark can be safely ignored in this context, since its lifetime is an order of magnitude shorter than typical QCD processes. As a consequence, the top quark does not undergo any hadronization effects (for instance, "toponium", i.e. $t\bar{t}$ bound states have never been observed), but rather decays weakly into a W-boson and a b-quark.

The mass of the b-quark is rather large (and to some extent this is also true for the charm quark), so that one may attempt to determine their values from perturbative expansions in α_s of some mass-dependent quantity. By contrast, in the light quark sector non-perturbative effects such as spontaneous chiral symmetry breaking dominate. As far as the determination of the masses of the u, d, s quarks is concerned, ChPT is of limited value, since only ratios of quark masses can be predicted, but not their absolute values. The reason is that although the light quark masses appear as parameters of ChPT, their values cannot be fixed by chiral symmetry (see Sect. 5.6.1 for more details). The absolute normalization must therefore be provided by non-perturbative methods such as lattice simulations or QCD sum rules.

Below we will focus on attempts to compute the values of the light quark masses in units of some hadronic quantity. As indicated in Sect. 5.5.1, this entails the knowledge of the renormalization factor that links lattice regularization to the chosen continuum scheme. Lattice simulations have maximum impact in the light quark sector, owing to the dominance of non-perturbative effects, which is in fact signified by the large uncertainties quoted for the values of the u, d and s quark masses in the particle data book [61].

The general procedure for the determination of light quark masses in lattice QCD starts from the PCAC relation, Eq. (5.116). Assuming exact isospin symmetry, $m_u = m_d$, one can consider a generic light flavour ℓ with mass $m_\ell \equiv \hat{m} = \frac{1}{2}(m_u + m_d)$. In order to determine, say, the combination $\hat{m} + m_s$, one must define a particular hadronic renormalization scheme, by specifying the lattice scale and the hadronic quantity that fixes the value of $\hat{m} + m_s$. Furthermore, the renormalization factor which connects hadronic and continuum schemes must be known. Equation (5.116) can then be rewritten such that it yields the sum of RG-invariant quark masses $\hat{M} + M_s$ in units of the quantity Q which sets the lattice spacing:

$$\frac{\hat{M} + M_s}{Q} = Z_M \times \left(\frac{f_{PS}^{bare} Q}{G_{PS}^{bare}} \right)\Bigg|_{m_{PS} = m_K} \times \left(\frac{m_K^2}{Q^2} \right)\Bigg|_{exp} + O(a^p). \qquad (5.156)$$

In this expression, the subscript "exp" denotes the experimental values for the respective quantities, while the matrix element G_{PS}^{bare} is given by

$$G_{PS}^{bare}\Big|_{m_{PS}=m_K} \equiv G_K^{bare} = \langle 0 \, | (\bar{\ell}\gamma_5 s)_{lat} | \, K \rangle. \tag{5.157}$$

The pseudoscalar decay constant f_{PS}^{bare} parameterizes the matrix element of the unrenormalized axial current, i.e.

$$f_{PS}^{bare} m_{PS}\Big|_{m_{PS}=m_K} \equiv f_K^{bare} m_K = \langle 0 \, | (\bar{\ell}\gamma_0\gamma_5 s)_{lat} | \, K \rangle. \tag{5.158}$$

The renormalization factor Z_M relates the bare current quark mass to the RG-invariant mass. Thus, the task for lattice calculations is to compute the ratio $f_{PS}^{bare} Q / G_{PS}^{bare}$ for a generic pseudoscalar state and tune the bare quark mass such that $m_{PS} = m_K$. By combining the result with the renormalization factor Z_M and the experimental value of m_K^2/Q^2, the RGI quark masses in units of Q are obtained up to lattice artefacts of order a^p, where p is characteristic of the details of the discretization. Since the RGI quark masses are scale- and scheme-independent quantities, the factor Z_M depends only on the bare coupling g_0. Using the Schrödinger functional as the intermediate renormalization scheme, non-perturbative estimates of Z_M computed for $O(a)$ improved Wilson fermions within a wide range of bare couplings, have been published in Refs. [75] and [78]. In this case, Z_M is given by

$$Z_M(g_0) = \frac{M}{\bar{m}_{SF}(\mu_0)} \frac{Z_A(g_0)}{Z_P(g_0, a\mu_0)}, \tag{5.159}$$

where the ratio $M/\bar{m}_{SF}(\mu_0)$ is computed via the finite-size scaling procedure. The transition between lattice regularization and the SF-scheme is accomplished by determining Z_P and the renormalization factor Z_A of the axial current.[14] Note that the dependence on the intermediate matching scale μ_0 drops out completely in this expression. Finally, the conversion to the $\overline{MS}$-scheme is performed by considering

$$Z_m(g_0, a\mu) \equiv \frac{\bar{m}_{\overline{MS}}(\mu)}{M} Z_M(g_0), \tag{5.160}$$

where the ratio $\bar{m}_{\overline{MS}}(\mu)/M$ can be computed through the numerical integration of the perturbative approximation of the anomalous dimension τ and the β-function at four loops. This yields [35, 78]

$$\frac{\bar{m}_{\overline{MS}}(2\,\text{GeV})}{M} = \begin{cases} 0.7208, & N_f = 0 \\ 0.7013, & N_f = 2. \end{cases} \tag{5.161}$$

---

[14]If the fermionic discretization preserves chiral symmetry $Z_A = 1$, while for Wilson fermions Z_A should be computed non-perturbatively. For the SF this was performed in Refs. [91, 92].

Table 5.2 Results for the strange quark mass in the $\overline{\mathrm{MS}}$-scheme at $\mu = 2\,\mathrm{GeV}$ and for the ratio $m_s/\hat{m}$, in the continuum limit of the quenched approximation

Collaboration	Action	Q	Renorm.	$m_s^{(Q)}$ [MeV]	$m_s/\hat{m}$	$m_s^{(r_0)}$ [MeV]
SPQ$_{\mathrm{cd}}R$ [93]	Wilson	m_{K^*}	RI	105(9)(6)	24.3(2)(6)	95(9)(5)
CP-PACS [47]	Wilson	m_ρ	pert.	114(2)($^6_3$)	26.5($^{5.1}_{3.4}$)	98(2)($^6_3$)
ALPHA/UKQCD [35]	Clover	f_K	SF	97(4)		99(4)
JLQCD [94]	Stagg.	m_ρ	RI	106(7)	25.1(2.4)	95(6)

The table includes information on the fermionic action employed in the simulations, the quantity Q that sets the scale, and the type of renormalization (RI/MOM, SF or tadpole improved perturbation theory. The right-most column contains the results for the strange quark mass when converted into a common hadronic scheme, in which the scale is set by $Q' = 1/r_0$, assuming that $r_0 = 0.5\,\mathrm{fm}$.

Estimates for the strange quark mass itself can be obtained in two ways: first, one combines $\hat{M} + M_s$ with the ratio $M_s/\hat{M} = 24.4 \pm 1.4$ estimated in ChPT [38]. Alternatively, one might attempt to compute $\hat{M}$ directly from lattice data, by considering Eq. (5.116) for a pion. In this case, however, one relies on chiral extrapolations, because of the difficulties involved when tuning the masses of the light quarks towards the values of the physical up- and down-quark masses.

In Table 5.2 we present a selection of results for the mass of the strange quark in the quenched approximation, normalized in the $\overline{\mathrm{MS}}$-scheme at $\mu = 2\,\mathrm{GeV}$, as well as the ratio $M_s/\hat{M}$. Two observations are worth mentioning: first, direct determinations of $M_s/\hat{M}$ via chiral extrapolations agree well with the estimate from ChPT, even though the chiral limit is ill-defined in the quenched approximation. Second, the different systematics in the simulations (lattice actions, renormalization of local operators) generate a spread of seemingly incompatible results for the mass of the strange quark. However, the spread can be traced to the particular choice of hadronic renormalization scheme. To this end one can compute the relation between quark masses computed for two different lattice scales, Q and Q'. From Eq. (5.156) one easily infers that the strange quark mass $m_s^{(Q')}$ estimated using Q', is related to its counterpart $m_s^{(Q)}$ via [37]

$$m_s^{(Q')} \,[\mathrm{MeV}] = \left(\frac{Q'}{Q}\right)_{\mathrm{lat}} \left(\frac{Q}{Q'}\right)_{\mathrm{exp}} m_s^{(Q)} \,[\mathrm{MeV}]. \qquad (5.162)$$

Here, the subscripts "lat" and "exp" refer to lattice and experimental estimates of the scale ratios. The ratio $(Q'/Q)_{\mathrm{lat}}$ can be determined in the continuum limit using published lattice data, and the deviation of the proportionality factor from unity is a measure of the relative quenching effects, when either Q or Q' is chosen to set the scale. Once the results have been converted to the common scale r_0, the estimates for m_s in the continuum limit show remarkable consistency, despite the very different systematic effects among the simulations included in this analysis (c.f. Table 5.2). This demonstrates that lattice artefacts and renormalization effects can be controlled at the level of a few percent with the available techniques.

Table 5.3 Selection of recent unquenched results for the light quark masses

Collaboration	N_f	Action	Q	Ren.	m_s [MeV]	$m_s/\hat{m}$	$\hat{m}$ [MeV]
CPPACS/ JLQCD [95]	2+1	Clover	m_ρ	pert.	$91.1(^{14.6}_{6.2})$		$3.54(^{0.64}_{0.35})$
HPQCD [96]	2+1	Stagg.	$\Delta_{\Upsilon'-\Upsilon}$	pert.	87(8)	27.4(4)	3.2(3)
QCDSF/ UKQCD [97]	2	Clover	m_N, r_0	RI	111(9)	27(3)	4.1(4)
SPQ$_{cd}$R [98]	2	Wilson	m_{K^*}	RI	$101(^{26}_{8})$		4.3(4)
ALPHA [78]	2	Clover	r_0	SF	97(22)		
CPPACS [99]	2	Clover	m_ρ	pert.	$88(^4_6)$	26(2)	$3.44(^{14}_{22})$
ETM [100]	2	tmQCD	f_π	RI	105(3)(9)	27.3(3)(1.2)	3.85(12)(40)

The challenge for current and future simulations is to eliminate the remaining uncertainty due to quenching. Several simulations with $N_f = 2$ or $2 + 1$ flavours of dynamical quarks[15] based on different fermionic discretizations have produced results for the light quark masses, which are shown in Table 5.3. Despite the enormous progress that has been made in simulating light dynamical quarks, it is important to realize that systematic effects such as lattice artefacts and/or renormalization effects are currently not as well controlled as in the quenched theory. The fact that affordable lattice spacings are still relatively large implies that extrapolations to the continuum limit are in general longer than in the quenched approximation, thereby leading to larger errors. In some cases it is not even clear whether the leading lattice artefacts in dynamical simulations have been isolated. Also, the quantity Q that sets the scale must be known at least as accurately as the quark mass itself, and hence the determination of these observables may prove just as costly. Finally, dynamical quark masses are still fairly large, especially in many simulations using Wilson fermions, and thus the long and potentially uncontrolled chiral extrapolations significantly affect estimates for the isospin-averaged light quark mass $\hat{m}$.

5.6   Spontaneous Chiral Symmetry Breaking

Chiral symmetry has already been mentioned in connection with the masses of the light quarks. Here we will extend the general framework and elaborate on effective descriptions of QCD at low energies, which can be treated analytically. As we shall see, much can be learnt via the interplay of such effective theories and lattice simulations of QCD.

---

[15]$N_f = 2$ usually denotes a degenerate doublet of light (u, d) quarks, while $N_f = 2 + 1$ denotes a degenerate doublet together with a heavier third flavour, i.e. the strange quark.

Massless QCD with N_f flavours is invariant under independent rotations of the left- and right-handed components of the quarks fields. If one defines the field Ψ as the vector of N_f Dirac spinors ψ_i via

$$\Psi = \left(\psi_1, \ldots, \psi_{N_f}\right)^T, \tag{5.163}$$

its left- and right-handed components are given by

$$\Psi_L := \left(\mathbb{1}_{N_f} \otimes P_-\right) \Psi, \quad \Psi_R := \left(\mathbb{1}_{N_f} \otimes P_+\right) \Psi, \quad P_\pm = \tfrac{1}{2}\left(1 \pm \gamma_5\right). \tag{5.164}$$

The action of the massless theory is then invariant under transformations like

$$\Psi \to \Psi' = \exp\left\{iP_-(\omega_L \cdot T) + iP_+(\omega_R \cdot T)\right\} \Psi, \tag{5.165}$$

where ω_L, ω_R are real vectors, and T denotes the generators of $SU(N_f)$, which satisfy

$$\left[T^a, T^b\right] = if^{abc}T^c, \qquad \mathrm{Tr}\left(T^a T^b\right) = \tfrac{1}{2}\delta^{ab}. \tag{5.166}$$

The above transformation can be rewritten in terms of vector and axial rotations, i.e.

$$\Psi \to \Psi' = \exp\left\{i\alpha_V \cdot T + i\alpha_A \cdot T\gamma_5\right\} \Psi, \tag{5.167}$$

where $\alpha_V \equiv \tfrac{1}{2}(\omega_R + \omega_L)$ and $\alpha_A \equiv \tfrac{1}{2}(\omega_R - \omega_L)$. Invariance under these transformation laws is what one usually means when one says that (massless) QCD is invariant under a global $SU(N_f)_L \otimes SU(N_f)_R$ symmetry.

Actually, QCD has even more global symmetries, namely a $U(1)_V$ symmetry, which corresponds to a common rotation of all quark flavours. The conserved charge derived from the Noether current, which is associated with this unbroken symmetry, is the quark number. The conservation of the axial current associated with the remaining axial $U(1)$ symmetry is, however, severely broken by an anomalous term, which gives rise to strong non-perturbative effects generated by instantons. Without going into further detail here, we refer to common textbooks.

Returning now to $SU(N_f)_L \otimes SU(N_f)_R$, we note that symmetries in sub-nuclear physics are usually deduced from the particle spectrum. That is, symmetries manifest themselves through the occurrence of mass-degenerate (or nearly degenerate) particle multiplets that can be grouped according to the irreducible representations of the symmetry group. Indeed, for $N_f = 3$ one finds that the light pseudoscalar mesons, i.e. the pions, kaons and η-mesons form an octet. The mass splittings among the members of the octet are small when viewed on typical hadronic scales, and arise due to the unequal, non-zero masses of the light quarks. However, if the pseudoscalar octet were interpreted as a manifestation of an (approximate) $SU(3)_L \otimes SU(3)_R$ chiral symmetry, one would expect that each member of the octet is accompanied by a parity partner, i.e. a scalar meson, whose mass is of the same order of magnitude. This is not observed in experiment, where the lightest scalar mesons are found to lie 600–700 MeV above the pseudoscalar octet. One therefore concludes that the symmetry must be spontaneously broken. The term "spontaneous breaking" refers to the fact that theories like QCD possess more internal symmetries than those that can be inferred from the particle spectrum. In general, spontaneously broken symmetries are not realized as symmetry transformations involving the physical states of the theory. In particular, the ground state, i.e. the vacuum, is not invariant under the transformation. As discussed in many textbooks, it is precisely the invariance of the vacuum under the symmetry transformation that is required to ensure the degeneracy of the particle spectrum. If the vacuum is not invariant, certain operators may acquire a non-vanishing expectation value. In fact, a sufficient condition for the spontaneous breaking of the physical $SU(3)_L \otimes SU(3)_R$ chiral symmetry is fulfilled if the expectation value of the scalar density, $\bar{\Psi}\Psi$, is non-zero, i.e.

$$\langle \bar{\Psi}\Psi \rangle \equiv \langle \bar{u}u + \bar{d}d + \bar{s}s \rangle \neq 0. \tag{5.168}$$

Furthermore, according to Goldstone's theorem [101], the generator of each broken symmetry is associated with a massless particle. Since the masses of the members of the pseudoscalar octet are rather small in comparison with the proton mass, they are identified as the Goldstone bosons of the spontaneously broken chiral symmetry.

Spontaneous chiral symmetry breaking is an entirely non-perturbative phenomenon. The task is then to explore the breaking mechanism and compute the value of the quark condensate $\langle \bar{\Psi}\Psi \rangle$. As shall be outlined below, this can be achieved through the interplay of lattice simulations and effective low-energy descriptions of QCD.

5.6.1 Chiral Perturbation Theory

Chiral Perturbation Theory (ChPT) has already been mentioned in connection with extrapolations of lattice data to the physical values of the up- and down-quark masses, and also in the context of lattice determinations of the strange quark mass. Here we present a brief introduction into the general formalism. More thorough reviews can be found in Refs. [102, 103].

Chiral Perturbation Theory is an effective theory, based on a systematic expansion of the low-energy dynamics of QCD in powers of the 4-momentum and the quark mass about the chiral limit [89, 90], i.e.

$$\mathcal{L}_{\text{eff}} = \mathcal{L}_{\text{eff}}^{(2)} + \mathcal{L}_{\text{eff}}^{(4)} + \dots, \tag{5.169}$$

where the superscripts label the order of the expansion in powers of p. In contrast to QCD, the basic degrees of freedom which appear in $\mathcal{L}_{\text{eff}}$ are the Goldstone bosons, rather than the fundamental quarks and gluons. ChPT is parameterized in terms of a set of empirical couplings, usually called "low-energy constants" (LECs). At lowest order, the effective chiral Lagrangian (in Euclidean space-time) reads

$$\mathcal{L}_{\text{eff}}^{(2)} = \tfrac{1}{2}F_0^2\left\{\tfrac{1}{2}\text{Tr}\left(\partial_\mu U^\dagger \partial_\mu U\right) - B_0\text{Tr}\left(\mathcal{M}(U + U^\dagger)\right)\right\}, \qquad (5.170)$$

where $\mathcal{M} = \text{diag}(m_u, m_d, m_s)$ is the quark mass matrix, and $U(x)$ collects the Goldstone boson fields, i.e.

$$U(x) = \exp\left(\frac{i}{F_0}\lambda \cdot \phi(x)\right); \quad \lambda \cdot \phi \equiv \sum_{a=1}^{8}\lambda^a\phi_a = \begin{pmatrix} \pi^0 + \frac{1}{\sqrt{3}}\eta & \sqrt{2}\pi^+ & \sqrt{2}K^+ \\ \sqrt{2}\pi^- & -\pi^0 + \frac{1}{\sqrt{3}}\eta & \sqrt{2}K^0 \\ \sqrt{2}K^- & \sqrt{2}\bar{K}^0 & -\frac{2}{\sqrt{3}}\eta \end{pmatrix}. \tag{5.171}$$

The λ^a's denote the Gell-Mann matrices which are normalized as $\text{Tr}(\lambda^a\lambda^b) = 2\delta^{ab}$. The LECs at leading order are B_0 and F_0, where the latter corresponds to the pion decay constant in the chiral limit.[16] The expression for $\mathcal{L}_{\text{eff}}^{(4)}$, i.e. the interaction terms at next-to-leading order in the chiral expansion, contains 12 additional interaction terms, multiplied by the LECs $L_1, \ldots, L_{10}$, H_1, H_2. The values of the LECs are usually determined by matching the expressions of ChPT for physical observables to experimental data. However, it turns out that the complete set of LECs cannot be obtained in this way. Rather, in order to fix the values of some LECs, one must resort to additional theoretical assumptions. One particular example is the value of B_0, which appears in the chiral expansion of the pion mass at lowest order (see also Eq. (5.80)):

$$m_\pi^2 = B_0(m_u + m_d). \tag{5.172}$$

From this expressions it is clear that B_0 can only be determined using m_π as input if the physical values of the quark masses are known in the first place. By the same token, the value of $\hat{m} = \frac{1}{2}(m_u + m_d)$ can only be inferred if an estimate for B_0 is available. However, the a priori unknown parameter B_0 drops out in suitably chosen ratios of $m_\pi^2, m_K^2, \ldots$ This explains why ChPT can be used to predict the ratios of the light quark masses but fails to provide an absolute mass scale. Another reason why the complete set LECs cannot be determined from chiral symmetry considerations alone is the fact that the effective Lagrangian beyond leading order is invariant under a symmetry transformation which involves the LECs and the mass matrix $\mathcal{M}$, but which is absent in QCD. This is the so-called "Kaplan-Manohar ambiguity" [104]. At this point it is clear that lattice simulations of QCD can provide valuable input for the determination of LECs. For instance, since the values of the

---

[16]We use capital symbols for decay constants whenever we refer to a normalization in which $F_\pi \simeq 93\,\text{MeV}$.

quark masses are input parameters in the simulations, lattice QCD allows to map out the quark mass dependence of the masses of Goldstone bosons and thus determine the LEC B_0. We shall see below that B_0 is related to the quark condensate $\langle \bar{\Psi} \Psi \rangle$ which can be considered as the order parameter for spontaneous chiral symmetry breaking. Furthermore, as we have already discussed in Sect. 5.5.6, absolute values of quark masses are accessible via lattice QCD.

We end our brief introduction to ChPT with the derivation of a few relations which will be useful for our discussion of chiral symmetry breaking below. In particular, we shall derive the leading-order mass formulae such as Eq. (5.80) and establish a link between the quark condensate and B_0. To this end we expand the field U in the chiral Lagrangian $\mathcal{L}_{\text{eff}}^{(2)}$ in powers of the Goldstone boson fields. Assuming exact isospin symmetry, $m_u = m_d$, one finds at lowest order in ϕ_a:

$$\mathcal{L}_{\text{eff}}^{(2)} = \frac{1}{2} \sum_{a=1}^{8} \partial_\mu \phi_a \partial_\mu \phi_a + \dots$$

$$+ \frac{1}{2}(m_u + m_d) B_0 \sum_{a=1}^{3} \phi_a^2 + \frac{1}{2}(\hat{m} + m_s) B_0 \sum_{a=4}^{7} \phi_a^2$$

$$+ \frac{1}{3}(\hat{m} + 2m_s) B_0 \phi_8^2 + \dots. \tag{5.173}$$

After identifying ϕ_1, ϕ_2, ϕ_3 with the pions, $\phi_4, \dots, \phi_7$ with the kaons, and $\phi_8 \equiv \eta$, one derives the leading-order relations between the quark masses and the masses of the Goldstone bosons, viz.

$$m_\pi^2 = 2B_0\hat{m}, \quad m_K^2 = B_0(\hat{m} + m_s), \quad m_\eta^2 = \tfrac{2}{3}B_0(\hat{m} + 2m_s). \tag{5.174}$$

Thus, the relation for a generic pseudoscalar Goldstone boson made up of quarks with masses m_1 and m_2 is precisely what was already shown in Eq. (5.80). We note that from Eq. (5.174) one easily derives the Gell-Mann–Okubo mass relation, i.e.

$$3m_\eta^2 + m_\pi^2 - 4m_K^2 = 0, \tag{5.175}$$

which is satisfied experimentally within a few percent. Furthermore, Eq. (5.174) yields the ratio $m_s/\hat{m}$ at lowest order, viz.

$$\frac{m_s}{\hat{m}} = \frac{2m_K^2 - m_\pi^2}{m_\pi^2} \simeq 24, \tag{5.176}$$

which is already close to the estimate at next-to-leading order of $m_s/\hat{m} = 24.4 \pm 1.5$ [38], quoted in Sect. 5.5.6.

For the discussion of spontaneous symmetry breaking, it is useful to establish a connection between the quark condensate in QCD, $\langle \bar{\Psi} \Psi \rangle$, and the LECs which parameterize the effective chiral Lagrangian. This link is provided by the so-called Gell-Mann–Oakes–Renner relation [105], which we are going to derive below. To this end we consider the QCD Lagrangian in the continuum:

$$\mathcal{L}_{\text{QCD}} = -\frac{1}{4}F^a_{\mu\nu}(x)F^a_{\mu\nu}(x) + \sum_f \bar{\psi}_f(x)\left(\gamma_\mu D_\mu + m_f\right)\psi_f(x). \tag{5.177}$$

The path integral is defined as

$$Z_{\text{QCD}} = \int D[A_\mu]D[\bar{\psi},\psi]\exp\left\{-\int d^4x\,\mathcal{L}_{\text{QCD}}\right\}, \tag{5.178}$$

and the expression for the quark condensate can be formally derived by taking derivatives with respect to the light quark masses, i.e.

$$\sum_{f=u,d,s}\frac{\partial \ln Z_{\text{QCD}}}{\partial m_f}\bigg|_{m_f=0} = -\left\langle \bar{u}u + \bar{d}d + \bar{s}s\right\rangle\big|_{m_f=0} \equiv -\left\langle \bar{\Psi}\Psi\right\rangle. \tag{5.179}$$

What is the analogue of this expression in the effective chiral theory? To answer this question one takes the lowest-order chiral Lagrangian of Eq. (5.170) and defines the corresponding path integral[17]

$$Z_{\text{ChPT}} = \int D[U]\exp\left\{-\int d^4x\,\mathcal{L}^{(2)}_{\text{eff}}\right\}. \tag{5.180}$$

Since $\mathcal{L}^{(2)}_{\text{eff}}$ contains the quark mass matrix one can consider similar derivatives, i.e.

$$\sum_{f=u,d,s}\frac{\partial \ln Z_{\text{ChPT}}}{\partial m_f}\bigg|_{m_f=0} = \frac{F_0^2 B_0}{2}\sum_{f=u,d,s}\frac{\partial}{\partial m_f}\left\langle \text{Tr}\,\mathcal{M}(U+U^\dagger)\right\rangle\bigg|_{m_f=0} = 3\cdot F_0^2 B_0 + \dots, \tag{5.181}$$

and comparison with Eq. (5.179) yields

$$-\frac{1}{3}\left\langle \bar{u}u + \bar{d}d + \bar{s}s\right\rangle \equiv \Sigma = F_0^2 B_0. \tag{5.182}$$

In other words, the quark condensate is related to the slope parameter in the lowest-order mass formulae and the pion decay constant in the chiral limit, F_0. This result is known as the Gell-Mann–Oakes–Renner relation.

---

[17]It should be obvious that the field U must not be confused with the link variable considered in previous sections.

5.6.2 Lattice Calculations of the Quark Condensate

The Gell-Mann–Oakes–Renner relation is the starting point for many lattice determinations of the quark condensate. For a generic pseudoscalar meson consisting of a mass-degenerate quark and antiquark, i.e. $m_1 = m_2 \equiv m$, the LEC Σ is given by

$$\Sigma = \lim_{m \to 0} \left(\frac{m_{PS}^2 F_{PS}^2}{2m} \right). \tag{5.183}$$

The technical drawback of this straightforward approach is that the chiral limit in the above expression is difficult to take in practice, as we have mentioned several times already. In the quenched approximation the situation is even worse: due to the appearance of quenched chiral logarithms (c.f. Eq. (5.82)) the ratio m_{PS}^2/m becomes singular at vanishing quark mass, and hence the chiral limit does not exist. Since the quenched approximation is being abandoned, this issue will gradually become irrelevant.

However, a more serious obstacle remains in the case of dynamical simulations with Wilson fermions: since this particular type of regularization breaks chiral symmetry explicitly, the matching of simulation data at non-zero lattice spacing to the expressions of ChPT is—strictly speaking—not permitted. Matching is certainly justified if a fermionic discretization is employed which preserves chiral symmetry, such as overlap or domain wall fermions, or if results obtained using Wilson fermions are extrapolated to the continuum limit before a comparison to ChPT is performed.

A complementary approach for determining the condensate on the lattice is based on the Banks–Casher relation [106]. It provides a link between the LEC Σ and the spectral properties of the Dirac operator, viz.

$$\Sigma = \lim_{\lambda \to 0} \lim_{m \to 0} \lim_{V \to \infty} \frac{\pi}{V} \rho(\lambda), \tag{5.184}$$

where V is the space-time volume. The spectral density $\rho(\lambda)$ is defined as follows: Let $\mathcal{D}$ denote the massless Dirac operator in the continuum, satisfying $\{\gamma_5, \mathcal{D}\} = 0$. Its eigenvalue equation reads

$$\mathcal{D}\psi_n = i\lambda_n \psi_n, \qquad \lambda_n \in \mathbb{R}, \tag{5.185}$$

where the eigenvalues and eigenfunctions depend on the gauge field. A suitable definition of the spectral density is then represented by

$$\rho(\lambda) := \sum_n \langle \delta(\lambda - \lambda_n) \rangle, \tag{5.186}$$

where the expectation value is taken with respect to the QCD functional integral.[18] Note that in Eq. (5.184) the ordering of limits must be obeyed. In particular, since the spontaneous breaking of a continuous symmetry cannot occur in finite volume, the limit $V \to \infty$ must be taken before the chiral limit and the spectrum in the deep infrared are considered.

The Banks–Casher relation provides not only a method to determine the condensate, but also suggests a mechanism how spontaneous chiral symmetry breaking comes about. Indeed, Eq. (5.184) implies that a non-zero value of the quark condensate is generated through a non-vanishing value of the spectral density in the deep infrared. In other words, spontaneous chiral symmetry breaking is driven by an accumulation of small eigenvalues. An immediate consequence of the Banks–Casher relation is that the level spacing $\Delta\lambda$ between the small eigenvalues is given by

$$\Delta\lambda \equiv \frac{1}{\rho(\lambda)} = \frac{\pi}{V\Sigma}. \qquad (5.187)$$

Hence, as $V \to \infty$ the level spacing becomes arbitrarily small. In the free theory, i.e. in the absence of a non-trivial gauge field one finds that $\rho(\lambda) \propto \lambda^3$, which vanishes as $\lambda \to 0$. The accumulation of eigenvalues near zero with a rate predicted by Eq. (5.187) must therefore arise through the interaction with the gauge field.

In order to test the Banks–Casher scenario, a possible strategy is to compute the spectral density and check whether it actually produces an arbitrarily dense spectrum near the origin. Analytic predictions for $\rho(\lambda)$ can be derived in the framework of effective theories of QCD at low energies, namely ChPT, as well as chiral Random Matrix Theory (RMT). The latter also yields predictions for the distributions of individual eigenvalues, in addition to the spectral density.

Chiral Random Matrix Theory goes back to an idea of Wigner who tried to utilize statistical properties for the theoretical description of systems with many degrees of freedom and complicated dynamics, such as nuclear resonances. Rather than trying to model the local interactions within such a system explicitly, all possible interactions that are consistent with the symmetries of the theory are equally likely. The Hamiltonian is then approximated by a matrix whose elements are uncorrelated but obey a particular probability distribution. The main guiding principle for the RMT description of QCD is the requirement that all global symmetries must be respected. The massless Dirac operator can then be represented by an $N \times N$ matrix $\hat{D}$ with an off-diagonal block structure which is characteristic for systems with chiral symmetry:

$$\hat{D} = \begin{pmatrix} 0 & W \\ -W^\dagger & 0 \end{pmatrix} \begin{matrix} \}N_+ \\ \}N_- \end{matrix}. \qquad (5.188)$$

---

[18]A normalization factor of V^{-1} must be included in Eq. (5.184) since $\rho(\lambda)$ is proportional to the volume.

As illustrated by the above expression, the matrix W is, in general, rectangular with N_+ rows and N_- columns, such that $N = N_+ + N_-$. For $N_+ \neq N_-$ the matrix $\hat{D}$ has $|N_+ - N_-|$ zero modes, and the index $\nu \equiv N_+ - N_-$ may be identified with the topological charge in QCD. With this definition, $\hat{D}$ is anti-hermitian and has purely imaginary eigenvalues which come in complex conjugate pairs:

$$\hat{D}\phi_n = i\mu_n\phi_n, \qquad \mu_n \in \mathbb{R}. \tag{5.189}$$

One can define the system's partition function in a sector of fixed topological charge ν via

$$\mathcal{Z}_\nu = \int D[W] \det\left(\hat{D} + m\right)^{N_f} e^{-\frac{1}{2}N\mathrm{Tr}\,(W^\dagger W)}, \tag{5.190}$$

where N_f is—as usual—the number of dynamical quark flavours. It makes sense to identify the matrix size N with the physical volume V of the theory (up to some proportionality constant).

In order to study the spectral properties of $\hat{D}$ in the deep infrared, it is useful to rescale the eigenvalues by the system size

$$z \equiv \mu_n N, \qquad N \propto V \tag{5.191}$$

since, according to Eq. (5.187), the level spacing of the scaled eigenvalues z is of order one. The so-called *microscopic spectral density* in the sector of topological charge ν is then defined as

$$\rho_s^{(\nu)}(z) := \lim_{N \to \infty} \sum_n \langle \delta(z - \mu_n N) \rangle_\nu, \tag{5.192}$$

where the expectation value $\langle \cdots \rangle_\nu$ is taken with respect to the partition function $\mathcal{Z}_\nu$. An explicit expression for $\rho_s^{(\nu)}(z)$ in terms of Bessel functions has been worked out by Verbaarschot and Zahed [107]

$$\rho_s^{(\nu)}(z) = \frac{z}{2} \left\{ \left[J_{\nu+N_f}(z) \right]^2 - J_{N_f+\nu+1}(z)\, J_{N_f+\nu-1}(z) \right\}. \tag{5.193}$$

The microscopic spectral density is the convolution of the distribution functions $p_k^{(\nu)}$ of the individual scaled eigenvalues, i.e.

$$\rho_s^{(\nu)}(z) = \sum_{k=1}^{\infty} p_k^{(\nu)}(z), \qquad \int_0^\infty dz\, p_k^{(\nu)}(z) = 1. \tag{5.194}$$

Fig. 5.16 RMT predictions for the microscopic spectral density and distributions for individual eigenvalues in the sector with topological charge $\nu = 0$

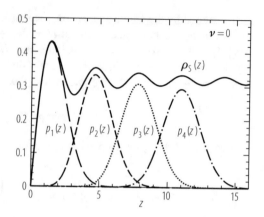

Chiral RMT yields predictions for these distributions. For instance, for the lowest eigenvalue in the sector with $\nu = 0$ one obtains for $N_f = 0$

$$p_1^{(0)}(z) = \frac{1}{2} z \, e^{-z^2/4}. \tag{5.195}$$

For further illustration the microscopic spectral density and the distribution functions for a few of the lowest eigenvalues are plotted in Fig. 5.16. The result for $\rho_s^{(\nu)}(z)$ indicates that an accumulation of small eigenvalues does indeed take place. Since one considers the simultaneous limits $\mu \to 0$ and $N \to \infty$ for fixed z, a non-zero value of $\rho_s^{(\nu)}(z)$ for finite z signals that the spectrum is packed more and more densely near the origin.

Can the predictions of RMT be verified from first principles in simulations of lattice QCD? The answer is 'yes', provided one considers a particular kinematical situation, commonly referred to as the "ϵ-regime" of QCD. It is based on the formulation of QCD in a large but finite volume of spatial size L and for arbitrarily small quark mass. The Compton wavelength of the pion then exceeds the spatial size, and thus the ϵ-regime is characterized by

$$m_\pi L \ll 1, \qquad F_\pi L \gg 1. \tag{5.196}$$

In this particular situation the path integral of the theory is dominated by zero momentum modes. In a symmetric finite box with volume $V = L^4$, the minimum non-zero momentum is given by $p_{min} \propto 1/L$. Let us recall the expression for the lowest-order effective chiral Lagrangian, i.e.

$$\mathcal{L}_{\text{eff}}^{(2)} = \frac{1}{2} F_0^2 \left\{ \frac{1}{2} \text{Tr} \left(\partial_\mu U^\dagger \partial_\mu U \right) - m \Sigma \, \text{Tr} \left(e^{i\theta/N_f} U + \text{h.c.} \right) \right\}, \tag{5.197}$$

where we have included the vacuum angle θ and assumed that $\mathcal{M} \equiv m\mathbb{1}$. If the quark mass m is tuned so that

$$m \Sigma \ll F_0^2 p_{\min}^2 \sim F_0^2/L^2, \tag{5.198}$$

the statistical weight of fields with $\partial_\mu U \neq 0$ will be strongly suppressed in the path integral. In other words, the mass term will dominate over the kinetic term, except for fields U with $\partial_\mu U = 0$. Since $2m\Sigma/F_0^2 = m_{PS}^2$, the conditions in Eq. (5.196), which define the kinematical situation of the ϵ-regime, are equivalent to

$$m \Sigma V \ll 1. \tag{5.199}$$

The zero-momentum part can be represented by a constant SU(3) matrix U_0 such that

$$U(x) = U_0\, e^{2i\xi(x)/F_0}, \qquad U_0 \in SU(3), \tag{5.200}$$

where the field ξ incorporates the fluctuations about the zero momentum mode. According to Leutwyler and Smilga [108], the path integral of the theory in topological sector ν can be written in the form

$$Z_\nu^{(0)} = \int D[U_0]\,(\det U_0)^\nu \exp\left(m\Sigma V\, \mathrm{ReTr}\, U_0\right). \tag{5.201}$$

After this somewhat lengthy preparatory discussion, the connection between QCD in the ϵ-regime and chiral RMT can finally be established. An important result derived by Shuryak and Verbaarschot [109] states that the path integral $Z_\nu^{(0)}$ can be mapped exactly onto the partition function $\mathcal{Z}_\nu$ of RMT. One therefore expects that the low-lying eigenvalues of QCD in the ϵ-regime are distributed in the same way as those in RMT. By computing the former in a lattice simulation and performing a comparison to the analytically known distributions in RMT, one may verify the Banks–Casher scenario of spontaneous chiral symmetry breaking.

The Neuberger-Dirac operator D_N of Eq. (5.47) is ideally suited for this task. Since it satisfies the Ginsparg-Wilson relation, chiral symmetry is preserved at the level of the discretized theory. Furthermore, D_N can be shown to satisfy an exact index theorem, so that it sustains $|\nu|$ exact zero modes on gauge configurations with topological charge ν. This allows for an unambiguous identification of topological sectors to which the path integral $Z_\nu^{(0)}$ is restricted [110]. Therefore, the investigation of spontaneous chiral symmetry breaking is a prime example where it is absolutely vital that the lattice-regularized theory obeys the same symmetries that are present in the continuum.

Before we proceed we must elucidate the relation of the spectra of the random matrix $\hat{D}$ and the Neuberger-Dirac operator. While the eigenvalues of $\hat{D}$ are purely imaginary, the operator D_N is unitary, and hence its eigenvalues lie on a circle with radius $1/\bar{a}$ in the complex plane, centered around the point $1/\bar{a}$ on the real axis.

Fig. 5.17 Comparison of
simulation results for ratios of
eigenvalues with Random
Matrix Theory (horizontal
bars) in the sectors with
topological charge
$\nu = 0, 1, 2$ [111]

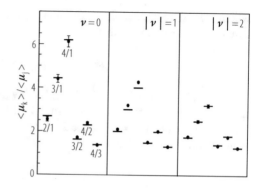

Thus, if γ denotes an eigenvalues of D_N, it can be parameterized as

$$\gamma = \frac{1}{a}\left(1 - e^{i\phi}\right), \qquad \bar{a} = \frac{a}{1+s}. \tag{5.202}$$

Since the radius of the circle diverges in the continuum limit, the low-lying part of
the spectrum satisfies $|\gamma| \ll 1/\bar{a}$, and hence $\mathrm{Re}\gamma \simeq 0$. One can then identify an
eigenvalue μ of $\hat{D}$ with $\mathrm{Im}\gamma$, i.e.

$$\mu \quad \leftrightarrow \quad \mathrm{Im}\gamma \simeq |\gamma| = \frac{1}{a}[2(1 - \cos\phi)]. \tag{5.203}$$

A simple but effective check of the RMT description of the low-lying spectrum can
be performed by comparing ratios of scaled eigenvalues. The combination $|\gamma_k|\Sigma V$
of the kth eigenvalue in QCD corresponds to $\mu_k N$ in RMT. If the low-lying spectra
in the two theories indeed coincide one expects the following equalities in a given
topological sector ν

$$\frac{\langle|\gamma_k|\rangle_\nu}{\langle|\gamma_j|\rangle_\nu} \stackrel{!}{=} \frac{\langle\mu_k\rangle_\nu}{\langle\mu_j\rangle_\nu} \equiv \int_0^\infty dz\, z\, p_k^{(\nu)}(z) \Big/ \int_0^\infty dz\, z\, p_j^{(\nu)}(z). \tag{5.204}$$

While the ratio $\langle|\gamma_k|\rangle_\nu/\langle|\gamma_j|\rangle_\nu$ is determined in the simulation, the two integrals on
the right-hand side can be evaluated analytically for the first few eigenvalues.[19]

In Refs. [111, 112] ratios for some of the lowest eigenvalues have been computed
in the quenched approximation. The results from [111] are shown in Fig. 5.17 for a
box size $L = 1.49$ fm. The agreement between lattice results and RMT is excellent.
By contrast, a smaller box size of about 1 fm yields significant discrepancies
between QCD and RMT, which can be as large as 10 standard deviations. This
is a reflection of the fact that the large volume limit must be taken before the RMT

---

[19]The expressions for the distributions $p_k^{(\nu)}(z)$ become rapidly more complicated as k increases,
so that one may have to resort to numerical evaluations of the integrals.

behaviour sets in. Similar findings have been reported for QCD with $N_f = 2$ flavours of dynamical overlap quarks [113].

The confirmation of the RMT prediction for the distribution of the low-lying eigenvalues supports the Banks–Casher scenario of spontaneous chiral symmetry breaking. In a subsequent step one may therefore extract the LEC Σ via the relation

$$\langle |\gamma_k| \rangle_\nu \Sigma V = \langle \mu_k N \rangle_\nu \equiv \int_0^\infty dz\, z\, p_k^{(\nu)}(z). \tag{5.205}$$

If Σ is identified with the expectation value of the scalar density, as suggested by the effective low-energy description of QCD, it must be related to a particular continuum scheme, like the $\overline{MS}$-scheme of dimensional regularization. If the regularization prescription obeys chiral symmetry, the corresponding renormalization factor, Z_S, satisfies

$$Z_S = Z_P = 1/Z_m. \tag{5.206}$$

where Z_m relates the bare quark mass to the chosen continuum scheme (for instance, $\overline{MS}$). Provided that Z_S, or equivalently, Z_m has been computed for a range of bare couplings, the lattice estimates for Σ can be used to determine the renormalized condensate in units of some scale, e.g.

$$r_0^3 \Sigma_{\overline{MS}}(\mu) = Z_S(g_0, a\mu) r_0^3 \Sigma + O(a^2). \tag{5.207}$$

For the Neuberger-Dirac operator, Z_S has been computed non-perturbatively in the quenched approximation [114], employing the technique outlined in Ref. [115]. The resulting values for Z_S could then be combined with the results for Σ extracted from the matching to RMT from [111]. A subsequent extrapolation to vanishing lattice spacing yields the results for the renormalized condensate in the continuum limit:

$$\Sigma_{\overline{MS}}(2\,\text{GeV}) = (285 \pm 9\,\text{MeV})^3, \qquad (\text{scale set by } f_K). \tag{5.208}$$

The quoted error represents the total uncertainty arising from statistics, the uncertainty in the renormalization factor, and the continuum extrapolation. If the nucleon mass is used to set the scale the central value drops to 261 MeV, as a consequence of the scale ambiguity encountered in the quenched approximation. We stress once more that the chiral condensate is ill-defined in the quenched theory, and thus great care must be taken when the results are interpreted in the context of the full theory. Nevertheless, it is encouraging that for $N_f = 2$ flavours of dynamical quarks, a similar calculation [113] finds $\Sigma_{\overline{MS}}(2\,\text{GeV}) = (251 \pm 7 \pm 11\,\text{MeV})^3$ at $a \simeq 0.11$ fm, in good agreement with the quenched result, given the inherent ambiguities and inconsistencies of the latter.

Lattice results for the condensate have been reported by many other authors (e.g. [116–125]), employing a variety of approaches. Although the various calculations are subject to different systematics, the overall picture is rather consistent, with values for the condensate centering around $(250 \, \text{MeV})^3$. As for many other quantities, the influence of lattice artefacts and renormalization effects must be studied in more detail, especially in the case of fully dynamical calculations. It is also important to mention that analytic non-perturbative approaches to the strong interaction, such QCD sum rules, also give broadly consistent results with lattice simulations within the quoted uncertainties (see e.g. [126–128] and references therein). This completes the consistent picture of chiral symmetry and its spontaneous breaking in QCD.

5.7　Hadronic Weak Matrix Elements

The experimental programme at the *B*-factories BaBar and Belle, as well as many other experiments at high-energy colliders, such as the Tevatron and LEP, have greatly enhanced the accuracy of many observables related to flavour physics and the Cabibbo–Kobayashi–Maskawa (CKM) matrix. The main motivation for studying flavour physics is to gain a proper understanding of CP violation and, in turn, the matter-antimatter asymmetry which is apparently manifest in the universe. CP violation is incorporated into the Standard Model via a complex phase in the CKM matrix, and therefore a precise knowledge of its elements is required to decide whether or not additional sources of CP violation must be considered.

In order to make these statements more precise we recall some basic definitions. As is well known, the CKM matrix V_{CKM} relates flavour to mass eigenstates. For flavour-changing charged current transitions between up- and down-type quarks this implies that, in addition to the dominant transitions like $u \leftrightarrow d$, $c \leftrightarrow s$ and $t \leftrightarrow b$, there are further transitions of lesser strength. The CKM matrix is therefore expected to possess a hierarchical structure, with the diagonal elements V_{ud}, V_{cs} and V_{tb} being of order one. An approximate parameterization that takes this into account is due to Wolfenstein [129]. By expanding V_{CKM} in powers of the Cabibbo-angle $|V_{us}| \equiv \lambda \simeq 0.22$ one obtains

$$V_{\text{CKM}} \equiv \begin{pmatrix} V_{ud} & V_{us} & V_{ub} \\ V_{cd} & V_{cs} & V_{cb} \\ V_{td} & V_{ts} & V_{tb} \end{pmatrix} \simeq \begin{pmatrix} 1 - \lambda^2/2 & \lambda & A\lambda^3(\rho - i\eta) \\ -\lambda - i A^2\lambda^5\eta & 1 - \lambda^2/2 & A\lambda^2 \\ A\lambda^3(1 - \bar{\rho} - i\bar{\eta}) & -A\lambda^2 - i A\lambda^4\eta & 1 \end{pmatrix},$$

$$(5.209)$$

with the remaining parameters A, $\bar{\rho}$ and $\bar{\eta}$ of order one.[20] In the standard model, V_{CKM} is unitary, and, provided that one can determine its elements with sufficient precision, any deviation from unitarity would be a signature of "new physics".

---

[20]The relation of rescaled parameter $\bar{\rho}$ to ρ is given by $\bar{\rho} = \rho(1 - \lambda^2/2 + O(\lambda^4))$, and a similar relation holds for $\bar{\eta}$ and η.

Fig. 5.18 Constraints on the apex of the unitarity triangle [130]

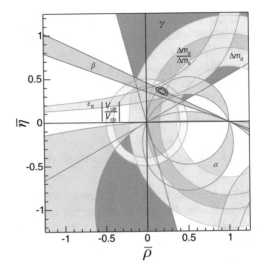

Unitarity gives rise to relations such as

$$V_{ud}V_{ub}^* + V_{cd}V_{cb}^* + V_{td}V_{tb}^* = 0, \qquad (5.210)$$

which can be represented by a triangle. The strategy that has been adopted in order to search for hints of new physics, is to use experimental and theoretical input to over-constrain the unitarity relations like those in Eq. (5.210). The current status is depicted in Fig. 5.18, where the unitarity triangle is plotted in the $(\bar{\rho}, \bar{\eta})$-plane [130].

The experimentally measured quantities, i.e. the mass differences ΔM_s, ΔM_d and ϵ_K, the latter of which parameterizes indirect CP violation in the kaon system, serve to constrain the apex of the unitarity triangle. They are proportional to the relevant CKM matrix elements, i.e.

$$\Delta M_d = \frac{G_F^2 M_W^2}{6\pi^2} \eta_B S\left(\frac{m_t}{M_W}\right) f_B^2 \hat{B}_B |V_{td}V_{tb}^*|^2, \qquad \frac{\Delta M_s}{\Delta M_d} = \frac{f_{B_s}^2 \hat{B}_{B_s}}{f_B^2 \hat{B}_B} \frac{m_{B_s}}{m_B} \frac{|V_{ts}|^2}{|V_{td}|^2},$$

$$\epsilon_K \propto \hat{B}_K \, \mathrm{Im}(V_{td}V_{ts}^*), \qquad (5.211)$$

where G_F is the Fermi constant, and M_W, m_t denote the masses of the W-boson and top quark, respectively. The proportionality factors in the above expressions involve the leptonic B-meson decay constants f_B and f_{B_s}, as well as the B-parameters $\hat{B}_B$, $\hat{B}_{B_s}$ and $\hat{B}_K$, which in turn parameterize the transition amplitudes for $B^0 - \bar{B}^0$, $B_s^0 - \bar{B}_s^0$, and $K^0 - \bar{K}^0$ mixing. While the decay constants are difficult to measure with sufficient accuracy, due to the fact that the leptonic decay rates are suppressed, the B-parameters are not at all accessible in experiment. One must therefore resort to theoretical estimates of these quantities. Since non-perturbative effects must inevitably be included, lattice simulations of QCD are ideally suited for this task.

Lattice calculations of weak hadronic matrix elements is a major activity within the lattice community, and a thorough coverage of all aspects would easily fill an entire chapter. We shall therefore concentrate on some of the most important quantities, and point out the main conceptual issues. It is strongly recommended that the reader consult the regular reviews of the topic at the annual lattice conferences, e.g. [131–135].

5.7.1 Weak Matrix Elements in the Kaon Sector

In the kaon sector, $K^0 - \bar{K}^0$ mixing is one of the most important processes. The B-parameter B_K parameterizes the non-perturbative contribution to indirect CP violation. It is defined by the ratio of the relevant operator matrix element to its value in the so-called "vacuum saturation approximation":

$$B_K(\mu) = \frac{\langle \bar{K}^0 | Q^{\Delta S=2}(\mu) | K^0 \rangle}{\frac{8}{3} f_K^2 m_K^2}. \tag{5.212}$$

Here, μ denotes the renormalization scale at which the $\Delta S = 2$ four-quark operator $Q^{\Delta S=2}$, defined by

$$Q^{\Delta S=2} = [\bar{s}\gamma_\mu(1-\gamma_5)d][\bar{s}\gamma_\mu(1-\gamma_5)d] \equiv O_{VV+AA} - O_{VA+AV}, \tag{5.213}$$

is considered. The relation between ϵ_K and the CKM matrix elements is provided by the RG-invariant B-parameter $\hat{B}_K$. In NLO perturbation theory $\hat{B}_K$ is related to $B_K(\mu)$ via

$$\hat{B}_K = \left(\frac{\bar{g}(\mu)^2}{4\pi}\right)^{\gamma_0/2b_0} \left\{1 + \bar{g}(\mu)^2 \left[\frac{b_0\gamma_1 - b_1\gamma_0}{2b_0^2}\right]\right\} B_K(\mu), \tag{5.214}$$

where γ_0, γ_1 denote the coefficients in the perturbative expansion of the anomalous dimension of $Q^{\Delta S=2}$. Since QCD is parity-conserving, the physically relevant operator in the above expression is the parity-even combination O_{VV+AA}. The typical left-handed chiral structure of this operator, which is characteristic for weak transitions, poses a problem for lattice calculations if Wilson fermions are employed. In this case the discretization breaks chiral symmetry explicitly, and thus O_{VV+AA} mixes under renormalization with operators involving the opposite chirality. Therefore, the general renormalization pattern is

$$O_{VV+AA}^R(\mu) = Z(g_0, a\mu) \left\{O_{VV+AA}^{bare} + \sum_{i=1}^{4} \Delta_i(g_0) O_i^{bare}\right\} \tag{5.215}$$

Thus, in order to determine the physical matrix element, one must not only determine the overall renormalization factor Z, but also the mixing coefficients Δ_i. Several techniques have been developed [136–138] to address this problem, which is merely an inconvenience rather than a serious obstacle. In a formulation based on staggered fermions the problem is absent, since the remnant $U(1) \otimes U(1)$ symmetry protects the operator from mixing with other chiralities. However, a drawback of the staggered formulation is the broken flavour ("taste") symmetry, which may lead to significant complications [139]. Fermionic discretizations based on the Ginsparg-Wilson relation, such as domain wall or overlap fermions do not suffer from the mixing problem, whilst preserving all flavour symmetries. Finally, the mixing problem can also be circumvented for Wilson-like discretizations in the context of twisted-mass QCD [140, 141]. With the help of a suitably chosen flavour rotation (see Eq. (5.51)), the matrix element of O_{VV+AA} in QCD can be mapped exactly onto that of the parity-odd operator O_{VA+AV} in the chirally twisted theory, viz.

$$\left\langle \bar{K}^0 \left| O_{VA+AV}^{bare} \right| K^0 \right\rangle_{tmQCD} = i \left\langle \bar{K}^0 \left| O_{VV+AA}^{bare} \right| K^0 \right\rangle_{QCD}. \qquad (5.216)$$

It has been shown that O_{VA+AV} renormalizes purely multiplicatively [142], i.e. all mixing coefficients vanish. The overall multiplicative, scale-dependent renormalization factor of O_{VA+AV} which yields the physical matrix element has been determined non-perturbatively [143], using the finite-size scaling procedure based on the Schrödinger functional formalism described in Sect. 5.5.2.

We now give a summary of the current status of B_K. Here, the calculation by the JLQCD Collaboration [154], based on staggered quarks in the quenched approximation, has served as a benchmark result for a long time. Their result, for which the perturbatively renormalized matrix element was extrapolated to the continuum limit, has since been confirmed by many other calculations employing different fermionic discretizations and different renormalization techniques. These include domain wall [148, 149] and overlap quarks [150, 151], as well as the Wilson formulation [153, 155]. Moreover, a calculation employing twisted mass QCD has been completed [152], which includes non-perturbative renormalization and a thorough investigation of the continuum limit.

Recently, results for B_K from simulations with dynamical quarks have become available, both for $N_f = 2$ [146, 147] and $N_f = 2 + 1$ flavours [144, 145]. A compilation of quenched and unquenched results is shown in Fig. 5.19. Although the figure suggests a trend in the data which points to slightly lower estimates for $\hat{B}_K$ if dynamical quarks are switched on (see Fig. 5.19), the quoted uncertainties are still too large to point to a significant deviation. In particular, a systematic study of the continuum limit in the unquenched case is not yet available. It is interesting to compare the results for $\hat{B}_K$ to the non-lattice determination in Ref. [130]. Here, the determinations of the angles of the unitarity triangle from experimental data in conjunction with direct measurements of ΔM_d, ΔM_s and ϵ_K allow to fit the values of several of the quantities in Eq. (5.211), which incorporate the hadronic

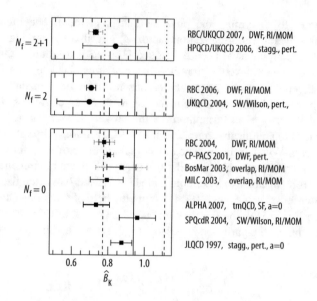

Fig. 5.19 Recent lattice results for the RGI kaon B-parameter $\hat{B}_K$. From top to bottom, the plotted values are taken from Refs. [144–154]. Dotted error bars (where shown) indicate the quoted systematic error. The labels include information on the fermionic discretization and the intermediate renormalization scheme, if non-perturbative renormalization was used. We also indicate whether or not the results have been extrapolated to the continuum limit. The vertical lines represent the (non-lattice) result from [130], with the quoted uncertainty (see text)

uncertainties. In this way one obtains $\hat{B}_K^{\text{non} - \text{lattice}} = 0.94 \pm 0.17$, which is shown as the vertical band in Fig. 5.19. Clearly, within the rather large error margins, this result is compatible with all lattice determinations, quenched or unquenched.

First Row Unitarity and the Value of $|V_{us}|$ In addition to Eq. (5.210), the unitarity of the CKM matrix implies many other constraints on its elements, such as those which appear in the first row:

$$|V_{ud}|^2 + |V_{us}|^2 + |V_{ub}|^2 = 1. \tag{5.217}$$

Owing to the smallness of $|V_{ub}|$, i.e. $|V_{ub}|^2 \simeq 2 \cdot 10^{-5}$, the direct verification of first row unitarity with the current experimental and theoretical accuracy rests on the precise knowledge of $|V_{ud}|$ and $|V_{us}|$. The value of $|V_{ud}|$ can be determined with high accuracy from super-allowed nuclear β-decays ($0^+ \rightarrow 0^+$ transitions), and in the current edition of the particle data book the best estimate is quoted as [61]

$$|V_{ud}| = 0.97377 \pm 0.00027. \tag{5.218}$$

The value of $|V_{us}|$ can be extracted from the decay rate of $K_{\ell 3}$ transitions, i.e.

$$\Gamma(K \to \pi \ell \nu_\ell) \propto \frac{G_F^2 m_K^5}{192 \pi^3} |V_{us}|^2 \left| f_+^{K\pi}(0) \right|^2, \qquad (5.219)$$

where $f_+^{K\pi}$ is one of the two form factors which parameterize the hadronic matrix element for semi-leptonic $K \to \pi \ell \nu_\ell$ transitions, i.e.

$$\langle \pi(\vec{p}_\pi) | (\bar{s} \gamma_\mu u)(0) | K(\vec{p}_K) \rangle = f_+^{K\pi}(q^2)(p_K + p_\pi)_\mu + f_-^{K\pi}(q^2)(p_K - p_\pi)_\mu,$$

$$q_\mu = (p_K - p_\pi)_\mu. \qquad (5.220)$$

In order to arrive at a precise estimate for $|V_{us}|$, $f_+^{K\pi}(q^2)$ must be determined with an accuracy at the level of 1%, since the decay rate and hence the combination $|V_{us}|^2 [f_+^{K\pi}]^2$ can be measured rather precisely. The form factor $f_+^{K\pi}$ admits a chiral expansion; At zero momentum transfer it reads

$$f_+^{K\pi}(0) = 1 + f_2 + f_4 + \ldots . \qquad (5.221)$$

While the leading chiral correction, $f_2 = -0.023$, has been computed long ago [156], knowledge on f_4 and the higher corrections is still fairly limited. The strategy pursued in lattice calculations [157] is based on computing the quantity

$$\Delta f \equiv f_+^{K\pi}(0) - (1 + f_2), \qquad (5.222)$$

which is a measure of the contributions beyond leading order. An old phenomenological estimate by Leutwyler and Roos [158] yields the value $\Delta f = -0.016(8)$. It is clearly desirable to check this result and ultimately replace it by a model-independent estimate based on QCD.

Semi-leptonic form factors can be determined in lattice simulations by computing suitable three-point correlation functions, in which the initial and final hadronic states are projected onto non-vanishing momentum. The main issues that must be addressed in order to judge the accuracy of the form factor determination are listed in the following:

- The dependence of the form factors on the momentum transfer q^2 must be modelled, in order to interpolate their values to $q^2 = 0$. Typical *ansätze* for the interpolation include linear or quadratic functions of q^2, as well as formulae based on pole dominance [159]. The freedom of choosing a particular *ansatz* introduces a certain ambiguity, since different model functions yield slightly different results. Via the introduction of so-called twisted boundary conditions [159–164], the q^2 resolution of form factors can be significantly improved;
- As for all quantities involving pions, a chiral extrapolation of lattice results must be performed. Clearly, in order to obtain $f_+^{K\pi}(0)$ and hence Δf with small controlled errors, a reliable chiral extrapolation is perhaps the single most

Table 5.4 Recently published lattice results for the quantity Δf

Collaboration	N_f	Action	$f_+^{K\pi}(0)$	Δf	m_π^{min} [MeV]
Bećirević et al. [157]	0	Clover	0.960(5)(6)	−0.017(5)(7)	490
RBC [165]	2	DWF	0.968(9)(6)	−0.009(9)(6)	490
UKQCD/RBC [166]	2 + 1	DWF	0.964(5)	−0.013(5)	330
Leutwyler and Roos [158]	./.	./.	0.961(8)	−0.016(8)	./.
Bijnens and Talavera [167]	./.	./.	0.976(10)	−0.001(10)	./.
Jamin et al. [168]	./.	./.	0.974(11)	−0.003(11)	./.
Cirigliano et al. [169]	./.	./.	0.984(12)	0.007(12)	./.

The minimum value of the pion mass used in the simulations is listed in the right-most column. The lower part of the table contains analytical estimates

important issue. Thus, the ability to simulate as deeply as possible in the chiral regime will be decisive for the final accuracy;

- Other systematic uncertainties include control over lattice artefacts, which is closely related to the renormalization of local operators, such as the vector current, which appears in Eq. (5.220). If chiral symmetry is broken explicitly, the (local) vector current is not conserved, and in order to guarantee a smooth approach to the continuum limit, its renormalization factor, Z_V, must be included. However, in all recent simulations the form factor has been extracted from suitably chosen ratios in which Z_V drops out.

A compilation of recent results for the form factor $f_+^{K\pi}(0)$ and the quantity Δf are presented in Table 5.4, where they are compared to analytical estimates. The agreement with the old result by Leutwyler and Roos is quite striking. Despite a tendency among the more recent analytical calculations to produce slightly larger estimates for Δf, all results are in good agreement within the quoted uncertainties.

An alternative method to determine $|V_{us}|$ from experimental data was proposed by Marciano [170]. Instead of considering semi-leptonic decays, it is based on the leptonic decay rates, i.e.

$$\frac{\Gamma(K \to \mu\bar{\nu}_\mu(\gamma))}{\Gamma(\pi \to e\bar{\nu}_e(\gamma))} \propto \frac{|V_{us}|^2}{|V_{ud}|^2} \frac{f_K^2 m_K}{f_\pi^2 m_\pi}. \tag{5.223}$$

Hence, the task is to provide an input value for the ratio of decay constants, f_K/f_π. This quantity is well-suited for lattice calculations in several respects: first, ratios of quantities can be computed with high statistical accuracy, owing to the fact that the fluctuations in the numerator and denominator are correlated. Second, the renormalization factor of the axial current, Z_A, drops out in the ratio f_K/f_π. However, since the quantity of interest involves a chiral extrapolation, the same caveats as in the case of the pion form factor, apply in this case. In particular, it is mandatory to go as close as possible to the physical mass of the pion. The quenched approximation is clearly of very limited value in this context, since the chiral behaviour and hence the actual value of f_K/f_π may strongly depend on the

Table 5.5 Recently published results for f_K/f_π in lattice QCD with dynamical quarks

Collaboration	N_f	Action	f_K/f_π	$m_\pi^{\min}$ [MeV]
CP-PACS [52]	2	Clover	1.19(3)	500
JLQCD [172]	2	Clover	$1.148(11)(^{12}_5)(^2_3)$	500
ETM [100]	2	tmQCD	1.227(9)(24)	290
MILC [173]	2 + 1	Stagg.	$1.208(2)(^7_{14})$	290
NPLQCD [174]	2 + 1	Stagg./DWF	$1.218(2)(^{11}_{24})$	290
RBC/UKQCD [175]	2 + 1	DWF	1.24(2)	330
HPQCD [176]	2 + 1	Stagg.	1.189(7)	250

number of active sea quarks. Furthermore, it is known that in the continuum limit of the quenched approximation the value f_K/f_π is underestimated by about 10% [171].

Recent results for f_K/f_π in lattice QCD with dynamical quarks are listed in Table 5.5. A caveat that applies to all such compilations is that systematic errors are not estimated in a uniform manner. For instance, none of the listed results (with the exception of [52]) is based on a systematic scaling study aimed at separating cutoff effects from the actual mass dependence, although the influence of lattice artefacts has been included in some error estimates by including cutoff effects into a generalized chiral fit. Moreover, not all of the listed values of f_K/f_π include finite-volume corrections, which can be computed in ChPT and incorporated into the *ansatz* for the chiral fit [177, 178]. Despite these caveats it appears, though, that the estimates for f_K/f_π based on fits including pion masses well below 500 MeV are compatible with each other.

5.7.2 Weak Matrix Elements in the Heavy Quark Sector

The main obstacle for calculations of weak matrix elements involving heavy quarks, and in particular the b-quark, is that one is faced with a multi-scale problem. In Sect. 5.2.5 we have already discussed systematic effects in lattice calculations that arise from finite-size effects and lattice artefacts. Translating the relations in (5.79) directly to the b-quark sector, one finds that the following inequalities cannot be satisfied simultaneously, at least not with the currently available computer power:

$$am_b \ll 1, \quad m_\pi L \gg 1, \quad L/a \lesssim 50. \tag{5.224}$$

Violation of the first relation implies the presence of large lattice artefacts, the second inequality must be satisfied if one wants to avoid uncontrolled finite-volume effects, and the third is dictated by memory capacities of current computers. With a b-quark mass of $m_b \approx 4$ GeV and typical inverse lattice spacings of $a^{-1} \lesssim 4.5$ GeV, it is evident that the b-quark cannot be studied directly, since its Compton wavelength is smaller or of the same order of magnitude than the lattice spacing itself.

Several strategies to deal with this problem have been applied over many years, among them the "static approximation" [179], the non-relativistic formulation (NRQCD) [180], the so-called "Fermilab-approach" [181] and finite-size scaling techniques [182, 183].

Since the charm quark is lighter than the b-quark by roughly a factor three, one may attempt to treat charm as a fully relativistic, propagating quark in simulations. Still, one can incur large lattice artefacts in this way, and a careful extrapolation to the continuum limit is then required. However, such an extrapolation may be spoilt if the leading lattice artefacts cannot be isolated in the results, due to the relatively large mass of the charm quark. Still, if one has reason to trust the results obtained for relativistic charm quarks, one may extrapolate them to the mass of b-quark, which is yet another way of circumventing the problem that the b-quark is too heavy to be treated relativistically. Typically, the *ansatz* for the extrapolation of a particular quantity to the mass of the b-quark is motivated by its expected quark mass dependence in Heavy Quark Effective Theory (HQET).

In the static approximation the b-quark is assumed to be infinitely heavy [179]. In this formalism it is convenient to represent the b-quark by a pair of spinors, $(\psi_h, \psi_{\bar{h}})$, which propagate forward and backward in time, respectively, and which satisfy

$$P_+\psi_h = \psi_h, \quad P_-\psi_{\bar{h}} = \psi_{\bar{h}}, \qquad P_\pm = \tfrac{1}{2}(1 \pm \gamma_0). \tag{5.225}$$

While the field ψ_h annihilates a heavy quark, $\psi_{\bar{h}}$ creates a heavy antiquark. The dynamics of these fields in the discretized version of the theory is described by the Eichten-Hill action [184]

$$S^{\text{stat}} = a^4 \sum_x \left\{ \mathcal{L}_h^{\text{stat}} + \mathcal{L}_{\bar{h}}^{\text{stat}} \right\}, \quad \mathcal{L}_h^{\text{stat}} = \bar{\psi}_h(x)\nabla_0^*\psi_h(x), \quad \mathcal{L}_{\bar{h}}^{\text{stat}} = -\bar{\psi}_{\bar{h}}(x)\nabla_0\psi_{\bar{h}}(x),$$
$$\tag{5.226}$$

where ∇_0, ∇_0^* denote the forward and backward covariant lattice derivatives in the temporal direction. Although the numerical computation of the quark propagator based on the Eichten-Hill action is relatively "cheap", simulation results in the static approximation typically suffer from relatively large statistical noise. Without going into detail we note that the signal-to-noise ratio can be significantly improved if one replaces the temporal link variables in ∇_0 and ∇_0^* by suitably chosen generalized parallel transporters. A full account can be found in Ref. [185].

Obviously, the static approximation represents only the leading term in an expansion of the quark action in inverse powers of the heavy quark mass, and thus one expects corrections in powers of $1/m_h$. As described in Ref. [182], one can set up a formalism in which the leading corrections to physical observables can be systematically computed as operator insertions in correlation functions defined with respect to the static action S^{stat}. Again, we refrain from describing any further details and refer the reader to the original literature [182, 186].

Higher-order corrections to the static approximation can also be incorporated into the theory by adding the appropriate $1/m_h$ terms to the action itself. In this way one obtains a non-relativistic version of QCD (NRQCD) [180], in which the mass of the heavy quark is imposed as a cutoff on relativistic momentum modes, i.e.

$$p \sim m_h v \ll m_h, \tag{5.227}$$

where v denotes the four-velocity of the heavy quark. Heuristically, the introduction of the cutoff is justified since the internal typical momentum modes of hadrons containing a heavy quark are much smaller than the mass of the latter. The loss of relativistic states can be compensated by adding new local interaction terms order by order in $p/m_h \sim v$ to $\mathcal{L}_h^{\text{stat}}$ and $\mathcal{L}_{\bar{h}}^{\text{stat}}$. In general, these additional interaction terms will generate mixing between quark and antiquark. However, by applying a Foldy-Wouthuysen transformation, the fields can be decoupled. At the level of the classical theory, the $1/m_h$ correction to the NRQCD Lagrangian for the forward propagating field reads

$$\mathcal{L}_h^{(1);\,\text{class}} = -\frac{1}{2m_h} \left\{ \bar{\psi}_h \boldsymbol{D} \cdot \boldsymbol{D} \psi_h + \bar{\psi}_h \boldsymbol{\sigma} \cdot \boldsymbol{B} \psi_h \right\}, \tag{5.228}$$

and $\boldsymbol{D}$ is the vector of the covariant derivatives in the spatial directions.

In the quantized version of the theory, the coefficients which multiply the fields in the above expression become dependent on the gauge coupling and must be appropriately tuned to guarantee the correct matching of the non-relativistic theory to standard QCD at order in $1/m_h$. Thus, the lattice-regularized version of the $1/m_h$ correction reads

$$\mathcal{L}_h^{(1)} = -\left\{ \omega_1 \bar{\psi}_h \boldsymbol{\nabla} \cdot \boldsymbol{\nabla} \psi_h + \omega_2 \bar{\psi}_h \boldsymbol{\sigma} \cdot \hat{\boldsymbol{B}} \psi_h \right\}, \tag{5.229}$$

where $\hat{\boldsymbol{B}}$ denotes a lattice representation of the magnetic field. The coefficients ω_1 and ω_2 are formally of order $1/m_h$ and are found to be linearly divergent in the lattice spacing a. Therefore, at a given order in the non-relativistic expansion of the action, a finite cutoff must be kept, and in this sense the effective theory is non-renormalizable. All this implies that in NRQCD the continuum limit, $a \to 0$, cannot be taken. Instead, one must argue that lattice artefacts are small in the range of lattice spacings where the calculations are performed.

Another approach can be based on the idea that the Wilson fermion action is suitably adapted for heavy quarks, such that the Wilson quark propagator does not deviate from the continuum behaviour even for quark masses $am \gtrsim 1$, i.e. for quark masses near or above the cutoff [181]. According to Ref. [181] this can be achieved by modifying the normalization of the quark fields (see Eq. (5.36)) in the discretized lattice theory, i.e.

$$\psi(x) \to \sqrt{2\kappa}\, e^{am_{\text{P}}/2} \psi(x), \qquad \bar{\psi}(x) \to \bar{\psi}(x)\, e^{am_{\text{P}}/2} \sqrt{2\kappa}, \tag{5.230}$$

where the "pole mass" am_P of the Wilson propagator is given by

$$am_P = \ln(1 + am), \qquad (5.231)$$

and am denotes the bare subtracted quark mass in the Wilson theory (see Eq. (5.39)). The factor $\sqrt{2\kappa}\, e^{am_P/2}$ is designed to interpolate smoothly between the relativistic and non-relativistic regimes. As a consequence, in order to cancel the effects of large quark masses in hadronic matrix elements involving b-quarks, the normalization of quark fields is modified according to the above prescription. The so-called "Fermilab approach" to heavy quark physics on the lattice is based on the normalization in Eq. (5.230). Essentially it amounts to formulating an effective theory for quarks, whose spatial momenta are small, $|a\vec{p}| \ll 1$, with mass-dependent coefficients. Like in the case of the static approximation, the formalism allows to take the continuum limit. Related approaches to the Fermilab method have been presented in Refs. [187, 188].

Finally, we briefly introduce another strategy to deal with heavy quarks on the lattice and the related multi-scale problem [182, 183]. Here the condition $m_\pi L \gg 1$ in Eq. (5.224) is sacrificed in favour of $am_b \ll 1$. In this way one is able to accommodate a fully relativistic b-quark at the expense of having to deal with strong finite-volume effects. The key observation is that the "distortion" due to unphysically small volumes can be computed in a series of finite-size scaling steps, which relate the results obtained on a sequence of lattice sizes $L_0, L_1, \ldots$. Like in the case of the non-perturbative determination of the RG running of the coupling and the quark mass discussed in Sect. 5.5.2, one can set up a recursive finite-size scaling procedure, which traces the volume dependence of observables. Here it is mostly sufficient to apply two or three steps in the scaling sequence.

In the remainder of this section we shall discuss some selected results. Regarding the vast number of individual results, we do not attempt to provide a complete review of the current status of lattice calculations of weak matrix elements in the heavy quark sector. Regular appraisals of the progress made in studying these systems can be found in the rapporteur talks on the subject at the annual conferences on lattice field theory [132, 133, 135]. Instead we shall discuss the relation between CKM matrix elements and the quantities that must be computed in order to extract the former from experimental data without resorting to model assumptions.

Heavy-Light Decay Constants From Eq. (5.211) and Fig. 5.18 one infers that the ratio $\xi \equiv f_{B_s}\sqrt{\hat{B}_{B_s}}/f_B\sqrt{\hat{B}_B}$ of decay constants and B-parameters is a key quantity, since it links $\Delta M_s/\Delta M_d$ to the ratio $|V_{ts}|^2/|V_{td}|^2$ of CKM matrix elements. Typically, one determines decay constants and B-parameters separately, since the former can be easily extracted from hadronic two-point functions, while the latter may undergo complicated mixing patterns, depending on the fermionic discretization. The decay constant of, say, a B^+ meson, is defined via the matrix element of the heavy-light axial current, i.e.

$$f_B m_B = \langle 0 | (\bar{u}\gamma_0\gamma_5 b) | B^+ \rangle. \qquad (5.232)$$

If the matrix element on the right-hand side is computed in a lattice simulation, then the axial current defined in the discretized theory must be matched to its counterpart in the continuum formulation. The details of the matching procedure depend on the type of fermionic discretization and the chosen treatment to represent the heavy-light axial current on the lattice (e.g. static approximation, NRQCD, etc.). If the b-quark is treated in the static approximation, the axial current has a non-vanishing anomalous dimension, and hence its running must be determined as well. Therefore, the various techniques which have been developed to compute the renormalization factors of local operators non-perturbatively, are of particular relevance also in the study of heavy-light decay constants [189]. In particular, non-perturbative estimates for the renormalization factor of the axial current, Z_A, are required to ensure a smooth convergence towards the continuum limit.

We now present results for f_B and f_{B_s}. From Chiral Perturbation Theory one expects that the bulk of the SU(3)-flavour breaking effect in ξ (i.e. the deviation of ξ from unity) is carried by the decay constants. The full expression at NLO for f_{B_s}/f_B reads [190]

$$\frac{f_{B_s}}{f_B} - 1 = (m_K^2 - m_\pi^2) f_2(\mu) - \frac{1+3g^2}{(4\pi f_\pi)^2} \left[\frac{1}{2} I_P(m_K) + \frac{1}{4} I_P(m_\eta) - \frac{3}{4} I_P(m_\pi) \right],$$
$$(5.233)$$

where $I_P(m_{PS}) = m_{PS}^2 \ln(m_{PS}^2/\mu^2)$ and f_2 is a low-energy constant, and g^2 is the strength of the $B^*B\pi$ vertex. As was pointed out by Kronfeld and Ryan [191], the contribution from the chiral logarithms can be sizeable, so that a naïve linear extrapolation of lattice data from the region of the strange quark mass tends to underestimate f_{B_s}/f_B. By contrast, the corresponding ratio B_{B_s}/B_B is expected to be close to one, since the coefficient of the chiral logarithm nearly vanishes. Since f_{B_s}/f_B enters directly into fits to the CKM parameters, many attempts were made to pin down its value precisely. As in the case of f_K/f_π discussed earlier, the main issue for lattice calculations is whether the quark masses employed in simulations are small enough to allow for a controlled chiral extrapolation based on the NLO formulae. The influence of the chiral logarithms has so far been detected only in simulations based on $N_f = 2+1$ flavours of rooted staggered quarks. Using NRQCD to treat the b-quark, the authors of [192] find

$$\frac{f_{B_s}}{f_B} = 1.20 \pm 0.03 \pm 0.01, \qquad N_f = 2+1, \qquad (5.234)$$

where the first error is statistical, while the second is an estimate of the systematic uncertainty. This result awaits confirmation from simulations with sea quark masses as small as those used in [192], but employing different fermionic discretizations, both in the sea and valence quark sectors. This is of particular relevance, since the typical spread among the recently published results is of the same order or even larger than the uncertainty quoted above. Further discussions and compilations of lattice data for f_{B_s}/f_B can be found in [133, 193].

Estimates for absolute values of heavy-light decay constants are also highly desirable, especially since f_B is hard to determine experimentally, even at the B-factories, since the $B \to \tau \nu_\tau$ decay rate is suppressed. For f_{B_s} the suppression is even stronger, and thus the prospects for an experimental determination of this quantity are extremely uncertain. The main issues facing lattice calculations are the influence of lattice artefacts in conjunction with the renormalization of the axial current, and the dependence of results on the number of dynamical quark flavours.

As an example for one of the most advanced quenched calculations for f_{B_s} we shall briefly discuss the result by the ALPHA collaboration [198], which also illustrates the interplay between various methods to treat the b-quark. In Ref. [198] the results obtained in the static approximation were combined with data computed around the mass of the charm quark. Provided that estimates for the decay constants in both datasets have been extrapolated to the continuum limit, a subsequent interpolation in the heavy quark mass yields the desired result for f_{B_s}. The *ansatz* for the interpolation is based on the expression

$$f_{PS}\sqrt{m_{PS}} = C_{PS}(M/\Lambda_{\overline{MS}})\, \gamma \left(1 + \frac{\delta}{m_{PS}}\right), \tag{5.235}$$

where f_{PS} is a generic heavy-light decay constant, γ, δ are real constants, and the factor C_{PS} arises from the matching between the static approximation and QCD with fully relativistic quarks. Thus, using the static approximation as the limiting case removes the systematic error due to the uncontrolled extrapolation to the mass of the b-quark. The resulting estimate for f_{B_s} is [198]

$$f_{B_s} = 193 \pm 6\,\text{MeV}, \qquad N_f = 0. \tag{5.236}$$

Non-perturbative renormalization has been employed in both the static approximation and the relativistic formulation. Except for the unknown systematic error due to quenching, the quoted error contains all uncertainties. The above result has been confirmed by the approach based on the finite-size scaling method [199].

Turning now to unquenched simulations, we compare the above value to the result by the HPQCD Collaboration [192], which was obtained using NRQCD for the b-quark, while $N_f = 2 + 1$ rooted staggered quarks were used as sea quarks. Here, the estimate for f_{B_s} results from a combination of the value for f_B and the ratio f_{B_s}/f_B already quoted in Eq. (5.234). In this way one obtains

$$f_{B_s} = 259 \pm 32\,\text{MeV}. \tag{5.237}$$

Thus, in spite of the large error, it appears that the inclusion of dynamical quark effects increases the estimate for heavy-light decay constants. This is also supported by other simulations. For instance, using their simulation results in quenched QCD

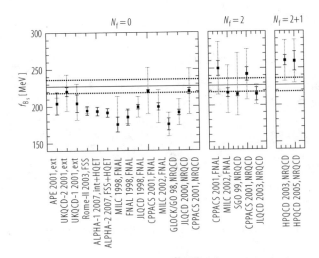

Fig. 5.20 Recent lattice results for f_{B_s}. From left to right the results are taken from the following papers. $N_f = 0$: [194–207]; $N_f = 2$: [203, 204, 207, 208], [209]; $N_f = 2 + 1$: [192, 210]. The labels indicate the method used to treat the b-quark in the simulation ("ext" and "int" stand for extrapolations and interpolations to the mass of the b-quark, respectively). The horizontal lines represent the (non-lattice) result from [130], with the quoted uncertainty

and with $N_f = 2$ flavour of dynamical Wilson quarks, the CP-PACS Collaboration find [203]

$$\frac{f_{B_s}^{N_f=2}}{f_{B_s}^{N_f=0}} = 1.14 \pm 0.05. \tag{5.238}$$

The "non-lattice" determination of f_{B_s} via fits using the experimental results for the angles of the unitarity triangle as input [130] also point to a larger value compared to the quenched theory, as can be seen from the horizontal band in the compilation in Fig. 5.20.

In current unquenched simulations, systematic effects such as lattice artefacts and the renormalization of local operators are not yet controlled at a similar level compared to the quenched approximation. Thus, despite the fact that these calculations are much more "realistic" in that they include sea quarks, the quoted overall uncertainties are still relatively large.

***B*-Parameters $\hat{B}_{B_d}$ and $\hat{B}_{B_s}$** Following the recent experimental determination of the mass difference ΔM_s at the Tevatron [211, 212], lattice determinations of the B-parameters $\hat{B}_{B_d}$ and $\hat{B}_{B_s}$ have received much attention. Although the first calculations date back to the 1990s, relatively few results are available, due to several specific technical difficulties. First, the complicated renormalization and mixing patterns of four-quark operators which afflict lattice calculations of the kaon B-parameter $\hat{B}_K$ are also encountered in the b-quark sector. Second, there is the

Table 5.6 Published lattice results for the B-parameter $B_{B_d}(m_b)$ in the $\overline{\text{MS}}$-scheme and the ratio B_{B_s}/B_{B_d}

Collaboration	N_f	Action	$B_{B_d}(m_b)$	B_{B_s}/B_{B_d}	
Gadiyak et al. [213]	2	DWF		1.06(6)(3)	static
JLQCD [209]	2	Clover	$0.836(27)\binom{56}{62}$	$1.017(16)\binom{56}{17}$	NRQCD
Gimenez et al. [214, 215]	0	Clover	0.81(5)(4)	1.01(1)	Static
UKQCD [215, 216]	0	Clover	0.79(4)(4)	1.02(2)	Static
Christensen et al. [217]	0	Wilson	0.98(4)	0.99(1)(1)	Static
SPQ$_{cd}$R [218, 219]	0	Clover	$0.87(4)\binom{5}{4}$	0.99(2)	rel./ext.
UKQCD [195]	0	Clover		$0.98(2)\binom{0}{2}$	rel./ext.

The method to treat the heavy quark is specified in the last column

added complication which arises from the fact that the b-quark cannot be simulated directly.

In Table 5.6 we list published results for $B_{B_d}(m_b)$ and the ratio B_{B_s}/B_{B_d} from a variety of methods to treat the heavy quark. The table shows that all results are broadly consistent with each other at the level of 10%, despite the different systematics. Moreover, none of the listed estimates is based on non-perturbative renormalization factors, and furthermore all entries have been computed for a fixed value of the lattice spacing, i.e. a systematic study of the continuum limit is lacking even in the quenched approximation. As for the ratio B_{B_s}/B_{B_d}, it should be mentioned that the quark masses in the simulations correspond to pion masses not much smaller than 500 MeV. However, in view of the fact that the bulk of the relevant SU(3)-flavour breaking effect in $\Delta M_s/\Delta M_d$ is expected to come from the ratio of decay constants, f_{B_s}/f_{B_d}, this may not be such a serious limitation. Results for B_{B_d} and B_{B_d} computed on dynamical gauge configurations with rooted staggered quarks should be published soon.

Another recent development is the implementation of non-perturbative renormalization for heavy-light four-quark operators in the static approximation [220, 221]. If the b-quark is treated in the static approximation, the $\Delta B = 2$ four-quark operator must be matched to its counterpart in the static theory, i.e.

$$Q^{\Delta B=2}(m_b) = C_{\text{L}}(m_b, \mu)\widetilde{Q}_1(\mu) + C_{\text{S}}(m_b, \mu)\widetilde{Q}_2(\mu), \qquad (5.239)$$

where

$$\widetilde{Q}_1 = \left(\bar{\psi}_h \gamma_\mu (1-\gamma_5)\ell\right)\left(\bar{\psi}_{\bar{h}} \gamma_\mu (1-\gamma_5)\ell\right) \equiv \widetilde{O}_{\text{VV+AA}} + \widetilde{O}_{\text{VA+AV}}$$

$$\widetilde{Q}_2 = \left(\bar{\psi}_h (1-\gamma_5)\ell\right)\left(\bar{\psi}_{\bar{h}} (1-\gamma_5)\ell\right) \equiv \widetilde{O}_{\text{SS+PP}} + \widetilde{O}_{\text{SP+PS}}, \qquad (5.240)$$

with ℓ denoting the light (d or s) flavour. For the physical matrix element only the parity-even operators $\widetilde{O}_{\text{VV+AA}}$ and $\widetilde{O}_{\text{SS+PP}}$ are relevant. If chiral symmetry is not preserved by the discretization, four-quark operators such as $\widetilde{O}_{\text{VV+AA}}$ undergo complicated mixing patterns under renormalization, which necessitate finite subtractions similar to those required for the operator O_{VV+AA} in Eq. (5.215). However, just as in the case of $K^0 - \bar{K}^0$ mixing, the parity-even operators $\tilde{O}_{VV+AA}$ and $\tilde{O}_{SS+PP}$ can be mapped onto their parity-odd counterparts $\tilde{O}_{VA+AV}$ and $\tilde{O}_{SP+PS}$ by a flavour rotation, which realizes the transition to tmQCD at maximal twist angle. Moreover, it can be shown [220] that the combinations

$$\tilde{O}_1' \equiv \tilde{O}_{VA+AV}, \qquad \tilde{O}_2' \equiv \tilde{O}_{VA+AV} + 4\tilde{O}_{SP+PS} \tag{5.241}$$

renormalize purely multiplicatively. The RG running of these operators, as well as the matching to hadronic schemes based on tmQCD have been determined non-perturbatively in the SF scheme for $N_f = 0$ [221] and $N_f = 2$ [222], which will eventually allow for a determination of $\hat{B}_{B_s}$ and $\hat{B}_B$ with full control over renormalization and discretization effects. Corrections of order $1/m_b$ can be taken into account through an interpolation between the results obtained in the static approximation and for relativistic heavy quarks with masses in the region of that of the charm quark.

Semi-Leptonic B-Decays The CKM elements $|V_{ub}|$ and $|V_{cb}|$, which appear in the unitarity triangle relation equation (5.210), can be extracted from both inclusive and exclusive B-meson decays. However, $|V_{ub}|$ is still one of the most poorly constrained CKM elements. Its value can be determined by combining lattice calculations of semi-leptonic form factors for exclusive decays such as $\bar{B}^0 \to \pi^+\ell^-\bar{\nu}_\ell$ with the experimentally measured decay rate. If the leptons are assumed to be massless, the latter yields the combination $[|V_{ub}| f_+(q^2)]^2$, while the form factor $f_+(q^2)$ can be extracted from the matrix element

$$\langle \pi(\vec{p}_\pi)|(\bar{b}\gamma_\mu u)(0)|B(\vec{p}_B)\rangle = \left[(p_B + p_\pi)_\mu - q_\mu \frac{m_B^2 - m_\pi^2}{q^2}\right] f_+(q^2) + q_\mu \frac{m_B^2 - m_\pi^2}{q^2} f_0(q^2). \tag{5.242}$$

Here, $q_\mu \equiv (p_B - p_\pi)_\mu$ denotes the momentum transfer. For a B-meson at rest one has

$$q^2 = m_B^2 + m_\pi^2 - 2m_B\sqrt{m_\pi^2 + \vec{p}_\pi^2}. \tag{5.243}$$

In order to avoid large lattice artefacts, typical values of the pion momentum in simulations are restricted to

$$|\vec{p}_\pi| \lesssim 1\,\text{GeV}. \tag{5.244}$$

Therefore, lattice calculations typically yield the form factors f_+ and f_0 near $q^2 = q_{max}^2$. By contrast, the bulk of the experimental data is recorded in bins with small values of q^2, since the decay rate is suppressed near q_{max}^2. Therefore, an extrapolation to small values of q^2 must be performed, which requires an

Fig. 5.21 Form factors f_+ (upper data set) and f_0 for $B \to \pi \ell \nu$ decays (taken from Ref. [224]). The data are taken from Refs. [225] (UKQCD), [226] (Abada et al.), [227] (El-Khadra et al.), [228] (JLQCD) and [229] (FNAL04). The unquenched results by HPQCD have been updated [230]

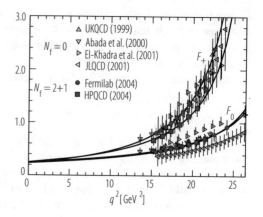

ansatz for the shape of the form factor. Although a parameterization of the q^2-dependence which goes beyond vector pole dominance and is also consistent with the expected heavy-quark scaling laws has been proposed [223], the extrapolation to small momentum transfers typically introduces some model dependence in the result for $|V_{ub}|$.

Figure 5.21 shows a compilation of lattice data for the form factors as a function of q^2 together with the curves which represent the extrapolations to $q^2 = 0$. The problem of the model dependence introduced by the extrapolation to small momentum transfer can be avoided by combining form factors from lattice simulations with the decay rate measured in restricted intervals of q^2, which overlap with the range of momentum transfers that are directly accessible in simulations. Such a procedure has been performed by the CLEO Collaboration [231]. The result for $|V_{ub}|$ obtained in this way is somewhat smaller compared to the standard method based on form factor extrapolations, but the uncertainties are still quite large. For the actual estimates of $|V_{ub}|$ obtained in this way, the reader may consult the original papers.

Semi-leptonic heavy-to-heavy decays such as $\bar{B} \to (D, D^*)\ell\bar{\nu}_\ell$ offer a way to determine $|V_{cb}|$. In this case it is convenient to use the four-velocities of the two mesons as the kinematical variables instead of the four-momenta. The decay amplitudes are then parameterized in terms of six form factors, i.e.

$$\frac{\langle D(v') |(\bar{c}\gamma^\mu b)| B(v)\rangle}{\sqrt{m_B m_D}} = (v + v')^\mu h_+(\omega) + (v - v')^\mu h_-(\omega)$$

$$\frac{\langle D^*(v', \epsilon) |(\bar{c}\gamma^\mu b)| B(v)\rangle}{\sqrt{m_B m_D^*}} = i\epsilon^{\mu\nu\alpha\beta}\epsilon_\nu^* v'_\alpha v_\beta h_V(\omega) \qquad (5.245)$$

$$\frac{\langle D^*(v', \epsilon) |(\bar{c}\gamma^\mu \gamma_5 b)| B(v)\rangle}{\sqrt{m_B m_D^*}} = (\omega + 1)\epsilon^{*\mu} h_{A_1}(\omega) - \epsilon^* \cdot v \left[v^\mu h_{A_2}(\omega) + v'^\mu h_{A_3}(\omega)\right],$$

where $\omega = v \cdot v'$. In the limit of infinite heavy quark mass, four out of these six form factors can be replaced by a single, universal form factor, $\xi(\omega)$, which is called the Isgur-Wise function [232]

$$m_b, m_c \to \infty \quad \Rightarrow \quad h_+(\omega) = h_{A_1}(\omega) = h_{A_3}(\omega) = h_V(\omega) = \xi(\omega), \qquad (5.246)$$

while $h_-(\omega)$ and $h_{A_2}(\omega)$ vanish as m_b, m_c become infinitely heavy. Outside the exact heavy-quark limit, the relation between the Isgur-Wise function and the form factors is modified. For instance,

$$h_+(\omega) = (1 + \beta_+(\omega) + \gamma_+(\omega))\, \xi(\omega), \qquad (5.247)$$

where β_+, γ_+ parameterize radiative corrections and corrections arising from operators of higher dimension, which are suppressed by additional inverse powers of the heavy quark mass. Similar relations hold for h_{A_1}, h_{A_3} and h_V. Another important result, known as Luke's Theorem [233], states that at zero recoil, $v = v'$, i.e. $\omega = 1$, the leading corrections to the form factors h_+ and h_{A_1} are quadratic in the inverse heavy quark mass.

With this setup one may devise a strategy to determine $|V_{cb}|$ by combining the experimentally determined decay rate with lattice calculations of the form factors. The differential decay rate for $\bar{B} \to D^* \ell \bar{\nu}_\ell$ in the limit of zero recoil reads

$$\lim_{\omega \to 1} \frac{1}{\sqrt{\omega^2 - 1}} \frac{d\Gamma(B \to D^* \ell \nu)}{d\omega} = |V_{cb}|^2 \frac{G_F^2}{4\pi^3}(m_B - m_{D^*})^2 m_{D^*}^3 [h_{A_1}(1)]^2,$$

$$(5.248)$$

which, owing to Luke's Theorem, receives corrections of order $1/m_c^2$ only. For $\omega > 1$ the single axial form factor h_{A_1} must be replaced by a linear combination of several form factors. Thus, the theoretical uncertainties appear to be controlled best at zero recoil. Since the rate is suppressed near $\omega = 1$, the measured decay rate must be extrapolated to that value to determine $|V_{cb}|$. Most of the published lattice calculations of the form factors and the Isgur-Wise function [234–238] are therefore focused on the determination of the slope of $\xi(\omega)$ at $\omega = 1$. The measured decay rate can then be extrapolated to zero recoil using a particular parameterization of $\xi(\omega)$, with its slope constrained via the lattice calculation. After taking radiative and power corrections into account, a value for $|V_{cb}|$ can be extracted.

A different but related strategy is to compute the form factors $h_+(1)$ and $h_{A_1}(1)$ directly via suitably chosen double ratios of hadronic matrix elements in which many systematic effects can be expected to cancel [239, 240]. Using the "Fermilab approach" for the heavy quarks in the quenched approximation, the authors of Ref. [240] find

$$h_{A_1}(1) = 0.913^{+0.024\ +0.017}_{-0.017\ -0.030}, \qquad (5.249)$$

where the first error is statistical, while the second represents an estimate of various systematic uncertainties added in quadrature. Again, this result can be combined with the experimental decay rate to determine $|V_{cb}|$. More details can be found in [240].

Most lattice studies of heavy-to-heavy semi-leptonic B-decays have been restricted to the quenched approximation. However, results for the form factors from dynamical simulations can be expected in the near future. Clearly, in order to have maximum impact on the determination of $|V_{cb}|$, systematic effects arising from lattice artefacts and the formulation used to treat the heavy quark must be controlled to a high degree.

5.8   Concluding Remarks

In this article we have introduced the lattice approach to QCD and discussed a variety of applications, which range from hadron spectroscopy, confinement, quark masses and the running coupling, to spontaneous chiral symmetry breaking and hadronic matrix elements for flavour physics. This illustrates not only the versatility of the lattice method, but also indicates that lattice calculations have become ever more important for making quantitative predictions in the notoriously difficult sector of non-perturbative QCD. Still, a great number of other applications have not even been covered here, including nucleon structure functions and form factors, calculations at finite temperature and/or chemical potential, or detailed investigations of the QCD vacuum structure.

That lattice calculations have reached this standing is owed to the enormous progress which been made in developing more efficient algorithms for dynamical fermions, better discretizations, as well as a number of new theoretical concepts such as non-perturbative renormalization. These developments, in conjunction with the availability of ever more powerful computers, shall allow for precise computations of many phenomenologically relevant quantities, which previously seemed virtually intractable.

5.9   Addendum: QCD on the Lattice

5.9.1   Introduction

Since the first edition of this article [241] the field of lattice QCD has undergone a huge transformation. While the actual methodology was well established at the time of writing (2007), few simulations employing dynamical quarks had produced results with controlled errors, having a direct impact on phenomenology and experiment. During the past ten years or so this has changed dramatically.

Simulations with light dynamical quarks, whose masses correspond to the physical value of the pion mass, have become the state of the art, and the effects of dynamical strange and charm quarks are now routinely included as well. In fact, lattice calculations of certain observables have reached (or are aiming for) a level of precision where the effects of the breaking of isospin symmetry can no longer be ignored. This necessitates that lattice QCD must account not only for the effects of unequal u and d quark masses but also for corrections due to electromagnetism, owing to the different electric charges of up- and down-type quarks.

In this context it is interesting to quote a remark by Ken Wilson, made at the 1989 International Conference on Lattice Field Theory [242]: "*I still believe that an extraordinary increase in computing power (10^8 is I think not enough) and equally powerful algorithmic advances will be necessary before a full interaction with experiment takes place.*" Given that, in 1989, the most powerful supercomputers could sustain 10 GFlops (i.e. 10^{10} floating point operations per second), Wilson's estimate was tantamount to requiring ExaFlops capabilities (10^{18} Flops) for lattice QCD to make an impact, a performance figure that has only been reached very recently by less than a handful of machines. The enormous progress that the field of lattice QCD has already seen over the past decade proves that Wilson's view was far too pessimistic.[21] For instance, results from lattice calculations for the decay constants and form factors of mesons and baryons containing heavy quarks are vital input for global analyses of observables in flavour physics, designed to constrain the elements of the Cabibbo–Kobayashi–Maskawa matrix. Furthermore, lattice QCD yields precise values for the masses of the light (u, d, s) quarks [244].

An impressive testimony to the importance of lattice QCD for the entire field of particle physics is the regular report provided by the Flavour Lattice Averaging Group (FLAG). Since its inception in 2007, FLAG has been charting the progress in lattice QCD, by collecting results for a range of phenomenologically relevant quantities. Taking inspiration from the Particle Data Group, FLAG assesses the quality of individual calculations and produces world averages by combining those results that satisfy a defined set of requirements regarding the overall control over systematic effects. Three editions of the FLAG report, published in 2010 [245], 2013 [246] and 2016 [247], have appeared until now, and a fourth one has been published in 2019 [248]. In fact, the current status of lattice calculations of many observables that have been reviewed in the first edition of this article can be found in these comprehensive reports.

This short review is organized as follows. In Sects. 5.9.2 and 5.9.3 we give an update of lattice calculations applied to hadron spectroscopy, weak hadronic matrix elements and the determination of Standard Model parameters such as quark masses and the strong coupling constant. These quantities were covered extensively in the original edition of [241]. Then, in Sect. 5.9.4 we extend the discussion to the determination of quantities that describe structural and other properties of the nucleon, such as form factors and the axial charge. Finally, in Sect. 5.9.5 we discuss lattice calculations of the hadronic contributions to the muon anomalous magnetic

---

[21] Even Wilson himself acknowledged, at least partially, that this was the case [243].

moment, which is a key quantity to study possible deviations from the Standard Model. The review concludes with a few remarks on the progress achieved over the past decade and an outlook for future calculations.

5.9.2 Hadron Spectroscopy

The calculation of the light hadron spectrum, i.e. the masses of the lowest-lying mesons and baryons has long been regarded a benchmark for lattice QCD. In the quenched approximation, i.e. in the absence of dynamical quarks, a significant deviation between the calculated spectrum and experiment at the level of 10–15% was observed. When the light hadron spectrum could eventually be accurately reproduced within the overall uncertainty after the inclusion of light dynamical quarks [249–252] (see Fig. 5.22), this was hailed as a major success of lattice QCD. Thanks to these milestone results, the credibility of lattice calculations was firmly established throughout the particle and hadron physics communities.

Calculations of the light hadron spectrum have since been further refined, by taking the effects of isospin breaking into account. Strong isospin breaking arises from the mass splitting between the u and d quarks, $m_u \neq m_d$. Since the electric charges of u and d quarks differ as well, electromagntic interactions are another source of isospin breaking. The formulation of QED on a lattice of finite volume poses considerable technical challenges since the photon is massless. There are several strategies to address the problem of the associated zero mode, and we refer the reader to recent reviews of the subject [253–255], which also serve as a guide to the literature.

After the inclusion of strong and electromagnetic isospin breaking effects, it became possible to perform another benchmark calculation, namely the accurate determination of the neutron-proton mass difference, as well as the mass splittings of other baryonic iso-multiplets [256–259]. The ability to determine isospin breaking effects arising from QED was also instrumental for calculations of the electromagnetic mass splittings of pions and kaons [260–265], which can be used to study violations of Dashen's theorem [266]. The latter states that the electromagnetic self-energies of the charged pions and kaons are identical, while those of their neutral

Fig. 5.22 The spectrum of the lowest-lying hadrons as computed in Ref. [250], to be compared to Fig. 5.5 of the original review [241]

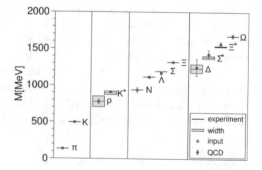

counterparts vanish. More details are found in section 3.1.1 of the FLAG report [247].

Another recent focus of lattice spectroscopy has been the determination of the excitation spectrum and the properties of hadronic resonances. This is a major refinement of previous calculations in which the masses of resonances (the simplest being the ρ-meson) were extracted naively from the exponential decay of the vector correlation function, thereby ignoring the fact that resonances are characterized both by a mass and a width. The general framework for the study of resonance properties in lattice QCD was developed by Lüscher already in the 1980s and 1990s [267–270], and it is only now that the potential of this elegant and powerful formalism can be fully exploited. The key idea that underlies the Lüscher method is the realization that computing the energy levels of multi-particle states in a finite volume gives access to the scattering phase shifts in infinite volume, provided that the spectrum (including excited states) can be determined sufficiently well for a range of kinematical situations. The latter are typically determined by the lattice volume and/or the total momentum of the multi-particle system in question.

To be more specific, let us consider the simplest resonance, the ρ-meson, whose properties can be accessed in p-wave $\pi\pi$ scattering. For energies below the inelastic threshold, the Lüscher condition reads

$$\phi(q) + \delta_1(k) = 0 \quad \mathrm{mod}\,\pi, \qquad q = \frac{kL}{2\pi}, \tag{5.250}$$

where $\phi(q)$ is a known kinematic function of the scaled scattering momentum in units of the box size, $q = kL/2\pi$ and δ_1 is the scattering phase shift. The scattering momentum k is determined from the nth energy level ω_n in a finite volume, according to

$$\omega_n = \sqrt{m_\pi^2 + k^2}, \tag{5.251}$$

where m_π is the pion mass. Figure 5.23 shows an example of a calculation of the p-wave scattering phase shift as a function of the centre-of-mass energy [271].

Fig. 5.23 The p-wave scattering phase shift of the ρ-meson, computed for $m_\pi = 280\,\mathrm{MeV}$ as a function of the centre-of-mass energy [271]. Data obtained for two different values of the lattice spacing (open and filled grey symbols) are shown. The solid line is obtained from a fit to a Breit-Wigner *ansatz* for the resonance

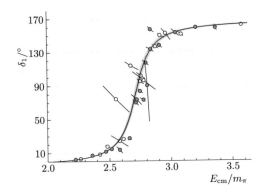

A crucial ingredient for the reliable determination of not just the energy level of the ground state but also the excitation spectrum is the use of correlator matrices computed using a suitable basis of interpolating operators (see Section 5.3 in Ref. [241]). The diagonalization of the correlator matrix can be achieved by solving a generalized eigenvalue problem from which the energy levels in a given channel can be determined [272–274]. The sometimes arduous task of constructing efficient interpolators for multi-particle states has been helped enormously by practical methods to compute "all-to-all" quark propagators [275] and, in particular, the so-called "distillation" technique [276, 277]. With these new developments it has been possible to perform lattice investigations of $\pi\pi$ scattering and the ρ resonance [278–291], as well as determinations of $K\pi$ [292, 293] and KK scattering lengths [294, 295]. The formalism has also been used to study meson-baryon [296–300] and baryon-baryon [301, 302] interactions.

While the original Lüscher formalism was derived for the case of elastic two-particle scattering, it has now been generalized to coupled-channel systems [303–307], including the treatment of three-particle thresholds [308–315]. It also opens the possibility to study weak non-leptonic kaon decays [316] and compute form factors for timelike momentum transfers [317–320].

Moreover, the experimental discovery of new charmonium-like resonances, commonly referred to as the X, Y and Z states, has kindled a new interest in hadron spectroscopy. A distinctive feature of the new resonances is their closeness to particle thresholds, and efforts are underway to gain a detailed understanding of the resonance structure in the charm sector. Using the formalism described above, there have been many calculations of a variety of charmonium-like resonances in lattice QCD. In view of the vast literature, we refer the reader to several recent reviews of the subject [321–323].

5.9.3 Parameters of the Standard Model

The Standard Model (SM) contains 19 parameters (excluding the neutrino sector) whose values are not predicted by the theory itself but must instead be fixed using experimental input. In many cases the relations between experimentally accessible observables and SM parameters involve quantities that encode the effects of the strong interactions. A well-known example is the kaon B-parameter B_K that enters the relation between the quantity ϵ_K, which is a measure of indirect CP violation, and a particular combination of Cabibbo–Kobayashi–Maskawa (CKM) matrix elements V_{td}, V_{ts}, i.e.

$$\epsilon_K \propto \hat{B}_K \, \text{Im} \, (V_{td} \, V_{ts}^*). \tag{5.252}$$

While ϵ_K can be determined experimentally from a ratio of decay amplitudes of long- and short-lived K-mesons, $K_{L,S} \rightarrow (\pi\pi)_{I=0}$, the parameter $\hat{B}_K$ must be extracted from the hadronic matrix element of a four-quark operator between

K^0 and $\bar{K}^0$ states. Obviously, such a calculation must be performed in the non-perturbative regime of QCD since it involves typical hadronic scales.

Other CKM matrix elements, such as V_{us}, V_{ub} and V_{cb} are related to weak processes involving kaons, D- and B-mesons, which are described by a variety of leptonic decay constants (and their ratios), form factors of semi-leptonic meson and baryon decays, as well as the B-parameters that encode strong interaction contributions to $B^0 - \bar{B}^0$ and $B_s^0 - \bar{B}_s^0$ mixing. All these quantities have been studied in lattice QCD for many years, and increasingly precise estimates with controlled systematic errors have appeared over the past decade. They have been instrumental for recent analyses of the unitarity of the CKM matrix [324–327].

Similar considerations apply to SM parameters such as the strong coupling constant α_s and the masses of the quarks. While the asymptotic scaling behaviour of α_s gives rise to the dimensionful Λ-parameter that encodes the intrinsic scale of QCD, the quark masses are external parameters. Providing the link between experimentally accessible quantities and quark masses, as well as expressing the Λ-parameter in units of some measurable low-energy quantity has been a primary task for lattice QCD. Lattice calculations have also be instrumental for determining the coupling constants of effective descriptions of QCD, such as the low-energy constants of Chiral Perturbation Theory.

The importance of accurate, model-independent determinations of SM parameters and input quantities for flavour physics has led to the foundation of the Flavour Lattice Averaging Group (FLAG). Updates of the FLAG report have appeared at regular intervals since the publication of its first edition in 2010 [245]. As part of its mission, FLAG issues global estimates and averages of lattice results, provided that they satisfy a set of defined quality criteria. FLAG estimates are quoted separately according to the sea quark content of the calculations that enter the global analyses, i.e. whether they have been obtained with a degenerate doublet of u, d quarks ($N_f = 2$) or with an additional dynamical strange ($N_f = 2 + 1$) and charm quark ($N_f = 2+1+1$). The current status of lattice QCD calculations of quark masses, the strong coupling, decay constants, form factors, mixing parameters and low-energy constants is summarized in Tables 1 and 2 of the 2016 FLAG report [247]. The FLAG webpage[22] contains additional updates. Below we comment on the current status of a few selected quantities.

Quark Masses According to FLAG, the strange quark mass is known to 1% precision, while the accuracy in the determination of the average u and d quark mass, $\hat{m} \equiv \frac{1}{2}(m_u + m_d)$, varies between 1–5%, depending on the sea quark content [328–332, 332–340]. Thanks to the recent progress in including the effects of isospin breaking in lattice QCD calculations, estimates for the masses of the individual u and d quarks could also be obtained, typically with $2 - 5\%$ precision [261, 262, 264, 330]. Furthermore, the masses of the heavy quarks have been determined with excellent precision [328, 330–332, 335, 337, 341–348].

---

[22] http://flag.unibe.ch/.

Running Coupling A milestone was achieved by the ALPHA collaboration, who published [349] an estimate for $\alpha_s(M_Z^2)$ obtained by tracing the scale evolution of the strong coupling non-perturbatively over several orders of magnitude into an energy range where the application of perturbation theory can be considered safe (at least as far as the quoted precision is concerned). Their main result is the determination of the Λ-parameter in three-flavour QCD, i.e. $\Lambda_{\overline{MS}}^{(3)} = 341(12)$ MeV, which can be matched to the Λ-parameter in the five-flavour theory using perturbation theory, giving $\Lambda_{\overline{MS}}^{(5)} = 215(10)(3)$ MeV. Finally, this is translated into the result for the strong coupling [349]:

$$\alpha_s^{\overline{MS}}(M_Z^2) = 0.11852(84). \qquad (5.253)$$

The quoted error is 30% smaller than that of the 2016 PDG estimate of $\alpha_s = 0.1181(11)$ [244]. The latter includes lattice results from Refs. [331, 335, 350–354].

Kaon Weak Matrix Elements The kaon B-parameter B_K is now known with an overall accuracy of 1.3% [336, 355–359]. Moreover, the calculations of matrix elements relevant for $K^0 - \bar{K}^0$ mixing have been extended to include operators that arise in extensions of the Standard Model [355, 358–363].

Lattice QCD results for kaon leptonic decay constants (more precisely: the ratio f_{K^+}/f_{π^+}) and the form factor $f_+(0)$ describing semi-leptonic $K \to \pi \ell \nu_\ell$ decays have now reached a level of precision that enables a competitive and model-independent determination of V_{us} (see Sect. 5.7.1 of the original review article). Moreover, it is possible to test the unitarity of the first row in the CKM matrix, i.e. the relation

$$|V_{ud}|^2 + |V_{us}|^2 + |V_{ub}|^2 = 1, \qquad (5.254)$$

by combining experimental information with lattice results for $f_+(0)$ and f_{K^+}/f_{π^+}. Neglecting the contribution from $|V_{ub}|^2 \approx 1.7 \cdot 10^{-5}$, one finds that $|V_{ud}|^2 + |V_{us}|^2$ can be determined with a total precision at the percent level, by combining the FLAG estimates[23] for $f_+(0)$ and f_{K^+}/f_{π^+} with the experimentally accessible combinations $|V_{us}|f_+(0) = 0.2165(4)$ and $|V_{us}/V_{ud}|f_{K^+}/f_{\pi^+} = 0.2760(4)$ [244, 364]. In QCD with dynamical light, strange and charm quarks ($N_f = 2+1+1$) the result is $|V_{ud}|^2 + |V_{us}|^2 = 0.9797(74)$, which signals a slight tension of 2.7 standard deviations with the Standard Model. The precision of the unitarity test can be sharpened considerably by replacing $|V_{ud}|$ with the value extracted from neutron β-decay, i.e. $|V_{ud}| = 0.97417(21)$ [365]. It is then sufficient to provide one additional constraint from lattice QCD, either in the form of $f_+(0)$ or the ratio f_{K^+}/f_{π^+}. Inserting the lattice result for $f_+(0)$ yields $|V_{ud}|^2 + |V_{us}|^2 = 0.99884(53)$, which again differs from unitarity by about 2σ. Using instead the

---

[23] See the web update at http://flag.unibe.ch/.

lattice result for f_{K^+}/f_{π^+} implies $|V_{ud}|^2 + |V_{us}|^2 = 0.99986(46)$. Thus, first-row unitarity can be probed with permil-level precision [247].

Heavy-Light Decay Constants and Form Factors The treatment of heavy quarks on the lattice presents additional significant challenges: since the mass of the charm quark is close to typical values of the inverse lattice spacing, which acts as the ultraviolet cutoff, lattice results are prone to suffering from large discretisation errors. Moreover, the mass of the bottom quark exceeds currently accessible values of a^{-1}, and specially designed methods are required for a consistent treatment. This has been discussed extensively in Sect. 5.7.2 of the original review.

The overall precision of lattice estimates for weak hadronic matrix elements involving charm and bottom quarks has vastly improved over the past decade. As shown in Table 2 of FLAG 2016 [247], the leptonic decay constants of the B and B_s mesons are now known at the level of 2%, while ratios such as f_{B_s}/f_B have been determined with even better accuracy [347, 366–373]. Since the 2016 edition of the FLAG report, new results obtained with $N_f = 2+1+1$ flavours of dynamical quarks [343, 374, 375] have pushed the overall precision to the sub-percent level, which is an impressive achievement. Also the estimates of the individual B-parameters $\hat{B}_B$ and $\hat{B}_{B_s}$, their ratios and combinations with the leptonic decay constants are now known with overall errors at the percent level [347, 370, 376, 377].

Results for form factors describing semi-leptonic decays of hadrons containing b-quarks, such as $B \rightarrow (D, D^*)\ell\nu$, or even $\Lambda_b \rightarrow p\ell\nu$ have reached a level of precision that is sufficient for competitive determinations of the CKM matrix elements V_{cb} and V_{ub} from exclusive processes. An extensive discussion is presented in the web update of the FLAG report.

5.9.4   Nucleon Matrix Elements

The understanding of the internal structure of the nucleon in terms of the fundamental interactions between its constituents, the quarks and gluons, has become a major activity within the field of lattice QCD. Structural information is encoded in quantities such as form factors, structure functions and (generalized) parton distribution functions (PDFs). An open problem in this context is the decomposition of the proton's spin in terms of the spins of quarks and gluons, as well as their angular momentum [378, 379]. Another important issue is the so-called "proton radius puzzle" [380], which arises due to the observed discrepancy between the proton radius extracted from the Lamb shift in muonic hydrogen [381, 382] compared to the more traditional determinations from electron-proton scattering [383] or the Lamb shift in electronic hydrogen [384]. Accurate knowledge of the electromagnetic form factors of the proton are indispensable in order to resolve—or corroborate—this puzzle.

The determination of quantities such as nucleon form factors in lattice QCD proceeds by calculating the corresponding hadronic matrix elements between nucleon initial and final states. A strong motivation for computing such quantities is provided by the fact that fundamental interactions are often probed in scattering experiments involving nuclear targets. For instance, probing the neutrino sector requires accurate knowledge of the scattering cross sections of neutrinos with nuclear targets. Similar considerations apply to the search for dark matter candidates. Therefore, precise determinations of the corresponding nucleon matrix elements are indispensable for exploring the limits of the SM.

The past decade has seen a huge rise in the number of publications describing lattice calculations of nucleon matrix elements. Quantities that have been studied include

- the electromagnetic form factors of the nucleon, $G_E(Q^2)$ and $G_M(Q^2)$, which give access to the electric and magnetic charge radii of the nucleon and its magnetic moment [385–397];
- the iso-vector axial charge of the nucleon, g_A, which is a measure of the strength of weak interaction in neutron β-decay [386, 387, 389, 392, 397–414], as well as the scalar and tensor charges, g_S and g_T [386, 393, 404–406, 411, 412, 414–419];
- axial and induced pseudoscalar form factors of the nucleon [397, 407, 409, 420, 421], as well as the strange electromagnetic and axial form factors [421–426, 529] which probe the quark sea inside the nucleon;
- the pion-nucleon σ-term $\sigma_{\pi N}$ [412, 427–438] and the strange content of the nucleon σ_s [412, 429–431, 435–444]. These σ-terms are proportional to nucleon matrix element of the flavour-diagonal scalar density, $\bar{q}q$, which parameterizes the rate of change in the nucleon mass due to a non-zero value of the corresponding quark mass.

Recent reviews, presented at the annual conference on lattice field theory, can be found in Refs. [445–447]. Some results on nucleon form factors and other matrix elements are reviewed in section 3.2.5 of [448], and a dedicated chapter has been prepared for the 2019 edition of the FLAG report. In addition, there has been a community effort in the form of a white paper [449] in which lattice results are used to reduce the overall uncertainties in polarized and unpolarized proton PDFs and their moments.

The relevant nucleon hadronic matrix elements are extracted from suitable three-point correlation functions of quark bilinears between interpolating operators representing the initial and final-state nucleons. Examples of the corresponding diagrams, with the initial-state nucleon placed at Euclidean time $t = 0$ (the source), the final-state nucleon at time t_s (the sink) and the operator insertion at time t, are shown in Fig. 5.24. In addition to the quark-connected diagram, in which the operator is inserted on a valence quark line, there are also quark-disconnected diagrams in which the operator probes the quark sea. The latter class of diagrams must be computed to determine, for instance, iso-scalar quantities, the strangeness form factors and the σ-terms.

Precise determinations of nucleon matrix elements with controlled statistical and systematic errors are particularly challenging. This is a consequence of the fact that the noise-to-signal ratio in three-point correlation functions corresponding to the

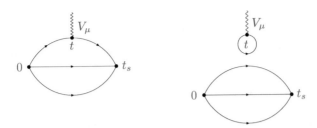

Fig. 5.24 Quark-connected (left) and disconnected (right) diagrams representing the interaction of the vector current with the nucleon

diagrams in Fig. 5.24 grows exponentially with a rate proportional to $\exp\{(m_N - \frac{3}{2}m_\pi)t_s\}$, where m_N and m_π denote the nucleon and pion masses, respectively, and t_s is the source-sink separation. Techniques designed to enhance the statistical signal at affordable numerical cost have been developed and applied, including the truncated solver method [450] and "all-mode-averaging" [451]. Furthermore, a technique to achieve an exponential error reduction via domain decomposition and multi-level integration has been proposed and tested in [452, 453]. So far, it has not been employed in actual calculations of nucleon matrix elements with dynamical quarks.

Quark-disconnected diagrams of the type shown on the right of Fig. 5.24 are intrinsically even noisier than their quark-connected counterparts and require special techniques that balance statistical accuracy against numerical cost. Commonly applied variance reduction techniques for quark-disconnected diagrams include hierarchical probing [454, 455], the coherent source sequential propagator method [389, 456] low-mode averaging [457, 458], the hopping parameter expansion [450, 459–461] and partitioning/dilution [275, 462]).

Despite these improvements, typical values of the source-sink separation t_s for which the signal has not yet disappeared into the noise are limited to $t_s \gtrsim 1.5$ fm. Since the correlation function is dominated by the ground state for $t, (t_s - t) \to \infty$, it is then not guaranteed that the matrix element of interest can be extracted without incurring a bias from unsuppressed excited state contributions, as long as one cannot probe the region $t_s > 1.5$ fm. Hence, in addition to "standard" systematic effects such as lattice artefacts or finite-volume effects, one must also ensure that the asymptotic regime of nucleon correlation functions has been correctly isolated. Indeed, controlling excited state effects has become perhaps the most important issue in current lattice calculations of nucleon matrix elements. The commonly used strategies include

- fits to three-point correlation functions or suitably defined ratios of correlators including sub-leading contributions from excited states [393, 394];
- calculations of three-point correlators summed over the operator insertion time t [463–467]. Contributions from excited states can be shown to be parametrically more strongly suppressed than in the standard case [468];

- increasing the projection of nucleon interpolators onto the ground state [404, 469], as well as the construction of an operator basis for the variational method, which allows for the projection onto the approximate ground state [456, 469, 470].

The first two approaches proceed by fitting data obtained in a finite interval of source-sink separations t_s to a function that describes the approach to the asymptotic behaviour. To be able to resolve the sub-leading contributions from excited states in such a fit obviously requires sufficiently precise input data.

Another challenge for lattice calculations of nucleon matrix elements is the accurate description of the pion mass dependence. Although simulations at or near the physical pion mass are now routinely performed, the result at the physical point is often obtained via an extrapolation in the pion mass. The fit *ansatz* for the pion mass dependence is usually derived from chiral effective theory. However, the convergence properties of baryonic chiral perturbation theory are not as well understood as in the mesonic sector, and it is still unclear whether the predicted functional form provides a good description in the pion mass range over which it is applied. It is thus mandatory to gather sufficiently precise results at small enough pion mass, in order to control the systematic uncertainty associated with the chiral extrapolation.

Instead of performing a detailed survey of a variety of nucleon observables, we single out one particular quantity—the iso-vector axial charge of the nucleon, g_A, which is perhaps the most widely studied of nucleon matrix elements in lattice QCD and serves to illustrate the current state of the art. The axial charge describes the coupling of the W boson to the nucleon. In Minkowski space notation it is defined by

$$\langle p(k, s') | \bar{u} \gamma^{\mu} \gamma_5 d | n(k, s) \rangle = g_A \, \bar{u}_p(k, s') \, \gamma^{\mu} \gamma_5 \, u_n(k, s), \qquad (5.255)$$

where $u_n(k, s)$ and $u_p(k, s')$ denote the Dirac spinors of the neutron and proton with four-momentum k and spins s and s', respectively. The axial charge has been measured experimentally in neutron β-decay, and the current world average quoted in the PDG is $g_A = 1.2724 \pm 0.0023$ [471]. Provided that the experimental sensitivity is sufficient, it may be possible to probe for scalar and tensor interactions that are generated by loop effects or arise due to new forces in extensions of the SM. The definitions of the associated scalar and tensor charges, g_S and g_T are derived from Eq. (5.255) by replacing the axial current $\bar{u} \gamma^{\mu} \gamma_5 d$ by the scalar density $\bar{u} d$ and the tensor current $\bar{u} \sigma^{\mu\nu} d$, respectively.

The calculation of g_A is facilitated by the fact that it is derived from a forward matrix element without any momentum transfer and, secondly, since the contributions from quark-disconnected diagrams cancel in the iso-vector combination, for mass-degenerate up and down quarks. Coupled with the fact that a precise experimental value is known, the iso-vector axial charge is a benchmark quantity for lattice calculations of nucleon matrix elements. Obviously, the ability of state-of-the-art lattice calculations to reproduce the experimental result will enhance the

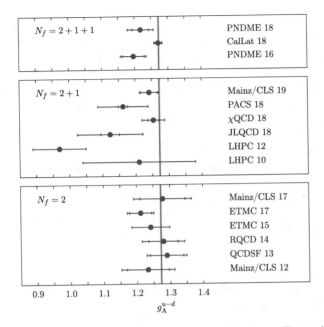

Fig. 5.25 Compilation of recent results for the isovector axial charge. The vertical red band indicates the PDG average [471]. Lattice results are labelled by PNDME 18 [411], CalLat 18 [410], PNDME 16 [406], Mainz/CLS 19 [414], PACS 18 [397], χQCD 18 [413], JLQCD 18 [412], LHPC 12 [392], LHPC 10 [389], Mainz/CLS 17 [409], ETMC 17 [407], ETMC 15 [405], RQCD 14 [404], QCDSF 13 [403] and Mainz/CLS 12 [402]

credibility of lattice predictions for the unmeasured charges g_S and g_T. Figure 5.25 shows a compilation of recent results for g_A, obtained in lattice QCD with $N_f = 2, 2+1$ and $2+1+1$ flavours of dynamical quarks. While most estimates agree with the experimental result within errors, it is clear that the overall precision of current lattice calculations does not match that of the experiments. To state this observation more precisely, we note that the typical total error of current lattice results is at the level of 1–3% while experiment is an order of magnitude more precise. It should also be mentioned that, more often than not, lattice results tend to be slightly lower that the PDG average. Whether this is due to a remnant bias from excited state contributions or indeed to any other systematic effect, must be investigated in future calculations able to realize larger source-sink separations.

The tendency to underestimate g_A in early lattice calculations of g_A has been attributed to unsuppressed excited state effects. In this context it is interesting to note that recent analyses of the contributions from $N\pi$ states to nucleon matrix elements based on chiral effective theory [472, 473] suggest that the asymptotic (physical) value of g_A is approached from above. The different conclusions drawn from numerical and analytic studies can only be reconciled if one succeeds in simulating significantly larger source-sink separations at affordable cost.

Given that lattice QCD calculations reproduce the experimental value of benchmark quantities such as the axial charge at the level of a few percent, it is interesting to look at quantities that have not been measured so far. Results for the (iso-vector) scalar and tensor charges have been reported in [386, 393, 404–406, 411, 412, 414–419]. For both quantities one obtains $g_S, g_T \approx 1$, and while the typical overall uncertainty in g_S is at the level of 10%, the tensor charge is determined with 3% precision, similar to that of g_A. The 2019 edition of the FLAG report contains a detailed compilation and comparison of results for the axial, scalar and tensor charges, as well as flavour-singlet charges and σ-terms. Calculations of these quantities have matured to a level which allows for global averages to be determined.

Lattice calculations of nucleon matrix elements is a rich subject, and while a comprehensive discussion of other quantities such as form factors and moments of PDFs is beyond the scope of this short review, we refer the reader to recent reviews [445–447], specific sections of [448] and the white paper on PDFs [449].

5.9.5 Hadronic Contributions to the Muon Anomalous Magnetic Moment

The SM describes with great accuracy and precision the properties of the constituents of the visible matter in the universe but leaves several profound questions unanswered. For instance, it cannot account for the matter-antimatter asymmetry and does not explain the vast hierarchy between the electroweak scale and the Planck mass. Most prominently, the SM cannot account for the presence of dark matter in the universe for which there is overwhelming observational evidence. Against this backdrop, the exploration of the limits of the SM and the search for "new physics" has become a major activity in particle physics. Traditionally, high-energy particle colliders have had the highest discovery potential. However, despite the fact that the LHC is the most powerful accelerator in the world, new particles that can, for instance, explain the dark matter puzzle have not been observed in the expected region. Therefore, additional search strategies must be pursued to detect evidence for physics beyond the SM.

Observables that can be measured with very high precision and for which similarly accurate theoretical predictions exist at the same time, play an increasingly important rôle for exploring the limits of the SM. One such quantity is the anomalous magnetic moment of the muon, $a_\mu \equiv \frac{1}{2}(g_\mu - 2)$, where g_μ denotes the muon's gyromagnetic ratio. There has been a persistent tension of about 3.5 standard deviations between the measured value and the SM prediction [244]:

$$a_\mu^{\text{exp}} - a_\mu^{\text{SM}} = (266 \pm 76) \cdot 10^{-11}. \tag{5.256}$$

As described in detail in the extensive reviews in Refs. [474] and [475], the SM estimate of the anomalous magnetic moment receives contributions from QED, the weak and the strong interactions, i.e.

$$a_\mu^{SM} = a_\mu^{QED} + a_\mu^{weak} + a_\mu^{strong}. \qquad (5.257)$$

While QED effects account for about 99.994% of the absolute value of a_μ^{SM}, its total uncertainty is completely dominated by the contribution from a_μ^{strong}. Since the latter is mostly due to hadronic effects that are intrinsically non-perturbative, it is clear that special attention must be paid to their reliable evaluation.

The most important quantum corrections to a_μ^{SM} arising from strong interaction physics are the leading hadronic vacuum polarization (HVP) and hadronic light-by-light scattering (HLbL) contributions. The HVP contribution, a_μ^{hvp}, which arises at order α^2 (where α is the fine structure constant), can be expressed in terms of a dispersion integral of the cross section ratio $R(s) = \sigma(e^+e^- \to \text{hadrons})/\sigma(e^+e^- \to \mu^+\mu^-)$, multiplied by a known kernel function. At small values of the centre-of-mass energy s, the dispersion integral is evaluated using experimental data for the R-ratio $R(s)$ as input [476–480]. For instance, the recent analysis of Ref. [479], which is based on the available data for $e^+e^- \to$ hadrons, produced an estimate of $a_\mu^{hvp} = (693.1 \pm 3.4) \cdot 10^{-10}$. While the total error is at the level of 0.5%, it is clear that experimental uncertainties enter the SM prediction for a_μ in this approach.

The HLbL contribution has been quantified mostly using hadronic models, although efforts are under way to formulate and apply a dispersive or data-driven framework to treat some of the dominant sub-processes [481–491]. The current SM estimate a_μ^{SM} is based on model calculations such as the "Glasgow consensus", i.e. $a_\mu^{hlbl} = (105 \pm 26) \cdot 10^{-11}$ [492]. Other studies, which have produced consistent results, can be found in Refs. [474, 478, 493].

Given the importance of a_μ for testing the limits of the SM, it is crucial to verify the current estimates of a_μ^{hvp} and a_μ^{hlbl} and possibly reduce their overall errors using an ab initio approach such as lattice QCD. Given that two new experiments (E989 at Fermilab and E34 at J-PARC) are set to improve the precision of the measurement of a_μ by a factor four, the importance of reliably estimating the hadronic contributions has become even higher. In order to make an impact, lattice QCD must be able to constrain a_μ^{hvp} with sub-percent accuracy, while an estimate of a_μ^{hlbl} at the level of 10% would already be a major step forward. Both tasks, however, present a considerable challenge to lattice QCD. The current status of lattice calculations of a_μ^{hvp} and a_μ^{hlbl} was reviewed extensively in Ref. [494], which can be consulted for details. Here we present merely an overview of the main issues and a guide to the literature.

The hadronic vacuum polarization contribution, a_μ^{hvp}, is accessible in lattice QCD via different integral representations involving the correlator of the electromagnetic current. The first possibility is to consider a convolution integral over Euclidean momenta Q^2 of the subtracted vacuum polarization function [500, 501]. The second possibility is the so-called time-momentum representation defined in Ref. [502], in which the product of the spatially summed vector correlator $G(x_0)$ and a kernel function is integrated over the Euclidean time x_0. A variant of the time-momentum representation uses the time moments of $G(x_0)$ [503]. Finally, there also exists a Lorentz-covariant formulation in coordinate space [504] involving the point-to-point vector correlator $G(x, y)$.

In order to meet the precision goal of sub-percent uncertainty, it is mandatory to have good control over the infrared regime which makes a sizeable contribution to a_μ^{hvp}. In the formulation of Refs. [500, 501] this implies that momenta corresponding to $Q^2 \lesssim m_\mu^2$ must be included, since this is where the convolution integral receives its dominant contribution. Instead, in the time-momentum representation or the Lorentz-covariant formulation one must constrain the long-distance regime of the correlator sufficiently well. The statistical accuracy that one can attain for a_μ^{hvp} is affected by the well-known noise problem encountered for the vector correlator, i.e. the fact that the signal-to-noise ratio increases exponentially at large distances.[24] Another limiting factor for the overall precision of a_μ^{hvp} in lattice QCD is the knowledge of the lattice scale [499, 505]. At first sight this may seem surprising, given that a_μ^{hvp} is a dimensionless quantity. However, employing the time-momentum representation, one easily sees that the lattice scale enters through the combination $(x_0 m_\mu)^2$ in the kernel function. Similar arguments exist for the other representations of a_μ^{hvp}. Furthermore, at the level of sub-percent precision, it is necessary to include the contributions from quark-disconnected diagrams and the effects from isospin breaking (see Sect. 5.9.2). All of this is explained in great detail in Ref. [494].

First exploratory calculations of a_μ^{hvp} in full QCD were published in 2008 [506], and in the following years several studies appeared [497, 507–509], employing a range of different discretisations of the quark action, which were mostly aimed at investigating systematic effects. The most recent calculations are focussed on reducing the overall uncertainties [495, 496, 498, 499, 510–515, 530]. A comparison of recent estimates for a_μ^{hvp} from lattice QCD to results obtained via the dispersive approach is shown in Fig. 5.26. As of now, current calculations cannot match the accuracy of the dispersive approach, but efforts are under way to reduce the uncertainties to a level that makes the lattice approach competitive with data-driven methods [494, 516].

In order to determine the hadronic light-by-light scattering contribution, it is necessary to formulate the problem in such a way that a_μ^{hlbl} is expressed in terms of quantities that can be computed on the lattice with affordable effort. Several different strategies have been proposed and are currently being pursued:

In a first method, the matrix element of the electromagnetic current between explicit muon initial and final states is computed is QCD+QED [517]. In order to isolate the desired light-by-light scattering contribution, one has to perform a non-perturbative subtraction. While the method has produced estimates in the expected

---

[24]This is similar to, but less severe, than the noise problem encountered in nucleon correlation functions discussed in Sect. 5.9.4 of this review.

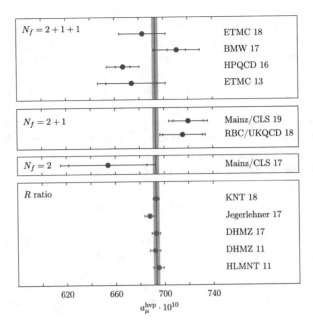

Fig. 5.26 Compilation of recent results for the hadronic vacuum polarisation contribution in units of 10^{-10}. The three panels represent calculations with different numbers of sea quarks. Lattice results are labelled by ETMC 18 [515], BMW 17 [495], HPQCD 16 [496], ETMC 13 [497], Mainz/CLS 19 [530], RBC/UKQCD 18 [498], and Mainz/CLS 17 [499]. The phenomenological determinations based on the R-ratio are labelled as HLMNT 11 [477], DHMZ 11 [476], Jegerlehner 17 [478] and KNT 18 [480]. The red vertical band denotes the estimate from dispersion theory quoted in KNT 18 [480]

range, statistical errors are large, as a result of the cancellation between two large numbers [518].

In another method proposed by the RBC/UKQCD Collaboration [519, 520], the light-by-light scattering diagram is evaluated by inserting three explicit photon propagators. The positions of the insertion of these propagators are then sampled stochastically. In this way, results for the quark-connected and the leading quark-disconnected contributions have been obtained, i.e.

$$(a_\mu^{\text{hlbl}})^{\text{conn}} = (116.0 \pm 9.6) \cdot 10^{-11}, \quad (a_\mu^{\text{hlbl}})^{\text{disc}} = (-62.5 \pm 8.0) \cdot 10^{-11}. \quad (5.258)$$

The sum of the two contributions gives $a_\mu^{\text{hlbl}} = (53.5 \pm 13.5) \cdot 10^{-11}$ which differs from the Glasgow consensus by a factor two. However, before jumping to conclusions one must take into account that systematic effects have not yet been fully quantified in these calculations.

The Mainz group has proposed a method in which the QED kernel function is computed semi-analytically in infinite volume [521–524]. This has the advantage that large finite-volume effects arising from the massless photon mode are absent.

The method has yet to produce explicit estimates for a_μ^{hlbl}. A variant was proposed by RBC/UKQCD in Ref. [525]. Another project of the Mainz group has focussed on the forward light-by-light scattering amplitude, which can be linked via the optical theorem and dispersive sum rules to models of the cross section for the process $\gamma^*\gamma^* \to$ hadrons [526, 527]. The results provide an important test for model estimates of a_μ^{hlbl}.

Finally, lattice QCD calculations can also be used to directly test model estimates of the expected dominant contribution to a_μ^{hlbl} from the pion pole, which requires knowledge of the transition form factor for $\pi^0 \to \gamma^*\gamma^*$. The calculation of Ref. [528], which was performed in two-flavour QCD, gives

$$(a_\mu^{\text{hlbl}})^{\pi^0} = (65.0 \pm 8.3) \cdot 10^{-11} \tag{5.259}$$

which is in very good agreement with model estimates [491]. It will be interesting to extend this calculation by including the corresponding contributions of the η and η' mesons.

This brief survey demonstrates that lattice QCD contributes in many different and complementary ways to constrain the hadronic contributions to the muon $g - 2$ more precisely.

5.9.6 Concluding Remarks

In this short review we have charted the progress of lattice QCD calculations over more than a decade, i.e. since the publication of the original review article. Back in 2007, lattice QCD was on the verge of providing estimates for hadronic observables from first principles, which were of immediate phenomenological relevance. In the meantime, lattice QCD has become an indispensable tool in particle and hadron physics: In addition to to providing accurate estimates of SM parameters and input quantities for analyses in flavour physics, lattice QCD is now also making inroads into field such as nucleon structure and precision observables. This underlines the important role of lattice calculations for exploring the limits of the SM and searches for new physics.

Furthermore, studying hadronic interactions, i.e. the physics of resonances and multi-hadron systems, has become a major activity in lattice QCD and also serves as a basis for the understanding of light nuclei from first principles. Other important applications of the lattice formulation that have not been covered in this article are studies of matter under extreme conditions. Indeed, many features of the QCD phase diagram and properties of the quark-gluon plasma that are otherwise inaccessible can nowadays be obtained reliably from lattice calculations. Perhaps the most significant development since Ken Wilson's 1989 remark, quoted in the introduction, is the fact that there is now a vigorous interaction between lattice QCD and experiment.

Acknowledgments I am grateful to Guido Altarelli, Andreas Jüttner, Stefan Scherer and Rainer Sommer for a careful reading of the original manuscript and many helpful suggestions.

References

1. M. Creutz, *Quarks, Gluons and Lattices*, Cambridge University Press (1983), Cambridge, UK.
2. I. Montvay and G. Münster, *Quantum fields on a lattice*, Cambridge University Press (1994), Cambridge, UK.
3. J. Smit, *Introduction to quantum fields on a lattice: A robust mate*, Cambridge Lect. Notes Phys. 15 (2002) 1.
4. H.J. Rothe, *Lattice gauge theories: An Introduction*, World Sci. Lect. Notes Phys. 74 (2005) 1.
5. K. Osterwalder and R. Schrader, Commun. Math. Phys. 31 (1973) 83; Commun. Math. Phys. 42 (1975) 281.
6. K.G. Wilson, Phys. Rev. D10 (1974) 2445.
7. P. Weisz, Nucl. Phys. B212 (1983) 1; P. Weisz and R. Wohlert, Nucl. Phys. B236 (1984) 397.
8. Y. Iwasaki, Nucl. Phys. B258 (1985) 141.
9. L. Susskind, Phys. Rev. D16 (1977) 3031.
10. D.B. Kaplan, Phys. Lett. B288 (1992) 342, hep-lat/9206013.
11. V. Furman and Y. Shamir, Nucl. Phys. B439 (1995) 54, hep-lat/9405004.
12. H. Neuberger, Phys. Lett. B417 (1998) 141, hep-lat/9707022; Phys. Lett. B427 (1998) 353, hep-lat/9801031.
13. P. Hasenfratz, Nucl. Phys. B (Proc. Suppl.) 63 (1998) 53, hep-lat/9709110.
14. K. Symanzik, Nucl. Phys. B226 (1983) 187; Nucl. Phys. B226 (1983) 205.
15. B. Sheikholeslami and R. Wohlert, Nucl. Phys. B259 (1985) 572.
16. M. Lüscher, S. Sint, R. Sommer and P. Weisz, Nucl. Phys. B478 (1996) 365, hep-lat/9605038.
17. M. Lüscher, S. Sint, R. Sommer, P. Weisz and U. Wolff, Nucl. Phys. B491 (1997) 323, hep-lat/9609035.
18. H. Kluberg-Stern, A. Morel, O. Napoly and B. Petersson, Nucl. Phys. B220 (1983) 447.
19. G.P. Lepage, Phys. Rev. D59 (1999) 074502, hep-lat/9809157.
20. MILC Collaboration, K. Orginos, D. Toussaint and R.L. Sugar, Phys. Rev. D60 (1999) 054503, hep-lat/9903032.
21. H.B. Nielsen and M. Ninomiya, Nucl. Phys. B185 (1981) 20; Nucl. Phys. B193 (1981) 173.
22. P.H. Ginsparg and K.G. Wilson, Phys. Rev. D25 (1982) 2649.
23. P. Hasenfratz, V. Laliena and F. Niedermayer, Phys. Lett. B427 (1998) 125, hep-lat/9801021.
24. M. Lüscher, Phys. Lett. B428 (1998) 342, hep-lat/9802011.
25. P. Hernández, K. Jansen and M. Lüscher, Nucl. Phys. B552 (1999) 363, hep-lat/9808010.
26. ALPHA Collaboration, R. Frezzotti, P.A. Grassi, S. Sint and P. Weisz, JHEP 08 (2001) 058, hep-lat/0101001.
27. R. Frezzotti and G.C. Rossi, JHEP 08 (2004) 007, hep-lat/0306014.
28. S. Sint, *Lattice QCD with a chiral twist*, (2007), hep-lat/0702008.
29. A. Shindler, *Twisted mass lattice QCD*, (2007), arXiv:0707.4093 [hep-lat].
30. N. Metropolis, A.W. Rosenbluth, M.N. Rosenbluth, A.H. Teller and E. Teller, J. Chem. Phys. 21 (1953) 1087.
31. S. Duane, A.D. Kennedy, B.J. Pendleton and D. Roweth, Phys. Lett. B195 (1987) 216.
32. M. Lüscher, Commun. Math. Phys. 54 (1977) 283.
33. R. Sommer, Nucl. Phys. B411 (1994) 839, hep-lat/9310022.
34. ALPHA Collaboration, M. Guagnelli, R. Sommer and H. Wittig, Nucl. Phys. B535 (1998) 389, hep-lat/9806005.
35. ALPHA & UKQCD Collaborations, J. Garden, J. Heitger, R. Sommer and H. Wittig, Nucl. Phys. B571 (2000) 237, hep-lat/9906013.

36. M. Lüscher, Commun. Math. Phys. 104 (1986) 177.
37. H. Wittig, Nucl. Phys. B (Proc. Suppl.) 119 (2003) 59, hep-lat/0210025.
38. H. Leutwyler, Phys. Lett. B378 (1996) 313, hep-ph/9602366.
39. S.R. Sharpe, PoS LAT2006 (2006) 022, hep-lat/0610094.
40. HPQCD Collaboration, C.T.H. Davies et al., Phys. Rev. Lett. 92 (2004) 022001, hep-lat/0304004.
41. ETM Collaboration, P. Boucaud et al., Phys. Lett. B650 (2007) 304, hep-lat/0701012.
42. M. Hasenbusch, Phys. Lett. B519 (2001) 177, hep-lat/0107019.
43. C. Urbach, K. Jansen, A. Shindler and U. Wenger, Comput. Phys. Commun. 174 (2006) 87, hep-lat/0506011.
44. M. Lüscher, Comput. Phys. Commun. 165 (2005) 199, hep-lat/0409106.
45. M.A. Clark and A.D. Kennedy, Phys. Rev. Lett. 98 (2007) 051601, hep-lat/0608015.
46. F. Butler, H. Chen, J. Sexton, A. Vaccarino and D. Weingarten, Nucl. Phys. B430 (1994) 179, hep-lat/9405003.
47. CP-PACS Collaboration, S. Aoki et al., Phys. Rev. Lett. 84 (2000) 238, hep-lat/9904012; Phys. Rev. D67 (2003) 034503, hep-lat/0206009.
48. UKQCD Collaboration, K.C. Bowler et al., Phys. Rev. D62 (2000) 054506, hep-lat/9910022.
49. MILC Collaboration, C.W. Bernard et al., Phys. Rev. Lett. 81 (1998) 3087, hep-lat/9805004.
50. C.W. Bernard et al., Phys. Rev. D64 (2001) 054506, hep-lat/0104002.
51. BGR Collaboration, C. Gattringer et al., Nucl. Phys. B677 (2004) 3, hep-lat/0307013.
52. CP-PACS Collaboration, A. Ali Khan et al., Phys. Rev. D65 (2002) 054505, hep-lat/0105015.
53. C.J. Morningstar and M.J. Peardon, Phys. Rev. D60 (1999) 034509, hep-lat/9901004.
54. APE Collaboration, M. Albanese et al., Phys. Lett. 192B (1987) 163.
55. M. Teper, Phys. Lett. B183 (1987) 345.
56. M. Lüscher and U. Wolff, Nucl. Phys. B339 (1990) 222.
57. Y. Chen et al., Phys. Rev. D73 (2006) 014516, hep-lat/0510074.
58. UKQCD Collaboration, G.S. Bali et al., Phys. Lett. B309 (1993) 378, hep-lat/9304012.
59. J. Sexton, A. Vaccarino and D. Weingarten, Phys. Rev. Lett. 75 (1995) 4563, hep-lat/9510022.
60. M.J. Teper, Glueball masses and other physical properties of SU(N) gauge theories in D = (3+1): A Review of lattice results for theorists, (1998), hep-th/9812187.
61. Particle Data Group, W.M. Yao et al., J. Phys. G33 (2006) 1.
62. E. Klempt and A. Zaitsev, Phys. Rept. 454 (2007) 1, arXiv:0708.4016 [hep-ph].
63. M. Lüscher, K. Symanzik and P. Weisz, Nucl. Phys. B173 (1980) 365; M. Lüscher, Nucl. Phys. B180 (1981) 317.
64. G. Parisi, R. Petronzio and F. Rapuano, Phys. Lett. 128B (1983) 418.
65. S. Necco and R. Sommer, Nucl. Phys. B622 (2002) 328, hep-lat/0108008.
66. G.S. Bali, Phys. Rept. 343 (2001) 1, hep-ph/0001312.
67. M. Lüscher and P. Weisz, JHEP 07 (2002) 049, hep-lat/0207003.
68. O. Philipsen and H. Wittig, Phys. Rev. Lett. 81 (1998) 4056, hep-lat/9807020.
69. ALPHA Collaboration, F. Knechtli and R. Sommer, Phys. Lett. B440 (1998) 345, hep-lat/9807022; Nucl. Phys. B590 (2000) 309, hep-lat/0005021.
70. SESAM Collaboration, G.S. Bali, H. Neff, T. Duessel, T. Lippert and K. Schilling, Phys. Rev. D71 (2005) 114513, hep-lat/0505012.
71. T. van Ritbergen, J.A.M. Vermaseren and S.A. Larin, Phys. Lett. B400 (1997) 379, hep-ph/9701390; K.G. Chetyrkin, Phys. Lett. B404 (1997) 161, hep-ph/9703278; J.A.M. Vermaseren, S.A. Larin and T. van Ritbergen, Phys. Lett. B405 (1997) 327, hep-ph/9703284.
72. M. Lüscher, R. Narayanan, P. Weisz and U. Wolff, Nucl. Phys. B384 (1992) 168, hep-lat/9207009.
73. S. Sint, Nucl. Phys. B421 (1994) 135, hep-lat/9312079.
74. S. Sint, Nucl. Phys. B451 (1995) 416, hep-lat/9504005.
75. ALPHA Collaboration, S. Capitani, M. Lüscher, R. Sommer and H. Wittig, Nucl. Phys. B544 (1999) 669, hep-lat/9810063.
76. G. Martinelli, C. Pittori, C.T. Sachrajda, M. Testa and A. Vladikas, Nucl. Phys. B445 (1995) 81, hep-lat/9411010.

77. ALPHA Collaboration, M. Della Morte et al., Nucl. Phys. B713 (2005) 378, hep-lat/0411025.
78. ALPHA Collaboration, M. Della Morte et al., Nucl. Phys. B729 (2005) 117, hep-lat/0507035.
79. Y. Taniguchi, JHEP 10 (2006) 027, hep-lat/0604002.
80. S. Sint, PoS LAT2005 (2006) 235, hep-lat/0511034.
81. M. Lüscher, JHEP 05 (2006) 042, hep-lat/0603029.
82. G. Parisi, Presented at 20th Int. Conf. on High Energy Physics, Madison, Wis., Jul 17–23, 1980.
83. G.P. Lepage and P.B. Mackenzie, Phys. Rev. D48 (1993) 2250, hep-lat/9209022.
84. M. Lüscher, R. Sommer, P. Weisz and U. Wolff, Nucl. Phys. B413 (1994) 481, hep-lat/9309005.
85. S. Sint and R. Sommer, Nucl. Phys. B465 (1996) 71, hep-lat/9508012.
86. HPQCD Collaboration, Q. Mason et al., Phys. Rev. Lett. 95 (2005) 052002, hep-lat/0503005.
87. Y. Schröder, Phys. Lett. B447 (1999) 321, hep-ph/9812205.
88. S. Bethke, Nucl. Phys. Proc. Suppl. 135 (2004) 345, hep-ex/0407021.
89. J. Gasser and H. Leutwyler, Ann. Phys. 158 (1984) 142.
90. J. Gasser and H. Leutwyler, Nucl. Phys. B250 (1985) 465.
91. M. Lüscher, S. Sint, R. Sommer and H. Wittig, Nucl. Phys. B491 (1997) 344, hep-lat/9611015.
92. M. Della Morte, R. Hoffmann, F. Knechtli, R. Sommer and U. Wolff, JHEP 07 (2005) 007, hep-lat/0505026.
93. SPQ$_{cd}$R Collaboration, D. Bećirević, V. Lubicz and C. Tarantino, Phys. Lett. B558 (2003) 69, hep-lat/0208003.
94. JLQCD Collaboration, S. Aoki et al., Phys. Rev. Lett. 82 (1999) 4392, hep-lat/9901019.
95. JLQCD Collaboration, T. Ishikawa et al., Phys. Rev. D78 (2008) 011502.
96. HPQCD Collaboration, Q. Mason, H.D. Trottier, R. Horgan, C.T.H. Davies and G.P. Lepage, Phys. Rev. D73 (2006) 114501, hep-ph/0511160.
97. M. Göckeler et al., Phys. Rev. D73 (2006) 054508, hep-lat/0601004.
98. D. Bećirević et al., Nucl. Phys. B734 (2006) 138, hep-lat/0510014.
99. CP-PACS Collaboration, A. Ali Khan et al., Phys. Rev. Lett. 85 (2000) 4674, hep-lat/0004010.
100. ETM Collaboration, B. Blossier et al., JHEP 04 (2008) 020.
101. J. Goldstone, Nuovo Cim. 19 (1961) 154.
102. S. Scherer, Adv. Nucl. Phys. 27 (2003) 277, hep-ph/0210398.
103. V. Bernard and U.G. Meißner, Chiral perturbation theory, (2006), hep-ph/0611231.
104. D.B. Kaplan and A.V. Manohar, Phys. Rev. Lett. 56 (1986) 2004.
105. M. Gell-Mann, R.J. Oakes and B. Renner, Phys. Rev. 175 (1968) 2195.
106. T. Banks and A. Casher, Nucl. Phys. B169 (1980) 103.
107. J.J.M. Verbaarschot and I. Zahed, Phys. Rev. Lett. 70 (1993) 3852, hep-th/9303012.
108. H. Leutwyler and A. Smilga, Phys. Rev. D46 (1992) 5607.
109. E.V. Shuryak and J.J.M. Verbaarschot, Nucl. Phys. A560 (1993) 306, hep-th/9212088.
110. L. Giusti, C. Hoelbling, M. Lüscher and H. Wittig, Comput. Phys. Commun. 153 (2003) 31, hep-lat/0212012.
111. L. Giusti, M. Lüscher, P. Weisz and H. Wittig, JHEP 11 (2003) 023, hep-lat/0309189.
112. W. Bietenholz, K. Jansen and S. Shcheredin, JHEP 07 (2003) 033, hep-lat/0306022.
113. H. Fukaya et al., Phys. Rev. D76 (2007) 054503, arXiv:0705.3322 [hep-lat].
114. J. Wennekers and H. Wittig, JHEP 09 (2005) 059, hep-lat/0507026.
115. P. Hernández, K. Jansen, L. Lellouch and H. Wittig, JHEP 07 (2001) 018, hep-lat/0106011.
116. L. Giusti, F. Rapuano, M. Talevi and A. Vladikas, Nucl. Phys. B538 (1999) 249, hep-lat/9807014.
117. P. Hernández, K. Jansen and L. Lellouch, Phys. Lett. B469 (1999) 198, hep-lat/9907022.
118. T. Blum et al., Phys. Rev. D69 (2004) 074502, hep-lat/0007038.
119. MILC Collaboration, T.A. DeGrand, Phys. Rev. D64 (2001) 117501, hep-lat/0107014.
120. L. Giusti, C. Hoelbling and C. Rebbi, Phys. Rev. D64 (2001) 114508, hep-lat/0108007.
121. P. Hernández, K. Jansen, L. Lellouch and H. Wittig, Nucl. Phys. B (Proc. Suppl.) 106 (2002) 766, hep-lat/0110199.

122. P. Hasenfratz, S. Hauswirth, T. Jörg, F. Niedermayer and K. Holland, Nucl. Phys. B643 (2002) 280, hep-lat/0205010.
123. D. Bećirević and V. Lubicz, Phys. Lett. B600 (2004) 83, hep-ph/0403044.
124. V. Gimenez, V. Lubicz, F. Mescia, V. Porretti and J. Reyes, Eur. Phys. J. C41 (2005) 535, hep-lat/0503001.
125. C. McNeile, Phys. Lett. B619 (2005) 124, hep-lat/0504006.
126. H.G. Dosch and S. Narison, Phys. Lett. B417 (1998) 173, hep-ph/9709215.
127. S. Narison, (2002), hep-ph/0202200.
128. M.R. Pennington, (2002), hep-ph/0207220.
129. L. Wolfenstein, Phys. Rev. Lett. 51 (1983) 1945.
130. UTfit Collaboration, M. Bona et al., JHEP 10 (2006) 081, hep-ph/0606167.
131. C. Dawson, PoS LAT2005 (2006) 007.
132. M. Okamoto, PoS LAT2005 (2006) 013, hep-lat/0510113.
133. T. Onogi, PoS LAT2006 (2006) 017, hep-lat/0610115.
134. A. Jüttner, PoS LAT2007 (2007) 014, arXiv:0711.1239 [hep-lat].
135. M. Della Morte, PoS LAT2007 (2007) 008, arXiv:0711.3160 [hep-lat].
136. L. Conti et al., Phys. Lett. B421 (1998) 273, hep-lat/9711053.
137. JLQCD Collaboration, S. Aoki et al., Phys. Rev. D60 (1999) 034511, hep-lat/9901018.
138. D. Bećirević et al., Phys. Lett. B487 (2000) 74, hep-lat/0005013.
139. R.S. Van de Water and S.R. Sharpe, Phys. Rev. D73 (2006) 014003, hep-lat/0507012.
140. M. Guagnelli, J. Heitger, C. Pena, S. Sint and A. Vladikas, Nucl. Phys. B (Proc. Suppl.) 106 (2002) 320, hep-lat/0110097.
141. R. Frezzotti and G.C. Rossi, JHEP 10 (2004) 070, hep-lat/0407002.
142. A. Donini, V. Gimenez, G. Martinelli, M. Talevi and A. Vladikas, Eur. Phys. J. C10 (1999) 121, hep-lat/9902030.
143. ALPHA Collaboration, M. Guagnelli, J. Heitger, C. Pena, S. Sint and A. Vladikas, JHEP 03 (2006) 088, hep-lat/0505002.
144. RBC and UKQCD Collaborations, D.J. Antonio et al., Phys. Rev. Lett. 100 (2008) 032001, hep-ph/0702042.
145. HPQCD and UKQCD Collaborations, E. Gamiz et al., Phys. Rev. D73 (2006) 114502, hep-lat/0603023.
146. Y. Aoki et al., Phys. Rev. D72 (2005) 114505, hep-lat/0411006.
147. UKQCD Collaboration, J.M. Flynn, F. Mescia and A.S.B. Tariq, JHEP 11 (2004) 049, hep-lat/0406013.
148. RBC Collaboration, T. Blum et al., Phys. Rev. D68 (2003) 114506, hep-lat/0110075.
149. CP-PACS Collaboration, A. Ali Khan et al., Phys. Rev. D64 (2001) 114506, hep-lat/0105020.
150. N. Garron, L. Giusti, C. Hoelbling, L. Lellouch and C. Rebbi, Phys. Rev. Lett. 92 (2004) 042001, hep-ph/0306295.
151. MILC Collaboration, T.A. DeGrand, Phys. Rev. D69 (2004) 014504, hep-lat/0309026.
152. ALPHA Collaboration, P. Dimopoulos et al., Nucl. Phys. B749 (2006) 69, hep-ph/0601002; Nucl. Phys. B776 (2007) 258, hep-lat/0702017.
153. D. Bećirević, P. Boucaud, V. Gimenez, V. Lubicz and M. Papinutto, Eur. Phys. J. C37 (2004) 315, hep-lat/0407004.
154. JLQCD Collaboration, S. Aoki et al., Phys. Rev. Lett. 80 (1998) 5271, hep-lat/9710073.
155. D. Bećirević, D. Meloni and A. Retico, JHEP 01 (2001) 012, hep-lat/0012009.
156. J. Gasser and H. Leutwyler, Nucl. Phys. B250 (1985) 517.
157. D. Bećirević et al., Nucl. Phys. B705 (2005) 339, hep-ph/0403217.
158. H. Leutwyler and M. Roos, Z. Phys. C25 (1984) 91.
159. UKQCD Collaboration, P.A. Boyle, J.M. Flynn, A. Jüttner, C.T. Sachrajda and J.M. Zanotti, JHEP 05 (2007) 016, hep-lat/0703005.
160. P.F. Bedaque, Phys. Lett. B593 (2004) 82, nucl-th/0402051.
161. G.M. de Divitiis, R. Petronzio and N. Tantalo, Phys. Lett. B595 (2004) 408, hep-lat/0405002.
162. C.T. Sachrajda and G. Villadoro, Phys. Lett. B609 (2005) 73, hep-lat/0411033.

163. UKQCD Collaboration, J.M. Flynn, A. Jüttner and C.T. Sachrajda, Phys. Lett. B632 (2006) 313, hep-lat/0506016.
164. F.J. Jiang and B.C. Tiburzi, Phys. Lett. B645 (2007) 314, hep-lat/0610103.
165. C. Dawson, T. Izubuchi, T. Kaneko, S. Sasaki and A. Soni, Phys. Rev. D74 (2006) 114502, hep-ph/0607162.
166. UKQCD and RBC Collaborations, D.J. Antonio et al., (2007), hep-lat/0702026; UKQCD and RBC Collaborations, P.A. Boyle et al., Phys. Rev. Lett. 100 (2008) 141601, arXiv:0710.5136 [hep-lat].
167. J. Bijnens and P. Talavera, Nucl. Phys. B669 (2003) 341, hep-ph/0303103.
168. M. Jamin, J.A. Oller and A. Pich, JHEP 02 (2004) 047, hep-ph/0401080.
169. V. Cirigliano et al., JHEP 04 (2005) 006, hep-ph/0503108.
170. W.J. Marciano, Phys. Rev. Lett. 93 (2004) 231803, hep-ph/0402299.
171. ALPHA Collaboration, J. Heitger, R. Sommer and H. Wittig, Nucl. Phys. B588 (2000) 377, hep-lat/0006026.
172. JLQCD Collaboration, S. Aoki et al., Phys. Rev. D68 (2003) 054502, hep-lat/0212039.
173. MILC Collaboration, C. Aubin et al., Phys. Rev. D70 (2004) 114501, hep-lat/0407028.
174. S.R. Beane, P.F. Bedaque, K. Orginos and M.J. Savage, Phys. Rev. D75 (2007) 094501, hep-lat/0606023.
175. RBC and UKQCD Collaborations, C. Allton et al., Phys. Rev. D76 (2007) 014504, hep-lat/0701013.
176. HPQCD Collaboration, E. Follana, C.T.H. Davies, G.P. Lepage and J. Shigemitsu, Phys. Rev. Lett. 100 (2008) 062002, arXiv:0706.1726 [hep-lat].
177. D. Bećirević and G. Villadoro, Phys. Rev. D69 (2004) 054010, hep-lat/0311028.
178. G. Colangelo, S. Durr and C. Haefeli, Nucl. Phys. B721 (2005) 136, hep-lat/0503014.
179. E. Eichten, Nucl. Phys. B (Proc. Suppl.) 4 (1988) 170.
180. B.A. Thacker and G.P. Lepage, Phys. Rev. D43 (1991) 196; G.P. Lepage, L. Magnea, C. Nakhleh, U. Magnea and K. Hornbostel, Phys. Rev. D46 (1992) 4052, hep-lat/9205007.
181. A.X. El-Khadra, A.S. Kronfeld and P.B. Mackenzie, Phys. Rev. D55 (1997) 3933, hep-lat/9604004.
182. ALPHA Collaboration, J. Heitger and R. Sommer, JHEP 02 (2004) 022, hep-lat/0310035.
183. M. Guagnelli, F. Palombi, R. Petronzio and N. Tantalo, Phys. Lett. B546 (2002) 237, hep-lat/0206023.
184. E. Eichten and B.R. Hill, Phys. Lett. B234 (1990) 511.
185. M. Della Morte, A. Shindler and R. Sommer, JHEP 08 (2005) 051, hep-lat/0506008.
186. R. Sommer, Non-perturbative QCD: Renormalization, O(a)-improvement and matching to heavy quark effective theory, (2006), hep-lat/0611020.
187. S. Aoki, Y. Kuramashi and S.i. Tominaga, Prog. Theor. Phys. 109 (2003) 383, hep-lat/0107009.
188. N.H. Christ, M. Li and H.W. Lin, Phys. Rev. D76 (2007) 074505, hep-lat/0608006.
189. ALPHA Collaboration, J. Heitger, M. Kurth and R. Sommer, Nucl. Phys. B669 (2003) 173, hep-lat/0302019.
190. B. Grinstein, E.E. Jenkins, A.V. Manohar, M.J. Savage and M.B. Wise, Nucl. Phys. B380 (1992) 369, hep-ph/9204207; J.L. Goity, Phys. Rev. D46 (1992) 3929, hep-ph/9206230; M.J. Booth, Phys. Rev. D51 (1995) 2338, hep-ph/9411433; S.R. Sharpe and Y. Zhang, Phys. Rev. D53 (1996) 5125, hep-lat/9510037.
191. A.S. Kronfeld and S.M. Ryan, Phys. Lett. B543 (2002) 59, hep-ph/0206058.
192. HPQCD Collaboration, A. Gray et al., Phys. Rev. Lett. 95 (2005) 212001, hep-lat/0507015.
193. N. Tantalo, Lattice calculations for B and K mixing, (2007), hep-ph/0703241.
194. D. Bećirević et al., Nucl. Phys. B618 (2001) 241, hep-lat/0002025.
195. UKQCD Collaboration, L. Lellouch and C.J.D. Lin, Phys. Rev. D64 (2001) 094501, hep-ph/0011086.
196. UKQCD Collaboration, K.C. Bowler et al., Nucl. Phys. B619 (2001) 507, hep-lat/0007020.
197. G.M. de Divitiis, M. Guagnelli, F. Palombi, R. Petronzio and N. Tantalo, Nucl. Phys. B672 (2003) 372, hep-lat/0307005.

198. M. Della Morte et al., JHEP 0802 (2008) 078, arXiv:0710.2201 [hep-lat].
199. D. Guazzini, R. Sommer and N. Tantalo, JHEP 0801 (2008) 076, arXiv:0710.2229 [hep-lat].
200. C.W. Bernard et al., Phys. Rev. Lett. 81 (1998) 4812, hep-ph/9806412.
201. A.X. El-Khadra, A.S. Kronfeld, P.B. Mackenzie, S.M. Ryan and J.N. Simone, Phys. Rev. D58 (1998) 014506, hep-ph/9711426.
202. JLQCD Collaboration, S. Aoki et al., Phys. Rev. Lett. 80 (1998) 5711.
203. CP-PACS Collaboration, A. Ali Khan et al., Phys. Rev. D64 (2001) 034505, hep-lat/0010009.
204. MILC Collaboration, C. Bernard et al., Phys. Rev. D66 (2002) 094501, hep-lat/0206016.
205. A. Ali Khan et al., Phys. Lett. B427 (1998) 132, hep-lat/9801038.
206. JLQCD Collaboration, K.I. Ishikawa et al., Phys. Rev. D61 (2000) 074501, hep-lat/9905036.
207. CP-PACS Collaboration, A. Ali Khan et al., Phys. Rev. D64 (2001) 054504, hep-lat/0103020.
208. S. Collins et al., Phys. Rev. D60 (1999) 074504, hep-lat/9901001.
209. JLQCD Collaboration, S. Aoki et al., Phys. Rev. Lett. 91 (2003) 212001, hep-ph/0307039.
210. M. Wingate, C.T.H. Davies, A. Gray, G.P. Lepage and J. Shigemitsu, Phys. Rev. Lett. 92 (2004) 162001, hep-ph/0311130.
211. D0 Collaboration, V.M. Abazov et al., Phys. Rev. Lett. 97 (2006) 021802, hep-ex/0603029.
212. CDF Collaboration, A. Abulencia et al., Phys. Rev. Lett. 97 (2006) 062003, hep-ex/0606027.
213. V. Gadiyak and O. Loktik, Phys. Rev. D72 (2005) 114504, hep-lat/0509075.
214. V. Gimenez and G. Martinelli, Phys. Lett. B398 (1997) 135, hep-lat/9610024.
215. V. Gimenez and J. Reyes, Nucl. Phys. B545 (1999) 576, hep-lat/9806023.
216. UKQCD Collaboration, A.K. Ewing et al., Phys. Rev. D54 (1996) 3526, hep-lat/9508030.
217. J.C. Christensen, T. Draper and C. McNeile, Phys. Rev. D56 (1997) 6993, hep-lat/9610026.
218. D. Bećirević et al., Nucl. Phys. B618 (2001) 241, hep-lat/0002025.
219. D. Bećirević, V. Gimenez, G. Martinelli, M. Papinutto and J. Reyes, JHEP 04 (2002) 025, hep-lat/0110091.
220. F. Palombi, M. Papinutto, C. Pena and H. Wittig, JHEP 08 (2006) 017, hep-lat/0604014.
221. F. Palombi, M. Papinutto, C. Pena and H. Wittig, JHEP 09 (2007) 062, arXiv:0706.4153 [hep-lat].
222. P. Dimopoulos et al., PoS LAT2007 (2007) 368, arXiv:0710.2862 [hep-lat]; ALPHA Collaboration, P. Dimopoulos et al., JHEP 0805 (2008) 065, arXiv:0712.2429 [hep-lat].
223. D. Bećirević and A.B. Kaidalov, Phys. Lett. B478 (2000) 417, hep-ph/9904490.
224. S. Hashimoto, Int. J. Mod. Phys. A20 (2005) 5133, hep-ph/0411126.
225. UKQCD Collaboration, K.C. Bowler et al., Phys. Lett. B486 (2000) 111, hep-lat/9911011.
226. A. Abada et al., Nucl. Phys. B619 (2001) 565, hep-lat/0011065.
227. A.X. El-Khadra, A.S. Kronfeld, P.B. Mackenzie, S.M. Ryan and J.N. Simone, Phys. Rev. D64 (2001) 014502, hep-ph/0101023.
228. JLQCD Collaboration, S. Aoki et al., Phys. Rev. D64 (2001) 114505, hep-lat/0106024.
229. M. Okamoto et al., Nucl. Phys. B (Proc. Suppl.) 140 (2005) 461, hep-lat/0409116.
230. HPQCD Collaboration, E. Dalgic et al., Phys. Rev. D73 (2006) 074502, hep-lat/0601021.
231. CLEO Collaboration, S.B. Athar et al., Phys. Rev. D68 (2003) 072003, hep-ex/0304019.
232. N. Isgur and M.B. Wise, Phys. Lett. B232 (1989) 113; Phys. Lett. B237 (1990) 527.
233. M.E. Luke, Phys. Lett. B252 (1990) 447.
234. C.W. Bernard, Y. Shen and A. Soni, Phys. Lett. B317 (1993) 164, hep-lat/9307005.
235. UKQCD Collaboration, S.P. Booth et al., Phys. Rev. Lett. 72 (1994) 462, hep-lat/9308019.
236. UKQCD Collaboration, K.C. Bowler et al., Phys. Rev. D52 (1995) 5067, hep-ph/9504231.
237. UKQCD Collaboration, K.C. Bowler, G. Douglas, R.D. Kenway, G.N. Lacagnina and C.M. Maynard, Nucl. Phys. B637 (2002) 293, hep-lat/0202029.
238. G.M. de Divitiis, E. Molinaro, R. Petronzio and N. Tantalo, Phys. Lett. B655 (2007) 45, arXiv:0707.0582 [hep-lat]; G.M. de Divitiis, R. Petronzio and N. Tantalo, JHEP 0710 (2007) 062, arXiv:0707.0587 [hep-lat].
239. S. Hashimoto et al., Phys. Rev. D61 (2000) 014502, hep-ph/9906376.
240. S. Hashimoto, A.S. Kronfeld, P.B. Mackenzie, S.M. Ryan and J.N. Simone, Phys. Rev. D66 (2002) 014503, hep-ph/0110253.
241. H. Wittig, (2008).

242. K.G. Wilson, Nucl. Phys. Proc. Suppl. 17 (1990) 82.
243. K.G. Wilson, Nucl. Phys. Proc. Suppl. 140 (2005) 3, hep-lat/0412043.
244. Particle Data Group, C. Patrignani et al., Chin. Phys. C40 (2016) 100001.
245. G. Colangelo et al., Eur. Phys. J. C71 (2011) 1695, 1011.4408.
246. S. Aoki et al., Eur. Phys. J. C74 (2014) 2890, 1310.8555.
247. S. Aoki et al., Eur. Phys. J. C77 (2017) 112, 1607.00299.
248. S. Aoki et al., Eur. Phys. J. C80 (2020) 113. https://doi.org/10.1140/epjc/s10052-019-7354-7
249. PACS-CS, S. Aoki et al., Phys. Rev. D79 (2009) 034503, 0807.1661.
250. S. Dürr et al., Science 322 (2008) 1224, 0906.3599.
251. ETM, C. Alexandrou et al., Phys. Rev. D78 (2008) 014509, 0803.3190.
252. MILC, A. Bazavov et al., Rev. Mod. Phys. 82 (2010) 1349, 0903.3598.
253. N. Tantalo, PoS LATTICE2013 (2014) 007, 1311.2797.
254. A. Portelli, PoS LATTICE2014 (2015) 013, 1505.07057.
255. A. Patella, PoS LATTICE2016 (2017) 020, 1702.03857.
256. T. Blum, R. Zhou, T. Doi, M. Hayakawa, T. Izubuchi, S. Uno and N. Yamada, Phys. Rev. D82 (2010) 094508, 1006.1311.
257. Budapest-Marseille-Wuppertal, S. Borsanyi et al., Phys. Rev. Lett. 111 (2013) 252001, 1306.2287.
258. S. Borsanyi et al., Science 347 (2015) 1452, 1406.4088.
259. R. Horsley et al., J. Phys. G43 (2016) 10LT02, 1508.06401.
260. T. Blum, T. Doi, M. Hayakawa, T. Izubuchi and N. Yamada, Phys. Rev. D76 (2007) 114508, 0708.0484.
261. RM123, G.M. de Divitiis et al., Phys. Rev. D87 (2013) 114505, 1303.4896.
262. D. Giusti, V. Lubicz, C. Tarantino, G. Martinelli, S. Sanfilippo, S. Simula and N. Tantalo, Phys. Rev. D95 (2017) 114504, 1704.06561.
263. R. Horsley et al., JHEP 04 (2016) 093, 1509.00799.
264. Z. Fodor et al., Phys. Rev. Lett. 117 (2016) 082001, 1604.07112.
265. MILC, S. Basak et al., Phys. Rev. D99 (2019) 034503, 1807.05556.
266. R.F. Dashen, Phys. Rev. 183 (1969) 1245.
267. M. Lüscher, Commun.Math.Phys. 104 (1986) 177.
268. M. Lüscher, Commun.Math.Phys. 105 (1986) 153.
269. M. Lüscher, Nucl. Phys. B354 (1991) 531.
270. M. Lüscher, Nucl. Phys. B364 (1991) 237.
271. C. Andersen, J. Bulava, B. Hörz and C. Morningstar, Nucl. Phys. B939 (2019) 145, 1808.05007.
272. C. Michael, Nucl. Phys. B259 (1985) 58.
273. M. Lüscher and U. Wolff, Nucl. Phys. B339 (1990) 222.
274. B. Blossier, M. Della Morte, G. von Hippel, T. Mendes and R. Sommer, JHEP 04 (2009) 094, 0902.1265.
275. J. Foley, K. Jimmy Juge, A. O'Cais, M. Peardon, S.M. Ryan and J.I. Skullerud, Comput. Phys. Commun. 172 (2005) 145, hep-lat/0505023.
276. Hadron Spectrum, M. Peardon et al., Phys. Rev. D80 (2009) 054506, 0905.2160.
277. C. Morningstar, J. Bulava, J. Foley, K.J. Juge, D. Lenkner, M. Peardon and C.H. Wong, Phys. Rev. D83 (2011) 114505, 1104.3870.
278. CP-PACS, T. Yamazaki et al., Phys. Rev. D70 (2004) 074513, hep-lat/0402025.
279. NPLQCD, S.R. Beane, P.F. Bedaque, K. Orginos and M.J. Savage, Phys. Rev. D73 (2006) 054503, hep-lat/0506013.
280. S.R. Beane, T.C. Luu, K. Orginos, A. Parreno, M.J. Savage, A. Torok and A. Walker-Loud, Phys. Rev. D77 (2008) 014505, 0706.3026.
281. X. Feng, K. Jansen and D.B. Renner, Phys. Lett. B684 (2010) 268, 0909.3255.
282. T. Yagi, S. Hashimoto, O. Morimatsu and M. Ohtani, (2011), 1108.2970.
283. Z. Fu, Phys. Rev. D87 (2013) 074501, 1303.0517.
284. PACS-CS, K. Sasaki, N. Ishizuka, M. Oka and T. Yamazaki, Phys. Rev. D89 (2014) 054502, 1311.7226.

285. D.J. Wilson, R.A. Briceño, J.J. Dudek, R.G. Edwards and C.E. Thomas, Phys. Rev. D92 (2015) 094502, 1507.02599.
286. ETM, C. Helmes et al., JHEP 09 (2015) 109, 1506.00408.
287. RQCD, G.S. Bali, S. Collins, A. Cox, G. Donald, M. Göckeler, C.B. Lang and A. Schäfer, Phys. Rev. D93 (2016) 054509, 1512.08678.
288. J. Bulava, B. Fahy, B. Hörz, K.J. Juge, C. Morningstar and C.H. Wong, Nucl. Phys. B910 (2016) 842, 1604.05593.
289. L. Liu et al., Phys. Rev. D96 (2017) 054516, 1612.02061.
290. D. Guo, A. Alexandru, R. Molina and M. Döring, Phys. Rev. D94 (2016) 034501, 1605.03993.
291. C. Morningstar, J. Bulava, B. Singha, R. Brett, J. Fallica, A. Hanlon and B. Hörz, Nucl. Phys. B924 (2017) 477, 1707.05817.
292. S. Prelovsek, L. Leskovec, C.B. Lang and D. Mohler, Phys. Rev. D88 (2013) 054508, 1307.0736.
293. R. Brett, J. Bulava, J. Fallica, A. Hanlon, B. Hörz and C. Morningstar, Nucl. Phys. B932 (2018) 29, 1802.03100.
294. C. Helmes, C. Jost, B. Knippschild, B. Kostrzewa, L. Liu, C. Urbach and M. Werner, PoS LATTICE2016 (2016) 135, 1611.09584.
295. C. Helmes, C. Jost, B. Knippschild, B. Kostrzewa, L. Liu, C. Urbach and M. Werner, Phys. Rev. D96 (2017) 034510, 1703.04737.
296. A. Torok, S.R. Beane, W. Detmold, T.C. Luu, K. Orginos, A. Parreno, M.J. Savage and A. Walker-Loud, Phys. Rev. D81 (2010) 074506, 0907.1913.
297. C.B. Lang and V. Verduci, Phys. Rev. D87 (2013) 054502, 1212.5055.
298. W. Detmold and A. Nicholson, Phys. Rev. D93 (2016) 114511, 1511.02275.
299. C.B. Lang, L. Leskovec, M. Padmanath and S. Prelovsek, Phys. Rev. D95 (2017) 014510, 1610.01422.
300. C.W. Andersen, J. Bulava, B. Hörz and C. Morningstar, Phys. Rev. D97 (2018) 014506, 1710.01557.
301. K. Orginos, A. Parreno, M.J. Savage, S.R. Beane, E. Chang and W. Detmold, Phys. Rev. D92 (2015) 114512, 1508.07583.
302. A. Francis, J.R. Green, P.M. Junnarkar, C. Miao, T.D. Rae and H. Wittig, (2018), 1805.03966.
303. S. He, X. Feng and C. Liu, JHEP 07 (2005) 011, hep-lat/0504019.
304. V. Bernard, M. Lage, U.G. Meissner and A. Rusetsky, JHEP 01 (2011) 019, 1010.6018.
305. M.T. Hansen and S.R. Sharpe, Phys. Rev. D86 (2012) 016007, 1204.0826.
306. R.A. Briceño and Z. Davoudi, Phys. Rev. D88 (2013) 094507, 1204.1110.
307. P. Guo, J. Dudek, R. Edwards and A.P. Szczepaniak, Phys. Rev. D88 (2013) 014501, 1211.0929.
308. L. Roca and E. Oset, Phys. Rev. D85 (2012) 054507, 1201.0438.
309. K. Polejaeva and A. Rusetsky, Eur. Phys. J. A48 (2012) 67, 1203.1241.
310. R.A. Briceño and Z. Davoudi, Phys. Rev. D87 (2013) 094507, 1212.3398.
311. M.T. Hansen and S.R. Sharpe, Phys. Rev. D90 (2014) 116003, 1408.5933.
312. M.T. Hansen and S.R. Sharpe, Phys. Rev. D93 (2016) 096006, 1602.00324, [Erratum: Phys. Rev. D96, no. 3, 039901(2017)].
313. M.T. Hansen and S.R. Sharpe, Phys. Rev. D95 (2017) 034501, 1609.04317.
314. R.A. Briceño, M.T. Hansen and S.R. Sharpe, Phys. Rev. D95 (2017) 074510, 1701.07465.
315. R.A. Briceño, M.T. Hansen and S.R. Sharpe, Phys. Rev. D98 (2018) 014506, 1803.04169.
316. L. Lellouch and M. Lüscher, Commun. Math. Phys. 219 (2001) 31, hep-lat/0003023.
317. H.B. Meyer, Phys. Rev. Lett. 107 (2011) 072002, 1105.1892.
318. X. Feng, S. Aoki, S. Hashimoto and T. Kaneko, Phys. Rev. D91 (2015) 054504, 1412.6319.
319. J. Bulava, B. Hörz, B. Fahy, K.J. Juge, C. Morningstar and C.H. Wong, PoS LATTICE2015 (2016) 069, 1511.02351.
320. F. Erben, J.R. Green, D. Mohler and H. Wittig, Phys. Rev. D101 (2020) 054504, 1910.01083.
321. S. Prelovsek, PoS LATTICE2014 (2014) 015, 1411.0405.
322. C. Liu, PoS LATTICE2016 (2017) 006, 1612.00103.
323. D. Mohler, EPJ Web Conf. 181 (2018) 01027.

324. CKMfitter Group, J. Charles et al., Eur. Phys. J. C41 (2005) 1, hep-ph/0406184.
325. J. Charles et al., Phys. Rev. D91 (2015) 073007, 1501.05013.
326. UTfit, M. Bona et al., JHEP 10 (2006) 081, hep-ph/0606167.
327. UTfit, M. Bona et al., JHEP 03 (2008) 049, 0707.0636.
328. Fermilab Lattice, MILC, TUMQCD, A. Bazavov et al., Phys. Rev. D98 (2018) 054517, 1802.04248.
329. Y. Maezawa and P. Petreczky, Phys. Rev. D94 (2016) 034507, 1606.08798.
330. ETM, N. Carrasco et al., Nucl. Phys. B887 (2014) 19, 1403.4504.
331. B. Chakraborty et al., Phys. Rev. D91 (2015) 054508, 1408.4169.
332. Fermilab Lattice, MILC, A. Bazavov et al., Phys. Rev. D90 (2014) 074509, 1407.3772.
333. S. Dürr et al., Phys. Lett. B701 (2011) 265, 1011.2403.
334. S. Dürr et al., JHEP 08 (2011) 148, 1011.2711.
335. C. McNeile, C.T.H. Davies, E. Follana, K. Hornbostel and G.P. Lepage, Phys. Rev. D82 (2010) 034512, 1004.4285.
336. RBC/UKQCD, T. Blum et al., Phys. Rev. D93 (2016) 074505, 1411.7017.
337. ETM, B. Blossier, P. Dimopoulos, R. Frezzotti, V. Lubicz, M. Petschlies, F. Sanfilippo, S. Simula and C. Tarantino, Phys. Rev. D82 (2010) 114513, 1010.3659.
338. P. Fritzsch, F. Knechtli, B. Leder, M. Marinkovic, S. Schaefer, R. Sommer and F. Virotta, Nucl. Phys. B865 (2012) 397, 1205.5380.
339. A. Bazavov et al., PoS LATTICE2010 (2010) 083, 1011.1792.
340. T. Burch et al., Phys. Rev. D81 (2010) 034508, 0912.2701.
341. P. Gambino, A. Melis and S. Simula, Phys. Rev. D96 (2017) 014511, 1704.06105.
342. K. Nakayama, B. Fahy and S. Hashimoto, Phys. Rev. D94 (2016) 054507, 1606.01002.
343. ETM, A. Bussone et al., Phys. Rev. D93 (2016) 114505, 1603.04306.
344. Y.B. Yang et al., Phys. Rev. D92 (2015) 034517, 1410.3343.
345. C.T.H. Davies et al., Phys. Rev. Lett. 104 (2010) 132003, 0910.3102.
346. B. Colquhoun, R.J. Dowdall, C.T.H. Davies, K. Hornbostel and G.P. Lepage, Phys. Rev. D91 (2015) 074514, 1408.5768.
347. ETM, N. Carrasco et al., JHEP 03 (2014) 016, 1308.1851.
348. F. Bernardoni et al., Phys. Lett. B730 (2014) 171, 1311.5498.
349. ALPHA, M. Bruno et al., Phys. Rev. Lett. 119 (2017) 102001, 1706.03821.
350. K. Maltman, D. Leinweber, P. Moran and A. Sternbeck, Phys. Rev. D78 (2008) 114504, 0807.2020.
351. PACS-CS, S. Aoki et al., JHEP 10 (2009) 053, 0906.3906.
352. B. Blossier et al., Phys. Rev. Lett. 108 (2012) 262002, 1201.5770.
353. ETM, B. Blossier, P. Boucaud, M. Brinet, F. De Soto, V. Morenas, O. Pene, K. Petrov and J. Rodriguez-Quintero, Phys. Rev. D89 (2014) 014507, 1310.3763.
354. A. Bazavov, N. Brambilla, X. Garcia i Tormo, P. Petreczky, J. Soto and A. Vairo, Phys. Rev. D90 (2014) 074038, 1407.8437.
355. ETM, N. Carrasco, P. Dimopoulos, R. Frezzotti, V. Lubicz, G.C. Rossi, S. Simula and C. Tarantino, Phys. Rev. D92 (2015) 034516, 1505.06639.
356. S. Dürr et al., Phys. Lett. B705 (2011) 477, 1106.3230.
357. J. Laiho and R.S. Van de Water, PoS LATTICE2011 (2011) 293, 1112.4861.
358. SWME, B.J. Choi et al., Phys. Rev. D93 (2016) 014511, 1509.00592.
359. ETM, V. Bertone et al., JHEP 03 (2013) 089, 1207.1287, [Erratum: JHEP07,143(2013)].
360. RBC/UKQCD, P.A. Boyle, N. Garron and R.J. Hudspith, Phys. Rev. D86 (2012) 054028, 1206.5737.
361. RBC/UKQCD, N. Garron, R.J. Hudspith and A.T. Lytle, JHEP 11 (2016) 001, 1609.03334.
362. SWME, T. Bae et al., Phys. Rev. D88 (2013) 071503, 1309.2040.
363. SWME, J. Leem et al., PoS LATTICE2014 (2014) 370, 1411.1501.
364. M. Moulson, PoS CKM2016 (2017) 033, 1704.04104.
365. J.C. Hardy and I.S. Towner, Phys. Rev. C91 (2015) 025501, 1411.5987.
366. ALPHA, F. Bernardoni et al., Phys. Lett. B735 (2014) 349, 1404.3590.
367. ETM, N. Carrasco et al., PoS LATTICE2013 (2014) 382, 1310.1851.

368. H. Na, C.J. Monahan, C.T.H. Davies, R. Horgan, G.P. Lepage and J. Shigemitsu, Phys. Rev. D86 (2012) 034506, 1202.4914.

369. N.H. Christ, J.M. Flynn, T. Izubuchi, T. Kawanai, C. Lehner, A. Soni, R.S. Van de Water and O. Witzel, Phys. Rev. D91 (2015) 054502, 1404.4670.

370. Y. Aoki, T. Ishikawa, T. Izubuchi, C. Lehner and A. Soni, Phys. Rev. D91 (2015) 114505, 1406.6192.

371. C. McNeile, C.T.H. Davies, E. Follana, K. Hornbostel and G.P. Lepage, Phys. Rev. D85 (2012) 031503, 1110.4510.

372. Fermilab Lattice, MILC, A. Bazavov et al., Phys. Rev. D85 (2012) 114506, 1112.3051.

373. HPQCD, R.J. Dowdall, C.T.H. Davies, R.R. Horgan, C.J. Monahan and J. Shigemitsu, Phys. Rev. Lett. 110 (2013) 222003, 1302.2644.

374. A. Bazavov et al., Phys. Rev. D98 (2018) 074512, 1712.09262.

375. C. Hughes, C.T.H. Davies and C.J. Monahan, Phys. Rev. D97 (2018) 054509, 1711.09981.

376. HPQCD, E. Gamiz, C.T.H. Davies, G.P. Lepage, J. Shigemitsu and M. Wingate, Phys. Rev. D80 (2009) 014503, 0902.1815.

377. Fermilab Lattice, MILC, A. Bazavov et al., Phys. Rev. D93 (2016) 113016, 1602.03560.

378. R.L. Jaffe and A. Manohar, Nucl. Phys. B337 (1990) 509.

379. X.D. Ji, Phys. Rev. Lett. 78 (1997) 610, hep-ph/9603249.

380. R. Pohl, R. Gilman, G.A. Miller and K. Pachucki, Ann. Rev. Nucl. Part. Sci. 63 (2013) 175, 1301.0905.

381. R. Pohl et al., Nature 466 (2010) 213.

382. R. Pohl et al., Science 353 (2016) 669.

383. A1, J.C. Bernauer et al., Phys. Rev. C90 (2014) 015206, 1307.6227.

384. P.J. Mohr, D.B. Newell and B.N. Taylor, Rev. Mod. Phys. 88 (2016) 035009, 1507.07956.

385. C. Alexandrou, G. Koutsou, J.W. Negele and A. Tsapalis, Phys. Rev. D74 (2006) 034508, hep-lat/0605017.

386. H.W. Lin, T. Blum, S. Ohta, S. Sasaki and T. Yamazaki, Phys. Rev. D78 (2008) 014505, 0802.0863.

387. T. Yamazaki, Y. Aoki, T. Blum, H.W. Lin, S. Ohta, S. Sasaki, R. Tweedie and J. Zanotti, Phys. Rev. D79 (2009) 114505, 0904.2039.

388. S. Syritsyn et al., Phys. Rev. D81 (2010) 034507, 0907.4194.

389. LHP, J. Bratt et al., Phys. Rev. D82 (2010) 094502, 1001.3620.

390. S. Collins et al., Phys. Rev. D84 (2011) 074507, 1106.3580.

391. C. Alexandrou et al., Phys. Rev. D83 (2011) 094502, 1102.2208.

392. J.R. Green, M. Engelhardt, S. Krieg, J.W. Negele, A.V. Pochinsky and S.N. Syritsyn, Phys. Lett. B734 (2014) 290, 1209.1687.

393. T. Bhattacharya, S.D. Cohen, R. Gupta, A. Joseph, H.W. Lin and B. Yoon, Phys. Rev. D89 (2014) 094502, 1306.5435.

394. S. Capitani et al., Phys. Rev. D92 (2015) 054511, 1504.04628.

395. QCDSF, UKQCD, CSSM, A.J. Chambers et al., Phys. Rev. D96 (2017) 114509, 1702.01513.

396. C. Alexandrou, M. Constantinou, K. Hadjiyiannakou, K. Jansen, C. Kallidonis, G. Koutsou and A. Vaquero Aviles-Casco, Phys. Rev. D96 (2017) 034503, 1706.00469.

397. PACS, K.I. Ishikawa, Y. Kuramashi, S. Sasaki, N. Tsukamoto, A. Ukawa and T. Yamazaki, Phys. Rev. D98 (2018) 074510, 1807.03974.

398. R.G. Edwards et al., Phys. Rev. Lett. 96 (2006) 052001, hep-lat/0510062.

399. A. Ali Khan et al., Phys. Rev. D74 (2006) 094508, hep-lat/0603028.

400. T. Yamazaki et al., Phys. Rev. Lett. 100 (2008) 171602, 0801.4016.

401. ETM, C. Alexandrou et al., Phys. Rev. D83 (2011) 045010, 1012.0857.

402. S. Capitani et al., Phys. Rev. D86 (2012) 074502, 1205.0180.

403. R. Horsley et al., Phys. Lett. B732 (2014) 41, 1302.2233.

404. G.S. Bali et al., Phys. Rev. D91 (2015) 054501, 1412.7336.

405. A. Abdel-Rehim et al., Phys. Rev. D92 (2015) 114513, 1507.04936, [Erratum: Phys. Rev.D93,no.3,039904(2016)].

406. T. Bhattacharya, V. Cirigliano, S. Cohen, R. Gupta, H.W. Lin and B. Yoon, Phys. Rev. D94 (2016) 054508, 1606.07049.
407. C. Alexandrou, M. Constantinou, K. Hadjiyiannakou, K. Jansen, C. Kallidonis, G. Koutsou and A. Vaquero Aviles-Casco, Phys. Rev. D96 (2017) 054507, 1705.03399.
408. E. Berkowitz et al., (2017), 1704.01114.
409. S. Capitani et al., Int. J. Mod. Phys. A34 (2019) 1950009, 1705.06186.
410. C.C. Chang et al., Nature 558 (2018) 91, 1805.12130.
411. R. Gupta, Y.C. Jang, B. Yoon, H.W. Lin, V. Cirigliano and T. Bhattacharya, (2018), 1806.09006.
412. JLQCD, N. Yamanaka, S. Hashimoto, T. Kaneko and H. Ohki, Phys. Rev. D98 (2018) 054516, 1805.10507.
413. J. Liang, Y.B. Yang, T. Draper, M. Gong and K.F. Liu, Phys. Rev. D98 (2018) 074505, 1806.08366.
414. T. Harris, G. von Hippel, P. Junnarkar, H.B. Meyer, K. Ottnad, J. Wilhelm, H. Wittig and L. Wrang, Phys. Rev. D100 (2019) 034513, 1905.01291.
415. Y. Aoki, T. Blum, H.W. Lin, S. Ohta, S. Sasaki, R. Tweedie, J. Zanotti and T. Yamazaki, Phys. Rev. D82 (2010) 014501, 1003.3387.
416. J.R. Green, J.W. Negele, A.V. Pochinsky, S.N. Syritsyn, M. Engelhardt and S. Krieg, Phys. Rev. D86 (2012) 114509, 1206.4527.
417. PNDME, T. Bhattacharya, V. Cirigliano, S. Cohen, R. Gupta, A. Joseph, H.W. Lin and B. Yoon, Phys. Rev. D92 (2015) 094511, 1506.06411.
418. T. Bhattacharya, V. Cirigliano, R. Gupta, H.W. Lin and B. Yoon, Phys. Rev. Lett. 115 (2015) 212002, 1506.04196.
419. C. Alexandrou et al., Phys. Rev. D95 (2017) 114514, 1703.08788, [erratum: Phys. Rev.D96,no.9,099906(2017)].
420. R. Gupta, Y.C. Jang, H.W. Lin, B. Yoon and T. Bhattacharya, Phys. Rev. D96 (2017) 114503, 1705.06834.
421. J. Green et al., Phys. Rev. D95 (2017) 114502, 1703.06703.
422. T. Doi, M. Deka, S.J. Dong, T. Draper, K.F. Liu, D. Mankame, N. Mathur and T. Streuer, Phys. Rev. D80 (2009) 094503, 0903.3232.
423. R. Babich, R.C. Brower, M.A. Clark, G.T. Fleming, J.C. Osborn, C. Rebbi and D. Schaich, Phys. Rev. D85 (2012) 054510, 1012.0562.
424. P.E. Shanahan et al., Phys. Rev. Lett. 114 (2015) 091802, 1403.6537.
425. J. Green et al., Phys. Rev. D92 (2015) 031501, 1505.01803.
426. R.S. Sufian, Y.B. Yang, A. Alexandru, T. Draper, J. Liang and K.F. Liu, Phys. Rev. Lett. 118 (2017) 042001, 1606.07075.
427. H. Ohki et al., Phys. Rev. D78 (2008) 054502, 0806.4744.
428. PACS-CS, K.I. Ishikawa et al., Phys. Rev. D80 (2009) 054502, 0905.0962.
429. J. Martin Camalich, L.S. Geng and M.J. Vicente Vacas, Phys. Rev. D82 (2010) 074504, 1003.1929.
430. S. Dürr et al., Phys. Rev. D85 (2012) 014509, 1109.4265, [Erratum: Phys. Rev.D93,no.3,039905(2016)].
431. QCDSF-UKQCD, R. Horsley et al., Phys. Rev. D85 (2012) 034506, 1110.4971.
432. G.S. Bali et al., Nucl. Phys. B866 (2013) 1, 1206.7034.
433. P.E. Shanahan, A.W. Thomas and R.D. Young, Phys. Rev. D87 (2013) 074503, 1205.5365.
434. C. Alexandrou, V. Drach, K. Jansen, C. Kallidonis and G. Koutsou, Phys. Rev. D90 (2014) 074501, 1406.4310.
435. S. Dürr et al., Phys. Rev. Lett. 116 (2016) 172001, 1510.08013.
436. χQCD, Y.B. Yang, A. Alexandru, T. Draper, J. Liang and K.F. Liu, Phys. Rev. D94 (2016) 054503, 1511.09089.
437. RQCD, G.S. Bali, S. Collins, D. Richtmann, A. Schäfer, W. Söldner and A. Sternbeck, Phys. Rev. D93 (2016) 094504, 1603.00827.

438. ETM, A. Abdel-Rehim, C. Alexandrou, M. Constantinou, K. Hadjiyiannakou, K. Jansen, C. Kallidonis, G. Koutsou and A. Vaquero Aviles-Casco, Phys. Rev. Lett. 116 (2016) 252001, 1601.01624.

439. MILC, D. Toussaint and W. Freeman, Phys. Rev. Lett. 103 (2009) 122002, 0905.2432.

440. JLQCD, H. Ohki, K. Takeda, S. Aoki, S. Hashimoto, T. Kaneko, H. Matsufuru, J. Noaki and T. Onogi, Phys. Rev. D87 (2013) 034509, 1208.4185.

441. M. Engelhardt, Phys. Rev. D86 (2012) 114510, 1210.0025.

442. MILC, W. Freeman and D. Toussaint, Phys. Rev. D88 (2013) 054503, 1204.3866.

443. χQCD, M. Gong et al., Phys. Rev. D88 (2013) 014503, 1304.1194.

444. P. Junnarkar and A. Walker-Loud, Phys. Rev. D87 (2013) 114510, 1301.1114.

445. S. Syritsyn, PoS LATTICE2013 (2014) 009, 1403.4686.

446. M. Constantinou, PoS LATTICE2014 (2015) 001, 1411.0078.

447. J. Green, 36th International Symposium on Lattice Field Theory (Lattice 2018) East Lansing, MI, United States, July 22–28, 2018, 2018, 1812.10574.

448. N. Brambilla et al., Eur. Phys. J. C74 (2014) 2981, 1404.3723.

449. H.W. Lin et al., Prog. Part. Nucl. Phys. 100 (2018) 107, 1711.07916.

450. G.S. Bali, S. Collins and A. Schafer, Comput. Phys. Commun. 181 (2010) 1570, 0910.3970.

451. T. Blum, T. Izubuchi and E. Shintani, Phys. Rev. D88 (2013) 094503, 1208.4349.

452. M. Cè, L. Giusti and S. Schaefer, Phys. Rev. D93 (2016) 094507, 1601.04587.

453. M. Cè, L. Giusti and S. Schaefer, Phys. Rev. D95 (2017) 034503, 1609.02419.

454. A. Stathopoulos, J. Laeuchli and K. Orginos, (2013), 1302.4018.

455. A.S. Gambhir, A. Stathopoulos, K. Orginos, B. Yoon, R. Gupta and S. Syritsyn, PoS LATTICE2016 (2016) 265, 1611.01193.

456. B. Yoon et al., Phys. Rev. D93 (2016) 114506, 1602.07737.

457. T.A. DeGrand and S. Schaefer, Comput. Phys. Commun. 159 (2004) 185, hep-lat/0401011.

458. L. Giusti, P. Hernandez, M. Laine, P. Weisz and H. Wittig, JHEP 04 (2004) 013, hep-lat/0402002.

459. R. Gupta, A. Patel, C.F. Baillie, G. Guralnik, G.W. Kilcup and S.R. Sharpe, Phys. Rev. D40 (1989) 2072.

460. C. Thron, S.J. Dong, K.F. Liu and H.P. Ying, Phys. Rev. D57 (1998) 1642, hep-lat/9707001.

461. V. Gülpers, G. von Hippel and H. Wittig, Phys. Rev. D89 (2014) 094503, 1309.2104.

462. S. Bernardson, P. McCarty and C. Thron, Comput. Phys. Commun. 78 (1993) 256.

463. L. Maiani, G. Martinelli, M.L. Paciello and B. Taglienti, Nucl. Phys. B293 (1987) 420.

464. S. Güsken, K. Schilling, R. Sommer, K.H. Mütter and A. Patel, Phys. Lett. B212 (1988) 216.

465. S. Güsken, U. Löw, K.H. Mütter, R. Sommer, A. Patel and K. Schilling, Phys. Lett. B227 (1989) 266.

466. S.J. Dong, K.F. Liu and A.G. Williams, Phys. Rev. D58 (1998) 074504, hep-ph/9712483.

467. S. Capitani, B. Knippschild, M. Della Morte and H. Wittig, PoS LATTICE2010 (2010) 147, 1011.1358.

468. J. Bulava, M. Donnellan and R. Sommer, JHEP 01 (2012) 140, 1108.3774.

469. B.J. Owen, J. Dragos, W. Kamleh, D.B. Leinweber, M.S. Mahbub, B.J. Menadue and J.M. Zanotti, Phys. Lett. B723 (2013) 217, 1212.4668.

470. J. Dragos et al., Phys. Rev. D94 (2016) 074505, 1606.03195.

471. Particle Data Group, M. Tanabashi et al., Phys. Rev. D98 (2018) 030001.

472. O. Bär, Int. J. Mod. Phys. A32 (2017) 1730011, 1705.02806.

473. O. Bär, EPJ Web Conf. 175 (2018) 01007, 1708.00380.

474. F. Jegerlehner and A. Nyffeler, Phys. Rept. 477 (2009) 1, 0902.3360.

475. F. Jegerlehner, Springer Tracts Mod. Phys. 274 (2017) pp.1.

476. M. Davier, A. Höcker, B. Malaescu and Z. Zhang, Eur. Phys. J. C71 (2011) 1515, 1010.4180.

477. K. Hagiwara, R. Liao, A.D. Martin, D. Nomura and T. Teubner, J. Phys. G38 (2011) 085003, 1105.3149.

478. F. Jegerlehner, EPJ Web Conf. 166 (2018) 00022, 1705.00263.

479. M. Davier, A. Hoecker, B. Malaescu and Z. Zhang, Eur. Phys. J. C77 (2017) 827, 1706.09436.

480. A. Keshavarzi, D. Nomura and T. Teubner, Phys. Rev. D97 (2018) 114025, 1802.02995.

481. G. Colangelo, M. Hoferichter, M. Procura and P. Stoffer, JHEP 09 (2014) 091, 1402.7081.
482. G. Colangelo, M. Hoferichter, B. Kubis, M. Procura and P. Stoffer, Phys. Lett. B738 (2014) 6, 1408.2517.
483. G. Colangelo, M. Hoferichter, M. Procura and P. Stoffer, JHEP 09 (2015) 074, 1506.01386.
484. G. Colangelo, M. Hoferichter, M. Procura and P. Stoffer, Phys. Rev. Lett. 118 (2017) 232001, 1701.06554.
485. G. Colangelo, M. Hoferichter, M. Procura and P. Stoffer, JHEP 04 (2017) 161, 1702.07347.
486. V. Pascalutsa and M. Vanderhaeghen, Phys. Rev. Lett. 105 (2010) 201603, 1008.1088.
487. V. Pascalutsa, V. Pauk and M. Vanderhaeghen, Phys. Rev. D85 (2012) 116001, 1204.0740.
488. V. Pauk and M. Vanderhaeghen, (2014), 1403.7503.
489. V. Pauk and M. Vanderhaeghen, Phys. Rev. D90 (2014) 113012, 1409.0819.
490. I. Danilkin and M. Vanderhaeghen, Phys. Rev. D95 (2017) 014019, 1611.04646.
491. A. Nyffeler, Phys. Rev. D94 (2016) 053006, 1602.03398.
492. J. Prades, E. de Rafael and A. Vainshtein, Adv. Ser. Direct. High Energy Phys. 20 (2009) 303, 0901.0306.
493. A. Nyffeler, Phys. Rev. D79 (2009) 073012, 0901.1172.
494. H.B. Meyer and H. Wittig, Prog. Part. Nucl. Phys. 104 (2019) 46, 1807.09370.
495. Budapest-Marseille-Wuppertal, S. Borsanyi et al., Phys. Rev. Lett. 121 (2018) 022002, 1711.04980.
496. B. Chakraborty, C.T.H. Davies, P.G. de Oliviera, J. Koponen, G.P. Lepage and R.S. Van de Water, Phys. Rev. D96 (2017) 034516, 1601.03071.
497. ETM, F. Burger, X. Feng, G. Hotzel, K. Jansen, M. Petschlies and D.B. Renner, JHEP 02 (2014) 099, 1308.4327.
498. RBC/UKQCD, T. Blum et al., Phys. Rev. Lett. 121 (2018) 022003, 1801.07224.
499. M. Della Morte et al., JHEP 10 (2017) 020, 1705.01775.
500. B.e. Lautrup, A. Peterman and E. de Rafael, Phys. Rept. 3 (1972) 193.
501. T. Blum, Phys. Rev. Lett. 91 (2003) 052001, hep-lat/0212018.
502. D. Bernecker and H.B. Meyer, Eur. Phys. J. A47 (2011) 148, 1107.4388.
503. HPQCD, B. Chakraborty, C.T.H. Davies, G.C. Donald, R.J. Dowdall, J. Koponen, G.P. Lepage and T. Teubner, Phys. Rev. D89 (2014) 114501, 1403.1778.
504. H.B. Meyer, Eur. Phys. J. C77 (2017) 616, 1706.01139.
505. M. Della Morte et al., EPJ Web Conf. 175 (2018) 06031, 1710.10072.
506. C. Aubin and T. Blum, Phys. Rev. D75 (2007) 114502, hep-lat/0608011.
507. X. Feng, K. Jansen, M. Petschlies and D.B. Renner, Phys. Rev. Lett. 107 (2011) 081802, 1103.4818.
508. P. Boyle, L. Del Debbio, E. Kerrane and J. Zanotti, Phys. Rev. D85 (2012) 074504, 1107.1497.
509. M. Della Morte, B. Jäger, A. Jüttner and H. Wittig, JHEP 1203 (2012) 055, 1112.2894.
510. T. Blum et al., Phys. Rev. Lett. 116 (2016) 232002, 1512.09054.
511. RBC/UKQCD, T. Blum et al., JHEP 04 (2016) 063, 1602.01767.
512. S. Borsanyi et al., Phys. Rev. D96 (2017) 074507, 1612.02364.
513. D. Giusti, V. Lubicz, G. Martinelli, F. Sanfilippo and S. Simula, JHEP 10 (2017) 157, 1707.03019.
514. Fermilab Lattice, LATTICE-HPQCD, MILC, B. Chakraborty et al., Phys. Rev. Lett. 120 (2018) 152001, 1710.11212.
515. D. Giusti, F. Sanfilippo and S. Simula, Phys. Rev. D98 (2018) 114504, 1808.00887.
516. K. Miura, PoS LATTICE2018 (2019) 010, 1901.09052.
517. M. Hayakawa, T. Blum, T. Izubuchi and N. Yamada, PoS LAT2005 (2006) 353, hep-lat/0509016.
518. T. Blum, S. Chowdhury, M. Hayakawa and T. Izubuchi, Phys. Rev. Lett. 114 (2015) 012001, 1407.2923.
519. T. Blum, N. Christ, M. Hayakawa, T. Izubuchi, L. Jin and C. Lehner, Phys. Rev. D93 (2016) 014503, 1510.07100.
520. T. Blum, N. Christ, M. Hayakawa, T. Izubuchi, L. Jin, C. Jung and C. Lehner, Phys. Rev. Lett. 118 (2017) 022005, 1610.04603.

521. J. Green, N. Asmussen, O. Gryniuk, G. von Hippel, H.B. Meyer, A. Nyffeler and V. Pascalutsa, PoS LATTICE2015 (2016) 109, 1510.08384.
522. N. Asmussen, J. Green, H.B. Meyer and A. Nyffeler, PoS LATTICE2016 (2016) 164, 1609.08454.
523. N. Asmussen, A. Gérardin, H.B. Meyer and A. Nyffeler, EPJ Web Conf. 175 (2018) 06023, 1711.02466.
524. N. Asmussen et al., EPJ Web Conf. 179 (2018) 01017, 1801.04238.
525. T. Blum, N. Christ, M. Hayakawa, T. Izubuchi, L. Jin, C. Jung and C. Lehner, Phys. Rev. D96 (2017) 034515, 1705.01067.
526. J. Green, O. Gryniuk, G. von Hippel, H.B. Meyer and V. Pascalutsa, Phys. Rev. Lett. 115 (2015) 222003, 1507.01577.
527. A. Gérardin, J. Green, O. Gryniuk, G. von Hippel, H.B. Meyer, V. Pascalutsa and H. Wittig, Phys. Rev. D98 (2018) 074501, 1712.00421.
528. A. Gérardin, H.B. Meyer and A. Nyffeler, Phys. Rev. D94 (2016) 074507, 1607.08174.
529. D. Djukanovic, K. Ottnad, J. Wilhelm and H. Wittig, Phys. Rev. Lett. 123, 21 (2019), 212001.
530. A. Gérardin, M. Cè, G. von Hippel, B. Hörz, H.B. Meyer, D. Mohler, K. Ottnad, J. Wilhelm and H. Wittig, Phys. Rev. D100 (2019) 014510, 1904.03120.

Chapter 6
The Discovery of the Higgs Boson at the LHC

Peter Jenni and Tejinder S. Virdee

6.1 Introduction and the Standard Model

The standard model of particle physics (SM) is a theory that is based upon principles of great beauty and simplicity. The theory comprises the building blocks of visible matter, the fundamental fermions: quarks and leptons, and the fundamental bosons that mediate three of the four fundamental interactions; photons for electromagnetism, the W and Z bosons for the weak interaction and gluons for the strong interaction (Fig. 6.1).

The SM provides a very successful description of the visible universe and has been verified in many experiments to a very high precision. It has an enormous range of applicability and validity. So far no significant deviations have been observed experimentally.

The possibility of installing a proton-proton accelerator in the LEP tunnel, after the e^+e^- programme, was being discussed in the 1980's. At the time there were many profound open questions in particle physics, and several are still present. In simple terms these are: what is the origin of mass i.e. how do fundamental particles acquire mass, and why do they have the masses that they have? Why is there more matter than anti-matter? What is dark matter? What is the path towards unification of all forces? Do we live in a world with more space-time dimensions than the familiar four? The LHC [1, 2] was conceived to address or shed light on these questions.

P. Jenni
CERN, Geneva, Switzerland

Albert-Ludwigs University Freiburg, Freiburg im Breisgau, Germany

T. S. Virdee (✉)
Imperial College London, London, UK
e-mail: Tejinder.Virdee@cern.ch

© The Author(s) 2020
H. Schopper (ed.), *Particle Physics Reference Library*,
https://doi.org/10.1007/978-3-030-38207-0_6

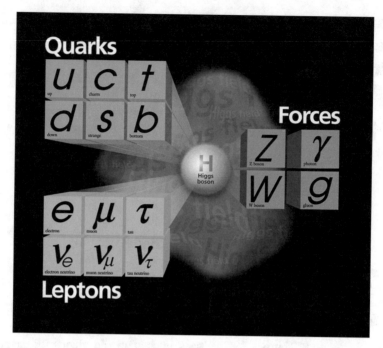

Fig. 6.1 Particle content of the SM, including the Higgs boson considered to be the keystone of the SM

The question of how fundamental particles acquire mass was first posed in the following form: how does the photon remain massless, giving the electromagnetic force an infinite range, whilst the W and Z bosons acquire a seemingly large mass, explaining the short-range of the weak nuclear force.

In 1964 three groups of physicists, Englert and Brout; Higgs; and Guralnik, Hagen, and Kibble [3–7], proposed that there exists an omnipresent field, pervading the universe, and fundamental particles can acquire mass by interacting with this field. At the heart of the mechanism endowing mass was spontaneous symmetry breaking of a local gauge symmetry, through the field's non-zero vacuum expectation value. The new field being a quantum field had an associated quantum, which became known as the Higgs boson.

Today, it seems remarkable that not much attention was paid to the papers [3–7], and even less to the associated Higgs boson. This was partly due to the fact that in the early 1960's most particle physicists were trying to make sense of a plethora of new particles being discovered.

In 1967 Kibble [8] generalized his earlier work with Guralnik and Hagen and brought the mechanism of spontaneous symmetry breaking closer to its application to the description of the real world, one in which the photon remains massless and the W and Z particles become massive [9]. This vein of work reached fruition in the seminal papers of Weinberg [10] and Salam [11], which raised the prospect of the unification of electromagnetism and weak interactions, now labeled the electro-weak theory. Earlier work on a similar model had been carried out by S. Glashow [12]. Weinberg and Salam assumed that W and Z bosons acquired mass by interacting with the field introduced in the earlier papers [3–7].

Both Weinberg and Salam conjectured that such a model would be renormalizable i.e. calculations would give finite answers. The key prediction of their theory was the existence of the Z^0 boson, in addition to the long-known charged W bosons. Again not much attention was paid to these papers.

The situation changed dramatically in 1971. t'Hooft in a *tour de force*, using methods developed by Veltman, outlined the proof that, indeed, the electro-weak theory would be renormalizable [13]. The electro-weak theory started being taken very seriously, so much so that Weinberg's paper [10] has now become the most cited paper in physics.

Experimentally, the 1973 discovery of weak neutral currents [14], mediated by the Z^0 boson, provided strong evidence for the verity of the electro-weak theory.

In parallel much progress had been made in understanding the particles that were being discovered in the 1950s and 1960s. Eventually, these were understood through an underlying gauge field theory, where the "charge" of strong interactions was labeled "colour", and the interactions of coloured quarks are mediated by gluons. The theory [15, 16] displayed two main properties: colour confinement, resulting in the hadrons being colourless, and asymptotic freedom, leading to a steady decrease in the strength of the interaction between quarks and gluons as the interaction energy scale increases. The latter enabled the use of perturbation theory for calculating strong interaction processes at high energies, which has been key to understanding the physics at the LHC.

Further major discoveries included those of new quarks and the gluon meant that the discovery, in 1983, of the W and Z bosons [17, 18] at CERN set the stage for the search for the Higgs boson. The Higgs boson, that earlier had been considered to be a minor and uninteresting feature of the spontaneous breaking mechanism, became to assume a role of central importance as the still missing key particle of the SM. The SM worked so well that the Higgs boson, or something else doing the same job, more or less had to be present.

In 1984, one year after the discovery of the W and Z bosons, a workshop was held in Lausanne where first ideas were discussed about a possible proton-proton collider and associated experiments to make a search for such a particle. The aim was to reuse the LEP tunnel after the end of the electron-positron programme. Amongst the leading protagonists were the scientists from UA1 and UA2 experiments. An exploratory machine was required to cover the wide range of mass values possible for the SM Higgs boson, its diverse decay signatures and production mechanisms and to discover any new high-mass particles at a centre-of-mass energy ten times higher than previously probed. A hadron (proton-proton) collider is such a machine as long as the proton energy is high enough and the instantaneous luminosity, $\mathscr{L}$, measured in $cm^{-2} s^{-1}$, is sufficiently large. The rate of production of a given particle is determined by $\mathscr{L} \times \sigma$ where σ is the cross section of the production reaction, measured in units of cm^2. The most interesting and easily detectable final states at a hadron collider involve charged leptons and photons and have a low $\sigma \times$ BR, where BR is the branching ratio into the decay mode of interest.

A major goal of the LHC thus became the elucidation of the mechanism for electroweak symmetry breaking. It also was clear that a search had to be made for new physics at the TeV energy scale as the SM is logically incomplete; it does not incorporate gravity. A promising avenue is the superstring theory, an attempt towards a unified theory with dramatic predictions of extra space dimensions and supersymmetry.

The LHC and its experiments were designed to find new particles, new forces and new symmetries amongst which could be the Higgs boson(s), supersymmetric particles, Z' bosons, or evidence of extra space dimensions. An experiment that could cover the detection of all these hypothesized but yet undiscovered particles would provide the best opportunity to discover whatever else might be produced at LHC energies.

In July 2012 the ATLAS and CMS collaborations discovered a Higgs boson [19, 20].

This paper is based on the previous articles [1, 21–23] written by the authors, some with M. Della Negra, using the recently published results from the ATLAS and CMS Collaborations on the measurements of the properties of the Higgs boson.

6.2   The SM Higgs Boson

In the early 1990's the search for the SM Higgs boson played a pivotal role in the design of the ATLAS and CMS experiments. The mass of the Higgs boson (m_H) is not predicted by theory, but from general considerations, $m_H < 1$ TeV. At the start of the LHC operation, direct searches for the Higgs boson carried out at the LEP collider led to a lower bound of $m_H > 114.4$ GeV at 95% CL [24], whilst precision electroweak constraints, including LEP data, implied that $m_H < 152$ GeV at 95% confidence level (CL) [25]. At time of the discovery at CERN, CDF and D0 experiments operating the Tevatron proton antiproton collider, detected an excess of events in the range 120–135 GeV [26].

It is known that quantum corrections make the mass of any fundamental scalar particle, such as the SM Higgs boson, float up to the next highest mass scale present in the theory, which in the absence of extensions to the SM, can be as high as 10^{15} GeV. Hence finding the scalar Higgs boson would immediately raise a more puzzling question: Why should it have a mass in the range between 100 GeV and 1 TeV? One appealing hypothesis, much discussed at the time, and still being investigated, predicts a new symmetry labeled *supersymmetry*. For every known SM particle there would be a partner with spin differing by half a unit; fermions would have boson superpartners and vice versa, thus doubling the number of fundamental particles. The contributions from the boson and fermion superpartners, and vice a versa, with amplitudes of opposite signs, would lead to their cancellation, and allow a low mass for the Higgs boson. In supersymmetry five Higgs bosons are predicted to exist with one resembling the SM Higgs boson with a mass below about 140 GeV. The lightest of this new species of superparticles could be the candidate for *dark matter* whose presence, by mass, in the universe is around five times more abundant than ordinary matter.

In 1975, physicists had already started to turn their attention to how a putative Higgs boson would manifest itself in experiments [27].

The search for the SM Higgs boson provided a stringent benchmark for evaluating the physics performance of various experiment designs under consideration in the early 1990s and heavily influenced the conceptual design of the general-purpose experiments, ATLAS and CMS.

6.2.1   Higgs Boson: Production and Decay

Although the mass of the Higgs boson is not predicted by theory, at a given mass all of its other properties are precisely predicted within the SM. The SM Higgs boson is short-lived (10^{-23} s) and hence the experiments only detect the decay products.

The cross sections for differing production mechanisms and the branching fractions for differing decay modes of the SM Higgs boson, as a function of mass, are illustrated in Fig. 6.2a, b, respectively [28], and the principal ones for $m_H = 125$ GeV and at $\sqrt{s} = 14$ TeV are tabulated in Table 6.1. The uncertainties on these numbers can be found in the twiki in Reference [28].

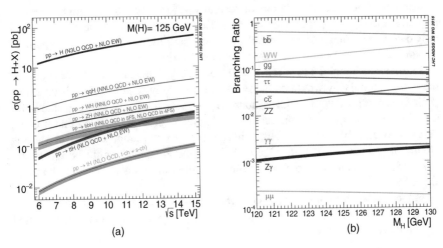

Fig. 6.2 (a) SM Higgs boson production cross sections as a function of the centre-of-mass energy, $\sqrt{s}$, for pp. collisions. The VBF process is indicated here as qqH [28]. The theoretical uncertainties are indicated as bands. (b) Branching ratios for the main decays of the SM Higgs boson near $m_H = 125$GeV [28]. The theoretical uncertainties are indicated as bands

Table 6.1 The Production cross sections and decay branching fractions at $\sqrt{s} = 14$ TeV

Production process	Cross section (pb)	Decay Mode H→	Branching fraction (%)
gg fusion (pp → H)	54.67	bb	58.24
qqH (VBF)	4.28	WW$^{(*)}$	21.37
WH (associated)	1.51	ττ	6.27
ZH (associated)	0.99	ZZ$^{(*)}$	2.62
ttH	0.61	γγ	2.27×10^{-1}
		μμ	2.18×10^{-2}

The dominant Higgs-boson production mechanism, labeled pp. → H in Fig. 6.2a (for masses up to ≈ 700 GeV) is gluon–gluon fusion.

The vector boson fusion (VBF) mechanism (WW$^{(*)}$ or ZZ$^{(*)}$), labeled pp. → qqH in Fig. 6.2a, becomes important for the production of higher-mass Higgs bosons. Here, the quarks that emit the W or Z bosons have transverse momenta of the order of W and Z masses. The detection of the resulting high-energy jets in the forward pseudorapidity[1] regions, $2.0 < |\eta| < 5.0$, can be used to tag the reaction, improving the signal-to-noise ratio. Tagging of forward jets from the VBF process has turned out to be very important in the measurements of many of the properties of the Higgs boson.

The production of the Higgs boson in association with W and Z boson, labeled pp. → W or Z H in Fig. 6.2a, or the production via the t-tbar fusion, has a much lower cross section, but nevertheless has been important for the final states with large backgrounds such as b-bbar, $\tau^+\tau^-$ or $\mu^+\mu^-$.

The Higgs boson decays in one of several ways (decay modes) into known SM particles, the types depending on its mass. Hence a search had to be envisaged not only over a large range of masses but also many possible decay modes: into pairs of photons, Z bosons, W bosons, τ leptons, and b quarks.

In the mass interval $110 < m_H < 150$ GeV, early detailed studies indicated that the two-photon decay would be the main channel likely to give a significant signal [29]. Detailed studies of another mode, H → ZZ$^{(*)}$ → $\ell\ell\ell\ell$, where ℓ stands for a charged electron or a muon, dubbed the "golden" mode, suggested that it could be used to cleanly detect the Higgs boson over a wide range of masses starting around $m_H = 130$ GeV [30]. One or both of the Z bosons would be virtual for $m_H < 180$ GeV, and the upper end of the detection range was indicated to be about $m_H < 600$ GeV.

In the region $700 < m_H < 1000$ GeV the cross-section decreases so Higgs boson decays via W and Z decays, where the W and Z decays are to channels with higher branching fractions, have to be employed.

---

[1] The pseudorapidity $\eta = -\ln[\tan(\theta/2)]$ where and θ is the polar angle measured from the positive z axis (along the beam direction).

6.3 The Large Hadron Collider

6.3.1 The Road to the LHC

With the prospect of ground-breaking physics at the LHC, several workshops and conferences followed, where the formidable experimental challenges started to appear manageable, provided that enough R&D work on detectors could be carried out. In 1987 the workshop in La Thuile of the so-called "Rubbia Long-Range Planning Committee" resulted in the recommendation of a proton-proton collider, labeled the Large Hadron Collider (LHC), as the next accelerator for CERN. Meetings of note were the ECFA LHC Workshop in Aachen in 1990 [31], and "Towards the LHC Experimental Programme" [32] which took place in Evian-les-Bains, France in March 1992. At Evian several proto-collaborations presented their designs in "Expressions of Interest". In addition, from the early 1990s, CERN's LHC Detector R&D Committee (DRDC), which reviewed and steered R&D groupings, greatly stimulated innovative developments in detector technology.

Table 6.2 lists the major steps on the long road to the discovery of the Higgs boson.

Table 6.2 The LHC Timeline

1984	Workshop on a Large Hadron Collider in the LEP tunnel, Lausanne, Switzerland.
1987	Workshop on the Physics at Future Accelerators, La Thuile, Italy. The Rubbia "Long-Range Planning Committee" recommends the Large Hadron Collider as the right choice for CERN's future.
1990	LHC Workshop, Aachen, Germany (discussion of physics, technologies and detector design concepts).
1992	General Meeting on LHC Physics and Detectors, Evian-les-Bains, France (with four general-purpose experiment designs presented).
1993	Three Letters of Intent evaluated by the CERN peer review committee LHCC. ATLAS and CMS selected to proceed to a detailed technical proposal.
1994	The LHC accelerator approved for construction, initially in two stages.
1996	ATLAS and CMS Technical Proposals approved.
1997	Formal approval for ATLAS and CMS to move to construction (materials cost ceiling of 475 MCHF).
1997	Construction commences (after approval of detailed Technical Design Reports of detector subsystems).
2000	Assembly of experiments commences, LEP accelerator is closed down to make way for the LHC.
2008	LHC experiments ready for pp. collisions. LHC starts operation. An incident stops LHC operation.
2009	LHC restarts operation, pp. collisions recorded by LHC detectors.
2010	LHC collides protons at high energy (centre of mass energy of 7 TeV).
2012	LHC operates at $\sqrt{s} = 8$ TeV: discovery of a Higgs boson.
2015	LHC operates at $\sqrt{s} = 13$ TeV.

6.3.2 The Challenges of the LHC Accelerator

In this section we outline some of the features and the technological challenges of the LHC [1].

Protons are accelerated by high electric fields generated in superconducting r.f. cavities and are guided around the accelerator by powerful superconducting dipole magnets. The dipole magnets are designed to operate at 8.3 Tesla, allowing the proton beams to be accelerated to 7 TeV, with the current carrying conductor cooled down to 1.9 K in a bath of superfluid helium. The beam pipe in which the protons circulate is under a better vacuum, and at a lower temperature, than that found in inter-planetary space.

The choices of two-in-one high-field superconducting dipole magnets operating at a temperature of 1.9 K, cooled by super-fluid helium were critical to a competitive and affordable design. The LHC could only be competitive with the Superconducting Super Collider (SSC), whose construction had started in the early 1990s in Texas, U.S.A, if the instantaneous luminosity could be an order of magnitude higher (at 10^{34} cm^{-2} s^{-1}). However, the SSC was later cancelled in October 1993.

The main challenges for the accelerator were to build more than one thousand two hundred 15 m long superconducting dipoles able to reach the required magnetic field, the large distributed cryogenic plant to cool the magnets and other superconducting accelerator structures, and the control of the beams, whose stored energy will reach, in design operation, a value of 350 MJ. This magnitude requires extraordinary precautions for beam handling, since if, for any reason this beam is lost in an uncontrolled way, it can do considerable damage to the machine elements, which would result in months of down time.

The counter-rotating LHC beams are organized in 2808 bunches, each of ~10^{11} protons per bunch separated by 25 ns, leading to a bunch crossing rate of ~40 MHz.

Proton beams were first circulated in the LHC in September 2008, and in the days that followed, rapid progress was made in getting a beam to circulate with very good lifetime. Soon after the start a technical incident occurred in the last of the eight sectors to be tested as it was being ramped up to the pre-agreed start-up energy of 5 TeV. The root cause was a failure of one of the 50,000 soldered joints. Substantial damage was done to a large part of the sector involved. After repairs lasting about a year, the LHC started operating again in November 2009. Collisions took place at the injection energy (450 GeV per beam), followed in 2010 and 2011, by a very successful operation at a centre-of-mass energy of 7 TeV. In 2012 the centre-of-mass energy was increased to 8 TeV. The performance surpassed expectations and an integrated luminosity of ~25 fb^{-1}, corresponding to 2×10^{15} proton-proton interactions, was delivered. This is labeled Run 1.

During the period 2015–2018 the LHC operated at a proton-proton centre-of-mass energy of 13 TeV and delivered a total of over 150 fb^{-1} of integrated luminosity. The collider performed close to, or beyond, its design values in many parameters, operating at 13 TeV and reaching peak luminosities of 2×10^{34}

Table 6.3 Some of the LHC parameters of attained/design performance for ATLAS and CMS

	Achieved	Design
Energy	13 TeV	14 TeV
Max. no. of bunches	2556	2808
Bunch spacing (ns)	25	25
Protons/bunch (10^{11})	1.1	1.15
β_* (cm)	30	55
Peak luminosity (10^{34} cm^{-2} s^{-1})	2.1	1.0
Total integrated luminosity (fb^{-1})		
$\sqrt{s} = 7$ TeV (Run 1)	5	
$\sqrt{s} = 8$ TeV (Run 1)	20	
$\sqrt{s} = 13$ TeV (Run 2)	140	

cm^{-2} s^{-1}, twice the design value. This period of operation is labeled Run 2. The achieved performance at the time of writing (2018) can be found in Table 6.3.

6.4 The ATLAS and CMS Experiments

Not only was the putative SM Higgs boson to be rarely produced in the proton collisions, but also it decays into particles (isolated photons, electrons, and muons) that are the best identifiable signatures of its production at the LHC also was expected to be rare. The rarity is illustrated by the fact that Higgs boson production and decay to one such distinguishable signature (H $\rightarrow$ ZZ$^{(*)}$ $\rightarrow$ 4 l) happens roughly once in 10^{13} proton-proton collisions. So a vast number of proton-proton collisions per second have to be delivered by the accelerator and examined by the experiments. At the end of 2018, the LHC was operating at a collision rate of around 10^9 per second. The ATLAS and CMS detectors operate in the harsh environment created by this huge rate of proton-proton collisions. The challenges posed are discussed in reference [33, 34].

6.4.1 The Challenges for ATLAS and CMS Experiments

At the Aachen workshop the physics case for the LHC was thoroughly examined. The experimental search for the Higgs boson across the entire possible range of mass was fully explored for the first time. There was a prevalent prejudice of the protagonists of supersymmetry that m_H should be smaller than 135 GeV. As the decay width of the SM Higgs boson is about 5.5 MeV at $m_H = 100$ GeV, and 8.3 MeV at 150 GeV, the width of the reconstructed invariant ($\gamma\gamma$ or 4 l) mass distribution, and hence the signal/background ratio, would be limited by the electron/photon energy resolution of the electromagnetic calorimeter, and the charged particle momentum resolution of the inner tracker and the muon spectrometer. This lower end of the remaining open mass range was considered to be especially difficult in hadron colliders. Hence the LHC experiments had to pay particular attention to the performance requirements imposed by the search for the Higgs boson in this low mass range. As a consequence much importance was placed on the tracking (inner and muon), as well as the magnetic field strength, and the electromagnetic calorimeters.

The search for the high-mass Higgs boson, particles predicted by SUSY, and other exotic states mentioned above, required excellent resolution for jets and missing transverse momentum (p_T^{miss}), requiring full solid angle calorimeter coverage.

A saying prevalent in the late 1980's and early 1990's captured the challenge: 'We think we know how to build a high energy, high luminosity hadron collider— but we don't have the technology to build a detector for it'. Making discoveries in the unprecedented high collision rate environment, generated by around one billion proton-proton interactions per second, with several tens of simultaneous collisions per bunch crossing, would require extraordinary detectors. Many technical, financial, industrial and human challenges lay ahead, which were all overcome, to yield experiments of unprecedented complexity and power. A flavour can be attained from articles in reference [35].

At the Evian meeting in 1992 four experiment designs were presented: two deploying toroids (one with a superconducting magnet in the barrel) and two deploying superconducting high-field solenoids. The choice of the magnetic field configuration determined the overall design of the experiments.

The collaborations deploying toroids merged to form the ATLAS Collaboration. The ATLAS design [35] was based on a very large superconducting air-core toroid for the measurement of muons, and supplemented by a superconducting 2 Tesla solenoid to provide the magnetic field for inner tracking and by a liquid-argon/lead electromagnetic calorimeter with a novel "accordion" geometry. The CMS design [36] was based on a single large-bore, long, high-field solenoid for analyzing muons, together with powerful microstrip-based inner tracking and an electromagnetic calorimeter comprising scintillating crystals.

On top of the selected event of interest, an average of up to around 40 other proton-proton events are superimposed. These superposed events are referred to as *minimum-bias* events, because no selection is made. Thus thousands of particles emerge from the interaction region every 25 ns where one nanosecond (ns) = 10^{-9} s. Hence the products of an interaction under study can be confused with those from other interactions in the same bunch crossing. This problem, known as *pileup*, clearly becomes more severe if the response time of a detector element and its electronic signal is longer than 25 ns. The effect of pileup can be reduced by using highly granular detectors with fast, short duration, signals, giving low *occupancy* (i.e., a low probability that a detector element will give a signal) at the expense of having large numbers of detector channels. The resulting millions of electronic channels require very good time synchronization.

The large flux of particles emanating from the interaction region creates a high-radiation environment requiring radiation-hard detectors and front-end electronics.

Access for maintenance is very difficult, time consuming, and highly restricted. Hence, a high degree of long-term operational reliability had to be attained, comparable to that which is usually associated with instruments flying on space missions.

The event selection process (called the *trigger*) must select among the billion interactions that occur each second since no more than a thousand events per second can be stored for subsequent analysis. The short time between bunch crossings, 25 ns, has major implications for the design of the readout and trigger systems. It takes a long time to make a trigger decision, yet new events occur in every crossing and a trigger decision must be made for every crossing; the selection process is split in several levels. The first of these is the Level-1 trigger decision, which takes about 3 μs and selects, on average, one crossing out of 400. During this time the data must be stored in *pipelines* integrated into the front-end electronics. In CMS, the data from these selected events are then moved into a commercial farm of CPUs to select and store about one thousand/s of the most interesting events for subsequent analysis.

It cannot be stressed enough how important were the many years of R&D and prototyping that preceded the start of detector construction. Technologies had to be developed far beyond what was the state-of-the-art in early 1990s, in terms of granularity, speed of readout, radiation tolerance, reliability, and very importantly cost. For many detector subsystems, there were initially several technologies considered, as it was far from certain which technologies would be able to attain the required performance. In many cases several variants were developed, prototyped and tested, before choosing the one best able to fulfill the stringent requirements. This involved building and testing increasingly more realistic and larger prototypes, in a process that involved industry from the outset. This took place over a number of years before construction commenced in the second half of the 1990s.

In the 1990's the two collaborations, ATLAS and CMS, grew rapidly in terms of people and institutes. Today each comprises over 3500 scientists and engineers, from over 150 institutions in more than 40 countries. The talents and resources of all these scientists were needed to build the experiments, which are now performing extraordinarily well at the LHC.

The single most important aspect of the experiment design and layout is the magnetic field configuration for the identification of muons and the measurement of their momentum. Large bending power is needed to measure precisely the momentum of charged particles. This forces a choice of superconducting technology for the magnets. The design configurations chosen by ATLAS and CMS are discussed below.

6.4.2   The ATLAS Detector

The design of the ATLAS detector [35], shown in Fig. 6.3 (top), is based on a novel superconducting air-core toroid magnet system, containing ~80 km of

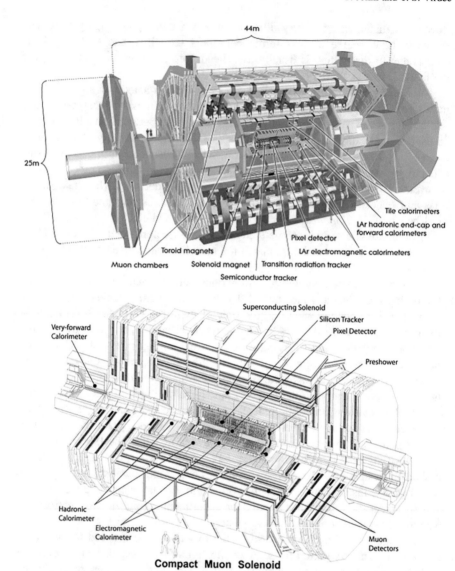

44m

25m

Tile calorimeters

LAr hadronic end-cap and forward calorimeters

Pixel detector

LAr electromagnetic calorimeters

Toroid magnets

Muon chambers   Solenoid magnet | Transition radiation tracker

Semiconductor tracker

Superconducting Solenoid

Silicon Tracker

Pixel Detector

Very-forward Calorimeter

Preshower

Hadronic Calorimeter

Electromagnetic Calorimeter

Muon Detectors

Compact Muon Solenoid

Fig. 6.3 Schematic longitudinal cut-away views of (top) the ATLAS and (bottom) the CMS detectors, showing the different layers around the LHC beam axis, with the collision point in the centre

superconductor cable, in eight separate barrel coils (each 25×5 m^2 in a 'racetrack' shape) and two matching endcap toroid systems. A field of ~0.5 Tesla is generated over a large volume. The toroids are complemented with a thin superconducting central solenoid (2.4 m diameter, 5.3 m length) that provides an axial magnetic field of 2 Tesla.

The electromagnetic calorimeter consists of a lead/liquid-argon sampling calorimeter in a novel 'accordion' geometry. A plastic scintillator—iron sampling hadron calorimeter, also with a novel geometry, is used in the barrel part of the experiment. Liquid-argon hadronic calorimeters are employed in the endcap regions near the beam axis. The electromagnetic and hadronic calorimeters have almost 200,000 and 20,000 cells, respectively, and are in an almost field-free region between the toroids and the solenoid.

The momentum of the muons is precisely measured after traversing the calorimeters in the air-core toroid field over a distance of ~5 m. About 1200 large muon chambers of various shapes, with a total area of 5000 m^2, measure the impact position with an accuracy of better than 0.1 mm. Another set of about 4200 fast chambers is used to provide the "trigger".

The reconstruction of all charged particles, and that of displaced vertices, is achieved in the inner detector, which combines highly granular pixel (50 × 400 μm^2 elements, leading to 80 million channels) and microstrip (13 cm × 80 μm elements, leading to six million channels) silicon semiconductor sensors placed close to the beam axis, and a 'straw tube' gaseous detector (350,000 channels) which provides about 30–40 signal hits per track. The latter also helps in the identification of electrons using information from the effects of transition radiation.

The air-core magnet system allows a relatively lightweight overall structure leading to a detector weighing 7000 tons. The muon spectrometer defines the overall diameter of 25 m and length of 44 m of the ATLAS detector.

6.4.3 The CMS Detector

The design of the CMS detector [36], shown in Fig. 6.3 (bottom), is based on a state-of-the-art superconducting high-field solenoid, which first reached the design field of 4 Tesla in 2006.

The solenoid generates a uniform magnetic field parallel to the direction of the LHC beams. The field is produced by a current of 20 kA flowing through a reinforced Nb-Ti superconducting coil built in four layers. Economic and transportation constraints limited the outer radius of the coil to 3 m and its length to 13 m. The field is returned through a 1.5 m thick iron yoke, which houses four muon stations to ensure robustness of identification and measurement and full geometric coverage.

The CMS design was first optimized to cleanly identify, trigger and measure muons, e.g. arising from processes such as H → ZZ$^{(*)}$ → 4 μ and few TeV mass Z' → 2 μ, over a wide range of momenta. The muons trace a spiral path in the magnetic field and are identified and reconstructed in ~3000 m^2 of gas chambers interleaved with the iron plates in the return yoke. Another ~500 fast chambers are used to provide a second system of detectors for the Level-1 muon trigger.

The next design priority was driven by the search for the decay of the SM Higgs boson into two photons. A new type of scintillating crystal was selected: lead-tungstate ($PbWO_4$) crystal.

The solution to charged particle tracking was to opt for a small number of precise position measurements of each charged track (~13 each with a position resolution of ~15 μm per measurement) leading to a large number of cells distributed inside a cylindrical volume 5.8 m long and 2.5 m in diameter: 66 million $100 \times 150 \mu m^2$ silicon pixels and 9.3 million silicon microstrips ranging from ~10 cm $\times$ 80 μm to ~20 cm $\times$ 180 μm. The 198 m^2 area of active silicon of the CMS tracker is by far the largest silicon tracker ever built.

Finally the hadron calorimeter, comprising ~3000 projective towers covering almost the full solid angle, is built from alternate plates of ~5 cm brass absorber and ~4 mm thick scintillator plates that sample the energy. The scintillation light is detected by photodetectors (hybrid photodiodes) that can operate in the strong magnetic field.

6.4.4 Installation and Commissioning

The two very different and complementary detector concepts, ATLAS and CMS, resulted in two different strategies for the underground installation of these experiments.

Given its size and its magnet structure, the ATLAS detector had to be assembled directly in the underground cavern. The installation process began in summer 2003 (after the completion of civil engineering work that started in 1998) and ended in summer 2008. Figure 6.4 (top) shows the completion of the barrel toroid magnet system with the insertion of the barrel calorimeters. Figure 6.4 (bottom) shows one end of the cylindrical barrel detector after 3.5 years of installation work, 1.5 years before completion. The ends of four of the barrel toroid coils are visible, illustrating the eightfold symmetry of the structure.

The iron yoke of the CMS detector is divided into five barrel-wheels and three endcap disks at each end, giving a total weight of 12,500 tons. This structure enabled the detector to be assembled and tested in a large surface hall while the underground cavern was being prepared. The sections, weighing between 350 tons and 2000 tons, were then lowered sequentially between October 2006 and January 2008, using a dedicated gantry system equipped with strand jacks: a pioneering use of this technology to simplify the underground assembly of large experiments. Figure 6.5 top shows the lowering of the heaviest and central section, supporting the superconducting coil. Figure 6.5 bottom shows the transverse section of the barrel part of CMS illustrating the successive layers of detection starting from the centre where the collisions occur: the inner tracker, the crystal calorimeter, the hadron calorimeter, the superconducting coil, and the iron yoke instrumented with the four muon stations. The last muon station is at a radius of 7.4 m.

Individual detector components (e.g. chambers) of both experiments were built and assembled in a distributed way all around the globe in the numerous participating institutes and were typically first tested at their production sites, then after delivery to CERN, and finally again after their installation in the underground

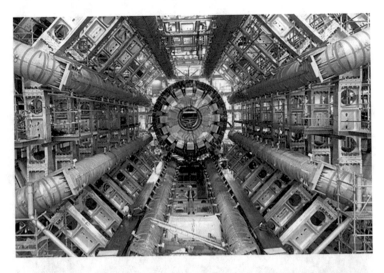

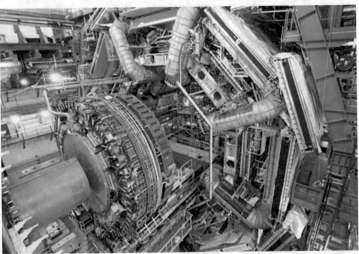

Fig. 6.4 (top) Photograph of the barrel toroid magnet system after the completion of the the installation of the eight coils, (bottom) Photograph of one end of the ATLAS detector barrel with the calorimeter end-cap still retracted before its insertion into the barrel toroid magnet structure (February 2007 during the installation phase)

caverns. The collaborations also invested enormous effort in testing representative samples of the detectors in test beams at CERN and other accelerator laboratories around the world. These test-beam campaigns not only verified that performance criteria were met over the several years of production of detector components, but also were used to prepare the calibration and alignment data for LHC operation. The so-called large combined test-beam setups, which represented whole 'slices' of the different detector layers of the final detectors, proved to be very important.

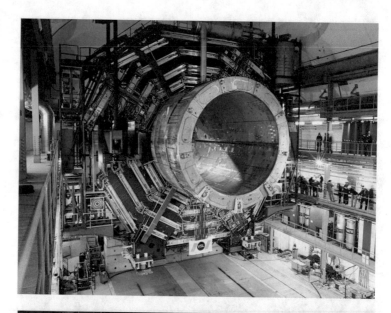

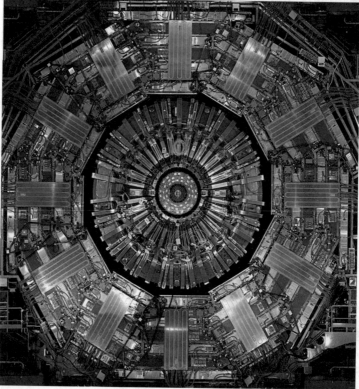

Fig. 6.5 (top) Photograph showing the lowering of the central barrel part and solenoid of the CMS detector during its installation in the cavern in 2007; (bottom) Photograph of transverse section of the barrel part of CMS illustrating the successive layers of detection

During the installation, the experiments made extensive use of the constant flow of cosmic rays impinging on Earth providing a reasonable flux of muons even at a depth of 100 m underground. Typically a few hundred per second traverse the detectors. These muons were used to check the whole chain from sub-detector hardware to analysis programs of the experiments, and to align the detector elements and calibrate their response prior to the proton-proton collisions. In particular, after the LHC incident on 19th September 2008 the experiments used the 15 months LHC down time, before the first collisions on 23rd November 2009, to run the full detectors in very extensive cosmic-ray campaigns, collecting many hundreds of millions of muon events. These runs allowed both ATLAS and CMS to be ready for physics operation, with pre-calibrated and pre-aligned detectors, by the time of the first pp collisions.

6.5   Experiment Software and LHC Worldwide Computing Grid

The experiment collaborations themselves develop the software that enables reconstruction, from raw data, of analyzable objects such as electrons, photons, jets, b jets, muons, and other charged tracks, and their energies or momenta. Algorithms have to be run to calibrate the energy deposits; align the hits from charged particles; and correct for changes in detector response arising from irradiation, variation in environmental parameters such as temperature, or changes in the position of detecting elements. The software packages must also simulate the response of the detectors to the passage of particles generated in simulated events occurring in bunch crossings that contain interesting physics processes, as well as simple backgrounds. These include processes such as the production of W or Z bosons, QCD jets, or Higgs bosons and their decays. Such simulations helped prepare, prior to the first collisions, the experiments' end-to-end processing and analysis chains, which were crucial for the rapid delivery of physics results of outstanding quality and quantity soon after the first collisions.

The LHC computing system, termed the LHC Worldwide Computing Grid (WLCG) [1], was conceived to make effective use of distributed resources, work on a large scale, and enable all the experiments' scientists, wherever they were based, to have access to LHC data, and without regard to the extent of the resources they themselves could afford. The WLCG provided the backbone for the analysis capabilities of the experiments. The global WLCG has continued to grow, now encompassing around 170 computing centers in 42 countries, with an infrastructure that provides access to some 600,000 computing cores, around 500 PB of storage (50% on disk and 50% on tape), and a network that frequently runs at 100 Gb s^{-1} between larger sites and at 10 Gb s^{-1} between smaller sites. The security of access have been instrumental for building a truly federated computing infrastructure for science. Although the individual computing tasks described above were already familiar in particle physics, the scope, scale, and geographical spread of the LHC computing and data analysis are unprecedented.

6.6   Operation of the LHC: The Start of Data Taking

On the tenth of September 2008 first beams circulated in the Large Hadron Collider. Nine days later, during the powering test of the last octant, alarms reached the LHC accelerator's control room and safety systems were activated to protect the accelerator. It turned out that one of the 50,000 soldered joints had malfunctioned. This led to an electrical arc that pierced the vacuum enclosure of a superconducting dipole bending magnet leading a massive escape of helium, the pressure wave of which caused considerable damage. The accelerator went offline for repairs. The ATLAS and CMS experiments continued to run round-the-clock for a few months recording billions of traversals of muons from cosmic rays. These data demonstrated that the experiments were in a good shape to take collision data. After a few tweaks the ATLAS and CMS experiments were even better prepared for first collisions, which came on 23rd November 2009. The first collision data were rapidly distributed, analysed and physics results produced.

Following a preliminary low-energy run in the autumn of 2009, the ATLAS and CMS experiments started recording high-energy proton-proton collisions in March 2010 at $\sqrt{s} = 7$ TeV. Some 45 pb^{-1} of data were recorded, sufficient to demonstrate that the experiments were working well, according to the ambitious design specifications and the results they were producing were consistent with the predictions from known SM physics. Many parameters were examined, including the efficiency of identification and reconstruction of physics objects, the measured energy and momentum resolutions, the resolution of peaks in invariant mass distribution, and more. An example of the performance from the CMS experiment is the comparison of the observed width of Y particle with the design mass resolution. The width is expected to be dominated by instrumental resolution. Figure 6.6 shows that the observed width is measured to be 70 MeV consistent with the design value. Also observed in such di-muon invariant mass distributions is a history of decades of particle physics indicating the excellent performance of the experiments. The next step was to see if known physics could be measured as per the predictions of the SM, extrapolated to the new energies.

6.6.1   Measurement of SM Processes to Verify Experiment Performance

Observation and accurate measurement of the production of known SM particles at the LHC collision energies is a pre-requisite for the exploration of new physics,

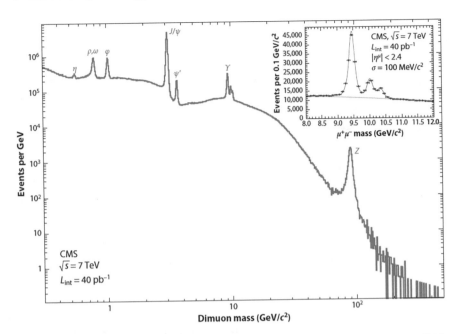

Fig. 6.6 The distribution of the invariant mass for di-muon events, shown here from CMS, displays the various well-known resonant states of the SM. The inset illustrates the excellent mass resolution for the three states of the Y family. The mass resolutions in the central region are; 28 MeV (0.9%) for J/ψ, 69 MeV (0.7%) for Y(1S), both dominated by instrumental resolution and $\Gamma = 2.5$ GeV for the Z dominated by its natural width, and are equal to the design values

including the search for the Higgs boson. The SM processes, such as W and Z production, are often considered to be 'standard candles' for the experiments.

In the ATLAS and CMS experiments, SM physics can be studied with unprecedented precision, allowing comparison with the predictions of the SM with small instrumental systematic errors. The data collected so far have enabled many precise measurements of SM processes, including the production of light quarks and gluons, bottom and top quarks, and W and Z bosons, singly and in pairs, and with varying numbers of jets resulting from higher order processes. A summary of such studies is shown in Fig. 6.7, where measurements of cross sections for various selected electroweak and QCD processes are compared with predictions from the SM. These very diverse measurements, probing cross-sections over a range of many orders of magnitude, established that the experiments were "physics commissioned" and ready for discoveries. The detector performance was well understood and known SM processes were correctly observed, crucially important as they often constitute large backgrounds to signatures of new physics, such as those expected for the Higgs boson.

The speed with which these measurements verified the SM predictions for known physics is a tribute to the large amount of work done by many groups, including

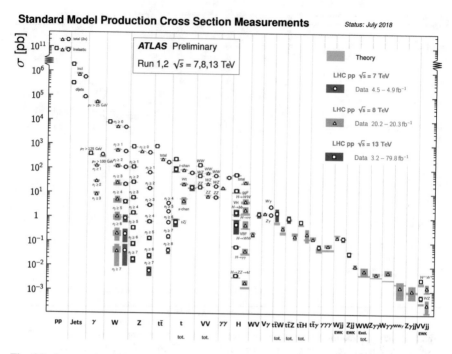

Fig. 6.7 A comparison of cross-section measurements for electroweak and QCD processes with theoretical predictions from the SM, shown here as example from the ATLAS experiment

theorists, other collider experiments at LEP, Tevatron, HERA, b-factories and to the good preparation of the ATLAS and CMS experiments.

In what follows the production of b-bar, $\tau^+\tau^-$, W^+W^-, etc. will be denoted by bb, $\tau\tau$, $WW^{(*)}$, etc.

Using all the data so far collected extensive searches for new physics, beyond the standard model, have been performed. No new physics beyond the SM has yet been discovered. Limits have been set on e.g. quark substructure, supersymmetric particles (e.g. disfavouring gluino masses below 1.5 TeV in simple models of supersymmetry), potential new bosons (e.g disfavouring new heavy SSM W′ and Z′ bosons with masses below 3 TeV for couplings similar to the ones for the known W and Z bosons) and semi-classical black holes in the context of large extra dimensions (with masses below 10 TeV).

6.7 The Discovery and Properties of a Higgs Boson

Undoubtedly, the most striking result to emerge from the ATLAS and CMS experiments is the discovery of the Higgs boson at a mass of ~125 GeV [19] and [20], respectively.

The SM Higgs boson couples to the different pairs of particles in a set proportion i.e. for fermions (f) proportional to m_f^2, and for bosons (V) proportional m_V^4/v^2 where v is the vacuum expectation value of the scalar field ($v = 246$ GeV). Once produced the Higgs boson disintegrates immediately into known SM particles. Both the production modes and decay modes and rates are precisely predicted in the SM. A search had to be made over a large range of masses and the many possible decay modes with differing branching ratios as shown in Fig. 6.2 and Table 6.1. For example, at $m_H = 125$ GeV the SM boson is predicted to decay into pairs of photons with branching fraction (BR) of 2.2×10^{-3}, into Z bosons and then four electrons or muons or two muons and two electrons with BR = 1.25×10^{-4}, into a pair of W bosons and then into $llvv$ with BR ~1%, etc.

For a given Higgs boson mass hypothesis, the sensitivity of the search depends on:

– the mass of the Higgs boson
– the Higgs boson production cross section (Fig. 6.2a and Table 6.1),
– the decay branching fraction into the selected final state (Fig. 6.2b and Table 6.1),
– the signal selection efficiency,
– the observed Higgs boson mass resolution, and
– the level of backgrounds with the same or a similar final state.

Comparisons with the expectations from the SM for the various combinations of production and decay modes are usually cast in terms of modifiers such as signal strength, μ, that is the ratio of the measured production $\times$ decay rate of the signal and the SM expectation i.e. $\mu = \sigma.BR/(\sigma.BR)_{SM}$. Signal strength of one would be indicative of the SM Higgs boson.

CMS and ATLAS increasingly use global event reconstruction algorithms, labeled particle-flow reconstruction, that attempt to identify, reconstruct and provide the measurement of the energy of particles by combining information from the inner tracker, the calorimeters and the muon system in an optimized manner. Hadronic jets are clustered from the reconstructed particles using the infrared- and collinear-safe anti-k_T algorithm with a distance parameter usually set at 0.4. The jet momenta are measured by summing vectorially the momenta of all particles in the jet. Jets originating from b-jets are identified by discriminants that include the presence of particles originating from vertices displaced from the primary interaction vertex. A typical b-jet efficiency of around 70% is attained for a 1% misidentification probability for light quarks and gluons. The missing transverse momentum vector is taken as the negative of the vector sum of the momenta of all reconstructed particles in the event; its magnitude is labeled p_T^{miss}.

By the end of 2012 (LHC Run 1) the total amount of data that had been examined corresponded to an integrated luminosities of ~5 fb^{-1} at $\sqrt{s} = 7$ TeV and ~ 20 fb^{-1} at $\sqrt{s} = 8$ TeV, equating to the examination of some 2×10^{15} proton-proton collisions.

By the end of the Run 2 (2018) the total amount of data that had been recorded corresponded to an integrated luminosities ~150 fb^{-1} at $\sqrt{s} = 13$ TeV, equating to some 1.5×10^{16} proton-proton collisions.

6.7.1 Event and Physics Objects Reconstruction and Analysis Techniques

It is convenient to subdivide the analysis of data relating to Higgs bosons according to the decay channel, using datasets where the data have been selected to contain a particular set of final state particles. To improve the sensitivity, the events in this dataset are usually separated into categories that are intended to reflect the expected signal-to-background ratio. Several multivariate methods are used in the analyses

- to improve event reconstruction, estimates of the energies/momenta of physics objects (e.g. photons and electrons, etc.),
- to identify physics objects (such as electrons, photons, b-quarks, tau leptons, etc.)
- to categorize events according to particular production (e.g. ggH, VBF, VH, ttH etc.), decay mode, or expected signal-to-background ratio.

The reader can find the exact description of the multivariate methods used within the individual papers referenced in the sections below.

Charged leptons and photons originating from the fundamental partonic processes tend to be "isolated" i.e. no other particles surround the one of interest. A relative isolation condition is applied on such particles. The sum of transverse momenta of accompanying particles, within an angular radius of approximately 0.3, around the particle of interest, is divided by the transverse momentum of the particle of interest. A cut on this ratio is made, the value of which is separately optimized for electrons, muons or photons. A correction to the accompanying energy is applied when the instantaneous luminosity is high and undesirable energy from pileup interactions is accidentally captured in the region.

6.7.2 The Discovery: Results from the 2011 and Partial 2012 Datasets

In the 2011 data-taking run the ATLAS and CMS experiments recorded data at $\sqrt{s} = 7$ TeV corresponding to an integrated luminosity of ~5 fb^{-1}. In December 2011, the first "tantalizing hints" of a new particle from both the CMS and ATLAS experiments were shown at CERN. The general conclusion was that both experiments were seeing an excess of unusual events at roughly the same place in mass (in the mass range 120–130 GeV) in two different decay channels. That set the stage for data taking in 2012.

In January 2012 it was decided to slightly increase the energy of the protons from 3.5 to 4 TeV, giving a centre of mass energy of 8 TeV. By June 2012 the number of high-energy collisions examined had doubled and both CMS and ATLAS had greatly improved their analyses so it was decided to look at the area which had shown the excess of events but only after all the algorithms and selection procedures had been agreed, in case a bias was inadvertently introduced. These data led to the discovery of a Higgs boson, independently in both the ATLAS and CMS experiments in July 2012.

In this section we shall concentrate on the region of low mass ($114.4 < m_H < 150$ GeV) where the two channels particularly suited for unambiguous discovery are the decays to two photons and to two Z bosons, where one or both of the Z bosons could be virtual, subsequently decaying into four electrons, four muons or two electrons and two muons. These two decay modes are particularly suited for discovery as the observed mass resolution (~1% of m_H) is the best and the backgrounds are manageable or small.

6.7.2.1 The H → γγ Decay Mode

In the H → $\gamma\gamma$ analysis a search is made for a narrow peak in the diphoton invariant mass distribution in the mass range 110–150 GeV, on a large irreducible background from QCD production of two photons (via quark-antiquark annihilation and the gluon-fusion or "box" diagrams). There is also a reducible background where one or more of the reconstructed photon candidates originate from misidentification of jet fragments, with the process of QCD Compton scattering dominating. The relative fractions of these backgrounds in the selected events are illustrated in Fig. 6.9a.

The event selection requires two "isolated" photon candidates satisfying p_T and photon identification criteria. As an example, CMS applies a threshold of $p_T = m_{\gamma\gamma}/3$ ($m_{\gamma\gamma}/4$) to the leading (sub-leading) photon in p_T, where $m_{\gamma\gamma}$ is the diphoton invariant mass. Scaling the p_T thresholds in this way avoids distortion of the shape of the $m_{\gamma\gamma}$ distribution. The background is estimated from data, without the use of MC simulation, by fitting the diphoton invariant mass distribution in a range ($100 < m_{\gamma\gamma} < 180$ GeV).

The results from the CMS experiments are shown in Fig. 6.8a [20]. A clear peak at a diphoton mass of around 125 GeV is seen. A similar result was obtained in the ATLAS experiment [19].

6.7.2.2 The H → ZZ$^{(*)}$ → 4 l Decay Mode

In the H → ZZ$^{(*)}$ → 4 l decay mode a search is made for a narrow four-charged lepton mass peak in the presence of a small continuum background. The background sources include an irreducible four-lepton contribution from direct ZZ$^{(*)}$ production via quark-antiquark and gluon–gluon processes. Reducible background contributions arise from Z + bb and tt production where the final states contain two isolated leptons and two b-quark jets producing secondary leptons.

The event selection requires two pairs of same-flavour, oppositely charged isolated leptons. Since there are differences in the reducible background rates and mass resolutions between the sub-channels 4e, 4 μ, and 2e2μ, they are analysed separately. Electrons are typically required to have $p_T > 7$ GeV. The corresponding requirements for muons are $p_T > 5$ GeV. Both electrons and muons are required to

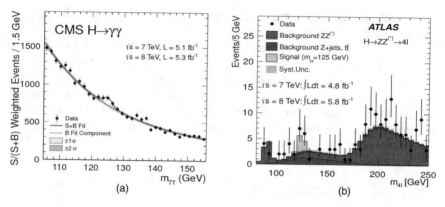

Fig. 6.8 (a) The two-photon invariant mass distribution of selected candidates in the CMS experiment, weighted by S/B of the category in which it falls. The lines represent the fitted background and the expected signal contribution ($m_H = 125$ GeV). (b) The four-lepton invariant mass distribution in the ATLAS experiment for selected candidates relative to the background expectation. The expected signal contribution ($m_H = 125$ GeV) is also shown

be isolated. The pair with invariant mass closest to the Z boson mass is required to have a mass in the range 40–120 GeV and the other pair is required to have a mass in the range 12–120 GeV. The $ZZ^{(*)}$ background, which is dominant, is evaluated from Monte Carlo simulation studies.

The m_{4l} distribution is shown in Fig. 6.8b for the ATLAS experiment [19]. A clear peak is observed at ~125 GeV in addition to the one at the Z mass. The latter is due to the conversion of an inner bremsstrahlung photon emitted simultaneously with the dilepton pair. A similar result was obtained by the CMS experiment [20].

6.7.2.3 Combinations

A search was also made in other decay modes of a possible Higgs boson and combined to yield the final results published in August 2012 by ATLAS [19] and CMS [20] experiments. Both ATLAS and CMS independently discovered a new heavy boson at approximately the same mass, clearly evident in the two different decay modes, $\gamma\gamma$ and $ZZ^{(*)}$. The observed (expected) local significances were 6.0σ (5.0σ) and 5.0σ (5.8σ) in ATLAS and CMS respectively, indicating that a new particle had been discovered.

The decay into two bosons (two photons; two Z bosons or two W bosons) implied that the new particle is a boson with spin different from one, and its decay into two photons that it carries either spin-0 or spin-2.

The results presented by both ATLAS and CMS collaborations were consistent, within uncertainties, with the expectations for a SM Higgs boson. Both noted that collection of more data would enable a more rigorous test of this conclusion and an investigation of whether the properties of the new particle imply physics beyond the SM.

6.7.3   Results from the Data Recorded Subsequent to the Discovery

The combined results from the ATLAS and CMS experiment from Run 1 on the Higgs boson production, decay rates and constraints on its couplings were published in 2016 [37]. These results have been superseded by the ones presented below. Results are presented from the most recently published papers (in journals or submitted to the hep arXiv) from the two collaborations. The integrated luminosity differs from one result to another and is indicated in the legends of the plots presented.

The LHC centre of mass energy was increased from $\sqrt{s} = 8$ TeV to $\sqrt{s} = 13$ TeV in 2015. At the higher value of $\sqrt{s}$ the predicted cross-sections for the dominant ggH production mode and the rare ttH production mode increased by factors of ~2.3 and ~ 3.8, respectively. This and the larger datasets from Run 2 allow a more precise comparison of the properties of the Higgs boson with respect to those predicted by the SM. In addition, since the discovery, the theoretical predictions have become more accurate with the inclusion of further (higher) order corrections. Details can be found below in the references included in the individual papers of the two collaborations.

The two collaborations have also improved the reconstruction of physics objects and the methods of analysis. Event categorization and machine learning methods are deployed to study almost all the different production and decay modes. The analyses described below divide events into multiple categories reflecting the different Higgs boson production channels to improve the sensitivity of the measurements. Associated production processes (WH and ZH), or the ttH production process, are tagged by requiring the presence of additional leptons or jets. The VBF process is tagged using distinctive kinematic properties such the presence of two jets with a large separation in pseudorapidity and a large invariant jet-jet mass. In some cases the kinematic characteristics of the whole event, such as large missing p_T, are used to preferentially select events e.g. arising from ZH production where the Z boson decays to neutrinos.

6.7.3.1   The H $\rightarrow \gamma\gamma$

As the H $\rightarrow \gamma\gamma$ decay proceeds via W-boson and top-quark loops, it is especially sensitive to the presence of any undiscovered heavy charged fermions and bosons. Any significant deviation from the precise SM prediction for the cross section would be indicative of new physics.

The H $\rightarrow$ $\gamma\gamma$ mode provides good sensitivity to almost all Higgs boson production processes. The interference between W-loop and top-loops provides sensitivity to the relative sign of the fermion and boson couplings.

It is common to use a dedicated boosted decision tree discriminator to select and categorize events; it is constrained using the diphoton kinematic variables, photon isolation and identification variables, and per-event estimated diphoton mass resolution for the pair of photons in the event.

ATLAS has measured the properties of the H $\rightarrow$ $\gamma\gamma$ mode using 79.8 fb^{-1} of collision data recorded at $\sqrt{s}$ = 13 TeV [38]. The properties measured include the signal strength, the cross section measurements for the production of a Higgs boson through gluon–gluon fusion, vector boson fusion, and in association with a vector boson or a top-quark pair. They are found to be compatible with the predictions of the SM. The signal strength is measured to be μ = 1.06 $\pm$ 0.08 (stat), $^{+0.08}_{-0.07}$ (exp), $^{+0.07}_{-0.06}$ (theo), improving on the precision of the previous ATLAS measurement at $\sqrt{s}$ = 7 and 8 TeV by over a factor of three. The cross section for the production of the Higgs boson decaying to two isolated photons in the fiducial region of the selection of photons is measured to be 60.4 $\pm$ 6.1 (stat) $\pm$ 6.0 (exp) $\pm$ 60.3 (theo) fb, in good agreement with the SM value of 63.5 $\pm$ 3.3 fb. The differential cross section, sensitive to higher order QCD corrections and properties of the Higgs boson, such as its spin and CP quantum numbers, is illustrated in Fig. 6.9b and no significant deviation from a wide array of SM predictions is observed.

CMS has reported results from the H $\rightarrow$ $\gamma\gamma$ decay channel based on data collected at $\sqrt{s}$ = 13 TeV corresponding to an integrated luminosity of 35.6 fb^{-1} [39]. The diphoton invariant mass distribution, observed in CMS, is shown in

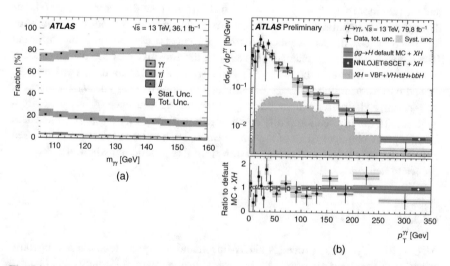

Fig. 6.9 (a) The data-driven determination of event fractions for $\gamma\gamma$, γ-jet, and jet-jet events as a function of m$_{\gamma\gamma}$ after the final selection. (b) The fiducial differential cross section for pp. $\rightarrow$ $\gamma\gamma$ as a function of p$_T$($\gamma\gamma$) compared to the SM expectations

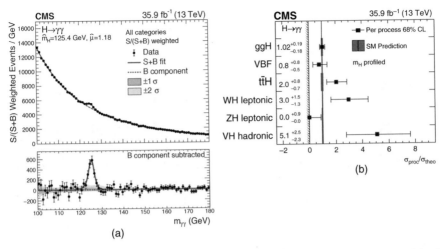

Fig. 6.10 (a) Data and signal-plus-background model fits for all categories summed, weighted by their sensitivity. The one (green) and two (yellow) standard deviation bands include the uncertainties in the background component of the fit. The lower panel shows the residuals after the background subtraction. (b) Cross section ratios measured for each process (black points) compared to the SM expectations and their uncertainties (blue band). The signal strength modifiers are constrained to be nonnegative, as indicated by the vertical line and hashed pattern at zero

Fig. 6.10a. The measured signal strength is found to be $1.18^{+0.17}_{-0.14}$, largely insensitive to the precise mass value assigned to the Higgs boson. Signal strengths associated with the different Higgs boson production mechanisms are shown in Fig. 6.10b and found to be compatible with the expectations from the SM.

6.7.3.2   H → ZZ$^{(*)}$ → 4 l Decay Mode

The Higgs boson decay H → ZZ$^{(*)}$ → 4 l is the most significant process in constraining the HZZ coupling. To study the differing production mechanisms involved, the events are categorized on the basis of the presence of jets, b-tagged jets, leptons, p_T^{miss}, and various matrix element discriminants that make use of the information about the additional objects: VBF (1- and 2-jet), VH hadronic, VH leptonic, ttH, VH p_T^{miss}, and untagged categories.

ATLAS has studied the coupling properties of the Higgs boson in the four-lepton (e,μ) decay channel using 36.1 fb^{-1} of pp. collision data recorded at $\sqrt{s} = 13$ TeV [40]. The four-lepton invariant mass distribution is illustrated in Fig. 6.11a. Cross sections are measured for the main production modes and the ratio of sigma.BR/(sigma.BR)$_{SM}$ are plotted in Fig. 6.11b. The inclusive cross section times branching fraction for H → ZZ$^{(*)}$ decay and for a Higgs boson absolute rapidity below 2.5 is measured to be $1.73^{+0.24}_{-0.23}$ pb, the statistical error dominating, compared to the SM prediction of 1.34 ± 0.09 pb.

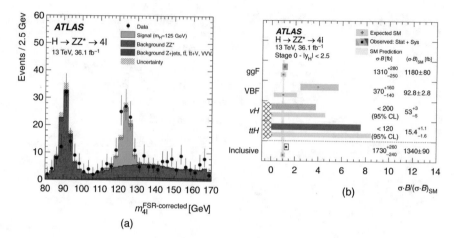

Fig. 6.11 (a) Observed and expected four-lepton invariant mass distribution for the selected Higgs boson candidates with a constrained Z boson mass. (b) Observed and expected SM values of the cross-section ratios σ.B normalized by the SM expectation (σ.B)$_{SM}$ for the inclusive production

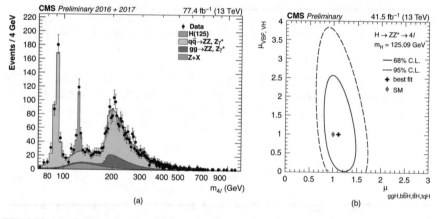

Fig. 6.12 (a) Distribution of the four-lepton reconstructed invariant mass m$_{4l}$ in the full mass range, combining data from 2016 an 2017. Points with error bars represent the data and stacked histograms represent expected distributions of the signal and background processes. The ZZ$^{(*)}$ backgrounds are normalized to the SM expectation, the Z + X background to the estimation from data. (b) Result of the 2D likelihood scan for the μ(ggH, ttH, bbH, tqH) and μ(VBF,VH) signal-strength modifiers

CMS has studied the coupling properties using 77.4 fb^{-1} of pp. collision data recorded at $\sqrt{s} = 13$ TeV [41]. The four-lepton invariant mass distribution is illustrated in Fig. 6.12a. The signal strength is measured to be $\mu = 1.06^{+0.15}_{-0.13}$ at m$_H = 125.09$ GeV, the combined ATLAS and CMS measurement of the Higgs boson mass [42]. The result of a 2D likelihood scan of the signal strengths for the individual Higgs boson production modes are also measured and are shown in Fig. 6.12b. All measurements are consistent with the expectations from the SM.

6.7.3.3   H → WW$^{(*)}$ → 2 l 2ν Decay Mode

The H → WW$^{(*)}$ decay mode has a large branching fraction (~20%) and a relatively low-background final state. The study of this final state in which both W bosons decay leptonically is based on the signature with two isolated, oppositely charged, high p_T leptons (electrons or muons) and large missing transverse momentum, E_T^{miss}, due to the undetected neutrinos. The signal sensitivity is improved by separating events according to lepton flavor; into e^+e^-, $\mu^+\mu^-$, and eμ samples and according to jet multiplicity into 0-jet and 1-jet samples. The dominant background arises from irreducible non-resonant WW$^{(*)}$ production, and the dominant uncertainties arise from the estimation, using the data themselves, of the backgrounds from top quark pair, WW$^{(*)}$ and DY production.

The final states are categorized according to the number of associated jets, with the 0-jet category dominating the overall sensitivity. Events are selected that contain two leptons of either different or the same flavour. The large background from tt production, the different and same flavour final states are further categorized with 0, 1 and 2 associated jets. In the different-flavour final state, dedicated 2-jet categories are included to enhance the sensitivity to VBF and VH production mechanisms.

ATLAS has presented measurements of the inclusive cross section of Higgs boson production via the gluon–gluon fusion (ggF) and vector-boson fusion (VBF) modes [43], based on an integrated luminosity of 36.1 fb^{-1} recorded at $\sqrt{s} = 13$ TeV in 2015–2016. The combined transverse mass distribution for $N_{jet} \leq 1$ is shown in Fig. 6.13a. The ggF and VBF cross-sections times the H → WW$^{(*)}$ branching ratio are measured to be 12.6 ± 1.0(stat) $^{+1.9}_{-1.8}$ (syst) pb and 0.50 ± 0.24 (stat) ± 0.18 (syst) pb, respectively, in agreement with the SM predictions, as illustrated in Fig. 6.13b.

CMS has published results on the decay mode H → WW$^{(*)}$ using data corresponding to an integrated luminosity of 35.9 fb^{-1}, collected at $\sqrt{s} = 13$ TeV during 2016 [44]. The expected relative fraction of different Higgs boson production mechanisms in each category is shown in Fig. 6.14a, together with the expected signal yield. Combining all channels, the observed cross section times branching fraction is $1.28^{+0.18}_{-0.17}$ times the SM prediction for the Higgs boson with a mass of 125.09 GeV. The ratio of the observed and the predicted cross sections for the main Higgs boson production modes is shown in Fig. 6.14b. All are consistent with the predictions from the SM.

6.7.3.4   The H → ττ Decay Mode

All of the decay modes discussed so far test the direct coupling of the Higgs boson to bosons, and only indirectly probe, through quantum loops, its coupling to fermions.

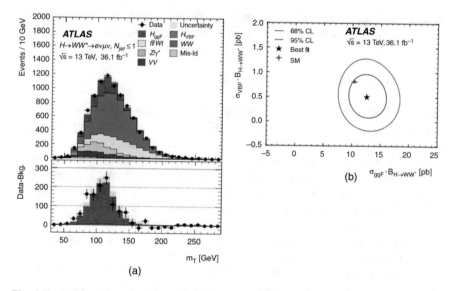

(a)

Fig. 6.13 (a) Post-fit combined transverse mass distribution for $N_{jet} \leq 1$. The bottom panel shows the difference between the data and the estimated background compared to the distribution for a SM Higgs boson with $m_H = 125$GeV. The H_{VBF} contribution is too small to be visible. (b) 68% and 95% confidence level two-dimensional likelihood contours of $\sigma_{ggF} \cdot B_H \rightarrow WW^{(*)}$ vs. $\sigma_{VBF} \cdot B_H \rightarrow WW^{(*)}$, compared to the SM prediction shown by the red marker

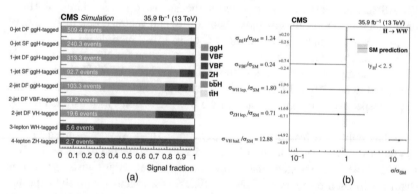

Fig. 6.14 (a) Expected relative fraction of different Higgs boson production mechanisms in each category included in the combination, together with the expected signal yield. (b) Observed cross sections for the main Higgs boson production modes, normalized to the SM predictions. The vertical line and band correspond to the SM prediction and associated theoretical uncertainty

The $H \rightarrow \tau\tau$ mode provides the best sensitivity for the direct measurement for Higgs boson coupling to fermions. It benefits from a relatively large branching fraction, a moderate mass resolution (~10–20%) and provides good sensitivity to both the ggH and VBF production processes.

The H $\rightarrow \tau\tau$ mode is studied via tau decays to $e\mu$, $\mu\mu$, $e\tau_h$, $\mu\tau_h$, $\tau_h\tau_h$, where electrons and muons arise from leptonic τ-decays and τ_h denotes a τ lepton decaying hadronically. Each of these categories is further divided into three sub-categories labeled 0-jet, boosted and VBF. The 0-jet category helps constrain background normalisation, identification efficiencies, and energy scales, and systematic uncertainties in the background model. The main irreducible background, $Z \rightarrow \tau\tau$ production, and the largest reducible backgrounds (W + jets, QCD multijet production, and top quark pair) are evaluated from control samples in data.

CMS has observed the H $\rightarrow \tau\tau$ mode using a data sample corresponding to an integrated luminosity of 35.9fb^{-1} at $\sqrt{s} = 13$ TeV [45]. Figure 6.15a shows the distribution of the decimal logarithm of the ratio of the expected signal and the sum of expected signal and expected background in each bin of the mass distributions used to extract the results, in all signal regions. The background contributions are separated by decay channel. The inset shows the corresponding difference between the observed data and expected background distributions divided by the background expectation, as well as the signal expectation divided by the background expectation. The best fit of the product of the observed H $\rightarrow \tau\tau$ signal production cross section and branching fraction is $1.09^{+0.27}_{-0.26}$ times the SM expectation. The combination with the corresponding measurement performed with data collected by the CMS experiment at center-of-mass energies of 7 and 8 TeV leads to an observed

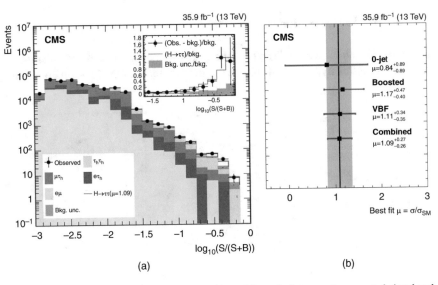

Fig. 6.15 (a) Distribution of the decimal logarithm of the ratio between the expected signal and the sum of expected signal and expected. The inset shows the corresponding difference between the observed data and expected background distributions divided by the background expectation, as well as the signal expectation divided by the background expectation. (b) Best fit signal strength per category

significance of 5.9 standard deviations, equal to the expected significance. Figure 6.15b right plots the signal strength per category for $m_H = 125.09$ GeV.

ATLAS has observed the H $\rightarrow \tau\tau$ mode using 36.1 fb^{-1} of data recorded at $\sqrt{s} = 13$ TeV [46]. All combinations of leptonic and hadronic tau decays were considered. Combining all data taken at $\sqrt{s} = 7$, 8 and 13 TeV, the observed (expected) significance is found to be 6.4 (5.4) standard deviations. Using the data taken at $\sqrt{s} = 13$ TeV, the total cross section, in the H $\rightarrow \tau\tau$ decay channel, is measured to be 3.71 ± 0.59 (stat) $^{+0.87}_{-0.74}$ (syst) pb, for $m_H = 125$ GeV, assuming the relative contributions of its production modes predicted by the SM. Total cross sections are determined separately for vector boson fusion production and gluon–gluon fusion production to be σ(VBF, H $\rightarrow \tau\tau$) $= 0.28 \pm 0.09$ (stat) $^{+0.11}_{-0.09}$ (syst) pb and σ(ggF, H $\rightarrow \tau\tau$) $= 3.0 \pm 1.0$ (stat.) $^{+1.6}_{-1.2}$ (syst) pb, respectively. The measured values for σ(H $\rightarrow \tau\tau$), when only the data of individual channels are used, are shown in Fig. 6.16a, along with the result from the combined fit. The theory uncertainty in the predicted signal cross section is shown by the yellow band. Figure 6.16b shows the likelihood contours in the variables (ggF, H $\rightarrow \tau\tau$) and (VBF, H $\rightarrow \tau\tau$) for the combination of all channels. The 68% and 95% CL contours are shown as dashed and solid lines, respectively, for $m_H = 125$ GeV. The SM expectation is indicated by a plus symbol and the best fit to the data is shown as a star. All measurements are in agreement with SM expectations.

6.7.3.5 H→b$\bar{b}$ Decay Mode

In the SM, fermions couple directly to the Higgs boson via the Yukawa interaction. A clear test of this hypothesis would be the measurement of the H→ $b\bar{b}$ coupling.

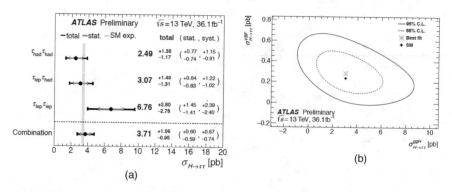

Fig. 6.16 (a) Measured values for σ(H $\rightarrow \tau\tau$) when only the data of individual channels are used. Also shown is the result from the combined fit. The total $\pm$ 1σ uncertainty in the measurement is indicated in black, with the individual contribution from the statistical uncertainty in blue. (b) Likelihood contours for the combination of all channels in the [(ggF, H $\rightarrow \tau\tau$) v/s (VBF, H $\rightarrow \tau\tau$)] plane. The SM expectation is indicated by a plus symbol and the best fit to the data is shown as a star

The H → bb decay mode has by far the largest branching ratio (~58%). However, this is the most difficult decay channel to observe, since bottom quark pairs are prolifically produced by QCD processes and give rise to a formidable background. The cross section for b-quark pair production, σ_{bb}(QCD), is $\sim 10^7 \times \sigma$(H → bb). Therefore the search concentrates on Higgs boson production in association with a W or Z boson using the following decay modes: W → $e\nu/\mu\nu$ and Z → ee or $\mu\mu$ or $\nu\nu$. The Z → $\nu\nu$ decay is identified by the requirement of a large missing transverse energy. The Higgs boson candidate is reconstructed by requiring two b-tagged jets.

Events are selected in 0-, 1- and 2-charged lepton (e or μ) channels, to explore the ZH → $\nu\nu$bb, WH → $l\nu$bb, ZH → llbb signatures, respectively. Both experiments introduced several improvements since the initial searches including more efficient identification of b-jets, better dijet mass resolution and use of multivariate discriminants that better separate signal from background. Multivariate discriminants, built from variables that describe the kinematics of the selected events, are used to maximise the sensitivity to the Higgs boson signal. The signal extraction method is validated with, for example, the diboson analysis where the nominal multivariate analysis is modified to extract the VZ, Z → bb diboson process.

ATLAS has observed the mode H → bb by analyzing the combined data from Run 1 and Run 2 [47], corresponding to an integrated luminosity of 80fb^{-1} yielding an observed (expected) significance of 5.4 (5.5) standard deviations, thus providing direct observation of the Higgs boson decay into b-quarks. The signal strength is measured to be 1.01 ± 0.12(stat) $^{+0.16}_{-0.15}$(syst). Figure 6.17a shows the distribution of m_{bb} in data after subtraction of all backgrounds except for the WZ and ZZ$^{(*)}$ diboson processes using data taken at $\sqrt{s} = 13$ TeV. The contributions from all lepton channels, p^V_T regions, and number-of-jets categories are summed and

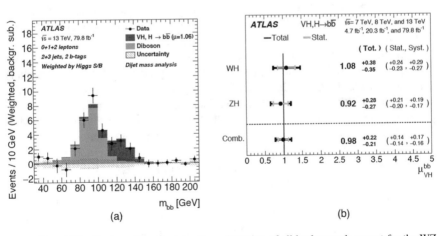

(a)   (b)

Fig. 6.17 (a) Distribution of m_{bb} in data after subtraction of all backgrounds except for the WZ and ZZ diboson processes, as obtained with the dijet-mass analysis. (b) Fitted values of the Higgs boson signal strength, μ, for $m_H = 125$ GeV for the WH and ZH processes and their combination, using the 7 TeV, 8 TeV and 13 TeV data

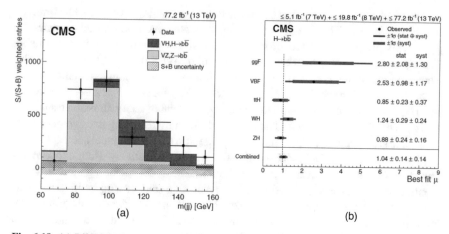

Fig. 6.18 (**a**) Dijet invariant mass distribution for events weighted by S/(S + B) in all channels combined in the 2016 and 2017 data sets. The error bar for each bin represents the presubtraction 1σ statistical uncertainty on the data, while the gray hatching indicates the 1σ total uncertainty on the signal and all background components. (**b**) Best-fit value of the H → bb signal strength with its 1σ systematic (red) and total (blue) uncertainties for the five individual production modes considered, as well as the overall combined result. The vertical dashed line indicates the SM expectation

weighted by their respective S/B, with S being the total fitted signal and B the total fitted background in each region. The expected contribution of the associated WH and ZH production of a SM Higgs boson with $m_H = 125$ GeV is shown, scaled by the measured signal strength ($\mu = 1.06$). The size of the combined statistical and systematic uncertainty for the fitted background is indicated by the hatched band. Figure 6.17b shows the fitted values of the Higgs boson signal strength, where $\mu(VHbb) = 0.98^{+0.22}_{-0.21}$ for $m_H = 125$ GeV for the WH and ZH processes and their combination, using the 7 TeV, 8 TeV and 13 TeV data.

CMS has observed the mode H → bb. Figure 6.18a shows the weighted dijet invariant mass distribution for events weighted by S/(S + B) in all channels combined in the 2016 and 2017 data sets [48]. The data (points), the fitted VH signal (red) and VZ background (grey) distributions, with all other fitted background processes subtracted, except that from dibosons are shown in Fig. 6.18a. Figure 6.18b shows the best-fit value of the H → bb signal strength for the five individual production modes considered, as well as the overall combined result. The vertical dashed line indicates the SM expectation. All results are extracted from a single fit with $m_H = 125.09$ GeV. CMS has made measurements, using data collected at $\sqrt{s} = 7$, 8, and 13 TeV, and observes an excess of events at $m_H = 125$ GeV with a significance of 5.6 standard deviations, where the expectation for the SM Higgs boson is 5.5, and a signal strength of 1.04 ± 0.14 (stat.) ± 0.14 (syst.).

6.7.3.6   H → μμ Decay Mode

The H → $\mu^+\mu^-$ decay mode extends the test of the Higgs boson's coupling to the second generation of fermions. Several scenarios beyond the SM predict a higher branching fraction than the one predicted in the SM (2.2×10^{-4} at $m_H = 125\,\text{GeV}$).

The dominant and irreducible background arises from the $Z/\gamma_* \to \mu\mu$ process that has a rate several orders of magnitude larger than that from the SM Higgs boson signal. However, due to the precise muon momentum measurement achieved by ATLAS and CMS, the dimuon mass resolution is excellent ($\approx$ 2–3%). A search is performed for a narrow peak over a large but smoothly falling background. For optimal search sensitivity, events are divided into several categories. Taking advantage of the superior muon momentum measurement in the central region events can be subdivided by the pseudorapidity of the muons, or by selections aiming at specific production processes. A category selecting the vector boson fusion process with its distinctive signature and relatively large cross section is particularly useful.

ATLAS has performed this search using data corresponding to an integrated luminosity of 36.1 fb^{-1} collected at $\sqrt{s} = 13\,\text{TeV}$ [49]. No significant excess is observed above the expected background. When combined with the data taken at $\sqrt{s} = 7$ and 8 TeV, the observed (expected) cross-section upper limit is 2.8 (2.9) times the SM prediction.

The search in CMS, using an integrated luminosity corresponding to 35.9 fb^{-1} recorded at $\sqrt{s} = 13\,\text{TeV}$ [50], and combining with data taken at $\sqrt{s} = 7$ and 8 TeV, yielded an observed (expected) cross-section upper limit is 2.92 (2.16) times the Standard Model prediction.

6.7.3.7   ttbar H Production Mode

As $m_t > m_H$ the Yukawa coupling of the Higgs boson to top quarks cannot be tested directly. However, it can be measured through the measurement in the pp. → ttH production process. The coupling of the Higgs boson to the top quark, the heaviest particle in the SM, could be very sensitive to the effects of physics beyond the SM.

Although the pp. → ttH production process only contributes around 1% of the total Higgs-boson production cross section, the top quarks in the final state offer a distinctive signature and allow many Higgs-boson decay modes to be accessed. Of these, the decay to two b-quarks, the Higgs boson decay mode with the largest branching fraction, also is sensitive to the b-quark's Yukawa coupling, the second largest in the SM.

A top quark decays almost exclusively to a bottom quark and a W boson, with the W boson subsequently decaying either to a quark and an antiquark or to a charged lepton and its associated neutrino. The Higgs boson has a rich spectrum of decay modes, and ttH production is studied using a wide variety of final state event topologies, with the Higgs boson decaying into bb, $WW^{(*)}$, $\tau\tau$, $\gamma\gamma$, and $ZZ^{(*)}$ pairs.

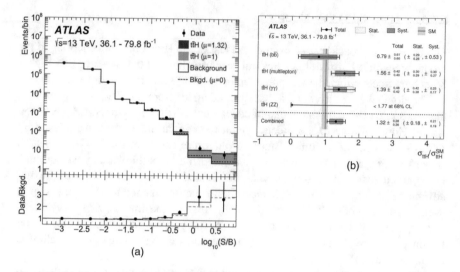

Fig. 6.19 (**a**) Observed event yields in all analysis categories in up to 79.8fb^{-1} of 13 TeV data. The lower panel shows the ratio of the data to the background estimated from the fit with freely floating signal, compared to the expected distribution including the signal assuming $\mu = 1.32$ (full red) and $\mu = 1$ (dashed orange). The error bars on the data are statistical. (**b**) Combined *ttH* production cross section, as well as cross sections measured in the individual analyses, divided by the SM prediction

ATLAS has observed this production mode using data taken at $\sqrt{s} = 7$ TeV, 8 TeV and 13 TeV corresponding to integrated luminosities up to 79.8 fb^{-1}. The Higgs boson decays included comprise bb, WW^*, $\tau + \tau-$, $\gamma\gamma$, and ZZ^*. The observed significance is 6.3σ, compared to an expectation of 5.1σ [51]. Assuming SM branching fractions, the total *ttH* production cross section at $\sqrt{s} = 13$ TeV is measured to be 670 ± 90(stat.) $^{+110}_{-100}$(syst.) fb, in agreement with the SM prediction. Figure 6.19a shows the observed event yields in all analysis categories. The background yields correspond to the observed fit results, and the signal yields are shown for both the observed results ($\mu = 1.32$) and the SM prediction ($\mu = 1$). The ranking of the discriminant bins is carried out by $\log_{10}(S/B)$, where S is the extracted signal yield and B the extracted background yield. Figure 6.19b shows the combined *ttH* production cross section, as well as cross sections measured in the individual analyses, divided by the SM prediction. The black lines show the total uncertainties, and the bands indicate the statistical and systematic uncertainties. The red vertical line indicates the SM cross-section prediction, and the grey band represents the PDF and α_S uncertainties and the uncertainties due to missing higher-order corrections.

CMS has observed ttH production in a combined analysis of data at $\sqrt{s} = 7$, 8, and 13 TeV, corresponding to integrated luminosities of up to 5.1, 19.7, and 35.9 fb^{-1}, respectively [52]. An excess of events is observed with an observed (expected) significance of 5.2 (4.2) standard deviations, over the expectation from

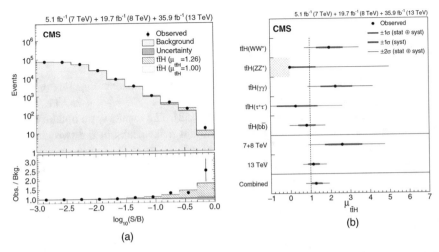

Fig. 6.20 (**a**) Distribution of events as a function of the decimal logarithm of S/B. The shaded histogram shows the expected background distribution. The two hatched histograms, each stacked on top of the background histogram, show the signal expectation for the SM ($\mu_{ttH} = 1$) and the observed ($\mu_{ttH} = 1.26$) signal strengths. The lower panel shows the ratios of the expected signal and observed results relative to the expected background. (**b**) Best fit value of the ttH signal strength modifier μ_{ttH}, with its 1σ and 2σ confidence intervals for upper section) the five individual decay channels considered, middle section) the combined result for 7 + 8 TeV alone and for 13 TeV alone, and lower section) the overall combined result. The SM expectation is shown as a dashed vertical line

the background-only hypothesis for $m_H = 125.09$. The combined best-fit signal strength normalized to the standard model prediction is $1.26^{+0.31}_{-0.26}$. Figure 6.20a shows the distribution of events as a function of the decimal logarithm of S/B, where S and B are the expected postfit signal (with $\mu_{ttH} = 1$) and background yields, respectively, in each bin of the distributions considered in this combination. The shaded histogram shows the expected background distribution. The two hatched histograms, each stacked on top of the background histogram, show the signal expectation for the SM ($\mu_{ttH} = 1$) and the observed ($\mu_{ttH} = 1.26$) signal strengths. The lower panel shows the ratios of the expected signal and observed results relative to the expected background. Figure 6.20b plots the ttH signal strength modifiers, μ_{ttH}, for the various selections and the overall combined result. The SM expectation is shown as a dashed vertical line.

6.7.4 Combining the Results

6.7.4.1 Mass of the Observed State

The mass of the Higgs boson is measured using the two decay channels that give the best mass resolutions namely $H \rightarrow \gamma\gamma$ and $H \rightarrow ZZ^{(*)} \rightarrow 4\,l$. ATLAS and CMS have combined their results from Run 1 [53]. The results were obtained from a simultaneous fit to the reconstructed invariant mass peaks in the two channels and for the two experiments. The measured masses from the individual channels and the two experiments were found to be consistent amongst themselves. The combined measured mass of the Higgs boson was found to be $m_H = 125.09 \pm 0.21$ (stat.) ± 0.11 (syst.) GeV, a value subsequently used in many of the analyses discussed above. The results of these measurements and more recent ones are shown in Fig. 6.21 [54].

The mass of the Higgs boson, combined with the measured top quark mass, has cosmological implications. The current measurement of m_H, along with that of the top quark mass [$m_t = 173.21 \pm 0.51$ (stat) ± 0.71 (syst)] indicate that our universe is in a metastable state, which eventually will tunnel through the potential barrier to the true vacuum in which space collapses, albeit over a period of time that is many orders of magnitude larger than the lifetime of the universe so far.

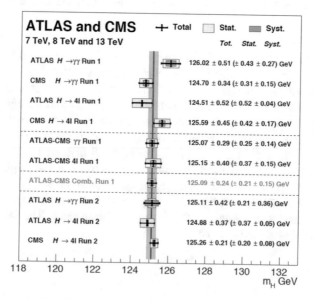

Fig. 6.21 Summary of the CMS and ATLAS mass measurements in the $\gamma\gamma$ and $ZZ^{(*)}$ channels in Run 1 and Run 2. Particle Data Group [54]

6.7.4.2   Compatibility of the Observed State with the SM Higgs Boson Hypothesis: Signal Strength

In the SM, the Higgs boson is a fundamental scalar particle with spin-parity $J^P = 0^+$, and couples to fundamental fermions as m_f^2/v^2 and to fundamental bosons as m_V^4/v^2 where $v = 246$ GeV. Several individual tests of compatibility with expectations from the SM have been discussed above.

Here we discuss the signal strength, μ, as determined by the combination of results from all channels. ATLAS and CMS combined their data from Run 1 [37] from the analysis of five production processes, namely gluon fusion, vector boson fusion, and associated production with a W or a Z boson or a pair of top quarks, and of the five decay modes $H \rightarrow \gamma\gamma$, $ZZ^{(*)}$, $WW^{(*)}$, bb, and $\tau\tau$. All results are reported assuming a value of 125.09 GeV for the Higgs boson mass. The Higgs boson production and decay rates measured by the two experiments are combined within the context of three generic parameterisations: two based on cross sections and branching fractions, and one on ratios of coupling modifiers. Several interpretations of the measurements with more model-dependent parameterisations are also given. The combined signal yield relative to the SM prediction is measured to be 1.09 ± 0.11. The error is broken down as ± 0.07 (statistical), ± 0.04 (experimental systematic), ± 0.03 (theoretical on background) and ± 0.07 (theoretical on signal).

The most recent measured values of μ, using the above-mentioned channels, are:

- ATLAS: $\mu = 1.13\ ^{+0.09}_{-0.08}$, using 79.8 fb^{-1} at $\sqrt{s} = 13$ TeV [55],
- CMS: $\mu = 1.17 \pm 0.1$, using 35.9 fb^{-1} at $\sqrt{s} = 13$ TeV [56].

The error in the measurement from ATLAS has the following breakdown: ± 0.05 (statistical), ± 0.05 (experimental systematic), ± 0.03 (theoretical on background) and $(+0.05, -0.04)$ (theoretical on signal). This should give a flavor of the possibilities for extrapolation into the future.

6.7.4.3   Compatibility of the Observed State with the SM Higgs Boson Hypothesis: Couplings

The 25 products, $\mu_i \times \mu_f$, where i (f) is the production (decay) index can also be considered as free parameters. This can be viewed as the measurements of cross sections times branching fractions, sigma×BR, by production mechanism and decay mode. The results from the ATLAS and CMS have been combined and are illustrated in Fig. 6.22 from the Particle Data Group [54] showing compatibility with the SM.

Figure 6.23a, from the ATLAS experiment [55], illustrates the dependence of the Higgs boson couplings on mass of the decay particles (μ, τ, b-quark, W, Z and t-quark). The plot is made for reduced coupling strength modifiers, κ, using $\kappa_F(m_F/v)$ for fermions and $(\sqrt{\kappa_V})(m_V/v)$ for vector bosons, where $v = 246.22$ GeV. The couplings modifiers κ_F and κ_V are measured assuming no BSM contributions

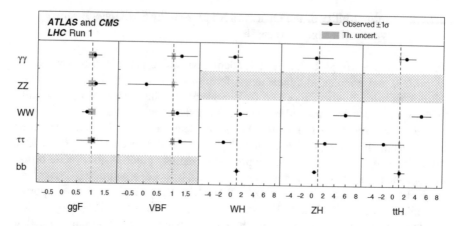

Fig. 6.22 Combined measurements of the products σ.BR for the five main production and five main decay modes. The hatched combinations require more data for a meaningful confidence interval to be provided

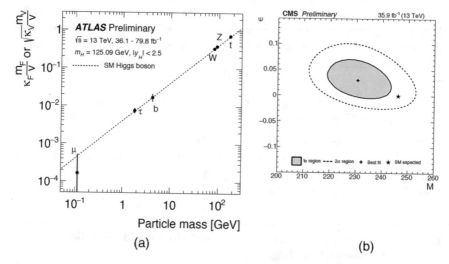

Fig. 6.23 (**a**) Reduced coupling strength modifiers $\kappa_F(m_F/v)$ for fermions (t, b, τ, μ) and $(\sqrt{\kappa_V})(m_V/v)$ for vector bosons (W, Z) as a function of their masses m_F and m_V, respectively. The SM prediction for both cases is also shown (dotted line). (**b**) Likelihood scan in the M-ε plane. The best-fit point and, 1σ, 2σ CL regions are shown, along with the SM prediction

to the Higgs boson decays, and the SM structure of loop processes. The line is the expectation from the SM.

CMS has made a similar fit, shown in Fig. 6.23b [56], using a phenomenological parameterization relating the masses of the fermions and vector bosons to the corresponding κ modifiers using two parameters, denoted M and ε. In such a model one can relate the coupling modifiers to M and ε as $\kappa_F = v \, m_f^\epsilon/M^{1+\epsilon}$ for fermions, and $\kappa_V = v\, m_V^{2\epsilon}/M^{1+2\epsilon}$ for vector bosons. The SM expectation, $\kappa_i = 1$, is recovered when $(M, \varepsilon) = (v, 0)$.

Both Fig. 6.23a and b demonstrate good compatibility with the SM within the errors on the measurements.

6.7.4.4 Compatibility of the Observed State with the SM Higgs Boson Hypothesis: Quantum Numbers

Ascertaining the quantum numbers of the Higgs boson is essential to the understanding of its nature and its coupling properties. According to the Landau-Yang theorem the observations made in the diphoton channel excludes the spin-1 hypothesis and restricts possibilities for the boson to have spin-0 or -2. The diphoton decay mode also implies that the boson has charge conjugation has C-even.

To identify the spin-parity of the Higgs boson the production and decay processes are examined in several analyses. The angular distributions of the decay particles can be used to test various spin hypotheses.

Much can be gleaned from the decay mode $H \rightarrow ZZ^{(*)} \rightarrow 4\, l$, where the full final state is reconstructed, including the angular variables sensitive to the spin-parity, along with a very favourable signal/background ratio. CMS has used the information from the five angles (see Fig. 6.24a) and the two dilepton pair masses combined to form a discriminant based on the 0^+ nature of the Higgs boson [57].

ATLAS has also tested various J^P hypotheses, and in particular 0^+ and 0^-.

In all scenarios investigated by both CMS [57] and ATLAS [58] experiments, the data are compatible with the 0^+ hypothesis, excluding a pseudoscalar nature at CLS levels of 99.8% and 98.0%, respectively. The expected distribution in ATLAS of the test statistic for the SM hypothesis (in blue) and several alternative spin and parity hypotheses is compared in Fig. 6.24b. The combination of the three decay processes allows the exclusion of all considered non-SM hypotheses at amore than 99.9% CL in favour of the SM spin-0 hypothesis.

6.8 Conclusions and Outlook

In July 2012 the ATLAS and CMS experiments announced the discovery of a Higgs boson, confirming the conjecture put forward in the 1960s. Further results from the two experiments show that, within the current measurement precision, the Higgs boson has the properties predicted by the SM. However, several theories of physics beyond the SM (BSM) predict the existence of more than one Higgs boson, and one of these would only be subtly different from that predicted in the SM one, with signal strengths differing by between 0.5–5%, depending on the model in question, indicative of the required level of sensitivity to distinguish it from a SM Higgs boson.

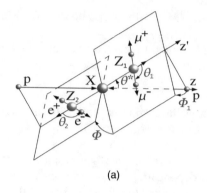

(a)

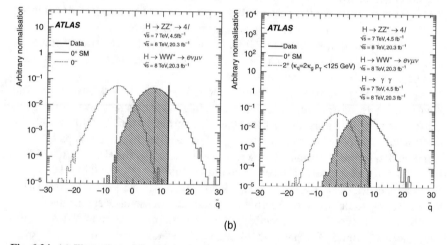

(b)

Fig. 6.24 (a) Illustration of the production and decay of a particle $X \to Z_1 Z_2 \to 4\,l$ with the two production angles θ_* and Φ_1 shown in the X rest frame and three decay angles θ_1, θ_2, and Φ shown in the Z_i and X rest frames, respectively. (b) Examples of distributions of the test statistic ($\tilde{q}$) defined in for the combination of decay channels left) 0^+ versus 0^- right) 0^+ versus the spin-2 model. The observed values are indicated by the vertical solid line and the expected medians by the dashed lines. The shaded areas correspond to the integrals of the expected distributions used for the rejection of each hypothesis

In Run 2 (2015–2018) the LHC provided proton-proton collisions at $\sqrt{s} = 13$ TeV with a peak instantaneous luminosity of 2×10^{34} cm^{-2} s^{-1}, a factor of two beyond the design value. It is intended to operate the accelerator at $\sqrt{s} = 14$ TeV after the next long shutdown (2019–2020) and to integrate a luminosity corresponding to some 300 fb^{-1} by the end of Run 3 (2021–2024). More precise measurements of the properties of the new boson will be made, as well as a more extensive exploration of physics beyond the SM, for which many possibilities are conjectured including supersymmetry, extra dimensions, unified theories, superstrings etc.

The results presented in Chap. 6 are still mostly dominated by statistical errors. The ATLAS and CMS experiments continually update their results that can be found through their websites quoted in references [36, 37]. Much more data need to be collected to enable rigorous testing of the compatibility of the Higgs boson with the SM and to get clues to physics lying beyond the SM in case of a significant deviation. This is one of the main motivations for the high luminosity LHC project, labeled the HL-LHC.

Europe's topmost priority in particle physics calls for the exploitation of the full potential of the LHC, including the high-luminosity upgrade of the accelerator and detectors with a view to collecting ten times more data than in the initial design. It is planned to increase the instantaneous luminosity of the LHC to 5×10^{34} cm^{-2} s^{-1}, and record, by around 2035, an integrated luminosity corresponding to ~3000 fb^{-1} (ten times larger than the original design value). Such an integrated luminosity also requires substantial upgrades of the ATLAS and CMS experiments, now underway, to allow a very precise measurement of the properties of the Higgs boson the study of its rare decay modes and self-coupling, in addition to the search for physics beyond the SM. Many theories beyond the SM make different predictions for the properties of one or more Higgs bosons.

Based on the currently analysed data ATLAS and CMS experiments have recently made projections of the attainable sensitivity for such measurements by the end of the HL-LHC phase [59, 60]. As around 150 million Higgs bosons will be produced a search can also be made for exotic and rare decays of the boson. Figures 6.25 and 6.26 show two sets of projections; Scenario 1 (S1) using the current theoretical errors or Scenario 2 (S2) where the theoretical errors are halved. The extrapolations show the possibility of measuring the individual signal strengths with a precision of between 5–10% for an integrated luminosity of 300 fb^{-1}, and a few percent for a dataset corresponding to 3000 fb^{-1} per experiment, dominated by theoretical erros. The per-production mode signal strength parameters are projected to be measurable with uncertainties of between 3–6% for a dataset corresponding to

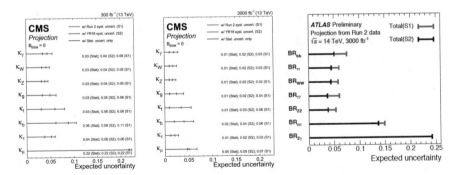

Fig. 6.25 Summary plot from CMS showing the total expected $\pm 1\sigma$ uncertainties in S1 and S2 on the per-decay mode signal strength parameters for 300 fb^{-1} (left) and 3000 fb^{-1} (centre). The statistical-only component of the uncertainty is also shown. (right) Expected uncertainty on the branching ratio measurements in ATLAS for the gg, ZZ$^{(*)}$, WW$^{(*)}$, tt, bb, $\mu\mu$ and Zγ decay channels normalized to their SM predictions assuming SM production cross section for scenarios S1 (red) and S2 (black)

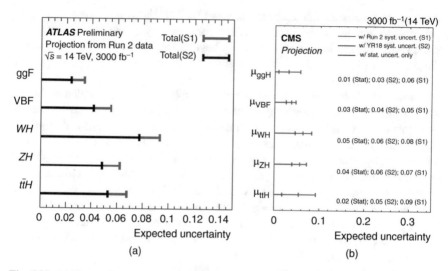

Fig. 6.26 (a) Expected uncertainty on the measurement of cross sections in ATLAS for the ggF, VBF, WH, ZH and ttH production modes normalized to their SM predictions assuming SM branching fractions for Scenarios S1 (red) and S2 (black). (b) Summary plot from CMS showing the total expected $\pm 1\sigma$ uncertainties in S1 and S2 on the cross section measurements for 300 fb^{-1} (left) and 3000 fb^{-1} (right). The statistical-only component of the uncertainty is also shown

3000 fb^{-1} per experiment. Of particular note, in view of a future electron-positron collider, is the projection for the measurement of the ttH coupling with a precision of ~5% per experiment.

The discovery of a Higgs boson implies the discovery of a fundamental scalar field that pervades the universe. Astronomical and astrophysical measurements point to the following composition of energy-matter in the universe: ~4% normal matter that "shines", ~23% dark matter, and the rest in the form of "dark energy." Dark matter is weakly and gravitationally interacting matter with no electromagnetic or strong interactions. These are the properties carried by the lightest supersymmetic particle. Hence the question: Is dark matter supersymmetric in nature? Fundamental scalar fields could well have played a critical role in the conjectured inflation of our universe immediately after the Big Bang, and in the recently observed accelerating expansion of the universe that, among other measurements, signals the presence of dark energy in our universe.

The discovery of the Higgs boson could turn out to be a portal to physics beyond the SM. Physicists at the LHC are eagerly looking forward to further running of the LHC, and the HL-LHC, and to establishing the true nature of the Higgs boson, to find clues or answers to some of the other fundamental open questions in particle physics and cosmology. The exploitation of the LHC is in its infancy, having recorded a small fraction of the finally anticipated integrated luminosity, and the expectations for other discoveries in the coming decades are high.

Acknowledgements The construction, and now the operation and exploitation, of the large and complex ATLAS and CMS experiments have required the talents, the resources, and the dedication of thousands of scientists, engineers and technicians worldwide. This paper is dedicated to all who have worked on these experiments. The superb construction, and efficient operation of the LHC accelerator and the WLCG computing are gratefully acknowledged.

The authors are greatly indebted to C. Seez for careful reading of this manuscript.

References

1. L. Evans (ed.), The Large Hadron Collider, a Marvel of Technology, EPFL Press, 2nd Edition, 2018;
2. L. Evans, P. Bryant (ed.) and LHC Machine, JINST 03 (2008) S08001.
3. F. Englert and R. Brout, Phys. Rev. Lett. **13** (1964) 321.
4. P.W. Higgs, Phys. Lett. **12** (1964) 132.
5. P.W. Higgs, Phys. Rev. Lett. **13** (1964) 508.
6. G.S. Guralnik, C.R. Hagen, and T.W.B. Kibble, Phys. Rev. Lett. **13** (1964) 585.
7. P.W. Higgs, Phys. Rev. 145 (1966) 1156.
8. T.W.B. Kibble, Phys. Rev. 155 (1967) 1554.
9. The Infinity Puzzle, Frank Close, Basic Books (2011).
10. S. Weinberg, Phys. Rev. Lett. 19 (1967) 1264.
11. A. Salam, Proceedings of the eighth Nobel symposium, ed. N. Svartholm, p. 367. Almqvist & Wiskell, 1968.
12. S.L. Glashow, Nucl. Phys. 22 (1961) 579.
13. G. t'Hooft and M. Veltman, Nucl. Phys. B44(1972)189.
14. F. J. Hasert et al., Phys. Lett. B46(1973)38.
15. D. Gross and F.Wilczek, Phys. Rev. Lett. 30(1973)1343.
16. D. H. Politzer, Phys. Rev. Lett. 30(1973)1346.
17. UA1 Collaboration, Experimental Observation of isolated large transverse energy electrons with associated missing energy at $\sqrt{s}$=540 GeV, Phys. Lett. B122 (1983) 103; Experimental Observation of Lepton Pairs of Invariant Mass Around 95 GeV/c2 at the CERN SPS Collider, Phys. Lett. 126B (1983) 398.
18. UA2 Collaboration, Observation of single isolated electrons of high transverse momentum in events with missing transverse energy at the CERN p-pbar collider, Phys. Lett. B122 (1983) 476; Evidence for $Z^0 \to e^+ e^-$ at the CERN anti-p $-$ p Collider, Phys. Lett. 129B (1983) 130.
19. ATLAS Collaboration, Observation of a new particle in the search for the Standard Model Higgs boson with the ATLAS detector at the LHC, Phys. Lett. B716 (2012) 1.
20. CMS Collaboration, Observation of a new boson at a mass of 125 GeV with the CMS experiment at the LHC, Phys. Lett. B716 (2012) 30.
21. M. Della Negra, P. Jenni and T.S. Virdee, Journey in the Search of the Higgs Boson: The ATLAS and CMS Experiments at eth Large Hadron Collider, Science Vol. 338 no. 6114 (2012) 1560.
22. P. Jenni and T.S. Virdee, The Discovery of the Higgs Boson at the LHC, in "60 Years of CERN Experiments", Ed. H. Schopper and L. Di Lella (World Scientific, 2015), 978-981-4644-14-3.
23. M. Della Negra, P. Jenni and T. S. Virdee, The Construction of ATLAS and CMS, Ann. Rev. Part. Sci. 68 (2018) 183.
24. ALEPH, DELPHI, L3, OPAL Collaborations, and LEP Working Group for Higgs Boson Searches, Phys. Lett. B 565 (2003) 61.

25. ALEPH, CDF, D0, DELPHI, L3, OPAL, SLD Collaborations, the LEP Electroweak Working Group, the Tevatron Electroweak Working Group, and the SLD Electroweak and Heavy Flavour Groups, Precision electroweak measurements and constraints on the standard model, CERN PH-EP-2010-095, http://lepewwg.web.cern.ch/LEPEWWG/plots/winter2012/, arXiv:1012.2367, 2010, http://cdsweb.cern.ch/record/1313716.
26. CDF and D0 Collaborations, Evidence for a Particle Produced in Association with Weak Bosons and Decaying to Bottom-Antibottom Quark Pair in Higgs Boson Searches at the Tevatron, Phys. Rev. Lett. 109, 071894 (2012).
27. J.R. Ellis, M.K. Gaillard, D.V. Nanopoulos, Nucl. Phys. B 106 (1976) 292.
28. LHC Higgs Cross Section Working Group, https://twiki.cern.ch/twiki/bin/view/LHCPhysics/ LHCHXSWG, http://arxiv.org/abs/1101.0593 (2011), http://arxiv.org/abs/1201.3084 (2012), http://arXiv:1307.1347 (2013), http://arXiv:1610.07922 (2016).
29. C.J. Seez, T.S. Virdee, L. Di Lella, R.H. Kleiss, Z. Kunszt, W.J. Stirling, in: G. Jarlskog, D. Rein (Eds.), Proceedings of the Large Hadron Collider Workshop, Aachen, Germany, 1990, p. 474, CERN 90-10-V-2/ECFA 90-133-Vol-2.
30. M. Della Negra, D. Froidevaux, K. Jakobs, R. Kinnunen, R. Kleiss, A. Nisati, T. Sjostrand, in: G. Jarlskog, D. Rein (Eds.), Proceedings of the Large Hadron Collider Workshop, Aachen, Germany, 1990, p. 509, CERN 90-10-V-2/ECFA 90-133-Vol-2.
31. Proceedings of the Large Hadron Collider Workshop, Aachen, Germany, 1990, CERN 90-10-V-2/ECFA 90-133.
32. G. Flugge, ed. Proceedings of the General Meeting on LHC Physics Detectors: Towards the LHC Experimental Programme, Geneva: ECFA/CERN (1992).
33. N. Ellis, T. S. Virdee, Ann. Rev. Nucl. Sci. 44 (1994) 609.
34. The LHC Detector Challenge, Physics World, Vol. 17, No. 9, (2004); Detectors at LHC, Phys. Rep. 403-404 (2004) 401.
35. ATLAS Collaboration, https://atlas.cern/discover/about, Letter of Intent, CERN-LHCC-92-004 (1992); Technical Proposal, CERN-LHCC-1994-043 (1994); JINST 3 (2008) S08003.
36. CMS Collaboration, https://cms.cern/, Letter of Intent, CERN-LHCC-92-003 (1992); Technical Proposal, CERN-LHCC-1994-038 (1994); JINST 3 (2008) S08004.
37. ATLAS and CMS Collaborations, Measurement of the Higgs boson production and decay rates and constraints on its couplings from a combined ATLAS and CMS analyss of the LHC pp collision data at $\sqrt{s}=7$ and 8 TeV, JHEP08 (2016) 045.
38. ATLAS Collaboration, Measurements of the Higgs boson properties in the diphoton decay channel using 80 fb^{-1} of pp collision data with the ATLAS detector, ATLAS-CONF-2018-028 (2018).
39. CMS Collaboration, Measurements of Higgs boson properties in the diphoton decay channel in proton-proton collisions at $\sqrt{s} = 13$ TeV, JHEP11 (2018) 185.
40. ATLAS Collaboration, Measurements of Higgs boson coupling properties in the H $\rightarrow$ ZZ* $\rightarrow$ 4l decay channel at $\sqrt{s} = 13$ TeV with the ATLAS detector, JHEP03 (2018) 095.
41. CMS Collaboration, Measurements of the properties of Higgs boson decaying in the four-lepton final state at $\sqrt{s} = 13$ TeV, CMS PAS HIG-18-001.
42. ATLAS and CMS Collaborations, Combined measurement of the Higgs Boson Mass in pp Collisions at $\sqrt{s}=7$ and 8 TeV with the ATLAS and CMS Experiments, Phys. Rev. Lett. 114 191803 (2015).
43. ATLAS Collaboration, Measurements of gluon–gluon fusion and vector-boson fusion Higgs boson production cross-sections in the H$\rightarrow$WW*$\rightarrow$e$\nu\mu\nu$ decay channel in pp collisions at $\sqrt{s}=13$TeV with the ATLAS detector, Phys. Lett. B 789 (2019) 508.
44. CMS Collaboration, Measurements of the properties of Higgs boson decaying to a W boson pair in pp collisions at $\sqrt{s} = 13$ TeV, Phys. Lett. B 791 (2019) 96.
45. CMS Collaboration, Observation of the Higgs boson decay to a pair of τ leptons with the CMS detector, Phys. Lett. B 779 (2018) 283.
46. ATLAS Collaboration, Cross-section measurements of the Higgs boson decaying to a pair of tau leptons in proton-proton collisions at $\sqrt{s}=13$ TeV with the ATLAS detector, Phys. Rev. D 99, 072001 (2019).

47. ATLAS Collaboration, Observation of H→ $b\bar{b}$ decays and VH production with the ATLAS detector, Phy. Lett. B786 (2018) 59.
48. CMS Collaboration, Observation of the Higgs Boson decay to Bottom Quarks, Phys. Rev. Lett. 121, 121801 (2018).
49. ATLAS Collaboration, Search for the Dimuon Decay of the Higgs Boson in pp Collisions at $\sqrt{s}$=13 TeV with the ATLAS detector, Phys. Rev. Lett. 119, 051802 (2017).
50. CMS Collaboration, Search for Higgs boson decaying to two muons in proton-proton collisions at $\sqrt{s}$=13 TeV, Phys. Rev. Lett. 122, 021801 (2019).
51. ATLAS Collaboration, Observation of Higgs boson production in association with a top quark pair at the LHC with the ATLAS detector, Phys. Lett. B 784 (2018) 173.
52. CMS Collaboration, Observation of ttH Production, Phys. Rev. Lett 120, 231801 (2018).
53. ATLAS and CMS Collaborations, Precise determination of the mass of the Higgs boson and tests of compatibility of its couplings with the standard model predictions using collisions at 7 and 8 TeV, Eur. Phys. J. C (2015) 75:212.
54. C. Patrignani et al., Particle Physics Group, Chin. Phys. C, 40 100001 (2016)
55. ATLAS Collaboration, Combined measurements of Higgs boson production and decay in proton-proton collision date at $\sqrt{s}$=13 TeV, ATLAS-CONF-2018-031 (2018).
56. CMS Collaboration, Combined measurements of Higgs boson couplings in proton-proton collisions at $\sqrt{s}$=13 TeV, Eur. Phys. J. C 79 (2019) 421.
57. CMS Collaboration, Study of the mass and spin-parity of the Higgs boson candidate via its decays to Z boson pairs, Phys. Rev. Lett., 110 (2013) 081803.
58. ATLAS Collaboration, Study of the spin of the Higgs boson in diboson decays with the ATLAS detector, Eur. Phys. J C75 (2016) 476.
59. ATLAS Collaboration, Projections for measurements of Higgs boson cross sections, branching fractions, coupling parameters and mass with the ATLAS detector, ATL-PHY-PUB-2018-054.
60. CMS Collaboration, Sensitivity projections for Higgs boson properties measurements at the HL-LHC, CMS PAS FTR-18-011 (2018).

Chapter 7
Relativistic Nucleus-Nucleus Collisions and the QCD Matter Phase Diagram

Reinhard Stock

7.1 Introduction

7.1.1 Overview

This review will be concerned with our knowledge of extended matter under the governance of strong interaction, in short: QCD matter. Strictly speaking, the hadrons are representing the first layer of extended QCD architecture. In fact we encounter the characteristic phenomena of confinement as distances grow to the scale of 1 fm (i.e. hadron size): loss of the chiral symmetry property of the elementary QCD Lagrangian via non-perturbative generation of "massive" quark and gluon condensates, that replace the bare QCD vacuum [1]. However, given such first experiences of transition from short range perturbative QCD phenomena (jet physics etc.), toward extended, non perturbative QCD hadron structure, we shall proceed here to systems with dimensions far exceeding the force range: matter in the interior of heavy nuclei, or in neutron stars, and primordial matter in the cosmological era from electro-weak decoupling (10^{-12} s) to hadron formation ($0.5 \cdot 10^{-5}$ s). This primordial matter, prior to hadronization, should be deconfined in its QCD sector, forming a plasma (i.e. color conducting) state of quarks and gluons [2]: the Quark Gluon Plasma (QGP).

In order to recreate matter at the corresponding high energy density in the terrestrial laboratory one collides heavy nuclei (also called "heavy ions") at ultrarelativistic energies. Quantum Chromodynamics predicts [2–4] a phase transformation

R. Stock (✉)
Goethe University Frankfurt, Frankfurt, Germany

Frankfurt Institute of Advanced Studies (FIAS), Frankfurt, Germany

Institut fuer Kernphysik, Goethe Universitaet, Frankfurt, Germany
e-mail: stock@ikf.uni-frankfurt.de

© The Author(s) 2020
H. Schopper (ed.), *Particle Physics Reference Library*,
https://doi.org/10.1007/978-3-030-38207-0_7

to occur between deconfined quarks and confined hadrons. At near-zero net baryon density (corresponding to big bang conditions) non-perturbative Lattice-QCD places this transition at an energy density of about 1 GeV/fm^3, and at a critical temperature, $T_{crit} \approx 170$ MeV [4–8] (see the article on Lattice QCD in this Volume). The ultimate goal of the physics with ultrarelativistic heavy ions is to locate this transition, elaborate its properties, and gain insight into the detailed nature of the deconfined QGP phase that should exist above. What is meant by the term "ultrarelativistic" is defined by the requirement that the reaction dynamics reaches or exceeds the critical density $\epsilon \approx 1$ GeV/fm^3. Required beam energies turn out [8] to be $\sqrt{s} \geq 10$ GeV, and various experimental programs have been carried out or are being prepared at the CERN SPS (up to about 20 GeV), at the BNL RHIC collider (up to 200 GeV) and finally reaching up to 5.5 TeV at the LHC of CERN.

QCD confinement-deconfinement is of course not limited to the domain that is relevant to cosmological expansion dynamics, at very small excess of baryon over anti-baryon number density and, thus, near zero baryo-chemical potential μ_B. In fact, modern QCD suggests [9–11] a detailed phase diagram of QCD matter and its states, in the plane of T and baryo-chemical potential μ_B. For a map of the QCD matter phase diagram we are thus employing the terminology of the grand canonical Gibbs ensemble that describes an extended volume V of partonic or hadronic matter at temperature T. In it, total particle number is not conserved at relativistic energy, due to particle production-annihilation processes occurring at the microscopic level. However, the probability distributions (partition functions) describing the relative particle species abundances have to respect the presence of certain, to be conserved net quantum numbers (i), notably non-zero net baryon number and zero net strangeness and charm. Their global conservation is achieved by a thermodynamic trick, adding to the system Lagrangian a so-called Lagrange multiplier term, for each of such quantum number conservation tasks. This procedure enters a "chemical potential" μ_i that modifies the partition function via an extra term $\exp\left(-\mu_i/T\right)$ occurring in the phase space integral (see Sect. 7.3 for detail). It modifies the canonical "punishment factor" ($\exp\left(-E/T\right)$), where E is the total particle energy in vacuum, to arrive at an analogous grand canonical factor for the extended medium, of $\exp\left(-E/T - \mu_i/T\right)$. This concept is of prime importance for a description of the state of matter created in heavy ion collisions, where net-baryon number (valence quarks) carrying objects are considered—extended "fireballs" of QCD matter. The same applies to the matter in the interior of neutron stars. The corresponding conservation of net baryon number is introduced into the grand canonical statistical model of QCD matter via the "baryo-chemical potential" μ_B.

We employ this terminology to draw a phase diagram of QCD matter in Fig. 7.1, in the variables T and μ_B. Note that μ_B is high at low energies of collisions creating a matter fireball. In a head-on collision of two mass 200 nuclei at $\sqrt{s} = 15$ GeV the fireball contains about equal numbers of newly created quark-antiquark pairs (of zero net baryon number), and of initial valence quarks. The accommodation of the latter, into created hadronic species, thus requires a formidable redistribution task of net baryon number, reflecting in a high value of μ_B. Conversely, at LHC

Fig. 7.1 Sketch of the QCD matter phase diagram in the plane of temperature T and baryo-chemical potential μ_B. The parton-hadron phase transition line from lattice QCD [8–11] ends in a critical point E. A cross-over transition occurs at smaller μ_B. Also shown are the points of hadro-chemical freeze-out from the grand canonical statistical model

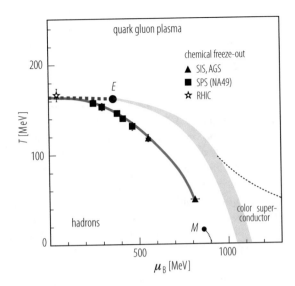

energy (5.5 TeV for Pb+Pb collisions), the initial valence quarks constitute a mere 5% fraction of the total quark density, correspondingly requiring a small value of μ_B. In the extreme, big bang matter evolves toward hadronization (at $T=170$ MeV) featuring a quark over antiquark density excess of 10^{-9} only, resulting in $\mu_B \approx 0$.

Note that the limits of existence of the hadronic phase are not only reached by temperature increase, to the so-called Hagedorn value T_H (which coincides with T_{crit} at $\mu_B \to 0$), but also by density increase to $\varrho > (5\text{–}10)\,\varrho_0$: "cold compression" beyond the nuclear matter ground state baryon density ϱ_0 of about 0.16 B/fm^3. We are talking about the deep interior sections of neutron stars or about neutron star mergers [12–14]. A sketch of the present view of the QCD phase diagram [9–11] is given in Fig. 7.1. It is dominated by the parton-hadron phase transition line that interpolates smoothly between the extremes of predominant matter heating (high T, low μ_B) and predominant matter compression ($T \to 0$, $\mu_B > 1$ GeV). Onward from the latter conditions, the transition is expected to be of first order [15] until the critical point of QCD matter is reached at $200 \leq \mu_B\,(E) \leq 500$ MeV. The relatively large position uncertainty reflects the preliminary character of Lattice QCD calculations at finite μ_B [9–11]. Onward from the critical point, E, the phase transformation at lower μ_B is a cross-over[11].

We note, however, that these estimates represent a major recent advance of lattice theory which was, for two decades, believed to be restricted to the $\mu_B = 0$ situation. Onward from the critical point, toward lower μ_B, the phase transformation should acquire the properties of a rapid cross-over [16], thus also including the case of primordial cosmological expansion. This would finally rule out former ideas, based on the picture of a violent first order "explosive" cosmological hadronization phase transition, that might have caused non-homogeneous conditions, prevailing during early nucleo-synthesis [17], and fluctuations of global matter distribution density that could have served as seedlings of galactic cluster formation [18]. However, it needs to be stressed that the conjectured order of phase transformation, occurring along the parton-hadron phase boundary line, has not been unambiguously confirmed by experiment, as of now.

On the other hand, the *position* of the QCD phase boundary at low μ_B has, in fact, been located by the hadronization points in the T, μ_B plane that are also illustrated in Fig. 7.1. They are obtained from statistical model analysis [19] of the various hadron multiplicities created in nucleus-nucleus collisions, which results in a $[T, \mu_B]$ determination at each incident energy, which ranges from SIS via AGS and SPS to RHIC energies, i.e. $3 \leq \sqrt{s} \leq 200$ GeV. Toward low μ_B these hadronic freeze-out points merge with the lattice QCD parton-hadron coexistence line: hadron formation coincides with hadronic species freeze-out. These points also indicate the μ_B domain of the phase diagram which is accessible to relativistic nuclear collisions. The domain at $\mu_B \geq 1.5$ GeV which is predicted to be in a further new phase of QCD featuring color-flavor locking and color superconductivity [20] will probably be accessible only to astrophysical observation.

One may wonder how states and phases of matter in thermodynamical equilibrium—as implied by a description in grand canonical variables—can be sampled via the dynamical evolution of relativistic nuclear collisions. Employing heavy nuclei, $A \approx 200$, as projectiles/targets or in colliding beams (RHIC, LHC), transverse dimensions of the primordial interaction volume do not exceed about 8 fm, and strong interaction ceases after about 20 fm/c. We shall devote an entire later section to the aspects of equilibrium (Sect. 7.2.5) but note, for now, that the time and dimension scale of primordial perturbative QCD interaction at the microscopic partonic level amounts to subfractions of 1 fm/c, the latter scale, however, being representative of non perturbative processes (confinement, "string" formation etc.). The A+A fireball size thus exceeds, by far, the elementary non perturbative scale. An equilibrium quark gluon plasma represents an extended non-perturbative QCD object, and the question whether its relaxation time scale can be provided by the expansion time scale of an A+A collision, needs careful examination. Reassuringly, however, the hadrons that are supposedly created from such a preceding non-perturbative QGP phase at top SPS and RHIC energy, do in fact exhibit perfect hadrochemical equilibrium, the derived $[T, \mu_B]$ values [19] thus legitimately appearing in the phase diagram, Fig. 7.1.

In the present review we will order the physics observables to be treated, in sequence of their origin from successive stages that characterize the overall dynamical evolution of a relativistic nucleus-nucleus collision. In rough outline this evolution can be seen to proceed in three major steps. An initial period of matter compression and heating occurs in the course of interpenetration of the projectile and target baryon density distributions. Inelastic processes occurring at the microscopic level convert initial beam longitudinal energy to new internal and transverse degrees of freedom, by breaking up the initial baryon structure functions. Their partons thus acquire virtual mass, populating transverse phase space in the course of inelastic perturbative QCD shower multiplication. This stage should be far from thermal equilibrium, initially. However, in step two, inelastic interaction between the two arising parton fields (opposing each other in longitudinal phase space) should lead to a pile-up of partonic energy density centered at mid-rapidity (the longitudinal coordinate of the overall center of mass). Due to this mutual stopping down of the initial target and projectile parton fragmentation showers, and from the concurrent decrease of parton virtuality (with decreasing average square momentum transfer Q^2) there results a slowdown of the time scales governing the dynamical evolution. Equilibrium could be approached here, the system "lands" on the T, μ plane of Fig. 7.1, at temperatures of about 300 and 200 MeV at top RHIC and top SPS energy, respectively. The third step, system expansion and decay, thus occurs from well above the QCD parton-hadron boundary line. Hadrons and hadronic resonances then form, which decouple swiftly from further inelastic transmutation so that their yield ratios become stationary ("frozen-out"). A final expansion period dilutes the system to a degree such that strong interaction ceases all together.

In order to verify in detail this qualitative overall model, and to ascertain the existence (and to study the properties) of the different states of QCD that are populated in sequence, one seeks observable physics quantities that convey information imprinted during distinct stages of the dynamical evolution, and "freezing-out" without significant obliteration by subsequent stages. Ordered in sequence of their formation in the course of the dynamics, the most relevant such observables are briefly characterized below:

1. Suppression of J/Ψ and Y production by Debye-screening in the QGP. These vector mesons result from primordial, pQCD production of $c\bar{c}$ and $b\bar{b}$ pairs that would hadronize unimpeded in elementary collisions but are broken up if immersed into a npQCD deconfined QGP, at certain characteristic temperature thresholds.
2. Suppression of dijets which arise from primordial $q\bar{q}$ pair production fragmenting into partonic showers (jets) in vacuum but being attenuated by QGP-medium induced gluonic bremsstrahlung: Jet quenching in A+A collisions.

   a. A variant of this: *any* primordial hard parton suffers a high, specific loss of energy when traversing a deconfined medium: High p_T suppression in A+A collisions.

3. Hydrodynamic collective motion develops with the onset of (local) thermal equilibrium. It is created by partonic pressure gradients that reflect the initial collisional impact geometry via non-isotropies in particle emission called "directed" and "elliptic" flow. The latter reveals properties of the QGP, seen here as an ideal partonic fluid.

   a. Radial hydrodynamical expansion flow ("Hubble expansion") is a variant of the above that occurs in central, head on collisions with cylinder symmetry, as a consequence of an isentropic expansion. It should be sensitive to the mixed phase conditions characteristic of a first order parton-hadron phase transition.

4. Hadronic "chemical" freeze-out fixes the abundance ratios of the hadronic species into an equilibrium distribution. Occurring very close to, or at hadronization, it reveals the dynamical evolution path in the $[T, \mu_B]$ plane and determines the critical temperature and density of QCD. The yield distributions in A+A collisions show a dramatic strangeness enhancement effect, characteristic of an extended QCD medium.
5. Fluctuations, from one collision event to another (and even within a single given event) can be quantified in A+A collisions due to the high charged hadron multiplicity density (of up to 600 per rapidity unit at top RHIC energy). Such event-by-event (Debye) fluctuations of pion rapidity density and mean transverse momentum (event "temperature"), as well as event-wise fluctuations of the strange to non-strange hadron abundance ratio (may) reflect the existence and position of the conjectured critical point of QCD (Fig. 7.1).
6. Two particle Bose-Einstein-Correlations are the analog of the Hanbury-Brown, Twiss (HBT) effect of quantum optics. They result from the last interaction experienced by hadrons, i.e. from the global decoupling stage. Owing to a near isentropic hadronic expansion they reveal information on the overall space-time-development of the "fireball" evolution.

In an overall view the first group of observables (1 to 2a) is anchored in established pQCD physics that is well known from theoretical and experimental analysis of elementary collisions (e^+e^- annihilation, pp and $p\bar{p}$ data). In fact, the first generation of high Q^2 baryon collisions, occurring at the microscopic level in A+A collisions, should closely resemble such processes. However, their primary partonic products do not escape into pQCD vacuum but get attenuated by interaction with the concurrently developing extended high density medium, thus serving as diagnostic tracer probes of that state. The remaining observables capture snapshots of the bulk matter medium itself. After initial equilibration we may confront elliptic flow data with QCD during the corresponding partonic phase of the dynamical evolution employing thermodynamic [21] and hydrodynamic [22] models of a high temperature parton plasma. The hydro-model stays applicable well into the hadronic phase. Hadron formation (confinement) occurs in between these phases (at about 5 μs time in the cosmological evolution). In fact relativistic nuclear collision data may help to finally pin down the mechanism(s) of this fascinating QCD process [23–25] as we can vary the conditions of its occurrence, along the parton-hadron phase separation line of Fig. 7.1, by proper choice of collisional energy $\sqrt{s}$, and system size A, while maintaining the overall conditions of an extended imbedding medium of high energy density within which various patterns [9–11, 15, 16] of the hadronization phase transition may establish. The remaining physics observables (3a, 5 and 6 above) essentially provide for auxiliary information about the bulk matter system as it traverses (and emerges from) the hadronization stage, with special emphasis placed on manifestations of the conjectured critical point.

The present review will briefly cover each of the above physics observables in a separate chapter, beginning with the phenomena of confinement and hadronization (Sect. 7.3), then to turn to the preceding primordial dynamics, e.g. to elliptical flow (Sect. 7.4), high p_T and jet quenching (Sect. 7.5) and quarkonium suppression (Sect. 7.6) as well as in-medium ϱ-meson "melting". We then turn to the late period, with correlation and fluctuation studies (Sect. 7.7). We conclude (Sect. 7.8) with a summary, including an outlook to the future of the research field.

However, before turning to such specific observables we shall continue this introductory chapter, with a look at the origin, and earlier development of the ideas that have shaped this field of research (Sect. 7.1.2). Then we turn to a detailed description of the overall dynamical evolution of relativistic nucleus-nucleus collisions, and to the typical overall patterns governing the final distributions in transverse and longitudinal (rapidity) phase space (Sect. 7.2). The aspects of an approach toward equilibrium, at various stages of the dynamical evolution (which are of key importance toward the intended elucidation of the QCD matter phase diagram), will be considered, in particular.

7.1.2 History

The search for the phase diagram of strongly interacting matter arose in the 1960s, from a coincidence of ideas developing—at first fairly independently—in nuclear and astrophysics. In fact, the nuclear proton-neutron matter, a quantum liquid at $T = 0$ and energy density $\epsilon = 0.15$ GeV/fm^3, represents the ground state of extended QCD matter. Of course, QCD was unknown during the development of traditional nuclear physics, and the extended matter aspects of nuclei—such as compressibility or the equation of state, in general—did not receive much attention until the advent, in the 1960s, of relativistic nuclear mean field theory, notably s-matrix theory by Brueckner [26] and the σ-model of Walecka [27]. These theories developed the novel view of "*infinite* nuclear matter" structure, based on in-medium properties of the constituent baryons that share parts of their vacuum mass and surface structure with the surrounding, continuous field of relativistic scalar and vector mesons. Most importantly, in the light of subsequent development, these theories allowed for a generalization away from ground state density and zero temperature. Such developments turned out to be of key relevance for acute nuclear astrophysics problems: the dynamics of type II super-novae and the stability of neutron stars, which both required the relation of pressure to density and temperature of hadronic matter, i.e. the hadronic matter equation of state (EOS). H.A. Bethe et al. [28] postulated that the final stages of supernova collapse should evolve through the density interval $0.1 \leq \varrho/\varrho_0 \leq 5$ where $\varrho_0 = 0.16$ (baryons per fm^3) is the nuclear matter ground state density, and a similar domain was expected for the neutron star density variation from surface to interior [29]. It was clear that, at the highest thus considered densities the EOS might soften due to strange hadron production caused by increasing Fermi energy. However the field theoretical models permitted no reliable extrapolation to such high densities (which, in retrospect, are perhaps not reached in supernova dynamics [30]), and the experimental information concerning the EOS from study of the giant monopole resonance—a collective density oscillation also called "breathing mode"—covered only the parabolic minimum occurring in the related function of energy vs. density at $T = 0$, $\varrho = \varrho_0$.

The situation changed around 1970 due to the prediction made by W. Greiner et al. [31] that nucleus-nucleus collisions, at relatively modest relativistic energies, would result in shock compression. This mechanism promised to reach matter densities far beyond those of a mere superposition (i.e. $\varrho/\varrho_0 \leq 2\gamma$) of initial target and projectile densities. Coinciding in time, the newly developed Bevalac accelerator at LBL Berkeley offered projectiles up to ^{38}Ar, at just the required energies, 100 MeV $\leq E_{\text{Lab}}$/nucleon $\leq$ 2 GeV. The field of "relativistic heavy ion physics" was born. The topic was confronted, at first, with experimental methods available both from nuclear and particle physics. It was shown that particle production (here still restricted to pions and kaons) could indeed be linked to the equation of state [32] and that, even more spectacularly, the entire "fireball" of mutually stopped hadrons developed decay modes very closely resembling the initial predictions of hydrodynamical shock flow modes [33] which directly link primordial pressure gradients with collective velocity fields of matter streaming out, again governed by the nuclear/hadronic matter EOS. Actually, both these statements do, in fact, apply (mutatis mutandis) up to the present ultra-relativistic energies (see Sects. 7.2–7.4). However it turned out soon that the equation of state at low or even zero temperature (as required in supernova and neutron star studies) could only be obtained in a semi-empirical manner [34]. The reason: compression can, in such collisions, be only accomplished along with temperature and entropy increase. In an ideal baryon gas $\exp(s/A) \propto T^{3/2}/\varrho$, i.e. $T^{3/2}$ will grow faster than ϱ in a non-isentropic compression. Thus the reaction dynamics will be sensitive to various *isothermes* of the ground state EOS $P = f(\varrho, T = 0)$, staying at $T \gg 0$, throughout, and, moreover, not at constant T. Thus a relativistic dynamical mean field model is required in order to interactively deduce the $T = 0$ EOS from data [34]. The EOS result thus remains model dependent.

The ideas concerning creation of a quark gluon plasma arose almost concurrent with the heavy ion shock compression proposal. In 1974 T.D. Lee formulated the idea that the non-perturbative vacuum condensates could be "melted down ... by distributing high energy or high nucleon density over a relatively large volume" [35]. Collins and Perry [36] realized that the asymptotic freedom property of QCD implies the existence of an ultra-hot form of matter with deconfined quarks and gluons, an idea that gained wide recognition when S. Weinberg [37] proposed an asymptotic freedom phase at the beginning of "The first Three minutes". In fact, this idea of deconfinement by asymptotic freedom (with implied temperature of several GeV) was correct, but somewhat besides the point, as everybody expected, likewise, that deconfinement sets in right above the limiting hadron temperature of R. Hagedorn [38], $T_{\text{H}} \approx 160$ MeV. A medium existing down to that temperature would, however, feature an average momentum square transfer $Q^2 < 1$ GeV^2, i.e. be far into the non perturbative domain, and very far from asymptotic freedom. Right above the hadron to parton transition the "quark gluon plasma" (as it was named by E. Shuryak [39]) is not a weakly coupled ideal pQCD gas as soon became obvious by Lattice QCD calculations for extended matter [40]. Seen in retrospect one obviously cannot defend a picture of point like quarks (with "current" masses) at $Q^2 \leq 0.2$ GeV^2 where size scales of 0.5 to 1 fm must play a dominating role.

An analytic QCD description of deconfinement does not exist. For heavy quarkonia, $c\bar{c}$ (J/Ψ) and $b\bar{b}$ (Y) deconfinement in partonic matter, Matsui and Satz proposed [41] a Debye screening mechanism, caused by the high spatial density of free color carriers, that removes the confining long range potential as T increases toward about $2\,T_c$, an effect reproduced by modern lattice QCD [42]. However, light hadron deconfinement cannot be understood with a non-relativistic potential model. Such critical remarks not withstanding, we shall demonstrate in Sects. 7.3–7.6 that the very existence, and also crucial properties of the QGP can in fact be inferred from experiment, and be confronted with corresponding predictions of recent lattice QCD theory.

Our present level of an initial understanding of the phase diagram of QCD matter (Fig. 7.1), is the result of a steady development of both experiment and theory, that began about three decades ago, deriving initial momentum from the Bevalac physics at LBL which motivated—along with the developing formulation of the quark gluon plasma research goals—a succession of experimental facilities progressing toward higher $\sqrt{s}$. Beginning with the AGS at BNL (^{28}Si and ^{197}Au beams with $\sqrt{s} \leq$ 5 GeV), the next steps were taken at the CERN SPS ($\sqrt{s}$ from 6 to 20 GeV; ^{16}O, ^{32}S, ^{208}Pb beams), and at the Relativistic Heavy Ion Collider RHIC (the first facility constructed explicitly for nuclear collisions) which offers beams of ^{64}Cu and ^{197}Au at $20 \leq \sqrt{s} \leq 200$ GeV. A final, gigantic step in energy will be taken 2008 with the CERN Large Hadron Collider: ^{208}Pb beams at $\sqrt{s} = 5.5$ TeV.

7.2 Bulk Hadron Production in A+A Collisions

In this section we take an overall look at bulk hadron production in nucleus-nucleus collisions. In view of the high total c.m. energies involved at e.g. top SPS ($E_{cm}^{tot} \approx 3.3$ TeV) and top RHIC (38 TeV) energies, in central Pb+Pb (SPS) and Au+Au (RHIC) collisions, one can expect an extraordinarily high spatial density of produced particles. Thus, as an overall idea of analysis, one will try to relate the observed flow of energy into transverse and longitudinal phase space and particle species to the high energy density contained in the primordial interaction volume, thus to infer about its contained matter. The typical experimental patterns of such collisions, both in collider mode at RHIC and in a fixed target configuration at the SPS, are illustrated in Fig. 7.2 which shows a fractional view of the total distribution of charged particles (about 4000 and 1600, respectively) within the tracking volume of the STAR and NA49 experiments.

Most of these tracks correspond to "thermal" pions (p_T up to 2 GeV) and, in general, such thermal hadrons make up for about 95% of the observed multiplicity: the bulk of hadron production. Their distributions in phase space will be illustrated in the subsections below. This will lead to a first insight into the overall reaction

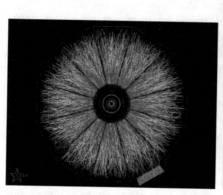
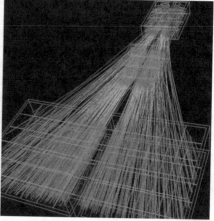

Fig. 7.2 Charged particle tracks in central Au+Au and Pb+Pb collision events, in collider geometry (top) from RHIC STAR TPC tracking at $\sqrt{s} = 200$ GeV, and in fixed target geometry (bottom) from NA49 at the SPS, $\sqrt{s} = 17.3$ GeV

dynamics, and also set the stage for consideration of the rare signals, imbedded in this thermal bulk production: correlations, jets, heavy flavors, fluctuations, which are the subject of later chapters.

7.2.1 Particle Multiplicity and Transverse Energy Density

Particle production can be assessed globally by the total created transverse energy, the overall result of the collisional creation of *transverse* momentum p_T or transverse mass ($m_T = \sqrt{p_T^2 + m_0^2}$), at the microscopic level. Figure 7.3 shows the distribution of total transverse energy $E_T = \sum_i E(\Theta_i) \cdot \sin \Theta$ resulting from a calorimetric measurement of energy flow into calorimeter cells centered at angle Θ_i relative to the beam [43], for ^{32}S + ^{197}Au collisions at $\sqrt{s} = 20$ GeV, and for ^{208}Pb + ^{208}Pb collisions at $\sqrt{s} = 17.3$ GeV.

The shape is characteristic of the impact parameter probability distribution (for equal size spheres in the Pb+Pb case). The turnoff at $E_T = 520$ GeV indicates the point where geometry runs out of steam, i.e. where $b \to 0$, a configuration generally referred to as a "central collision". The adjacent shoulder results from genuine event by event fluctuations of the actual number of participant nucleons from target and projectile (recall the diffuse Woods-Saxon nuclear density profiles), and from experimental factors like calorimeter resolution and limited acceptance. The latter covers 1.3 units of pseudo-rapidity and contains mid-rapidity $\eta_{mid} = 2.9$. Re-normalizing [43] to $\Delta \eta = 1$ leads to $dE_T/d\eta$ (mid) $= 400$ GeV, in agreement with the corresponding WA80 result [44]. Also, the total transverse energy of central

Fig. 7.3 Minimum bias
distribution of total transverse
energy in Pb+Pb collisions at
$\sqrt{s} = 17.3$ GeV, and S+Au
collisions at $\sqrt{s} = 20$ GeV, in
the rapidity interval
$2.1 < y < 3.4$, from [43]

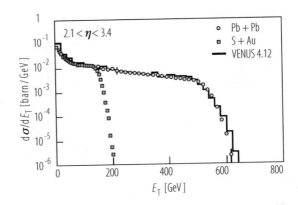

Pb+Pb collisions at $\sqrt{s} = 17.3$ GeV turns out to be about 1.2 TeV. As the definition of a central collision, indicated in Fig. 7.3, can be shown [42] to correspond to an average nucleon participant number of $N_{part} = 370$ one finds an average total transverse energy per nucleon pair, of $E_T/(0.5\,N_{part}) = 6.5$ GeV. After proper consideration of the baryon pair rest mass (not contained in the calorimetric E_T response but in the corresponding $\sqrt{s}$) one concludes [43] that the observed total E_T corresponds to about $0.6\,E_T^{max}$, the maximal E_T derived from a situation of "complete stopping" in which the incident $\sqrt{s}$ gets fully transformed into internal excitation of a single, ideal isotropic fireball located at mid-rapidity. The remaining fraction of E_T^{max} thus stays in longitudinal motion, reflecting the onset, at SPS energy, of a transition from a central fireball to a longitudinally extended "fire-tube", i.e. a cylindrical volume of high primordial energy density. In the limit of much higher $\sqrt{s}$ one may extrapolate to the idealization of a boost invariant primordial interaction volume, introduced by Bjorken [45].

We shall show below (Sect. 7.2.2) that the charged particle rapidity distributions, from top SPS to top RHIC energies, do in fact substantiate a development toward a boost-invariant situation. One may thus employ the Bjorken model for an estimate of the primordial spatial energy density ϵ, related to the energy density in rapidity space via the relation [45]

$$\epsilon(\tau_0) = \frac{1}{\pi R^2} \frac{1}{\tau_0} \frac{dE_T}{dy} \qquad (7.1)$$

where the initially produced collision volume is considered as a cylinder of length $dz = \tau_0\,dy$ and transverse radius $R \propto A^{1/3}$. Inserting for πR^2 the longitudinally projected overlap area of Pb nuclei colliding near head-on ("centrally"), and assuming that the evolution of primordial pQCD shower multiplication (i.e. the energy transformation into internal degrees of freedom) proceeds at a time scale $\tau_0 \leq 1$ fm/c, the above average transverse energy density, of $dE_T/dy = 400$ GeV at top SPS energy [43, 44] leads to the estimate

$$\epsilon(\tau_0 = 1 \text{ fm}) = 3.0 \pm 0.6 \text{ GeV/fm}^3, \qquad (7.2)$$

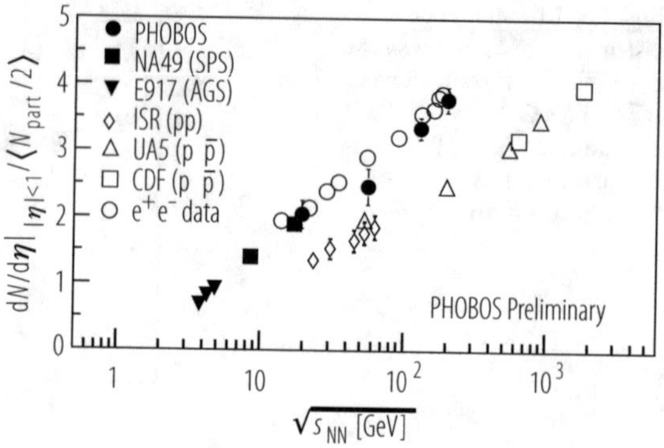

Fig. 7.4 Charged hadron rapidity density at mid-rapidity vs. $\sqrt{s}$, compiled from e^+e^-, pp, $p\bar{p}$ and A+A collisions [53]

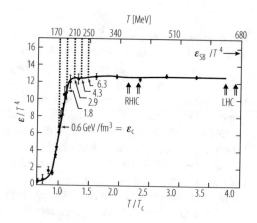

Fig. 7.5 Lattice QCD results at zero baryon potential for energy density ϵ/T^4 versus T/T_c with three light quark flavors, compared to the Stefan-Boltzmann-limit ϵ_{SB} of an ideal quark-gluon gas [48]

thus exceeding, by far, the estimate of the critical energy density ϵ_0 obtained from lattice QCD (see below), of about 1.0 GeV/fm^3. Increasing the collision energy to $\sqrt{s} = 200$ GeV for Au+Au at RHIC, and keeping the same formation time, $\tau_0 = 1$ fm/c (a conservative estimate as we shall show in Sect. 7.2.4), the Bjorken estimate grows to $\epsilon \approx 6.0 \pm 1$ GeV/fm^3. This statement is based on the increase of charged particle multiplicity density at mid-rapidity with $\sqrt{s}$, as illustrated in Fig. 7.4. From top SPS to top RHIC energy [46] the density per participant nucleon pair almost doubles. However, at $\sqrt{s} = 200$ GeV the formation or thermalization time τ_0, employed in the Bjorken model [45], was argued [47] to be shorter by a factor of about 4. We will return to such estimates of τ_0 in Sect. 7.2.5 but note, for now, that the above choice of $\tau_0 = 1$ fm/c represents a conservative upper limit at RHIC energy.

These Bjorken-estimates of spatial transverse energy density are confronted in Fig. 7.5 with lattice QCD results obtained for three dynamical light quark flavors [48], and for zero baryo-chemical potential (as is realistic for RHIC energy and beyond but still remains a fair approximation at top SPS energy where $\mu_B \approx$ 250 MeV). The energy density of an ideal, relativistic parton gas scales with the fourth power of the temperature,

$$\epsilon = gT^4 \tag{7.3}$$

where g is related to the number of degrees of freedom. For an ideal gluon gas, $g = 16\,\pi^2/30$; in an interacting system the effective g is smaller. The results of Fig. 7.5 show, in fact, that the Stefan-Boltzmann limit ϵ_{SB} is not reached, due to non perturbative effects, even at four times the critical temperature $T_c = 170$ MeV. The density $\epsilon/T^4 = g$ is seen to ascend steeply, within the interval $T_c \pm 25$ MeV. At T_c the critical QCD energy density $\epsilon = 0.6$–1.0 GeV/fm^3. Relating the thermal energy density with the Bjorken estimates discussed above, one arrives at an estimate of the initial temperatures reached in nucleus-nucleus collisions, thus implying thermal partonic equilibrium to be accomplished at time scale τ_0 (see Sect. 7.2.5). For the SPS, RHIC and LHC energy domains this gives an initial temperature in the range $190 \leq T^{\mathrm{SPS}} \leq 220$ MeV, $220 \leq T^{\mathrm{RHIC}} \leq 400$ MeV (assuming [47] that τ_0 decreases to about 0.3 fm/c here) and $T^{\mathrm{LHC}} \geq 600$ MeV, respectively. From such estimates one tends to conclude that the immediate vicinity of the phase transformation is sampled at SPS energy, whereas the dynamical evolution at RHIC and LHC energies dives deeply into the "quark-gluon-plasma" domain of QCD. We shall return to a more critical discussion of such ascertations in Sect. 7.2.5.

One further aspect of the mid-rapidity charged particle densities per participant pair requires attention: the comparison with data from elementary collisions. Figure 7.4 shows a compilation of pp, $p\overline{p}$ and e^+e^- data covering the range from ISR to LEP and Tevatron energies.

The data from e^+e^- represent dN_{ch}/dy, the rapidity density along the event thrust axis, calculated assuming the pion mass [49] (the difference between dN/dy and $dN/d\eta$ can be ignored here). Remarkably, they superimpose with the central A+A collision data, whereas pp and $p\overline{p}$ show similar slope but amount to only about 60% of the AA and e^+e^- values. This difference between e^+e^- annihilation to hadrons, and pp or $p\overline{p}$ hadro-production has been ascribed [50] to the characteristic leading particle effect of minimum bias hadron-hadron collisions which is absent in e^+e^-. It thus appears to be reduced in AA collisions due to subsequent interaction of the leading parton with the oncoming thickness of the remaining target/projectile density distribution. This naturally leads to the scaling of total particle production with N_{part} that is illustrated in Fig. 7.6, for three RHIC energies and minimum bias Au+Au collisions; the close agreement with e^+e^- annihilation data is obvious again. One might conclude that, analogously, the participating nucleons get "annihilated" at high $\sqrt{s}$, their net quantum number content being spread out over phase space (as we shall show in the next section).

Fig. 7.6 The total number of charged hadrons per participant pair shown as a function of N_{part} in Au+Au collisions at three RHIC energies [53]

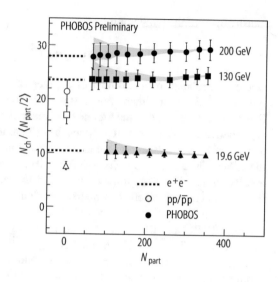

7.2.2 Rapidity Distributions

Particle production number in A+A collisions depends globally on $\sqrt{s}$ and collision centrality, and differentially on p_T and rapidity y, for each particle species i. Integrating over p_T results in the rapidity distribution dN_i/dy. Particle rapidity,[1] $y = \sinh^{-1} p_L/M_T$ (where $M_T = \sqrt{m^2 + p_T^2}$), requires mass identification. If that is unknown one employs pseudo-rapidity ($\eta = -\ln[\tan(\Theta/2)]$) instead. This is also chosen if the joint rapidity distribution of several unresolved particle species is considered: notably the charged hadron distribution. We show two examples in Fig. 7.7. The left panel illustrates charged particle production in $p\bar{p}$ collisions studied by UA1 at $\sqrt{s} = 540$ GeV [51]. Whereas the minimum bias distribution (dots) exhibits the required symmetry about the center of mass coordinate, $\eta = 0$, the rapidity distribution corresponding to events in which a W boson was produced (histogram) features, both, a higher average charged particle yield, and an asymmetric shape. The former effect can be seen to reflect the expectation that the W production rate increases with the "centrality" of $p\bar{p}$ collisions, involving more primordial partons as the collisional overlap of the partonic density profiles gets larger, thus also increasing the overall, softer hadro-production rate. The asymmetry should result from a detector bias favoring W identification at negative rapidity: the transverse W energy, of about 100 GeV would *locally* deplete the energy store

---

[1] The rapidity variable represents a compact (logarithmic) description of longitudinal phase space. It is based on longitudinal particle velocity (derived from p_{long} and m), $y = 1/2\ln((1 + \beta_L)/(1 - \beta_L))$. The rapidity distribution dN/dy is shape invariant under longitudinal Lorentz transformation, and centered at "mid-rapidity" $y_{mid} = y_{CM}$, for all produced particle species; see Figs. 7.7, 7.8, 7.9, and 7.10.

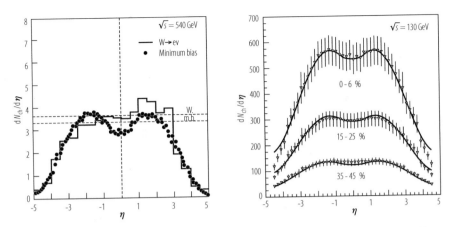

Fig. 7.7 Left panel: charged particle pseudo-rapidity distribution in $p\bar{p}$ collisions at $\sqrt{s} = 540$ GeV [51]. Right panel: same in RHIC Au+Au collisions at $\sqrt{s} = 130$ GeV at different centralities [52]. Closed lines represent fits with the color glass condensate model [64]

available for associated soft production. If correct, this interpretation suggests that the wide rapidity gap between target and projectile, arising at such high $\sqrt{s}$, of width $\Delta y \approx 2 \ln (2\gamma_{CM})$, makes it possible to define local sub-intervals of rapidity within which the species composition of produced particles varies.

The right panel of Fig. 7.7 shows charged particle pseudo-rapidity density distributions for Au+Au collisions at $\sqrt{s} = 130$ GeV measured by RHIC experiment PHOBOS [52] at three different collision centralities, from "central" (the 6% highest charged particle multiplicity events) to semi-peripheral (the corresponding 35–45% cut). We will turn to centrality selection in more detail below. Let us first remark that the slight dip at mid-rapidity and, moreover, the distribution shape in general, are common to $p\bar{p}$ and Au+Au. This is also the case for e^+e^- annihilation as is shown in Fig. 7.8 which compares the ALEPH rapidity distribution along the mean p_T ("thrust") axis of jet production in e^+e^- at $\sqrt{s} = 200$ GeV [49] with the scaled PHOBOS-RHIC distribution of central Au+Au at the same $\sqrt{s}$ [53]. Note that the mid-rapidity values contained in Figs. 7.7 and 7.8 have been employed already in Fig. 7.4, which showed the overall $\sqrt{s}$ dependence of mid-rapidity charged particle production. What we concluded there was a perfect scaling of A+A with e^+e^- data at $\sqrt{s} \geq 20$ GeV and a 40% suppression of the corresponding pp, $p\bar{p}$ yields. We see here that this observation holds, semi-quantitatively, for the entire rapidity distributions. These are not ideally boost invariant at the energies considered here but one sees in dN_{ch}/dy a relatively smooth "plateau" region extending over $|y| \leq 1.5$–2.5.

The production spectrum of charged hadrons is, by far, dominated by soft pions ($p_T \leq 1$ GeV/c) which contribute about 85% of the total yield, both in elementary and nuclear collisions. The evolution of the π^- rapidity distribution with $\sqrt{s}$ is illustrated in Fig. 7.9 for central Au+Au and Pb+Pb collisions from AGS via SPS to RHIC energy, $2.7 \leq \sqrt{s} \leq 200$ GeV [54].

Fig. 7.8 Pseudo-rapidity
distribution of charged
hadrons produced in central
Au+Au collisions at
$\sqrt{s} = 200$ GeV compared
with e^+e^- data at similar
energy. The former data
normalized by $N_{part}/2$. From
ref. [53]

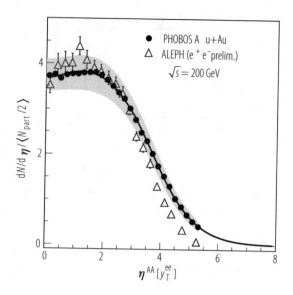

At lower $\sqrt{s}$ the distributions are well described by single Gaussian fits [54] with $\sigma(y)$ nearly linearly proportional to the total rapidity gap $\Delta y \propto \ln \sqrt{s}$ as shown in the right hand panel of Fig. 7.9. Also illustrated is the prediction of the schematic hydrodynamical model proposed by Landau [55],

$$\sigma^2 \propto \ln \left(\frac{\sqrt{s}}{2m_p} \right) \tag{7.4}$$

which pictures hadron production in high $\sqrt{s}$ pp collisions to proceed via a dynamics of initial complete "stopping down" of the reactants matter/energy content in a mid-rapidity fireball that would then expand via 1-dimensional ideal hydrodynamics. Remarkably, this model that has always been considered a wildly extremal proposal falls rather close to the lower $\sqrt{s}$ data for central A+A collisions but, as longitudinal phase space widens approaching boost invariance we expect that the (non-Gaussian) width of the rapidity distribution grows linearly with the rapidity gap Δy. LHC data will finally confirm this expectation, but Figs. 7.7, 7.8, and 7.9 clearly show the advent of boost invariance, already at $\sqrt{s} = 200$ GeV.

A short didactic aside: At low $\sqrt{s}$ the total rapidity gap $\Delta y = 2$–3 does closely resemble the total rapidity width obtained for a thermal pion velocity distribution at temperature $T = 120$–150 MeV, of a single mid-rapidity fireball, the y-distribution of which represents the longitudinal component according to the relation [19]

$$\frac{dN}{dy} \propto \left(m^2 T + \frac{2mT^2}{\cosh y} + \frac{2T^2}{\cosh^2 y} \right) \exp \left[-m \, \cosh y / T \right] \tag{7.5}$$

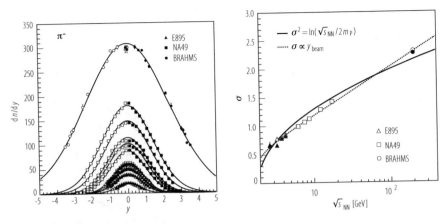

Fig. 7.9 Left panel: negative pion rapidity distributions in central Au+Au and Pb+Pb collisions from AGS via SPS to RHIC energies [54]. Right panel: the Gaussian rapidity width of pions versus $\sqrt{s}$, confronted by Landau model predictions (solid line) [54]

where m is the pion mass. *Any* model of preferentially longitudinal expansion of the pion emitting source, away from a trivial single central "completely stopped" fireball, can be significantly tested only once $\Delta y > 3$ which occurs upward from SPS energy. The agreement of the Landau model prediction with the data in Fig. 7.9 is thus fortuitous, below $\sqrt{s} \approx 10$ GeV, as *any* created fireball occupies the entire rapidity gap with pions.

The Landau model offers an extreme view of the mechanism of "stopping", by which the initial longitudinal energy of the projectile partons or nucleons is inelastically transferred to produced particles and redistributed in transverse and longitudinal phase space, of which we saw the total transverse fraction in Fig. 7.3. Obviously e^+e^- annihilation to hadrons represents the extreme stopping situation. Hadronic and nuclear collisions offer the possibility to analyze the final distribution in phase space of their non-zero net quantum numbers, notably net baryon number. Figure 7.10 shows the net-proton rapidity distribution (i.e. the proton rapidity distribution subtracted by the antiproton distribution) for central Pb+Pb/Au+Au collisions at AGS ($\sqrt{s} = 5.5$ GeV), SPS ($\sqrt{s} \leq 17.3$ GeV) and RHIC ($\sqrt{s} = 200$ GeV) [56]. With increasing energy we see a central (but non-Gaussian) peak developing into a double-hump structure that widens toward RHIC leaving a plateau about mid-rapidity. The RHIC-BRAHMS experiment acceptance for p, $\overline{p}$ identification does unfortunately not reach up to the beam fragmentation domain at $y_p = 5.4$ (nor does any other RHIC experiment) but only to $y \approx 3.2$, with the consequence that the major fraction of p^{net} is not accounted for. However the mid-rapidity region is by no means net baryon free. At SPS energy the NA49 acceptance covers the major part of the total rapidity gap, and we observe in detail a net p distribution shifted down from $y_p = 2.9$ by an average rapidity shift [56] of $\langle \delta y \rangle = 1.7$. From Fig. 7.10 we infer that $\langle \delta y \rangle$ cannot scale linearly

Fig. 7.10 Net proton rapidity distributions in central Au+Au/Pb+Pb collisions at AGS, SPS and RHIC energies [56, 57]

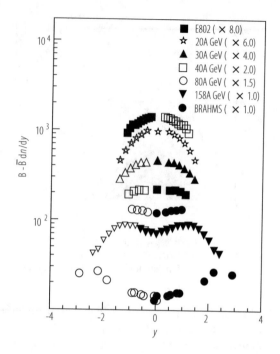

with $y_p \approx \ln(2\gamma_{CM}) \approx \ln\sqrt{s}$ for ever—as it does up to top SPS energy where $\langle\delta y\rangle = 0.58\, y_p$ [56]. Because extrapolating this relation to $\sqrt{s} = 200$ GeV would result in $\langle\delta y\rangle = 3.1$, and with $y_p \approx 5.4$ at this energy we would expect to observe a major fraction of net proton yield in the vicinity of $y = 2.3$ which is not the case. A saturation must thus occur in the $\langle\delta y\rangle$ vs. $\sqrt{s}$ dependence.

The re-distribution of net baryon density over longitudinal phase space is, of course, only partially captured by the net proton yield but a recent study [57] has shown that proper inclusion of neutron[2] and hyperon production data at SPS and RHIC energy scales up, of course, the dN/dy distributions of Fig. 7.10 but leaves the peculiarities of their shapes essentially unchanged. As the net baryon rapidity density distribution should resemble the final valence quark distribution the Landau model is ruled out as the valence quarks are seen to be streaming from their initial position at beam rapidity toward mid-rapidity (not vice versa). It is remarkable, however, to see that some fraction gets transported very far, during the primordial partonic non-equilibrium phase. We shall turn to its theoretical description in Sect. 7.2.4 but note, for now, that pp collisions studied at the CERN ISR [58] lead to a qualitatively similar net baryon rapidity distribution, albeit characterized by a smaller $\langle\delta y\rangle$.

---

[2]Neutrons are not directly measured in the SPS and RHIC experiments but their production rate, relative to protons, reflects in the ratio of tritium to ^{3}He production measured by NA49 [57], applying the isospin mirror symmetry of the corresponding nuclear wave functions.

The data described above suggest that the stopping mechanism universally resides in the primordial, first generation of collisions at the microscopic level. The rapidity distributions of charged particle multiplicity, transverse energy and valence quark exhibit qualitatively similar shapes (which also evolve similarly with $\sqrt{s}$) in pp, $p\bar{p}$, e^+e^- reactions, on the one hand, and in central or semi-peripheral collisions of $A \approx 200$ nuclei, on the other. Comparing in detail we formulate a nuclear modification factor for the bulk hadron rapidity distributions,

$$R_y^{AA} \equiv \frac{dN^{ch}/dy\,(y) \quad \text{in A+A}}{0.5\,N_{part}\,dN^{ch}/dy \quad \text{in pp}} \tag{7.6}$$

where $N_{part} < 2A$ is the mean number of "participating nucleons" (which undergo at least one inelastic collision with another nucleon) which increases with collision centrality. For identical nuclei colliding $\langle N_{part}^{proj} \rangle \simeq \langle N_{part}^{targ} \rangle$ and thus $0.5\,N_{part}$ gives the number of opposing nucleon pairs. $R^{AA} = 1$ if each such "opposing" pair contributes the same fraction to the total A+A yield as is produced in minimum bias pp at similar $\sqrt{s}$. From Figs. 7.4 and 7.6 we infer that for $|\eta| < 1$, $R^{AA} = 1.5$ at top RHIC energy, and for the pseudo-rapidity integrated total N^{ch} we find $R^{AA} = 1.36$, in central Au+Au collisions. AA collisions thus provide for a higher stopping power than pp (which is also reflected in the higher rapidity shift $\langle \delta y \rangle$ of Fig. 7.10). The observation that their stopping power resembles the e^+e^- inelasticity suggests a substantially reduced leading particle effect in central collisions of heavy nuclei. This might not be surprising. In a Glauber-view of successive minimum bias nucleon collisions occurring during interpenetration, each participating nucleon is struck $\nu > 3$ times on average, which might saturate the possible inelasticity, removing the leading fragment.

This view naturally leads to the scaling of the total particle production in nuclear collisions with N_{part}, as seen clearly in Fig. 7.6, reminiscent of the "wounded nucleon model" [59] but with the scaling factor determined by e^+e^- rather than pp [60]. Overall we conclude from the still rather close similarity between nuclear and elementary collisions that the mechanisms of longitudinal phase space population occur primordially, during interpenetration which is over after 0.15 fm/c at RHIC, and after 1.5 fm/c at SPS energy. I.e. it is the primordial non-equilibrium pQCD shower evolution that accounts for stopping, and its time extent should be a lower limit to the formation time τ_0 employed in the Bjorken model [45], Eq. (7.1). Equilibration at the partonic level might begin at $t > \tau_0$ only (the development toward a quark-gluon-plasma phase), but the primordial parton redistribution processes set the stage for this phase, and control the relaxation time scales involved in equilibration [61]. More about this in Sect. 7.2.5. We infer the existence of a saturation scale [62] controlling the total inelasticity: with ever higher reactant thickness, proportional to $A^{1/3}$, one does not get a total rapidity or energy density proportional to $A^{4/3}$ (the number of "successive binary collisions") but to $A^{1.08}$ only [63]. Note that the lines shown in Fig. 7.7 (right panel) refer to such a saturation theory: the color glass condensate (CGC) model [64] developed by McLerran and

Venugopulan. The success of these models demonstrates that "successive binary baryon scattering" is not an appropriate picture at high $\sqrt{s}$. One can free the partons from the nucleonic parton density distributions only *once*, and their corresponding transverse areal density sets the stage for the ensuing QCD parton shower evolution [62]. Moreover, an additional saturation effect appears to modify this evolution at high transverse areal parton density (see Sect. 7.2.4).

7.2.3 Dependence on System Size

We have discussed above a first attempt toward a variable (N_{part}) that scales the system size dependence in A+A collisions. Note that one can vary the size either by centrally colliding a sequence of nuclei, $A_1 + A_1$, $A_2 + A_2$ etc., or by selecting different windows in N_{part} out of minimum bias collision ensembles obtained for heavy nuclei for which BNL employs ^{197}Au and CERN ^{208}Pb. The third alternative, scattering a relatively light projectile, such as ^{32}S, from increasing A nuclear targets, has been employed initially both at the AGS and SPS but got disfavored in view of numerous disadvantages, of both experimental (the need to measure the entire rapidity distribution, i.e. lab momenta from about 0.3–100 GeV/c, with uniform efficiency) and theoretical nature (different density distributions of projectile and target; occurrence of an "effective" center of mass, different for hard and soft collisions, and depending on impact parameter).

The determination of N_{part} is of central interest, and thus we need to look at technicalities, briefly. The approximate linear scaling with N_{part} that we observed in the total transverse energy and the total charged particle number (Figs. 7.3 and 7.6) is a reflection of the primordial redistribution of partons and energy. Whereas all observable properties that refer to the system evolution at later times, which are of interest as potential signals from the equilibrium, QCD plasma "matter" phase, have different specific dependences on N_{part}, be it suppressions (high p_T signals, jets, quarkonia production) or enhancements (collective hydrodynamic flow, strangeness production). N_{part} thus emerges as a suitable common reference scale.

N_{part} captures the number of potentially directly hit nucleons. It is estimated from an eikonal straight trajectory Glauber model as applied to the overlap region arising, in dependence of impact parameter b, from the superposition along beam direction of the two initial Woods-Saxon density distributions of the interacting nuclei. To account for the dilute surfaces of these distributions (within which the intersecting nucleons might not find an interaction partner) each incident nucleon trajectory gets equipped with a transverse radius that represents the total inelastic NN cross section at the corresponding $\sqrt{s}$. The formalism is imbedded into a Monte Carlo simulation (for detail see [66]) starting from random microscopic nucleon positions within the transversely projected initial Woods-Saxon density profiles. Overlapping cross sectional tubes of target and projectile nucleons are counted as a participant nucleon pair. Owing to the statistics of nucleon initial position sampling each considered impact parameter geometry thus results in a probability distribution of derived N_{part}.

Its width σ defines the resolution $\Delta(b)$ of impact parameter b determination within this scheme via the relation

$$\frac{1}{\Delta(b)}\,\sigma(b) \approx \frac{d\left\langle N_{\text{part}}(b)\right\rangle}{db} \qquad (7.7)$$

which, at $A = 200$, leads to the expectation to determine b with about 1.5 fm resolution [66], by measuring N_{part}.

How to measure N_{part}? In fixed target experiments one can calorimetrically count all particles with beam momentum per nucleon and superimposed Fermi momentum distributions of nucleons, i.e. one looks for particles in the beam fragmentation domain $y_{\text{beam}} \pm 0.5$, $p_T \leq 0.25$ GeV/c. These are identified as spectator nucleons, and $N_{\text{part}}^{\text{proj}} = A - N_{\text{spec}}^{\text{proj}}$. For identical nuclear collision systems $\left\langle N_{\text{part}}^{\text{proj}}\right\rangle = \left\langle N_{\text{part}}^{\text{targ}}\right\rangle$, and thus N_{part} gets approximated by $2N_{\text{part}}^{\text{proj}}$. This scheme was employed in the CERN experiments NA49 and WA80, and generalized [67] in a way that is illustrated in Fig. 7.11.

The top panel shows the minimum bias distribution of total energy registered in a forward calorimeter that covers the beam fragment domain in Pb+Pb collisions at lab. energy of 158 GeV per projectile nucleon, $\sqrt{s} = 17.3$ GeV. The energy spectrum extends from about 3 TeV which corresponds to about 20 projectile spectators (indicating a "central" collision), to about 32 TeV which is close to the total beam energy and thus corresponds to extremely peripheral collisions. Note that the shape of this forward energy spectrum is the mirror image of the minimum bias transverse energy distribution of Fig. 7.3, both recorded by NA49. From both figures we see that the *ideal* head-on, $b \to 0$ collision cannot be selected from these (or any other) data, owing to the facts that $b = 0$ carries zero geometrical weight,

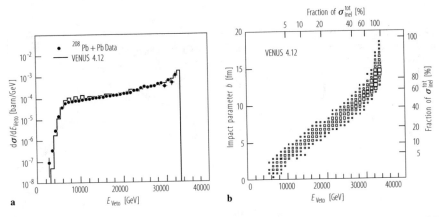

Fig. 7.11 (a) Energy spectrum of the forward calorimeter in Pb+Pb collisions at 158A GeV; (b) impact parameter and fraction of total inelastic cross section related to forward energy from the VENUS model [67, 68]

and that the diffuse Woods-Saxon nuclear density profiles lead to a fluctuation of participant nucleon number at given finite b. Thus the N_{part} fluctuation at finite weight impact parameters overshadows the genuinely small contribution of near zero impact parameters. Selecting "central" collisions, either by an on-line trigger cut on minimal forward energy or maximal total transverse energy or charged particle rapidity density, or by corresponding off-line selection, one thus faces a compromise between event statistics and selectivity for impact parameters near zero. In the example of Fig. 7.11 these considerations suggest a cut at about 8 TeV which selects the 5% most inelastic events, from among the overall minimum bias distribution, then to be labeled as "central" collisions. This selection corresponds to a soft cutoff at $b \leq 3$ fm.

The selectivity of this, or of other less stringent cuts on collision centrality is then established by comparison to a Glauber or cascade model. The bottom panel of Fig. 7.11 employs the VENUS hadron/string cascade model [68] which starts from a Monte Carlo position sampling of the nucleons imbedded in Woods-Saxon nuclear density profiles but (unlike in a Glauber scheme with straight trajectory overlap projection) following the cascade of inelastic hadron/string multiplication, again by Monte Carlo sampling. It reproduces the forward energy data reasonably well and one can thus read off the average impact parameter and participant nucleon number corresponding to any desired cut on the percent fraction of the total minimum bias cross section. Moreover, it is clear that this procedure can also be based on the total minimum bias transverse energy distribution, Fig. 7.3, which is the mirror image of the forward energy distribution in Fig. 7.11, or on the total, and even the mid-rapidity charged particle density (Fig. 7.6). The latter method is employed by the RHIC experiments STAR and PHENIX.

How well this machinery works is illustrated in Fig. 7.12 by RHIC-PHOBOS results at $\sqrt{s} = 200$ GeV [52]. The charged particle pseudo-rapidity density distributions are shown for central (3–6% highest N_{ch} cut) Cu+Cu collisions, with

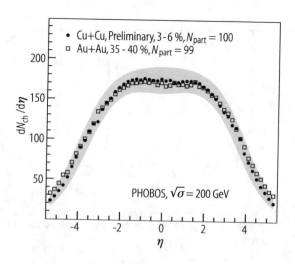

Fig. 7.12 Charged hadron pseudo-rapidity distributions in Cu+Cu and Au+Au collisions at $\sqrt{s} = 200$ GeV, with similar $N_{part} \approx 100$ [52]

Fig. 7.13 Charged pion multiplicity normalized by N_W vs. centrality in p+p, C+C, Si+Si and Pb+Pb collisions at $\sqrt{s} = 17.3$ GeV [67, 69]

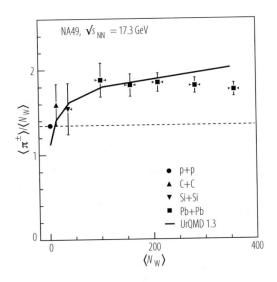

$\langle N_{part} \rangle = 100$, and semi-peripheral Au+Au collisions selecting the cut window (35–40%) such that the same $\langle N_{part} \rangle$ emerges. The distributions are nearly identical. In extrapolation to $N_{part} = 2$ one would expect to find agreement between min. bias p+p, and "super-peripheral" A+A collisions, at least at high energy where the nuclear Fermi momentum plays no large role. Figure 7.13 shows that this expectation is correct [69]. As it is technically difficult to select $N_{part} = 2$ from $A = 200$ nuclei colliding, NA49 fragmented the incident SPS Pb beam to study $^{12}C+^{12}C$ and $^{28}Si+^{28}Si$ collisions [67]. These systems are isospin symmetric, and Fig. 7.13 thus plots $0.5 \ (\langle \pi^+ \rangle + \langle \pi^- \rangle)/ \langle N_W \rangle$ including p+p where $N_W = 2$ by definition. We see that the pion multiplicity of A+A collisions interpolates to the p+p data point.

Note that NA49 employs the term "wounded nucleon" number (N_W) to count the nucleons that underwent at least one inelastic *nucleon-nucleon* collision. This is what the RHIC experiments (that follow a Glauber model) call N_{part} whereas NA49 reserves this term for nucleons that underwent *any* inelastic collision. Thus N_W in Fig. 7.13 has the same definition as N_{part} in Figs. 7.4, 7.6, 7.8, and 7.12. We see that a smooth increase joins the p+p data, via the light A+A central collisions, to a saturation setting in with semi-peripheral Pb+Pb collisions, the overall, relative increase amounting to about 40% (as we saw in Fig. 7.4).

There is nothing like an $N_{part}^{1/3}$ increase (the thickness of the reactants) observed here, pointing to the saturation mechanism(s) mentioned in the previous section, which are seen from Fig. 7.13 to dampen the initial, fast increase once the primordial interaction volume contains about 80 nucleons. In the Glauber model view of successive collisions (to which we attach only symbolical significance at high $\sqrt{s}$) this volume corresponds to $\langle \nu \rangle \approx 3$, and within the terminology of such models we might thus argue, intuitively, that the initial *geometrical* cross section, attached to the nucleon structure function as a whole, has disappeared at $\langle \nu \rangle \approx 3$, all constituent partons being freed.

7.2.4 Gluon Saturation in A+A Collisions

We will now take a closer look at the saturation phenomena of high energy QCD scattering, and apply results obtained for deep inelastic electron-proton reactions to nuclear collisions, a procedure that relies on a universality of high energy hadron scattering. This arises at high $\sqrt{s}$, and at relatively low momentum transfer squared Q^2 (the condition governing bulk charged particle production near mid-rapidity at RHIC, where Feynman $x \approx 0.01$ and $Q^2 \leq 5 \text{ GeV}^2$). Universality comes about as the transverse resolution becomes higher and higher, with Q^2, so that within the small area tested by the collision there is no difference whether the partons sampled there belong to the transverse gluon and quark density projection of any hadron species, or even of a nucleus. And saturation arises once the areal transverse parton density exceeds the resolution, leading to interfering QCD sub-amplitudes that do not reflect in the total cross section in a manner similar to the mere summation of separate, resolved color charges [61–65, 70, 71].

The ideas of saturation and universality are motivated by HERA deep inelastic scattering (DIS) data [72] on the gluon distribution function shown in Fig. 7.14 (left side). The gluon rapidity density, $xG(x, Q^2) = (\mathrm{d}N^{\text{gluon}})/(\mathrm{d}y)$ rises rapidly as a function of decreasing fractional momentum, x, or increasing resolution, Q^2. The origin of this rise in the gluon density is, ultimately, the non-abelian nature of

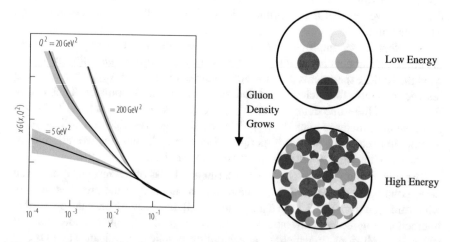

Fig. 7.14 (Left) HERA data for the gluon distribution function as a function of fractional momentum x and square momentum transfer Q^2 [72]. (Right) Saturation of gluons in a hadron; a head on view as x decreases [75]

QCD. Due to the intrinsic non-linearity of QCD [70, 71], gluon showers generate more gluon showers, producing an avalanche toward small x. As a consequence of this exponential growth the spatial density of gluons (per unit transverse area per unit rapidity) of any hadron or nucleus must increase as x decreases [65]. This follows because the transverse size, as seen via the total cross section, rises more slowly toward higher energy than the number of gluons. This is illustrated in Fig. 7.14 (right side). In a head-on view of a hadronic projectile more and more partons (mostly gluons) appear as x decreases. This picture reflects a representation of the hadron in the "infinite momentum frame" where it has a large light-cone longitudinal momentum $P^+ \gg M$. In this frame one can describe the hadron wave function as a collection of constituents carrying a fraction $p^+ = xP^+$, $0 \leq x < 1$, of the total longitudinal momentum [73] ("light cone quantization" method [74]). In DIS at large $sqrts$ and Q^2 one measures the quark distributions dN_q/dx at small x, deriving from this the gluon distributions $xG(x, Q^2)$ of Fig. 7.14.

It is useful [75] to consider the rapidity distribution implied by the parton distributions, in this picture. Defining $y = y_{hadron} - \ln(1/x)$ as the rapidity of the potentially struck parton, the invariant rapidity distribution results as

$$dN/dy = x\, dN/dx = xG(x, Q^2). \tag{7.8}$$

At high Q^2 the measured quark and gluon structure functions are thus simply related to the number of partons per unit rapidity, resolved in the hadronic wave function.

The above textbook level [74, 75] recapitulation leads, however, to an important application: the dN/dy distribution of constituent partons of a hadron (or nucleus), determined by the DIS experiments, is similar to the rapidity distribution of produced particles in hadron-hadron or A+A collisions as we expect the initial gluon rapidity density to be represented in the finally observed, produced hadrons, at high $\sqrt{s}$. Due to the longitudinal boost invariance of the rapidity distribution, we can apply the above conclusions to hadron-hadron or A+A collisions at high $\sqrt{s}$, by replacing the infinite momentum frame hadron rapidity by the center of mass frame projectile rapidity, y_{proj}, while retaining the result that the rapidity density of potentially interacting partons grows with increasing distance from y_{proj} like

$$\Delta y \equiv y_{proj} - y = \ln(1/x). \tag{7.9}$$

At RHIC energy, $\sqrt{s} = 200$ GeV, Δy at mid-rapidity thus corresponds to $x < 10^{-2}$ (well into the domain of growing structure function gluon density, Fig. 7.14), and the two intersecting partonic transverse density distributions thus attempt to resolve each other given the densely packed situation that is depicted in the lower circle of Fig. 7.14 (right panel). At given Q^2 (which is modest, $Q^2 \leq 5$ GeV^2, for bulk hadron production at mid-rapidity) the packing density at mid-rapidity will increase toward higher $\sqrt{s}$ as

$$\Delta y^{midrap} \approx \ln(\sqrt{s}/M), \text{ i.e. } 1/x \approx \sqrt{s}/M \tag{7.10}$$

thus sampling smaller x domains in Fig. 7.14 according to Eq. (7.9). It will further increase in proceeding from hadronic to nuclear reaction partners A+A. Will it be in proportion to $A^{4/3}$? We know from the previous sections (Sects. 7.2.2 and 7.2.3) that this is not the case, the data indicating an increase with $A^{1.08}$. This observation is, in fact caused by the parton saturation effect, to which we turn now.

For given transverse resolution Q^2 and increasing $1/x$ the parton density of Fig. 7.14 becomes so large that one cannot neglect their mutual interactions any longer. One expects such interactions to produce "*shadowing*", a decrease of the scattering cross section relative to incoherent independent scattering [70, 71]. As an effect of such shadowed interactions there occurs [75] a *saturation* [61–65, 70, 71, 75] of the cross section at each given Q^2, slowing the increase with $1/x$ to become logarithmic once $1/x$ exceeds a certain critical value $x_s(Q^2)$. Conversely, for fixed x, saturation occurs for transverse momenta below some critical $Q^2(x)$,

$$Q_s^2(x) = \alpha_s N_c \, \frac{1}{\pi R^2} \, \frac{dN}{dy} \tag{7.11}$$

where dN/dy is the x-dependent gluon density (at $y = y_{proj} - \ln(1/x)$). Q_s^2 is called the *saturation scale*. In Eq. (7.11) πR^2 is the hadron area (in transverse projection), and $\alpha_s N_c$ is the color charge squared of a single gluon. More intuitively, $Q_s^2(x)$ defines an inversely proportional resolution area $F_s(x)$ and at each x we have to choose $F_s(x)$ such that the ratio of total area πR^2 to $F_s(x)$ (the number of resolved areal pixels) equals the number of single gluon charge sources featured by the total hadron area. As a consequence the saturation scale $Q_s^2(x)$ defines a critical areal resolution, with two different types of QCD scattering theory defined, at each x, for $Q^2 > Q_s^2$ and $Q^2 < Q_s^2$, respectively [62, 65, 75].

As one expects a soft transition between such theories, to occur along the transition line implied by $Q_s^2(x)$, the two types of QCD scattering are best studied with processes featuring typical Q^2 well above, or below $Q_s^2(x)$. Jet production at $\sqrt{s} \geq 200$ GeV in $p\bar{p}$ or AA collisions with typical Q^2 above about 10^3 GeV^2, clearly falls into the former class, to be described e.g. by QCD DGLAP evolution of partonic showers [76]. The acronym DGLAP refers to the inventors of the perturbative QCD evolution of parton scattering with the "running" strong coupling constant $\alpha_s(Q^2)$, Dokshitzer, Gribov, Levine, Altarelli and Parisi. On the other hand, mid-rapidity bulk hadron production at the upcoming CERN LHC facility ($\sqrt{s} = 14$ TeV for pp, and 5.5 TeV for A+A), with typical $Q^2 \leq 5$ GeV^2 at $x \leq 10^{-3}$, will present a clear case for QCD saturation physics, as formulated e.g. in the "Color Glass Condensate (CGC)" formalism developed by McLerran, Venugopalan and collaborators [64, 65, 75, 77]. This model develops a classical gluon field theory for the limiting case of a high areal occupation number density, i.e. for the conceivable limit of the situation depicted in Fig. 7.14 (right hand panel) where the amalgamating small x gluons would overlap completely, within any finite resolution area at modest Q^2. Classical field theory captures, by construction, the effects of color charge coherence, absent in DGLAP parton cascade evolution theories [75]. This model appears to work well already at $\sqrt{s}$ as "low" as at

RHIC, as far as small Q^2 bulk charged particle production is concerned. We have illustrated this by the CGC model fits [64] to the PHOBOS charged particle rapidity distributions, shown in Fig. 7.7.

Conversely, QCD processes falling in the transition region between such limiting conditions, such that typical $Q^2 \approx Q_s^2(x)$, should present observables that are functions of the ratio between the transferred momentum Q^2 and the appropriate saturation scale, expressed by $Q_s^2(x)$. As Q^2 defines the effective transverse sampling area, and $Q_s^2(x)$ the characteristic areal size at which saturation is expected to set in, a characteristic behavior of cross sections, namely that they are universal functions of Q^2/Q_s^2, is called "geometric scaling". The HERA ep scattering data obey this scaling law closely [78], and the idea arises to apply the universality principle that we mentioned above: at small enough x, all hadrons or nuclei are similar, their specific properties only coming in via the appropriate saturation scales $Q_s^2(x, h)$ or $Q_s^2(x, A)$. Knowing the latter for RHIC conditions we will understand the systematics of charged particle production illustrated in the previous chapter, and thus also be able to extrapolate toward LHC conditions in pp and AA collisions.

All data for the virtual photo-absorption cross section $\sigma^{\gamma p}(x, Q^2)$ in deep inelastic ep scattering with $x \leq 0.01$ (which is also the RHIC mid-rapidity x-domain) have been found [78] to lie on a single curve when plotted against Q^2/Q_s^2, with

$$Q_s^2(x) \sim \left(\frac{x_0}{x}\right)^\lambda 1 \text{ GeV}^2 \qquad (7.12)$$

with $\lambda \simeq 0.3$ and $x_0 \simeq 10^{-4}$. This scaling [79] with $\tau = Q^2/Q_s^2$ is shown in Fig. 7.15 (top panel) to interpolate all data. A chain of arguments, proposed by Armesto et al. [63] connects a fit to these data with photo-absorption data for (virtual) photon-A interactions [80] via the geometrical scaling ansatz

$$\frac{\sigma^{\gamma A}(\tau_A)}{\pi R_A^2} = \frac{\sigma^{\gamma p}(\tau_p = \tau_A)}{\pi R_p^2} \qquad (7.13)$$

assuming that the scale in the nucleus grows with the ratio of the transverse parton densities, raised to the power $1/\delta$ (a free parameter),

$$Q_{s,A}^2 = Q_{s,p}^2 \left(\frac{A\pi R_p^2}{\pi R_A^2}\right)^{1/\delta}, \quad \tau_A = \tau_h \left(\frac{\pi R_A^2}{A\pi R_h^2}\right)^{1/\delta}. \qquad (7.14)$$

Figure 7.15 (middle and bottom panels) shows their fit to the nuclear photo-absorption data which fixes $\delta = 0.79$ and $\pi R_p^2 = 1.57$ fm^2 (see ref. [63] for detail). The essential step in transforming these findings to the case of A+A collisions is then taken by the empirical ansatz

$$\frac{dN^{AA}}{dy} \text{ (at } y \simeq 0) \propto Q_{s,A}^2(x)\pi R_A^2 \qquad (7.15)$$

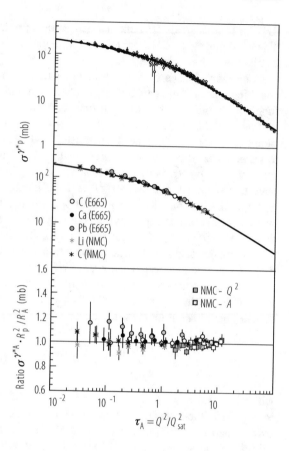

Fig. 7.15 (Top) Geometric scaling of the virtual photo-absorption cross section $\sigma^{\gamma p}$ on protons; (middle) cross sections for nuclei normalized according to Eq. (7.13); (bottom) the ratio of $\sigma^{\gamma A}$ to a fit of $\sigma^{\gamma p}$ (see [63] for data reference)

by which the mid-rapidity parton (gluon) density dN/dy in Eq. (7.11) gets related to the charged particle mid-rapidity density at $y \approx 0$ [70, 81], measured in nucleus-nucleus collisions. Replacing, further, the total nucleon number $2A$ in a collision of identical nuclei of mass A by the number N_{part} of participating nucleons, the final result is [63]

$$\frac{1}{N_{part}} \frac{dN^{AA}}{dy} \text{ (at } y \approx 0) = N_0(\sqrt{s})^\lambda N_{part}^\alpha \tag{7.16}$$

where the exponent $\alpha \equiv (1 - \delta)/3 \, \delta = 0.089$, and $N_0 = 0.47$. The exponent α is *far* smaller than 1/3, a value that represents the thickness of the reactants, and would be our naive guess in a picture of "successive" independent nucleon participant collisions, whose average number $\langle \nu \rangle \propto (N_{part}/2)^{1/3}$. The observational fact (see Fig. 7.13) that $\alpha < 1/3$ for mid-rapidity low Q^2 bulk hadron production in A+A collisions illustrates the importance of the QCD saturation effect. This is shown [63] in Fig. 7.16 where Eq. (7.16) is applied to the RHIC PHOBOS data for mid-rapidity charged particle rapidity density per participant pair, in Au+Au collisions at

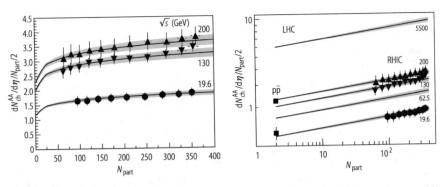

Fig. 7.16 Saturation model fit [63] applied to RHIC charged hadron multiplicity data at mid-rapidity normalized by number of participant pairs, at various energies [82]. Also shown is an extrapolation to $p\overline{p}$ data and a prediction for minimum bias Pb+Pb collisions at LHC energy, $\sqrt{s} = 5500$ GeV

$\sqrt{s} = 19.6$, 130 and 200 GeV [82], also including a *prediction* for LHC energy. Note that the *factorization of energy and centrality dependence*, implied by the RHIC data [52], is well captured by Eq. (7.11) and the resulting fits in Fig. 7.16. Furthermore, the steeper slope, predicted for $N_{part} \leq 60$ (not covered by the employed data set), interpolates to the corresponding pp and $p\overline{p}$ data, at $N_{part} = 2$. It resembles the pattern observed in the NA49 data (Fig. 7.13) for small N_{part} collisions of light A+A systems, at $\sqrt{s} = 17$–20 GeV, and may be seen, to reflect the onset of QCD saturation. Finally we note that the conclusions of the above, partially heuristic approach [63], represented by Eqs. (7.13)–(7.16), have been backed up by the CGC theory of McLerran and Venugopulan [64, 65, 75], predictions of which we have illustrated in Fig. 7.7.

Bulk hadron production in AA collisions at high $\sqrt{s}$ can be related, via the assumption of universality of high energy QCD scattering, to the phenomenon of geometric scaling first observed in HERA deep inelastic ep cross sections. The underlying feature is a QCD saturation effect arising from the diverging areal parton density, as confronted with the limited areal resolution Q^2, inherent in the considered scattering process. The "saturation scale" $Q_s^2(x, A)$ captures the condition that a single partonic charge source within the transverse partonic density profile can just be resolved by a sufficiently high Q^2. Bulk hadron production in A+A collisions falls below this scale.

7.2.5 Transverse Phase Space: Equilibrium and the QGP State

At RHIC energy, $\sqrt{s} = 200$ GeV, the Au+Au collision reactants are longitudinally contracted discs. At a nuclear radius $R \approx A^{1/3}$ fm and Lorentz $\gamma \approx 100$ their primordial interpenetration phase ends at time $\tau_0 \leq 0.15$ fm/c. This time scale is absent in e^+e^- annihilation at similar $\sqrt{s}$ where $\tau_0 \approx 0.1$ fm/c marks the end of the primordial pQCD partonic shower evolution [83] during which the initially created $q\bar{q}$ pair, of "virtually" $Q = \sqrt{s}/2$ each, multiplies in the course of the QCD DGLAP evolution in perturbative vacuum, giving rise to daughter partons of far lower virtuality, of a few GeV. In A+A collisions this shower era should last longer, due to the interpenetrational spread of primordial collision time. It should be over by about 0.25 fm/c. The shower partons in e^+e^- annihilation are localized within back to back cone geometry reflecting the directions of the primordial quark pair. The eventually observed "jet" signal, created by an initial Q^2 of 10^4 GeV^2, is established by then. Upon a slow-down of the dynamical evolution time scale to $\tau \approx$ 1 fm/c the shower partons fragment further, acquiring transverse momentum and yet lower virtuality, then to enter a non perturbative QCD phase of color neutralization during which hadron-like singlet parton clusters are formed. Their net initial pQCD virtuality, in pQCD vacuum, is recast in terms of non-perturbative vacuum hadron mass. The evolution ends with on-shell, observed jet-hadrons after about 3 fm/c of overall reaction time.

Remarkably, even in this, somehow most elementary process of QCD evolution, an aspect of equilibrium formation is observed, not in the narrowly focused final dijet momentum topology but in the relative production rates of the various created hadronic species. This so-called "hadrochemical" equilibrium among the hadronic species is documented in Fig. 7.17. The hadron multiplicities per e^+e^- annihilation event at $\sqrt{s} = 91.2$ GeV [38] are confronted with a Hagedorn [38] canonical statistical Gibbs ensemble prediction [84] which reveals that the apparent species equilibrium was fixed at a temperature of $T = 165$ MeV, which turns out to be the universal hadronization temperature of all elementary and nuclear collisions at high $\sqrt{s}$ (Hagedorns limiting temperature of the hadronic phase of matter). We shall return to this topic in Sect. 7.3 but note, for now, that reactions with as few as 20 charged particles exhibit such statistical equilibrium properties, a pre-requisite for application of thermodynamic or hydrodynamic concepts.

What happens with parton (and hadron) dynamics in A+A collisions after τ_0? There will not be a QCD evolution in vacuum (which would be over after 3 fm/c) as the transverse radius of the interacting system is large. It may grow to about twice the nuclear radius, i.e. to about 15 fm before interactions cease; i.e. the system needs about 15 fm/c to decouple. This simple fact is the key to our expectation that the expansive evolution of the initial high energy density deposited in a cylinder of considerable diameter (about 10 fm), may create certain equilibrium properties that allow us to treat the contained particles and energy in terms of thermodynamic phases of matter, such as a partonic QGP liquid, or a hadronic liquid or gas, etc. Such that the expansion dynamics makes contact to the phase diagram illustrated in Fig. 7.1. This expectation turns out to be justified as we shall describe in Sects. 7.3 and 7.4. What results for the evolution after τ_0 in a central A+A collision is sketched in Fig. 7.18 by means of a schematic 2-dimensional light cone diagram, which is entered by the two reactant nuclei along $z = t$ trajectories where z is the beam direction and Lorentz contraction has been taken to an extreme, such that there

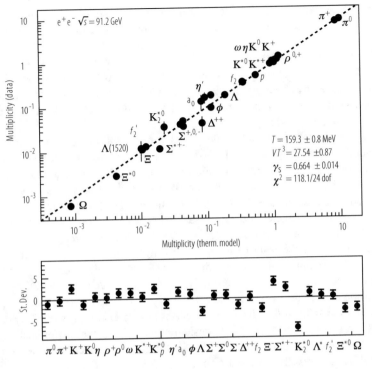

Fig. 7.17 Hadron multiplicities in LEP e^+e^- annihilation at $\sqrt{s} = 91.2$ GeV confronted with the predictions of the canonical statistical hadronization model [84]

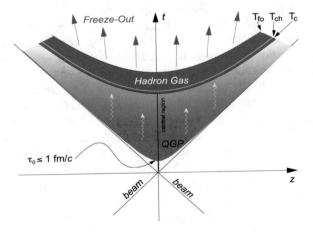

Fig. 7.18 Schematic light cone diagram of the evolution of a high energy heavy ion collision, indicating a formation phase τ_0 (see text)

occurs an idealized $t = z = 0$ interaction "point". Toward positive t the light cone proper time profiles of progressing parton-hadron matter evolution are illustrated. The first profile illustrated here corresponds to the end of formation time τ_0. From our above discussion of the e^+e^- annihilation process one obtains a first estimate, $\tau_0 \geq 0.25$ fm/c (including interpenetration time of 0.15 fm/c at RHIC) which refers to processes of very high $Q^2 \geq 10^3$ GeV^2, far above the saturation scale Q_s^2 discussed in the previous section. The latter scale has to be taken into account for low p_T hadron production.

It is the specific resolution scale Q^2 of a QCD sub-process, as enveloped in the overall collision dynamics of two slabs of given transverse partonic structure function density, that determines which fraction of the constituent partons enters interaction. In the simple case of extremely high Q^2 processes the answer is that all constituents are resolved. However, at modest Q^2 (dominating bulk hadron production) the characteristic QCD saturation scale $Q_s^2(x)$ gains prominence, defined such that processes with $Q^2 < Q_s^2$ do not exploit the initial transverse parton densities at the level of independent single constituent color field sources (see Eq. (7.11)). For such processes the proper formation time scale, τ_0, is of order of the inverse saturation momentum [61], $1/Q_s \sim 0.2$ fm/c at $\sqrt{s} = 200$ GeV. The first profile of the time evolution, sketched in Fig. 7.18, should correspond to proper time $t = \tau_0 = 0.25$ fm/c at RHIC energy. At top SPS energy, $\sqrt{s} = 17.3$ GeV, we cannot refer to such detailed QCD considerations. A pragmatic approach suggests to take the interpenetration time, at $\gamma \approx 8.5$, for guidance concerning the formation time, which thus results as $\tau_0 \approx 1.5$ fm/c.

In summary of the above considerations we assume that the initial partonic color sources, as contained in the structure functions (Fig. 7.14), are spread out in longitudinal phase space after light cone proper time $t = \tau_0 \approx 0.2$ fm/c, at top RHIC energy, and after $\tau_0 \approx 1.5$ fm/c at top SPS energy. No significant transverse expansion has occurred at this early stage, in a central collision of $A \approx 200$ nuclei with transverse diameter of about 12 fm. The Bjorken estimate [45] of initial energy density ϵ (Eq. (7.1)) refers to exactly this condition, after formation time τ_0. In order to account for the finite longitudinal source size and interpenetration time, at RHIC, we finally put the average $\tau_0 \approx 0.3$ fm, at $\sqrt{s} = 200$ GeV, indicating the "initialization time" after which all partons that have been resolved from the structure functions are engaged in shower multiplication. As is apparent from Fig. 7.18, this time scale is Lorentz dilated for partons with a large longitudinal momentum, or rapidity. This means that the slow particles are produced first toward the center of the collision region, and the fast (large rapidity) particles are produced later, away from the collision region. This Bjorken "inside-out" correlation [45] between coordinate- and momentum-space is similar to the Hubble expansion pattern in cosmology: more distant galaxies have higher outward velocities. This means that the matter created in A+A collisions at high $\sqrt{s}$ is also born expanding, however with the difference that the Hubble flow is initially one dimensional along the collision axis. This pattern will continue, at $\sqrt{s} = 200$ GeV, until the system begins to feel the effects of finite size in the transverse direction which will occur at some time t_0 in the vicinity of 1 fm/c. However, the tight correlation between position and momentum initially imprinted on the system will survive all further expansive evolution of the initial "firetube", and is well recovered in the expansion pattern of the finally released hadrons of modest p_T as we shall show when discussing radial flow and pion pair Bose-Einstein momentum correlation (see Sects. 7.2.6 and 7.7).

In order to proceed to a more quantitative description of the primordial dynamics (that occurs onward from τ_0 for as long the time period of predominantly longitudinal expansion might extend) we return to the Bjorken estimate of energy density, corresponding to this picture [45], as implied by Eq. (7.1), which we now recast as

$$\epsilon = \left(\frac{\mathrm{d}N_h}{\mathrm{d}y}\right)\left\langle E_h^T\right\rangle (\pi\, R_A^2\, t_0)^{-1} \qquad (7.17)$$

where the first term is the (average) total hadron multiplicity per unit rapidity which, multiplied with the average hadron transverse energy, equals the total transverse energy recorded in the calorimetric study shown in Fig. 7.3, as employed in Eq. (7.1). The quantity R_A is, strictly speaking, *not* the radius parameter of the spherical Woods-Saxon nuclear density profile but the *rms* of the reactant overlap profiles as projected onto the transverse plane (and thus slightly smaller than $R_A \approx A^{1/3}$ fm). Employing $A^{1/3}$ here (as is done throughout) leads to a conservative estimate of ϵ, a minor concern. However, the basic assumption in Eq. (7.17) is to identify the primordial transverse energy "radiation", of an interactional cylindric source of radius R_A and length t_0 (where $\tau_0 \leq t_0 \leq 1$ fm/c, not Lorentz dilated at midrapidity), with the finally emerging bulk hadronic transverse energy. We justify this assumption by the two observations, made above, that

1. the bulk hadron multiplicity density per unit rapidity $(\mathrm{d}N_h)/(\mathrm{d}y)$ resembles the parton density, primordially released at saturation scale τ_0 (Figs. 7.7 and 7.16) at $\sqrt{s} = 200$ GeV, and that
2. the global emission pattern of bulk hadrons (in rapidity and p_T) closely reflects the initial correlation between coordinate and momentum space, characteristic of a primordial period of a predominantly longitudinal expansion, as implied in the Bjorken model.

Both these observations are surprising, at first sight. The Bjorken model was conceived for elementary hadron collisions where the expansion proceeds into vacuum, i.e. directly toward observation. Figure 7.18 proposes that, to the contrary, primordially produced partons have to transform through further, successive stages of partonic and hadronic matter, at decreasing but still substantial energy density, in central A+A collisions. The very fact of high energy density, with implied short mean free path of the constituent particles, invites a hydrodynamic description of the expansive evolution. With initial conditions fixed between τ_0 and t_0, an ensuing 3-dimensional hydrodynamic expansion would preserve the primordial Bjorken-type correlation between position and momentum space, up to lower density conditions and, thus, close to emission of the eventually observed hadrons. We thus feel

justified to employ Eq. (7.1) or (7.17) for the initial conditions at RHIC, obtaining
[61, 84]

$$6 \text{ GeV/fm}^3 \leq \epsilon \leq 20 \text{ GeV/fm}^3 \tag{7.18}$$

for the interval $0.3 \text{ fm}/c \leq t_0 \leq 1 \text{ fm}/c$, in central Au+Au collisions at $y \approx 0$
and $\sqrt{s} = 200$ GeV. The energy density at top SPS energy, $\sqrt{s} = 17.3$ GeV, can
similarly be estimated [43, 44] to amount to about 3 GeV/fm^3 at a t_0 of 1 fm/c but
we cannot identify conditions at $\tau_0 < t_0$ in this case as the mere interpenetration of
two Pb nuclei takes 1.4 fm/c. Thus the commonly accepted $t_0 = 1$ fm/c may lead
to a high estimate. An application of the parton-hadron transport model of Ellis and
Geiger [85, 86] to this collision finds $\epsilon = 3.3$ GeV/fm^3 at $t = 1$ fm/c. A primordial
energy density of about 3 GeV/fm^3 is 20 times $\rho_0 \approx 0.15$ GeV/fm^3, the average
energy density of ground state nuclear matter, and it also exceeds, by far, the critical
QCD energy density, of $0.6 \leq \epsilon_c \leq 1$ GeV/fm^3 according to lattice QCD [48]. The
initial dynamics thus clearly proceeds in a deconfined QCD system also at top SPS
energy, and similarly so with strikingly higher energy density, at RHIC, where time
scales below 1 fm/c can be resolved.

However, in order now to clarify the key question as to whether, and when conditions of partonic dynamical equilibrium may arise under such initial conditions,
we need estimates both of the proper relaxation time scale (which will, obviously,
depend on energy density and related collision frequency), and of the expansion
time scale as governed by the overall evolution of the collision volume. Only if
τ (relax.) $< \tau$ (expans.) one may conclude that the "deconfined partonic system"
can be identified with a "deconfined QGP *state* of QCD matter" as described e.g. by
lattice QCD, and implied in the phase diagram of QCD matter suggested in Fig. 7.1.

For guidance concerning the overall time-order of the system evolution we
consider information [87] obtained from Bose-Einstein correlation analysis of pion
pair emission in momentum space (for detail see Sect. 7.7). Note that pions should
be emitted at *any* stage of the evolution, after formation time, from the surface
regions of the evolving "fire-tube". Bulk emission of pions occurs, of course, after
hadronization (the latest stages illustrated in the evolution sketch given in Fig. 7.18).
The dynamical pion source expansion models by Heinz [88] and Sinyukov [89]
elaborate a Gaussian emission time profile, with mean τ_f (the decoupling time) and
width $\Delta \tau$ (the duration of emission).

Figure 7.19 shows an application of this analysis to central Pb+Pb collision
negative pion pair correlation data obtained by NA49 at top SPS energy, $\sqrt{s} =$
17.3 GeV [90], where $\tau_f \approx 8$ fm/c and $\Delta \tau \approx 4$ fm/c (note that $\tau = 0$ in Fig. 7.19
corresponds, not to interaction time $t = 0$ but to $t \approx 1.4$ fm/c, the end of the
interpenetration phase). We see, first of all, that the overall dynamical evolution
of a central Pb+Pb collision at $\sqrt{s} = 17.3$ GeV is ending at about 15 fm/c;
the proper time defines the position of the last, decoupling profile illustrated in
Fig. 7.18, for the SPS collisions considered here. While the details of Fig. 7.19
will turn out to be relevant to our later discussion of hadronization (Sect. 7.3) and
hadronic expansion (Sect. 7.4), we are concerned here with the average proper time

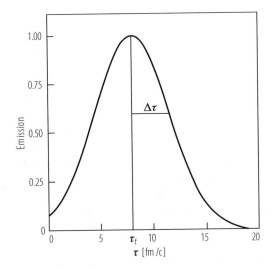

Fig. 7.19 Time profile of pion decoupling rate from the fireball in a central Pb+Pb collision, with $\tau = 0$ the end of the formation phase. Bose-Einstein correlation of $\pi^-\pi^-$ pairs yields an average Gaussian decoupling profile with $\tau_f = 8$ fm/c and duration of emission parameter $\Delta\tau = 4$ fm/c [87, 88]

at which the partonic phase ends. After consideration of the duration widths of these latter expansion phases [86, 87] one arrives at an estimate for the average time, spent before hadronization, of $\Delta t = 3$–4 fm/c, again in agreement with the parton cascade model mentioned above [86]. This model also leads to the conclusion that parton thermal equilibrium is, at least, closely approached locally in these central Pb+Pb collisions as far as mid-rapidity hadron production is concerned (at forward-backward rapidity the cascade re-scattering processes do not suffice, however).

This finding agrees with earlier predictions of $\tau_{relax} = 1$–2 fm/c at top SPS energy [91]. However we note that all such calculations employ perturbative QCD methods, implying the paradoxical consequence that equilibrium is closely approached only *at the end* of the partonic phase, at such low $\sqrt{s}$, i.e. in a QGP state at about $T = 200$ MeV which is, by definition, of non-perturbative nature. We shall return to the question of partonic equilibrium attainment at SPS energy in the discussion of the hadronization process in nuclear collisions (Sect. 7.3).

Equilibrium conditions should set in earlier at top RHIC energy. As transverse partonic expansion should set in after the proper time interval 0.3 fm/c $\leq t_0 \leq 1$ fm/c (which is now resolved by the early dynamics, unlike at top SPS energy), we take guidance from the Bjorken estimate of primordial energy density which is based on transverse energy production *data*. Conservatively interpreting the result in Eq. (7.18) we conclude that ϵ is about four times higher than at $\sqrt{s} = 17.3$ GeV in the above proper time interval. As the binary partonic collision frequency scales with the square of with the square of the density ρ (related to the energy density ϵ via the relation $\epsilon = \langle E \rangle \rho = T\rho$), and is inversely proportional to the relaxation time τ_{relax} we expect

$$\tau_{relax} \propto (1/\rho)^2 \approx (T/\epsilon)^2 \tag{7.19}$$

which implies that $\tau_{\text{relax}}(\text{RHIC}) \approx 0.25\ \tau_{\text{relax}}(\text{SPS}) \approx 0.5$ fm/c if we employ the estimate $T(\text{RHIC}) = 2T(\text{SPS})$. This crude estimate is, however, confirmed by the parton transport model of Molar and Gyulassy [92].

Partonic equilibration at $\sqrt{s} = 200$ GeV should thus set in at a time scale commensurate to the (slightly smaller) formation time scale, at which the to be participant partons are resolved from the initial nucleon structure functions and enter shower multiplication. Extrapolating to the conditions expected at LHC energy ($\sqrt{s} = 5.5$ TeV for A+A collisions), where the initial parton density of the structure functions in Fig. 7.14 is even higher ($x \approx 10^{-3}$ at mid-rapidity), and so is the initial energy density, we may expect conditions at which the resolved partons are almost "born into equilibrium".

Early dynamical local equilibrium at RHIC is required to understand the observations concerning elliptic flow, with which we shall deal, in detail, in Sect. 7.4. This term refers to a collective anisotropic azimuthal emission pattern of bulk hadrons in semi-peripheral collisions, a hydrodynamical phenomenon that originates from the initial geometrical non-isotropy of the primordial interaction zone [93, 94]. A detailed hydrodynamic model analysis of the corresponding elliptic flow signal at RHIC [95] leads to the conclusion that local equilibrium (a prerequisite to the hydrodynamic description) sets in at $t_0 \approx 0.6$ fm/c. This conclusion agrees with the estimate via Eq. (7.19) above, based on Bjorken energy density and corresponding parton collisions frequency.

We note that the concept of a hydrodynamic evolution appears to be, almost necessarily ingrained in the physics of a system born into (Hubble-type) expansion, with a primordial correlation between coordinate and momentum space, and at extreme initial parton density at which the partonic mean free path length λ is close to the overall spatial resolution resulting from the saturation scale, i.e. $\lambda \approx 1/Q_s$.

The above considerations suggest that a quark-gluon plasma state should be created early in the expansion dynamics at $\sqrt{s} = 200$ GeV, at about $T = 300$ MeV, that expands hydrodynamically until hadronization is reached, at $T \approx 165-170$ MeV. Its manifestations will be considered in Sects. 7.3–7.6. At the lower SPS energy, up to 17.3 GeV, we can conclude, with some caution, that a deconfined hadronic matter system should exist at $T \approx 200$ MeV, in the closer vicinity of the hadronization transition. It may closely resemble the QGP state of lattice QCD, near T_c.

7.2.6 Bulk Hadron Transverse Spectra and Radial Expansion Flow

In this chapter we analyze bulk hadron transverse momentum spectra obtained at SPS and RHIC energy, confronting the data with predictions of the hydrodynamical model of collective expansion matter flow that we have suggested in the previous section, to arise, almost necessarily, from the primordial Hubble-type coupling between coordinate and momentum space that prevails at the onset of the dynamical evolution in A+A collisions at high $\sqrt{s}$. As all hadronic transverse momentum spectra initially follow an approximately exponential fall-off (see below) the bulk hadronic output is represented by thermal transverse spectra at $p_T \leq 2$ GeV/c. We shall turn to high p_T information in later sections.

Furthermore we shall focus here on mid-rapidity production in near central A+A collisions, because hydrodynamic models refer to an initialization period characterized by Bjorken-type longitudinal boost invariance, which we have seen in Figs. 7.7 and 7.9 to be restricted to a relatively narrow interval centered at mid-rapidity. Central collisions are selected to exploit the azimuthal symmetry of emission, in an ideal impact parameter $b \to 0$ geometry. We thus select the predominant, relevant hydrodynamic "radial flow" expansion mode, from among other, azimuthally oriented (directed) flow patterns that arise once this cylindrical symmetry (with respect to the beam direction) is broken in finite impact parameter geometries.

In order to define, quantitatively, the flow phenomena mentioned above, we rewrite the invariant cross section for production of hadron species i in terms of transverse momentum, rapidity, impact parameter b and azimuthal emission angle φ_p (relative to the reaction plane),

$$\frac{dN_i(b)}{p_T\, dp_T\, dy\, d\varphi_p} = \frac{1}{2\pi} \frac{dN_i(b)}{p_T\, dp_T\, dy} \left[1 + 2v_1^i(p_T, b)\cos\varphi_p + 2v_2^i(p_T, b)\cos(2\varphi_p) + \ldots \right]$$

(7.20)

where we have expanded the dependence on φ_p into a Fourier series. Due to reflection symmetry with respect to the reaction plane in collisions of identical nuclei, only cosine terms appear. Restricting to mid-rapidity production all odd harmonics vanish, in particular the "directed flow" coefficient v_1^i, and we have dropped the y-dependence in the flow coefficients v_1^i and v_2^i. The latter quantifies the amount of "elliptic flow", to which we turn in Sect. 7.4. In the following, we will restrict to central collisions which we shall idealize as near-zero impact parameter processes governed by cylinder symmetry, whence all azimuthal dependence (expressed by the v_1^i, v_2^i, ... terms) vanishes, and the invariant cross section reduces to the first term in Eq. (7.20), which by definition also corresponds to all measurements in which the orientation of the reaction plane is not observed.

Typical transverse momentum spectra of the latter type are shown in Fig. 7.20, for charged hadron production in Au+Au collisions at $\sqrt{s} = 200$ GeV, exhibiting mid-rapidity data at various collision centralities [97]. We observe a clear-cut transition, from bulk hadron emission at $p_T \leq 2$ GeV/c featuring a near-exponential cross section (i.e. a thermal spectrum), to a high p_T power-law spectral pattern. Within the context of our previous discussion (Sect. 7.2.4) we tentatively identify the low p_T region with the QCD physics near saturation scale. Hadron production at $p_T \to 10$ GeV/c should, on the other hand, be the consequence of primordial leading parton fragmentation originating from "hard", high Q^2 perturbative QCD processes.

Fig. 7.20 Transverse
momentum spectra of
charged hadrons in Au+Au
collisions at $\sqrt{s} = 200$ GeV,
in dependence of collision
centrality [97] (offset as
indicated), featuring
transition from exponential to
power law shape

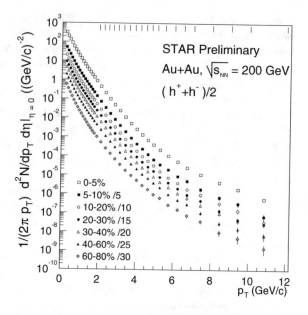

We thus identify bulk hadron production at low p_T as the emergence of
the initial parton saturation conditions that give rise to high energy density and
small equilibration time scale, leading to a hydrodynamical bulk matter expansion
evolution. Conversely, the initially produced hard partons, from high Q^2 processes,
are not thermalized into the bulk but traverse it, as tracers, while being attenuated
by medium-induced rescattering and gluon radiation, the combined effects being
reflected in the high p_T inclusive hadron yield, and in jet correlations of hadron
emission. We shall turn to the latter physics observables in Sect. 7.5, while staying
here with low p_T physics, related to hydrodynamical expansion modes, focusing on
radially symmetric expansion.

In order to infer from the spectral shapes of the hadronic species about the
expansion mechanism, we first transform to the transverse mass variable, $m_T = (p_T^2 + m^2)^{1/2}$, via

$$\frac{1}{2\pi} \frac{dN_i}{p_T \, dp_T \, dy} = \frac{1}{2\pi} \frac{dN_i}{m_T \, dm_T \, dy} \tag{7.21}$$

because it has been shown in p+p collisions [98] near RHIC energy that the m_T
distributions of various hadronic species exhibit a universal pattern ("m_T scaling")
at low m_T:

$$\frac{1}{2\pi} \frac{dN_i}{m_T \, dm_T \, dy} = A_i \exp(-m_T^i / T) \tag{7.22}$$

with a universal inverse slope parameter T and a species dependent normalization factor A. Hagedorn showed [99] that this scaling is characteristic of an adiabatic expansion of a fireball at temperature T. We recall that, on the other hand, an ideal hydrodynamical expansion is isentropic.

Figure 7.21 shows the $\sqrt{s}$ dependence of the average transverse kinetic energy $\langle m_T^i \rangle - m^i$ for pions, kaons and protons observed at mid-rapidity in central Au+Au/Pb+Pb collisions [54]. Similarly, the inverse slope parameter T resulting from a fit of Eq. (7.22) to K^+ and K^- transverse mass spectra (at $p_T \leq 2$ GeV/c) is shown in Fig. 7.22, both for nuclear and p+p collisions [100]. We see, first of all, that m_T scaling does not apply in A+A collisions, and that the kaon inverse slope

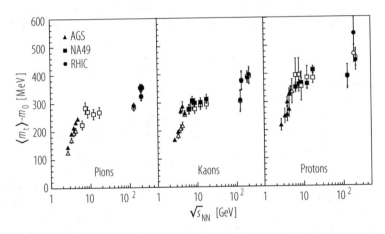

Fig. 7.21 The average transverse kinetic energy $\langle m_T \rangle - m_0$ for pions, kaons and protons vs. $\sqrt{s}$ in central Au+Au/Pb+Pb collisions [54]. Open symbols represent negative hadrons

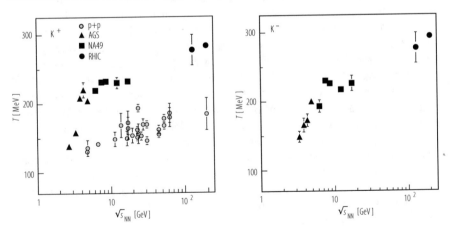

Fig. 7.22 The inverse slope parameter T of Eq. (7.22) for K^+ and K^- transverse mass spectra at $p_T < 2$ GeV/c and mid-rapidity in central A+A, and in minimum bias p+p collisions [100]

parameter, $T \approx 230$ MeV over the SPS energy regime, cannot be identified with the fireball temperature at hadron formation which is $T_h \approx 165$ MeV from Fig. 7.1. The latter is seen, however, to be well represented by the p+p spectral data exhibited in the left panel of Fig. 7.22. There is, thus, not only thermal energy present in A+A transverse expansion, but also hydrodynamical radial flow.

We note that the indications in Figs. 7.21 and 7.22, of a plateau in both $\langle m_T \rangle$ and T, extending over the domain of SPS energies, $6 \leq \sqrt{s} \leq 17$ GeV, have not yet been explained by any fundamental expansive evolution model, including hydrodynamics. Within the framework of the latter model, this is a consequence of the *initialization problem* [96] which requires a detailed modeling, both of primordial energy density vs. equilibration time scale, and of the appropriate partonic matter equation of state (EOS) which relates expansion pressure to energy density. At top RHIC energy, this initialization of hydro-flow occurs, both, at a time scale $t_0 \approx 0.6$ fm/c which is far smaller than the time scale of eventual bulk hadronization ($t \approx 3$ fm/c), and at a primordial energy density far in excess of the critical QCD confinement density. After initialization, the partonic plasma phase thus dominates the overall expansive evolution, over a time interval far exceeding the formation and relaxation time scale.

Thus, at RHIC energy, parton transport [92] and relativistic hydrodynamic [95, 96] models establish a well developed expansion mode that survives the subsequent stages of hadronization and hadronic expansion. This is reflected in their success in describing elliptic flow. On the other hand, the hydrodynamical model far overestimates elliptic flow at SPS energy [96] at which, as we have shown in Sect. 7.2.5, the initialization period may be not well separated from the confinement (hadronization) stage. Thus, whereas the expansion evolution at $\sqrt{s} = 200$ GeV (occurring at near-zero baryo-chemical potential in Fig. 7.1) "races" across the parton-hadron phase boundary with fully established flow patterns, near $\mu_B = 0$ where lattice QCD predicts the phase transformation to be merely a soft cross-over [16], the dynamics at $\sqrt{s} = 10$–20 GeV may originate from only slightly above, or even at the phase boundary, thus sampling the domain $200 \leq \mu_B \leq 500$ MeV where the equation of state might exhibit a "softest point" [96]. The hydrodynamic model thus faces formidable uncertainties regarding initialization at SPS energy.

The plateau in Figs. 7.21 and 7.22 may be the consequence of the fact that not much flow is generated in, or transmitted from the partonic phase, at SPS energies, because it is initialized close to the phase boundary [100] where the expected critical point [9, 10] (Fig. 7.1), and the corresponding adjacent first order phase transition might focus [101] or stall [96] the expansion trajectory, such that the observed radial flow stems almost exclusively from the hadronic expansion phase. The SPS plateau, which we shall subsequently encounter in other bulk hadron variables (elliptic flow, HBT radii) might thus emerge as a consequence of the critical point or, in general, of the flatness of the parton-hadron coexistence line. RHIC dynamics, on the other hand, originates from far above this line.

Hadronic expansion is known to proceed isentropically [102]: commensurate to expansive volume increase the momentum space volume must decrease, from a random isotropic thermal distribution to a restricted momentum orientation preferentially perpendicular to the fireball surface, i.e. radial. The initial thermal energy, implied by the hadron formation temperature $T_H = 165$ MeV, will thus fall down to a residual T_F at hadronic decoupling from the flow field ("thermal freeze-out") plus a radial transverse kinetic energy term $m_i \langle \beta_T \rangle^2$ where m_i is the mass of the considered hadron species and $\langle \beta_T \rangle$ the average radial velocity. We thus expect [103] for the slope of equation (7.22):

$$T = T_F + m_i \langle \beta_T \rangle^2 \,, \quad p_T \leq 2 \text{ GeV}/c \tag{7.23}$$

and

$$T = T_F \left(\frac{1 + \langle v_T \rangle}{1 - \langle v_T \rangle} \right)^{1/2} \,, \quad p_T \gg m_i \tag{7.24}$$

the latter expression valid at p_T larger than hadron mass scale (T then is the "blue-shifted temperature" at decoupling [104] and $\langle v_T \rangle$ the average transverse velocity). The assumption that radial flow mostly originates from the hadronic expansion phase is underlined by the proportionality of flow energy to hadron mass (Eq. (7.23)).

Figure 7.23 illustrates this proportionality, by a recent compilation [103] of RHIC results for central Au+Au collisions at $\sqrt{s} = 200$ GeV, and SPS results for central Pb+Pb collisions at top SPS energy, $\sqrt{s} = 17.3$ GeV. At the latter energy the slope parameter of the Φ meson is seen to be close to that of the similar mass baryons p and Λ, emphasizing the occurrence of m_i scaling as opposed to valence quark number scaling that we will encounter in RHIC elliptic flow data [94]. As is obvious from Fig. 7.23 the multi-strange hyperons and charmonia exhibit a slope saturation which is usually explained [103] as a consequence of their small total cross sections of rescattering from other hadrons, leading to an early decoupling from the bulk hadron radial flow field, such that $\langle \beta_T \rangle_\Omega < \langle \beta_T \rangle_p$.

Fig. 7.23 Hadron slope parameters T at mid-rapidity as a function of mass. For Pb+Pb at $\sqrt{s} = 17.3$ GeV (triangles) and Au+Au at $\sqrt{s} = 200$ GeV (circles); from [103]

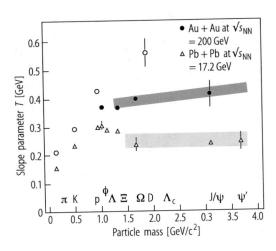

According to our observations with Eq. (7.23) a hydrodynamical ansatz for the transverse mass spectrum of hadrons should thus contain the variables "true temperature" T_F at decoupling from the flow field, and its average velocity $\langle \beta_T \rangle$, common to all hadrons. This is the case for the blast wave model [104] developed as an approximation to the full hydrodynamic formalism [96], assuming a common decoupling or "freeze-out" from flow, for all hadronic species, and a boost-invariant longitudinal expansion:

$$\frac{dN_i}{m_T \, dm_T \, dy} = A_i \, m_T \, K_1 \left(\frac{m_T \cosh \rho}{T_F} \right) I_0 \left(\frac{p_T \sinh \rho}{T_F} \right) \qquad (7.25)$$

where $\rho = \tanh^{-1} \beta_T$. In an extended version of this model a function is included that describes the radial profile of the transverse velocity field, $\beta_T(r) = \beta_T^{\max} \, r/R$, instead of employing a fixed β_T at decoupling [106]. Figure 7.24 flow shows [54] the resulting energy dependence of T_F and $\langle \beta_T \rangle$, for the same set of data as implied already in Figs. 7.21 and 7.22. The "true" decoupling temperature rises steeply at the AGS and less so at SPS energy (as does $\langle \beta_T \rangle$), to a value of about 95 MeV at top SPS energy, which is considerably lower than the chemical freeze-out temperature, $T_H = 165$ MeV, at which the hadronic species relative yield composition of the hadronic phase becomes stationary (see Sect. 7.3, and Fig. 7.1). Chemical decoupling thus occurs early, near the parton-hadron phase boundary, whereas hadronic radial flow ceases after significant further expansion and cooling, whence the surface radial velocity (its average value given by $\langle \beta_T \rangle$ in Fig. 7.24) approaches $\beta_T \approx 0.65$. Both data sets again exhibit an indication of saturation, over the interval toward top SPS energy: the SPS plateau. This supports our above conjecture that radial flow is, predominantly, a consequence of isentropic bulk hadronic expansion in this energy domain, which sets in at T_H. At RHIC energy, both parameters exhibit a further rise,

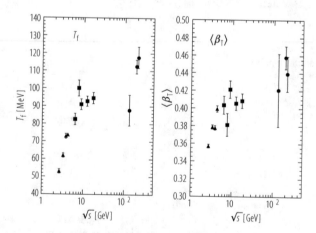

Fig. 7.24 Hadron decoupling temperature T_f, and average radial flow velocity $\langle \beta_T \rangle$ extracted from blast wave model (see Eq. (7.25)) fits of m_T spectra vs. $\sqrt{s}$ [54]

suggesting that primordial partonic flow begins to contribute significantly to radial flow.

7.3 Hadronization and Hadronic Freeze-Out in A+A Collisions

Within the course of the global expansion of the primordial reaction volume the local flow "cells" will hit the parton-hadron phase boundary as their energy density approaches $\epsilon_{crit} \approx 1$ GeV/fm^3. Hadronization will thus occur, not at an instant over the entire interaction volume, but within a finite overall time interval [86] that results from the spread of proper time at which individual cells, or coherent clusters of such cells (as developed during expansion) arrive at the phase boundary. However, irrespective of such a local-temporal occurrence, the hadronization process (which is governed by non perturbative QCD at the low Q^2 corresponding to bulk hadronization) universally results in a novel, *global* equilibrium property that concerns the relative abundance of produced hadrons and resonances. This so-called "hadrochemical equilibrium state" is directly observable, in contrast to the stages of primordial parton equilibration that are only indirectly assessed, via dynamical model studies.

This equilibrium population of species occurs both in elementary and nuclear collisions [107]. We have seen in Fig. 7.17 a first illustration, by e^+e^- annihilation data at $\sqrt{s} = 91.2$ GeV LEP energy, that are well reproduced by the partition functions of the statistical hadronization model (SHM) in its canonical form [84]. The derived hadronization temperature, $T_H = 165$ MeV, turns out to be universal to all elementary and nuclear collision processes at $\sqrt{s} \geq 20$ GeV, and it agrees with the limiting temperature predicted by Hagedorn [38] to occur in any multi-hadronic equilibrium system once the energy density approaches about 0.6 GeV/fm^3. Thus, the upper limit of hadronic equilibrium density corresponds, closely, to the lower limit, $\epsilon_{crit} = 0.6$–1.0 GeV/fm^3 of partonic equilibrium matter, according to lattice QCD [48]. In elementary collisions only about 20 partons or hadrons participate: there should be no chance to approach thermodynamic equilibrium of species by rescattering cascades, neither in the partonic nor in the hadronic phase. The fact that, nevertheless, the hadron formation temperature T_H coincides with the Hagedorn limiting temperature and with the QCD confinement temperature, is a consequence of the non-perturbative QCD hadronization process itself [85], which "gives birth" to hadrons/resonances in canonical equilibrium, at high $\sqrt{s}$, as we shall see below. This process also governs A+A collisions but, as it occurs here under conditions of high energy density extended over considerable volume, the SHM description now requires a *grand* canonical ensemble, with important consequences for production of strange hadrons (strangeness enhancement).

The grand canonical order of hadron/resonance production in central A+A collisions, and its characteristic strangeness enhancement shows that a state of extended matter that is quantum mechanically coherent must exist at hadronization [87, 88, 107]. Whether or not it also reflects partonic equilibrium properties (including flavor equilibrium), that would allow us to claim the direct observation of a quark gluon plasma state near T_c, cannot be decided on the basis of this observation alone, as the hadronization process somehow generates, by itself, the observed hadronic equilibrium. This conclusion, however, is still the subject of controversy [107].

Two typical examples of grand canonical SHM application are illustrated in Figs. 7.25 and 7.26, the first showing total hadron multiplicities in central Pb+Pb collisions at $\sqrt{s} = 17.3$ GeV by NA49 [100] confronted with SHM predictions by Becattini et al. [19]. This plot is similar to Fig. 7.17 in which e^+e^- annihilation to hadrons is confronted with a SHM prediction derived from the *canonical* ensemble [84]. Central Au+Au collision data at $\sqrt{s} = 200$ GeV from several RHIC experiments are compared to grand canonical model predictions by Braun-Munzinger et al. [108] in Fig. 7.26. The key model parameters, T_H and the baryo-chemical potential μ_B result as 159 MeV (160 MeV), and 247 MeV (20 MeV) at $\sqrt{s} = 17.3$ (200) GeV, respectively. The universality of the hadronization temperature is obvious from comparison of the present values with the results of the canonical

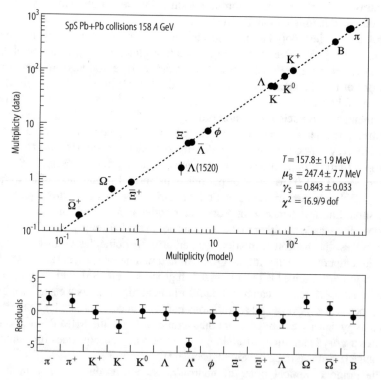

Fig. 7.25 Total hadron multiplicities in central Pb+Pb collisions at $\sqrt{s} = 17.3$ GeV [100] versus prediction of the grand canonical statistical hadronization model [19]

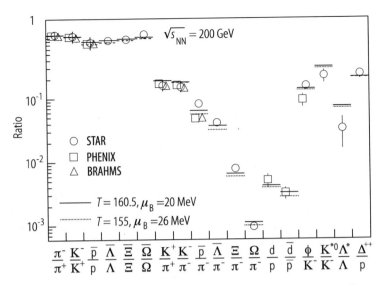

Fig. 7.26 Hadron multiplicity ratios at mid-rapidity in central Au+Au collisions at $\sqrt{s} = 200$ GeV from RHIC experiments STAR, PHENIX and BRAHMS, compared to predictions of the grand canonical statistical model [108]

procedure employed in e^+e^- annihilation to hadrons at $\sqrt{s} = 91.2$ GeV (Fig. 7.17), and in canonical SHM fits [109] to p+p collision data at $\sqrt{s} = 27.4$ GeV where $T_H = 159$ and 169 MeV, respectively.

Figures 7.25 and 7.26 illustrate two different approaches employed in grand canonical SHM application, the former addressing the values of the hadronic multiplicities as obtained in approximate full 4π acceptance (within limitations implied by detector performance), the latter employing a set of multiplicity *ratios* obtained in the vicinity of mid-rapidity as invited, at RHIC energy, by the limited acceptance of the STAR and PHENIX experiments. The latter approach is appropriate, clearly, in the limit of boost-invariant rapidity distributions where hadron production ratios would not depend on the choice of the observational rapidity interval. We have shown in Sect. 7.2.2 that such conditions do, in fact, set in at top RHIC energy, as referred to in Fig. 7.26. However, at low $\sqrt{s}$ the y-distributions are far from boost-invariant, and the total rapidity gap Δy may become comparable, in the extreme case, to the natural rapidity widths of hadrons emitted in the idealized situation of a single, isotropically decaying fireball positioned at mid-rapidity. Its rapidity spectra, Eq. (7.5), resemble Gaussians with widths $\Gamma_i \approx 2.35 (T/m_i)^{1/2}$ for hadron masses m_i. Clearly, the particle ratios $(dN_i/dy)/(dN_j/dy)$ then depend strongly on the position of the rapidity interval dy: away from $y = 0$ heavy hadrons will be strongly suppressed, and particle yields in narrow rapidity intervals are useless for a statistical model analysis unless it is known a priori that the radiator is a single stationary spherical fireball [110]. This is not the case toward top SPS energy (see Fig. 7.10), due to significant primordial longitudinal expansion of the hadron emitting source. Given such conditions, the total multiplicity per collision event (the invariant yield divided by the total overall inelastic cross section) should be employed in the SHM analysis, as is exemplified in Fig. 7.25.

7.3.1 Hadronic Freeze-Out from Expansion Flow

The hadronic multiplicities result from integration of the invariant triple differential cross section over p_T and y. Instrumental, experiment-specific conditions tend to result in incomplete p_T and/or y acceptances. It is important to ascertain that the effects of hydrodynamic transverse and longitudinal flow do not blast a significant part of the total hadron yield to outside the acceptance, and that they, more generally, do not change the relative hadron yield composition, thus basically affecting the SHM analysis. To see that hadronization incorporates only the internal energy in the co-moving frame [110], we first assume that hadrochemical freeze-out occurs on a sharp hypersurface Σ, and write the total yield of particle species i as

$$N_i = \int \frac{d^3 p}{E} \int_\Sigma p^\mu \, d^3 \sigma_\mu(x) \, f_i(x, p) = \int_\Sigma d^3 \sigma_\mu(x) j_i^\mu(x) \qquad (7.26)$$

where $d^3 \sigma$ is the outward normal vector on the surface, and

$$j_i^\mu(x) = g_i \int d^4 p \, 2\Theta(p^0)\delta(p^2 - m_i^2) \, p^\mu (\exp[p \cdot u(x) - \mu_i]/T \pm 1)^{-1} \qquad (7.27)$$

is the grand canonical number current density of species i, μ_i the chemical potential, $u(x)$ the local flow velocity, and g_i the degeneracy factor. In thermal equilibrium it is given by

$$j_i^\mu(x) = \rho_i(x)u^\mu(x) \text{ with}$$

$$\rho_i(x) = u_\mu(x)j_i^\mu(x) = \int d^4 p \, 2\Theta(p^0)\delta(p^2 - m_i^2) \, p \cdot u(x) \, f_i(p \cdot u(x); T; \mu_i)$$

$$= \int d^3 p' \, f_i(E_{p'}; T, \mu_i) = \rho_i(T, \mu_i). \qquad (7.28)$$

Here $E_{p'}$ is the energy in the local rest frame at point x. The total particle yield of species i is therefore

$$N_i = \rho_i(T, \mu_i) \int_\Sigma d^3 \sigma_\mu(x)u^\mu(x) = \rho_i(T, \mu_i) \, V_\Sigma(u^\mu) \qquad (7.29)$$

where only the total comoving volume V_Σ of the freeze-out hypersurface Σ depends on the flow profile u^μ. V is thus a common total volume factor at hadronization (to be determined separately), and the flow pattern drops out from the yield distribution over species in 4π acceptance [110]. For nuclear collisions at SPS energies and below one thus should perform a SHM analysis of the total, 4π-integrated hadronic multiplicities, as was done in Fig. 7.25.

We note that the derivation above illustrates the termination problem of the hydrodynamic description of A+A collisions, the validity of which depends on conditions of a short mean free path, $\lambda < 1$ fm. A precise argumentation suggests that two different free paths are relevant here, concerning hadron occupation number and hadron spectral freeze-out, respectively. As hadrochemical freeze-out occurs in the immediate vicinity of T_c (and $T_H \approx 160$–165 MeV from Figs. 7.25 and 7.26), the hadron species distribution stays constant throughout the ensuing hadronic phase, i.e. the "chemical" mean free path abruptly becomes infinite at T_H, whereas elastic and resonant rescattering may well extend far into the hadronic phase, and so does collective pressure and flow. In fact we have seen in Sect. 7.2.6 that the decoupling from flow occurs at T_F as low as 90–100 MeV (Fig. 7.24). Thus the hydrodynamic evolution of high $\sqrt{s}$ collisions has to be, somehow artificially, stopped at the parton-hadron boundary in order to get the correct hadron multiplicities N_i, of Eqs. (7.26)–(7.29), which then stay frozen-out during the subsequent hadronic expansion.

Equations (7.26)–(7.29) demonstrate the application of the Cooper-Frye prescription [111] for termination of the hydrodynamic evolution. The hyper-surface Σ describes the space-time location at which individual flow cells arrive at the freeze-out conditions, $\epsilon = \epsilon_c$ and $T = T_c$, of hadronization. At this point, the resulting hadron/resonance spectra (for species i) are then given by the Cooper-Frye formula

$$E \frac{dN_i}{d^3p} = \frac{dN_i}{dy\, p_T\, dp_T} = \frac{g_i}{(2\pi)^3} \int_\Sigma f_i(p \cdot u(x), x)\, p \cdot d^3\sigma(x), \qquad (7.30)$$

where $p^\mu f_i\, d^3\sigma_\mu$ is the local flux of particle i with momentum p through the surface Σ. For the phase space distribution f in this formula one takes the local equilibrium distribution at hadronic species freeze-out from the grand canonical SHM

$$f_i(E, x) = [\exp\{(E_i - \mu_i(x))/T\} \pm 1]^{-1} \qquad (7.31)$$

boosted with the local flow velocity $u^\mu(x)$ to the global reference frame by the substitution $E \to p \cdot u(x)$. Fixing $T = T_c$ (taken e.g. from lattice QCD) the hadron multiplicities N_i then follow from Eq. (7.29), and one compares to experiment, as in Figs. 7.25 and 7.26. In order now to follow the further evolution, throughout the hadronic rescattering phase, and to finally compare predictions of Eq. (7.30) to the observed flow data as represented by the various Fourier-terms of Eq. (7.20) one has to re-initialize (with hadronic EOS) the expansion from $\Sigma(T_c) = 165$ MeV) until final decoupling [96], at $T \approx 100$ MeV, thus describing e.g. radial and elliptic flow.

Alternatively, one might end the hydrodynamic description at $T = T_c$ and match the thus obtained phase space distribution of Eq. (7.30) to a microscopic hadron transport model of the hadronic expansion phase [95, 112]. This procedure

Fig. 7.27 Modification of mid-rapidity hadron multiplicities in central Au+Au collisions at $\sqrt{s} = 200$ GeV after chemical freeze-out at $T = T_c$. Squares show a hydrodynamic model prediction at $T = T_c$ (without further interaction); circles show the result of an attached UrQMD hadronic cascade expansion calculation [114]

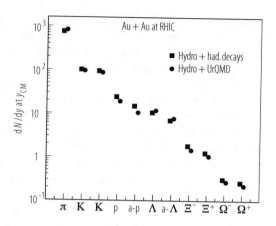

is illustrated in Fig. 7.27 by an UrQMD [113] calculation of Bass and Dumitru [114] for central Au+Au collisions at top RHIC energy. We select here the results concerning the survival of the hadronic multiplicities N_i throughout the dynamics of the hadronic expansion phase, which we have postulated above, based on the equality of the hadronization temperatures, $T_H \approx 160$ MeV, observed in e^+e^- annihilation (Fig. 7.17), where no hadronic expansion phase exists, and in central collisions of $A \approx 200$ nuclei (Figs. 7.25 and reffig:Figure26). In fact, Fig. 7.27 shows that the $\{N_i\}$ observed at the end of the hadronic cascade evolution agree, closely, with the initial $\{N_i\}$ as derived from a Cooper-Frye procedure (Eq. (7.29)) directly at hadronization. On the other hand, p_T spectra and radial flow observables change, drastically, during the hadronic cascade expansion phase.

The hadronic multiplicity distribution $\{N_i\}$, arising from the hadronization process at high $\sqrt{s}$, freezes-out instantaneously also in A+A collisions, and is thus preserved throughout the (isentropic) hadronic expansion phase. *It is thus directly measurable* and, moreover, its hadrochemical equilibrium features lend themselves to an analysis within the framework of Hagedorn-type statistical, grand canonical models. As we shall show below, the outcome of this analysis is contained in a $[T_H, \mu_B]$ parameter pair that reflects the conditions of QCD matter prevailing at hadronization, at each considered $\sqrt{s}$. In fact, the $[T, \mu]$ points resulting from the SHM analysis exhibited in Figs. 7.25 and 7.26 (at $\sqrt{s} = 17.3$ and 200 GeV, respectively) have been shown in the QCD matter phase diagram of Fig. 7.1 to approach, closely, the parton-hadron phase coexistence line predicted by lattice QCD. Thus, $T_H \approx T_c$ at high $\sqrt{s}$: hadrochemical freeze-out occurs in the immediate vicinity of QCD hadronization, thus providing for a location of the QCD phase boundary.

7.3.2 Grand Canonical Strangeness Enhancement

The statistical model analysis [19, 107, 108] of the hadronization species distribution N_i in A+A collisions is based on the grand canonical partition function for species i,

$$\ln Z_i = \frac{g_i V}{6\pi^2 T} \int_0^\infty \frac{k^4 \, dk}{E_i(k) \exp\{(E_i(k) - \mu_i)/T\} \pm 1} \tag{7.32}$$

where $E_i^2 = k^2 + m_i^2$, and $\mu_i \equiv \mu_B B_i + \mu_s S_i + \mu_I I_3^i$ is the total chemical potential for baryon number B, strangeness S and isospin 3-component I_3. Its role in Eq. (7.32) is to enforce, *on average* over the entire hadron source volume, the conservation of these quantum numbers. In fact, making use of overall strangeness neutrality ($\sum_i N_i S_i = 0$) as well as of conserved baryon number (participant $Z+N$) and isospin (participant $(N-Z)/Z$) one can reduce μ_i to a single effective potential μ_B. Hadronic freeze-out is thus captured in three parameters, T, V and μ_B. The density of hadron/resonance species i then results as

$$n_i = \frac{T}{V} \frac{\delta}{\delta_\mu} \ln Z_i \tag{7.33}$$

which gives

$$N_i = V n_i = \frac{g_i V}{(2\pi)^2} \int_0^\infty \frac{k^2 \, dk}{\exp\{(E_i(k) - \mu_i)/T\} \pm 1}. \tag{7.34}$$

We see that the common freeze-out volume parameter is canceled if one considers hadron multiplicity ratios, N_i/N_j, as was done in Fig. 7.26. Integration over momentum yields the one-particle function

$$N_i = \frac{V T g_i}{2\pi^2} m_i^2 \sum_{n=1}^\infty \frac{(\pm 1)^{n+1}}{n} K_2\left(\frac{nm_i}{T}\right) \exp\left(\frac{n\mu_i}{T}\right) \tag{7.35}$$

where K_2 is the modified Bessel function. At high T the effects of Bose or Fermi statistics (represented by the ± 1 term in the denominators of Eqs. (7.32) and (7.34)) may be ignored, finally leading to the Boltzmann approximation

$$N_i = \frac{V T g i}{2\pi^2} m_i^2 K_2\left(\frac{m_i}{T}\right) \exp\left(\frac{\mu_i}{T}\right) \tag{7.36}$$

which is the first term of Eq. (7.35). This approximation is employed throughout the SHM analysis. It describes the *primary* yield of hadron species i, directly at hadronization. The abundance of hadronic resonance states is obtained convoluting equation (7.34) with a relativistic Breit-Wigner distribution [19]. Finally, the overall multiplicity, to be compared to the data, is determined as the sum of the primary multiplicity equation (7.36) and the contributions arising from the unresolved decay of heavier hadrons and resonances:

$$N_i^{\text{observed}} = N_i^{\text{primary}} + \sum_j \text{Br}(j \to i)\, N_j. \tag{7.37}$$

After having exposed the formal gear of grand canonical ensemble analysis we note that Eq. (7.36) permits a simple, first orientation concerning the relation of T to μ_B in A+A collisions by considering, e.g., the antiproton to proton production ratio. From Eq. (7.36) we infer the simple expression

$$N(\overline{p})/N(p) = \exp(-2\mu_B/T). \tag{7.38}$$

Taking the mid-rapidity value 0.8 for $\overline{p}/p$ (from Fig. 7.26) at top RHIC energy, and assuming that hadronization occurs directly at the QCD phase boundary, and hence $T \approx T_c \approx 165$ MeV, we get $\mu_B \simeq 18$ MeV from Eq. (7.38), in close agreement with the result, $\mu_B = 20$ MeV, obtained [108] from the full SHM analysis. Equation (7.38) illustrates the role played by μ_B in the grand canonical ensemble. It logarithmically depends on the ratio of newly created quark-antiquark pairs (the latter represented by the $\overline{p}$ yield), to the total number of quarks including the net baryon number-carrying valence quarks (represented by the p yield).

The most outstanding property of the hadronic multiplicities observed in central A+A collisions is the enhancement of all strange hadron species, by factors ranging from about 2 to 20, as compared to the corresponding production rates in elementary hadron-hadron (and e^+e^- annihilation) reactions at the same $\sqrt{s}$. I.e. the nuclear collision modifies the relative strangeness output by a "nuclear modification factor", $R_s^{\text{AA}} = N_s^{\text{AA}}/0.5\, N_{\text{part}} \cdot N_s^{pp}$, which depends on $\sqrt{s}$ and N_{part} and features a hierarchy with regard to the strangeness number $s = 1, 2, 3$ of the considered species, $R_{s=1}^{\text{AA}} < R_{s=2}^{\text{AA}} < R_{s=3}^{\text{AA}}$. These properties are illustrated in Figs. 7.28 and 7.29. The former shows the ratio of total K^+ to positive pion multiplicities in central Au+Au/Pb+Pb collisions, from lower AGS to top RHIC energies, in comparison to corresponding ratios from minimum bias p+p collisions [100]. We have chosen this ratio, instead of $\langle K^+ \rangle / N_{\text{part}}$, because it reflects, rather directly, the "Wroblewski ratio" of produced strange to non-strange quarks [107], contained in the produced hadrons,

$$\lambda_s \equiv \frac{2(\langle s \rangle + \langle \overline{s} \rangle)}{\langle u \rangle + \langle d \rangle + \langle \overline{u} \rangle + \langle \overline{d} \rangle} \approx \begin{cases} 0.2 \text{ in pp} \\ 0.45 \text{ in AA.} \end{cases} \tag{7.39}$$

The low value of λ_s in pp (and all other elementary) collisions reflects a quark population far away from u, d, s flavor equilibrium, indicating *strangeness suppression* [109].

The so-called *strangeness enhancement* property of A+A collisions (obvious from Figs. 7.28 and 7.29) is, thus, seen as the removal of strangeness suppression;

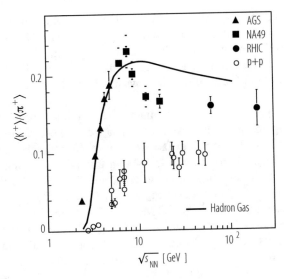

Fig. 7.28 The ratio of total K^+ to total π^+ multiplicity as a function of $\sqrt{s}$, in central Au+Au and Pb+Pb collisions and in p+p minimum bias collisions [100]

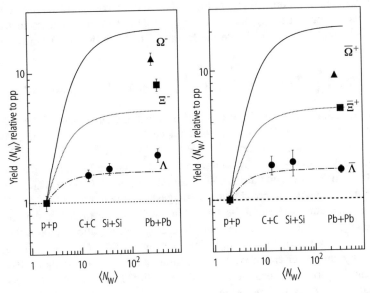

Fig. 7.29 The nuclear modification factors $R^{\text{AA}}_{s=1,2,3}$ for hyperon and anti-hyperon production in nucleus-nucleus collisions at $\sqrt{s} = 17.3$ GeV, relative to the p+p reference at the same energy scaled by $N_W (= N_{\text{part}})$. The NA49 data refer to total 4π yields [116]. Closed lines represent the inverse strangeness suppression factors from ref. [119], at this energy

it is also referred to as a *strangeness saturation*, in SHM analysis [107, 108], for the reason that $\lambda_s \approx 0.45$ corresponds to the grand canonical limit of strangeness production, implicit in the analysis illustrated in Figs. 7.25 and 7.26. The average $R_{s=1}^{AA}$ at $\sqrt{s} \geq 10$ GeV thus is about 2.2, both in the data of Fig. 7.28 and in the statistical model. It increases (Fig. 7.29) toward about 10 in $s = 3$ production of Ω hyperons.

In order to provide for a first guidance concerning the above facts and terminology regarding strangeness production we propose an extremely naive argument, based on the empirical fact of a universal hadronization temperature (Figs. 7.17, 7.25, and 7.26) at high $\sqrt{s}$. Noting that $\langle s \rangle = \langle \bar{s} \rangle$ and $\langle u \rangle \approx \langle \bar{u} \rangle \approx \langle d \rangle \approx \langle \bar{d} \rangle$ in a QGP system at μ_B near zero, and $T = 165$ MeV, just prior to hadronization, $\lambda_s \approx \langle s \rangle / \langle u \rangle \approx \exp\{(m_u - m_s)/T\} = 0.45$ at $p_T \to 0$ if we take current quark masses, $m_s - m_u \approx 130$ MeV. I.e. the value of λ_s in A+A collisions at high $\sqrt{s}$ resembles that of a grand canonical QGP at $\mu_B \to 0$, as was indeed shown in a 3 flavor lattice QCD calculation [115] at $T \approx T_c$. On the contrary, a p+p collision features no QGP but a small fireball volume, at $T \approx T_c$, within which local strangeness neutrality, $\langle s \rangle = \langle \bar{s} \rangle$ has to be strictly enforced, implying a canonical treatment [109]. In our naive model the exponential penalty factor thus contains twice the strangeness quark mass in the exponent, λ_s in pp collisions $\approx \exp\{2(m_u - m_s)/T\} \approx 0.2$, in agreement with the observations concerning strangeness suppression, which are thus referred to as canonical suppression. In a further extension of our toy model, now ignoring the u, d masses in comparison to $m_s \approx 135$ MeV, we can estimate the hierarchy of hyperon enhancement in A+A collisions,

$$R_s^{AA} \propto N_s^{AA}/N_s^{pp} \cdot 0.5 \, N_{part} \approx \exp\{(-sm_s + 2sm_s)/T\} = 2.2, 5.1, 11.6 \quad (7.40)$$

for $s = 1, 2, 3$, respectively. Figure 7.29 shows that these estimates correspond well with the data [116] for R^{AA} derived in 4π acceptance for Λ, Ξ and Ω as well as for their antiparticles, from central Pb+Pb collisions at $\sqrt{s} = 17.3$ GeV. The p+p reference data, and C+C, Si+Si central collisions (obtained by fragmentation of the SPS Pb beam) refer to separate NA49 measurements at the same energy.

The above, qualitative considerations suggest that the relative strangeness yields reflect a transition concerning the fireball volume (that is formed in the course of a preceding dynamical evolution) once it enters hadronization. Within the small volumes, featured by elementary collisions (see Sect. 7.3.3), phase space is severely reduced by the requirement of *local* quantum number conservation [109, 117] including, in particular, local strangeness neutrality. These constraints are seen to be removed in A+A collisions, in which extended volumes of high primordial energy density are formed. Entering the hadronization stage, after an evolution of expansive cooling, these extended volumes will decay to hadrons under conditions of global quantum mechanical coherence, resulting in quantum number conservation occurring, non-locally, and *on average* over the entire decaying volume. This large coherent volume decay mode removes the restrictions, implied by local quantum number balancing. In the above, naive model we have thus assumed that the hadronization of an Omega hyperon in A+A collisions faces the phase space penalty factor of only three s quarks to be gathered, the corresponding three $\bar{s}$ quarks being taken care of elsewhere in the extended volume by global strangeness conservation. In the framework of the SHM this situation is represented by the grand canonical ensemble (Eqs. (7.34), (7.36)); the global chemical potential μ_B expresses quantum number conservation *on average*. Strict, local conservation is represented by the canonical ensemble.

The grand canonical (GC) situation can be shown to be the large collision volume limit (with high multiplicities $\{N_i\}$) of the canonical (C) formulation [118, 119], with a continuous transition concerning the degree of canonical strangeness suppression [119]. To see this one starts from a system that is already in the GC limit with respect to baryon number and charge conservation whereas strangeness is treated canonically. Restricting to $s = 1$ and -1 the GC strange particle densities can be written (from Eq. (7.36)) as

$$n^{GC}_{s=\pm 1} = \frac{Z_{s=\pm 1}}{V} \lambda^{\pm 1}_s \tag{7.41}$$

with

$$Z_{s=\pm 1} = \frac{V g_s}{2\pi^2} m^2_s K_2 \left(\frac{m_s}{T}\right) \exp\left\{(B_s \mu_B + Q_s \mu_Q)/T\right\} \tag{7.42}$$

and a "fugacity factor" $\lambda^{\pm 1}_s = \exp(\mu_s/T)$. The canonical strange particle density can be written as [119]

$$n^C_s = n^{GC}_s \cdot (\tilde{\lambda}_s) \tag{7.43}$$

with an effective fugacity factor

$$\tilde{\lambda}_s = \frac{S_{\pm 1}}{\sqrt{S_1 S_{-1}}} \frac{I_1(x)}{I_0(x)} \tag{7.44}$$

where $S_{\pm 1} = \sum_{s=\pm 1} Z_{s=\pm 1}$ is the sum over all created hadrons and resonances with $s = \pm 1$, the $I_n(x)$ are modified Bessel functions, and $x = 2\sqrt{S_1 S_{-1}}$ is proportional to the total fireball volume V. In the limit $x \approx V \to \infty$ the suppression factor $I_1(x)/I_0(x) \to 1$, and the ratio $S_{\pm 1}/\sqrt{S_1 S_{-1}}$ corresponds exactly to the fugacity λ_s in the GC formulation (see Eq. (7.41)). Thus the C and GC formulations are equivalent in this limit, and the canonical strangeness suppression effect disappears. Upon generalization to the complete strange hadron spectrum, with $s = \pm 1, \pm 2, \pm 3$, the strangeness suppression factor results [119] as

$$\eta(s) = I_s(x)/I_0(x). \tag{7.45}$$

In particular for small x (volume), $\eta(s) \to (x/2)^s$, and one expects that the larger the strangeness content of the particle the smaller the suppression factor, and hence the larger the enhancement in going from elementary to central A+A collisions. This explains the hierarchy addressed in Eq. (7.40), and apparent from the data shown in Fig. 7.29. In fact, the curves shown in this figure represent the results obtained from Eq. (7.45), for $s = 1, 2, 3$ hyperon production at $\sqrt{s} = 17.3$ GeV [119]. They are seen to be in qualitative agreement with the data. However the scarcity of data, existing at top SPS energy for total hyperon yields, obtained in 4π acceptance (recall the arguments in Sect. 7.3.1) both for A+A and p+p collisions does not yet permit to cover the SHM strangeness saturation curves in detail, for $s > 1$.

This saturation is seen in Fig. 7.29, to set in already at modest system sizes, but sequentially so, for ascending hyperon strangeness. Note that SHM saturation is sequentially approached, from Eq. (7.45), with increasing fireball *volume V*. In order to make contact to the experimental size scaling with centrality, e.g. N_{part}, the model of ref. [119], which is illustrated in Fig. 7.29, has converted the genuine volume scale to the N_{part} scale by assuming a universal eigenvolume of 7 fm³ per participant nucleon. I.e. $N_{part} = 10$ really means a coherent fireball volume of 70 fm³, in Fig. 7.29. Within this definition, saturation of $s = 1, 2, 3$ sets in at fireball volumes at hadronization of about 60, 240 and 600 fm³, respectively: this is the *real* message of the SHM curves in Fig. 7.29.

The above direct translation of coherent fireball volume to participant number is problematic [120] as it assumes that all participating nucleons enter into a single primordially coherent fireball. This is, however, not the case [120] particularly in the relative small scattering systems that cover the initial, steep increase of $\eta(s)$, where several local high density clusters are formed, each containing a fraction of N_{part}. This is revealed by a percolation model [120] of cluster overlap attached to a Glauber calculation of the collision/energy density. At each N_{part} an average cluster volume distribution results which can be transformed by Eq. (7.45) to an average $\{\eta(s, V)\}$ distribution whose weighted mean is the appropriate effective canonical suppression factor corresponding to N_{part}. On the latter scale, the SHM suppression curve thus shifts to higher N_{part}, as is shown in Fig. 7.30 for the K^+/π^+ ratio vs. N_{part}, measured at mid-rapidity by PHENIX in Au+Au collisions at $\sqrt{s} = 200$ GeV,

Fig. 7.30 The mid-rapidity K^+ to π^+ ratio vs. N_{part} in minimum bias Au+Au collisions at $\sqrt{s} = 200$ GeV, compared to the percolation model [120] (solid line); a prediction of which for Cu+Cu at similar energy is given by the long dashed line (see text for detail)

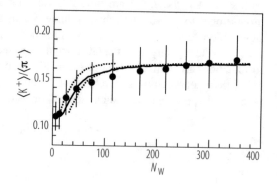

which is reproduced by the percolation model [120]. Also included is a prediction for Cu+Cu at this energy which rises more steeply on the common N_{part} scale because the collision and energy density reached in central Cu+Cu collisions, at $N_{part} \approx 100$, exceeds that in peripheral Au+Au collisions (at the same N_{part}) which share a more prominent contribution from the dilute surface regions of the nuclear density profile. We note, finally, that this deviation from universal N_{part} scaling does not contradict the observations of a perfect such scaling as far as overall charged particle multiplicity densities are concerned (recall Fig. 7.12) which are dominated by pions, not subject to size dependent canonical suppression.

7.3.3 Origin of Hadro-Chemical Equilibrium

The statistical hadronization model (SHM) is *not* a model of the QCD confinement process leading to hadrons, which occurs once the dynamical cooling evolution of the system arrives at T_c. At this stage the partonic reaction volume, small in elementary collisions but extended in A+A collisions, will decay (by whatever elementary QCD process) to on-shell hadrons and resonances. This coherent quantum mechanical decay results in a de-coherent quasi-classical, primordial on-shell hadron-resonance population which, at the instant of its formation, lends itself to a quasi-classical Gibbs ensemble description. Its detailed modalities (canonical for small decaying systems, grand canonical for extended fireballs in A+A collisions), and its derived parameters [T, μ_B] merely recast the conditions, prevailing at hadronization. The success of SHM analysis thus implies that the QCD hadronization process ends in statistical equilibrium concerning the hadron-resonance species population.

In order to identify mechanisms in QCD hadronization that introduce the hadro-chemical equilibrium we refer to jet hadronization in e^+e^- annihilation reactions, which we showed in Fig. 7.17 to be well described by the canonical SHM. In di-jet formation at LEP energy, $\sqrt{s} = 92$ GeV, we find a charged particle multiplicity of about 10 per jet, and we estimate that, likewise, about 10 primordial partons participate on either side of the back-to-back di-jet [85]. There is thus no chance for either partonic or hadronic, extensive rescattering toward chemical equilibrium. However, in the jet hadronization models developed by Amati and Veneziano [83], Webber [121] and Ellis and Geiger [85] the period of QCD DGLAP parton shower evolution (and of perturbative QCD, in general) ends with local color neutralization, by formation of spatial partonic singlet clusters. This QCD "color pre-confinement" [83] process reminds of a coalescence mechanism, in which the momenta and the initial virtual masses of the individual clustering partons get converted to internal, invariant virtual mass of color neutral, *spatially extended* objects. Their mass spectrum [121] extends from about 0.5 to 10 GeV. This cluster mass distribution, shown in Fig. 7.31, represents the first stochastic element in this hadronization model.

Fig. 7.31 Invariant mass spectrum of color neutralization clusters in the Veneziano-Webber hadronization model [83, 121]

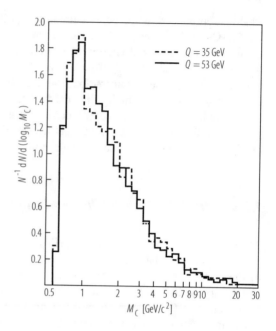

The clusters are then re-interpreted within non-perturbative QCD: their internal, initially perturbative QCD vacuum energy gets replaced by non-perturbative quark and gluon condensates, making the clusters appear like hadronic resonances. Their subsequent quantum mechanical decay to on-shell hadrons is governed by the phase space weights given by the hadron and resonance spectrum [85, 121]. I.e. the clusters decay under "phase space dominance" [85], the outcome being a micro-canonical or a canonical hadron and resonance ensemble [84, 107]. The apparent hadro-chemical equilibrium thus is the consequence of QCD color neutralization to clusters, and their quantum mechanical decay under local quantum number conservation and phase space weights. We note that the alternative description of hadronization, by string decay [122], contains a quantum mechanical tunneling mechanism, leading to a similar phase space dominance [123].

Hadronization in e^+e^- annihilation thus occurs from local clusters (or strings), isolated in vacuum, of different mass but similar energy density corresponding to QCD confinement. These clusters are boosted with respect to each other but it was shown [124] that for a Lorentz invariant scalar, such as multiplicity, the contributions of each cluster (at similar T) can be represented by a single canonical system with volume equal to the sum of clusters. In the fit of Fig. 7.17 this volume sum amounts to about 45 fm^3 [84]; the individual cluster volumes are thus quite small, of magnitude a few fm^3 [85]. This implies maximum canonical strangeness suppression but may, in fact, require a micro-canonical treatment of strangeness [109], implying a further suppression. These MC effects are oftentimes included [125] in the canonical partition functions by an extra strangeness fugacity parameter $\gamma_s < 1$ which suppresses $s = 1, 2, 3$ in a hierarchical manner, $\langle N_i(s) \rangle \approx (\gamma_s)^{s_i}$. The fit of Fig. 7.17 requires $\gamma_s = 0.66$, a value typical of canonical multiplicity analysis in p+p, p+$\bar{p}$ and e^+e^- annihilation collisions [109] at $\sqrt{s} \geq 30$ GeV.

The above picture, of hadrochemical equilibrium resulting from the combined stochastic features of QCD color neutralization by cluster formation, and subsequent quantum mechanical decay to the on-shell hadron and resonance spectrum (under phase space governance) lends itself to a straight forward extension to A+A collisions. The essential new features, of grand canonical hadronization including strangeness enhancement, should result from the fact that extended space-time volumes of $\epsilon > \epsilon_{crit}$ are formed in the course of primordial partonic shower evolution, an overlap effect increasing both with $\sqrt{s}$ and with the size of the primordial interaction volume. As the volume of the elementary hadronization clusters amounts to several fm^3 it is inevitable that the clusters coalesce, to form extended "super-cluster" volumes prior to hadronization [120]. As these super-clusters develop toward hadronization via non perturbative QCD dynamics, it is *plausible* to assume an overall quantum mechanical coherence to arise over the entire extended volume, which will thus decay to hadrons under global quantum number conservation, the decay products thus modeled by the GC ensemble.

Our expectation that space-time coalescence of individual hadronization clusters will lead to a global, quantum mechanically coherent extended super-cluster volume, that decays under phase space dominance, appears as an analogy to the dynamics and quantum mechanics governing low energy nuclear fission from a preceding "compound nucleus" [126]. Note that the observation of a smooth transition from canonical strangeness suppression to grand canonical saturation (Figs. 7.29, 7.30) lends further support to the above picture of a percolative growth [120] of the volume that is about to undergo hadronization.

An extended, coherent quark gluon plasma state would, of course, represent an ideal example of such a volume [127] and, in fact, we could imagine that the spatial extension of the plasma state results from a percolative overlap of primordial zones of high energy density, which becomes more prominent with increasing $\sqrt{s}$ and N_{part}. A QGP state preceding hadronization will thus lead to all the observed features. However, to be precise: the hadronizing QCD system of extended matter decaying quantum coherently, could still be a non-equilibrium precursor of the ideal equilibrium QGP, because we have seen above that hadrochemical equilibrium also occurs in e^+e^- annihilation, where no partonic equilibrium exists. It gets established in the course of hadronization, irrespective of the degree of equilibrium prevailing in the preceding partonic phase.

7.3.4 Hadronization vs. Rapidity and $\sqrt{s}$

We have argued in Sect. 7.3.1 that, at relatively low $\sqrt{s}$, the total rapidity gap Δy does not significantly exceed the natural thermal rapidity spreading width $\Gamma_i \approx 2.35 \, (T/m_i)^{1/2}$ of a single, isotropically decaying fireball, centered at mid-rapidity and emitting hadrons of mass m_i [110]. However, this procedure involves an idealization because in the real Pb+Pb collision the intersecting dilute surface sections of the nuclear density profiles will lead to a significant contribution of single-scattering NN collisions, outside the central high density fireball. The leading hadron properties of such "corona collisions" result in wider proper rapidity distributions, quite different from those of the central fireball decay hadrons. Their contribution will thus be prominent near target/projectile rapidity, and will feature a canonically suppressed strangeness. The one-fireball assumption, although inevitable at small Δy, does not quite resemble the physical reality. This may explain the need for an additional strangeness suppression factor in the GC one-particle partition function (Eq. (7.32)) that has, unfortunately, also been labeled γ_s but expresses physics reasons quite different from the extra suppression factor that reflects micro-canonical phase space constraints in elementary collisions. It turns out that all GC analysis of central A+A collisions at low $\sqrt{s}$, and addressed to total 4π multiplicities, requires a γ_s of 0.7–0.85 [19]; in the fit of Fig. 7.25 $\gamma_s = 0.84$.

At RHIC, $\Delta y \approx 11 \gg \Gamma_i$, and such difficulties disappear: $\gamma_s \approx 1$ at midrapidity and, moreover, the wide gap permits a SHM analysis which is differential in y. Figure 7.32 shows the y-dependence of the ratios π^-/π^+, K^-/K^+ and $\bar{p}/p$ as obtained by BRAHMS [128] in central Au+Au collisions at $\sqrt{s} = 200$ GeV. The figure shows a dramatic dependence of the $\bar{p}/p$ ratio, which reflects the local baryochemical potential according to Eq. (7.38). At $y_{CM} > 1$ the $\bar{p}/p$ ratio drops down steeply, to about 0.2 at $y \approx 3.5$, thus making close contact to the top SPS energy value obtained by NA49 [129]. The K^-/K^+ ratio follows a similar but weaker drop-off pattern, to about 0.65 again matching with the top SPS energy value of about 0.6 [130]. The deviation from unity of these ratios reflects the rapidity densities of initial valence u, d quarks, relative to the densities of newly created light and strange quark-antiquark pairs, i.e. the y distribution of the net baryon number density, and of the related baryo-chemical potential of the GC ensemble.

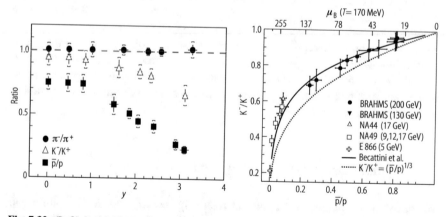

Fig. 7.32 (Left) Anti-hadron to hadron ratios as a function of rapidity in central Au+Au collisions at $\sqrt{s} = 200$ GeV. (Right) Interpretation of the correlation between $\bar{p}/p$ and K^-/K^+ in terms of baryo-chemical potential μ_B variation in the grand canonical statistical model. From [128]

Thus, in analyzing successive bins of the rapidity distributions in Fig. 7.32, the major variation in the GC fit concerns the baryo-chemical potential $\mu_B(y)$ which increases from about 20 MeV (Fig. 7.26) at mid-rapidity, to about 150 MeV at $y \geq 3$ while the hadronization temperature stays constant, at $T = 160$ MeV. This interplay between K^-/K^+, $\overline{p}/p$ and μ_B is illustrated [128] in the right hand panel of Fig. 7.32, and shown to be well accounted for by the GC statistical model [131].

These considerations imply that hadronization at RHIC (and LHC) energy occurs *local* in *y*-space and *late* in time. The density distribution of net baryon number results from the primordial pQCD shower evolution (c.f. Sect. 7.2.4), and is thus fixed at formation time, $t_0 \leq 0.6$ fm/c at RHIC. Hadronization of the bulk partonic matter occurs later, at $t \geq 3$ fm/c [86, 95], and transmits the local conditions in rapidity space by preserving the local net baryon quantum number density. Most importantly we conclude that hadronization occurs, not from a single longitudinally boosted fireball but from a succession of "super-clusters", of different partonic composition depending on *y*, and decaying at different time due to the Lorentz-boost that increases with *y*, in an "inside-outside" pattern (c.f. Fig. 7.18). We are thus witnessing at hadronization a Hubble expanding system of local fireballs. The detailed implications of this picture have not been analyzed yet. Note that a central RHIC collision thus does not correspond to a single hadronization "point" in the $[T, \mu]$ plane of Fig. 7.1 but samples $\{T, \mu\}$ along the QCD parton-hadron coexistence line [132].

Throughout this chapter we have discussed hadronic freeze-out at high $\sqrt{s}$ only (top SPS to RHIC energy), because of the proximity of the chemical freeze-out parameters $[T, \mu_B]$ to the QCD phase boundary from lattice QCD, which suggests an overall picture of hadronization, to occur directly from a partonic cluster or super-cluster. Our discussion of the GC statistical hadronization model has been explicitly or implicitly based on the assumption that hadronic freeze-out coincides with hadronization. However, the GC model has also been applied successfully to hadro-chemical freeze-out at $\sqrt{s}$ down to a few GeV [19, 107, 108] where it is not expected that the dynamical evolution traverses the phase boundary at all, but grand canonical multiplicity distributions, and their characteristic strangeness enhancement pattern, are observed throughout. Toward lower $\sqrt{s}$, T decreases while μ_B increases, as is shown in Fig. 7.33 which presents a compilation of all reported freeze-out parameters [108].

These points have also been included in the phase diagram of Fig. 7.1 which shows that they are gradually branching away from the phase separation boundary line that could recently be predicted by lattice QCD once new methods had been developed to extend the theory to finite μ_B [9, 10]. At $\sqrt{s} \geq 20$ GeV we see that

$$\epsilon_c(QCD) \approx \epsilon_H \approx \epsilon_{GC} \tag{7.46}$$

where ϵ_{GC} is the freeze-out density inferred from GC analysis [19, 107, 108].

In turn, the GC hadronic freeze-out points drop below the lattice QCD coexistence line at lower $\sqrt{s}$, implying that chemical freeze-out now occurs within the hadronic expansion phase. This requires a model of freeze-out, now governed by

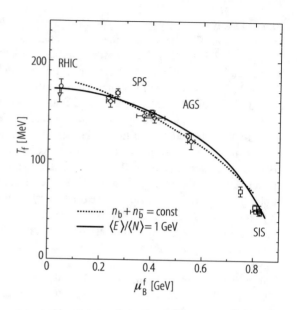

Fig. 7.33 Energy dependence of the hadro-chemical freeze-out points obtained by grand canonical statistical model analysis in the plane [T, μ_B], with interpolating curve at fixed energy per particle of about 1 GeV [107, 139]

the properties of a high density hadronic medium, upon expansive cooling and dilution. Holding on to the model of a quantum mechanical de-coherence decay to on-shell hadrons that we discussed in Sect. 7.3.3, we argue that an initial, extended high density hadronic fireball, given sufficient life-time at T smaller, but not far below T_c, could also be seen as a quantum mechanically coherent super-cluster, as governed by effective mean fields [133]. In such a medium hadrons, at T near T_c, acquire effective masses and/or decay widths far off their corresponding properties in vacuum: they are off-shell, approaching conditions of QCD chiral symmetry restoration as $T \rightarrow T_c$ [134]. This symmetry is inherent in the elementary QCD Lagrangian, and "softly" broken within the light quark sector by the small non-zero current quark masses, but severely broken at $T \rightarrow 0$ by the high effective constituent quark masses that get dressed by non perturbative QCD vacuum condensates. Pictorially speaking, hadrons gradually loose this dressing as $T \rightarrow T_c$ [135], introducing a change, away from in vacuum properties, in the hadronic mass and width spectrum. Such in-medium chiral restoration effects have, in fact, been observed in relativistic A+A collisions, by means of reconstructing the in-medium decay of the ρ vector meson to an observed e^+e^- pair [136] (see Sect. 7.6.3).

A dense, high T hadronic system, with mean-field induced off-shell constituents is also, clearly, quantum mechanically coherent. At a certain characteristic density, $\epsilon < \epsilon_c$, and temperature $T < T_c$, as reached in the course of overall hadronic expansion, this extended medium will undergo a decoherence transition to classical on-shell hadrons. Its frozen-out hadronic multiplicity distribution should be, again, characterized by the phase space weights of a grand canonical ensemble at $T < T_c$. Theoretical studies of such a mean field hadronic expansion mode [137] have also shown that such mechanisms play essentially no role at $\sqrt{s} \geq 20$ GeV because the expanding system is already in rapid flow once it traverses the phase boundary, with an expansion time scale shorter than the formation time scale of mean field phenomena. At lower energies, on the other hand, the system might not even dive into the deconfined phase but spend a comparatively long time in its direct vicinity, at the turning point between compression and re-expansion where all dynamical time constants are large, and the hadron density is high, such that the inelastic hadronic transmutation rate becomes high (particularly in collisions of more than two hadronic reactants, with reaction rates [138] proportional to ϵ^n), and sufficiently so for maintaining hadronic chemical equilibrium after it is first established at maximum hadron density, in low $\sqrt{s}$ systems that do not cross the phase boundary at all.

The GC freeze-out parameters $[T, \mu]$ at various $\sqrt{s}$ in Fig. 7.33 permit a smooth interpolation in the T, μ plane [139], which, in turn, allows for GC model predictions which are continuous in $\sqrt{s}$. Such a curve is shown in Fig. 7.28 compared to the 4π data points for the K^+/π^+ multiplicity ratio in central collisions Au+Au/Pb+Pb, at all $\sqrt{s}$ investigated thus far. It exhibits a smooth maximum, due to the interplay of T saturation and μ_B fall-off to zero, but does not account for the sharp peak structure seen in the data at $\sqrt{s} \approx 7$ GeV and $\mu_B \approx 480$ MeV. This behavior is not a peculiarity of the K^+ channel only; it also is reflected in an unusually high Wroblewski ratio (see Eq. (7.39)) obtained at $\sqrt{s} = 7.6$ GeV, of $\lambda_s = 0.60$ [19]. This sharp strangeness maximum is unexplained as of yet. It implies that hadron formation at this $\sqrt{s}$ reflects influences that are less prominent above and below, and most attempts to understand the effect [141–143] are centered at the assumption that at this particular $\sqrt{s}$ the overall bulk dynamics will settle directly at the phase boundary where, moreover, finite μ_B lattice theory also expects a QCD critical point [9–11]. This would cause a softest point to occur in the equation of state, i.e. a minimum in the relation of expansion pressure vs. energy density, slowing down the dynamical evolution [144, 145], and thus increasing the sensitivity to expansion modes characteristic of a first order phase transition [143], which occurs at $\mu_B \geq \mu_B^{crit}$. Such conditions may modify the K/π ratio (Fig. 7.28) [143].

It thus appears that the interval from top AGS to lower SPS energy, $5 \leq \sqrt{s} \leq 10$ GeV, promises highly interesting information regarding the QCD phase diagram (Fig. 7.1) in the direct vicinity of the parton-hadron coexistence line. In particular, the physics of a critical point of QCD matter deserves further study. Observable consequences also comprise so-called "critical fluctuations" [146, 147] of multiplicity density, mean transverse momentum and hadron-chemical composition [148], the latter in fact being observed near $\sqrt{s} = 7$ GeV in an event by event study of the K/π ratio in central Pb+Pb collisions [149]. We shall return to critical point physics in Sect. 7.7.

7.4 Elliptic Flow

We have up to now mostly stressed the importance of central collisions and mid-rapidity data because they provide for the highest primordial energy density and avoid problems caused by emission anisotropy and the presence of cold spectator sections of target and projectile nuclei. On the other hand, a fundamentally new window of observation is opened by non-central collisions as the finite impact parameter breaks cylinder symmetry, defining emission anisotropies with respect to the orientation of the impact vector $\vec{b}$ as we have shown in Eq. (7.20). In a strongly interacting fireball collision dynamics, the initial geometric anisotropy of the reaction volume gets transferred to the final momentum spectra and thus becomes experimentally accessible. Furthermore, the high charged particle multiplicity allows for an event-by-event determination of the reaction plane (direction of $\vec{b}$), enabling the study of observables at azimuth φ, relative to the known reaction plane. We shall show that this opens a window into the very early stages of A+A collisions onward from the end of nuclear interpenetration, at $\tau \approx 2\,R(A)/\gamma_{CM}$. Our observation thus begins at the extreme energy densities prevailing right at formation time (Sects. 7.2.4 and 7.2.5), i.e. concurrent with the initialization phase of relativistic hydrodynamic expansion. We access the phase diagram of Fig. 7.1 in regions *far above* the QCD phase boundary.

Before turning to the details of elliptic flow data we wish to illustrate [96] the above statements. Figure 7.34 exhibits the transverse projection of primordial energy density, assumed to be proportional to the number density of participant nucleons in the overlap volume arising from a Au+Au collision at impact parameter $b = 7$ fm. The nuclear density profiles (assumed to be of Woods-Saxon type) intersect in an ellipsoidal fireball, with minor axis along the direction of $\vec{b}$ which

Fig. 7.34 Transverse projection of primordial binary collision density in an Au+Au collision at impact parameter 7 fm, exhibiting participant parton spatial excentricity [96]

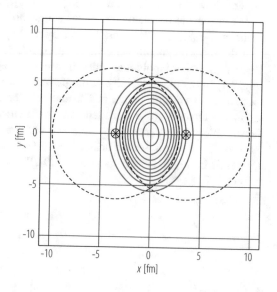

is positioned at $y = 0$. The obvious geometrical deformation can be quantified by the spatial excentricity (unfortunately also labeled ϵ in the literature)

$$\epsilon_x(b) = \frac{\langle y^2 - x^2 \rangle}{\langle y^2 + x^2 \rangle} \tag{7.47}$$

where the averages are taken with respect to the transverse density profiles of Fig. 7.34. ϵ_x is zero for $b = 0$, reaching a value of about 0.3 in the case $b = 7$ fm illustrated in Fig. 7.34.

Translated into the initialization of the hydrodynamic expansion the density anisotropy implies a corresponding pressure anisotropy. The pressure is higher in x than in y direction, and thus is the initial acceleration, leading to an increasing momentum anisotropy,

$$\epsilon_p(\tau) = \frac{\int dx\, dy\, (T^{xx} - T^{yy})}{\int dx\, dy\, (T^{xx} + T^{yy})} \tag{7.48}$$

where $T^{\mu x}_{(x)}$ is the fluid's energy-momentum tensor. Figure 7.35 shows [96, 150] the time evolution of the spatial and momentum anisotropies for the collision considered in Fig. 7.34, implementing two different equations of state which are modeled with (without) implication of a first order phase transition in "RHIC" ("EOS1"). A steep initial rise is observed for ϵ_p, in both cases: momentum anisotropy builds up during the early partonic phase at RHIC, while the spatial deformation disappears. I.e. the initial source geometry, which is washed out later on, imprints a flow anisotropy which is preserved, and observable as "elliptic flow". A first order phase transition essentially stalls the buildup of ϵ_p at about $\tau = 3$ fm/c when the system enters the

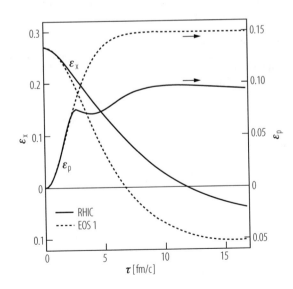

Fig. 7.35 Time evolution of the spatial excentricity ϵ_x and the momentum space anisotropy ϵ_p (Eqs. (7.47) and (7.48)) in the hydrodynamic model of an Au+Au collision at $b = 7$ fm, occurring at $\sqrt{s} = 200$ GeV [96]. The dynamics is illustrated with two equations of state

mixed phase, such that the emerging signal is almost entirely due to the partonic phase. We have to note here that the "ideal fluid" (zero viscosity) hydrodynamics [150] employed in Fig. 7.35 is, at first, a mere hypothesis, in view also of the fact that microscopic transport models have predicted a significant viscosity, both for a perturbative QCD parton gas [92, 151] and for a hadron gas [95, 152]. Proven to be correct by the data, the ideal fluid description of elliptic flow tells us that the QGP is a non-perturbative liquid [153, 154].

Elliptic flow is quantified by the coefficient v_2 of the second harmonic term in the Fourier expansion (see Eq. (7.20)) of the invariant cross section; it depends on $\sqrt{s}$, b, y and p_T. Figure 7.36 shows the $\sqrt{s}$ dependence of v_2 at mid-rapidity and averaged over p_T, in Au+Au/Pb+Pb semi-peripheral collisions [93, 155]. We see that the momentum space anisotropy is relatively small overall, but exhibits a steep rise toward top RHIC energy.

Figure 7.37 shows the (pseudo)-rapidity dependence of v_2 at $\sqrt{s} = 130$ and 200 GeV as obtained by PHOBOS [156] for charged particles in minimum bias Au+Au collisions. It resembles the corresponding charged particle rapidity density distribution of Fig. 7.8, suggesting that prominent elliptic flow arises only at the highest attainable primordial energy density.

That such conditions are just reached at top RHIC energy is shown in Figs. 7.38 and 7.39. The former combines STAR [157] and PHENIX [158] data for the p_T dependence of elliptic flow, observed for various identified hadron species $\pi^{\pm}$, $K^{\pm}$, p, K^0 and Λ, $\overline{\Lambda}$ in Au+Au at 200 GeV. The predicted hydrodynamic flow pattern [96, 159] agrees well with observations in the bulk $p_T < 2$ GeV/c domain. Figure 7.39 (from [155]) unifies average v_2 data from AGS to top RHIC energies in a scaled representation [93] where v_2 divided by the initial spatial anisotropy ϵ_x

Fig. 7.36 Energy dependence of the elliptic flow parameter v_2 at mid-rapidity and averaged over p_T, in Au+Au and Pb+Pb semi-peripheral collisions [155]

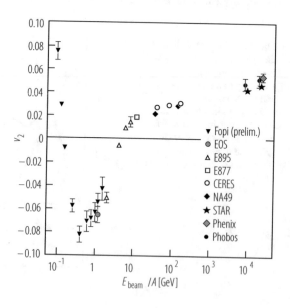

Fig. 7.37 Pseudo-rapidity dependence of the p_T-averaged elliptic flow coefficient v_2 for charged hadrons at $\sqrt{s} = 130$ and 200 GeV [156]

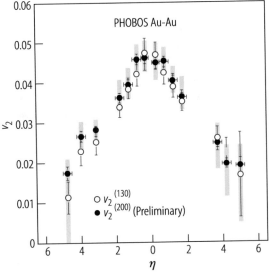

Fig. 7.38 Transverse momentum dependence of elliptic flow v_2 for mesons and baryons in Au+Au collisions at $\sqrt{s} = 200$ GeV. The hydrodynamic model [96, 159] describes the mass dependence at $p_T \leq 2$ GeV/c

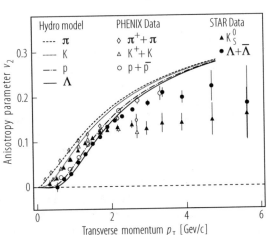

is plotted versus charged particle mid-rapidity density per unit transverse area S, the latter giving the density weighted transverse surface area of primordial overlap, Fig. 7.34. Figure 7.39 includes the hydrodynamic predictions [95, 96, 150, 159, 161] for various primordial participant or energy densities as implied by the quantity $(1/S)\, dn_{ch}/dy$ [93]. Scaling v_2 by ϵ_x enhances the elliptic flow effect of near-central collisions where ϵ_x is small, and we see that only such collisions at top RHIC energy reach the hydrodynamical ideal flow limit in models that include an EOS ansatz which incorporates [96] the effect of a first order phase transition, which reduces the primordial flow signal as was shown in Fig. 7.35.

At top RHIC energy, the interval between $t_0 \approx 0.6$ fm/c, and hadronization time, $t_H \approx 3$ fm/c, is long enough to establish dynamical consequences of an early

Fig. 7.39 Elliptic flow v_2 scaled by spatial excentricity ϵ as a function of charged particle density per unit transverse area S, from AGS to top RHIC energy. The hydrodynamic limit is only attained at RHIC [155]

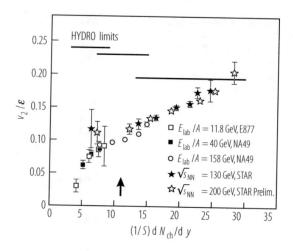

approach toward local equilibrium. The "lucky coincidence" of such a primordial resolution of dynamical time scale, with the extreme primordial density, offered by semi-central collisions of heavy nuclei, results in an extremely short mean free path of the primordial matter constituents, thus inviting a hydrodynamic description of the expansive evolution. Consistent application of this model reveals a low viscosity: the primordial matter resembles an ideal fluid, quite different from earlier concepts, of a weakly interacting partonic gas plasma state (QGP) governed by perturbative QCD screening conditions [36, 41].

A further, characteristic scaling property of elliptic flow is derived from the p_T dependence of v_2, observed for the different hadronic species. In Fig. 7.38 one observes a hadron mass dependence, the v_2 signal of pions and charged kaons rising faster with p_T than that of baryons. Clearly, within a hydrodynamic flow *velocity* field entering hadronization, heavier hadronic species will capture a higher p_T, at a given flow velocity. However, unlike in hadronic radial expansion flow phenomena (c.f. Sect. 7.2.6) it is not the hadronic *mass* that sets the scale for the total p_T derived, per particle species, from the elliptic flow field, but the hadronic valence quark content. This conclusion is elaborated [94] in Fig. 7.40.

The left panel shows measurements of the p_T dependence of v_2 for several hadronic species, in minimum bias Au+Au collisions at $\sqrt{s} = 200$ GeV [161]. The middle panel bears out the hydrodynamically expected [162] particle mass scaling when v_2 is plotted vs. the relativistic transverse kinetic energy $K E_T \equiv m_T - m$ where $m_T = (p_T^2 + m^2)^{1/2}$. For $K E_T \geq 1$ GeV, clear splitting into a meson branch (lower v_2) and a baryon branch (higher v_2) occurs. However, both of these branches show good scaling separately. The right panel shows the result obtained after scaling both v_2 and $K E_T$ (i.e. the data in the middle panel) by the constituent quark number, $n_q = 2$ for mesons and $n_q = 3$ for baryons. The resulting perfect, universal scaling is an indication of the inherent quark degrees of freedom in the flowing matter as it approaches hadronization. We thus assert that the bulk of the elliptic flow signal develops in the pre-hadronization phase.

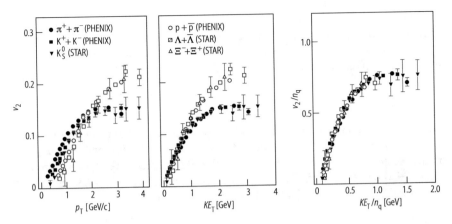

Fig. 7.40 v_2 vs. p_T (left panel) and transverse kinetic energy $KE_T = m_T - m_0$ (middle) for several hadronic species in min. bias Au+Au collisions at $\sqrt{s} = 200$ GeV, showing separate meson and baryon branches. Scaling (right panel) is obtained by valence quark number n_q, dividing v_2 and KE_T [94]

The above scaling analysis has been extended to Φ meson and Ω hyperon production, and also to first PHENIX results [163] concerning elliptic flow of the charmed D meson [94], with perfect agreement to the observations made in Fig. 7.40, of a separation into meson/hadron branches on the KE_T scale, which merge into a universal v_2 scaling once both v_2 and KE_T per valence quark are considered. The observation that the D meson charm quark apparently shares in the universal flow pattern is remarkable as its proper relaxation time is, in principle, lengthened by a factor M/T [164]. A high partonic rescattering cross section σ is thus required in the primordial QGP fireball, to reduce the partonic mean free path $\lambda = 1/n\sigma$ (where n is the partonic density), such that $\lambda \ll A^{1/3}$ (the overall system size) and, simultaneously, $\lambda < 1$ fm in order to conform with the near-zero mean free path implication of the hydrodynamic description of the elliptic flow, which reproduces the data gathered at RHIC energy. The presence of a high partonic rescattering cross section was born out in a parton transport model study [92] of the steep linear rise of the elliptic flow signal with p_T (Fig. 7.38). In such a classical Boltzmann equation approach the cross sections required for the system to translate the initial spatial into momentum space anisotropy, fast enough before the initial deformation gets washed out during expansion (c.f. Fig. 7.35), turn out to exceed by almost an order of magnitude the values expected in a perturbative QCD quark-gluon gas [92].

The non-perturbative quark-gluon plasma is thus a strongly coupled state (which has been labeled sQGP [165]). At RHIC energy this reduces the partonic mean free path to a degree that makes hydrodynamics applicable. A Navier-Stokes analysis [166] of RHIC flow data indicates that the viscosity of the QGP must be about ten times smaller than expected if the QGP were a weakly interacting pQCD Debye screened plasma. This justifies the use of perfect fluid dynamics.

Considering first attempts to derive a quantitative estimate of the dimensionless ratio of (shear) viscosity to entropy, we note that η/s is a good way to characterize the intrinsic ability of a substance to relax toward equilibrium [167]. It can be estimated from the expression [94]

$$\eta/s \approx T \lambda_f c_s \tag{7.49}$$

where T is the temperature, λ_f the mean free path, and c_s is the sound speed derived from the partonic matter EOS. A fit by the perfect fluid "Buda-Lund" model [168] to the scaled v_2 data shown in Fig. 7.40 yields $T = 165 \pm 3$ MeV; c_s is estimated as 0.35 ± 0.05 [94, 162], and $\lambda_f \approx 0.30$ fm taken from a parton cascade calculation including $2 \leftrightarrow 3$ scattering [169]. The overall result is [94]

$$\eta/s = 0.09 \pm 0.02 \tag{7.50}$$

in agreement with former estimates of Teaney and Gavin [170]. This value is very small and, in fact, close to the universal lower bound of $\eta/s = 1/4\pi$ recently derived field theoretically [171].

Elliptic flow measurements thus confirm that the quark-gluon matter produced as $\sqrt{s} \rightarrow 200$ GeV is to a good approximation in local thermal equilibrium up to about 3–4 fm/c. In addition, the final hadron mass dependence of the flow pattern is consistent with a universal scaling appropriate for a nearly non-viscous hydrodynamic flow of partons, and the observed v_2 signal reflects a primordial equation of state that is consistent with first numerical QCD computations [153, 154, 165] of a strongly coupled quark-gluon plasma (sQGP) state. First estimates of its proper shear viscosity to entropy ratio, η/s, are emerging from systematic analysis of the elliptic flow signal. At lower $\sqrt{s}$ precursor elliptic flow phenomena are observed, as well, but are more difficult to analyze as the crucial, new feature offered by top RHIC energies is missing here: a clear cut separation in time, between primordial formation of local partonic equilibrium conditions, and hadronization. At RHIC (and at future LHC) energy elliptic flow systematics thus captures the emerging quark-gluon state of QCD at energy densities in the vicinity of $\epsilon = 6$–15 GeV/fm^3, at temperature $T \approx 300$ MeV, and $\mu_B \rightarrow 0$, describing it as a strongly coupled, low viscosity liquid medium.

7.5 In-medium Attenuation of High p_T Hadronand Jet Production

In the preceding sections we have followed the dynamical evolution of bulk matter in A+A collisions (at $p_T \leq 2$ GeV which covers about 95% of the hadronic output), from initial formation time of partonic matter which reflects in charged particle transverse energy and multiplicity density, also giving birth to hadrons, and to the elliptic expansion flow signal.

An alternative approach toward QCD plasma diagnostics exploits the idea [41, 172] of implanting partonic products of primordial high Q^2 processes into the evolving bulk medium, that could serve as "tracers" of the surrounding, co-traveling matter. The ideal situation, of being able to scatter well defined partons, or electrons, from a plasma fireball, is approximated by employing primordially formed charm-anticharm quark pairs [41], or leading partons from primordial di-jet production [172]. Both processes are firmly anchored in perturbative QCD and well studied in elementary collisions (where such partons are directly released into vacuum), which thus serve as a reference in an analysis that quantifies the in-medium modification of such tracer partons. Not a surprise, in view of our above inferences, from elliptic flow, of a high temperature, strongly coupled primordial medium: these in-medium modifications are quite dramatic, leading to a suppression of J/Ψ production from primordial $c\bar{c}$ pairs (Sect. 7.6), and to high p_T hadron and jet quenching, the subject of this chapter.

7.5.1 High p_T Inclusive Hadron Production Quenching

At top RHIC energy, $\sqrt{s} = 200$ GeV, di-jet production from primordial hard pQCD parton-parton collisions (of partons from the initial baryonic structure functions) is the source of "leading" partons, with E_T up to about 30 GeV. They are derived from the inclusive cross section arising if the A+A collision is considered, first, as an *incoherent* superposition of independent nucleon-nucleon collisions, as enveloped within the target-projectile nucleon densities. In this framework, the pQCD cross section for producing an E_T parton in A+B takes the form of "factorization" [173]

$$\frac{d\sigma}{dE_T \, dy} = \sum_{a,b} \int_{x_a} dx_a \int_{x_b} dx_b \, f_{a/A}(x_a) f_{b/B}(x_b) \frac{d\sigma_{ab}}{dE_T \, dy} \qquad (7.51)$$

where the $f(x)$ are the parton distributions inside projectile A and target B nuclei, and the last term is the pQCD hard scattering cross section. This equation describes the primordial production rate of hard partons, leading to the conclusion [172] that, at RHIC energy, all hadrons at $p_T \geq 6$–10 GeV should arise from initial pQCD parton production.

As partons are effectively frozen during the hard scattering, one can treat each nucleus as a collection of free partons. Thus, with regard to high p_T production, the density of partons within the parton distribution function of an atomic number A nucleus should be equivalent to the superposition of A independent nucleons N:

$$f_{a/A}(x, Q^2) = A \, f_{a/N}(x, Q^2). \qquad (7.52)$$

From Eqs. (7.51) and (7.52) it is clear that the primordial high Q^2 inclusive parton cross section in a minimum bias A+B reaction scales as $A \cdot B$ times the corresponding (N+N or) p+p cross section. Furthermore, as each leading parton ends up in an observed high p_T hadron h, we thus write the invariant hard hadron cross section as

$$E \, d\sigma_{AB \to h}/d^3 p = A \cdot B \cdot E \, d\sigma_{pp \to h}/d^3 p. \qquad (7.53)$$

Since nucleus-nucleus experiments usually measure invariant yields N_h for a given centrality bin, corresponding to an average impact parameter b, one writes instead:

$$E \, dN_{AB \to h}(b)/d^3 p = \langle T_{AB}(b) \rangle \, E \, d\sigma_{pp \to h}/d^3 p, \qquad (7.54)$$

where $T_{AB}(b)$ is the Glauber geometrical overlap function of nuclei A, B at impact parameter b, which accounts for the average number of participant parton collisions at given impact geometry [174], $\langle N_{coll}(b) \rangle$. One can thus quantify the attenuating medium effects, as experienced by the primordially produced tracer parton on its way toward hadronization, by the so-called *nuclear modification factor* for hard collisions (analogous to Eq. (7.6), that refers to soft, bulk hadron production):

$$R_{AB}(p_T, y, b) = \frac{d^2 N_{AB}/dy \, dp_T}{\langle T_{AB}(b) \rangle \, d^2 \sigma_{pp}/dy \, dp_T}. \qquad (7.55)$$

Obviously, this concept of assessing the in-medium modification of hadron production at high p_T requires corresponding p+p collision data, as a reference basis. Such data have been, in fact, gathered at top RHIC, and top SPS energies, $\sqrt{s} = 200$ and 17.3 GeV, respectively. Alternatively, in situations where the relevant reference data are not known, one considers the production ratio of hadronic species h, observed in central relative to peripheral collisions:

$$R_{CP}(p_T, y) = \frac{d^2 N_h(b_1)/dy \, dp_T}{d^2 N_h(b_2)/dy \, dp_T} \times \frac{\langle T_{AB}(b_2) \rangle}{\langle T_{AB}(b_1) \rangle} \qquad (7.56)$$

where $b_1 \ll b_2$ are the average impact parameters corresponding to the employed trigger criteria for "central" and "peripheral" A+A collisions, respectively. This ratio recasts, to a certain extent, the in-medium attenuation analysis, offered by R_{AB}, insofar as peripheral collisions approach the limiting conditions, of a few single nucleon-nucleon collisions occurring in the dilute surface sections of the nuclear density profiles, i.e. essentially in vacuum.

Employing the above analysis schemes, the RHIC experiments have, in fact, demonstrated a dramatic in-medium suppression of the high p_T yield, for virtually all hadronic species. Figure 7.41 shows R_{AA} for neutral pions produced in min. bias Cu+Cu and Au+Au collisions at $\sqrt{s} = 200$ GeV where PHENIX extended the p_T range up to 18 GeV/c [175]; the nuclear modification factor refers to the range

Fig. 7.41 The nuclear modification factor R_{AA} for π^0 in min. bias Cu+Cu and Au+Au collisions at $\sqrt{s} = 200$ GeV, in the range $p_T > 7$ GeV/c [175], plotted vs. centrality as measured by participant nucleon number N_{part}

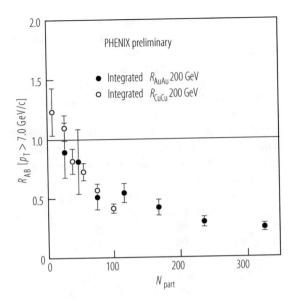

Fig. 7.42 R_{AA} for π^0 production in central Au+Au collisions at $\sqrt{s} = 200$ GeV [175], compared to a hydrodynamic calculation for different opacities (transport coefficients [195]) of the plasma [204]

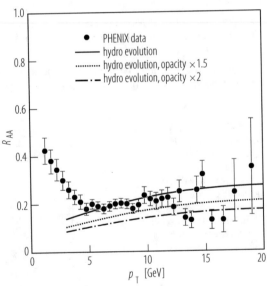

$p_T > 7.0$ GeV/c and is shown as a function of centrality, reflected by N_{part}. We infer a drastic suppression, by an $R_{AA} \, approx \, 0.25$ in near central collisions. Figure 7.42 shows the p_T dependence [175] of neutral pion R_{AA} in central Au+Au collisions ($N_{part} = 350$), with a suppression to below 0.2 at $p_T \geq 4$ GeV/c.

Note that R_{AA} *cannot reach arbitrarily small values* because of the unattenuated contribution of quasi-in-vacuum surface "corona" single nucleon-nucleon collisions, closely resembling the $p + p \rightarrow \pi^0 + X$ inclusive yield employed in the denominator of R_{AA}, Eq. (7.55). Even in a central trigger configuration this suggests a lower bound, $R_{AA} \approx 0.15$. Nuclear attenuation of high p_T pions thus appears to be almost "maximal", implying a situation in which the interior sections of the high energy density fireball feature a very high opacity, i.e. they are almost "black" to the high p_T partons contained in pions.

We remark here, briefly, on a confusing feature that occurs in Fig. 7.42 (and in the majority of other R_{AA} vs. p_T plots emerging from the RHIC experiments): the pQCD number of collision scaling employed in R_{AA} does *not* describe pion production at low p_T as we have demonstrated in Sect. 7.2. The entries at $p_T \leq$ 3 GeV/c are thus besides the point, and so are pQCD guided model predictions, as shown here [176].

Nuclear modification analysis at RHIC covers, by now, the high p_T production of a multitude of mesonic and baryonic species [177], most remarkably even including the charmed D meson which is measured via electrons from semi-leptonic heavy flavor decay by PHENIX [175] and STAR [178]. Figure 7.43 illustrates the first PHENIX results from central Au+Au at $\sqrt{s} = 200$ GeV, R_{AA} falling to about 0.3 at 5 GeV/c. Heavy flavor attenuation thus resembles that of light quarks, as is also attested by the predictions of in-medium parton transport models [179, 180] included in Fig. 7.43, which cast the medium opacity into an effective parton transport coefficient $\hat{q}$ which is seen here to approach a value 14 GeV²/c at high p_T, again corresponding to a highly opaque medium. We shall describe this approach in more detail below.

A most fundamental cross-check of the in-medium attenuation picture of color charged partons consists in measuring R_{AA} for primordial, "direct" photons.

Fig. 7.43 The nuclear modification factor R_{AA} vs. p_T for electrons from semi-leptonic decays of heavy flavor (mostly D) mesons in central Au+Au collisions at $\sqrt{s} = 200$ GeV [175]; with calculations of in medium energy loss using different attenuation models [179, 180]

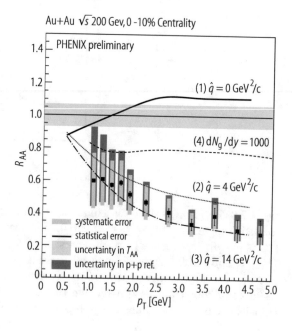

Figure 7.44 shows first PHENIX results [181] for direct photon R_{AA} vs. p_T in central Au+Au at $\sqrt{s} = 200\,\text{GeV}$. *Photons obey pQCD number of collisions scaling,* $R_{AA} \approx 1$! Also included in Fig. 7.44 are the attenuation ratios for neutral pions (already shown in Fig. 7.42), and for η mesons that follow the pattern of extreme suppression. In essence, the PHENIX results in Figs. 7.43 and 7.44 wrap up all one needs to know for a theoretical analysis of fireball medium opacity, for various flavors, and indicate transparency for photons.

We turn, briefly, to R_{CP} as opposed to R_{AA} analysis, in order to ascertain similar resulting conclusions at RHIC energy. Figure 7.45 illustrates an R_{CP} analysis of π, p and charged hadron high p_T production in Au+Au at $\sqrt{s} = 200\,\text{GeV}$ by STAR

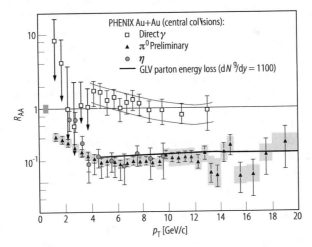

Fig. 7.44 Modification factor R_{AA} vs. p_T for direct photons in central Au+Au at 200 GeV (squares). Also shown are R_{AA} for π^0 (triangles) and η (circles) [181] fitted by the attenuation model [176, 180]

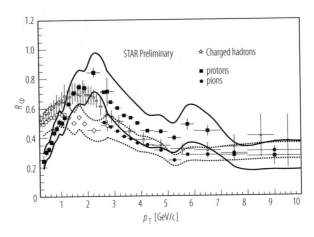

Fig. 7.45 R_{CP}, the ratio of scaled π, p and charged hadron yields vs. p_T in central (5%) and peripheral (60–80%) Au+Au collisions at $\sqrt{s} = 200\,\text{GeV}$ [182]

[182], the central to peripheral yield ratio referring to a 5%, and a 60–80% cut of minimum bias data. We showed in Eq. (7.56) that the R_{CP} measure also refers to a picture of pQCD number of binary collision scaling, inappropriate at low p_T. Thus ignoring the features of Fig. 7.45 at $p_T \leq 3$ GeV/c we conclude that the high p_T data again suggest a suppression by about 0.3, common to pions and protons, thus approaching the ratio, of about 0.2, observed in Figs. 7.41 and 7.42 which employ the "ideal" in-vacuum $p + p \rightarrow$ hadron $+ X$ reference.

At top SPS energy, $\sqrt{s} = 17.3$ GeV, the experimentally covered p_T range is fairly limited [183], $p_T < 4$ GeV/c. Figure 7.46 shows NA49 results, R_{CP} for p and charged pions. Contrary to former expectations that such data would be overwhelmed by Croonin-enhancement [184] the same systematic behavior as at RHIC is observed, qualitatively: R_{CP} (baryon)$>$ R_{CP} (meson) at $p_T > 3$ GeV/c. Note that, again, the data do not approach unity at $p_T \rightarrow 0$ because of the employed binary scaling, and that the strong rise of the proton signal at $p_T < 2$ GeV/c is largely the result of strong radial flow in central Pb+Pb collisions, an effect much less prominent in pion p_T spectra. The high p_T suppression is much weaker than at RHIC but it is strong enough that the expected Croonin enhancement of high p_T mesons is not observed. These SPS data, as well as first results obtained at the intermediate RHIC energy of $\sqrt{s} = 62.4$ GeV [185] are reproduced by an attenuation model based on the primordial gluon density $d N_g/d y$ that scales as the charged particle midrapidity density $d N_{ch}/d y$ [176], and was also employed in Figs. 7.43 and 7.44.

Fig. 7.46 R_{CP} results from SPS Pb+Pb collisions at $\sqrt{s} = 17.3$ GeV, for pions and protons [183], with attenuation model fits [176]

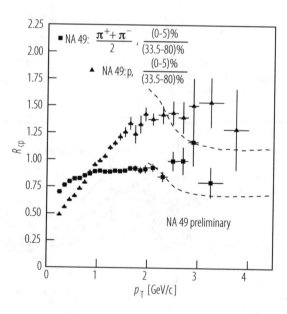

7.5.2 Energy Loss in a QCD Medium

The attenuation model that we have hinted at consists of a gluon radiative energy loss theory of the primordially produced leading, high p_T parton as it traverses a medium of color charges, by means of emission of gluon bremsstrahlung. We expect that the resulting partonic specific energy loss, per unit pathlength (i.e. its dE/dx) should reflect characteristic properties of the traversed medium, most prominently the spatial density of color charges [176] but also the average momentum transfer, per in-medium collision of the considered parton or, more general, per unit pathlength at constant density. Most importantly, an aspect of non-abelian QCD leads to a characteristic difference from the corresponding QED situation: the radiated gluon is itself color-charged, and its emission probability is influenced, again, by its subsequent interaction in the medium [186] which, in turn, is proportional to medium color charge density and traversed pathlength L. Thus, the traversed path-length L in-medium occurs, both in the probability to emit a bremsstrahlung gluon, and in its subsequent rescattering trajectory, also of length L, until the gluon finally decoheres. Quantum mechanical coherence thus leads to the conclusion that non-abelian dE/dx is not proportional to pathlength L (as in QED) but to L^2 [187].

This phenomenon occurs at intermediate values of the radiated gluon energy, ω, in between the limits known as the Bethe-Heitler, and the factorization regimes [186],

$$\omega_{BH} \approx \lambda\, q_T^2 \ll \omega \ll \omega_{fact} \approx L^2\, q_T^2 / \lambda \le E \tag{7.57}$$

where λ is the in-medium mean free path, q_T^2 the (average) parton transverse momentum square, created per collision, and E the total cm energy of the traveling charge. In the BDMPSZ model [186–188] the properties of the medium are encoded in the transport coefficient, defined as the average induced transverse momentum squared per unit mean free path,

$$\hat{q} = \left\langle q_T^2 \right\rangle / \lambda. \tag{7.58}$$

The scale of the radiated gluon energy distribution $\omega\, dN/d\omega$ is set by the characteristic gluon energy [186, 187]

$$\omega_c = \frac{1}{2}\, \hat{q}\, L^2. \tag{7.59}$$

To see more explicitly how the various properties of the color charged medium enter in $\hat{q}$ we rewrite it as

$$\hat{q} = \rho \int q_T^2\, dq_T^2\, \frac{d\sigma}{dq_T^2} \equiv \rho\sigma \left\langle q_T^2 \right\rangle = \lambda^{-1} \left\langle q_T^2 \right\rangle \tag{7.60}$$

where ρ is the color charge density of scattering centers, σ the effective binary cross section for interaction of the considered leading parton at scale q^2 (which may depend on quark flavor), and $\langle q_T^2 \rangle$ as above. Obviously, both σ and $\langle q_T^2 \rangle$ refer to detailed, intrinsic properties of the QCD medium, globally described by density ρ. The leading parton cross section with in-medium color charges should depend on the implied resolution scale, $Q^2 = \langle q_T^2 \rangle$, and can thus be obtained from perturbative QCD [186–188] only if $Q^2 > Q_{sat}^2$, the saturation scale that we discussed in Sect. 7.2. Likewise, $\langle q_T^2 \rangle$ itself reflects a medium property, the effective range of the color force, which is different in confined and deconfined media. Hadron size limits the force range in the former case, such that $\hat{q}$ is minimal in ground state hadronic matter also, of course, due to the small energy density $\rho = \rho_0 = 0.15$ GeV/fm^3 [189]. This was, in fact, confirmed by a RHIC run with deuteron-gold collisions, in which mid-rapidity hadrons traverse sections of cold Au nucleus spectator matter. Figure 7.47 shows results obtained for R_{dA} dependence on p_T, for π^0 from PHENIX [190], and for charged hadrons from STAR [191]. For comparison, both panels also include the corresponding R_{AA} data for central Au+Au collisions (all at $\sqrt{s} = 200$ GeV/c), exhibiting the typical, drastic high p_T quenching of inclusive hadron production, clearly absent in d+Au collisions.

We have shown a first application of the BDMPSZ model, to RHIC inclusive D meson production [179], in Fig. 7.43. Before engaging in further model application we note, first, that Eqs. (7.57)–(7.60) above refer to the idealized conditions of an infinitely extended medium of uniform composition. In reality, the fireball medium expands, at time scale concurrent with the proper time incurred in the leading partons propagation over distance L, such that all ingredients in $\hat{q}$, exhibited in Eq. (7.60), vary with expansion time [192]. However, before turning to adaption to reality of the infinite matter model, we wish to expose its prediction for the final connection of the specific average partonic energy loss $\langle \Delta E \rangle$ and in-medium path

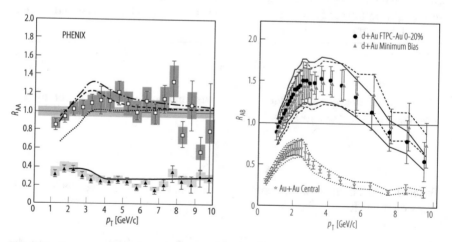

Fig. 7.47 R_{AA} vs. p_T for d+Au collisions at $\sqrt{s} = 200$ GeV, compared to central Au+Au results, for π^0 (left) and charged hadrons (right). From [190, 191]

length L traversed:

$$\langle \Delta E \rangle = \int \omega \frac{dN}{d\omega} \, d\omega \propto \alpha_s \, C_R \, \hat{q} \, L^2. \tag{7.61}$$

This relation [193] represents the eikonal approximation limit of extended medium and high leading parton initial energy $E > \omega_c$ (Eq. (7.59)). The average energy loss is thus proportional to the appropriate strong coupling constant α_s, to the Casimir factor corresponding to the leading parton (with value 4/3 for quarks, and 3 for gluons), as well as to $\hat{q}$ and L^2.

In order to verify the non-abelian nature of radiative parton energy loss in a partonic QCD medium it would, of course, be most convincing if a direct, explicit L^2 dependence of $\langle \Delta E \rangle$ could be demonstrated. Such a demonstration is lacking thus far, chiefly because of the obvious difficulty of simultaneously knowing the primordial parton energy E, the transport coefficient $\hat{q}$ and—in finite nuclear collision geometry—its variation along the actual traversed path L as the surrounding medium expands with propagation time. Moreover, the partonic medium induced energy loss ΔE of the primordial parton is not directly observable. Even if we assume that high p_T partons evolve into the observed hadrons only *after* leaving the fireball medium [176], their ensuing "fragmentation" to hadrons (which is known from $p + p$ jet physics) results in several hadrons usually comprising a "leading" hadron which transports a major fraction $\langle z \rangle \equiv \langle E^h/E^p \rangle$ of the fragmenting parton energy E^p, which, in turn, equals E^p (primordial)—ΔE, with ΔE sampled from a probability distribution with mean $\langle \Delta E \rangle$ according to Eq. (7.61). The observed leading hadron energy or transverse momentum is thus subject to sampling, both, z from the fragmentation function, and ΔE from in-medium energy loss. Finally, inclusive high p_T leading hadron observation in A+A collisions involves an average over all potential initial parton production points, within the primordially produced density profile. A specific distribution of in medium path lengths $f(L)$ arises, for each such production point, which, moreover, depends on a model of space-time fireball expansion. The final inclusive yield thus requires a further, weighted volume average over $f(L)$ per production point. Thus, typical of an inclusive mode of observation, the "ideal" relationship of Eq. (7.61), between radiative in-medium energy loss ΔE and traversed path length L gets shrouded by double averages, independently occurring at either side of the equation [176, 179, 189, 194–196]. A detailed L^2 law verification cannot be expected from inclusive central collision data *alone* (see next section).

However, the unmistakably clear signal of a strong, in-medium high p_T parton quenching effect, gathered at RHIC by R_{AA} measurement for a multitude of hadronic species (Figs. 7.20, 7.41, 7.42, 7.43, 7.45, and 7.47), in Au+Au collisions at $\sqrt{s} = 200$ GeV, has resulted in first estimates of the transport coefficient $\hat{q}$, the medium—specific quantity entering Eq. (7.61), in addition to the geometry—specific path length L. In fact, the transport coefficient can, to some extent, be analyzed independently, owing to the fact that $\hat{q} \propto \varrho$ from Eq. (7.60). The density ϱ falls down rapidly during expansion, but it is initially rather well constrained by the conditions of one-dimensional Bjorken expansion that we have described in Sects. 7.2 and 7.4. *The major contribution to partonic ΔE arises in the early expansion phase (via a high $\hat{q}$), in close analogy to the formation of the elliptic flow signal.* These two signals are, thus, closely correlated: the primordial hydrodynamic expansion phase of bulk matter evolution sets the stage for the attenuation, during this stage of QCD matter, of primordially produced "tracer" partons, traversing the bulk matter medium as test particles.

The bias in partonic ΔE to the primordial expansion period is borne out in an expression [193, 195] which replaces the $\hat{q}$ coefficient, appropriate to an infinitely extended static medium considered in the original BDMPSZ model, by an effective, expansion time averaged

$$\hat{q}_{\text{eff}} = \frac{2}{L^2} \int_{t_0}^{L} dt \, (t - t_0) \, \hat{q} \, (t) \qquad (7.62)$$

to be employed in the realistic case of an expanding fireball source. Due to the rapid fall-off of ϱ, in $\hat{q} = \varrho \sigma \langle q_T^2 \rangle$ from Eq. (7.60), the integral depends, far more strongly, on $\hat{q}$ ($t \approx t_0$) than on total path length L. Furthermore, inserting $\hat{q}_{\text{eff}}$ into the BDMPSZ formula [193, 195] for the transverse momentum downward shift, occurring in leading parton or hadron p_T spectra (of power law form $p_T^{-\nu}$, see Fig. 7.20)

$$\Delta p_T \approx -\alpha_s \sqrt{\pi \hat{q} L^2 p_T / \nu}, \qquad (7.63)$$

we see that the first order proportionality to L^2 is removed. The downward p_T shift is thus, primarily, a consequence of $\hat{q}_{\text{eff}}$ which, in turn, is biased to reflect the "ideal" transport coefficient $\hat{q}$ at early evolution time. Within this terminology, the p_T shift (see Eq. (7.63)) determines the experimentally measured ratio $R_{AA}(p_T)$ which quantifies the effective transport coefficient $\hat{q}_{\text{eff}}$ for the p_T domain considered. It can be related, as a cross check, to the initial gluon rapidity density if the collision region expands according to Bjorken scaling [187, 197]:

$$\hat{q} = \alpha_s \frac{2}{L} R_A^{-2} \frac{dN^g}{dy}. \qquad (7.64)$$

A typical result of application of the model described above [195] is shown in Fig. 7.48. Analogous to Fig. 7.41, R_{AA} for neutral pions and charged hadrons is averaged over the range $4.5 \leq p_T \leq 10$ GeV/c, and shown as a function of centrality (assessed by N_{part}) in minimum bias Au+Au collision at $\sqrt{s} = 200$ GeV [190, 191, 198]. A path-averaged $\hat{q}_{\text{eff}}$ of 14 GeV^2/fm is inferred from the fit, in close agreement to the value found in [195].

A more recent study [199] of the PHENIX R_{AA} data for π^0 in central Au+Au collisions (Fig. 7.42) is shown in Fig. 7.49. The analysis is carried out in the framework of the WHDG model [200], which replaces the (effective) transport coefficient $\hat{q}$ (employed in the BDMPSZ model [186–188], and turned into the data analysis

Fig. 7.48 The effective transport coefficient $\hat{q} = 14$ GeV^2/fm in the parton quenching model (PQM) of [195] determined from the centrality dependence of R_{AA} for π^0 and charged hadrons, averaged over 4.5 $\leq p_T \leq 10$ GeV/c, in Au+Au at $\sqrt{s} = 200$ GeV

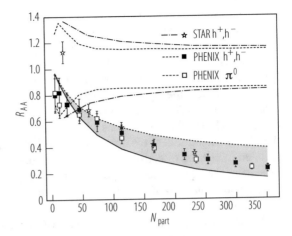

Fig. 7.49 Application of the WDHG transport model [200] based on Eq. (7.64) to PHENIX R_{AA} data for π^0 [175], indicating primordial $1000 \leq (\mathrm{d}\,N^g)/(\mathrm{d}\,y) \leq 2000$

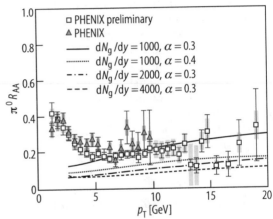

formalism of [193, 195]) by the primordial gluon mid-rapidity density $\mathrm{d}N^g/\mathrm{d}y$, as the fundamental parameter, via Eq. (7.64). The initial gluon density, in turn, being related to the charged hadron mid-rapidity density [61]. Figure 7.49 shows that, within the still preliminary statistics at $p_T > 10$ GeV/c, the "conservative" estimate of $\alpha_s = 0.3$ and $\mathrm{d}N^g/\mathrm{d}y = 1000$ does not appear to be the most appropriate choice, the data rather requiring $1000 < \mathrm{d}N^g/\mathrm{d}y < 2000$. Overall, Fig. 7.49 demonstrates a certain insensitivity of the data, to the basic parameters of theoretical high p_T quenching models of inclusive hadron production at RHIC energy, that we have already inferred from Fig. 7.43, concerning choice of $\hat{q}$.

Radiative in-medium energy loss of primordially produced partons, traversing the evolving bulk medium as "tracers", must be extremely strong. This is indicated by the inclusive attenuation ratios R_{AA}, which fall down to about 0.2 in central collisions thus almost reaching the absolute lower limit of about 0.15 that arises from the unavoidable fraction of un-attenuated primordial surface "corona" nucleon-nucleon interaction products [201].

Not expected from early estimates of medium opacity based on the picture of a weakly coupled perturbative QCD medium [172, 202], the interior sections of the central Au+Au interaction volume must be almost "black" toward high p_T parton transport [194, 195] at $\sqrt{s} = 200$ GeV, also including charm quark propagation (Fig. 7.43). The remaining signal should thus stem, primarily, from the dilute surface sections, and from the finite fraction of partons traversing the interior with small, or zero radiative energy loss, as a consequence of the finite width of the ΔE probability distribution [193, 194]. Seen in this light, the smooth decrease of R_{AA} with centrality (Fig. 7.48) should reflect the combined effects, of a decreasing surface to volume ratio, an increasing effective $\hat{q}$ (due to interior density increase) that confronts the increasing average geometrical potential path length $\langle L \rangle$ (essentially enhancing its effect), and a thus diminishing fraction of primordial high p_T partons experiencing a small ΔE.

Not surprisingly, the ideal non abelian QCD relationship of ΔE proportional to in-medium high p_T parton path length L^2 can, thus, not be established from inclusive high p_T quenching data alone. We shall show in the next section that di-jet primordial production can offer a mechanism of higher resolution. The inclusive R_{AA} attenuation data, obtained at RHIC, are seen to establish an unexpected, high opacity of the primordial interaction volume, extending to high p_T parton propagation. The required, high transport coefficient $\hat{q} = 14$ GeV^2/fm from Fig. 7.48, confirms and extends the picture derived from elliptic flow data [167]: at top RHIC energy the plasma is non-perturbatively, strongly coupled, a new challenge to lattice QCD [61]. The QGP may be largely composed of colored, string-like partonic aggregates [203].

7.5.3 Di-jet Production and Attenuation in A+A Collisions

In order to analyze leading parton attenuation in a more constrained situation [204], one investigates parton tracer attenuation under *further geometrical constraints* concerning the in-medium path length L, by means of di-jet analysis, and/or by varying the primordial parton density that enters $\hat{q}$ via Eq. (7.60) in studies at different $\sqrt{s}$ while maintaining the observational geometrical constraints.

We shall concentrate here on di-jet attenuation data obtained in Au+Au collisions at top RHIC energy, $\sqrt{s} = 200$ GeV. At this relatively modest energy the initial pQCD production cross section of leading partons (as described in Eq. (7.51)) reaches up to $p_T = 25$ GeV/c. The ensuing DGLAP shower multiplication initiates "parton fragmentation" to hadrons [83, 85, 121], each carrying a momentum fraction $z_T = p_T/p_T$ (primord. parton). The created ensemble of hadrons h belonging to the observed hadronic jet can be wrapped up by the total *fragmentation function*

$$F_h(z, \sqrt{s}) = \sum_i \int \frac{dz}{z} C_i(z, \sqrt{s}) \, D_{\text{part} \rightarrow h}(z, \sqrt{s}) \tag{7.65}$$

which summarizes the contributions arising from the different shower partons i. Here, C_i are the weight coefficients of the particular process, and $D_{\text{part}(i)\to h}$ are the individual fragmentation functions (FFs) for turning parton i into hadron h. Similar to the parton distribution functions (PDFs) in Eq. (7.51), derived from deep inelastic electron-parton scattering (DIS) and shown in Fig. 7.14, the FFs are semi-empirical non-perturbative QCD functions that have an intuitive probabilistic interpretation. They quantify the probability that the primordial parton produced at short distance $1/Q$ fragments into i shower partons, forming a jet that includes the hadron h [205, 206].

At Fermilab energy, $\sqrt{s} = 1.8$ TeV, the jet spectrum reaches up to $E_T \approx 400$ GeV, and a typical 100 GeV jet comprises about 10 hadrons which can be identified above background by jet-cone reconstruction algorithms [205]. This allows for a complete determination of the corresponding fragmentation function, and for a rather accurate reconstruction of the p_T and E_T of the primordial parton that initiated the jet. Similar conditions will prevail in jet spectroscopy of Pb+Pb collisions at LHC energy, $\sqrt{s} = 5.5$ TeV.

However, at RHIC energy a typical jet at $15 \leq E_T \leq 25$ GeV features a fragmentation function comprised of a few hadrons with E_T in the 2–15 GeV range. Considering the high background, arising in the lower fraction of this energy domain from concurrent, unrelated high p_T hadron production processes, a complete jet-cone analysis cannot succeed. The RHIC experiments thus confront back-to-back di-jet production with an analysis of the azimuthal correlation between high p_T hadrons. Defining the observational geometry by selecting a high p_T "trigger" hadron observed at azimuthal emission angle φ_{trig}, the associated production of high p_T hadrons is inspected as a function of $\Delta\varphi = \varphi_{\text{ass}} - \varphi_{\text{trig}}$. If the trigger has caught a leading jet hadron one expects the hadrons of the balancing back-to-back jet to occur at the side opposite to the trigger, $\Delta\varphi \approx \pi$. The trigger condition thus imposes the definition of a "near-side" and an "away side" azimuthal domain. Furthermore, the relatively narrow rapidity acceptance of the STAR and PHENIX experiments (centered at $y = 0$) selects di-jets with axis perpendicular to the beam direction.

Originating from a uniform distribution of primordial back-to-back di-parton production vertices, throughout the primordial reaction volume, the trigger selected di-jet partons thus experience an (anti-)correlated average path length $\langle L \rangle$ to arrive at the surface while experiencing medium-specific attenuation, with $\langle L_{\text{trig}} \rangle \approx 2R - \langle L_{\text{away}} \rangle$, R being the transverse medium radius. No such geometric constraint exists in the study of inclusive high p_T hadron production. We thus expect information different from the inclusive $R_{\text{AA}}(p_T)$ signal. The geometrical selectivity can be even further constrained by fixing the direction of the impact parameter (i.e. the reaction plane) in semi-central collisions (recall Sect. 7.4), and observing the di-jet correlation signal in dependence of the di-jet axis orientation relative to the reaction plane.

The very first di-hadron correlation measurements confirmed the existence of strong in-medium attenuation. Figure 7.50 shows the azimuthal yield distributions, per trigger hadron, as observed by STAR at $\sqrt{s} = 200$ GeV [207]. The left panel

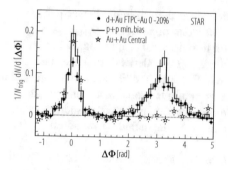 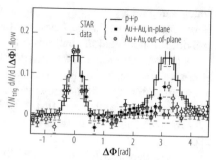

Fig. 7.50 Di-hadron correlation from back-to-back di-jet production in Au+Au collisions at $\sqrt{s} =$ 200 GeV. The trigger particle is at azimuth $\Phi = 0$, with a $p_T > 4$ GeV threshold. The away side peak at $\Delta\Phi = \pi$ is observed (left panel) in p+p, d+A but absent in central Au+Au. Right panel shows the correlation in Au+Au for different orientations of the trigger direction relative to the reaction plane [207]

shows the distribution of hadrons with $p_T \geq 2$ GeV/c relative to a trigger hadron with $p_T^{trig} \geq 4$ GeV/c. Data for p+p, d+Au and central Au+Au are illustrated. At the "near side" (the trigger position is $\Phi = 0$) all three reactions exhibit a similar narrow distribution of hadrons associated with the trigger hadron, typical of a jet cone fragmentation mechanism. Note that the associated near-side central Au+Au signal thus exhibits no signs of an attenuation softened fragmentation function, indicating that the trigger imposed high p_T hadron should predominantly stem from primordial jet production vertex points located near to the surface of the reaction volume, in azimuthal trigger direction. Thus, conversely, the balancing opposite jet has to develop while traversing almost the entire transverse diameter of the interaction volume. I.e. $\langle L_{oppos} \rangle \approx 2\,R$ thus emphasizing the expectation that di-jet spectroscopy should allow for stricter constraints on path length L in comparison to single high p_T hadron R_{AA} analysis. In fact, no trigger related away side signal of $p_T > 2$ GeV/c hadrons is observed in Fig. 7.50 for central Au+Au collisions, whereas p+p and central d+Au collisions exhibit a clear away-side di-jet signal.

We conclude that the trigger bias, selecting a single near side hadron of $p_T \geq 4$ GeV/c in central Au+Au collisions, responds to a primordial di-jet of about 10 GeV per back-to-back parton. After traversal of in medium average path length $L \to 2\,R$ the fragmentation function of the opposite side parton contains on average no hadron at $p_T > 2$ GeV/c, indicating that it should have lost a fraction $\langle \Delta E_T \rangle \geq 5$ GeV. The medium is thus highly opaque, but the total disappearance of the opposite side signal can only provide for a lower limit estimate of $\langle \Delta E_T \rangle$, within the trigger conditions employed here. We shall show below that the situation changes with more recent RHIC data [208] that extend the trigger hadron p_T range toward 20 GeV/c.

However, the right hand panel of Fig. 7.50 shows that an improved constraint on partonic in-medium path length can already be obtained by studying the di-jet back-to-back production geometry in correlation with the orientation of the reaction plane that arises from non-zero impact parameter in semi-central A+A collisions. We have seen in Fig. 7.34 that such collisions exhibit an elliptical primordial transverse density profile, with minor axis along the impact vector $\vec{b}$, defining the reaction plane. Di-jets with axis "in-plane" thus traverse a shorter in-medium path length as orthogonal "out-of-plane" jets, the difference in average path length being quantified by the spatial excentricity $\epsilon(b)$, Eq. (7.47). Figure 7.50 (right) shows the di-hadron correlation results in semi-peripheral Au+Au collisions, as compared to the in-vacuum p+p reference. At the 20–60% centrality window employed here, out of plane jet emission occurs along the major axis, the reaction volume diameter still of magnitude $2R$, as in central collisions. The trigger condition thus again selects opposite side path lengths $L \rightarrow 2\,R$, but the energy density should be lower than in central collisions. Even so, the average opacity along the away side parton path appears to be high enough to wipe out the correlation signal. In-plane geometry, however, shows a partially attenuated signal (as compared to the global p+p reference) at the opposite side, corresponding to path lengths $L \approx R$. These data thus provide for first information concerning the relation of $\langle \Delta E \rangle$ and average traversed path length [176].

Obviously, one wants to extend the above study to higher di-jet energies, i.e. to measurement of di-hadron correlations at hadron trigger $p_T \rightarrow 20$ GeV/c, conditions sampling the very low primordial cross section, at $\sqrt{s} = 200$ GeV, of jet production at primordial $E_T \rightarrow 30$ GeV [208, 209]. Figure 7.51 shows the corresponding jet correlations selected with high p_T trigger, $8 < p_T^{\text{trig}} < 15$ GeV/c, and high p_T associated hadrons, $p_T > 6$ GeV/c, in minimum bias d+Au, semi-central Au+Au and central Au+Au at $\sqrt{s} = 200$ GeV. Very clear and almost background-free back-to-back jets are observed *in all three cases*, in sharp contrast with the away side jet total disappearance in central Au+Au at lower jet energy, Fig. 7.50 (left panel). The near side trigger associated hadron yield decreases only slightly from d+A to central Au+Au, while the away side yield drops down by an

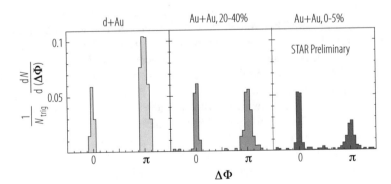

Fig. 7.51 Di-hadron correlation at high p_T in central Au+Au collisions at $\sqrt{s} = 200$ GeV, compared to d+Au and peripheral Au+Au; for $8 \leq p_T^{\text{trig}} \leq 15$ GeV, and $p_T^{\text{assoc}} > 6$ GeV. From [208]

attenuation factor of about 0.2, the signal thus *not being completely extinguished.* We infer $R_{AA}(L \rightarrow 2R) \approx 0.2$.

In order to show how such data can be evaluated in a picture of in-medium leading parton attenuation (as finally reflected in leading hadron production observed in the above di-hadron correlation data) we briefly consult the pQCD factorization [210] prediction for the inclusive production of a high p_T hadron at central rapidity, in the nuclear collision $A + B \rightarrow h + x$ [196],

$$\frac{d^3\sigma_{AB \rightarrow hx}}{d^2 p_T \, dy} = K_{NLO} \sum_{abc} \int d\vec{r} \, dx_a \, dx_b \, dz_c \, F_{a/A}(x_a, Q^2, \vec{r})$$

$$\times F_{b/B}(x_b, Q^2, \vec{b} - \vec{r}) \frac{d^3\sigma_{ab-c}}{d^2 p_{T(c)} \, dy_c}(x_a, x_b, Q^2)$$

$$\times \frac{1}{z_c^2} D_{h/c}(z_c, Q^2) \tag{7.66}$$

where the parton (a, b) distribution functions F in nucleus A, B and the elementary pQCD cross section for $a + b \rightarrow c + x$ have been already implied in Eq. (7.51). Their spatial integral gets convoluted with the fragmentation function D that describes the conversion of the leading parton c to a hadron carrying a fraction $0 < z_c < 1$ of its transverse momentum. K is a factor introduced as a phenomenological correction for "next to leading order" (NLO) QCD effects. Within the (further) approximation that the leading parton c suffers medium induced gluon bremsstrahlung energy loss but hadronizes outside the interaction volume (in vacuum), the in-medium quenching leads, merely, to a re-scaling of the fragmentation function,

$$D_{h/c}^{med} = \int d\epsilon \, P(\epsilon) \frac{1}{1 - \epsilon} D_{h/c}^{vac}\left(\frac{z_c}{1 - \epsilon}, Q^2\right), \tag{7.67}$$

where the primary parton is implied to lose an energy fraction $\epsilon = \Delta E/E_c$ with probability $P(\epsilon)$ [196]. Therefore the leading hadron is a fragment of a parton with reduced energy $(1 - \epsilon)E_c$, and accordingly must carry a larger fraction of the parton energy, $z_c/(1 - \epsilon)$. If no final state quenching is considered, $P(\epsilon)$ reduces to $\delta(\epsilon)$. The entire effect of medium attenuation on the leading parton is thus contained in the shift of the fragmentation function.

An application of this formalism [196] is shown in Fig. 7.52. The in-medium modification of the hadron-triggered fragmentation function (see Eq. (7.67)) is evaluated for central Au+Au collisions at $\sqrt{s} = 200$ GeV. In adaptation to the modalities of RHIC di-hadron correlation data, the opposite side fragmentation function (for observation of trigger-related hadrons with $p_T > 2$ GeV/c) is studied in dependence of the trigger selected p_T window. Its attenuation is quantified by the ratio D (with quenching) to D (without quenching), as a function of the fraction z_T, of opposite side hadron p_T to trigger hadron p_T. Referring to the observational conditions implied in Fig. 7.50 (left) $p_T^{trig} \approx 4$–6 GeV/c and opposite side $p_T >$

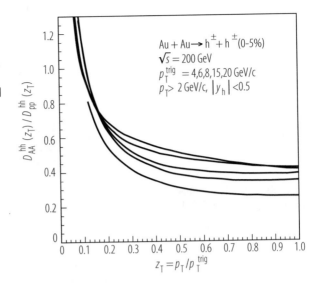

Fig. 7.52 The ratio of the hadron-triggered fragmentation function, Eq. (7.67), in central Au+Au and in p+p collisions, for different values of p_T^{trig} [196]

2 GeV, the predicted suppression amounts to a factor of about 0.35, whereas the data imply a factor smaller than 0.2. Likewise, comparing to the data of Fig. 7.51, the (harder) trigger conditions should lead to a suppression of about 0.45 from Fig. 7.52 but the observed value is close to 0.2. The predictions appear to fall short of the actually observed suppression. This calculation employs an ansatz for the transport coefficient $\hat{q}$ similar to Eq. (7.64), recurring to the primordial gluon density at $\tau_0 \leq 0.6$ fm/c which, in turn, is estimated by the charged hadron mid-rapidity density [197]. However, this pQCD based argument can also not reproduce the magnitude of the effective transport coefficient (Eq. (7.62)), $\hat{q}_{eff} \approx 10$–15 GeV^2/fm, shown in refs. [194, 195, 199] to be *required* by the large observed suppression (see Fig. 7.43).

It has been argued [199, 211] that these models need refinement by introducing more realistic dynamics, and/or by completely abandoning the pQCD ansatz for hard parton in-medium transport [212, 213]. We shall return to these novel suggestions, of how to treat dynamical, non equilibrium quantities of the "strongly coupled" parton plasma of non perturbative QCD (toward which equilibrium lattice theory can only give hints), in our final conclusion (Sect. 7.8). In the meanwhile, we note that the expected non abelian behavior, $\Delta E \propto L^2$, could not be verified quantitatively, as of yet [211], because of the unexpectedly high opacity of the fireball interior sections at $\sqrt{s} = 200$ GeV, in combination with the limited jet energy range that is available at RHIC energy. It appears possible, however, to extend the analysis of the "back side jet re-appearance" data [208, 209] (Fig. 7.51) toward this goal. The situation should improve at LHC energy, $\sqrt{s} = 5.5$ TeV, where primordial 100–200 GeV jets are abundant, such that the opposite side jet can be reconstructed with explicit use of the complete Fermilab jet cone recognition algorithms [205] even if their in-medium E_T loss ranges up to 50 GeV.

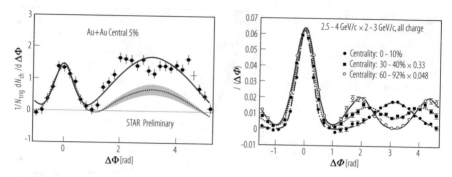

Fig. 7.53 Di-hadron correlation: away side emission pattern in central Au+Au collisions, compared to pp data by STAR [214] (left panel) and to peripheral Au+Au (right panel) by PHENIX [215]

A further prediction of the model employed in Fig. 7.52 has been confirmed by the RHIC experiments. The high medium opacity at top RHIC energy leads to an intriguing emission pattern of low p_T opposite side hadrons. Clearly, the trigger-selected E_T flux, of up to 20 GeV, toward the away-side, cannot remain unnoticeable. Inspection of the attenuated fragmentation functions [196] in Fig. 7.52 reveals an *enhanced* emission of bremsstrahlung gluon hadronization products at $z_T \leq 0.1$ This fraction of in-medium jet-degradation products has in fact been observed, as is shown in Fig. 7.53. The left panel shows STAR results [214] for the di-hadron correlation in central Au+Au at $\sqrt{s} = 200$ GeV, with near-side hadron trigger $4 < p_T < 6$ GeV/c, and opposite side observation extended to soft hadrons, $0.15 < p_T < 4$ GeV/c. A prominent double-peak structure is indicated, symmetric about $\Delta\Phi = \pi$. The right panel shows high resolution PHENIX results [215] for Au+Au at three centralities, from peripheral to central collisions. For the former, the typical p+p-like away side peak (c.f. Fig. 7.47) is recovered, while a double peak appears in the latter case, shifted from $\Delta\Phi = \pi$ by $\pm\delta\Phi \approx 70^0$. A hypothetical mechanism comes to mind [213], of sideward matter acceleration in a "Mach-cone" directed mechanism of compressional shock waves initiated by the in-medium energy loss of the opposite side leading jet parton, which traverses the medium at "super-sonic" velocity, i.e. at $v > v_s$, the appropriate speed of sound in a parton plasma.

If confirmed by pending studies of the away-side multi-hadron correlation, that might ascertain the implied conical shape of the soft hadron emission pattern (about the direction $\Delta\Phi \approx \pi$ of the leading parton), this mechanism might lead to the *determination of the sound (or shock wave) velocity* of a strongly coupled parton plasma: a third characteristic QGP matter property, in addition to viscosity η (from elliptic flow) and $\hat{q}$ (from high p_T parton attenuation). We note that the implied concept, of measuring the shock wave transport velocity of a strongly interacting medium, dates back to the 1959 idea of Glassgold et al. [216], to study shock wave emission in central p+A collisions at AGS and PS proton energy, of about 30 GeV. In hindsight we can understand the inconclusiveness of such searches: the "low energy" incident proton does not preserve its identity, breaking up into relatively soft partons in the course of traversing the target nucleus. Their coupling to the surrounding cold baryonic target matter is weak, and possible collective transverse compressional waves are swiftly dissipated by the cold, highly viscous [61] hadronic medium. The shock compression and Mach-cone emission mechanism was then revived by Greiner et al. [217] for central ^{20}Ne + U collisions at Bevalac energy [218]. The recent RHIC observations in Fig. 7.53 have been called *"conical flow"* [219]; they demonstrate, again, the strong coupling to the medium, of even a 20 GeV/c leading parton [220].

7.6   Vector Meson and Direct Photon Production: Penetrating Probes

This chapter is devoted to three observables that have been profoundly studied already at the SPS, as tracers of the created fireball medium:

1. J/Ψ "charmonium" production, as suppressed in a high T QGP medium,
2. "direct" photons as black body T sensors of the QGP state;
3. ρ meson in medium production, studied by di-lepton decay.

These three observables represent an internally connected set of ideas of high density QCD matter diagnostics: all serve as medium tracers but in a complementary way. Matsui and Satz [41] realized that at the modest top SPS energy, $17.3 < \sqrt{s} < 20$ GeV, the production rate of $c\bar{c}$ pairs (that would in part develop toward charmonium production, J/Ψ, Ψ', etc.) was so low that only the primordial, *first generation* nucleon-nucleon collisions in an A+A event would create a measurable yield (nothing after thermalization). I.e. the primordial $c\bar{c}$ yield would be well-estimated by $A^{4/3} \sigma_{pp}(c\bar{c})$. Initially produced in a superposition of states including color triplets and octets [221] the emerging $c\bar{c}$ pairs thus co-travel with the developing high energy density fireball medium, as tracers, on their way toward J/Ψ or Ψ' and D, $\bar{D}$ formation. Attenuation, characteristic of medium properties, will break up leading $c\bar{c}$ pairs resulting in a suppression of the eventually observed J/Ψ yield: another "quenching" observable.

As the suppression of J/ψ is related to medium temperature and density, the extent of charmonium quenching should also be related to the rate of black body thermal fireball radiation, by photon emission via elementary $q\bar{q} \rightarrow g\gamma$ and $qg \rightarrow q\gamma$ processes in the plasma [222, 223]. Photons leave the interaction volume essentially un-rescattered, and their radiative intensity, proportional to T^4, makes them an ideal probe of the initial fireball temperature. *This thus could be an ideal diagnostics of the early deconfined matter*, but it is difficult to disentangle from a multitude of concurrent low p_T photon sources [224], most prominently $\pi^0 \rightarrow 2\gamma$ decay, with cross sections higher by several orders of magnitude. The thermal photon signal thus becomes more promising the higher the initial temperature T_i which might reach up to 500–600 MeV at LHC energy.

Similar to photons, in medium created lepton pairs [225] escape essentially unattenuated as was shown for the Drell-Yan process, $q\bar{q} \rightarrow L\bar{L}$ by CERN experiment NA38/NA50 [226]. Thermal di-lepton production in the mass region ≤ 1 GeV is largely mediated by light vector mesons. Among these, the ρ meson is of particular interest due to its short lifetime (1.3 fm/c), making it an ideal tracer (via its in-medium di-lepton decay) for the modification of hadrons composed of *light quarks*, in the vicinity of $T = T_c$. This modification signal is thus complementary to the *heavy quark* charmonium J/Ψ suppression (break up) effect that sets in at $T \geq 1.5$–$2T_c$ (see below). Moreover, in addition to the deconfinement breakup mechanism by QCD plasma Debye screening of the color force potential acting on the $c\bar{c}$ pair [41], the QCD chiral symmetry restoration mechanism [227] can be studied via in-medium modification of the ρ spectral function as $T \rightarrow T_c$. Note that the in vacuum ρ mass and width properties are owed to non-perturbative QCD condensate structures [1, 228] which spontaneously break the chiral symmetry of the QCD Lagrangian, at $T \rightarrow 0$. These properties should change, in the vicinity of T_c, and be reflected in modifications of the di-electron or di-myon decay spectra—unlike the suppression effect on the J/Ψ which simply dissolves primordial $c\bar{c}$ pairs before they can hadronize, a yes-no-effect whose onset with $\sqrt{s}$ or centrality serves as a plasma thermometer, by observing R_{AA} or $R_{CP} < 1$.

7.6.1 Charmonium Suppression

Due to the high charm and bottom quark masses, the "quarkonium" states of $c\bar{c}$ and $b\bar{b}$ can be described in non-relativistic potential theory [229, 230], using

$$V(r) = \sigma r - \frac{\alpha}{r} \tag{7.68}$$

as the confining potential [231], with string tension $\sigma = 0.2$ GeV^2 and gauge coupling $\alpha = \pi/12$. We are interested in the states J/Ψ (3.097), χ_c (3.53) and Ψ' (3.685) which are the 1S, 1P and 2S levels. The decay of the latter two feeds into the J/Ψ, accounting for about 40% of its yield. The radii derived from Eq. (7.68) are 0.25, 0.36 and 0.45 fm, respectively, well below hadron size at least for the J/Ψ and χ_c states.

With increasing temperature, $\sigma(T)$ decreases, and at deconfinement $\sigma(T_c) = 0$. For $T \geq T_c$ we thus expect

$$V(r) = -\frac{\alpha}{r} \exp\left[-r/r_D(T)\right] \tag{7.69}$$

where $r_D(T)$ is the QCD Debye screening radius. It was initially estimated from a SU (2) gauge theory of thermal gluons [41], to amount to about 0.2–0.3 fm at

$T/T_c = 1.5$. In this picture, the screened potential (Eq. (7.69)) can still give rise to bound $c\bar{c}$ states provided their radius is smaller than r_D. The pioneering study of Matsui and Satz [41] concluded that screening in a QGP system would dissolve the J/Ψ, or its $c\bar{c}$ precursor, at $T \geq 1.3 \, T_c$ whereas the χ_c and Ψ' states would be suppressed already directly above T_c.

The corresponding energy density for J/Ψ suppression, employing

$$\epsilon/\epsilon_c \approx (T/T_c)^4 \approx 2.9 \tag{7.70}$$

(obtained with $\epsilon_c \approx 1 \, \text{GeV/fm}^3$ from lattice QCD), would thus amount to about $2.9 \, \text{GeV/fm}^3$. This motivated an extensive experimental effort at the CERN SPS Pb beam at $\sqrt{s} = 17.3$ GeV. We have seen in Sect. 7.2 that the Bjorken estimate [45] of average transverse energy density reached in central Pb+Pb collisions [43, 44] amounts to $\epsilon = (3.0 \pm 0.6) \, \text{GeV/fm}^3$, with higher ϵ to be expected in the interior fireball sections: encouraging conditions.

However, the above approach assumes the validity of a two-body potential treatment at *finite* T, near a conjectured critical point of QCD. More recently the quarkonium spectrum was calculated *directly* in finite temperature lattice QCD [232], with the striking result that the J/Ψ dissociation temperature in a realistic non-perturbative treatment of the QCD plasma state moves up to about $T = 2 \, T_c$, whereas χ_c and Ψ' dissociation is expected [230] to occur at $T = (1.1–1.2)T_c$.

In addition to high T breakup of $c\bar{c}$ or J/Ψ, we have to take account of the so-called "normal suppression" of charmonium yields, observed in proton-nucleus collisions [233]. This effect is due to a re-scattering dissociation of the primordially produced, pre-hadronic $c\bar{c}$ system upon traversal of (cold) hadronic matter [234]. It can be studied in p+A collisions where the data on J/Ψ production relative to pp collisions can be described by the survival probability

$$S_{pA} \equiv \frac{\sigma_{pA}}{A\sigma_{pp}} = \int d^2b \int dz \, \rho_A(b,z) \exp\left\{-(A-1)\int_z^\infty dz' \, \rho_A(b,z')\sigma_{abs}\right\} \tag{7.71}$$

where σ_{abs} is the effective cross section for the "absorption" (break-up) of the $c\bar{c}$ in cold nuclear matter, and ρ_A is the transverse nuclear density profile. The data [233] suggest $\sigma_{abs} = 4.2 \, mb$. The generalization of Eq. (7.71) to the nucleus-nucleus case [235] gives a good description of the J/Ψ suppression (relative to binary pp scaling) in S+U and peripheral Pb+Pb collisions at top SPS energy [226]. It has thus become customary to quantify the J/Ψ suppression in central A+A collisions by relating the observed yield, not directly to the scaled pp yield (thus obtaining R_{AA}), but to a hypothetical "normal absorption" yield baseline, established by Eq. (7.71). All *further* absorption is called "anomalous suppression".

Figure 7.54 shows the results gathered at $\sqrt{s} = 17.3$ GeV by the NA38–NA50–NA60 di-muon spectrometer [236], for minimum bias S+U ($\sqrt{s} = 20$ GeV), In+In and Pb+Pb. Up to $N_{part} \approx 100$ all yields gather at the "normal absorption"

Fig. 7.54 J/Ψ production measured in minimum bias collisions of S+U at $\sqrt{s} = 20$ GeV and Pb+Pb and In+In at $\sqrt{s} = 17.3$ GeV. The yield is scaled by "normal nuclear absorption", Eq. (7.71) [236]

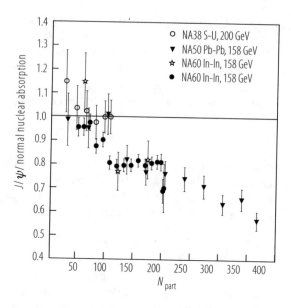

expectation from p+A scaling. A plateau at 0.8 follows for intermediate N_{part} values up to about 200 (which corresponds to central In+In collisions, so the NA60 data end here), and a final falloff toward 0.6 for central Pb+Pb collisions. It appears natural to interpret the former step as an indication of Ψ' suppression, the final step as χ_c dissociation. No genuine J/Ψ suppression is indicated. The expectation from lattice calculations [232] that J/Ψ dissociation does not occur until $T \approx 2\,T_c$ (and, thus, $\epsilon \approx 16\,\epsilon_c$) is thus compatible with Fig. 7.54. We know from Eq. (7.70) that $T \leq 1.3\,T_c$ at top SPS energy. The data are thus compatible with no break-up of the J/Ψ at the SPS, unlike at top RHIC energy where one expects $T \approx 2T_c$ [61, 96].

The RHIC data obtained by PHENIX [237] are shown in Fig. 7.55. Minimum bias Au+Au collisions are covered at mid-rapidity, as well as at $1.2 < y < 2.2$, and plotted vs. N_{part}. Due to a parallel measurement of J/Ψ production in p+p collisions [238] the PHENIX experiment is in the position to show R_{AA}, but this is done without re-normalization to p-A absorption. J/Ψ is suppressed in central collisions, $R_{AA} \leq 0.2$. Note that R_{AA} *cannot drop down* below about 0.15, due to unsuppressed surface contributions. The suppression is thus stronger than at top SPS energy[3]—in fact it is almost maximal. We conclude that in central Au+Au at $\sqrt{s} = 200$ GeV the charmonium signal gets significantly quenched, in accord with the inferences about the primordial temperature that we presented in Sects. 7.2.5 and 7.4, to amount to about 300 MeV, i.e. $T/T_c \approx 2$ as implied for J/Ψ dissociation by the lattice calculations [232].

---

[3]Dropping the unfortunate distinction between normal and anomalous absorption one gets $R_{AA} = 0.35$ for the central Pb+Pb collisions in Fig. 7.54, almost a factor 2 above the RHIC value.

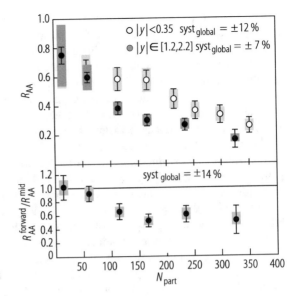

Fig. 7.55 R_{AA} for J/Ψ production in minimum bias Au+Au collisions at $\sqrt{s} = 200$ GeV, by PHENIX [237] at mid-rapidity and at $1.2 < y < 2.2$

The above interpretation is still a matter of controversy [239]. It remains unclear whether successive, distinct stages of normal and anomalous J/Ψ suppression are compatible with the dynamical evolution in central A+A collisions. A further open question refers to the difference in Fig. 7.55 of the $\langle y \rangle = 0$ and $\langle y \rangle = 1.7$ data at intermediate centrality [239] interpretation of which should be suspended until the advent of higher statistics data.

A different aspect of charmonium production in A+A collisions requires attention: at the stage of universal hadronization a certain fraction of bound $c\bar{c}$ mesons results, as the outcome of the density of uncorrelated c and $\bar{c}$ quarks in the QGP medium as it enters hadronization [240, 241]. The stage of statistical hadronization (recall Sect. 7.3) is omitted in the charmonium suppression models [41, 230, 234, 235], which proceed in two steps (only): primordial nucleon-nucleon collisions produce an initial input of $c\bar{c}$ pairs, proportional to $\sigma_{c\bar{c}}^{NN} \times N_{coll}^{AA}$ (b). In vacuum, the low relative momentum fraction of these $c\bar{c}$ pairs (about 1%) would evolve into charmonium hadrons, after a formation time of several fm/c. In the concurrently developing QGP medium, however, the initial $c\bar{c}$ correlated pairs may dissociate owing to the absence (due to color screening [41]) of vacuum-like bound states [232], at high T. At $T \geq 2\,T_c$ all fireball charmonium production would thus cease, all $c\bar{c}$ pairs ending up in singly charmed hadrons. This picture [230] is incomplete.

Even if all $c\bar{c}$ pairs break up at RHIC during the early phase of central collisions, the single charm and anti-charm quarks flow along in the expanding medium of equilibrated light and strange quarks. This picture is supported [94] by the observed elliptic flow of charm merging into the universal trend. Charm cannot, however, be chemically (flavor) equilibrated at the low plasma temperature, where it is essentially neither newly created nor annihilated [241]. The initially produced c and

$\bar{c}$ quarks (a few per central Au+Au collision at RHIC, a few tens at LHC) finally undergo statistical hadronization at $T = T_c$ along with all other light and strange quarks. To deal with the non-equilibrium overall charm abundance an extra charm fugacity factor γ_c is introduced into the statistical model [108] calculation (for details see [241]). J/Ψ and Ψ' are thus created in non-perturbative hadronization, with multiplicities proportional to γ_c^2 and phase space weights, along with all other charmed hadrons. This "regeneration" model also agrees with the RHIC data of Fig. 7.55, albeit within a large systematic uncertainty [241].

We note that the term regeneration is, in fact, *misleading*. The statistical hadronization process does *not* recover the initial, small fraction of correlated $c\bar{c}$ pairs that would end up in J/Ψ in vacuum. It arises from the *total density* of primordially produced c and $\bar{c}$, uncorrelated in the hadronizing fireball volume.

The statistical hadronization J/Ψ production process, sketched above, thus has the unfortunate property of providing a trivial background charmonium yield, unrelated to the deconfinement signal [41] referring to the primordial J/Ψ yield. Only about 1% of the primordial $c\bar{c}$ yield results in charmonia, in vacuum. The in-medium deconfinement process breaking up the $c\bar{c}$ correlation on its way to charmonia, thus constitutes a mere 1% fraction of the total charmed quark and anti-quark number. The regeneration process is insensitive to this 1% fraction, deconfined or not. At T_c, charm hadronization reacts only to the total abundance of c and $\bar{c}$, as imprinted into the dynamical evolution by the perturbative QCD $c\bar{c}$ production rate of initial nucleon-nucleon collisions. At RHIC, it turns out [241] that the c and $\bar{c}$ density is low, giving rise to substantial canonical suppression (recalling Eqs (7.38)–(7.42) in Sect. 7.3) of the two charm quark charmonia, relative to D mesons, during hadronization. With a tenfold c, $\bar{c}$ density at LHC, grand canonical charmonium production will set in, thus probably overshooting the primordial yield reference, $\sigma_{NN}^{J/\Psi} \times N_{coll}$. Thus we expect $R_{AA} > 1$ at the LHC. The role of a critical deconfinement "thermometer" is lost for J/Ψ at LHC, but the bottonium Y states can take over, being deconfined well above $T = 300$ MeV [242].

The RHIC result [237] for J/Ψ in central Au+Au collisions (Fig. 7.55), namely that $R_{AA} \rightarrow 0.2$, represents the lucky coincidence that the initial temperature, $T \approx 300$ MeV, is high enough to dissolve the correlated $c\bar{c}$ charmonium precursor states, while the J/Ψ suppression is not yet overshadowed by the trivial hadronization yield of J/Ψ.

7.6.2   Direct Photons

Photons are produced during all stages of the dynamical evolution in A+A collisions. About 98% stem from final electromagnetic hadron decays, not of interest in the present context, other then by noting that their rate has to be painstakingly measured experimentally, in order to obtain "direct" photon spectra at low p_T by

Fig. 7.56 The WA98 direct photon transverse momentum spectrum for central Pb+Pb collisions at $\sqrt{s} = 17.3$ GeV. Also indicated are scaled pA results above 2.0 GeV/c and pQCD estimates [243]. From [224]

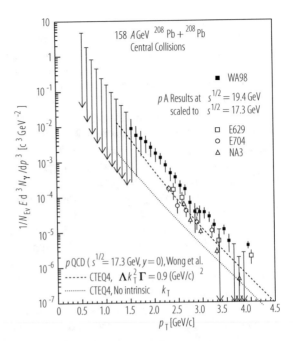

subtracting the decay fraction from the total. This was done [224] by WA98 in central Pb+Pb collisions at top SPS energy; we show their result in Fig. 7.56.

Only upper limits could be obtained at $p_T \leq 1.5$ GeV/c due to overwhelming background from π^0 and η decay, thus also obscuring the major spectral domain in which to look for a direct photon QCD plasma black body radiation source, i.e. for thermal photons of a fireball at T between T_c and about 250 MeV [223]. Several data from p+p and p+A collisions at nearby $\sqrt{s}$ and scaled up to central Pb+Pb are included in Fig. 7.56, at $p_T \geq 2$ GeV/c, and one sees the Pb+Pb data in clear excess of such contributions from primordial bremsstrahlung and hard, pQCD initial partonic collisions [243]. This excess ranges up to about 3.5 GeV/c, in Fig. 7.56. Above, the hard pQCD initial collision yield dominates [244] over thermal production.

In contrast to all other primordial high p_T pQCD yields (e.g. J/Ψ, charm, jet leading partons) this photon yield is *not* attenuated in the medium of A+A collisions. We have shown in Fig. 7.44 the R_{AA} for the PHENIX central Au+Au direct photon results [181] at RHIC $\sqrt{s} = 200$ GeV, obtained in a background substraction procedure [245] similar to the one undertaken by WA98. This procedure gives reliable data at $p_T > 4.0$ GeV/c, at RHIC, and we observe $R_{AA} = 1$. Hard initial photons are not attenuated, and there is no sign of any other direct photon contribution besides the primordial pQCD yield which, in fact, is shown (by $R_{AA} = 1$) to obey binary scaling. However, there is no hint to plasma thermal radiation (except for a trend at the very limit of statistical significance, at $p_T < 4.5$ GeV/c) in this high p_T window.

The WA98 SPS data, with thermal radiation enhancement indicated in the interval $1.5 < p_T < 3.5$ GeV/c, thus remained as the sole evidence until, more recently, the PHENIX experiment gained low p_T data [246] exploiting the fact that any source of real photons emits also virtual photons γ^* leading to internal conversion to an e^+e^- pair (the Dalitz effect). To identify this yield the invariant mass distribution of e^+e^- pairs is analyzed outside the phase space limits of π^0 Dalitz decay; the decay pairs of all remaining hadron sources (η, Δ) is subtracted as a "cocktail". The remaining pair yield is then converted assuming $\gamma^*_{dir}/\gamma^*_{inclusive} = \gamma_{dir}/\gamma_{inclusive}$ (see ref. [246] for detail), thus finally obtaining data representative of γ_{dir} in this approach. Figure 7.57 shows the corresponding p_T distribution which covers the interval $1.3 \leq p_T \leq 4.5$ GeV/c, within which the conventional direct photon extraction method did not give significant results [181]. The PHENIX experiment has also obtained direct photon spectra in p+p and d+Au at $\sqrt{s} = 200$ GeV [247] which are both well accounted for [246] by a next to leading order (NLO) pQCD photon production model [248]. These data were already employed in deriving $R_{AA} = 1$ for central Au+Au collisions, as shown in Fig. 7.44 and referred to, above. The pQCD fits derived from p+p and d+A are shown in Fig. 7.57 after binary scaling to Au+Au (pQCD $\times T_{AA}$). They merge with the yield at $p_T \geq 4$ GeV/c but demonstrate a large excess yield below 3 GeV/c. That excess is well described by adding a thermal photon component resulting from the hydrodynamic model of d'Enterria and Peressounko [249]. It traces the dynamical evolution during the early stages of equilibrium attainment,

Fig. 7.57 Internal conversion measurement of direct photons in central Au+Au collisions at 200 GeV [246]. Predictions by pQCD [248] and thermal hydrodynamic [249] models are included

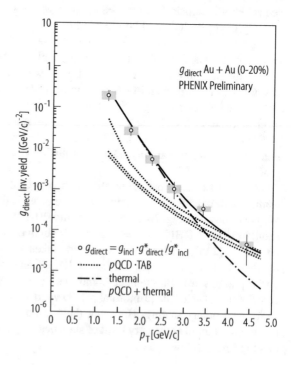

in which the photon luminosity of the emerging QGP matter phase is maximal. The hydrodynamic model provides for the space-time evolution of the local photon emission rate [223] which includes hard thermal loop diagrams to all orders, and Landau-Migdal-Pomeranchuk (LPM) in-medium interference effects. Similar, in outline, to previous models that combined algorithms of plasma photon radiation luminosity with hydrodynamic expansion [250], the model [249] fits the data in Fig. 7.57. It implies a picture in which the early stage of approach toward thermal equilibrium at RHIC is governed by a symbolic, initial, effective "temperature" of about 550 MeV which, after equilibration at $t \approx 0.6$ fm/c, *corresponds to* $T \approx 360\ MeV$ *in the primordial plasma* [249]: close to the consensus about initial T as derived from J/Ψ suppression, jet attenuation, elliptic flow and transverse energy coupled with the 1-dimensional Bjorken expansion model.

However, we note that the employed theoretical framework, hard thermal loop (HTL) QCD perturbation theory of a weakly coupled plasma state, as combined with hydrodynamics, has a tendency to call for rather high initial T values. This applies both to the above analysis [249] of RHIC data which is based on the thermal field theory model of Arnold, Moore and Yaffe [223], and to previous analysis [250] of the WA98 SPS data of Fig. 7.56. Direct photon production is even more strongly biased toward the primordial, high T evolution than jet attenuation (that may be proportional to T^3 [212]). Thus, by implication of the model [249] that fits the low p_T RHIC data of Fig. 7.57, the yield is highly sensitive to the (pre-equilibrium) formation period, $0.15 < t < 0.6$ fm/c, where the HTL model might not be fully applicable. This illustrates the present state of the art. The model(s) based on perturbative QCD require extreme initial "temperatures" to produce the high photon yield, indicated by the RHIC experiment. The strongly coupled nature [213] of the non perturbative local equilibrium QGP state at RHIC, $T \approx 300$ MeV, may provide for an alternative approach to plasma photon production.

7.6.3 Low Mass Dilepton Spectra: Vector Mesons In-medium

We have dealt with dilepton spectra throughout the above discussion of J/Ψ and direct photon production, as messenger processes sensitive to the energy density prevailing (during and) at the end of the primordial equilibration phase. The third tracer observable, low mass vector meson dilepton decay in-medium, samples—on the contrary—the conditions and modalities of hadron deconfinement in the vicinity of $T = T_c$. SPS energy is ideally suited for such studies as the QGP fireball is prepared near T_c whereas, at RHIC, it races through the T_c domain with developed expansion flow. The major relevant data thus stem from the CERN SPS experiments NA45 [251] and NA60 [136], which have analyzed e^+e^- and $\mu^+\mu^-$ production at invariant mass from 0.2 to 1.4 GeV.

Figure 7.58 shows NA45 data [251] for e^+e^- production in central Pb+Au collisions at $\sqrt{s} = 17.3$ GeV. In searching for modifications of ϱ properties and of $\pi^+\pi^-$ annihilation via virtual intermediate ϱ decay to e^+e^-, in the high

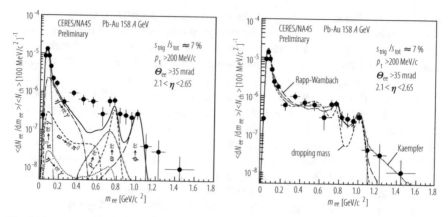

Fig. 7.58 Di-electron mass spectrum for central Pb+Au collisions at $\sqrt{s} = 17.3$ GeV with the hadron decay cocktail (left) and further in medium contributions (right) [251]; see text for detail

density environment near the hadron-parton coexistence line at $T = T_c$, the various background sources have to be under firm control. The Dalitz decays of π^0, η and η' and the in vacuo e^+e^- decays of ϱ, ω and Φ, which occur after hadronic freeze-out to on-shell particles, form a hadronic "cocktail" (left panel in Fig. 7.58) that is generated from yields provided by the grand canonical statistical model [108]. Within detector resolution, π, ω and Φ leave distinct peaks but the observed invariant mass distribution is not accounted for.

One needs also to account for background from Drell-Yan lepton pair production and open charm decay, both scaling with "number of collisions" $A^{4/3}$. The latter contribution arises from primordial $c\bar{c}$ charm production, $\langle c \rangle = \langle \bar{c} \rangle$ which leads to synchronous formation of $\langle D \rangle = \langle \overline{D} \rangle$ at hadronization; subsequent decays $D \rightarrow$ lepton + X, $\overline{D} \rightarrow$ antilepton + Y create $L\overline{L}$ pairs. This procedure is straight forward as no significant medium attenuation occurs besides the statistical charm redistribution conditions at hadronization (Sect. 7.6.1), governing $D, \overline{D}$ production [241].

Onward to non-trivial backgrounds in the invariant mass plot of Fig. 7.58, we recall the presence of thermal lepton pairs from virtual photon production in the early plasma phase, in parallel to real "direct" photon emission [252]. The spectrum of such pairs contains an average decay factor $\exp(M_{ll}/T)$, with T the (initial) plasma temperature. With $T \geq 220$ MeV assumed for top SPS energy [43, 44], this contribution is a background candidate over the entire invariant mass interval covered by the data. In general Drell-Yan, open charm decay and plasma radiation contributions are smooth, partially closing the "holes" in the hadronic cocktail undershoot of the data. This smoothing effect is helped, finally, by consideration of modifications concerning the ϱ meson spectral function near $T = T_c$, which [133–135] both affects the immediate $\varrho \rightarrow e^+e^-$ decay invariant mass region (through the fraction of in-medium ϱ decays vs. the in vacuo decay fraction after hadronic freeze-out) and, even more importantly, the contribution of in-medium

$\pi^+\pi^-$ annihilation to dileptons. The latter contribution accounts for the most obvious deviation between cocktail and data at $0.3 \leq m_{ll} \leq 0.7$ GeV in Fig. 7.58 (left panel).

The right hand panel of Fig. 7.58 shows the results of various theoretical models which address the sources of the significant dilepton excess over hadronic (in vacuum) and Drell-Yan cocktails, labeled "Rapp-Wambach" [133, 135], "dropping mass" (Brown-Rho [134]) and "Kaempfer" [252]. We shall return to these models below but note, for now, that the extra yield in central A+A collisions chiefly derives from $\pi^+\pi^-$ annihilation via the (in medium modified) ϱ resonance, and from modification of the ϱ peak itself.

With improved statistics and background control, the A+A specific extra dilepton yield below $M \approx 1.4$ GeV/c^2 can be represented by itself, after cocktail subtraction. This has been first accomplished by NA60 [136, 253] and, more recently, also by NA45 [254]. We show the former results in Fig. 7.59. The left panel shows the di-muon invariant mass spectrum in semi-central Indium-Indium collisions at top SPS energy $\sqrt{s} = 17.3$ GeV, compared to the hadronic cocktail, qualitatively in agreement with the left hand panel of Fig. 7.58 but with superior resolution and statistics. The cocktail subtraction procedure (see [136] for details) leads to an invariant mass spectrum of di-muon excess in In+In, shown in the right side panel of Fig. 7.59: an experimental landmark accomplishment. The ϱ vacuum decay contribution to the hadronic cocktail has been retained and is shown (thin solid line) to be a small fraction of the excess mass spectrum, which exhibits a broad distribution (that is almost structureless if the cocktail ϱ is also subtracted out),

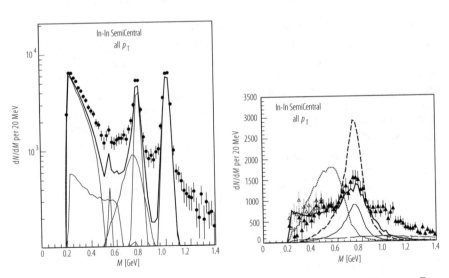

Fig. 7.59 (Left) Di-muon invariant mass spectrum in semi-central In+In collisions at $\sqrt{s} = 17.3$ GeV, with hadron final state decay cocktail. (Right) Excess mass spectrum after cocktail subtraction, confronted with fits from the broadening [133, 135, 255] and the dropping mass [228] models. From [136]

widening with collision centrality [136]. The best theoretical representation of the excess yield again results (like in Fig. 7.58, right panel) from the broadening model [133, 135, 255] where the ϱ spectral function is smeared due to various coupling mechanisms within the medium via the vector dominance model, prior to hadronic freeze-out. In the VDM ϱ couples, both, to pion pair annihilation, and to excited baryon states like the N^* (1520), via their $N\pi\pi$ decay branches.

The above data still imply serious conceptual questions. It is unclear to exactly which stage of the dynamical evolution (in the vicinity of $T = T_c$) the excess dilepton yield should correspond. As the observations appear to be coupled to in-medium ϱ meson "metabolism" we need to identify a period of temporal extension above the (in vacuo) ϱ half life (of 1.3 fm/c), safely 2 fm/c. This period should be located *in the vicinity of hadro-chemical freeze-out*. From top SPS to RHIC energy, hadro-chemical freeze-out should closely coincide with hadron formation, i.e. it should occur near the parton-hadron coexistence line, at $T = T_c$. The microscopic parton cascade model of reference [85] implements the Webber [121] phenomenological, non-perturbative QCD hadronization model (recall Sect. 7.3.3) which proposes pre-hadronization clusters of color neutralization (Fig. 7.31) as the central hadronization step. In it, so one might speculate, the transition from pQCD to non perturbative QCD creates the chiral condensates $\langle q\bar{q} \rangle$, spontaneously breaking chiral symmetry [256] (see below), and creating hadronic mass. The overall process, from pQCD color neutralization to on-shell hadrons, takes about 2.5 fm/c [85, 86] at top SPS energy. This could, thus, be the period of excess dilepton yield creation. However, the relation of the models employed above [133–135, 255, 256] to this primordial spontaneous creation of chiral condensates is still essentially unknown [256].

Thus, at present, the 1990s paradigm of a direct observation of the chiral phase transition in QCD has been lost. The Brown-Rho model [134] predicted the ϱ mass to drop to zero at $T = T_c$, occurring as a certain power of the ratio $\langle q\bar{q} \rangle^{\text{med}} / \langle q\bar{q} \rangle^{\text{vac}}$ of the chiral condensate in medium and in vacuum which approaches zero at the chiral phase transition temperature, then expected to coincidence with the deconfinement temperature. This "dropping mass" model is ruled out by the data in Fig. 7.58 and 7.59. This is, perhaps, a further manifestation of the fact that the deconfined QGP state at $T \geq T_c$ is not a simple pQCD gas of quarks and gluons [213]. In fact, lattice calculations [232, 257] find indications of surviving light $q\bar{q}$ pair correlations in the vector channel at $T \geq T_c$. Thus the two most prominent symmetries of the QCD Lagrangian, non abelian gauge invariance (related to confinement) and chiral invariance (related to mass) might exhibit *different* critical patterns at $T = T_c$ and low baryo-chemical potential. This conjecture is best illustrated by the observation that the broad, *structureless* NA60 excess dilepton spectrum of Fig. 7.59 (after cocktail ϱ subtraction) is equally well reproduced by a $T \approx$ 160–170 MeV calculation in hadronic (equilibrium) matter [133, 253, 254], and by a thermal QGP fireball of $q\bar{q}$ annihilation at this average temperature [252], as illustrated here by the model curve labeled "Kaempfer" in Fig. 7.58 (right panel). This observation has invited the concept of *parton-hadron duality* near T_c [258], which might be provocatively translated as "the QCD chiral transition properties cannot be unambiguously disentangled from the deconfinement transition effects at T_c." [256].

We may be looking at the wrong domain of $[T, \mu_B]$ space: too high $\sqrt{s}$ and, thus, too high T, too low μ_B. Already at top SPS energy the medium is dominated by the deconfinement, not by the chiral QCD transition. After hadronization the medium is still at T close to T_c but the density drops off within a few fm/c, not allowing for an equilibrium mean field state of chirally restored hadrons. It is thus perhaps not surprising that the data are seen to be dominated by simple broadening and lack of structure: perhaps that is all that happens to hadrons at T_c.

The chiral restoration transition should thus be studied at higher μ_B and lower T such that the dynamics achieves high baryon densities but still merely touches the critical (deconfinement) temperature. In fact, at $\mu_B \rightarrow 1$ GeV and $T < 100$ MeV the *chiral transition* should be of first order [259]. Here, in fact, the chiral condensate mass plays the role of the order parameter (in analogy to the magnetization in a spin system), which approaches zero as $T \rightarrow T_c$. We might thus speculate that, unlike at top SPS to LHC energy (where deconfinement dominates), the chiral QCD first order phase transition will dominate the phenomena occurring near the hadron-parton borderline, in the vicinity of $\sqrt{s} = 4$–6 GeV [142]. This requires a new experimental program, with low energy running at the SPS [260], at RHIC [261] and at the GSI FAIR project [262].

7.7   Fluctuation and Correlation Signals

Fluctuation and correlation signals in A+A collisions can be evaluated in single events, due to the high multiplicity of produced particles. Depending on the physics context we may be interested to see either a small, or a large nonstatistical fluctuation effect. For example in view of the universality of hadronization (Sect. 7.3) we would have difficulty with an event by event pion to baryon ratio (essentially μ_B^{-1}) fluctuating by, say, 50%. Conversely, searching for critical fluctuations in the vicinity of a predicted critical point of QCD [146, 147] we would be frustrated if event-wise $\langle p_T \rangle$, dN/dy (low p_T pion) [263] or strange to non-strange ratios like K/π [148, 264] would not exhibit any significant fluctuation beyond statistics. Overall, event by event fluctuation observables also carry a message concerning the robustness of our assumptions about equilibrium attainment. It turns out that equilibrium properties are not, merely, central limit consequences of ensemble averaging. Rather to the contrary, each $A \approx 200$ central collision event at $\sqrt{s} \geq 10$ GeV appears to offer, to the dynamical evolution of bulk properties, a sufficiently complete macroscopic limit. Such that we can view the event-wise bulk dynamics as "self analyzing".

Fig. 7.60 Initial transverse energy density distribution of a single central Au+Au collision event at $\sqrt{s} = 200$ GeV [265]

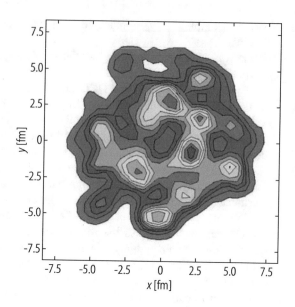

7.7.1 Elliptic Flow Fluctuation

That this ascertations is far from trivial is illustrated in Fig. 7.60. It shows [265] the primordial transverse energy density projection (the stage basic to all primordial observables) at $t \approx 0.2$ fm/c, in a central Au+Au collision at $\sqrt{s} = 200$ GeV, exhibiting an extremely clumpy and nonhomogeneous structure, apparently far away from equilibrium.

The imaging of this initial geometrical pattern of the collision volume by a hydrodynamic evolution even applies at the level of individual events, such as illustrated in Fig. 7.60. The PHOBOS Collaboration has shown [266–268] that an event by event analysis of the elliptic flow coefficient v_2 is possible (see ref. [266] for detail), by means of a maximum likelihood method. For initialization they sample the seemingly random initial spatial density distribution of single Au+Au collision events by the "participant excentricity" of individual Monte Carlo events,

$$\epsilon_{\text{part}} = \frac{\sqrt{(\sigma_y^2 - \sigma_x^2)^2 + 4\sigma_{xy}^2}}{\sigma_y^2 + \sigma_x^2} \qquad (7.72)$$

where $\sigma_{xy} = \langle xy \rangle - \langle x \rangle \langle y \rangle$. The *average* values of ϵ_{part} turn out to be similar to ϵ_x from Eq. (7.47), as expected, but the relative fluctuation width $\sigma(\epsilon)/\langle \epsilon \rangle_{\text{part}}$ turns out to be considerable. It is the point of this investigation [267, 268] to show that the observed relative event by event flow fluctuation equals the magnitude of the relative excentricity fluctuation. This is shown [268] in Fig. 7.61. The left panel demonstrates that the *average* $\langle v_2 \rangle$ obtained vs. N_{part} from the event-wise analysis agrees with the previously published [156] event-averaged PHOBOS data. The right

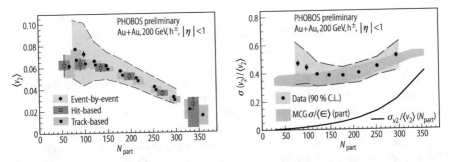

Fig. 7.61 (Left) Event by event elliptic flow analysis by PHOBOS gives an average $\langle v_2 \rangle$ that agrees with the result of ensemble analysis, in Au+Au at 200 GeV, for charged hadrons. (Right) Event-wise relative v_2 fluctuation vs. N_{part}, compared to the event-wise relative fluctuation of the participant excentricity, $\sigma(\epsilon_{part})/\langle \epsilon_{part} \rangle$. The closed line gives v_2 variation due to N_{part} number fluctuation. From [268]

panel shows that the event-wise relative fluctuation of v_2 is large: it amounts to about 0.45 and is equal to the relative fluctuation of ϵ_{part}, i.e.

$$\sigma(v_2)/\langle v_2 \rangle \approx \sigma(\epsilon_{part})/\langle \epsilon_{part} \rangle. \qquad (7.73)$$

The initial geometry appears to drive the hydrodynamic evolution of the system, not only on average but event-by-event [268], thus providing for an example of the self-analyzing property mentioned above. The v_2 signal thus emerges as the most sensitive and specific diagnostic instrument for the primordial conditions and their subsequent evolution: it reveals even the random (Fig. 7.60) initial fluctuations. In comparison the analysis with thermal photons is only sensitive to the primordial temperature [249], and restricted by very small cross sections and significant background from other sources and evolution times. It also does not give viscosity information.

7.7.2 Critical Point: Fluctuations from Diverging Susceptibilities

Recalling the goal defined in the introduction we seek observables that help to elaborate points or regions of the QCD phase diagram, Fig. 7.1. We have seen several observables that refer to the QCD plasma environment at $T \geq 300$ MeV, $\mu \approx 0$ (elliptic flow, jet attenuation, J/Ψ suppression, direct photon production), which could be first resolved within the primordial time interval $\tau \leq 1$ fm/c accessible at RHIC energy. The LHC will extend the reach of such observables toward $T \approx 600$ MeV, $x_F \leq 10^{-3}$, at $\mu_B = 0$. On the other hand, relativistic A+A collisions at lower energy permit a focus on the hypothetical QCD parton-hadron coexistence line, $T = T_c(\mu_B)$, with the domain $\mu_B \to 500$ MeV being accessible at the SPS. Characteristic observables are radial flow, hadro-chemical freeze-out, and chiral symmetry restoration effects in dilepton vector meson spectra. Focusing on this domain, we discuss fluctuations potentially associated with the existence of a critical point [8–11, 146, 147].

At the end of Sect. 7.6.3 we mentioned the conclusion from chiral symmetry restoration models [15, 259] that at high μ_B the phase transformation occurring at $T_c(\mu_B)$ should be a *chiral first order phase transition*. On the other hand, lattice QCD has characterized [16] the phase transformation at $\mu_B \to 0$, to be merely a rapid cross-over. Thus, the first order nature of the phase coexistence line in Fig. 7.1 has to end, with decreasing μ_B, in a QCD critical point, tentatively located by recent lattice QCD calculations [9–11] in the interval $\mu_B = 300$–500 MeV. The existence of such a point in the $[T, \mu_B]$ plane would imply fluctuations analogous to critical opalescence in QED [146, 147, 263]. Beyond this second order phase transition point the coexistence line would be the site of a rapid cross-over [16]. This overall theoretical proposal places potential observations related to the critical point itself, and/or to the onset of first order phase transition conditions at higher μ_B, within the domain of the lower SPS energies, $\sqrt{s} \leq 10$ GeV. Note that, at such low energies, the initialization of thermal equilibrium conditions should occur in the vicinity of T_c, unlike at RHIC and LHC, and that the central fireball spends considerable time near the coexistence line, at $300 \leq \mu_B \leq 500$ MeV.

To analyze potential observable effects of a critical point, we recall briefly the procedure in finite μ_B lattice theory that led to its discovery. One method to compute thermodynamic functions at $\mu_B > 0$ from the grand canonical partition function $Z(V, T, \mu_q)$ at $\mu_q = 0$ is to employ a Taylor expansion with respect to the chemical quark potential [10, 11, 270], defined by the derivatives of Z at $\mu = 0$. Of particular interest is the quark number density susceptibility,

$$\chi_{u,d} = T^2 \left(\frac{\delta^2}{\delta(\mu/T)^2} \frac{p}{T^4} \right) \tag{7.74}$$

which can also be written as

$$\chi_q = T^2 \left(\frac{\delta}{\delta(\mu_u/T)} + \frac{\delta}{\delta(\mu_d/T)} \right) \frac{n_u + n_d}{T^3} \tag{7.75}$$

with $\chi_q = (\chi_u + \chi_d)/2$ and quark number densities n_u, n_d. We see that the susceptibility refers to the quark number density fluctuation. The lattice result [10, 270] is shown in Fig. 7.62, a calculation with two dynamical flavors assuming $T_c = 150$ MeV and three choices of chemical quark potential, $\mu_q = 0, 75$ and 150 MeV, respectively, corresponding to $\mu_B = 3\,\mu_q = 0, 225$ and 450 MeV. These choices correspond to LHC/RHIC energy, top SPS energy and $\sqrt{s} \approx 6.5$ GeV, respectively. At $\mu_B = 0$ one sees a typical smooth cross-over transition at $T = T_c$ whereas a steep maximum of susceptibility occurs with $\mu_B = 450$ MeV. This suggests the presence of a critical point in the (T, μ_B) plane [270] in the vicinity of $(150$ MeV, 450 MeV$)$. For final confirmation one would like to see this maximum

Fig. 7.62 Quark number density susceptibility vs. temperature for light quarks in 2 flavor lattice QCD at finite μ_B. The calculation refers to $T_c = 150$ MeV and quark chemical potential $\mu_q/T_c = 0$, 0.5 and 1.0, respectively [270]. Smooth lines interpolate the calculated points; error bars indicate lattice statistics

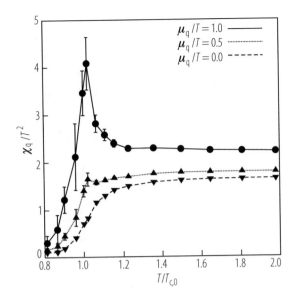

disappear again, toward $\mu_q > T_c$, but this is beyond the convergence domain of the employed Taylor expansion (see ref. [9] for alternative approaches).

From Fig. 7.62 we also expect a divergence of the strangeness susceptibility, for which no results from a 3 dynamical flavors calculation at finite μ_B exist to date. A lattice calculation at $\mu_B = 0$, $T \approx 1.5\, T_c$ suggests [271] that the u, d, s quark flavors densities fluctuate *uncorrelated* (but we do not know whether that is also true at $\mu_B = \mu_B^{crit}$). This could thus be observed in event by event analysis, in particular as a fluctuation of the Wroblewski ratio $\lambda_s = 2(s + \bar{s})/(u + \bar{u} + d + \bar{d})$ which is approximated by the event-wise ratio $(K^+ + K^-)/(\pi^+ + \pi^-)$. This was first measured by NA49 in central collisions of Pb+Pb at top SPS energy; the result [272] is shown in Fig. 7.63. The data result from a maximum likelihood analysis of track-wise specific ionization in the domain $3.5 \leq y \leq 5$ slightly forward of mid-rapidity. The width σ_{data} is almost perfectly reproduced by the mixed event reference, such that the difference,

$$\sigma_{dyn} = \sqrt{(\sigma_{data}^2 - \sigma_{mix}^2)} \qquad (7.76)$$

amounts to about 3% of σ_{data} only, at $\sqrt{s} = 17.3$ GeV. This analysis has more recently been extended to all energies available thus far, at the SPS [273] and at RHIC [274]. Figure 7.64 shows that σ_{dyn} stays constant from top SPS to top RHIC energy but exhibits a steep rise toward lower energies that persists down to the lowest SPS energy, $\sqrt{s} = 6.2$ GeV. Figure 7.33 shows [107] that at this energy $\mu_B = 450$ MeV, thus corresponding to the susceptibility peak in Fig. 7.62. Again, as we noted about the peak in Fig. 7.62: if these data indicate a critical point effect in the vicinity of $\mu_B = 450$ MeV the relative fluctuation should *decrease* again, toward

Fig. 7.63 Event by event fluctuation of the K^{+-}/π^{+-} ratio in central collisions of Pb+Pb at $\sqrt{s} = 17.3$ GeV, relative to mixed event background (histogram) [272]

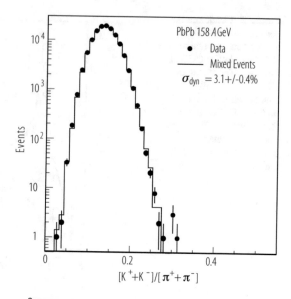

Fig. 7.64 The relative deviation of the event by event K/π fluctuation width from the mixed event background width, σ_{dyn} (see Eq. (7.76)); at SPS [273] and RHIC [274] energies

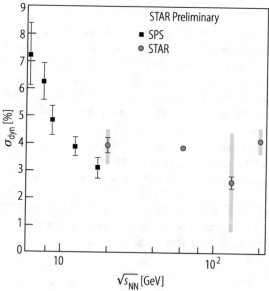

yet higher μ_B and lower $\sqrt{s}$. These data will, hence, be re-measured and extended to lower $\sqrt{s}$ by experiments in preparation [260–262] at CERN, RHIC and GSI-FAIR. This will also help to evaluate alternative tentative explanations invoking fluctuating canonical suppression [120], or strangeness trapping [275]. Finally, the position of the critical point needs to be ascertained by lattice theory.

7.7.3 Critical Fluctuation of the Sigma-Field, and Related Pionic Observables

Earlier investigations of critical QCD phenomena that might occur in high energy nuclear collisions were based on QCD chiral field theory [276]. The QCD critical point is associated with the chiral phase transition in so far as it appears as a remnant of a tri-critical point [147] corresponding to the "ideal" chiral limit that would occur if $m_u = m_d = 0$. Therefore the existence of a second-order critical point, at $\mu_B > 0$, is a fundamental property of QCD with small but non-zero quark masses [277]. The magnitude of the quark condensate, which plays the role of an order parameter of the spontaneously broken symmetry (generating hadronic mass), has the thermal expectation value

$$\langle \overline{q}q \rangle = \frac{1}{Z} \sum_n \langle n \mid \overline{q}q \mid n \rangle \exp(-E_n/T) \qquad (7.77)$$

with the partition function of hadronic states E_n

$$Z = \sum_n \exp(-E_n/T). \qquad (7.78)$$

The low energy behavior of the matrix elements $\langle n \mid \overline{q}q \mid n \rangle$ can be worked out in chiral perturbation theory [277]. At the QCD critical point the order parameter fluctuates strongly. Its magnitude $\langle \overline{q}q \rangle$ is identified with an isoscalar quantity, the so-called σ-field. The critical point communicates to the hadronic population via the $\sigma \leftrightarrow \pi\pi$ reaction, generating fluctuating fractions of the direct pion yield present near $T = T_c$, which thus gets imprinted with a fluctuation of transverse momentum (in the low p_T domain) stemming from σ mass fluctuation, downward toward the critical point. At it the isoscalar field ideally approaches zero mass, in order to provide for the long wavelength mode required by the divergence of the correlation length [147].

Note the relatively fragile structure of the argument. In an ideal, stationary infinite volume situation the sigma field would really become massless, or at least fall below the $\pi^+\pi^-$ threshold; thus its coupling to $\pi^+\pi^-$ becomes weak, and restricted to very small p_T. Furthermore, such primary soft pions, already small in number, are subject to intense subsequent re-absorption and re-scattering in the final hadronic cascade evolution [114]. In fact, experimental investigations of event by event p_T fluctuations in central A+A collisions, covering the entire $\sqrt{s}$ domain from low SPS to top RHIC energy have not found significant dynamic effects [278–281]. Figure 7.65 illustrates the first such measurement by NA49 [278] in central Pb+Pb collisions at $\sqrt{s} = 17.3$ GeV, at forward rapidity $4 < y < 5.5$, showing the distribution of event-wise charged particle average transverse momentum, a perfect Gaussian. It is very closely approximated by the mixed event distribution, ruling out a significant value of σ_{dyn} from Eq. (7.76). More sensitive measures of changes,

Fig. 7.65 Event by event fluctuation of average charged hadron p_T in the interval $4.0 < y < 5.5$, in central Pb+Pb collisions at $\sqrt{s} = 17.3$ GeV. Mixed event background given by histogram [278]

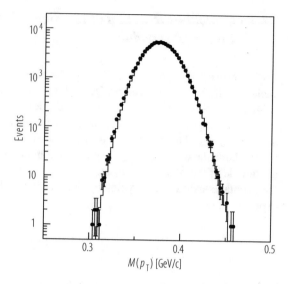

event by event, in the parent distribution in transverse momentum space, have been developed [282, 283]. NA49 has employed [278, 279] the measure $\Phi(p_T)$, defined as [282]

$$\Phi(p_T) = \sqrt{\frac{\langle Z^2 \rangle}{\langle N \rangle}} - \sqrt{\overline{z^2}} \tag{7.79}$$

where $z_i = p_{Ti} - \overline{p}_T$ for each particle, with $\overline{p}_T$ the overall inclusive average, and for each event $Z = \sum_N z_i$ is calculated. With the second term the trivial independent particle emission fluctuation is subtracted out, i.e. Φ vanishes if this is all. Indeed, the data of Fig. 7.65 lead to Φ compatible with zero. Furthermore, a recent NA49 study [279] at mid-rapidity, covers the range from $\sqrt{s} = 17.3$ to 6.3 GeV (where the K/π ratio fluctuation in Fig. 7.64 exhibits the much-discussed rise, and even the ensemble average in Fig. 7.28 shows the unexplained sharp peak) but finds no significant Φ signal.

Alternatively, one can base a signal of dynamical p_T fluctuation on the binary *correlation* of particle transverse momenta in a given event, i.e. on the co-variance $\langle p_{Ti} \, p_{Tj} \rangle$ [281, 283] of particles i, j in one event. Of course, the co-variance receives contributions from sources beyond our present concern, i.e. Bose-Einstein correlation, flow and jets (the jet activity becomes prominent at high $\sqrt{s}$, and will dominate the p_T fluctuation signal at the LHC). In co-variance analysis, the dynamical p_T fluctuation (of whatever origin) is recovered via its effect on correlations among the transverse momentum of particles. Such correlations can be quantified employing the two-particle p_T correlator [281, 284]

$$\langle \Delta p_{Ti} \, \Delta p_{Tj} \rangle = \frac{1}{M_{\text{pairs}}} \sum_{k=1}^{n} \sum_{i=1}^{N(k)} \sum_{j=i+1}^{N(k)} \Delta p_{Ti} \Delta p_{Tj} \tag{7.80}$$

where M_{pairs} is the total number of track pairs of the events k contained in the entire ensemble of n events, $N(k)$ is the number of tracks in event k, and $\Delta p_{Ti} = p_{Ti} - \overline{p}_T$ where $\overline{p}_T$ is the global ensemble mean p_T. The normalized dynamical fluctuation is then expressed [281] as

$$\sigma(p_T)_{dyn} = \sqrt{\langle \Delta p_{Ti} \, \Delta p_{Tj} \rangle} / \overline{p_T}. \tag{7.81}$$

It is zero for uncorrelated particle emission.

Figure 7.66 shows the analysis of p_T fluctuations based on the p_T correlator, for central Pb+Au SPS collisions by CERES [280] and for central Au+Au at four RHIC energies by STAR [281]. The signal is at the 1% level at all $\sqrt{s}$, with no hint at critical point phenomena. Its small but finite size could arise from a multitude of sources, e.g. Bose-Einstein statistics, Coulomb or flow effects, mini-jet-formation, but also from experimental conditions such as two-track resolution limits [284]. We note that even if a critical opalescence effect, related to a fluctuating chiral condensate at $T = T_{crit}$, couples to the primordial, low p_T pion pair population [147, 285], this signal might be dissipated away, and thus "thermalized" to the thermal freeze-out scale of about 90–110 MeV, as a pion experiences about 6 re-scatterings during the hadronic cascade [114]. On the other hand the hadro-chemical K/π ratio fluctuation (Fig. 7.64) would be preserved throughout the cascade (Sect. 7.3).

Fig. 7.66 Dynamical p_T event by event fluctuation analysis by $\sigma(p_T)_{dyn}$ of Eq. (7.81), vs. $\sqrt{s}$, showing SPS [280] and RHIC [281] data

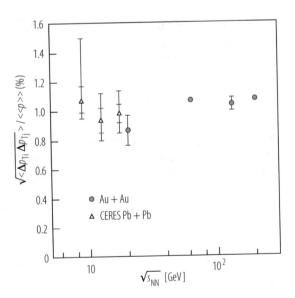

7.7.4 Bose-Einstein-Correlation

Identical boson pairs exhibit a positive correlation enhancement when $\Delta \vec{p}_{ij} \to 0$, an intensity correlation analogous to the historical Hanbury-Brown and Twiss effect (HBT) of two photon interferometry [286] employed in astrophysics. In nucleus-nucleus collisions this is an aspect of the symmetry of the N-pion wave function that describes pion pairs at the instant of their decoupling from strong interaction. The momentum space correlation function $C(q, K)$, $q = p_1 - p_2$, $K = \frac{1}{2}(p_1 + p_2)$ is the Fourier transform of the spatial emission function $S(x, K)$ which can be viewed as the probability that a meson pair with momentum K is emitted from the space-time point x in the freezing-out fireball density distribution [96, 287].

The aim of HBT two particle interferometry is to extract from the measured correlator $C(q, K)$ as much information about $S(x, K)$ as possible. In addition to the traditional HBT focus on geometrical properties of the source distribution ("HBT radii") there occurs information on time and duration of emission, as we have already employed in Fig. 7.19, illustrating τ_f and $\Delta \tau$ in central Pb+Pb collisions at top SPS energy [90]. Moreover, even more of dynamical information is accessible via the effects of collective flow on the emission function. In this respect, the observations make close contact to the hydrodynamic model freeze-out hypersurface (Sect. 7.3.1) and the contained flow profiles. HBT thus helps to visualize the *end* of the collective radial and elliptic flow evolution. This implies that we expect to gather evidence for the, perhaps, most consequential property of A+A collisions at high $\sqrt{s}$, namely the primordially imprinted Hubble expansion.

We do not engage here in a detailed exposition of HBT formalism and results as comprehensive recent reviews are available [96, 287]. Briefly, the measured quantity in $\pi^+\pi^+$, $\pi^-\pi^-$ or K^+K^+ interferometry is the correlator

$$C(q, K) = \frac{d^6 N / dp_1^3 \, dp_2^3}{d^3 N / dp_1^3 \ \ d^3 N / dp_2^3} \tag{7.82}$$

of pair yield normalized to the inclusive yield product. The correlator is related to the emission function which is interpreted as the Wigner phase space density of the emitting source [288]:

$$C(q, K) \approx 1 + \frac{| \int d^4 x \, S(x, K) \, e^{iqx} |^2}{| \int d^4 x \, S(x, K) |^2}. \tag{7.83}$$

Due to the experimental on-shell requirement $K_0 = \sqrt{K^2 + m^2}$ the 4-vector components of K are constrained on the left hand side. Thus, relation (7.83) cannot simply be inverted.

To proceed one employs a Gaussian ansatz on either side of Eq. (7.83). The experimental data are parametrized by source "radius" parameters $R_{ij}(K)$,

$$C(q, K) = 1 + \lambda(K) \exp\left[-\sum_{ij} R_{ij}^2(K) q_i q_j\right] \qquad (7.84)$$

employing $\lambda(K)$ *essentially as a fudge factor* (related nominally to possible coherence of emission effects in the source, which would dilute or remove the Bose-Einstein statistics effect of $C \rightarrow 2$ for $q \rightarrow 0$ but have never been seen in nuclear collisions). In Eq. (7.84) the sum runs over three of the four components of q, due again to the on-shell requirements [287]. For the emission function $S(x, K)$ a Gaussian profile is assumed about an "effective source center" $\overline{x}(K)$, thus

$$S(x, K) \rightarrow S(\overline{x}(K), K) \times G \qquad (7.85)$$

where G is a Gaussian in coordinates $\tilde{x}^\mu(K)$ relative to the center coordinates $\overline{x}^\mu(K)$. Inserting Eq. (7.85) into (7.83) one finally obtains

$$C(q, K) = 1 + \lambda(K) \exp\left[-q_\mu q_\nu \langle \tilde{x}^\mu \tilde{x}^\nu \rangle\right] \qquad (7.86)$$

where $\langle \tilde{x}^\mu \tilde{x}^\nu \rangle$ are the elements of the space-time variance of the correlation function, which re-interpret the "radii" R_{ij}^2 in Eq. (7.84). Assuming azimuthal symmetry (central collisions), cartesian parametrizations of the pair relative momentum q coordinates (corresponding to fixation of the space-time variance in Eq. (7.84)) have been introduced by Yano, Koonin and Podgoretskii [289], and, alternatively, by Pratt [290]. The latter, *out-side-longitudinal* coordinate system has the "long" direction along the beam axis. In the transverse plane, the "out" direction is chosen parallel to $K_T = (p_{1T} + p_{2T})/2$, the transverse component of the pair momentum K. The "side" direction is then orthogonal to the out- and long-direction but, moreover, it has the simplest geometrical interpretation (see ref. [287] for detail), to essentially reflect the transverse system size [288]. The parameters of Eq. (7.86) are thus defined; as an example we quote, from identification of Eq. (7.84) with (7.86), the resulting geometrical definition of the "side" radius,

$$R_{\text{side}}^2(K) = \langle \tilde{y}(K)^2 \rangle. \qquad (7.87)$$

Overall, this model of Fourier related correlators, $C(q, K)$ the experimentally accessible quantity (see Eq. (7.82)), and $S(x, K)$ the to-be-inferred spatial freeze-out fireball configuration, leads to the Gaussian ansatz [287]

$$C(q, K) = 1 + \lambda(K) \exp[-R_{\text{out}}^2(K) q_{\text{out}}^2 - R_{\text{side}}^2(K) q_{\text{side}}^2 - $$
$$R_{\text{long}}^2(K) q_{\text{long}}^2 + \text{cross terms}] \qquad (7.88)$$

which is fitted to the experimental correlation function (Eq. (7.82)). The experiment thus determines the variances R_{out}, R_{side} and R_{long}. In the so-called "local co-moving system" (LCMS), defined as the frame in which $p_{z,1} = -p_{z,2}$, i.e. $\beta_{long} = 0$, we obtain in addition to Eq. (7.87)

$$R_{out}(K)^2 = \left\langle (\tilde{x}(K) - \beta_T \tilde{t}(K))^2 \right\rangle$$

$$R_{long}^2(K) = \left\langle \tilde{z}(K)^2 \right\rangle \tag{7.89}$$

and finally, for azimuthal symmetry in central collisions with $\langle \tilde{x}^2 \rangle \approx \langle \tilde{y}^2 \rangle$ we find the "duration of emission" parameter illustrated in Fig. 7.19 [90]:

$$\left\langle \tilde{t}^2 \right\rangle \approx \frac{1}{\beta_T} \left(R_{out}^2 - R_{side}^2 \right). \tag{7.90}$$

The resulting reduction of the initial 8 coordinates of the meson pair stems, in summary, from the on-shell requirement, from azimuthal symmetry and from approximate Bjorken invariance in the LCMS system [287, 288]. One has to be aware of the latter idealizations. Note that in an expanding source all HBT parameters depend on K, the pair mean momentum (see below).

We make first use of the above parametrization in Fig. 7.67. From the purely spatial radii R_{side} and R_{long} one can define a mid-rapidity volume at pionic decoupling, $V_f = (2\pi)^{2/3} R_{side}^2 R_{long}$ which is shown [291] at $K = 0.2$ GeV for central Pb+Pb and Au+Au collisions from AGS [292] via SPS [90, 293] to RHIC [294] energy. The upper panel shows the $\sqrt{s}$ dependence, the lower illustrates

Fig. 7.67 Coherence freeze-out volume V_f from π^- pair Bose-Einstein correlation analysis in central Au+Au and Pb+Pb collisions, (upper panel) plotted vs. $\sqrt{s}$, (lower panel) vs. mid-rapidity charged particle density dN^{ch}/dy [291]

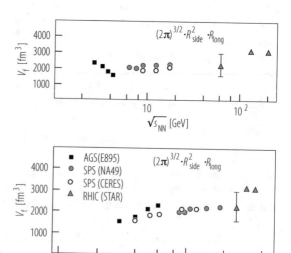

the dependence on the charged particle rapidity density dN_{ch}/dy which one might intuitively expect to be related to the freeze-out volume [295]. We see, firstly, that the plot vs. $\sqrt{s}$ exhibits a non-monotonous pattern at the transition from AGS to SPS energies [287, 295], whereas the plot vs. dN/dy rather features a rise toward a plateau that ends in a steep increase at RHIC energies. Second, the tenfold increase in charged particle rapidity density is reflected in only a doubling of the "volume" V_f.

The latter observation reminds us of the fact that the geometrical parameters obtained via the above analysis do not refer to the global source volume if that volume undergoes a collective Hubble expansion [96, 287]. A pion pair emitted with small relative momentum into the azimuthal direction $\vec{K}_T$ is likely to stem (only) from the fraction of the expanding source that also moves into this direction. This coupling between position and momentum in the source becomes more pronounced, both, with increasing K_T and increasing sources transverse velocity β_T from radial expansion. We have seen in Fig. 7.24 that the latter increases dramatically with $\sqrt{s}$, such that the coherence volume V_f comprises a decreasing fraction of the total fireball. It should thus rise much more slowly than proportional to the global dN/dy [96, 287].

A striking experimental confirmation of the Hubble expansion pattern in central A+A collisions is shown in Fig. 7.68. The illustrated HBT analysis of NA49 [90] at $\sqrt{s} = 17.3$ GeV, and of PHOBOS [294] at $\sqrt{s} = 200$ GeV, employs the alternative parametrization of the correlation function $C(q, K)$ introduced [289] by Yano, Koonin and Podgoretskii (YKP). Without describing the detail we note that the YKP correlation function contains the "YK velocity" β_{YK} describing the

Fig. 7.68 Emitting source rapidity (Y_{YKP}) as a function of pion pair rapidity ($Y_{\pi\pi}$). From π^- pair correlation analysis in central Pb+Pb collisions at $\sqrt{s} = 17.3$ GeV [90], and in central Au+Au collisions at $\sqrt{s} = 200$ GeV [294]

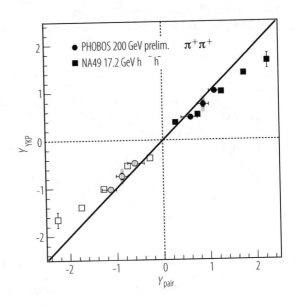

source's longitudinal collective motion in each interval of pion pair rapidity,

$$Y_{\pi\pi} = \frac{1}{2} \ln \left(\frac{E_1 + E_2 + p_{z1} + p_{z2}}{E_1 + E_2 - p_{z1} - p_{z2}} \right). \tag{7.91}$$

Defining the "YKP rapidity" corresponding to β_{YK} by

$$Y_{YKP} = \frac{1}{2} \ln \left(\frac{1 + \beta_{YK}}{1 - \beta_{YK}} \right) + y_{cm} \tag{7.92}$$

leads to the results in Fig. 7.68, indicating a strong correlation of the collective longitudinal source rapidity Y_{YKP} with the position of the pion pair in rapidity space.

We are interested in a possible non-monotonous $\sqrt{s}$ dependence of freeze-out parameters such as V_f because of recent predictions of the relativistic hydrodynamic model that a QCD critical point should act as an attractor of the isentropic expansion trajectories in the $[T, \mu_B]$ plane [101, 296]. Figure 7.69 shows such trajectories characterized by the ratio of entropy to net baryon number, S/n_B, which is conserved in each hydro-fluid cell throughout the expansion. Note that the S/n_B ratio is determined from the EOS during the primordial initialization stage in the partonic phase; the relationship between S/n_B and $\sqrt{s}$ is thus not affected by the presence or absence of a critical end point (CEP) at $T = T_c$ which, however, drastically influences the trajectory at later times as is obvious from comparing the left and right panels, the latter obtained with a first order phase transition at all μ_B but no CEP. In this case, the S/n_B = const. trajectories in the partonic phase all point to the origin in the $[T, \mu_q]$ plane because $\mu_q/T \propto \ln(S/n_B)$ in an ideal parton gas; whereas they point oppositely below $T = T_c$ because $T^{3/2}/\varrho \propto \ln(S/n_B)$ in a hadron gas. This organization is dramatically upset for the cases $S/n = 100$, 50, and 33 by the assumption of a CEP, tentatively placed here at $T = 155$ MeV, $\mu_B = 368$ MeV.

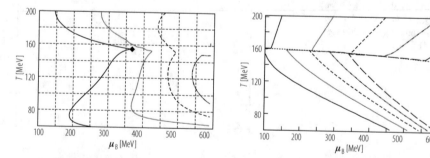

Fig. 7.69 (Left) Influence of a critical point on hydrodynamic model isentropic expansion trajectories characterized by various values of entropy to net baryon number s/n_B. (Right) The same trajectories without a critical point but a first order transition all along the phase boundary [296]

A first conclusion from this model is that it should not be essential to fine-tune $\sqrt{s}$ to make the system pass near the CEP because the attractor character would dominate over a substantial domain, e.g. at hadro-chemical freeze-out conditions $250 \leq \mu_B \leq 500$ MeV in the example of Fig. 7.69. Please note that hadro-chemical freeze-out is not treated correctly in this model, to occur at T_H which is $160 \leq T_H \leq 130$ MeV from Figs. 7.1 and 7.33. The trajectories shown here *below* T_H are thus not very realistic. The expected pattern could cause the "plateau" behavior of several observables at hadronic freeze-out over the range of SPS energies *that corresponds to the above interval of* μ_B and T_H, e.g. $\langle m_T \rangle$ and T in Figs. 7.21 and 7.22, elliptic flow v_2 at mid-rapidity in Fig. 7.36, and coherent hadronic freeze-out volume V_f from HBT in Fig. 7.67.

A consistent search for critical point effects implies, at first, a correct treatment of the hadronic expansion phase in hydro-models as above [101, 296], properly distinguishing chemical composition freeze-out, and eventual "thermal" freeze-out at hadronic decoupling. From such a model predictions for the $\sqrt{s}$ or S/n_B systematics could be provided for the HBT source parametrization implied by Eqs. (7.84)–(7.86). I.e. the hydrodynamic model provides the correlator $S(x, K)$ in cases with, and without a critical point which, as Fig. 7.69 shows, leads to considerable changes in the system trajectory in the domain near hadronization and onward to hadronic thermal freeze-out, at each given S/n or $\sqrt{s}$. On the experimental side, this trajectory is documented by $[T, \mu_B]$ at hadronic freeze-out from grand canonical analysis [140] (Sect. 7.3) which also yields S/n_B. Furthermore, as we shall show below, HBT in combination with the analysis of p_T or m_T spectra will describe the situation at thermal hadron freeze-out yielding T_f, β_T at the surface and the "true" transverse radius R_{geom} of the source, in addition to the coherence volume V_f illustrated in Fig. 7.67 (which also documents the present, insufficient data situation). Of course, we have to note that the lattice results concerning the critical point are not yet final [270], and neither is the hydrodynamic treatment [296] concerning the EOS in the vicinity of the CEP.

We turn to combined analysis of the two final processes of expansive evolution, formation of p_T spectra and decoupling to Bose-Einstein pair correlation, in order to show how the experimental information mentioned above can be gathered. At the level of "hydrodynamically inspired" analytic models for the emission function $S(x, K)$ several approaches [106, 297] have established the connection, arising from radial flow, between hadronic m_T spectra and the K_T dependence of the HBT radii R_{side}, R_{out} and R_{long}, Eqs. (7.87), (7.92), which fall down steeply with increasing K_T [291–294]. A combined analysis lifts continuous ambiguities between the thermal freeze-out temperature T_f and the radial expansion velocity β_T (for which a radial profile $\beta_T = (r/R_{side}) \beta_0$ is assumed), that exist in the blast wave model derivation of both the p_T spectra, Eq. (7.25), and the K_T dependence of R_s.

This was first demonstrated in a NA49 study of pion correlation in central Pb+Pb collisions at $\sqrt{s} = 17.3$ GeV [90] that is shown in Fig. 7.70. The ambiguous anticorrelation of fit parameters T_f and β_T^2 can be somewhat constrained if several hadronic species are fit concurrently. This is obvious in Fig. 7.70 from the overlap of parametrization regions for negative hadron, and for deuterium m_T spectra. An

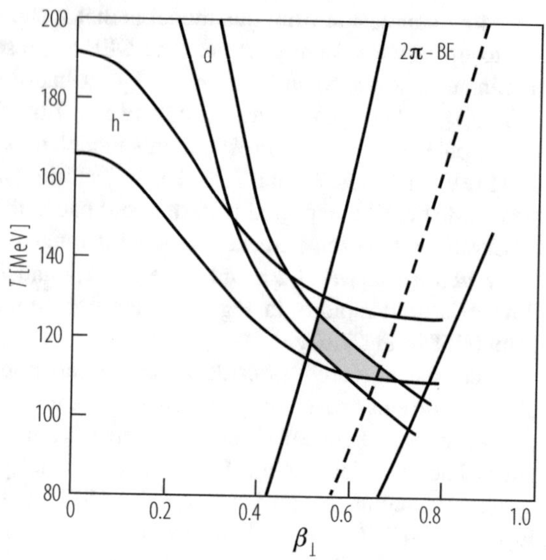

Fig. 7.70 Allowed regions of freeze-out temperature vs. radial expansion velocity for central Pb+Pb collisions at $\sqrt{s} = 17.3$ GeV and mid-rapidity, combining negative hadron and deuterium spectral data analysis with BE π^- correlation results on K_T dependence of $R_\perp \approx R_s$ [90]

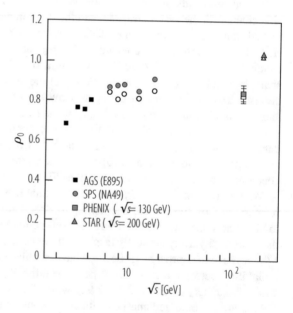

Fig. 7.71 The surface velocity (here denoted as ρ_0) of radial expansion at decoupling in central Au+Au and Pb+Pb collisions vs. $\sqrt{s}$. From combined analysis of hadron p_T spectra and π^- pair correlation [291, 297]

orthogonal constraint on β_T^2/T_f results from a blast wave model fit of the HBT K_T dependence of $R_T \approx R_{side}$ [298], employed here. We see that the three independent 1σ fit domains pin down T_f and β_T rather sharply, to [115 MeV, 0.55]. A relativistic correction [297] leads to $T_f = 105$ MeV, $\beta_T = 0.60$. The $\sqrt{s}$ dependence of β_T at the freeze-out surface, from such an analysis [291, 297] is shown in Fig. 7.71. The data again exhibit a plateau at SPS energies, which remains to be understood [299].

In the light of the considerations above, the plateau might turn out to reflect a critical point focusing effect.

7.8 Summary

We have seen that many physics observables in relativistic nucleus-nucleus collisions can, at RHIC energy $\sqrt{s} = 200$ GeV, be related to the primordial dynamical phase, from initial QCD parton shower formation to local momentum space equilibration. The time interval is 0.35 to 0.65 fm/c. This domain can be investigated at RHIC due to the short interpenetration time, of 0.15 fm/c. From among the bulk hadron production signals, total and midrapidity charged particle rapidity densities, dN_{ch}/dy, reflect the primordial parton saturation phenomenon [72]. It leads to an unexpectedly slow increase of multiplicity with fireball participant number N_{part}, and with $\sqrt{s}$. This observation signals the onset of nonperturbative QCD, a coherent shower multiplication by multigluon coherence [62, 65, 70, 71, 75]. It is expected to be even more dominant in LHC Pb+Pb collisions at 5.5 TeV. Furthermore, elliptic flow, a collective bulk hadron emission anisotropy, also originates from the primordial, nonisotropic spatial density profile of shower-produced partons [93, 96]. A hydrodynamic evolution sets in at $t < 1$ fm/c, implying the existence of a primordial equation of state (EOS) for partonic matter in local equilibrium. Moreover, the experimental elliptic flow data at RHIC are well described by "ideal fluid" hydrodynamics, i.e. by a very small shear viscosity η and a small mean free path λ [94, 95].

These observations indicate the existence of a (local) equilibrium state at very early times in the bulk parton matter stage of the dynamical evolution. This quark-gluon plasma (QGP) state of nonperturbative QCD was predicted by QCD lattice theory [11]. In Au+Au collisions at RHIC energy this state appears to be realized under the following conditions [61].

The energy density ϵ amounts to 6–15 GeV/fm^3, far above the parton-hadron QCD confinement region at $\epsilon = 1$ GeV/fm^3, and at about 55 times the density of nuclear ground state matter, $\rho_0 = 0.14$ GeV/fm^3. Translating to macroscopic units we are dealing with a matter density of about $1.3 \cdot 10^{19}$ kg/m^3, the density prevailing in the picosecond era of the cosmological evolution. The corresponding temperature amounts to $T = 300$–330 MeV (about $3.6 \cdot 10^{12}$ K), far above the "Hagedorn limit" [38] for the hadronic phase ($T_H = 165$ MeV). From an analysis of the production ratios of the various hadronic species (from pions to Omega hyperons) arising from this primordial partonic state at hadronization, the statistical hadronization model (SHM) determines its baryo-chemical potential [108], $\mu_B = 20$ MeV at RHIC energy. This value indicates that one is already close to the near-zero net baryon number conditions, $\mu_B \approx 0$, prevailing in the corresponding big bang evolution, where the density of particles exceeds that of antiparticles by a fraction of about 10^{-9} only.

Overall we thus obtain entries in the QCD phase diagram of Fig. 7.1. RHIC creates a parton plasma at about $T = 300$ MeV and $\mu_B = 20$ MeV. It turns out to behave like an ideal fluid and features an extremely short mean free path λ, thus inviting a description by relativistic hydrodynamics. The small shear viscosity η (or the small viscosity to entropy ratio η/s) are highlighted by the striking observation that even the fluctuations of primordial parton density in individual, single events, are preserved in the single event variation of elliptic flow [267, 268]. Moreover, the observed scaling of elliptic flow with hadron valence quark number [94]—and thus not with hadronic mass as in radial flow [103]—confirms the implied origin of the elliptic flow signal from the partonic phase.

At the LHC the phase of early QCD plasma formation is expected to shift to yet higher energy density, in the vicinity of $T = 600$ MeV and $\mu_B = 5$ MeV. One is thus getting nearer to the domain of QCD asymptotic freedom, and might expect a falloff of the partonic cross section which is extremely high at RHIC [61], as reflected by the small η/s and λ values.

The observed features of the QCD plasma produced at RHIC energy have invited the terminology of a "strongly coupled" quark-gluon plasma (sQGP [165]). Further evidence for the strongly coupled, non perturbative nature of the primordial partonic state stems from the various observed, strong in-medium attenuation effects on initially produced high p_T partons. In Au+Au collisions at $\sqrt{s} = 200$ GeV this high medium opacity leads to a universal quenching of the high p_T hadron yield [175] including, most remarkably, heavy charm quark propagation to D mesons [175, 178]. We have shown in Sect. 7.5 that the interior of the collisional fireball at midrapidity is almost "black" at $t < 1$ fm/c. This is also reflected in a strong suppression of the back-to-back correlation of hadrons from primordially produced di-jets [207, 208], and in a similarly strong suppression of the J/Ψ yield [237] which we have shown in Sect. 7.6 to be ascribed to an in-medium dissolution of the primordially produced $c\bar{c}$ pairs [41] at T about 300 MeV.

The underlying high medium opacity can be formally expressed [188, 193, 195] by an effective parton transport coefficient $\hat{q}$ (Eqs. (7.58) and (7.62)) which quantifies the medium induced transverse momentum squared per unit mean free path λ. The value of $\hat{q}$ derived from analysis of the various attenuation phenomena turns out to be almost an order of magnitude higher than what was expected from former, perturbative QCD based models [196]. Analogously, η/s has turned out to be much smaller than the previous perturbative QCD expectation [92]. The two quantities may be related [300] via the heuristic expression

$$\frac{\eta}{s} \approx 3.75\, C\, \frac{T^3}{\hat{q}} \tag{7.93}$$

with C a to be determined constant; $C = 1/3$ from [300]. This relation shows that a larger value of $\hat{q}$ implies a small value for the ratio η/s. The latter has a lower bound by the general quantum gauge field theory limit $\eta/s \geq (4\pi)^{-1}$ [171], a value not too far from the estimate $\eta/s = 0.09 \pm 0.02$ derived from relativistic hydrodynamics applied to elliptic flow v_2 [94, 170]. As a consequence, $\hat{q}$ cannot grow beyond a certain upper bound that should be established at LHC energy.

These considerations are an example of the recent intense theoretical search for alternative methods of real-time strong coupling calculations, complementary to lattice QCD. In this regime, lattice QCD has to date been the prime non-perturbative calculational tool. However, understanding collective flow, jet quenching and primordial photon radiation requires real time dynamics, on which lattice QCD information is at present both scarce and indirect. Complementary methods for real-time strong coupling calculations at high temperature are therefore being explored. For a class of non-abelian thermal gauge field theories, the conjecture of a correspondence between anti-de Sitter space-time theory and conformal field theory (the so-called AdS/CFT conjecture) has been shown [301] to present such an alternative. It maps nonperturbative problems at strong coupling onto calculable problems of classical gravity in a five-dimensional anti-de Sitter (ADS_5) black hole space-time theory [302]. In fact, this formalism has been recently applied [212] in a calculation of the transport coefficient $\hat{q}$ that governs in-medium jet attenuation, resulting in an effective, expansion time averaged $\hat{q}_{eff} = 5$ GeV^2/fm at $T = 300$ MeV corresponding to top RHIC energy, rather close to the experimental estimates (c.f. Figs. 7.41 and 7.48).

Thus, it does not seem to be too far-fetched to imagine [301] that the quark-gluon plasma of QCD, as explored at RHIC, and soon at the LHC (and theoretically in lattice QCD), and the thermal plasma of certain supersymmetric conformal gauge field theories (for example N = 4 "Super-Yang-Mills" (SYM) theory as employed in [212, 301]) share certain fundamental common properties.

The focus on early time in the dynamical evolution of matter in nucleus-nucleus collisions is specific to RHIC energy as the initial interpenetration period of two Lorentz contracted mass 200 nuclei amounts to 0.15 fm/c only. The subsequent evolution is thus reflected in a multitude of observables. It is, at first, described as a one-dimensional Hubble expansion [61], setting the stage for the emergence of the medium specific quantities addressed above (gluon saturation, direct photon production, hydrodynamic elliptic flow, jet quenching and J/Ψ suppression). These observables tend to settle toward their eventually observed patterns at $t \leq 1.0$–1.5 fm/c, owing to the fact that they are most prominently determined under primordial conditions of high temperature and density. For example, photon production reflects T^4, and the transport coefficient $\hat{q}$ falls with T^3 [212].

On the contrary, at the energy of former studies at the CERN SPS, $6 \leq \sqrt{s} \leq 20$ GeV, such early times stay essentially unresolved as the initial interpenetration takes upward of 1.5 fm/c. A fireball system in local (or global) equilibrium thus develops toward $t = 3$ fm/c, at T about 220 MeV, closer to the onset of hadronization [85, 86]. Also the baryo-chemical potential is much higher than in RHIC collisions, $250 \leq \mu_B \leq 450$ MeV. However, we thus gain insight into the QCD physics of the hadronization, and high μ_B domain of the phase diagram sketched in Fig. 7.1, in the vicinity of the conjectured parton-hadron coexistence line of QCD.

For reference of such data, e.g. statistical species equilibrium (Sect. 7.3), dilepton studies of light vector meson "melting" at the phase boundary (Sect. 7.6.3), and hadronic event-by-event fluctuations (Sects. 7.7.2 and 7.7.3), to theoretical QCD predictions, a recent progress of lattice QCD [8–10] is of foremost importance. The technical limitation of lattice QCD to the singular case of vanishing chemical potential, $\mu_B = 0$ (which arises from the Fermion determinant in the lattice formulation of the grand canonical partition function), has been overcome recently. Three different approaches have been outlined, the respective inherent approximation schemes touching upon the limits of both the mathematical and conceptual framework of lattice theory, and of present day computation power even with multi-teraflop machines. First results comprise the prediction of the parton-hadron phase boundary line, which interpolates between the well studied limits of the crossover-region at $\mu_B \to 0$, $T \geq T_c$ and the high net baryon density, low T region for which a first order character of the phase transition has been predicted by chiral QCD models [15]. We have illustrated this line in Fig. 7.1, and we have shown in Sect. 7.3 that hadronic freeze-out occurs at, or near this line at $\sqrt{s} \geq 17.3$ GeV (top SPS energy). The coexistence line includes an intermediate (T, μ_B) domain featuring a critical point of QCD at which the first order line at higher μ_B terminates, in a critical domain of (T, μ_B) in which the transition is of second order. One thus expects the nature of the confining hadronization transition—an open QCD question—to change from a crossover to a second order, and onward to a first order characteristics in a relatively small interval of μ_B that is accessible to nuclear collision dynamics at the relatively modest $\sqrt{s}$ of about 5 to 15 GeV. This domain has as of yet only received a first experimental glance, but the top AGS and low SPS energy experiments exhibit phenomena that can be connected to the occurrence of a critical point and/or a first order phase transition, notably the "SPS plateau" in $\langle m_T \rangle$, the non-monotonous K^+/π^+ excitation function, and the eventwise fluctuations of this ratio (Sects. 7.2.6, 7.3, 7.7.2 and 7.7.3). A renewed effort is underway at RHIC, at the CERN SPS and at the future GSI FAIR facility to study hadronization, in-medium meson modification induced by the onset of QCD chiral restoration, as well as critical fluctuations and flow, in the low $\sqrt{s}$ domain.

7.9   Postscript

Reinhard Stock

7.9.1   Progress of the Field

Returning, after a decade, to the topic of "Relativistic Nucleus-Nucleus Collisions and the QCD Phase Diagram", one is struck by the impressive progress that has transported many of the crucial themes, and physics observables, from initial, qualitative consideration to oftentimes quantitative comprehension. We shall briefly revisit some of these topics below, sketching the present state of the art, and mentioning some completely new developments (such as the study of light system collisions and, more spectacularly, the possible advent of Equation of State(EOS) information from neutron star mergers). This will, however, not be a real review but merely a narrative of recent progress.

The development of the field took a decisive turn by the startup of the CERN LHC collider which did not only move the energy frontier up to truly asymptotic values, 2.76 and 5.02 TeV per projectile nucleon pair in $Pb + Pb$ collisions, but also introduced a completely new generation of nuclear collision experiments, offering an extended reach of physics observables and, moreover, a vastly increased event statistics capability ranging upward well into the 10^9 domain. On the other hand, the "blockbuster" innovations from the late RHIC period, minimally viscous hydrodynamics with specific viscosity η/s tantalizing close to the fundamental field theoretical limit (establishing the QGP as a near-ideal liquid), transport coefficients of the QGP from in-medium-jet quenching, statistical production of charmonia and the impression that charmed quarks are also thermalized in the deconfined QCD phase, the saturation of the QCD hadronization temperature near 160 MeV, just to mention the highlights, have remained the cornerstones of LHC physics to a large degree. This holds, also, for the notoriously evasive gluon saturation and Colour Glass Condensate physics, and to the inconclusive critical point searches. In retrospect, one may state that one very central and important element of the RHIC progress (handed on to the LHC physics) consisted of a clear separation of the collisional initialization time period (the first fm/c interval) from the ensuing hydrodynamical evolution and its in-medium effects. This provided for a much clearer relationship between data and theory: it is the simple feature that projectile-target interpenetration which took several fm/c at SPS energies is now shrunk to subfractions of a fm/c thus providing for a sharp, global synchronization of the successive eras of collisional dynamics: initialization, flow expansion, hadronization, final hadron/resonance "afterburning". This is essential as one wants to tie the formation of the physics observables to a specific stage of the dynamical evolution, thus making them clearcut diagnostic tools.

One last thing. Neither RHIC nor the LHC experiments were built to explore phenomena typical of a large baryochemical potential, such as the critical point of QCD or the existence of a first order phase transition toward yet lower energies and, quite generally, the QCD phase diagram at high μ_B. However, at second thought a vigorous development at RHIC, the Beam Energy Scan(BES) program, as well as a re-vitalization of SPS experiments(the NA61/Shine experiment), and a concurrent brilliant extension of lattice QCD technique concerning ("critical") fluctuation observables, were undertaken. Also, the low energy, high μ_B domain will come under renewed focus with new facilities, FAIR at GSI Darmstadt and NICA at JINR Dubna, both under construction.

In the following we will present brief sketches of the recent and concurrent work, theoretical and experimental, concerning a couple of (subjectively selected) observables: reaction dynamics decomposed into 4 characteristic, separate stages, initialization, hydrodynamic expansion, hadronization and final hadron/resonance "gas" expansion. We shall focus on the two characteristic parameters of the QCD plasma medium, the specific viscosity η/s and the transport coefficient $\hat{q}$, as revealed by anisotropic flow and hadron/jet suppression, respectively. Then we turn to charmonium and bottonium suppression in the QCD medium, paying special attention to charmed meson statistical equilibrium hadronization. This leads us to hadronization and the information we have on the QCD phase diagram. These topics are all in full development since RHIC's first decade of operation, and have been introduced in the preceding review article. Then we turn to a couple of more recent topics, e.g. small systems analysis, formation of light nuclei and antinuclei, critical point searches and the possible role of (now observable) neutron star mergers to unravel the EOS of cold hadronic matter.

7.9.2 Reaction Dynamics

Whereas, at the SPS energies, a central $Pb + Pb$ collision featured some fraction of the collisional system already expanding toward hadronization at the time where the last participant nucleons just experienced their first collisions, the various evolutional stages are clearly separated from top RHIC energy onward. With Lorentz contraction factors in the hundreds to thousands domain, the primordial collisional system longitudinal size is less than a nucleon radius such that all primordial interactions occur "at once" so that after about 0.5 fm/c the initialization processes can be expected to settle down toward local equilibrium. This creates a primordial fireball that still features a highly clumpy energy density distribution. This distribution fluctuates from event to event due to a number of influences stemming from remaining impact parameter fluctuation, instantaneous density fluctuations within the average projectile-target Woods Saxon nucleon position/density profile (which are thus getting "photographed"), and by the presence or absence of colour saturation processes [303]. The latter can go along with the formation of a Colour Glass Condensate State [304] which decays to a parton system at the end of the initialization period, thus creating a particular topology of the primordial energy density distribution [305]. A model calculation of a zero impact parameter $Au + Au$ collision transverse energy density distribution was already shown in Fig. 7.60, exhibiting pronounced clumpiness. This distribution is then, approximately, translated into the energy-momentum tensor that starts the next stage: hydrodynamic expansion. The most remarkable feature of this evolution (see below) is the occurrence of a nearly ideal hydro-flow dynamics which approximatively preserves information, the better the lower dissipative processes like shear viscosity turn out to be. Thus, most remarkably, the eventwise fluctuating primordial fireball profile becomes measurable after flow decoupling, modulo the strength of shear viscosity. Two of the objects of desire in this physics [306]!

The hydrodynamic evolution stage is thus, theoretically, well separable from the initialization period. The transition is, in itself, of course a matter of theoretical model building. We mention here, in particular, investigations of the AdS/CFT correspondence method [307] with respect to primordial equilibration [308]. Collective flow then transports the density distribution toward decoupling; this stage is analyzed under the general idea to follow the Fourier components ϵ_n of the primordial energy density distribution eccentricity as they get delivered to observation in the form of spatial flow anisotropics quantified [309] by their Fourier harmonics decomposition coefficients v_n (see ref. [310] and bibliography therein). Strictly speaking, the hydro-phase does not directly "deliver" the observables of viscous, anisotropic flow to observation. This phase ends at hadronization because the mean free path in the ensuing hadron-resonance(HR) expansion stage is too long for hydrodynamics to apply. One thus introduces a transition from hydro to HR gas by means of a hadronization model that translates hydro matter flow into hadron-resonance propagation (by the Cooper-Frye formalism [311]). The latter is then described by a hadron transport model acting as an "afterburner" [312, 313]. Such so-called hybrid models thus consist of three stages that incorporate stage specific dynamical models for initialization, hydro flow and hadron-resonance afterburning. This allows for a very wide variety of theoretical choices concerning the overall dynamics. Add to this that the experimental data from RHIC and LHC can be analyzed with various techniques, delivering the flow coefficients v_n (reaction plane method, cumulant method etc.). Thus an almost excessive wealth of data to model comparisons have been undertaken, with the highest focus on elliptical flow v_2. Data exist for up to v_6. In general v_1—directed flow—and v_2 are predominantly related to impact geometry variations of mass 200 collisions, creating anisotropic primordial energy density gradients. Whereas v_3, in particular, appears to be more dependent on the primordial energy density profile and its even by event fluctuations, potentially related to Colour Glass Condensate(CGC) formation [314]. Very briefly summarizing the results one observes an anticorrelation between the initialization-, and the hydro flow-effects. The specific shear viscosity η/s (that quantifies the speed with which the system approaches equilibrium) is tantalizingly close to the Kovtun, Son and Stariets [315] minimum of $1/4\pi$, throughout, but its deduced value at RHIC goes up with changing the initialization model from a "trivial" Glauber trajectory choice, to a CGC model, as is illustrated in Fig. 7.72 with RHIC STAR data [314] for v_2 vs. theory [315]. Employing the Glauber initialization one deduces η/s to be near 0.08 (the KSS limit) whereas the CGC model requires twice the shear velocity, about 0.16. The Glauber initial fireball is wider, spatially, than the one from the CGC model. That ambiguity has not finally been settled yet, it persists at the LHC, where the specific viscosity is slightly larger. The overall present result, within the CGC type of hybrid dynamical models, is given by Eskola and coworkers [316] and illustrated in Fig. 7.73. Similar results for η/s have been given by other theory groups [313, 317]. In summary, an unambiguous conclusion about the QGP state as a near ideal liquid has been arrived at, whereas the existence of the CGC still remains to be finally established.

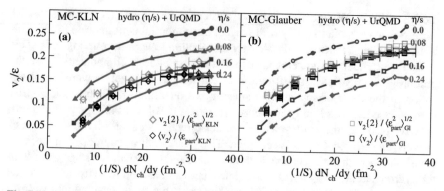

Fig. 7.72 STAR v_2 data [314] confronted with viscous hydrodynamics [315] for various choices of η/s, employing a Glauber- and a Colour Glass type initialization

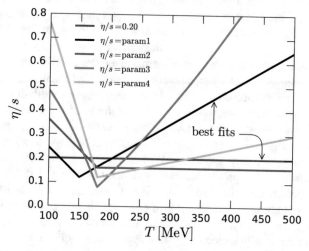

Fig. 7.73 Energy dependence of η/s from a simultaneous fit of RHIC and LHC data [316], with CGC initialization

7.9.3 Energy Loss in a QCD Medium: Hadron Suppression and Jet Quenching

The hydrodynamic model of the plasma evolution employs quantities that could be derived from QCD, for example the equation of state and the specific shear viscosity η/s. A different aspect of the QGP is seen by an individual hard parton, or a jet, traversing it. In general, the overall medium effect will be energy loss and momentum broadening, as well as a re-appearance of the lost energy in the form of soft emission. A major obstacle in the interpretation of single parton(seen as hard hadrons) and jet attenuation is the medium expansion, i.e. one needs, not a hydrodynamic but a genuinely microscopic QCD transport model. Then the key ingredient in a theoretical description will be the parton transport parameter $\hat{q}$ that we introduced in Sects. 7.5.2 and 7.5.3. We recall that it represents the average transverse momentum broadening square per unit pathlength, $\langle p_T^2 \rangle / \lambda$ (cf. [359–362, 366]). It is proportional to the local gluon number density. Recall further that the average in-medium energy loss is proportional to $\hat{q}$ [363]. This parameter can be systematically deduced from identified hadron p_T spectra in $A + A$ collisions as compared to the same spectra in minimum bias $p + p$ at similar energy, resulting in the nuclear modification factor R_{AA} as defined in Sect. 7.5.1. For jets the standard method is to estimate the jet energy loss by comparing the leading hadron (or fully reconstructed jet) energy on the trigger side with that of the away-side jet, where geometry has made sure that the trajectory traversed the QGP medium. Note, by the way, that $\hat{q}$ is inversely proportional to η/s from Eq. (7.93) (from old article!). Thus if the latter is in a sense "near-minimal" the transport coefficient has to be near-maximal. So the medium opaqueness might completely wipe out the jet at lower energies, as was indeed observed at RHIC (see Figs. 7.50 and 7.53): the black interior situation. At the LHC jet energies are always sufficient to observe both jet sides. We illustrate that in Fig. 7.74 [318] where we see very significant suppression in the "low" energy domain of recent LHC $Pb + Pb$ data, gradually weakening toward high p_T. This indicates that the energy loss is not growing in proportion to p_T. Corresponding results for the $p + Pb$ jet production at the LHC indicates essentially zero medium effect, asserting that the suppression in $Pb + Pb$ is indeed a final state, QGP effect. We mentioned above that all data concerning heavy flavour charm quark attenuation point to a behaviour similar to that of the

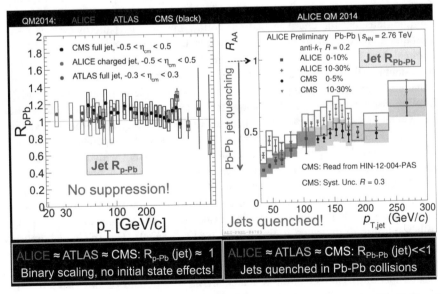

Fig. 7.74 Nuclear Modification Factors for jet production, obtained by the ALICE, CMS and ATLAS LHC collaborations, for p + Pb and Pb + Pb collisions[318]

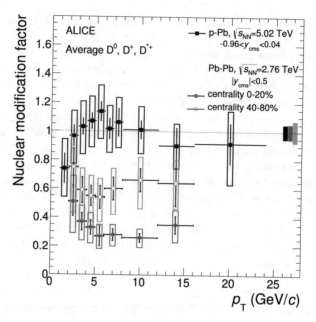

Fig. 7.75 R_{AA} factors for D meson production in $p + Pb$ and $Pb + Pb$, by ALICE at LHC [319]

light quarks. This is illustrated in Fig. 7.75 by ALICE data [319] for D meson production in $p + Pb$ and $Pb + Pb$, central $Pb + Pb$ collisions again showing drastic suppression whereas this is absent in the $p + Pb$ collision. We should also mention here that the charmed hadrons follow the flow v_n patterns observed for light quark hadrons. In the end, all $A + A$ data, both about the shear viscosity, near minimal, and the transport coefficient, have been shown to be semi-quantitatively consistent [320]. Figure 7.76 shows the result of current state of the art theory; quite remarkably, the theory groups have followed the example of the experimentalists, forming collaborations. Here we show the transport coefficient results from the JET Collaboration [320], extracted from fits to the combined data from RHIC and LHC, as of 2014. Indeed, the transport coefficient falls with temperature, consistent with the inversely proportional specific shear velocity rising with T. Note that the temperature scale reflects the energy density in the center of the fireball at an initial time of 0.6 fm/c, that we have ascribed above to the end of the primordial initialization period. This would be the medium in which the hard partons born by perturbative QCD interactions start embedding. The results of many theory collaborations are indicated in this Figure (see ref. [320]). The JET collaboration employs perturbative QCD technique to generate the jets in the primordial $A + A$ environment, then follows them through the co-travelling non-perturbative QCD plasma medium. This situation is characteristic of all "hard probes" studies: one has to combine the short range pQCD scale with the long range scale inherent in the structure(s) of the plasma QCD medium: the leading parton does not interact

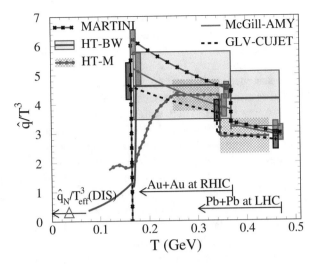

Fig. 7.76 Temperature dependence of the jet transport coefficient $\hat{q}/T^3$, extracted with different approaches for jet quenching at an initial jet energy of 10 GeV, in central A+A collisions (see ref. [320])

with a free gas of partons! The diagnostic of this state is the goal of all had probes physics. The present estimate [320] of the transport coefficient amounts to about 1.5 GeV^2/fm for a 10 GeV quark jet. The lost energy gets radiated by soft gluon emission from the plasma and is indeed recovered experimentally in hadrons populating the vicinity of the jet cone. The question of the elementary degrees of freedom in a high T QCD plasma is coming within reach by such studies.

7.9.4 Charmonium

The J/Ψ signal entered relativistic heavy ion physics as the "Holy Grail" observable of QGP formation by the work of Matsui and Satz [321], at the time the CERN SPS $A + A$ program took shape, in the late 1980s. From among all observables predicted then to provide evidence of QCD deconfinement and QGP formation, the proposed in-medium colour Debye-screening mechanism preventing the primordially produced $c\bar{c}$ pair to hadronize as a J/Ψ hidden charm meson was presenting a direct link between deconfined partons and the suppressed cross section of observed J/Ψ, which was relatively well known from elementary $p + p$ and $e^+ + e^-$ collisions. We have reported in Sect. 7.6.1 on the success of the charmonium suppression idea at the SPS (Fig. 7.54) and at RHIC (Fig. 7.55). However, we recall that it was predicted early on that if free charm and anticharm quarks were thermalized in the plasma medium (as has been verified meanwhile) they would hadronize statistically forming open charm mesons and charmonia [322] in relative hadro-chemical equilibrium, along with all other hadrons, albeit with a separate fugacity factor for c and $\bar{c}$ because these stem from the early hard collision phase unlike the majority of the lighter quarks [323]. We note that this mechanism is oftentimes called recombination, erroneously so as the thermal charm quarks do not really stem from Debey-screened J/Ψ breakup which covers a mere 1% fraction of the total primordial charm-antiquark production. The statistical $c\bar{c}$ hadronisation mechanism and its traces in RHIC data were soon to be fully substantiated [324]. At LHC energy we expect a significantly higher number density of thermalized c and $\bar{c}$ quarks as the system enters hadronization and, hence, a higher fraction of chemical equilibrium charmonia (for a recent review see ref [325]). Indeed, Fig. 7.77 (from [325]) shows that the R_{AA} ratios for J/Ψ drop down far more steeply with midrapidity multiplicity (centrality) at RHIC than at the LHC where, indeed, not much of the low energy charmonium suppression remains visible, being overshadowed here by the statistical production. We might remark that nobody nowadays insists on a proof of deconfinement anymore (as was the case in 1986) and, furthermore, that the universality of statistical hadron production in $A + A$ collisions, of which we see evidence here, is in itself a strong manifestation of deconfinement. To this topic we turn our attention next, but remark, in passing, that

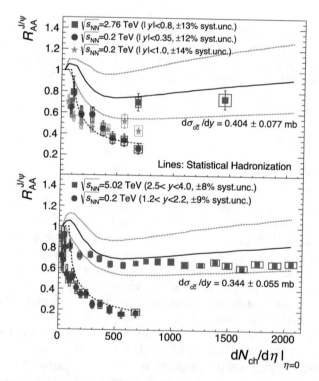

Fig. 7.77 Multiplicity(centrality) dependence of the Nuclear Modification Factor for J/Ψ production at mid-rapidity at top RHIC, and at LHC energy [325]

the suppression theme is now turning over to the bottonomium production at the LHC [326], visible there because of the much lower thermal bottom quark density.

7.9.5 Hadronization and the QCD Phase Diagram

This chapter will have two main topics, first the analysis of hadron production multiplicities with the Statistical Hadronization Model (SHM) that incorporates the canonical or grandcanonical partition functions of a hadron-resonance gas (HRG) model. This will be an update on Sect. 7.3. Second, however, we turn to new developments of lattice QCD theory, extension to finite baryochemical potential and determination of the hadronization parameters T vs. μ_B from, first, a determination of the overlap between Lattice QCD and the HRG model [327], to reveal the "hadronization point" and, second, from fitting new data concerning fluctuation of conserved charges to higher order Lattice susceptibilities [328]. Some remarks are also necessary with regard to the Lattice conclusion that the parton to hadron phase transition is "merely" a cross-over at small μ_B, top RHIC and LHC energy.

We have shown in Sect. 7.3.4 that, ideally, the Statistical Hadronization Model(SHM) freeze-out curve should reveal the QCD parton-hadron phase transformation line in the (T, μ_B) plane, i.e. the most prominent feature of the QCD phase diagram. This follows from the assumption that inelastic reactions (except resonance decays) among hadrons cease directly at the instant of hadronization: chemical hadron freeze-out determines the hadronization "point" corresponding to the chosen $A + A$ collision energy. Add, for correctness: as long as freeze-out occurs from the QCD hadronization transition, i.e. from a QGP. About freeze-out from hot, dense hadronic matter in expansion, we do not yet know the final interpretation of the hadronic multiplicities. Now, the above ideal picture, absence of final state effects from the afterburning phase, may require certain corrections. It has been shown [329] that, on the one hand, the bulk hadrons from a relativistic $A + A$ collision (i.e. pions and kaons that carry about 95% of the total cm energy output at LHC energy) do indeed pass the afterburner stage essentially unchanged. At the relatively low SPS energies, this applies also to protons and Lambdas. Whereas the antibaryon yields get reduced, away from chemical equilibrium, by annihilation processes that occur throughout the hadron/resonance expansion phase. At top RHIC and LHC energies, baryons and antibaryons are similarly affected, with the exception of the Omega hyperons which suffer little annihilation. These afterburner effects were quantified [330] by the microscopic transport model UrQMD [312] which is employed in many of the current multi-stage "hybrid" models [313, 316] for the afterburner stage. The thus obtained modification factors for each species were applied to the grand canonical partition functions of the SHM [330]. We show the results in Fig. 7.78, which gives (T, μ_B) points for AGS, SPS and LHC energies. The overall result of the annihilation corrections is seen to be an upward shift of the hadronization curve(see also [331]), above the curve resulting from the uncorrected SHM procedure which agrees with other SHM work [332]. The

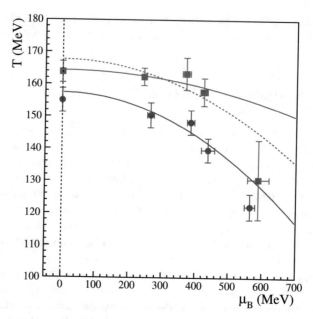

Fig. 7.78 Freeze-out points in the (T, μ_B) plane extracted from hadron multiplicity data at LHC, SPS and AGS energies, obtained with the standard Statistical Hadronization Model, and with the SHM corrected for baryon-antibaryon annihilation during the afterburner phase [330]

modified approach also explains the so-called "nonthermal proton to pion ratio puzzle" at the LHC [332], much discussed but simply resulting, in this model, from proton annihilation going to pions, thus decreasing the p/π ratio; of course a non-thermal effect [333]. The pseudo-critical temperature at $\mu_B=0$ turns out to be 164 ± 5 MeV, as compared to about 155 MeV in the standard approach. This new development is still vigorously contested by the ALICE community [325, 332]; it appears highly desirable that groups employing a different afterburner model turn their attention to the final state effects.

The reader will have noticed that entries from RHIC are missing in Fig. 7.78. This is the result of another tension in the community. By far the most comprehensive set of hadron multiplicity data stems from the STAR experiment at RHIC [334] which, up until now, has published baryon/antibaryon multiplicities without correction for feeddown from weak hyperon decays which are misidentified as primary particles (this omission is now disposed of, for subsequent data taking, by the new STAR vertex tracking detector [335]). Accidentally, the extra particles from feeddown do closely compensate for the losses by annihilation, and a picture of perfect equilibrium thus emerges from SHM analysis [334]. Then, apparently, there arises no p/π puzzle at top RHIC energy, the two missing corrections cancelling. We are at μ_B about 20 MeV, very close to the value of about 1 MeV encountered at LHC. A kink in the phase diagram? Clearly an inconsistency. In ALICE at the LHC, the primary vertex is very precisely reconstructed, eliminating secondary decays, and

NA49 at the SPS has employed a formidable correction simulation to determine the feeddown fractions, of order 50%. Once all this dust settles, one can still expect an interesting freeze-out curve [323], which should reflect influences of a critical point, perhaps even adjacent to a new, further form of QCD phase transition, from quarkyonic [337] to hadronic matter, that interpolates between the partonic and the hadronic phases at high μ_B [323]. This region in the QCD phase diagram will receive the necessary attention by NA61(SPS) and by the FAIR and NICA projects.

We turn to a second, very much elaborated source of information about the parton-hadron phase boundary, which was made possible by recent developments of Lattice QCD. In a way, this progress was a generalization of the attempt to extend Lattice QCD from $\mu_B=0$ to finite baryochemical potential [336], by a Taylor expansion of the reduced thermodynamic pressure in terms its derivatives with respect to μ_q/T :

$$\frac{p}{T^4} = \frac{1}{VT^3} \ln \mathcal{Z}. \tag{7.94}$$

$$\frac{p}{T^4}\bigg|_{T,\mu_q} = \frac{p}{T^4}\bigg|_{T,0} + \frac{1}{2!}\frac{\mu_q^2}{T^2}\frac{\partial^2(p/T^4)}{\partial(\mu_q/T)^2} + \frac{1}{4!}\frac{\mu_q^4}{T^4}\frac{\partial^4(p/T^4)}{\partial(\mu_q/T)^4} + \cdots \tag{7.95}$$

where derivatives are taken at $\mu_q = 0$. The derivatives of the reduced pressure, of order n on the r.h.side of equ. [304], with respect to a chemical potential, are called susceptibilities. In general, fluctuations of conserved quantum numbers are obtained as derivatives of the pressure to various chemical potentials μ_X/T, where X stands for net baryon number B, charge Q, strangeness S or charm C (see [328, 337, 338] and references therein). Also one can formulate correlations of net charges X and Y by mixed susceptibilities [339] which contain derivatives jointly to μ_X and μ_Y. Now, all such susceptibilities could be obtained from recent Lattice calculations, up to order six, as well as from the Hadron Resonance Gas (HRG) model and, most significantly, some of them can be related to experimentally accessible quantities. For example, net baryon number fluctuations (one can approximate them by net proton number fluctuations) are expressed by properties of the event by event multiplicity distribution, e.g. mean (M_B), variance (σ_B^2), skewness (S_B) and kurtosis (κ_B):

$$R_{12}^B(T,\mu_B) \equiv \frac{\chi_1^B(T,\mu_B)}{\chi_2^B(T,\mu_B)} \equiv \frac{M_B}{\sigma_B^2},$$

$$R_{31}^B(T,\mu_B) \equiv \frac{\chi_3^B(T,\mu_B)}{\chi_1^B(T,\mu_B)} \equiv \frac{S_B\sigma_B^3}{M_B},$$

$$R_{42}^B(T,\mu_B) \equiv \frac{\chi_4^B(T,\mu_B)}{\chi_2^B(T,\mu_B)} \equiv \kappa_B\sigma_B^2. \tag{7.96}$$

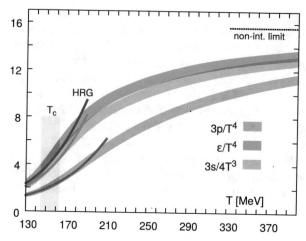

Fig. 7.79 The overlap between Lattice QCD and Hadron Resonance Gas(HRG) model calculations of the normalized pressure, energy density and entropy density, as a function of temperature [341]. It is seen that the results match over a relatively broad temperature domain, characteristic of a cross-over transition, and slightly different for the three quantities

where the n-th order susceptibilities χ_n^B are obtained from the corresponding partial derivatives of the Lattice or HRG pressure $P(T, \mu_X)$ with respect to the baryon chemical potential, and the mean, variance, skewness and curtosis from the data on net proton number fluctuations e.g. by the STAR Collaboration at a set of RHIC energies [340]. We cannot do justice to this very wide body of recent work here but merely present two typical results. Figure 7.79 illustrates the Lattice-HRG overlap analysis [341] technique showing the normalized pressure, energy density and entropy density as a function of temperature, at $\mu_B = 0$. The HRG curves start seriously departing from the Lattice results above about 170 MeV. Recall that the parton-hadron-transition is a relatively broad cross-over at μ_B=0; so we would conclude that the crossover domain ends here. The center of the pseudo-critical region might thus be at about 160 MeV, at $\mu_B = 0$; the authors also show the region center and width estimate from other analyses [337]: not really fully supported by the present analysis. The most recent results in the comparison of Lattice with the STAR kurtosis data are shown in Fig. 7.80 (taken from ref [338] which gives a comprehensive review of the topic). It shows the ratio of the two susceptibilities χ_4^B and χ_2^B, equal to kurtosis times variance, vs. temperature, from various state of the art Lattice calculations, and a band about a freeze-out temperature of $T = 153$ MeV indicated by the data. We cannot exhaust the topic which is under vigorous development but wish to enumerate points of concern that are presently under investigation:

1. It is unclear whether various hadronization observables like hadron multiplicities and higher order multiplicity fluctuations should refer to an identical freeze-out temperature within the relatively broad pseudocritical temperature band. The latter might freeze out later (sequential freeze-out).

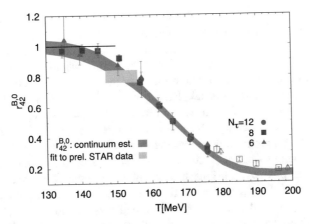

Fig. 7.80 The ratio of the two susceptibilities χ_4^B and χ_2^B, equal to kurtosis times variance, vs. temperature, from various state of the art Lattice calculations [338], and a band about a freeze-out temperature of $T = 153$ MeV indicated by the data

2. Net proton multiplicity is not a conserved quantity but taken here as a proxy for net baryon number: an approximation sensitive to the experimental energy and acceptance [342, 343].
3. Higher order fluctuations and their ratios receive sizeable contributions from the fluctuation, event by event, of the participant nucleon number [343].
4. Higher order fluctuations can be dampened in the course of the afterburner expansion [344].
5. The standard Hadron-Resonance Gas model employed in the Lattice-HRG overlap study may need amendments due to Van der Waals-Type hadron-hadron repulsion [345], or to high lying strange resonances that are not yet experimentally known [346].

Clearly, these questions need to be addressed before we can finally conclude on the QCD phase diagram and the parton-hadron phase boundary.

7.9.6 New Topics

7.9.6.1 Proton Induced Collisions

In defining the nuclear modification factor R_{AA} (Sect. 7.9.3) we have employed, in the denominator, the experimental minimum bias $p + p$ momentum distribution at the same cm energy as the energy per participant nucleon pair of the $A + A$ data forming the numerator. Thus, the $p + p$ cross section is employed as a reference, to incorporate the "no QGP physics" situation that is captured in the Glauber model which describes the $A + A$ collision as an superposition of independent, minimum bias nucleon-nucleon collisions occurring at the microscopic level. Of course this cannot be a realistic model but for the hard, partonic collisions occurring during the primordial interpenetration we may still employ it as a standard of comparison. Nevertheless it was tempting to check whether $p + p$ collisions revealed any of the effects ascribed to the QGP medium in $A + A$ collisions, such as grand canonical hadrochemical equilibrium at 160 MeV apparent temperature, collective hydrodynamic flow in its multifold Fourier harmonics phenomena, jet attenuation etc., thus making it useless as a "no new physics" standard reference. This experimental program occupied the first half of this decade, both at RHIC and LHC. Lacking space for detail, we briefly summarize the main results. Indeed, minimum bias $p + p$ collisions turned out to be essentially free of the new physics: ok to employ them in R_{AA}. However, onset behaviour of grand canonical hadronization [347], of elliptic flow [348], and new modification patterns of jet production [349] were observed in high multiplicity selections of $p + p$ data at LHC where the midrapidity charged particle density increases from about 8 in minimum bias mode, up to well beyond 20 in the ALICE study [347], with extremes ranging up to about 80. No surprise then: these are geometrically small but energy rich systems which lead to equilibrium conditions due to extreme density of degrees of freedom. Such that the bulk phenomena, advent of grandcanonical hadronization [347] and development of anisotropic hydro flow, exhibit onset patterns. Whereas the hard probes which depend on pathlength might well still stay unaffected of the medium. No big surprise, but initially the community was quite smitten by the surfacing of typical QGP signals in an elementary collision. A different, completely new aspect concerning jet production: the selection of extreme multiplicity density at midrapidity clearly represents a strong bias as to the primordial partonic collision generation, possibly selecting for changes in jet observables [349]. An interesting, unexplored aspect of jet production.

Along with $p + p$ came a wave of $p + Au/Pb$ collision studies. Here we would expect that the projectile side of rapidity space shows essentially $p + p$ properties whereas the heavy nucleus side shows, firstly, a strong dependence of impact parameter and participant nucleon number variation. Second, the resulting primordial fireballs should grow in proportion to N_{part}, leading to clear collective signals in the bulk production observables, emerging with increasing centrality (multiplicity), as was indeed observed [347, 350]. However, again the hard probes did not show significant attenuation as we have seen in Figs. 7.74 and 7.75 for LHC collisions [318, 319]. It is reassuring to confirm that extended path length in hot QGP matter is required in order to produce attenuation, whereas cold nuclear spectator matter has a very low transport parameter [320]. In general $p + Pb$ collisions share the complicating feature, common to all light-on-heavy collisions, that the effective center of mass for soft production moves away from the corresponding $p + p$ cm frame that controls hard processes, as N_{part} increases. At the LHC this shift goes up to two rapidity units for very central collisions, potentially causing more or less trivial effects in the attenuation of hard partons traversing the soft-produced QGP medium [351]. Also the effective midrapidity position for soft, bulk production is now outside the acceptance for ALICE.

7.9.6.2 Lambda Polarization and Fireball Vorticity

Huge global angular momenta are generated in non-central $A + A$ collisions. In fact they get converted to the vorticity of a QGP [352] and get transmitted to particle polarization at the stage of hadron formation and particlization whence hadrons are emitted from the hypersurface of a hydrodynamically expanding fireball (51, 52). The global angular momentum can thus lead to the local polarization of hadrons. The polarization, in turn, can be measured by Λ and $\bar{\Lambda}$ hyperon weak decay into a proton and a pion which is "self-analyzing" since the proton is emitted preferentially along the direction of the Lambda spin in its rest frame. The global polarization (the net polarization of the local ones in an event which is aligned in the direction of the event plane, i.e. along the direction of the angular momentum of the plasma) of Λ and $\bar{\Lambda}$ has recently been measured by the RHIC STAR Collaboration [355] at collisional energies below 62.4 GeV. At higher energies, including LHC, the falling of the polarization with energy [355] still precludes measurement. The measurement determines the event average of $\sin(\Phi_p - \Psi_{RP})$, where Φ_p and Ψ_{RP} are the azimuthal angle of the proton momentum in the Lambda rest frame, and that of the reaction plane. Its orientation cannot be directly measured but is approximated by the event plane determined from the hadronic directed flow; this is accounted for by a reaction plane resolution factor [355]. Alternative methods have been proposed recently [356]. For an overview see ref. [357].

From the data one can estimate [353, 354] the local vorticity in the plasma, the result implying that the QCD matter created in such collisions is the most vortical fluid known as of yet. Moreover, this new observable can shed a new, independent light on the equilibrium properties encountered (or not) in the course of (relativistic hydro) expansion toward hadronization [358].

7.9.6.3 EOS from Neutron Star Mergers

The recent observation of gravitational waves from a neutron star merger has, in fact, been anticipated with regard to its possible sensitivity to the EOS of dense hadronic matter at low temperature and high baryochemical potential. This is a topic of nuclear collision research already since the early BEVALAC studies at LBL when it was realized that the EOS should be of importance, both, to the dynamics of supernova explosions and to the radial structure of neutron stars, also affecting their maximum mass with regard to black hole formation. The hope was to derive the EOS from the data concerning hydrodynamic directed sidewards expansion, and from Kaon production in $A + A$ "compressive collisions" [359] at the low BEVALAC energies, in the GeV per nucleon energy range. One of the key problems arose from the fact that even at these energies the collisional fireball temperature ranges up to about 100 MeV whereas the above astrophysical phenomena require the EOS of cold, compressed matter [360]. The newly accessible Neutron star mergers promise to open up a window to the low T EOS [361], and to crucial transport parameters [362], once more experimental information becomes available, as can be expected.

7.9.6.4 Production of Light Nuclei in $A + A$ Collisions

At RHIC and LHC, a large number of light nuclei and antinuclei have been measured [325, 363, 364], from deuterium to anti-alpha. This continues a tradition dating back to BEVALAC time [365] and SPS [366]. Dating back is also the controversy concerning model description or, alternatively stated, the lack of a comprehensive understanding [325, 364, 367]. The Statistical Hadronization Model(SHM) gives a very good description(hadro-chemical freeze-out) of the LHC yields [325], along with all other hadron multiplicities. The coalescence model [365, 367] addresses the end of the hadron/resonance expansion stage (kinetic freeze-out), whence the produced nucleons, spread out in phase space, overlap, to a certain degree, with the internal wave functions of the various clusters. There are difficulties in both views if one takes them literally. The clusters cannot have existed at 160 MeV, the hadronization temperature. On the other hand, the coalescence cannot proceed on-shell because of the cluster binding energies (that are ignored in the model). All this points to a deeper flaw in, either the models, or their conventional understanding. The role of quantum mechanics seems to be missing, as pointed out long ago by E. Remler [368]. Hadronization does not directly produce a decoherent, on-shell state, in $A + A$ collisions. If hadronization occurs via the quantum mechanical decay of initially produced colour singlet clusters (see Sect. 5.3.3), then Fermis Golden Rule refers to the phase space weights AFTER decoherence. And, in fact, in the SHM partition functions we employ the set of in-vacuo free masses of the produced particles, also for the clusters, with apparent success. One might then speculate that the number densities and their ratios are fixed by the entropy at hadronization, but we have to wait until after the freeze-outs to have them decohere. This is a speculation, of course.

References

1. W. Weise, Proc. Intern. School Enrico Fermi, Course 43, edit. A. Molinari and L. Riccati, IOS Press Amsterdam 2003, p. 473;
   A. W. Thomas and W. Weise, The Structure of the Nucleon, Wiley-VCH publ. 2001
2. E. V. Shuryak, Phys. Rept. 61 (1980) 71;
   B. Müller, The Physics of the Quark-Gluon-Plasma, Springer Lecture Notes in Physics Vol.225, 1985;
   L. McLerran, The Physics of the Quark-Gluon Plasma, Rev. Mod. Phys.58 (1986) 1012
3. B. A. Freedman and L. McLerran, Phys. Rev. D16 (1977) 1169;
   E. V. Shuryak, Zh. Eksp. Teor. Fiz.74 (1978) 408;
   J. I. Kapusta, Nucl. Phys. B148 (1979) 461
4. L. Susskind, Phys. Rev. D20 (1979) 2610
5. J. Kuti, J. Polonyi and K. Szalachanyi, Phys. Lett. B98 (1981) 199

6. J. Engels, F. Karsch. I. Montvay and H. Satz, Phys. Lett. B101 (1981) 89
7. F. R. Brown, F. P. Butler, H. Chen, N. H. Christ, Z. Dong, W. Schaffer, L. I. Unger and A. Vaccarino, Phys. Rev. Lett. 65 (1990) 2491
8. F. Karsch, Nucl. Phys. A590 (1995) 372; R. Stock, hep-ph/9901415
9. Z. Fodor and S. D. Katz, JHEP 0203 (2002) 014;
   Ph. de Forcrand and O. Philipsen, Nucl. Phys. B642 (2002) 290
10. C. R. Allton et al., Phys. Rev. D68 (2003) 014507
11. F. Karsch and E. Laerman, in Quark-Gluon Plasma 3, eds. R. C. Hwa and X. N. Wang, World Scientific 2004, p.1
12. S. L. Shapiro and S. Teukolsky, Black Holes, White Dwarfs and Neutron Stars, Wiley Publ., 1983
13. F. Weber, Pulsars as Astrophysical Laboratories, Inst. of Physics Publ., 1999
14. R. Oechslin, H. T. Janka and A. Marek, astro-ph/0611047
15. R. Rapp, T. Schäfer and E. V. Shuryak, Annals Phys. 280 (2000) 35
16. R. D. Pisarski and F. Wilczek, Phys. Rev. D29 (1984) 338
17. E. Witten, Phys. Rev. D30 (1984) 272;
    H. Kurki-Suonio, R. A. Matzner, K.A. Olive and D. N. Schramm, Astrophys. J. 353 (1990) 406
18. K. Kajantie and H. Kurki-Suonio, Phys. Rev. D34 (1986) 1719
19. F. Becattini et al., Phys. Rev. C69 (2004) 024905
20. M. G. Alford, K. Rajagopal and F. Wilczek, Nucl. Phys. B537 (1999) 443;
    D. H. Rischke, Progr. Part. Nucl. Phys. 52 (2004) 197
21. J. P. Blaizot, E. Iancu and A. Rebhan, in Quark-Gluon Plasma 3, eds. R. C. Hwa and X. N. Wang, World Scientific 2004, p.60
22. G. Baym, B. L. Friman, J. P. Blaizot, M. Soyeur and W. Czyz, Nucl. Phys. A407 (1983) 541;
    H. Stöcker and W. Greiner, Phys. Rept. 137 (1986) 277;
    L. P. Csernai, Introduction to relativistic heavy ion collisions, Wiley 1994;
    P. Huovinen, P. F. Kolb and U. Heinz, Nucl. Phys. A698 (2002) 475
23. N. Cabibbo and G. Parisi, Phys. Lett. B59 (1975) 67
24. G. Veneziano, Nucl. Phys. B159 (1979) 213
25. G. T'Hooft, Nucl. Phys. B138 (1978) 1, and B190 (1981) 455;
    A. Di Giacomo, Proc. Intern. School Enrico Fermi, Course 43, edit. A. Molinari and L. Riccati, IOS Press Amsterdam 2003, p.401
26. K. Brückner, Phys. Rev. 96 (1954) 508
27. J. D. Walecka, Theoretical Nuclear and Subnuclear Physics, Oxford Univ. Press 1995
28. H. A. Bethe, G. E. Brown, J. Cooperstein and S. Kahana, Nucl. Phys. A324 (1979) 487;
    H. A. Bethe, G. E. Brown, Sci. Am. 252 (1985) 40
29. N. K. Glendenning, Compact Stars, Nucl. Astronomy and Astrophysics Library, Springer 1997
30. R. Buras, M. Rampp, H. Th. Janka and K. Kifonidis, Phys. Rev. Lett. 90 (2003) 241101;
    M. Prakash, S. Ratkovic and J. M. Lattimer, J. Phys. G30 (2004) 1279
31. W. Scheid, H. Müller and W. Greiner, Phys. Rev. Lett. 32 (1974) 741
32. C. F. Chapline, M. H. Johnson, E. Teller and M. S. Weiss, Phys. Rev. D8 (1973) 4302
33. H. A. Gustafsson et al., Plastic Ball Coll., Phys. Rev. Lett. 52 (1984) 1590;
    G. Buchwald, G. Graebner, J. Theis, J. Maruhn, W. Greiner and H. Stöcker, Phys. Rev. Lett. 52 (1984) 1594
34. C. Fuchs et al., Phys. Rev. Lett. 86 (2001) 1974
35. quoted after G. Baym, Nucl. Phys. A689 (2002) 23
36. J. C. Collins and M. J. Perry, Phys. Rev. Lett. 34 (1975) 1353
37. S. Weinberg, The first three minutes, Basic Books Publ. 1977
38. R. Hagedorn, Suppl. Nuovo Cimento 3 (1965) 147
39. E. V. Shuryak, Phys. Lett. B78 (1978) 150
40. F. Karsch, Nucl. Phys. A698 (2002) 199
41. T. Matsui and H. Satz, Phys. Lett. B178 (1986) 416

42. F. Karsch, E. Laermann and A. Peikert, Nucl. Phys. B605 (2001) 579
43. T. Alber et al., NA49 Coll., Phys. Rev. Lett 75 (1995) 3814
44. M. Agarwal et al., WA98 Coll., Eur. Phys. J. C18 (2001) 651
45. J. D. Bjorken, Phys. Rev. D27 (1983) 140
46. J. Adams et al., STAR Coll., Nucl. Phys. A757 (2005) 102
47. K. J. Eskola, K. Kajantie, P. Ruuskanen and K. Tuominen, Nucl. Phys. B570 (2000) 379;
    E. V. Shuryak and L. Xiong, Phys. Rev. Lett. 70 (1993) 2241
48. F. Karsch, E. Laerman and A. Peikert, Phys. Lett. B478 (2000) 447;
    S. Hands, Contemp. Phys. 42 (2001) 209
49. P. Abreu et al., Phys. Lett. B459 (1999) 597
50. M. Basile et al., Phys. Lett. B92 (1980) 367 and Phys. Lett. B95 (1980) 311
51. G. Panduri and C. Rubbia UA1 Coll., Nucl. Phys. A418 (1984) 117
52. G. Roland et al., PHOBOS Coll., Nucl. Phys. A774 (2006) 113
53. M. D. Baker et al., PHOBOS Coll., Nucl. Phys. A715 (2003) 65; P. A. Steinberg et al.,
    PHOBOS Coll., ibidem p.490
54. Ch. Blume, J. Phys. G31 (2005) 57, and references therein
55. L. D. Landau, Izv. Akad. Nauk. SSSR 17 (1953) 52;
    P. Carruthers and M. Duong-Van, Phys. Rev. D8 (1973) 859
56. I. G. Baerden et al., BRAHMS Coll., Phys. Rev. Lett. 93 (2004) 102301, and references to
    the other data therein
57. Ch. Blume et al., NA49 Coll., nucl-ex/0701042
58. P. Drijard et al., Nucl. Phys. B155 (1979) 269
59. A. Bialas, M. Bleszynski and W. Czyz, Nucl. Phys. B111 (1976) 461
60. A. Bialas and W. Czyz, Acta Phys. Pol. 36 (2005) 905
61. M. Gyulassy and L. McLerran, Nucl. Phys. A750 (2005) 30
62. A. H. Müller, Nucl. Phys. A715 (2003) 20, and references therein
63. N. Armestro, C. A. Salgado and U. A. Wiedemann, Phys. Rev. Lett. 94 (2005) 022002
64. J. Jalilian-Marian, J. Phys. G30 (2004) 751
65. L. D. McLerran and R. Venugopalan, Phys. Rev. D49 (1994) 2233 and ibid. 3352
66. D. Kharzeev, E. Levin and M. Nardi, hep-ph/0408050
67. S. Afanasiev et al., NA49 Coll., Nucl. Instr. Meth. A430 (1999) 210
68. K. Werner, Phys. Rept. 232 (1993) 87
69. B. Lungwitz, NA49 Coll., Frankfurt Thesis 2004
70. U. V. Gribov, E. M. Levin and M. G. Ryskin, Phys. Rept. 100 (1983) 1
71. A. H. Müller J. Qiu, Nucl. Phys. B268 (1986) 427;
    J. P. Blaizot and A. H. Müller, Nucl. Phys. B289 (1987) 847
72. J. Breitweg et al., Eur. Phys. J. 67 (1999) 609, and references therein
73. S. J. Brodsky, H. C. Pauli and S. S. Pinsky, Phys. Rept. 301 (1998) 299
74. R. K. Ellis, W. J. Stirling and B. R. Webber, QCD and Collider Physics, Cambridge
    Monographs 1996
75. For a review see E. Iancu and R. Venugopalan in Quark-Gluon Plasma 3, eds. R. C. Hwa and
    N. X. Wang, p. 249, World Scientific 2004
76. Y. L. Dokshitzer, Sov. Phys. JETP 46 (1977) 641;
    G. Altarelli and G. Parisi, Nucl. Phys. B126 (1977) 298
77. D. Kharzeev, E. Levin and M. Nardi, hep-ph/0111315
78. A. M. Stasto, K. Golec-Biernat and J. Kwiecinski, Phys. Rev. Lett. 86 (2001) 596;
    K. Golec-Biernat and M. Wüsthoff, Phys. Rev. D59 (1999) 3006
79. C. Adloff et al., Eur. Phys. J. C21 (2001) 33;
    J. Breitweg et al., Phys. Lett. B487 (2000) 53
80. M. R. Adams et al., Z. Phys. C67 (1995) 403;
    M. Arneodo et al., Nucl. Phys. B481 (1996) 3 and 23
81. R. Baier, A. H. Müller and D. Schiff, hep-ph/0403201, and references therein
82. B. B. Back et al., PHOBOS Coll., Phys. Rev. C65 (2002) 061901; nucl-ex/0405027
83. D. Amati and G. Veneziano, Phys. Lett. B83 (1979) 87

84. F. Becattini, Nucl. Phys. A702 (2002) 336
85. J. Ellis and K. Geiger, Phys. Rev. D54 (1996) 1967
86. K. Geiger and B. Müller, Nucl. Phys. B369 (1992) 600;
    K. Geiger and D. K. Shrivastava, Nucl. Phys. A661 (1999) 592
87. R. Stock, Phys. Lett. B456 (1999) 277
88. U. Heinz, Nucl. Phys. A610 (1996) 264
89. Y. M. Sinyukov, Nucl. Phys. A498 (1989) 151
90. H. Appelshäuser et al., NA49 Coll., Eur. Phys. J. C2 (1998) 611
91. J. Rafelski and B. Müller, Phys. Rev. Lett. 48 (1982) 1066
92. D. Molnar and M. Gyulassy, Nucl. Phys. A697 (2002) 495, and Nucl. Phys. A698 (2002) 379
93. S. A. Voloshin and A. M. Poskanzer, Phys. Lett. B474 (2000) 27;
    S. A. Voloshin, Nucl. Phys. A715 (2003) 379
94. R. A. Lacey and A. Taranenko, nucl-ex/0610029
95. D. Teaney, J. Laurent and E. V. Shuryak, Nucl. Phys. A698 (2002) 479, and nucl-th/0110037
96. P. F. Kolb and U. Heinz, in Quark-Gluon Plasma 3, edit. R. C. Hwa and Y. N. Wang, World
    Scientific 2004, p. 634, and references therein
97. J. L. Klay et al., STAR Coll., Nucl. Phys. A715 (2003) 733
98. K. Alpgard et al., Phys. Lett. B107 (1981) 310
99. R. Hagedorn and J. Ranft, Nuovo Cim. Suppl. 6 (1968) 169
100. M. Gazdzicki et al., NA49 Coll., J. Phys. G30 (2004) 701;
     M. Gazdzicki et al., Braz. Journal of Physics 34 (2004) 322
101. C. Nonaka and M. Asakawa, Phys. Rev. C71 (2005) 044904
102. G. Bertsch and P. Cuguon, Phys. Rev. C24 (1981) 269
103. N. Xu, J. Phys. G32 (2006) 123, and references therein
104. E. Schnedermann, J. Sollfrank and U. Heinz, Phys. Rev. C48 (1993) 2462
105. P. J. Siemens and J. Rassmussen, Phys. Rev. Lett. 43 (1979) 1486
106. U. Wiedemann and U. Heinz, Phys. Rev. C56 (1997) 3265
107. R. Stock, nucl-th/0703050;
     P. Braun-Munzinger, K. Redlich and J. Stachel, in Quark-Gluon Plasma 3, eds. R. C. Hwa
     and X. N. Wang, World Scientific 2004, p. 491
108. A. Andronic, P. Braun-Munzinger and J. Stachel, nucl-th/0511071
109. F. Becattini and U. Heinz, Z. Phys. C76 (1997) 269
110. U. Heinz, nucl-th/9810056
111. F. Cooper and G. Frye, Phys. Rev. D10 (1974) 186
112. H. Sorge, Z. Phys. C67 (1995) 479
113. M. Bleicher et al., J. Phys. G25 (1999) 1859
114. S. A. Bass and A. Dimitru, Phys. Rev. C61 (2000) 064909
115. R. V. Gavai and S. Gupta, J. Phys. G30 (2004) 1333
116. M. Mitrovski et al., NA49 Coll., J. Phys. G32 (2006) 43
117. E. V. Shuryak, Phys. Lett. B42 (1972) 357
118. J. Rafelski and M. Danos, Phys. Lett. B97 (1980) 279
119. A. Tounsi and K. Redlich, J. Phys. G28 (2002) 2095
120. C. Hoehne, F. Puehlhofer and R. Stock, Phys. Lett. B640 (2006) 96
121. B. R. Webber, Nucl. Phys. B238 (1984) 492
122. B. Andersson, G. Gustafsson, G. Ingelman and T. Sjöstrand, Phys. Rept. 97 (1983) 33
123. H. Satz, hep-ph/0612151; P. Castorina, D. Kharzeev and H. Satz, hep-ph/0704.1426
124. F. Becattini and G. Passaleva, Eur. Phys. Journ. C23 (2002) 551
125. J. Rafelski, Phys. Lett. B262 (1991) 333
126. W. S. C. Williams, Nuclear and Particle Physics, Oxford Univ. Press 1991
127. B. Müller and J. Rafelski, Phys. Rev. Lett. 48 (1982) 1066
128. I. G. Baerden et al., BRAHMS Coll., Phys. Rev. Lett. 90 (2003)102301
129. C. Alt et al., NA49 Coll., Phys. Rev. C73 (2006) 044910
130. S. V. Afanasiev et al., NA49 Coll., Phys. Rev. C69 (2004) 024902
131. F. Becattini et al., Phys. Rev. C64 (2001) 024901

132. D. Röhrich, Florenz
133. R. Rapp, J. Phys. G31 (2005) 217
134. G. E. Brown and M. Rho, Phys. Rept. 269 (1996) 333
135. R. Rapp and J. Wambach, Adv. Nucl. Phys. 25 (2000) 1
136. S. Damjanovic et al., NA60 Coll., Nucl. Phys. A774 (2006) 715
137. D. Zschiesche et al., Nucl. Phys. A681 (2001) 34
138. C. Greiner et al., J. Phys. G31 (2005) 61
139. P. Braun-Munzinger, J. Cleymans, H. Oeschler and K. Redlich, Nucl. Phys. A697 (2002) 902
140. F. Becattini, M. Gazdzicki and J. Manninen, hep-ph/0511092
141. M. Gazdzicki and M. I. Gorenstein, Acta Phys. Polon B30 (1999) 2705
142. R. Stock, J. Phys. G30 (2004) 633
143. V. Koch, A. Majumder and J. Randrup, nucl-th/0509030
144. E. G. Nikonow, A. A. Shanenko and V. D. Toneev, Heavy Ion Physics 8 (1998) 89
145. C. M. Hung and E. V. Shuryak, Phys. Rev. C57 (1998) 1891;
    M. Bleicher, hep-ph/0509314
146. M. Stephanov, K. Rajagopal and E. V. Shuryak, Phys. Rev. Lett. 81 (1998) 4816
147. M. Stephanov, K. Rajagopal and E. V. Shuryak, Phys. Rev. D60 (1999) 114028
148. V. Koch, A. Majumder and J. Randrup, nucl-th/0505052
149. Ch. Roland et al., NA49 Coll., J. Phys. G30 (2004) 1371
150. P. F. Kolb and U. Heinz, Nucl. Phys. A715 (2003) 653
151. P. Arnold, G. D. Moore and L. G. Yaffe, JHEP 0011 (2000) 001
152. A. Muronga, Phys. Rev. C69 (2004) 044901
153. T. Hirano and M. Gyulassy, Nucl. Phys. A769 (2006) 71
154. B. A. Gelman, E. V. Shuryak and I. Zahed, nucl-th/ 0601029
155. C. Alt et al., NA49 Coll., Phys. Rev. C68 (2003) 034903
156. B. B. Back et al., PHOBOS Coll., Nucl Phys. A715 (2003) 65
157. C. Adler et al., STAR Coll., Phys. Rev. C66 (2002) 034905;
    J. Adams et al., STAR Coll, Phys. Rev. Lett. 92 (2004) 052302 and 062301
158. S. S. Adler et al., PHENIX Coll., Phys. Rev. Lett. 91 (2003) 182301
159. P. Huovinen, P. F. Kolb, U. W. Heinz, P. V. Ruuskanen and S. A. Voloshin, Phys. Lett. B503 (2001) 58;
    P. Huovinen in Quark-Gluon Plasma 3, eds. R. C. Hwa and X. N. Wang, World Scientific 2004, p. 600
160. T. Hirano and Y. Nara, Phys. Rev. Lett. 91 (2003) 082301, and Phys. Rev. C68 (2003) 064902
161. A. Adare et al., PHENIX Coll., nucl-ex/0607021
162. M. Csanad, T. Csörgö, R. A. Lacey and B. Lorstadt, nucl-th/0605044;
    A. Adare et al., PHENIX Coll., nucl-ex/0608033
163. S. Sakai, PHENIX Coll., nucl-ex/0510027
164. G. D. Moore and D. Teaney, Phys. Rev. C71 (2005) 064904
165. E. V. Shuryak, Nucl. Phys. A774 (2006) 387, and references therein
166. D. Teaney, Phys. Rev. C68 (2003) 034913
167. L. P. Csernai, J. I. Kapusta and L. McLerran, nucl-th/0604032
168. T. Csörgö, L. P. Cernai, Y. Hama and T. Kodama, Heavy Ion Physics A21 (2004) 73;
    M. Csanad, T. Csörgö and B. Lörstadt, Nucl. Phys. A742 (2004) 80
169. Z. Xu and C. Greiner, Phys. Rev. C71 (2005) 064901
170. S. Gavin and M. Abdel-Aziz, nucl-th/0606061
171. P. K. Kovtun, D. T. Son and A. O. Starinets, Phys. Rev. Lett. 94 (2005) 111601
172. M. Gyulassy and M. Plumer, Nucl. Phys. A527 (1991) 641;
    M. Gyulassy, M. Plumer, M. Thoma and X. N. Wang, Nucl. Phys. A538 (1992) 37;
    X. N. Wang and M. Gyulassy, Phys. Rev. Lett. 68 (1992) 1480
173. J. C. Collins, D. E. Soper and G. Sterman, Nucl. Phys. B261 (1985) 104
174. D. d'Enterria, J. Phys. G30 (2004) 767
175. S. V. Greene et al., PHENIX Coll., Nucl. Phys. A774 (2006) 93
176. X. N. Wang, Phys. Lett. B595 (2004) 165

177. P. Jacobs and M. van Leeuwen, Nucl. Phys. A774 (2006) 237, and references therein
178. J. C. Dunlop et al., STAR Coll., Nucl. Phys. A774 (2006) 139
179. N. Armesto, S. Dainese, C. Salgado und U. Wiedemann, Phys. Rev. D71 (2005) 054027
180. M. Djordjevic, M. Gyulassy, R. Vogt and T. Wicks, nucl-th/0507019
181. Y. Akiba et al., PHENIX Coll., Nucl. Phys. A774 (2006) 403
182. O. Barannikova et al., STAR Coll., Nucl. Phys. A774 (2006) 465
183. M. M. Aggarwal et al., WA98 Coll., Eur. Phys. J. C23 (2002) 225;
     D. d'Enterria, nucl-ex/0403055;
     A. Laszlo and T. Schuster, NA49 Coll., Nucl. Phys. A774 (2006) 473;
     A. Dainese et al., NA57 Coll., Nucl. Phys. A774 (2006) 51
184. A. Accardi, hep-ph/0212148
185. S. Salur et al., STAR Coll., Nucl Phys. A774 (2006) 657
186. R. Baier, Y. L. Dokshitzer, A. H. Mueller, S. Peigne and D. Schiff, Nucl. Phys. B483 (1997) 291
187. R. Baier, Y. L. Dokshitzer, A. H. Mueller, S. Peigne and D. Schiff, Nucl. Phys. B484 (1997) 265
188. B. G. Zakharov, JETP Lett. 63 (1996) 952;
     ibid. 65 (1997) 615;
     ibid. 70 (1999) 176
189. I. Vitev, J. Phys. G30 (2004) 791
190. S. S. Adler et al., PHENIX Coll., Phys. Rev. C69 (2004) 034910
191. J. Adams et al., STAR Coll., Phys. Rev. Lett. 91 (2003) 072302
192. R. Baier, Y. L. Dokshitzer, A. H. Mueller, and D. Schiff, Phys. Rev. C58 (1998) 1706
193. C. A. Salgado and U. A. Wiedemann, Phys. Rev. D68 (2003) 014008
194. K. J. Eskola, A. Honkanen, C. A. Salgado and U. Wiedemann, Nucl. Phys. A747 (2005) 511
195. A. Dainese, C. Loizides and G. Paic, Eur. Phys. J. C38 (2005) 461
196. X. N. Wang, Phys. Rept. 280 (1997) 287
197. M. Gyulassy, I. Vitev and X. N. Wang, Phys. Rev. Lett. 86 (2001) 2537
198. S. S. Adler et al., PHENIX Coll., Phys. Rev. Lett. 91 (2003) 072301
199. W. A. Horowitz, nucl-th/0702084
200. S. Wicks, W. A. Horowitz, M. Djordjevic and M. Gyulassy, nucl-th/0512076
201. R. Stock, Phys. Rept. 135 (1986) 259
202. R. Baier, Nucl. Phys. A715 (2003) 209
203. J. Liao and E. V. Shuryak, hep-ph/0508035
204. Th. Renk and K. J. Eskola, hep-ph/0610059
205. T. Affolder et al., CDF Coll., Phys. Rev. D64 (2001) 032001
206. B. A. Kniehl, G. Kramer and B. Potter, Nucl. Phys. B597 (2001) 337
207. C. Adler et al., STAR Coll., Phys. Rev. Lett. 90 (2003) 082302;
     J. Adams et al., STAR Coll., Phys. Rev. Lett. 91 (2003) 072304;
     J. Adams et al., STAR Coll., Phys. Rev. Lett. 93 (2004) 252301
208. D. Magestro et al., STAR Coll., Nucl. Phys. A774 (2006) 573
209. T. Dietel et al., STAR. Coll., Nucl. Phys. A774 (2006) 569
210. J. C. Collins, D. E. Soper and G. Sterman, Adv. Ser. Direct. High Energy Phys. 5 (1988) 1
211. Th. Renk and K. Eskola, hep-ph/0610059
212. H. Liu, K. Rajagopal and U. A. Wiedemann, hep-ph/0605178
213. E. V. Shuryak, hep-ph/0703208
214. F. Wang et al., STAR Coll., J. Phys. G30 (2004) 1299
215. H. Buesching et al., PHENIX Coll., Nucl. Phys. A774 (2006) 103;
     Y. Akiba, ibidem p. 403
216. A. E. Glassgold, W. Heckrotte and K. M. Watson, Ann. Phys. (New York) 6 (1959) 1
217. H. Stöcker, J. Maruhn and W. Greiner, Z. Phys. A290 (1979) 297, and A293 (1979) 173
218. R. Stock et al., GSI-LBL Coll., Phys. Lett. 44 (1980) 1243
219. for critical comments concerning conical emission see C. A. Gagliardi et al., STAR Coll., Nucl. Phys. A774 (2006) 409

220. H. Stöcker, Nucl. Phys. A750 (2005) 121;
     J. Ruppert, Nucl. Phys. A774 (2006) 397;
     J. Csalderry-Solana, E. V. Shuryak and D. Teaney, hep-ph/0602183
221. M. Beneke and I. Z. Rothstein, Phys. Rev. D54 (1996) 2005
222. E. L. Feinberg, Nuovo Cim. A34 (1976) 391
223. J. Kapusta, P. Lichard and D. Seibert, Phys. Rev. D44 (1991) 2774;
     P. Arnold, G. D. Moore and L. Yaffe, J. High Energy Phys. 0112 (2001) 9;
     G. D. Moore, J. Phys. G30 (2004) 775
224. M. M. Aggarwal et al., WA98 Coll., Phys. Rev. Lett. 85 (2000) 3595
225. K. Kajantie and H. I. Miettinen, Z. Phys. C14 (1982) 357
226. G. Borges et al., NA50 Coll., J. Phys. G30 (2004) 1351;
     L. Ramello et al., NA50 Coll., Nucl. Phys. A715 (2003) 243
227. R. D. Pisarski, Phys. Lett. B110 (1982) 155
228. G. E. Brown and M. Rho, Phys. Rept. 363 (2002) 85
229. S. Jacobs, M. G. Olsson and C. Suchyta, Phys. Rev. D33 (1986) 3338
230. H. Satz, J. Phys. G32 (2006) 25, and hep-ph/0609197
231. E. Eichten et al., Phys. Rev. D21 (1980) 203
232. M. Asakawa and T. Hatsuda, Phys. Rev. Lett. 92 (2004) 012001;
     S. Datta et al., Phys. Rev. D69 (2004) 094507; G. Aarts et al., hep-lat/0603002
233. B. Alessandro et al., NA50 Coll., Eur. Phys. J. C39 (2005) 335
234. M. Nardi, Nucl. Phys. A774 (2006) 353
235. D. Kharzeev, C. Lourenco, M. Nardi and H. Satz, Z. Phys. C74 (1997) 307
236. E. Scomparin et al., NA60 Coll., Nucl. Phys. A774 (2006) 67
237. A. Adare et al., PHENIX Coll., nucl-ex/0611020
238. A. Adare et al., PHENIX Coll., hep-ex/0611020
239. M. J. Leitch, nucl-ex/0701021
240. M. Gazdzicki and M. I. Gorenstein, Phys. Rev. Lett. 83 (1999) 4009
241. A. Andronic, P. Braun-Munzinger, K. Redlich, J. Stachel, nucl-th/0611023, and references
     therein
242. S. Datta et al., hep-lat/0603002
243. For recent theory review, see C. Gale and K. L. Haglin, in Quark-Gluon-Plasma 3, R. Hwa
     and X. N. Wang eds., World Scientific 2004, p. 364;
     F. Gelis, Nucl. Phys. A715 (2003) 329
244. A. Dumitru et al., Phys. Rev. C64 (2001) 054909
245. S. S. Adler et al., PHENIX Coll., Phys. Rev. Lett. 94 (2005) 232301
246. S. Bathe et al., PHENIX Coll., Nucl. Phys. A774 (2006) 731
247. K. Okada et al., PHENIX Coll., hep-ex/0501066
248. L. E. Gordon and W. Vogelsang, Phys. Rev. D48 (1993) 3136
249. D. d'Enterria and D. Peressounko, nucl-th/0503054, and references therein
250. D. Huovinen, P. V. Ruskaanen, and S. S. Räsänen, Phys. Lett. B535 (2002) 109;
     D. K. Srivastava and B. Sinha, Phys. Rev. C64 (2001) 034902
251. D. Miskowiec et al., NA45 Coll., Nucl. Phys. A774 (2006) 43
252. K. Gallmeister, B. Kaempfer, O. P. Pavlenko and C. Gale, Nucl. Phys. A688 (2001) 939
253. R. Arnaldi et al., NA60 Coll., Phys. Rev. Lett. 96 (2006) 162302
254. D. Adamova et al., NA45 Coll., nucl-ex/0611022
255. H. van Hees and R. Rapp, Phys. Rev. Lett. 97 (2006) 102301
256. J. Wambach, Nucl. Phys. A715 (2003) 422
257. M. Asakawa, T. Hatsuda and Y. Nakahara, Nucl. Phys. A715 (2003) 863
258. R. Rapp and E. V. Shuryak, Phys. Lett. B473 (2000) 13
259. S. P. Klevansky, Rev. Mod. Phys. 64 (1992) 649;
     M. A. Stephanov, Phys. Rev. Lett. 76 (1996) 4472;
     M. A. Halasz et al., Phys. Rev. D58 (1998) 096007
260. M. Gazdzicki et al., NA61 Coll., nucl-ex/0612007
261. A. Cho, Science 312 (2006) 190

262. V. Friese, Nucl. Phys. A774 (2006) 377
263. N. G. Antoniou, F. K. Diakonos and A. S. Kapoyannis, hep-ph/0502131
264. R. Stock, hep-ph/0404125
265. T. Hirano, J. Phys. G30 (2004) 845
266. B. Alver et al., PHOBOS Coll., nucl-ex/0608025
267. B. Alver et al., PHOBOS Coll., nucl-ex/0702036
268. C. Loizides et al., PHOBOS Coll., nucl-ex/0701049
269. S. Mrowczynski and E. Shuryak, Acta Phys. Polon. B34 (2003) 4241;
     M. Miller and R. Snellings, nucl-ex/0312008
270. S. Ejiri et al., Nucl. Phys. A774 (2006) 837
271. R. V. Gavai and S. Gupta, Phys. Rev. D67 (2003) 034501
272. S. V. Afanasiev et al., NA49 Coll., Phys. Rev. Lett. 86 (2001) 1965
273. C. Roland et al., NA49 Coll., J. Phys. G30 (2004) 1381
274. S. Das et al., STAR Coll., SQM06
275. V. Koch, A. Majumder and J. Randrup, Nucl. Phys. A774 (2006) 643
276. E. V. Shuryak, The QCD Vacuum, Hadrons and Superdense Matter, Lecture Notes in Physics 8, World Scientific 1988
277. H. Leutwyler, Bern University preprint BUTP-91/43;
     N. G. Antoniou, Nucl. Phys. B92 (2001) 26
278. H. Appelshäuser et al., NA49 Coll., Phys. Lett. B459 (1999) 679
279. K. Grebieszkow et al., NA49 Coll., arXiv: 0707.4608
280. H. Sako and H. Appelshäuser, CERES Coll., J. Phys. G30 (2004) 1371
281. C. A. Pruneau et al., STAR Coll., Nucl. Phys. A774 (2006) 651
282. M. Gazdzicki and St. Mrowczynski, Z. Phys. C54 (1992) 127
283. S. A. Voloshin, nucl-ex/0109006
284. D. Adamova et al., CERES Coll., to be published
285. N. G. Antoniou et al., Nucl. Phys. A693 (2001) 799
286. R. Hanbury-Brown and R. Q. Twiss, Nature 178 (1956) 1046
287. B. Tomasik and U. A. Wiedemann, in Quark-Gluon-Plasma 3, R. C. Hwa and X. N. Wang (eds), World Scientific 2004, p.715
288. U. A. Wiedemann and U. Heinz, Phys. Rept. 319 (1999) 145
289. F. B. Yano and S. E. Koonin, Phys. Lett. B78 (1978) 556;
     M. I. Podgorestkii, Sov. J. Nucl. Phys. 37 (1983) 272
290. S. Pratt, Phys. Rev. D33 (1986) 72
291. C. Alt et al., NA49 Coll., to be published
292. M. A. Lisa et al., E895 Coll., Phys. Rev. Lett. 84 (2000) 2798
293. D. Adamova et al., CERES Coll., Nucl. Phys. A714 (2003) 124
294. B. B. Back et al., PHOBOS Coll., J. Phys. G30 (2004) 1053;
     J. Adams et al., STAR Coll., Phys. Rev. C71 (2005) 044906;
     S. S. Adler et al., PHENIX Coll., Phys. Rev. Lett 93 (2004) 152302
295. D. Adamova et al., CERES Coll., Phys. Rev. Lett. 90 (2003) 022301
296. M. Asakawa and Ch. Nonaka, Nucl. Phys. A774 (2006) 753
297. F. Retiere and M. A. Lisa, Phys. Rev. C70 (2004) 044907
298. U. Heinz et al., Phys. Lett. B382 (1996) 181
299. Y. Hama et al., Acta Phys. Polon 35 (2004) 179
300. A. Majumder, B. Müller and X. N. Wang, hep-ph/0703082
301. H. Liu, hep-ph/0702210, and references therein
302. J. M. Maldacena, Adv. Theor. Math. Phys. 2 (1998) 231, and Int. J. Theor. Phys. 38 (1999) 1113;
     E. Shuryak, S. J. Sin and I. Zahed, hep-th/0511199
303. L.V.Gribov, E.M.Levin, M.G.Ryskin, Phys. Pept.100(1983)1;
     L.D.Mclerran, R.Venugopalan, Phys.Rev.D49(1994)2233
304. T. Lappi and L.D.McLerran, Nucl.Phys. A772(2006)200
305. F.Gelis, Nucl.Phys.A931(2014)73

306. F.Antinori et al., arXiv:1409.2981
307. J.Casalderry, H.Liu, D.Mateos, K.Rajagopal, A.U.Wiedemann, arXiv:1101.0618,
     and references therein
308. R.A.Janik, Nucl.Phys.A931(2014)176
309. A.M.Poskanzer, S.A.Voloshin, Phys.Rev.C58(1998)1671
310. H.G.Ritter, R.Stock, arXiv:1408.4296
311. F.Cooper, G.Frye, Phys.Rev.D10(1974)186
312. H.Petersen et al., UrQMD Coll., Phys.Rev.C78(2008)044901
313. H.Song, S.S.Bass, U.Heinz, arXiv:1311.0157
314. J.Adams et al., STAR Coll., Phys.Rev.C72(2005)014904
315. U.Heinz, R.Snellings, arXiv:1301.2826
316. J.Eskola, H.Niemi, P.Paatelainen, K.Tuominen, arXiv:1704.04060
317. Ch.Gale, S.Jeon, B.Schenke, P.Tribedy, R.Venugopalan, Phys.Rev.Lett.110(2013)012302
318. J.Harris, talk at ERICE Nucl.Phys.School 2016
319. B.Abelev et al., ALICE Coll., Phys.Rev.Lett.113(2014)232301
320. X.N.Wang et al., Jet Coll., arXiv:1408.3519
321. T.Matsui, H.Satz, Phys.Lett.B178(1968)416
322. M.Gazdzicki, M.Gorenstein, Phys.Rev.Lett.83(1999)4009
323. A.Andronic, P.Braun-Munzinger, K.Redlich, J.Stachel, Nucl.Phys.A798(2007)334
324. P.Braun-Munzinger, J.Stachel, Landolt-Boernstein 23(2010)424, arXiv:0901.2500
325. A.Andronic, P.Braun-Munzinger, K.Redlich, J.Stachel, arXiv:1710.09425,
     and references therein
326. J.Adam et al., ALICE Coll.,Phys.Lett.B766(2017)212,
     and references therein
327. F.Karsch, K.Redlich, A.Tawfik, Phys.Lett.B571(2003)67;
     A.Bazavov et al., HotQCD Coll., Phys.Rev.D86(2012)034509;
     F.Karsch, arXiv:1312.2659
328. A.Bazavov et al., Phys.Rev.Lett.109(2012)192302;
     P.Alba et al., arXiv:1483.4903 and arXiv:1504.03262;
     F.Karsch, arXiv:1611.01973
329. F.Becattini, E.Grossi, M.Bleicher, J.Steinheimer, R.Stock, Phys.Rev.C90(2014)054907
330. F.Becattini, J.Steinheimer, R.Stock, M.Bleicher, Phys.Lett.B764(2017)241
331. J.Steinheimer, J.Aichelin, M.Bleicher, Phys.Rev.Lett.110(2013)042501
332. A.Andronic, P.Braun-Munzinger, K.Redlich, J.Stachel, Nucl.Phys.A904/905(2013)535;
     M.Floris et al., ALICE Coll. Nucl.Phys.A931(2014)103
333. F.Becattini, M.Bleicher, J.Steinheimer, R.Stock, arXiv:1712.03748
334. L.Adamczyk et al., STAR Coll., Phys.Rev.C96(2017)044904
335. L.Adamczyk et al., STAR Coll., Phys.Rev.Lett.118(2017)212301
336. C.R.Allton et al., Phys.Rev.D68(2003)014507
337. H.T.Ding, Nucl.Phys.A931(2014)52
338. A.Bazavov et al., Phys.Rev.D96(2017)074510
339. P.Braun-Munzinger, A.Kalweit, K.Redlich, J.Stachel, Phys.Lett.B747(2015)292
340. M.M.Aggarwal et al., STAR Coll., Phys.Rev.Lett.105(2010)022302;
     X.Luo et al., STAR Coll., PoS CPOD 2014,019, arXiv:1503.02558
341. A.Bazavov et al., HotQCD Coll., Phys.Rev.D90(2014)094503
342. M.Nahrgang et al., J.Phys.G38(2011)124150
343. P.Braun-Munzinger, H.Rustamov, J.Stachel, Nucl.Phys.A960(2017)114, arXiv:1622.00702
344. J.Steinheimer et al., Phys.Lett.B776(2018)32
345. P.Alba et al., Phys.Rev.C92(2015)064910
346. A.Bazavov et al., Phys.Rev.Lett. 113(2014)072001
347. J.Adam et al., ALICE Coll., Nature Phys.13(2017)535
348. C.Loizides, Nucl.Phys.A956(2016)200
349. S.Acharya et al., ALICE Coll., arXiv:1712.05603
350. K.Dusling, W.Li, B.Schenke, arXiv:1509.07939

351. P.Steinberg, arXiv:nucl-ex/0703002
352. F.Becattini, V.Chandra, L.DelZanna, E.Grossi, Annals Phys.338(2013)32
353. Z.T.Liang, X.N.Wang, Phys.Rev.Lett.94(2005)102301;
     J.H.Gao et al., Phys.Rev.Lett.109(2012)232301
354. F.Becattini, E.Grossi, Phys.Rev.D92(2015)045037
355. L.Adamczyk et al., STAR Coll., Nature548(2017)62
356. F.Becattini, arXiv:1711.08780;
     I.Siddique et al., arXiv:1710.00134
357. Q.Wang, arXiv:1704.04022
358. F.Becattini and Iu.Karpenko, arXiv:1707.07984
359. W.Scheid, H.Mueller, W.Greiner, Phys.Rev.Lett.32(1974)741
360. R.Stock, Phys.Rept.135(1986)259
361. M.Hanauske et al., J.Phys.Conf.Ser.878(2017)012031;
     Phys.Rev.D96(2017)043004
362. M.Alford et al., arXiv:1707.09475
363. Th.Kollegger and R.Stock, DOI:10.1007/978-3-319-00047-3
364. J.Schukraft, arXiv:1705.02646;
     S.Cho et al., arXiv:1702.00486
365. K.G.R.Doss et al., Plastic Ball Coll., Phys.Rev.C32(1985)116
366. T.Anticic et al., NA49Coll., Phys.Rev.C94(2016)044906
367. S.Mrowczynski, arXiv:1607.02267
368. M.Gyulassi, K.Frankel and E.A.Remler, Nucl.Phys.A402(1983)596

Chapter 8
Beyond the Standard Model

Eliezer Rabinovici

8.1 Introduction

Starting sometime in 2008/2009 one expects to be able to take a glimpse at physics at the TeV scale. This will be done through the Large Hadronic Collider (LHC) at CERN, Geneva. It will be a result of an unprecedented coordinated international scientific effort. This chapter is written in 2007. It is essentially inviting disaster to spell out in full detail what the current various theoretical speculations on the physics are, as well motivated as they may seem at this time. What I find of more value is to elaborate on some of the ideas and the motivations behind them. Some may stay with us, some may evolve and some may be discarded as the results of the experiments unfold. When the proton antiproton collider was turned on in the early eighties of the last century at Cern the theoretical ideas were ready to face the experimental results in confidence, a confidence which actually had prevailed. The emphasis was on the tremendous experimental challenges that needed to be overcome in both the production and the detection of the new particles. As far as theory was concerned this was about the physics of the standard model and not about the physics beyond it. The latter part was left safely unchallenged. That situation started changing when the large electron positron (LEP) collider experiments also at Cern were turned on as well the experiments at the Tevatron at Fermilab. Today it is with rather little, scientifically based, theoretical confidence that one is anticipating the outcome of the experiments. It is less the method and foundations that are tested and more the prejudices. It is these which are at the center of this chapter. Some claim to detect over the years an oscilatory behavior in the amount of conservatism expressed by leaders in physics. The generation in whose life time relativity and

E. Rabinovici (✉)
Hebrew University of Jerusalem, Jerusalem, Israel
e-mail: eliezer@vms.huji.ac.il

© The Author(s) 2020
H. Schopper (ed.), *Particle Physics Reference Library*,
https://doi.org/10.1007/978-3-030-38207-0_8

quantum mechanics were discovered remained non-conservative throughout their life. Some of the latter developed eventually such adventurous ideas as to form as a reaction a much more conservative following generation. The conservative generation perfected the inherited tools and has uncovered and constructed the Standard Model. They themselves were followed by a less conservative generation. The new generation was presented with a seemingly complete description of the known forces. In order to go outside the severe constraints of the Standard Model the new generation has drawn upon some of the more adventurous ideas of the older generation as well as created it own ideas. In a way almost all accepted notions were challenged. In the past such an attitude has led to major discoveries such as relativity and quantum mechanics. In some cases it was carried too far, the discovery of the neutrino was initially missed as energy conservation was temporarily given up.

The standard model is overall a very significant scientific achievement. It is a rather concise framework encompassing all the known properties of the known basic interactions. It is arguably the most impressive theoretical understanding of a large body of experimental information existing. An understanding backed by precise predictions all verified by high quality experiments. In this context it may seem surprising that one is searching for anything beyond the Standard Model. There are however diverse scientific reasons for the search of the beyond.

In 2007 the scientific community was aware of quite a few gaps in the understanding of the particle interactions. One class of observations posed obvious pressing problems:

- There is a large body of evidence that the so called dark matter should be composed mostly of different particle(s) than those that serve as the building blocks of the standard model. What are they? More recently also what is called a dark energy was needed to explain the data. Its possible origin(s) is under active study.
- A standard model for cosmology is forming and it includes in most cases versions of inflation. Such models seem to require the existence of a new heavy particle(s). What are they? On a more speculative note, models require more detailed understanding of how physical systems behave in big bang/crunch like circumstances and of how and if universes may form.

Another set of observations could be either defined as posing problems requiring an explanation or as pointing to new directions only after being combined with a certain amount of theoretical prejudice (*TP*). In the past some major advances were driven by such combinations.

- Three known interactions, the colored, the weak and the electromagnetic interactions all obey the dictum of quantum mechanics and are all well described by gauge theories but have otherwise very different properties and strengths at the energy scales probed till 2007. A *TP*, a strong and deeply ingrained one, suggests that all interactions, those known and those yet to be discovered should unify at a certain higher energy scale. This is the idea of a Grand Unified Theories (GUTS). Eventually it was found that to realize a particular aspect of this, the convergence of the all the couplings to a single value at an appropriate energy scale, there should be physics beyond the standard model. In Super Symmetric (SUSY) systems this indeed may occur. In fact the unification is achieved at a distance scale rather close to the Planck scale, the scale at which gravity becomes strong and its quantum effects, if there, should become noted. Thus the fourth force Gravity is naturally added to the unification scheme. An older realization of this TP was that all known interactions are but the low energy descriptions of a system containing only the gravitational force and residing in a higher number of space-time dimensions. This had led to Kaluza Klein theories and their variants. This idea has been revisited with the advent of SUSY and string theory.

- The observation of what is called dark energy and its very possible confirmation of a very small, but not vanishing, cosmological constant is viewed by many as a major problem lacking an explanation. The TP behind this is that any fundamental quantitative property of nature should be explained and not fine tuned. The absence of any significant amount of CP violation in the strong interactions is another such problem. The discovery of a new particle, the Axion, could be a signal of the solution of the latter one. The discovery of another particle, the Dilaton, could indicate a resolution of the former. This issue has led also to the reexamination of the so called enthropic principle.

- Many predictions of Classical General Relativity were confirmed experimentally over the years. The TP in 2007 is that there should be a quantum theory of gravity. Such a four dimensional quantum theory of gravity is not well defined within only field theory. This has been a driving force in the study of the properties of a theory were the basic constituents of nature are not particles but extended objects, including strings.

- A very successful framework to explain the basic interactions in particle physics is the so called Wilsonian one. It is very powerful when the laws of physics are such that different largely separated scales are essentially decoupled from each other. The physics beneath any energy scale (cutoff) is well described by operators whose scaling dimensions, in d spacetime dimensions, is not much larger than the same d. In many cases there is only a finite number of such operators, i.e. only a finite amount of terms in the Lagrangians describing the Physics below the energy scale. This TP has been tested successfully time and again but may eventually be falsified, perhaps in a theory of gravity. But given the validity of this method physical quantities should be only slightly dependent on the cutoff scale, and thus on the unknown physics extending beyond it. Generically the mass of scalar particles is strongly dependent on the cutoff. In particular instead of the scale of such masses being set by the weak interactions scale they will depend strongly on the cutoff. The only known, in 2007, physical scale beyond that of the weak interactions is of the order of the Planck scale. The discovery of Higgs particles whose masses is in the TeV range or lighter will thus require a large amount of fine tuning. The TP does not accept that. This is one manifestation of the so called hierarchy problem. This had led to ideas such as technicolor and supersymmetry as properties of beyond the standard model.

- It seems very likely in 2007 that eventually the neutrino masses will be added without caveats to the pantheon of particles, the particle data tables. Some *TPs* point to new physics at a rather high mass scale as the origin of small neutrino masses.

It is interesting that very few of the theoretical ideas and visions used to the address the ample set of problems mentioned above were originally created directly for that purpose. They were more tools whose applications were found only well after they were formed.

The theoretical structures discovered and created where an outcome of an urge to question, generalize and unify almost anything.

- In addition to the attempts to unify all symmetries and considering extra dimensions it was suggested that the topology of the extra dimensions may determine the low mass particle spectrum. These extra dimensions were assumed for year to be small but it turned out they could also be large, leading to theories in which large extra dimensions play a key role. A variant of this idea is that there exists a hidden sector where many desired things occur, SUSY is broken, the cosmological constant is (nearly) cancelled to name some. These effects are then communicated to the Standard Model particles by messengers. In many cases the weak gravitational force is designated that role.
- A natural direction of generalization is to question the point particle nature of the basic constituents of nature. Is there a consistent theory of elementary higher dimensional objects such as strings or membranes, this direction of research is developed under the name of string theory. It has led to much more satisfactory theory of gravity. It has success, faces many challenges not least of which is the present lack of experimental evidence and has a touch of magic. This magic has had already a significant impact on Mathematical Physics.
- The never satiated desire for simplicity may suggest to remove even the concept of a point particle which propagates in space time and for that matter to remove even the concept of space time. The idea is that space time is but an emergent, long distance, phenomena. The search for "true" underlying picture is still on.
- The quest for knowledge includes a succession of Copernican like revolutions. After each such step the researcher finds herself even further removed from the Center. A possible prospect for such a revolution is that our universe is but one of many universes. The idea emerges in quite a few contexts not necessarily unrelated. These include the many world interpretation of quantum mechanics, third quantization in quantum gravity, the mutliverse in models for inflation and in string theory. In string theory the idea has been refined in the brane world picture, our universe reflects an underlying structure in which different particles reside on different multi dimensional subspaces.
- In the process of quantization it was assumed that the space coordinates commute with each other as do their conjugate momenta. Mathematically one may construct noncommutative manifolds, manifolds whose coordinates do not commute. This had led to the study of physics on such manifolds. This an the

idea that the underlying theory of nature depends only on topology have both been extensively studied but at this stage still more in the realm of mathematics.

The various theoretical ideas mentioned above were extensively studied in the second part of the twentieth century. The respective developments are documented in many books covering thousands of pages. In this review we have made the following choices. The emphasis is on various aspects of SUSY. There has been significant progress in understanding the dynamics of systems which have in some form or another supersymmetric features. These systems have been researched in the weak coupling and the strong coupling regimes, perturbatively and non-perturbatively. The properties uncovered are remarkable in some cases. We also review, more briefly, the attempts to lay out a framework suitable for extracting supersymmetric signatures of nature. This rich area of research will better receive its due rewards after the accumulation of actual experimental data. The more "senior" areas of research such as examining the possibility of the existence of extra dimensions, grand unified theories and superstring theory will be reviewed in a much more descriptive manner. This choice was influenced by the possibility that the LHC will shed some concrete light on what is beyond the standard model.

8.2 Super Symmetry [1]

Super Symmetry (SUSY) embodies several forms of unification and generalization. It joins space time and internal symmetries and it generalizes the meaning of space time by adding fermionic components to the canonical space time bosonic coordinates. An original stated motivation for SUSY in field theory was to have a symmetry which was able to relate the self couplings of bosons, the self coupling of fermions and the Yukawa couplings of fermions to bosons. Eventually SUSY while indeed unifying such couplings has given rise to a multitude of possible non-equivalent ground states. Such a degeneracy of ground states was of a magnitude unknown before. In string theory, SUSY stabilized superstring theories by removing the tachyions which plagued the bosonic string theories. Over the years the motivations have varied and evolved. SUSY was called upon to emeliorate the so called hierarchy problem which will be reviewed below and it was perceived as an omen that in a SUSY theory one was able to arrange that the colored, weak and electromagnetic couplings unify at a single high energy scale. That scale not far from the Planck scale. In addition it was discovered that this enriched structure allows to obtain exact results in situations were only approximations were available before. In particular for four and higher dimensional systems. In its presence a very rich dynamics has been unraveled.

8.2.1 Elementary Particles in SUSY Models: Algebraic Structure

Many successes resulted from the application of symmetry principles to systems such as atoms, nuclei and elementary particles. They have been obtained even in the face of a rather incomplete understanding of the dynamics of these systems. Consequentially evermore higher symmetries were searched for, and in particular unifying ones. Such was the search for a symmetry that would contain in a non-trivial way both the Poincar'e space time symmetry and internal symmetries such as Isospin and flavor SU(3). It was shown that that was impossible, only a trivial product symmetry is allowed. Theorems are proved under assumptions, time and again new important directions emerge once significant loop holes in the assumptions are uncovered. Such was the case for obtaining consistently massive spin one particles and such was the case here. Allowing the algebra of the symmetry generators to be graded, i.e. to include both commutators and anti commutators a new structure containing both Poincar'e and internal symmetries was discovered.

A simple version of the SUSY algebra is given by the following anticommutation relations which obey the following commutation relations:

$$\{Q_\alpha, \overline{Q}_{\dot\alpha}\} = 2\sigma^\mu_{\alpha\dot\beta} P_\mu, \quad \{Q_\alpha, Q_\beta\} = \left\{\overline{Q}_{\dot\alpha}, \overline{Q}_{\dot\beta}\right\} = 0. \tag{8.1}$$

Where the Q_α are fermionic generators of supersymmetry. P_μ are the generators of space-time translations and the σ matrices are the Pauli matrices.

$$\left[P_\mu, Q_\alpha\right] = \left[P_\mu, \overline{Q}_{\dot\alpha}\right] = \left[P_\mu, P_\nu\right] = 0 \tag{8.2}$$

This is called the $N = 1$ supersymmetry algebra.

It can be generalized to include a higher number of supersymmetries. For example in four space-time dimensions there are also $N = 2$ and $N = 4$ supersymmetries:

$$\left\{Q^i_\alpha, \overline{Q}^j_{\dot\alpha}\right\} = 2\delta^{ij}\sigma^\mu_{\alpha\dot\beta} P_\mu + \delta_{\alpha\dot\beta} U_{ij} + (\gamma_5)_{\alpha\dot\beta} V_{ij}, \tag{8.3}$$

i and j run over the number of supersymmetries, U and V are the central charges i.e. they commute with all other charges, (they are antisymmetric in ij). When they do not vanish they are associated with what are called BPS states such as monopoles. The d = 4 realisations have as, $\mu, \nu = 0, 1, 2, 3$ the space-time indices. In four dimensions one has two component Weyl Fermions. Those with α or β indices transform under the $(0, \frac{1}{2})$ representation of the Lorentz group; and those with dotted indices, $\dot\alpha$ or $\dot\beta$ transform under the $(\frac{1}{2}, 0)$ representation.

The possible particle content of supersymmetric (SUSY) theories is determined by the SUSY algebra.

Consider first the massless representations of $N = 1$ supersymmetry.

The simplest is the called the chiral multiplet. It contains two real scalars and one Weyl Fermion:

$$\left(-\frac{1}{2}, 0, 0, \frac{1}{2}\right) \quad (\varphi, \psi) \quad (2, 2) \tag{8.4}$$

In the above table, first are written the helicities; then the associated component fields, φ denotes a complex scalar and ψ a Weyl Fermion; and finally are the number of physical degrees of freedom carried by the Bosons and Fermions. The massless multiplet containing a spin one boson and a spin one half Fermion is called the vector multiplet. Its content in the case of $N = 1$ is:

$$\left(-1, -\frac{1}{2}, \frac{1}{2}, 1\right) \quad (\lambda_\alpha, A_\mu) \quad (2, 2) \tag{8.5}$$

λ is a Weyl Fermion and A_μ is a vector field.

For $N = 2$ supersymmetry, there is a massless vector multiplet:

$$\left(-1 \begin{array}{ccc} -\frac{1}{2} & 0 & \frac{1}{2} \\ -\frac{1}{2} & 0 & \frac{1}{2} \end{array} 1\right) (0) \quad (\varphi, \psi) + (\lambda_\alpha, A_\mu) \quad (4, 4) \tag{8.6}$$

and a massless hypermultiplet which is given by:

$$\left(\begin{array}{ccc} & 0 & \\ -\frac{1}{2} & 0 & \frac{1}{2} \\ -\frac{1}{2}, & 0, & \frac{1}{2} \\ & 0 & \end{array}\right) \quad (\varphi_1, \psi_1) + (\varphi_2, \psi_2) \quad (4, 4). \tag{8.7}$$

For Massive multiplets, in $N = 1$, there is again the chiral multiplet which is the same as the massless multiplet but with now massive fields. The massive vector multiplet becomes:

$$\left(-1 \begin{array}{ccc} -\frac{1}{2} & 0 & \frac{1}{2} \\ -\frac{1}{2} & 0 & \frac{1}{2} \end{array} 1\right) (h, \psi_\alpha, \lambda_\alpha, A_\mu) \quad (4, 4) \tag{8.8}$$

Where h is a real scalar field. The massive vector multiplet has a different field content than the massless vector multiplet because a massive vector field has an additional physical degree of freedom. One sees that the massive vector multiplet is composed out of a massless chiral plus massless vector multiplet. This can occur dynamically; massive vector multiplets may appear by a supersymmetric analogue of the Higgs mechanism. With $N = 4$ supersymmetry, the massless vector multiplet is:

$$\begin{pmatrix} & & 0 & & \\ & -\frac{1}{2} & 0 & \frac{1}{2} & \\ & -\frac{1}{2} & 0 & \frac{1}{2} & \\ -1, & -\frac{1}{2} & 0 & \frac{1}{2} & , 1 \\ & -\frac{1}{2} & 0 & \frac{1}{2} & \\ & & 0 & & \end{pmatrix} \left(\lambda^a, \phi^I, A_\mu \right) \ (8, 8) \tag{8.9}$$

where $I = 1..6$, $a = 1..4$.

These are the unitary representations of the Super Symmetry algebra whose particle content allows them to participate in renormalizable interactions. Any higher supersymmetry in four dimensions would have to involve non-renormalizable terms. Mostly for particles with spin higher than one.

8.2.2 Supersymmetric Lagrangians

The task of writing down explicit supersymmetric Lagrangians was quite laborious. Originally all the interaction terms had to be written down explicitly. In some cases this had turned out to be much simpler by the introduction what are called superfields. These will be described below and make use of the anticommuting Grassman variables suitable to describe fermions. This is called the Super Space notation. In the spirit of generalization, the mathematical book keeping device has been elevated by some to a generalzation of regular space whose coordinates are denoted by communting numbers to a superspace in which some variables are Grassman variables. This superspace has its own geometircal properties and it was suggested to give it also a life of its own. It is not clear yet how fundamental the superspace description is but adopting this notation leads to considerable simplification. We note here that another generalization of space has been suggested. In regular space the coordinates commute also quantum mechanically, it was suggested to explore the situation that space coordinates not commute in quantizing the theory. This has experimental consequences, as of now this suggestion has no experimental backing.

8.2.2.1 Superspace, Chiral Fields and Lagrangians for Spin Zero and One-half Particles

Returning to superspace, spacetime can be extended to include Grassmann spinor coordinates, $\bar{\theta}_{\dot{\alpha}}, \theta_\alpha$. Superfields are functions of the superspace coordinates. Constructing a Lagrangian out of special types of superfields provides a useful way to construct explicitly supersymmetric Lagrangians. The integration formulas for Grassmann variables are:

$$\int d\theta_\alpha \theta_\alpha = \frac{\partial}{\partial \theta_\alpha} = 1, \quad \int d\theta_\alpha = 0 \tag{8.10}$$

Which results also in:

$$\int d^2\theta d^2\bar{\theta} \mathcal{L} = \int d^2\theta \frac{\partial^2 \mathcal{L}}{\partial \bar{\theta}_1 \bar{\theta}_2} \tag{8.11}$$

The supercharges can be realized in superspace by generators of supertranslations:

$$Q_\alpha = \frac{\partial}{\partial \theta_\alpha} - i\sigma^\mu_{\alpha\dot{\alpha}} \bar{\theta}^{\dot{\alpha}} \partial_\mu, \quad \overline{Q}_{\dot{\alpha}} = -\frac{\partial}{\partial \bar{\theta}_{\dot{\alpha}}} + i\theta^\alpha \sigma^\mu_{\alpha\dot{\alpha}} \partial_\mu. \tag{8.12}$$

To define the concept of (anti) chiral fields one defines a supercovariant derivative:

$$D_\alpha = \frac{\partial}{\partial \theta_\alpha} + i\sigma^\mu_{\alpha\dot{\alpha}} \bar{\theta}^{\dot{\alpha}} \partial_\mu, \quad \overline{D}_{\dot{\alpha}} = -\frac{\partial}{\partial \bar{\theta}_{\dot{\alpha}}} - i\theta^\alpha \sigma^\mu_{\alpha\dot{\alpha}} \partial_\mu. \tag{8.13}$$

A superfield Φ is called "chiral" if:

$$\overline{D}_{\dot{\alpha}} \, \Phi = 0. \tag{8.14}$$

Anti-chiral fields are defined by reversing the role of the θ and $\bar{\theta}$.
One introduces the variable,

$$y^\mu = x^\mu + i\theta\sigma^\mu\bar{\theta} \tag{8.15}$$

in terms of which the expansion of a chiral field is,

$$\Phi(y) = A(y) + \sqrt{2}\theta\psi(y) + \theta\theta F(y) \tag{8.16}$$

The Taylor expansion terminates after just a few terms because of the anticommuting property of the Grassmann coordinates. As a function of the coordinate x the expansion may be written as follows:

$$\begin{aligned} \Phi(x) \;=\;& A(x) + i\theta\sigma^\mu\bar{\theta}\,\partial_\mu A(x) + \tfrac{1}{4}\theta\theta\bar{\theta}\bar{\theta} * A(x) + \sqrt{2}\theta\psi(x) \\ &- \tfrac{i}{\sqrt{2}}\theta\theta\,\partial_\mu\psi(x)\sigma^\mu\bar{\theta} + \theta\theta F(x) \end{aligned} \tag{8.17}$$

The key point is that

$$\mathcal{L} = \int d^2\theta \; \Phi(x) \tag{8.18}$$

is a invariant under supersymmetric transformations (up to a total derivative).

After the integration some terms will disappear from the expansion of $\Phi(x)$ leaving only:

$$\Phi(x) = A(x) + \sqrt{2}\theta\psi(x) + \theta\theta F(x) \tag{8.19}$$

$A(x)$ will be associated with a complex Boson; $\psi(x)$ will be associated with a Weyl Fermion and $F(x)$ acts as an auxiliary field that carries no physical degrees of freedom. These are called the component fields of the superfield. The product of two chiral fields also produces a chiral field. Therefore, any polynomial, $W(\Phi)$, can be used to construct a supersymmetry invariant as

$$\mathcal{L} = \int d^2\theta W(\Phi) = F_{W(\Phi)} \tag{8.20}$$

is a supersymmetry invariant. This is used to provide a potential for the chiral field. The kinetic terms are described by:

$$\int d^2\theta d^2\bar{\theta} \; \overline{\Phi}_i \Phi_j = \overline{\Phi}_i \Phi_j \Big|_{\theta\theta\bar{\theta}\bar{\theta}} \tag{8.21}$$

Φ is an anti chiral field. After expanding and extracting the $\theta\theta\bar{\theta}\bar{\theta}$ term one obtains (up to total derivatives):

$$F_i^* F_f - |\partial_\mu A|^2 + \frac{i}{2}\partial_\mu \overline{\psi} \bar{\sigma}^\mu \psi \tag{8.22}$$

One has thus constructed the following Lagrangian:

$$\mathcal{L} = \overline{\Phi}_i \Phi_i \Big|_{\theta\theta\bar{\theta}\bar{\theta}} + \left[\lambda_i \Phi_i + \frac{1}{2}m_{ij}\Phi_i\Phi_j + \frac{1}{3}g_{ijk}\Phi_i\Phi_j\Phi_k \right]_{\theta\theta} \tag{8.23}$$

$$\begin{aligned} &= i\partial\overline{\psi}_i\bar{\sigma}\psi_i + A_i^* * A_i + F_i^* F_i + \lambda_i F_i + m_{ij}\left(A_i F_j - \tfrac{1}{2}\psi_i\psi_j\right) \\ &+ g_{ijk}\left(A_i A_j F_k - \psi_i\psi_j A_k\right) + h.c. \end{aligned} \tag{8.24}$$

One can eliminate the auxiliary fields F_i, F_i^* in favor of the fields carrying quantum degrees of freedom. The equation of motion for F_k^* is as follows:

$$F_k = \lambda_k^* + m_{ij}A_i^* + g_{ijk}^* A_i^* A_j^* \tag{8.25}$$

This gives:

$$
\begin{aligned}
\mathcal{L} &= i\partial\overline{\psi}_i\,\overline{\sigma}\psi_i + A_i^* * A_i - \tfrac{1}{2}m_{ij}\psi_i\psi_j - \tfrac{1}{2}m_{ij}^*\psi_i^*\psi_j^* \\
&- g_{ijk}\psi_i\psi_j A_k - g_{ijk}^*\overline{\psi}_i\overline{\psi}_j - A_k^* - F_i^* F_i
\end{aligned}
\tag{8.26}
$$

where the last term leads after integration on the F_is to a potential for the fields A, A^*; these are known as the F terms, $V_F(A^*, A)$. (Note, $V_F \geq 0$). It turns out as explained later that at the ground state this must vanish i.e. $V_F(A^*, A) = 0$ if SUSY is not to be spontaneously broken. This in turn implies that $F_i = 0$ for such a symmetric ground state. Although this is a classical analysis so far, in fact it is true to all orders in perturbation theory as there exists a non-renormalization theorem for such an effective potential in SUSY theories. The Lagrangian described above is called the Wess-Zumino Lagrangian (WZ).

So far the scalar fields have been defined over simple flat manifolds. To describe the kinetic terms of supersymmetric Lagrangians of systems containing scalar fields spanning complicated manifolds it is convenient to introduce the following supersymmetry invariant:

$$
\int d^4\theta\, K\left(\Phi, \overline{\Phi}\right).
\tag{8.27}
$$

$K\left(\Phi, \overline{\Phi}\right)$ is called the Kähler potential, unlike the potential $W(\Phi)$ but similar to the kinetic term introduced above, the Kähler potential depends on both Φ and $\overline{\Phi}$. One may add any function of Φ or $\overline{\Phi}$ to the integrand since these terms will vanish after integration. For the usual kinetic terms, K is taken to be given by $K = \Phi\overline{\Phi}$ which produces the $-\delta_{ij}\partial_\mu A^{*i}\partial^\mu A^j$ kinetic terms for the scalars. For the case of a sigma model with a target space whose metric is g_{ij}; this metric is related to the Kähler potential by:

$$
g_{ij} = \frac{\partial^2 K}{\partial\overline{\Phi}_i\partial\Phi_j}.
\tag{8.28}
$$

The above supersymmetry invariant (8.27) which previously gave the usual kinetic terms in the action, produces for general K the action of a supersymmetric sigma model, with the target space metric given by Eq. (8.28).

8.2.2.2 Global Symmetries

It is possible to construct Lagrangians which have a global symmetry which does not commute with supersymmetry and thus assigns different quantum numbers to particles in the same supermultiplet. This symmetry already in its discrete form forbids unwanted interaction terms which strongly violate baryon and lepton conservation laws. Such interactions arise due to the bosonic superpartners to standard model particles carrying Baryonic and Leptonic numbers. The symmetry also in its continuous U(1) version turns out to play a possible role in the possibilities to spontaneously break SUSY. It is called R symmetry.

R-symmetry is a global U(1) symmetry that does not commute with the super-symmetry. Its discrete version is called R parity. The action of the R-symmetry on a superfield Φ with R-character n as follows.

$$R\Phi\,(\theta, x) \;=\; \exp(2in\alpha)\,\Phi\,(\exp(-i\alpha\theta), x) \tag{8.29}$$

$$R\overline{\Phi}\,(\overline{\theta}, x) \;=\; \exp(-2in\alpha)\,\Phi\left(\exp\left(i\alpha\overline{\theta}\right), x\right) \tag{8.30}$$

Since the R-charge does not commute with the supersymmetry, the component fields of the chiral field have different R-charges. For a superfield Φ with R-character n, the R-charges of the component fields may be read off as follows:

$$R\text{ (lowest component of } \Phi) = R(A) \equiv n, \, R\,(\psi) = n - 1, \, R(F) = n - 2 \tag{8.31}$$

The R-charge of the Grassmann variables is given by:

$$R\,(\theta_\alpha) = 1, \, R\,(d\theta_\alpha) = -1 \tag{8.32}$$

with, barred variables having opposite R charge. The kinetic term $\overline{\Phi}\Phi$ is an R invariant. ($\overline{\theta\theta}\theta\theta$ is an invariant.) For the potential term,

$$\int d^2\theta\, W \tag{8.33}$$

to have zero R charge requires that $R(W) = 2$. For the resulting mass term from $W = \frac{1}{2}m\Phi^2$,

$$m\psi\psi + m^2|A|^2, \tag{8.34}$$

to have vanishing R-charge requires

$$R\,(\Phi) = R(A) = 1, \, R\,(\psi) = 0 \tag{8.35}$$

Adding the cubic term:

$$W_3 = \frac{\lambda}{3}\Phi^3 \tag{8.36}$$

produces

$$V = |\lambda|^2 |A|^4 + \lambda A \psi \psi. \tag{8.37}$$

This term is not R-invariant with the R-charges given by (8.35). To restore R-invariance requires λ is assigned an R-charge of -1. This can be viewed as simply a book keeping device or more physically one can view the coupling as the vacuum expectation value of some field. The expectation value inherits the quantum numbers of the field. This is how one treats for example the mass parameters of Fermions in the standard model. There is also one other global U(1) symmetry, one that commutes with the supersymmetry. All component fields are charged the same with respect to this U(1) symmetry. Demanding that the terms in the action maintain this symmetry requires an assignment of U(1) charges to λ, and m.

The charges are summarized in the following table:

	$U(1)$	$U(1)_R$
Φ	1	1
m	-2	0
λ	-3	-1
W	0	2

These symmetries can be used to prove important nonrenormalisation theorems. In particular it can be shown that the potential:

$$W = \frac{1}{2} m \, \Phi^2 + \frac{1}{3} \lambda \Phi^3. \tag{8.39}$$

does not change under renormalization.

These non-renormalization theorems play an important role in analyzing the dynamics of supersymmetric systems and in addressing the so called hierarchy problem.

8.2.2.3 Lagrangians for SUSY Gauge Theories

A vector superfield contains spin 1 and spin $\frac{1}{2}$ component fields. It obeys a reality condition $V = \overline{V}$.

$$\begin{aligned} V = {} & B + \theta \chi \overline{\theta} \overline{\chi} + \theta^2 C + \overline{\theta}^2 \overline{C} - \theta \sigma^\mu \overline{\theta} A_\mu \\ & + i \theta^2 \overline{\theta} \left(\overline{\lambda} + \tfrac{1}{2} \overline{\sigma}^\mu \partial_\mu \chi \right) - i \overline{\theta}^2 \theta \left(\lambda - \tfrac{1}{2} \sigma^\mu \partial_\mu \overline{\chi} \right) \\ & + \tfrac{1}{2} \theta^2 \overline{\theta}^2 \left(D^2 + \partial^2 B \right) \end{aligned} \tag{8.40}$$

B, D, A_μ are real and C is complex. The Lagrangian has a local U(1) symmetry with a gauge parameter, Λ an arbitrary chiral field:

$$V \rightarrow V + i\left(\Lambda - \overline{\Lambda}\right) \tag{8.41}$$

B, χ, and C are gauge artifacts and can be gauged away. The symmetry is actually U(1)$_C$ as opposed to the usual U(1)$_R$ because although the vector field transforms with a real gauge parameter, the other fields transform with gauge parameters that depend on the imaginary part of Λ.

It is possible to construct a chiral superfield, W_α, from V as follows

$$W_\alpha = -\frac{1}{4}\overline{DD}D_\alpha V, \quad \overline{D}_{\dot{\beta}} W_\alpha = 0 \tag{8.42}$$

One may choose a gauge (called the Wess Zumino gauge) in which B, C and χ vanish and then expand in terms of component fields,

$$
\begin{aligned}
V(y) &= -\theta\sigma^\mu\overline{\theta}A_\mu + i\theta^2\overline{\theta}\,\overline{\lambda} - i\overline{\theta}^2\theta\lambda + \tfrac{1}{2}\theta^2\overline{\theta}^2 D \\
W_\alpha(y) &= -i\lambda_\alpha + \left(\delta_\alpha^\beta D - \tfrac{i}{2}(\sigma^\mu\overline{\sigma}^\nu)_\beta^\alpha F_{\mu\nu}\right)\theta_\beta + \left(\sigma^\mu\partial_\mu\overline{\lambda}\right)_\alpha\theta^2
\end{aligned} \tag{8.43}
$$

Where A_μ is the vector field, $F_{\mu\nu}$ its field strength, λ is the spin $\frac{1}{2}$ field and D is an auxiliary scalar field. Under the symmetry (8.41), the component fields transform under a now U(1)$_R$ symmetry as:

$$A_\mu \rightarrow A_\mu - i\partial_\mu\left(B - B^*\right), \lambda \rightarrow \lambda, D \rightarrow D \tag{8.44}$$

Note, W is gauge invariant. The following supersymmetric gauge invariant Lagrangian is then constructed:

$$\mathcal{L} = \int d^2\theta \left(\frac{-i\tau}{16\pi}\right) W^\alpha W_\alpha + h.c. \tag{8.45}$$

where the coupling constant τ is now complex,

$$\tau = \frac{\theta}{2\pi} + i\frac{4\pi}{g^2.} \tag{8.46}$$

Expanding this in component fields produces,

$$\mathcal{L} = \frac{1}{4g^2}F_{\mu\nu}F^{\mu\nu} + \frac{1}{2g^2}D^2 - \frac{i}{g^2}\lambda\sigma D\overline{\lambda} + \frac{\theta}{32\pi^2}(*F)^{\mu\nu}F_{\mu\nu}. \tag{8.47}$$

D is a non-propagating field. The θ term couples to the instanton number density (this vanishes for abelian fields in a non-compact space). A monopole in the presence of such a coupling will obtain an electric charge. The supersymmetries acting on the component fields are, (up to total derivatives):

$$
\begin{aligned}
\delta_\epsilon A &= \sqrt{2}\epsilon\psi \\
\delta_\epsilon\psi &= i\sqrt{2}\sigma^\mu\bar{\epsilon}\partial_\mu A + \sqrt{2}\epsilon F \\
\delta_\epsilon F &= i\sqrt{2}\bar{\epsilon}\bar{\sigma}^\mu\partial_\mu\psi \\
\delta_\epsilon F_{\mu\nu} &= i\left(\epsilon\sigma_\mu\partial_\nu\bar{\lambda} + \bar{\epsilon}\bar{\sigma}_\mu\partial_\nu\lambda\right) - (\mu \leftrightarrow \nu) \\
\delta_\epsilon\lambda &= i\epsilon D + \sigma^{\mu\nu}\epsilon F_{\mu\nu} \\
\delta_\epsilon D &= \bar{\epsilon}\bar{\sigma}^\mu\partial_\mu\lambda - \epsilon\sigma^\mu\partial_\mu\bar{\lambda}.
\end{aligned}
\tag{8.48}
$$

One may also add to the action a term linear in the vector field V, known as a Fayet-Iliopoulos term:

$$
2K\int d^2\theta d^2\bar{\theta}V = KD = \int d\theta^\alpha W_\alpha + h.c.
\tag{8.49}
$$

It plays a possible role in the spontaneous breaking of SUSY. The U(1) gauge fields couple to charged chiral matter through the following term

$$
\mathcal{L} = \sum_i \int d^2\theta d^2\bar{\theta}\bar{\Phi}_i \exp\left(q_i V\right)\Phi_i
\tag{8.50}
$$

Under the gauge transformation

$$
V \to V + i\left(\Lambda - \bar{\Lambda}\right), \Phi_i \to \exp\left(-iq_i\Lambda\right)\Phi_i
\tag{8.51}
$$

Since there are chiral Fermions there is the possibility for chiral anomalies. In order that the theory is free from chiral anomalies one requires:

$$
\sum q_i = \sum q_i^3 = 0.
\tag{8.52}
$$

Writing out the term (8.50) in components produces:

$$
\mathcal{L} = F^*F - \left|\partial_\mu\phi + \frac{iq}{2}A_\mu\phi\right|^2 - i\bar{\phi}\bar{\sigma}\left(\partial_\mu + \frac{iq}{2}qA_\mu\right)\psi - \frac{iq}{\sqrt{2}}\left(\phi\bar{\lambda}\bar{\psi} - \bar{\phi}\lambda\psi\right)\frac{1}{2}qD\bar{\phi}\phi.
\tag{8.53}
$$

There are two auxiliary fields, the D and F fields.

Adding the kinetic term (8.47) for the vector field and a potential, $\tilde{W}(\Phi)$ for the matter, gives the total Lagrangian,

$$
\mathcal{L} = \int d^2\theta\left(W^\alpha W_\alpha + \int d^2\bar{\theta}\,\bar{\Phi}^i \exp\left(q_i V\right)\Phi_i + \tilde{W}(\Phi)\right)
\tag{8.54}
$$

this produces the following potential,

$$V = \sum_i \left| \frac{\partial \tilde{W}}{\partial \phi^i} \right|^2 + \frac{q^2}{4} \left(\left(2K + \sum |\phi_i|^2 \right) \right)^2 \tag{8.55}$$

So far only systems with U(1) vector fields were discussed. One can also consider non-Abelian gauge groups. The fields are in an adjoint representation of the group, $A^a{}_\mu, \lambda^a, D^a$, the index a is the group index, $(a = 1 \dots \text{dim(group)})$ and $D^a = \sum_i \bar{\phi}_i T^a_{R(\Phi_i)} \Big) \Phi_i$.

8.2.3 Supersymmetrical Particle Spectrum in Nature?

The classification of the particles was naturally followed by an attempt to correlate the known particles with the algebraic results. The photon is massless to a very good approximation, the only known fermionic particle that at the time was considered massless as well was the neutrino. It was found that in the standard models the two cannot be members of the same multiplet. As SUSY is broken, it was at the time expected that the breaking be spontaneous and in case of a global symmetry this would lead to a massless spin one half Goldstone particle, a spin half fermion in the case of broken SUSY. It would have also been nice and simple if the neutrino would be at least the Goldstone fermion or as it has become termed a Goldstino. This turned out not to be consistent with experiment as well. Eventually one got resigned to the situation that all known particles, be they bosons or fermions, have supersymmetric partners which are yet to be discovered. The yet to be confirmed spin zero elementary particle, the Higgs, has a spin one half superpartner—the Higgsino. In fact in a supersymmetric model extension of the standard model at least two Higgs fields are required. The SUSY interaction terms can each be composed only of chiral, or only of antichiral fields. For such interaction terms a single Higgs field would not permit to construct the Yukawa interaction terms needed to provide masses to all quarks and leptons. Also with only one field the theory would not be consistent as it would suffer what is called an anomaly. The particles carry a U(1) gauge charge, and in the presence of a single Higgsino that gauge symmetry would become invalid once quantum corrections are taken into account. An extra Higgsino, and thus an extra Higgs, is required to restore the gauge invariance at the quantum level. The two mentioned problems get to be resolved by adding the one extra Higgs supermultiplet. The superpartners of the known spin one half quarks and leptons are denoted squarks and sleptons and are required to be spin zero bosons. The superpartners of the various spin one known gauge particles are termed the photino, wino, zino and gluino. They are assagined spin one half and are in the adjoint representation of the gauge group. In the LHC a major effort is planned for observing these particles.

8.2.4 Spontaneous SUSY Breaking: Perturbative Analysis

Any attempt to relate supersymmetry to nature at the level of the known particles requires the symmetry to be broken. It could have been broken explicitly, in this section the currently known mechanisms for its spontaneous breaking are described. This includes the breaking at the classical level in a class of gauge theories (by D terms) and in theories with no gauge particles (by F terms). Also are described the dynamical breaking of Supersymmetry as well as its effective breaking in a metastable vacuum. For SUSY not to be spontaneously broken all the SUSY generators $Q^i{}_\alpha$ need to annihilate the ground state. As the Hamiltonian is constructed out of positive pairings of the SUSY generators, SUSY preservation occurs if and only if the energy of the ground state vanishes. Conversely, SUSY is broken if and only if the energy of the ground state does not vanish, $Q^i_\alpha \mid$ G.S. $> \neq$ 0 iff $E_{\text{G.S.}} \neq 0$. As the Hamiltonian of the SUSY system is non-negative, the non-vanishing ground state energy in the case spontaneous SUSY breaking is positive.

As for the actual mechanism for the spontaneous breaking, it turned out that the breaking of supersymmetry requires a somewhat elaborate structure.

8.2.4.1 F-terms

Consider first a system which contains only spin zero and spin one-half particles. In that case the condition for SUSY not to be broken is the vanishing of the potential generated by the F terms. Super Symmetry is thus unbroken when the following equations have a solution:

$$V_F = 0 \iff F_i = 0 \ \forall i, \tag{8.56}$$

These are n (complex) equations with n (complex) unknowns. Generically, they have a solution. Take the example of the one component WZ model, where

$$F_1 = -\lambda - mA + gA^2. \tag{8.57}$$

This has a solution. There is no supersymmetry breaking classically. The non-renormalization theorem for the F terms ensures this result to be correct to all orders in perturbation theory. The solution is:

$$V = A^*A|(gA - m)|^2 \tag{8.58}$$

and hence there are actually two supersymmetric vacuum: either at $< A > = m/g$ or at $< A> = 0$.

Let us examine now how supersymmetry may be spontaneously broken. The following anecdote may be of some pedagogical value. It turns out that a short time after supersymmetry was introduced arguments were published which claimed to prove that supersymmetry cannot be broken spontaneouly at all. Supersymmetry resisted breaking attmepts for both theories of scalars and gauge theories. One could be surprised that the breaking was first achieved in the gauge systems. This was done by Fayet and Illiopoulos. The presence in the collaboration of a student who paid little respect to the general counter arguements made the discovery possible. Fayet went on to discover the breaking mechanism also in supersymmetric scalar theories as did O'Raighfeartaigh.

We will describe four examples of mechanisms of breaking Super Symmetry. The first will be for theories not containing gauge particles. The second for systems containing gauge particles, the third is of a dynamical breaking of Super Symmetry and the fourth occurs if our universe happens to be in a long lived metastable vacuum of positive kinetic energy.

8.2.4.2 SUSY Breaking in Theories with Scalars and Spin One Half Particles by F Terms

The Fayet-O'Raifeartaigh potential contains three fields this is the minimal number needed in order to break supersymmetry. It is:

$$\mathcal{L}_{\text{Potential}} = \lambda \Phi_0 + m \Phi_1 \Phi_2 + g \Phi_0 \Phi_1 \Phi_1 + \text{h.c.} \tag{8.59}$$

Minimizing the potential leads to the following equations:

$$
\begin{aligned}
0 &= \lambda + g \Phi_1 \Phi_1 \\
0 &= m \Phi_2 + 2 g \Phi_0 \Phi_1 \\
0 &= m \Phi_1
\end{aligned}
\tag{8.60}
$$

These cannot be consistently solved so there cannot be a zero energy ground state and supersymmetry must be spontaneously broken. To find the ground state one must write out the full Lagrangian including the kinetic terms in component fields and then minimize. Doing so one discovers that in the ground state $A_1 = A_2 = 0$ and A_0 is arbitrary. The arbitrariness of A_0 is a flat direction in the potential, the field along the flat direction is called a moduli. Computing the masses by examining the quadratic terms for component fields gives the following spectrum: the six real scalars have masses: $0, 0, m^2 m^2, m^2 \pm 2g\lambda$; and the Fermions have masses: $0, 2\ m$. The zero mass Fermion is the Goldstino. We turn next to breaking of supersymmetry theories that are gauge invariant.

8.2.4.3 SUSY Breaking in Supersymmetric Gauge Theories

The potential for the supersymmetric gauge theory was obtained to be:

$$V = \sum_i \left| \frac{\partial \tilde{W}}{\partial \phi^i} \right|^2 + \frac{q^2}{4} \left(2K + \sum |\phi_i|^2 \right)^2 \tag{8.61}$$

The first term is the F-term and the second is the D-term. Both these terms need to vanish for supersymmetry to remain unbroken.

Some remarks about this potential are in order:

Generically, the F-terms should vanish since there are indeed n equations for n unknowns. If $< \varphi_i > = 0$, that is if the U(1) is not spontaneously broken then supersymmetry is broken if and only if $K_{F. I.} \neq 0$. When $K = 0$ and the F-terms have a vanishing solution then so will the D-term and there will be no supersymmetry breaking.

These ideas are demonstrated by the following example. Consider fields Φ_1, Φ_2 with opposite U(1) charges and Lagrangian given by:

$$\mathcal{L} = \frac{1}{4} \left(W^\alpha W_\alpha + h.c. \right) + \overline{\Phi}_1 \exp(eV)\Phi_1 + \overline{\Phi}_2 \exp(-eV)\Phi_2 + m\,\Phi_1\Phi_2 + h.c. + 2KV \tag{8.62}$$

This leads to the potential:

$$V = \frac{1}{2}D^2 + F_1 F_1^* + F_2 F_2^* \tag{8.63}$$

where

$$\begin{aligned} D + K + \tfrac{e}{2}\left(A_1^* A_1 - A_2^* A_2 \right) &= 0 \\ F_1 + m A_{2*} &= 0 \\ F_2 + m A_1^* &= 0 \end{aligned} \tag{8.64}$$

Leading to the following expression for the potential:

$$V = \frac{1}{2}K^2 + \left(m^2 + \frac{1}{2}eK \right) A_1^* A_1 + \left(m^2 - \frac{1}{2}eK \right) A_2^* A_2 + \frac{1}{8}e^2 \left(A_1^* A_1 - A_2^* A_2 \right)^2 \tag{8.65}$$

Consider the case, $m^2 > \frac{1}{2}eK$. The scalars have mass, $\sqrt{m^2 + \frac{1}{2}eK}$ and $\sqrt{m^2 - \frac{1}{2}eK}$. The vector field has zero mass. Two Fermions have mass m and one Fermion is massless. Since the vector field remains massless then the U(1) symmetry remains unbroken. For $K \neq 0$, supersymmetry is broken as the Bosons and Fermions have different masses. (For $K = 0$ though the symmetry is restored.) The massless Fermion (the Photino) is now a Goldstino. Note that a trace of the underlying supersymmetry survives as one still has $\text{TrM}^2{}_B = \text{TrM}^2{}_F$ even after the breaking of supersymmetry. M_B and M_F are the bosonic and fermionic mass matrices respectively.

Next, consider the case, $m^2 < \frac{1}{2}eK$; at the minimum, $A_1 = 0, A_2 = v$ where $v^2 \equiv 4\frac{\frac{1}{2}eK - m^2}{e^2}$.

The potential expanded around this minimum becomes, with $A \equiv A_1$ and $\tilde{A} \equiv A_2 - v$:

$$
\begin{aligned}
V = {} & \tfrac{2m^2}{e^2}\left(eK - m^2\right) + \tfrac{1}{2}\left(\tfrac{1}{2}e^2v^2\right)A_\mu A^\mu \\
& + 2m^2 A^* A = \tfrac{1}{2}\left(\tfrac{1}{2}e^2v^2\right)\left(\tfrac{1}{\sqrt{2}}\left(\tilde{A} + \tilde{A}^*\right)\right)^2 \\
& + \sqrt{m^2 + \tfrac{1}{2}e^2v^2}\left(\psi\tilde{\psi} + \overline{\psi\tilde{\psi}}\right) + 0 \times \lambda\bar{\lambda}
\end{aligned}
\tag{8.66}
$$

The first term implies that supersymmetry is broken for $m > 0$. The photon is massive, $m_\gamma^2 = \tfrac{1}{2}e^2v^2$ implying that the U(1) symmetry is broken as well. The Higgs field, $\tfrac{1}{\sqrt{2}}\left(\tilde{A} + \tilde{A}^*\right)^2$ has the same mass as the photon. Two Fermions have non-zero mass and there is one massless Fermion, the Goldstino.

In the above example there is both supersymmetry breaking and U(1) symmetry breaking except when $m = 0$ in which case the supersymmetry remains unbroken.

Next consider a more generic example where there is U(1) breaking but no supersymmetry breaking, Φ is neutral under the U(1) while Φ_+ has charge $+1$ and $\Phi-$ has charge -1. The potential is given by:

$$
\mathcal{L} = \frac{1}{2}m\Phi^2 + \mu\Phi_+\Phi_- + \lambda\Phi\Phi_+\Phi_- + h.c.
\tag{8.67}
$$

There are two branches of solutions to the vacuum equations (a denotes the vacuum expectation value of A, etc.):

$$
a_+a_- = 0, \quad a = -\frac{\lambda}{m}
\tag{8.68}
$$

which leads to no U(1) breaking and

$$
a_+a_- = -\frac{1}{8}\left(\lambda - \frac{m\mu}{g}\right), \quad a = -\frac{\mu}{g}
\tag{8.69}
$$

which breaks the U(1) symmetry.

Note, the presence of a flat direction:

$$
a_+ \to e^\alpha a_+, a_- \to e^{-\alpha}a_-
\tag{8.70}
$$

leaves $a-a_+$ fixed and the vacuum equations are still satisfied.

Typical generic supersymmetry breaking requires that some of the equations for the vanishing of the potential be redundant. Such could be the case if the system had an extra symmetry such as an R symmetry mentioned before. In the examples above containing only scalar fields this was indeed the case. The absence of a zero energy solution led also to a spontaneous breaking of the R-symmetry. This should lead to the presence of a Goldstone Boson corresponding to the broken

$U(1)_R$. Inverting this argument leads to the conclusion that supersymmetric breaking in nature is not easy to obtain since we do not observe even such a particle. This argument was revisited in recent years.

So far only systems with $U(1)$ vector fields were discussed. To have a non-vanishing FayetIliopoulos term there needs to be a $U(1)$ factor in the gauge group. To obtain a breaking of SUSY in a gauge system with does not have a $U(1)$ factor one needs to consider dynamical SUSY breaking. This is a more complex problem and new tools were developed to explore this possibility. The functional form of such a breaking allows to obtain a scale of SUSY breaking which is naturally much smaller than the relevant cutoff of the problem. This is reflected by the relation:

$$M_{\text{SUSY-breaking}} = M_{\text{cutoff}} \exp\left(-c/g\left(M_{\text{cutoff}}\right)\right). \tag{8.71}$$

8.2.5 Dynamics of SUSY Gauge Theories and SUSY Breaking

The description of the mechanism of dynamical breaking of SUSY is preceded by a survey of the general possible phase structure of gauge theories as well as its concrete realizations in SUSY gauge theories.

8.2.5.1 Phases of Gauge Theories

The phase structure of gauge theories can be introduced by analyzing them as statistical mechanical systems regulated by a finite lattice. The basics can be illustrated by considering $D = 4$ lattice gauge theories, in particular those for which the gauge fields which are Z_N valued. The system has a coupling g.

The effective "temperature" of the system is given by, $T = \frac{Ng^2}{2\pi}$.

For a given theory there is a lattice of electric and magnetically charged operators. The electric charge is denoted by n and the magnetic charge by m. An operator with charges (n,m) is *perturbative*, i.e., it is an irrelevant operator and weakly coupled to system, so long as the free energy, $F > 0$, i.e.,

$$n^2 T + \frac{m^2}{T} > \frac{C}{N}, \tag{8.72}$$

however, when the free energy is negative for the operator (n,m), it condenses indicating the presence of a relevant operator and hence an infra-red instability, this occur when,

$$n^2 T + \frac{m^2}{T} < \frac{C}{N}. \tag{8.73}$$

Keeping N,C fixed and vary T.

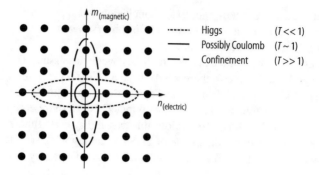

Fig. 8.1 Different possible phases

The system has three phases depending which operators condense. At small T, there is electric condensation which implies that there is electric charge screening, magnetic charges are confined, and the log of the Wilson loop is proportional to the length of the perimeter of the loop. (This is called the Higgs phase).

At high T, magnetic condensation occurs, this is the dual of electric condensation. Magnetic charges are screened, electric charges are confined and the log of the Wilson loop is proportional to the area. (This is called the confinement phase.) For intermediate values of T it is possible that there is neither screening of charges nor confinement, this is the Coulomb phase (Fig. 8.1).

In the presence of a θ parameter, an electric charge picks up a magnetic charge and becomes dyonic.

$$n' = n + \frac{\theta}{2\pi}m \tag{8.74}$$

This lead to a tilted lattice of dyonic charges and one may condense dyons with charges (n_0, m_0). This leads to what is called oblique confinement with the charges commensurate with (n_0, m_0) being screened and all other charges being confined.

These ideas relate to the gauge theories of the standard model. For QCD it was suggested that confinement occurs due the condensation of QCD monopoles. This is a magnetic superconductor dual to the electric one which describes the weak interactions. It is difficult to study this phenomenon directly. The Dirac monopole in a U(1) gauge theory is a singular object; however by embedding the monopole in a spontaneously broken non-Abelian theory with an additional scalar field one may smooth out the core of the monopole and remove the singularity. One may proceed analogously, by enriching QCD; adding scalars and making the theory supersymmetric one can calculate the condensation of monopoles in a four dimensional gauge theory.

In general the possible phase structure of gauge theories and their actual realizations were obtained by using various approximation schemes. While supersymmetry has not yet disclosed if it part of nature in some form or another, in its presence, the

Fig. 8.2 Possible phases of gauge theories (g_1 and g_2 are some relevant couplings)

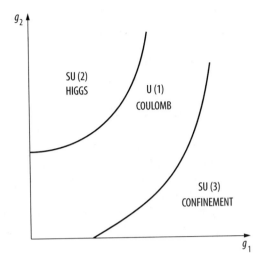

phase structure of gauge theories was exactly obtained in some cases. Most earlier exact important results in field theories, such as asymptotic freedom, were obtained in circumstances in which the couplings are weak. SUSY enables to obtain also results in the strong coupling regime. The analytic control seems to become larger the more supersymmetries the system possesses. This has been achieved in four dimensions for supersymmetric gauge theories with N = 1,2,4 supersymmetries. There are many new methods that have been utilized and the phase structures of these theories have been well investigated. Novel properties of these theories have been discovered such as new types of conformal field theories and new sorts of infra-red duality. To these we turn next (Fig. 8.2).

8.2.5.2 SUSY QCD: The Setup

The goal will be to examine theories that are simple supersymmetric extensions of QCD. Consider the case of a N = 1 vector multiplet with gauge group SU(N_C), and N_F chiral multiplets in the fundamental representation of SU(N_C), and N_F chiral multiplets in the antifundamental representation. The Lagrangian is:

$$\mathcal{L} = \int (-i\tau)\, \mathrm{Tr} W^\alpha W_\alpha d^2\theta + h.c. \\
+ Q_F^+ \exp(-2V)\, Q_F + \tilde{Q} \exp(2V)\, Q_F^+\big|_{\bar{\theta}\bar{\theta}\theta\theta} + m_F \tilde{Q}_F\, Q\big|_{\theta\theta} \tag{8.75}$$

where the coupling is:

$$\tau = \frac{\theta}{2\pi} + i\frac{4\pi}{g^2}. \tag{8.76}$$

Apart from the local SU(N_C) gauge symmetry, the fields are charged under the following global symmetries.

$$SU(N_F)_L \times SU(N_F)_R \times U(1)_V \times U(1)_A \times U(1)_{RC}$$

	SU(N_F)$_L$	SU(N_F)$_R$	U(1)$_V$	U(1)$_A$	U(1)$_{RC}$
Q_a^i	N_F	1	1	1	1
$\tilde{Q}_i^a$	1	$\overline{N}_F$	-1	1	1
W_α	1	1	0	0	1

When $N_C = 2$, because $2 \sim \bar{2}$, the global flavor symmetry is enhanced to SO($2N_F$)$_L$ × SO($2N_F$)$_R$.

There is an anomaly of the U(1)$_A$ × U(1)$_R$ symmetry. A single U(1) symmetry survives the anomaly. This is denoted as U(1)$_R$ and is a full quantum symmetry. The adjoint Fermion contributes $2N_C \times R(\lambda)$ to the anomaly. The Chiral Fermions contribute, $2N_F \times R_F$. $R(\lambda) = 1$, while R_F is now chosen so that the total anomaly vanishes,

$$2N_C + 2R_F N_F = 0. \tag{8.77}$$

This leads to

$$R_F = -\frac{N_C}{N_F}. \tag{8.78}$$

The Bosons in the chiral multiplet have an R-charge one greater than the Fermions in the multiplet. Thus,

$$R_B = 1 - \frac{N_C}{N_F} = \frac{N_F - N_C}{N_F}. \tag{8.79}$$

The non-anomalous R-charge, is given by:

$$R = R_C - \frac{N_C}{N_F} Q_A, \tag{8.80}$$

where R_C is the classical R-charge and A is the classical U(1)$_{rmA}$ charge. Following non-anomalous global charges: This leads to the

$$SU(N_F)_L \times SU(N_F)_R \times U(1)_V \times U(1)_R$$

	SU(N_F)$_L$	SU(N_F)$_R$	U(1)$_V$	U(1)$_R$
Q_a^i	N_F	1	1	$\frac{N_F - N_C}{N_F}$
$\tilde{Q}_i^a$	1	$\overline{N}_F$	-1	$\frac{N_F - N_C}{N_F}$
W_α	1	1	0	1

One is now ready to identify the classical moduli space.

8.2.5.3 The Moduli Space

The classical moduli space is given by solving the D-term and F-term equations:

$$
\begin{aligned}
D^a &= Q_F^\dagger T^a \, Q_F - \tilde{Q}_F T^a \tilde{Q}_F^\dagger \\
\overline{F}_{Q_F} &= -m_F \tilde{Q} \\
\overline{F}_{\tilde{Q}_F} &= -m_F Q
\end{aligned}
\tag{8.81}
$$

For $N_F = 0$ or for $N_F \neq 0$ and $m_F \neq 0$, there is no moduli space. Note, the vacuum structure is an infra-red property of the system hence having $m_F \neq 0$ is equivalent to setting $N_F = 0$ in the deep infrared.

Consider the quantum moduli space of the case where $N_F = 0$. One can show that the number of zero energy states of the system is no smaller than The Witten index, $\mathrm{Tr}(-1)^F = N_C$ i.e. the rank of the group $+1$. This number is larger than zero and thus there is no supersymmetry breaking in these systems. There are $2N_C$ Fermionic zero modes (from the vector multiplet). These Fermionic zero modes break through instanton effects the original $U(1)_R$ down to Z_{2N_C}. Further breaking occurs because the gluino two point function acquires a vacuum expectation value which breaks the symmetry down to Z_2. This indeed leaves N_C vacua. The gluino condensate is:

$$
< \lambda\lambda > = \exp\left(\frac{2\pi i k}{N_C}\right) \Lambda_{N_C}^3
\tag{8.82}
$$

where Λ_{NC} is the dynamically generated scale of the gauge theory and $k = 1, \ldots, N_C - 1$ label the vacua. Chiral symmetry breaking produces a mass gap. Note, because chiral symmetry is discrete there are no Goldstone Bosons. Further details of quantum moduli spaces will be discussed later.

Consider the case where $m_F = 0$ and $0 < N_F < N_C$. The classical moduli space is determined by the following solutions to the D-term equations:

$$
Q = \tilde{Q} =
\begin{pmatrix}
a_1 & 0 & & 0 & \ldots & 0 \\
0 & a_2 & & 0 & \ldots & 0 \\
 & & . & 0 & \ldots & 0 \\
 & & . & 0 & \ldots & 0 \\
 & & a_{NF} & 0 & \ldots & 0
\end{pmatrix}_{N^F \times N^C}
\tag{8.83}
$$

Where the row indicates the flavor and the column indicates the colour. There are N_F diagonal non-zero real entries, a_i. (To validate this classical analysis the vacuum expectation values must be much larger than any dynamically generated scale, i.e., $a_i \gg \Lambda$. The gauge symmetry is partially broken:

$$
SU(N_C) \rightarrow SU(N_C - N_F).
\tag{8.84}
$$

This is for generic values of a_i. By setting some subset of a_i to zero one may break to a subgroup of $SU(N_C)$ that is larger than $SU(N_C - N_F)$. Also, if $N_F = N_C - 1$ then the gauge group is complete broken. This is called the Higgs phase.

The number of massless vector Bosons becomes

$$N_C{}^2 - \left((N_C - N_F)^2 - 1\right) = 2N_C N_F - N_F{}^2, \tag{8.85}$$

the number of massless scalar fields becomes,

$$2N_C N_F - \left(2N_C N_F - N_F^2\right) = N_F^2. \tag{8.86}$$

The matrix

$$M_{\tilde{i}j} \equiv \tilde{Q}_{\tilde{i}} Q_j \tag{8.87}$$

forms a gauge invariant basis. The Kahler potential is then,

$$K = 2\mathrm{Tr}\sqrt{(M\overline{M})}. \tag{8.88}$$

When singularities appear, i.e. $\det M = 0$, it signals the presence of massless particles as well as of enhanced symmetries.

When $N_F \geq N_C$, one has the following classical moduli space,

$$
Q = \begin{pmatrix}
a_1 & 0 & & & \\
0 & a_2 & & & \\
& & \cdot & & \\
& & & \cdot & \\
& & & & a_{N_C} \\
0 & 0 & \cdots & \cdots & 0 \\
\cdot & \cdot & \cdots & \cdots & 0 \\
0 & 0 & \cdots & \cdots & 0
\end{pmatrix}_{N_F \times N_C}
\,,\quad
\tilde{Q} = \begin{pmatrix}
\tilde{a}_1 & 0 & & & \\
0 & \tilde{a}_2 & & & \\
& & \cdot & & \\
& & & \cdot & \\
& & & & \tilde{a}_{N_C} \\
0 & 0 & \cdots & \cdots & 0 \\
\cdot & \cdot & \cdots & \cdots & 0 \\
0 & 0 & \cdots & \cdots & 0
\end{pmatrix}_{N_F \times N_C}
\tag{8.89}
$$

with the constraint that

$$|a_i|^2 - |\tilde{a}_i|^2 = \rho. \tag{8.90}$$

Generically the $SU(N_C)$ symmetry is completely broken. However, when $a_i = \tilde{a}_i = 0$ then a subgroup of the $SU(N_C)$ can remain.

We will now consider some special cases, first the classical moduli space for $N_F = N_C$. The dimension of the moduli space is given by:

$$2N_C{}^2 - \left(N_C{}^2 - 1\right) = N_C{}^2 + 1 = N_F{}^2 + 1 \qquad (8.91)$$

There are $N_F{}^2$ degrees of freedom from M_{ij} and naively one would have two further degrees of freedom from:

$$B = \epsilon_{i_1 \ldots i_{N_C}} Q_{j_1}^{i_1} \cdots Q_{j_{N_F}}^{i_{N_C}}, \quad \tilde{B} = \epsilon_{i_1 \ldots i_{N_C}} \tilde{Q}_{j_1}^{i_1} \cdots \tilde{Q}_{j_{N_F}}^{i_{N_C}}. \qquad (8.92)$$

There is, however, a classical constraint:

$$\det M - B\tilde{B} = 0 \qquad (8.93)$$

which means M, B and $\tilde{B}$ are classically dependent. This leaves only $N_F^2 + 1$ independent moduli.

Generically, as well as the gauge symmetry being completely broken, the global flavor symmetry is also broken. There is a singular point in the moduli space where $M = 0 = B = \tilde{B}$.

Next, consider the case, $N_f = N_C + 1$, again there are N_F^2 moduli from $M_{\bar{i}j}$. There are also, $2(N_C + 1)$ degrees of freedom given by:

$$B_i = \epsilon_{ii_1 \ldots i_{N_C}} Q_{j_1}^{i_1} \cdots Q_{j_{N_F}}^{i_{N_C}}, \quad \tilde{B}_{\bar{i}} = \epsilon_{\bar{i}_1 \ldots i_{N_C}} \tilde{Q}_{j_1}^{i_1} \cdots \tilde{Q}_{j_{N_F}}^{i_{N_C}}. \qquad (8.94)$$

However, there are again the classical constraints:

$$\begin{aligned} \det M & - & M_{\bar{i}j} B^i B^{\bar{j}} = 0 \\ M_{\bar{j}i} B_i & = & M_{\bar{i}j} B_{\bar{j}} = 0 \end{aligned} \qquad (8.95)$$

giving again an $N_F^2 + 1$ dimension moduli space. (The moduli space is not smooth). There is a generic breaking of gauge symmetry. In all these cases the potential has flat directions of zero energy, SUSY is unbroken. The non-renormalization theorem extends this result to all order in perturbation theory, it will be the non-perturbative effects which will lift the vacuum degeneracies in some cases and will lead to dynamical breaking of SUSY.

8.2.5.4 Quantum Moduli Spaces/Dynamical SUSY Breaking

A rich structure emerges. One is required to examine on a case by case basis the role that quantum effects play in determining the exact moduli space. Quantum effects

Fig. 8.3 Potential for $1 < N_F < N_C$, it has no ground state

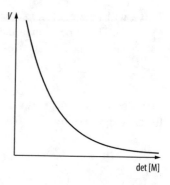

det [M]

both perturbative and nonperturbative can lift moduli. In what follows the quantum moduli space is examined for the separate cases: 1

$$\leq N_F \leq N_C - 1,\ N_F = N_C,\ N_F,\ N_C + 1,\ N_C + 1 < N_F \leq \frac{3N_C}{2},$$

$$\frac{3N_C}{2} < N_F < 3N_C,\ N_F = 3_C N,\ \text{and}\quad N_F > 3N_C.$$

We start with the study of the quantum moduli space for $0 < N_F < N_C$.

Classically, the dimension of the moduli space is N_F^2 from Q, $\tilde{Q}$. The following table summarizes the charges under the various groups.

	$SU(N_C)$	$SU(N_F)_L$	$SU(N_F)_R$	$U(1)_V$	$U(1)_A$	$U(1)_{R_{cl}}$	$U(1)_R$
Q^i_a	N_C	N_F	1	1	1	1	$\frac{N_F - N_C}{N_F}$
$\tilde{Q}^a_i$	$\overline{N_C}$	1	$\overline{N_F}$	-1	1	1	$\frac{N_F - N_C}{N_F}$
$\Lambda^{3N_C - N_F}$	1	1	1	0	$2N_F$	$2N_C$	0
M	1	N_F	$\overline{N_F}$	0	2	2	$2 - \frac{2N_C}{N_F}$
$\det M$	1	1	1	0	$2N_F$	$2N_F$	$2(N_F - N_C)$

Λ, the dynamically generated QCD scale is assigned charges as m and g were before. The power $3N_C - N_F$ is the coefficient in the one loop beta function. There is no Coulomb phase so W_α does not appear.

The symmetries imply, the superpotential, W, has the following form:

$$W = \left(\Lambda^{3N_C - N_F}\right)^a (\det M)^b c \tag{8.96}$$

a, b are to be determined. c is a numerical coefficient, If c does not vanish, the classical moduli space gets completely lifted by these nonperturbative effects (Fig. 8.3).

One examines the charges of W under the various symmetries. Automatically, the charges of W for the flavor symmetries, $SU(N_F)_L \times SU(N_F)_R$ and the $U(1)_V$ vanish.

If one requires the $U(1)_A$ charge to vanish then this implies $a = -b$. Requiring the $U(1)_R$ charge to vanish implies that $b = \frac{1}{N_F - N_C}$. These restrictions fix:

$$W = c \left(\frac{\Lambda^{3N_C - N_F}}{\det M} \right)^{\frac{1}{N_C - N_F}} \tag{8.97}$$

For non-vanishing c, all the moduli are now lifted and there is no ground state.

What is the value of c? This is a difficult to calculate directly unless there is complete higgsing. For $N_F = N_C - 1$ there is complete symmetry breaking and one can turn to weak coupling. From instant on calculations one calculates that $c \neq 0$ and the prepotential for the matter fields is

$$W \sim \left(\frac{\Lambda^{2N_C + 1}}{\det M} \right). \tag{8.98}$$

One may now go to $N_F < N_C - 1$ by adding masses and integrating out the heavy degrees of freedom. This produces

$$< M^i{}_j >_{\min} = \left(m^{-1} \right)^i{}_j \left(\Lambda^{3N_C - N_F} \det m \right)^{\frac{1}{N_C}}. \tag{8.99}$$

Thus in this region the dynamical effects remove the supersymmetric vaccum. In fact the system has no ground state and supersymmetry is broken dynamically. One can go one step further and consider the possibility that the effective potential above is modified so as to have a local minimum with positive vacuum energy. It is possible to construct many such examples and consider that supersymmetry seems broken as a result of the universe being for the (long) time being in the metastable state. Eventually the system may tunnel to another lower energy vacuum and perhaps eventually will reach a SUSY vacuum. In the analysis of such systems once needs also to take into account gravity. In the cases which follow supersymmetry is not broken, they are described to illustrate the rich structure which emerges once one can treat the system when it strongly interacts. Perhaps also structures like those of the dyonic condensates will yet play a role somewhere in nature (Fig. 8.4).

We now turn to study the dynamics and quantum moduli space for the $N_F \geq N_C$ case. In this situation there is a surviving moduli space. In the presence of a mass matrix, m_{ij} for matter one obtains

$$< M^i{}_j >= \left(m^{-1} \right)^i{}_j \left(\Lambda^{3N_C - N_F} \det m \right)^{\frac{1}{N_C}}. \tag{8.100}$$

Fig. 8.4 Potential with finite masses has a ground state. $M_{min} \to \infty$ as $m \to 0$

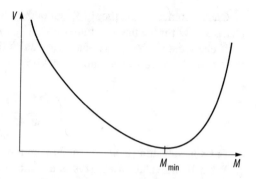

Previously, for the case of $N_F < N_C$, it turned out that $m \to 0$ implied $< M^i_j > \to \infty$ thus explicitly lifting the classical moduli space. For $N_F \geq N_C$ it is possible to have $m \to 0$ while keeping $< M^i_j >$ fixed.

Consider the case where $N_F = N_C$. Quantum effects alter the classical constraint to be:

$$\det M - B\tilde{B} = \Lambda^{2N_C}. \tag{8.101}$$

This has the effect of resolving the singularity in moduli space. The absence of a singularity means there will not be additional massless particles.

The physics of this theory depends on the position in moduli space of the vacuum. For large, $M/B/\tilde{B}$ one is sitting in the Higgs regime; however, for small $M/B/\tilde{B}$ one is in the confining regime. Note that M cannot be taken smaller than Λ. As the system has particles in the fundamental representations there is no actual phase transition between these two regimes, the transition is of a quantitative nature. In addition global symmetries need to be broken in order to satisfy the modified constraint equation.

Consider some examples: with the following expectation value,

$$< M^i_j >= \delta^i_j \Lambda^2, \quad < B\tilde{B} >= 0, \tag{8.102}$$

the global symmetries are broken to:

$$SU(N_F)_V \times U_B(1) \times U_R(1), \tag{8.103}$$

and there is chiral symmetry breaking. When,

$$< M^i_j >= 0, \quad < B\tilde{B} > \neq 0 \tag{8.104}$$

Fig. 8.5 Phases of super
QCD

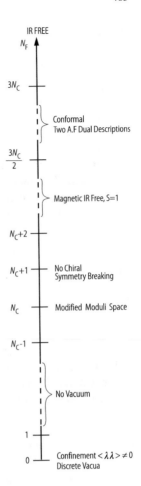

then the group is broken to:

$$SU(N_F)_L \times SU(N_F)_R \times U_R(1) \qquad (8.105)$$

which has chiral symmetry and also has confinement. This is an interesting situation because there is a dogma that as soon as a system has a bound state there will be chiral symmetry breaking (Fig. 8.5).

The dynamics for the case $N_F = N_C + 1$ brings about some different dynamics. The moduli space remains unchanged. The classical and quantum moduli spaces are the same and hence the singularity when $M = B = \tilde{B} = 0$ remains. This is not a theory of massless gluons but a theory of massless mesons and baryons. When, $M, B, B \neq 0$ then one is in a Higgs/confining phase. At the singular point when, $M = B = \tilde{B} = 0$ there is no global symmetry breaking but there is "confinement" with light baryons.

Fig. 8.6 Two systems with a different ultra-violet behavior flowing to the same infra-red fixed point.

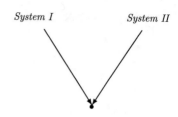

There is a suggestion that in this situation, $M, B, \tilde{B}$ become dynamically independent. The analogy is from the nonlinear sigma model, where because of strong infrared fluctuations there are n independent fields even though there is a classical constraint. The effective potential is:

$$W_{\text{eff}} = \frac{1}{\Lambda^{2N_C - 1}} \left(M^i{}_j B_i \tilde{B}^j - \det M \right) \tag{8.106}$$

the classical limit is taken by:

$$\Lambda \to 0 \tag{8.107}$$

which in turn imposes the classical constraint.

For higher values of the number of flavors N_F the plot thickens even more as new duality emerge. Infrared dualities which in some circumstances uncover new types of dynamical structures.

8.2.5.5 Infra-red Duality

Two systems are called infra-red dual if, when observed at longer and longer length scales, they become more and more similar (Fig. 8.6).

It has been observed that the following set of $N = 1$ supersymmetric gauge theories are pairwise infra-red dual called Seiberg dualities.

System		Dual System		
Gauge Group	#flavors	Gauge Group	#flavor	#singlets
$SU(N_C) SO(N_C) Sp(N_C)$	$N_F N_F 2N_F$	$SU(N_F - N_C)$ $SO(N_F - N_C + 4)$ $Sp(N_F - N_C - 2)$	$N_F N_F 2N_F$	$N_F^2 N_F^2 N_F^2$

For a given number of colors, N_C, the number of flavors, N_F, for which the infrared duality holds is always large enough so as to make the entries in the table meaningful. Note that the rank of the dual pairs is usually different. Lets explain why this result is so powerful. In general, it has been known for quite a long time that two systems which differ by irrelevant operator have the same infra-red behavior.

In these cases the UV structure of the Infrared dual theories is very different, the dual systems have a different numbers of colors. The common wisdom in hadronic physics has already identified very important cases of infra-red duality. For example, QCD, whose gauge group is $SU(N_C)$ and whose flavor group is $SU(N_F) \times SU(N_F) \times U(1)$, is expected to be infra-red dual to a theory of massless pions which are all color singlets. The pions, being the spin-0 Goldstone Bosons of the spontaneously broken chiral symmetry, are actually infra-red free in four dimensions. We have thus relearned that free spin-0 massless particles can actually be the infra-red ashes of a strongly-interacting theory, QCD whose ultraviolet behavior is described by other particles. By using supersymmetry, one can realize a situation where free massless spin-$\frac{1}{2}$ particles are also the infra-red resolution of another theory. This duality allows for the first time to ascribe a similar role to massless infra-red free spin-1 particles. Massless spin-1 particles play a very special role in our understanding of the basic interactions. This comes about in the following way: Consider, for example, the $N = 1$ supersymmetric model with N_C colors and N_F flavors. It is infra-red dual to a theory with $N_F - N_C$ colors and N_F flavors and N_F^2 color singlets. For a given N_C, if the number of flavors, N_F, is in the interval $N_C + 1 < N_F < \frac{3N_C}{2}$, the original theory is strongly coupled in the infra-red, while the dual theory has such a large number of flavors that it becomes infrared free. Thus the infra-red behavior of the strongly-coupled system is described by infrared free spin-1 massless fields (as well as its superpartners), i.e. infrared free massless spin-1 particles (for example photons in a SUSY system) could be, under certain circumstances, just the infra-red limit of a much more complicated ultraviolet theory. This is the first example of a weakly interacting theory in which spin one particles that in the infra-red may be viewed as bound states of the dual theory. The duality has passed a large number of consistency checks under many circumstances. The infrared duality relates two disconnected systems. From the point of view of string theory the two systems are embedded in a larger space of models, such that a continuous trajectory relates them. An additional new consequence of this duality follows for the case

$$\frac{3N_C}{2} < N_F < 3N_C$$

the two dual theories are both asymptotically free in the UV and are describe by the same nontrivial conformal field theory in the infrared. The panorama of these structures is given in Fig. 8.8. Finally a class of examples was found among $N = 2$ SUSY conformal systems for which there are two very different Lagrangian descriptions as far as the local symmetries are involved which are actually identical at all distance scales.

8.2.5.6 More General Matter Composition of SUSY Gauge Theories

One can enrich the structure of the theory by adding N_a particles in the adjoint representation. At first one has no matter in the fundamental representation and scalar multiplets which are adjoint valued. The potential for the scalars, φ_i is given by:

$$V = ([\varphi, \varphi])^2. \tag{8.108}$$

This potential obviously has a flat direction for diagonal φ. The gauge invariant macroscopic moduli would be $\mathrm{Tr}\varphi^k$. Consider the non-generic example of $N_C = 2$ and $N_a = 1$, the supersymmetry is now increased to $N = 2$. There is a single complex modulus, $\mathrm{Tr}\varphi^2$. Classically, SU(2) is broken to U(1) for $\mathrm{Tr}\phi^2 \neq 0$. One would expect a singularity at $\mathrm{Tr}\varphi^2 = 0$. The exact quantum potential vanishes in this case.

Naively, one could have expected that when $\mathrm{Tr}\varphi^2$ is of order Λ or smaller, one would expect that the strong infra-red fluctuations would wash away the expectation value for $\mathrm{Tr}\varphi^2$ and the theory would be confining. The surprising thing is that when SU(2) breaks down to U(1), because of the very strong constraints that supersymmetry imposes on the system, there are only two special points in moduli space and even there the theory is only on the verge of confinement. Everywhere else the theory is in the Coulomb phase. At the special points in the moduli space, new particles will become massless.

One can examine the effective theory at a generic point in moduli space where the theory is broken down to U(1). The Lagrangian is given by,

$$\mathcal{L} = \int d^2\theta \, \mathrm{Im} \left(\tau_{\mathrm{eff}} \left(\mathrm{Tr}\phi^2, g, \Lambda \right) W_\alpha W^\alpha \right) \tag{8.109}$$

The τ_{eff} is the effective complex coupling which is a function of the modulus, $\mathrm{Tr}\varphi^2$, the original couplings and the scale, Λ. This theory has an SL(2,Z) duality symmetry. The generators of the SL(2,Z) act on τ, defined by (8.76), as follows:

$$\tau \to -\frac{1}{\tau}, \quad \tau \to \tau + 1 \tag{8.110}$$

This is a generalization of the usual U(1) duality that occurs with electromagnetism to the case of a complex coupling. Recall the usual electromagnetic duality for Maxwell theory in the presence of charged matter is:

$$E \to B, B \to -E, e \to m, m \to -e. \tag{8.111}$$

This generalizes to a U(1) symmetry by defining:

$$E + iB, e + im. \tag{8.112}$$

The duality symmetry now acts by:

$$E + iB \rightarrow \exp(i\alpha)(E + iB), e + im \rightarrow \exp(i\alpha)(e + im). \qquad (8.113)$$

For the SU(2) case the moduli are given by $u = \mathrm{Tr}\varphi^2$, for SU(N_C) the moduli are given by $u_k = \mathrm{Tr}\varphi^k$, $k = 2, \ldots, N_C$. The classical moduli space is singular at times, there are no perturbative or nonperturbative corrections.

The dependence of τ on the moduli coordinate u was found. The vast number of results and literature on this will not be described here. Briefly:

The following complex equation,

$$y^2 = ax^3 + bx^2 + cx + d \qquad (8.114)$$

determines a torus. The complex structure of the torus, τ_{torus} will be identified with the complex coupling τ_{eff}. a,b,c,d are known holomorphic functions of the moduli, couplings, and scale, and so will implicitly determine τ_{torus}.

When $y(x)$ and $y'(x)$ vanish, for some value of x, τ is singular. Therefore,

$$\tau_{\mathrm{eff}} = i\infty, \, g^2_{eff} = 0 \qquad (8.115)$$

and the effective coupling vanishes. This reflects the presence of massless charged objects. This occurs for definite values of u in the moduli space. These new massless particles are monopoles or dyons. The theory is on the verge of confinement. For N $= 2$ supersymmetry that is the best one can do. The monopoles are massless but they have not condensed. For condensation to occur the monopoles should become tachyonic indicating an instability that produces a condensation. One can push this to confinement by adding a mass term: $\tilde{m}\mathrm{Tr}\varphi^2$, or generally for SU($N_C$) the term:

$$\delta W = g_k u_k. \qquad (8.116)$$

This breaks N $= 2$ supersymmetry down to N $= 1$. The effective prepotential is now:

$$W = M(u_k) q\tilde{q} + g_k u_k \qquad (8.117)$$

then

$$\frac{\partial W}{\partial u_k} = 0, \quad \frac{\partial W}{\partial (q\tilde{q})} = 0 \Rightarrow M(<u_k>) = 0, \quad \partial_{u_k} M(<u_k>) <q\tilde{q}> = -g_k \qquad (8.118)$$

Since generically,

$$\partial_{u_k} M(<u_k>) \neq 0 \qquad (8.119)$$

There will be condensation of the magnetic charge, confinement has been demonstrated to be indeed driven by monopole condensation. A monopole is usually a very heavy collective excitation. It is only the large amount of SUSY which allows one to follow the monople as it becomes massless and even condenses.

8.2.6   Dynamics of SUSY Gauge Theories with N > 1 SUSY

8.2.6.1   N = 4 Supersymmetry

In the presence of this large amount of supersymmetry in four space-time dimensions the particle content was described in an earlier section. It consists of spin one, spin one half and spin zero particles. The particles are all in the adjoint representation of the gauge group. They fall into representations of the SU(4) global symmetry group as well. The full Lagrangian is fully dictated by the symmetry. The large symmetry leads to several properties which can be demonstrated.

- The theory is scale invariant quantum mechanically. This was shown to all orders in perturbation theory as well as non-perturbatively. This served as an example of non-trivial four dimensional scale invariant theories.
- The theory has thus a meaningful coupling constant on which the physics truly depends (unlike massless QCD in which dimensionless quantities do not depend on the coupling). Moreover the coupling can be complexified by adding the θ parameter. (In the absence of a chiral anomaly the theory truly depends also on θ even though massless fermions are present). The theory is invariant under the modular group SL(2,Z), this group relates in particular small and large values of the coupling.
- The theory has flat directions along the scalar fields for any value of the coupling. The different points along the flat directions are not generically related by any symmetry. Each point characterizes a different vacuum choice for the system. These different vacua are called moduli. There is a special point in the moduli space and that is the point at the origin of field space where all scalar fields obtain a zero expectation value. At that point the theory is realized in a scale invariant manner. The massless fields rendering the theory with a rather complex analytic structure. Choosing different vacua along the moduli space leads to different residual gauge symmetries, as the scalar fields are in the adjoint representations the residual gauge group is at least $U(1)^r$ where r is the rank of the group. In each of these vacua the scale symmetry is spontaneously broken leading to a presence of a dilaton, the Goldstone Boson of broken scale invariance, in the spectrum. The vacuum energy is the same in all the phases associated with the different vacua choice. It has no dependence on any of the expectation values of the scalar fields. The spectrum includes massless and massive gauge particles.
- In the broken phases of the theory the conditions are appropriate for the presence of various solitons in the system. In general they contain particles particles which

have both electric and magnetic charges, called dyons. The massess of many of these particles, called BPS states is protected by the large SUSY to the extent that the mass dependence on the coupling is exactly known. This has applications in the counting microscopically the entropy of some black holes.

- Moreover, a hidden wish of theoreticians is that not only will one theory describe all physical phenomena but that that theory be exactly solvable. Given the complexity of four dimensional field theory this hope was suppressed. It had resurfaced when it was uncovered that quite a few properties of the $N = 4$ theory are exactly calculable. The system seems to have a large number of conserved quantities and this results in many so called integrability features.

- As will be discussed in the section on string theory, there is a large body of evidence that $N = 4$ theory encodes in it the information of special string theories which include gravity and black holes. These are string theories for strings propagation on a manifold part of which has a negative curvature and a negative cosmological constant.

To conclude, supersymmetric gauge theories have a very rich phase structure and many outstanding dynamical issues can be discussed reliably in the supersymmetric arena that are hard to address elsewhere.

8.2.7   Gauging Supersymmetry

During most of the second part of the twentieth century local symmetries were at center stage. The model unifying the electromagnetic and weak interactions actually united them mainly by using the concept that gauge theories do describe both. The stronger unification using only one non-semi simple group is yet to be achieved. The gauge theories have allowed to make precise calculable predictions for experimentally measurable quantities. Yet in the last year of that century some scientists suggested to move on and accept that gauge symmetries are simply redundent descriptions (choosing not to emphasize that this description allows for a local description of the theory). Be that as it may, at the time it was natural to promote global symmetry into a local symmetry. The result was very rewarding, it turned out that the "gauge particle" of local supersymmetry is a masselss fermion of the spin 3/2, called the gravitino, whose partner in the same supermultiplet is a massless spin two particle, the graviton. Local supersymmetry led to general coordinate invariance and the presence of gravity. This is called supergravity (SUGRA).

Lagrangians invariant under this local symmetry were found. Writing them down required even more efforts than those needed for the global supersymmetry case. Nevertheless this was achieved and a superfield notation was discovered as well. The Lagrangians are rather lengthy and we do not display them here. The system had additional interesting features.

- There is a Higgs like feature for such systems. The massless gravitino became massive when spontaneous breaking of supersymmetry occurred. The would-be goldstino became part of the massive gravitino. This may resolve the issue of the missing Goldstino.
- SUSY may persist in the presence of a negatively valued cosmological constant. Thus the spontaneous breaking of supersymmetry may be fine tuned so as to obtain a zero value or very small value for the cosmological constant. This is done by balancing the positive vacuum energy resulting from breaking supersymmetry against the value of the negative cosmological constant.
- The presence of gravity renders the Lagrangians to be superficially non-renormalizable. In fact in these theories the ultra violet divergences are much less severe than what would be expected by power counting. In the presence of local SUSY the supersymmetry can be enhanced up to $N = 8$ in four dimensions. Such a theory is intimately related to a ten dimensional theory with $N = 1$ supersymmetry in ten dimensions. Some scientists have the hope that this very special theory is in fact finite. Time will tell.
- Once global SUSY is embedded in supergravity one can imagine also terms which softly break supersymmetry and do not result from spontaneous breaking, i.e. one can add relevant terms to a Lagrangian describing for low energies a systems of particles containing superparticles with non-supersymmetric masses and interactions.

8.2.8   The Hierarchy Problem

SUSY was uncovered on the route of circumventing the no go theorem concerning the unification of a internal and space time symmetries into a unified group as well as attempting to find a symmetry which relates the different couplings in a rather general Lagrangian involving both bosons and fermions. It was taken up again when it was decided to declare the very large value of the ratio of the planck scale and the weak interaction scale as a problem. On a more technical level the problem was stated as follows. Consider a theory with an interacting scalar particle. Notice that to this date no fundamental spin zero elementary particles were observed, the discovery of an elementary spin zero Higgs particle would change this unfamiliar situation, a situation in which the simplest realization of a symmetry is not manifested in nature on an elementary level. No matter what the original mass of the scalar is, the interactions shift the mass. The mass shift is divergent as is the case for renormalizable theories. However according to the renormalization group ideas one always encodes the present knowledge valid up to an energy scale Λ in an effective low energy theory. In that case the mass shift is proportional to the cutoff Λ. As long as there is no physical scale near that of the weak interaction's scale of 1 TeV one needs to fine tune the input initial mass to obtain a Higgs mass of the weak interaction scale. This is to be contrasted with the case of a fermion mass. In that case the fermion mass is shifted by the interactions by an amount which is proportional only to the logarithem of the cut off and is proportional to the initial mass of the fermion. The shift vanishes when the initial fermion mass vanishes. The reason behind the vanishing of the mass of the fermion is that for a massless fermion the system obtains a new symmetry, chiral symmetry. This symmetry should be restored in the zero mass limit. As long as that symmetry is not broken the mass remains zero. If the symmetry is dynamically broken the mass shift can be very small. In the case of supersymmetrical systems it is the fermionic nature which prevails and thus the bosonic mass shifts are small, in a supersymmetric theory one may imagine that the hierarchy problem is solved. (The softer divergences of the supersymmetric systems are of similar origin as those responsible for the no renormalization theorems described before) The problem ab initio is a problem which involves in some manner theoretical taste and in any case it seems from the LEP data that if one insists on a hierarchy problem it is already present. SUSY has not come to the rescue in time and if it is a symmetry of nature, solving the hierarchy problem may well not be its main purpose.

An earlier attempt to solve the hierarchy problem involved the introduction of a new gauge symmetry similar in many ways to color called technicolor. This is a specific realization of the idea that the Higgs particle in not a fundamental particle. Indeed both in describing aspects of superconductivity and superfluidity the Higgs scalar is only an effective degree of freedom. As of 2007 these line of ideas were not consistent with some of the experimental data. An often quoted problem is that such an interaction induces flavor changing neutral interactions at a too high rate.

8.2.9 Effective Theories

One may wonder why the Lagrangians describing the basic interactions are expressed in terms of a finite number of terms rather then by an infinite one. As long as the laws of physics allow the decoupling of far away scales this can be explained. The theory is written down in full generality in the presence of a short distance/high energy cutoff. The cutoff Λ may reflect the scale below which one has no knowledge on the interactions. The terms are constrained only by possible symmetries, their number is a priori infinite. The physics beneath any lower energy scale, Λ' is obtained by integrating out all the degrees of freedom which are heavier than Λ'. A new set of terms replaces the original set, it contains generically less terms. This process can be repeated till a theory approaches a critical surface. The theory on the critical surface is scale invariant. The resulting theory near the critical surface is well described by operators whose scaling dimensions, in d space-time dimensions, can be determined near the surface and happen to be smaller and at most not much larger than d. In many cases there is only a finite number of such operators, i.e., only a finite amount of terms in the Lagrangian describe the physics near the critical surface. This not only explains the concise form of the Lagrangian but provides one with a systematic method to classify the allowed terms to appear, the power of the method is further enhanced when symmetries are present, as those constrain further the allowed terms. This can also sometimes be turned around. The collection of marginal and relevant operators may exhibit a symmetry of its own. This offers a proof that the symmetry should be present at low energies but may well disappear at higher energies if the irrelevant operators do not respect it. The resulting Lagrangian is called the effective low energy Lagrangian. It should contain all the light particles of the system and all the symmetries, may they be realized linearly or nonlinearly, (when the symmetry at question is spontaneously broken). Integrating out heavy particles results in a local Lagrangian consisting of a finite number of terms. Integrating out light degrees of freedom is likely to lead to non-local effects. The terms whose quantum scaling dimensions are smaller than d are called relevant terms. They become very large as the system is probed at lower energies and disappear at high energies, their number is usually finite and in many cases small. Examples of such terms are mass terms for both bosons and fermions in general and the non-Abelian Maxwell term for asymptotically free systems such as QCD. The terms whose scaling dimensions are exactly d are called marginal operators, they include the full Lagrangian for exactly scale invariant systems. Their number is also generically finite. They are equally important at all scales. Terms whose dimension is larger than d are called irrelevant operators. There is an infinite number of them. Each term on its own becomes insignificant in the low energy region and renders the theory non-renormalizable for high energies. (One can imagine examples where an infinite number of such terms collaborate to become relevant but that would most likely mean that the expansion around the critical surface should be modified). Those terms which have scaling dimensions not very much larger than d (such as five and six in four dimensions), can be useful hints for the scale at which the physics needs to be modified. For example the original four Fermi interactions are dimension six operators in four space-time dimensions. Their coefficient, the so called Fermi coupling, has dimension (-2) and hints at the nearly Tev scale of the weak interactions. Indeed at higher energies this irrelevant term is replaced by the classically marginal gauge interactions of the standard model. The replacement of an irrelevant operator by a relevant (or marginal) one at high energy leads to a well defined theory and it called UV completion. The UV completion is not unique, but its simplest version seems to do the job in the weak interactions case. This is clearly to be done when, like for the weak interactions in the 1950s, the leading term in the effective Lagrangian is irrelevant. When the irrelevant term appears in addition to marginal and relevant ones it may or may not indicate new physics.

8.2.10 MSSM Lagrangian

All this said one can write down the simplest low energy theory that by definition contains only marginal and relevant operators. For the case of systems with broken SUSY which still do not have a hierarchy problem one may allow in addition only those of the above operators which retain the at most logarithmic divergence structure of the theory. This requires that the classical marginal operators are all

Table 8.1 Internal quantum numbers of the Higgs superfields and one generation of matter superfields comprising the MSSM model.

Field	SU(3)$_c$	SU(2)$_L$	U(1)$_Y$
$\hat{L} = \begin{pmatrix} \hat{\nu}_{eL} \\ \hat{e}_L \end{pmatrix}$	1	2	-1
$\hat{E}^c$	1	1	2
$\hat{Q} = \begin{pmatrix} \hat{u}_L \\ \hat{d}_L \end{pmatrix}$	3	2	$\frac{1}{3}$
$\hat{U}^c$	3_*	1	$-\frac{4}{3}$
$\hat{D}^c$	3_*	1	$\frac{2}{3}$
$\hat{H}_u = \begin{pmatrix} \hat{h}_u^+ \\ \hat{h}_u^0 \end{pmatrix}$	1	2	1
$\hat{H}_d = \begin{pmatrix} \hat{h}_d^- \\ \hat{h}_d^0 \end{pmatrix}$	1	2_*	-1

supersymmetric as well as some of the classical dimension three operators. The dimension two relevant operators, namely the bosonic mass terms may break SUSY as long as the classically marginal terms are supersymmetric. The, so far, simplest model imposes minimality. It contains the Standard Model particles and interactions and their minimal extensions. Each Standard model particle is accompanied by a superpartner. Only one new supersymmetric multiplet is added in which neither the bosons nor the fermions are part of the Standard Model, this is a second Higgs field. The interactions are minimally extended to be N = 1 superymmetric. The system resulting for this construction is called the Minimal Super Symmetric Model (MSSM). The number of resulting terms may be minimal but it can hardly be considered as small, in fact it has 178 parameters. The superpotential in the MSSM contains the following terms:

$$\hat{f} = \mu \hat{H}_u^a \hat{H}_{da} + \sum_{i,j=1,3} \left[(f_u)_{ij} \epsilon_{ab} \hat{Q}_i^a \hat{H}_u^b \hat{U}_j^c + (f_d)_{ij} \hat{Q}_i^a \hat{H}_{da} \hat{D}_j^c + (f_e)_{ij} \hat{L}_i^a \hat{H}_{da} \hat{E}_j^c \right].$$

$$(8.120)$$

The indices of SU(2)$_L$ doublets are denoted above as a and b. The $(f_k)_{ij}$ are the appropriate Yukawa couplings. The Quarks, Leptons and Higgs particles are all chosen to be chiral fields and the notation, for one generation, is made explicit in Table 8.1.

The terms above were constructed to be invariant under the standard model symmetries, they turn out to conserve also both baryon (B) and lepton (L) numbers. However, the terms listed below are marginal and relevant terms which also conserve the standard model symmetries but do not conserve either B or L. The generic amount of violation induced by these terms is not consistent with the experimental data. Thus in MSSM one requires both B and L conservation. By the no renormalization theorems the symmetry will be respected to all orders in perturbation theory. Such symmetries which are preserved quantum mechanically in perturbation theory, once imposed classically, are denoted as natural in a "technical" sense.

$$\hat{f}_L = \sum_{ijk} \left[\lambda_{ijk} \epsilon_{ab} \hat{L}_i^a \hat{L}_j^b \hat{E}_k^c + \lambda'_{ijk} \epsilon_{ab} \hat{L}_i^a \hat{Q}_j^b \hat{D}_k^c \right] + \sum_i \mu'_i \epsilon_{ab} \hat{L}_i^a \hat{H}_u^b. \qquad (8.121)$$

$$\hat{f}_B = \sum_{ijk} \lambda''_{ijk} \hat{U}_i^c \hat{D}_j^c \hat{D}_k^c. \qquad (8.122)$$

The B and L symmetries are in any case not respected by non-perturbative effects. The above mentioned terms can also be forbidden by imposing a different global symmetry called R-parity which was already mentioned in the context of SUSY breaking. It is defined here as:

$$R = (-)^{3(B-L)+2s}, \qquad (8.123)$$

s denoting the spin of the particle. The standard model particles are even under the R-parity, while their superpartners are odd under it. In MSSM the bosonic partners of the standard model matter fermions carry non-zero L and B quantum numbers. The (non)conservation of B and L symmetries are not in general correlated. They happen to coincide on the terms disallowed above. For example, the terms below both respect R-parity but the first violates L number conservation and the second does not respect B number conservation.

$$\epsilon_{ab} \hat{L}^a \hat{H}_u^b \epsilon_{cd} \hat{L}^c \hat{H}_u^d \qquad (8.124)$$

and

$$\hat{U}_c \hat{U}_c \hat{D}_c \hat{E}_c \qquad (8.125)$$

For the record we write down all the dimension three and dimension two operators which softly break SUSY in the MSSM model. i and j run over the generations and are summed over as are the doublet SU(2) indices a and b.

$$\begin{aligned}
\mathcal{L}_{soft} = &- \left[\tilde{Q}_i^\dagger m_{Q_{ij}}^2 \, \tilde{Q}_j + \tilde{d}_{R_i}^\dagger m_{D_{ij}}^2 \tilde{d}_{R_j} + \tilde{u}_{R_i}^\dagger m_{U_{ij}}^2 \tilde{u}_{R_j} \right. \\
&+ \left. \tilde{L}_i^\dagger m_{L_{ij}}^2 \tilde{L}_j + \tilde{e}_{R_i}^\dagger m_{E_{ij}}^2 \tilde{e}_{R_j} + m_{H_u}^2 |H_u|^2 + m_{H_d}^2 |H_d|^2 \right] \\
&- \frac{1}{2} \left[M_1 \bar{\lambda}_0 \lambda_0 + M_2 \bar{\lambda}_A \lambda_A + M_3 \bar{\tilde{g}}_B \tilde{g}_B \right] \\
&- \frac{i}{2} \left[M_1' \bar{\lambda}_0 \gamma_5 \lambda_0 + M_2' \bar{\lambda}_A \gamma_5 \lambda_A + M_3' \bar{\tilde{g}}_B \gamma_5 \tilde{g}_B \right] \\
&+ \left[(a_u)_{ij} \epsilon_{ab} \tilde{Q}_i^a H_u^b \tilde{u}_{R_j}^\dagger + (a_d)_{ij} \tilde{Q}_i^a H_{da} \tilde{d}_{R_j}^\dagger + (a_e)_{ij} \tilde{L}_i^a H_{da} \tilde{e}_{R_j}^\dagger + h.c. \right] \\
&+ \left[(c_u)_{ij} \epsilon_{ab} \tilde{Q}_i^a H_d^{*b} \tilde{u}_{R_j}^\dagger + (c_d)_{ij} \tilde{Q}_i^a H_{ua}^* \tilde{d}_{R_j}^\dagger + (c_e)_{ij} \tilde{L}_i^a H_{ua}^* \tilde{e}_{R_j}^\dagger + h.c. \right] \\
&+ \left[b H_d^a H_{da} + h.c. \right].
\end{aligned}$$

$$(8.126)$$

The model, as written, has 178 independent parameters. Their number could be reduced if they had their origin in an underlying microscopic theory. In some present versions the matrices of parameters denoted above by c_{ij} are set to zero thus reducing by fiat the number of free parameters to 124. This is but the tip of an iceberg, one proceeds from versions of the MSSM and derives the mass spectrum of the supersymmetric particles while ensuring to preserve the known properties of the standard model particles as well as a variety of experimental bounds. The bounds range from cosmological ones to bounds on rare decays. We hope that in the not too distant future the fog will disperse and one will be able to write a rather concise item in a physics encyclopedia which would select the relevant physical components of this effort. As mentioned any model for SUSY breaking should respect present experimental constraints. Many models of spontaneous SUSY breaking fail to do this as a result it was suggested to add to the yet to be seen superpartners of the standard model particles also a hidden sector. In such models SUSY is to be broken in the hidden sector and its effects are supposed to be mediated to the "seen" sector. The agents or messengers which couple the two sectors vary. In some models the coupling is by the gravitational force at tree level, in others the coupling occurs first only at the one loop level and is called anomaly (Weyl anomaly) mediated. There are models in which the coupling is done through gauge interactions, they have a lower energy scale than the gravity mediated interactions. Each of these models has some advantages as well as disadvantages and are at this stage an active area of research.

We will end the section with a short glossary of terms currently used in describing MSSM features which were not mentioned above.

Short Glossary:

- Chargino—Charged supersymmetric partner of a charged standard model particle.
- Gaugino—Spin one half supersymmetric partner of a standard model gauge particle.
- LSP—Lightest supersymmetric partner of a standard model particle. It is long lived in many models.
- μ term—Term coupling the two different Higgs chiral supermultiplets.
- Neutralino—Neutral supersymmetric partner of a neutral standard model particle.
- squark—Spin zero supersymmetric partner of a standard model quark.
- slepton—Spin zero supersymmetric partner of a standard model lepton.
- $\tan(\beta)$—Ratio of the expectation values of the two Higgs fields.

8.3 Unification

8.3.1 Gauge Group Unification [2]

The standard model is unified along the lines that all of its components, the colour, weak and electromagnetic interactions are described by gauge theories. They are SU(3) $*$ SU(2) $*$ U(1). The gluons and the photons, the electrons and the quarks belong to different representations of the product gauge group, it is natural to attempt to have all of the elementary particles and interactions as a single representation of a single gauge group. Intermediate algebraic solutions to this problem have been found. For example the gauge group SU(5) does very economically unify the known gauge groups. Moreover it predicts that quarks and leptons can transmute into each other, in particular violating Baryon number conservation. The proton could decay in such models for example into an positron and a neutral pion. The original estimates of the half life time of the proton placed it possibly around 10^{31} years. That prediction was within experimental reached and initiated the construction of very ingenous experiments. Proton decay at that rate was not found, the lower bound on the proton life time was improved to be 10^{35}, thus invalidating the simplest version of the grand unified group SU(5). In that version the particles did not all belong to the same representation as one may have wished based on esthetics. There have been many other attempts since to find an appropriate unifying group these included SO(10) (which incorporated naturally a right handed SU(2) singlet neutrino) and exceptional groups such as E(6). This was done with and without SUSY. Using the renormalization group it was found that in the presence of SUSY the in teractions may indeed unify in magnitude at a high energy not much below the planck scale of 10^{19} GeV. At such scales it becomes difficult to ignore quantum gravity. In any case we will not review this vast subject further here.

8.3.2 Extra Dimensions and Unification [3]

The first ideas of unification by increasing the number of dimensions were suggested in classical field theory by Kaluza and Klein (KK) in the period of the 1920s. At the time there were two well known interactions, gravity and electromagnetism. KK suggested that space time is five dimensional rather than four dimensional and that there is but one fundamental interaction-gravity. In order to be consistent with observations it was suggested that the five dimensional space time is composed out of a fifth dimension which is a spatial circle of inverse radius m, and the usual four dimensional Minkowski component. The radius $1/m$ should be small enough to have not been observed yet. The resulting low energy (i.e. energies much lower than m) five dimensional Lagrangian decomposes into several four dimensional Lagrangians. They describe, four dimensional gravity, four dimensional Maxwell electrodynamics and the coupling of a neutral spin zero additional particle. The symmetry group of the low energy Lagrangian consists of four dimensional general coordinate invariance as well as a U(1) four dimensional gauge symmetry. This result is obtained in the following manner [3].

The basic five dimensional gravitational action is thus given by:

$$\hat{S} = \frac{1}{2\hat{\kappa}^2} \int d^5\hat{x}\sqrt{-\hat{g}}\,\hat{R} \qquad (8.127)$$

With $\hat{\kappa}^2$ being the five dimensional Newton constant. The action $\hat{S}$ is invariant under the fivedimensional general coordinate transformations

$$\delta\hat{g}_{\hat{\mu}\hat{\nu}} = \partial_{\hat{\mu}}\hat{\xi}^{\hat{\rho}}\hat{g}_{\hat{\rho}\hat{\nu}} + \partial_{\hat{\nu}}\hat{\xi}^{\hat{\rho}}\hat{g}_{\hat{\rho}\hat{\mu}} + \hat{\xi}^{\hat{\rho}}\partial_{\hat{\rho}}\hat{g}_{\hat{\mu}\hat{\nu}} \qquad (8.128)$$

A useful $4 + 1$ dimensional ansatz was:

$$\hat{g}_{\hat{\mu}\hat{\nu}} = e^{\phi/\sqrt{3}}\begin{pmatrix} g_{\mu\nu} + e^{-\sqrt{3}\phi}A_\mu A_\nu & e^{-\sqrt{3}\phi}A_\mu \\ e^{-\sqrt{3}\phi}A_\nu & e^{-\sqrt{3}\phi} \end{pmatrix}. \qquad (8.129)$$

The fields depend on the five dimensional coordinate $\hat{x}^{\hat{\mu}}$, which were written as: $\hat{x}^{\hat{\mu}} = (x^\mu, y)$, $\mu = 0,1,2,3$, and all unhatted quantities are four-dimensional. The fields $g_{\mu\nu}(x)$, $A_\mu(x)$ and $\varphi(x)$ are the spin 2 graviton, the spin 1 photon and the spin 0 dilaton respectively. The fields $g_{\mu\nu}(x,y)$, $A_\mu(x,y)$ and $\varphi(x,y)$ may be expanded in the form

$$g_{\mu\nu}(x, y) = \sum_{n=-\infty}^{n=\infty} g_{\mu\nu n}(x)e^{inmy},$$

$$A_\mu(x, y) = \sum_{n=-\infty}^{n=\infty} A_{\mu n}(x)e^{inmy}, \qquad (8.130)$$

$$\phi(x, y) = \sum_{n=-\infty}^{n=\infty} \phi_n e^{inmy}$$

Recall that m is the inverse radius in the fifth dimension.

In general the five dimensional theory can be described in terms of an infinite number of four dimensional fields. It also has an infinite number of four-dimensional symmetries appearing the Fourier expansion of the five dimensional general coordinate parameter $\hat{\xi}^{\hat{\mu}}(x, y)$

$$\hat{\xi}^\mu(x, y) = \sum_{n=-\infty}^{n=\infty} \xi^\mu{}_n(x)e^{inmy},$$

$$\hat{\xi}^4(x, y) = \sum_{n=-\infty}^{n=\infty} \xi^4{}_n(x)e^{inmy}. \qquad (8.131)$$

However, at energy scales much smaller than m only the $n = 0$ modes in the above sums enter the low effective low energy action.

The $n = 0$ modes in (8.130) are just the four dimensional graviton, photon and dilaton. Substituting (8.129) and (8.130) in the action (8.127), integrating over y and retaining just the $n = 0$ terms one obtains (dropping the 0 subscripts)

$$S = \frac{1}{2\kappa^2} \int d^4x \sqrt{-g} \left[R - \frac{1}{2} \partial_\mu \phi \partial^\mu \phi - \frac{1}{4} e^{-\sqrt{3}\phi} F_{\mu\nu} F^{\mu\nu} \right] \tag{8.132}$$

where $2\pi\kappa^2 = m\kappa^{\wedge 2}$ and $F_{\mu\nu} = \partial_\mu A_\nu - \partial_\nu A_\mu$. From (8.128), this action is invariant under general.

coordinate transformations with parameter $\xi^\mu{}_0$, i.e. (again dropping the 0 subscripts)

$$\begin{aligned}
\delta g_{\mu\nu} &= \partial_\mu \xi^\rho g_{\rho\nu} + \partial_\nu \xi^\rho g_{\mu\rho} + \xi^\rho \partial_\rho g_{\mu\nu} \\
\delta A_\mu &= \partial_\mu \xi^\rho A_\rho + \xi^\rho \partial_\rho A_\mu \\
\delta \phi &= \xi^\rho \partial_\rho \varphi,
\end{aligned} \tag{8.133}$$

local gauge transformations with parameter $\xi^4{}_0$

$$\delta A_\mu = \partial_\mu \xi^4 \tag{8.134}$$

and global scale transformations with parameter λ

$$\delta A_\mu = \lambda A_\mu, \delta \varphi = -2\lambda/\sqrt{3} \tag{8.135}$$

The symmetry of a vacuum, determined by the VEVs

$$< g_{\mu\nu} >= \eta_{\mu\nu}, < A_\mu >= 0, < \varphi >= \varphi_0 \tag{8.136}$$

is the four-dimensional Poincare group $\times R$. These expectation values have not been determined dynamically but let us settle here for them. At this level of the analysis the masslessness of the graviton is due to unbroken four dimensional general covariance, the masslessness of the photon is consistent with the four dimensional gauge invariance and the dilaton seems massless because it is the Goldstone boson associated with the spontaneous breakdown of the global scale invariance. In fact taking into account the actual periodicity in y which is not manifested for only the $n = 0$ low energy modes the symmetry acting on the scalar field is U(1) rather than R and is thus not a true scale symmetry. The field φ_0 is a pseudo-Goldstone boson and does not really deserve to called a dilaton. Further analysis uncovers an infinite tower of charged, massive spin 2 particles with charges e_n and masses m_n given by

$$e_n = n\sqrt{2}km, m_n = |n| m \tag{8.137}$$

Thus the KK ideas provide an explanation of the quantization of electric charge. (Note also that charge conjugation is the parity tranformation $y \to -y$ in the fifth dimesion.) If one indeed identifies the fundamental unit of charge $e = \sqrt{2}km$ with the charge on the electron, then one is forced to take m to be very large: the Planck mass 10^{19} GeV. Such extra scalar particles seem to be present abundantly in string theories as well. The experimental question was mainly to set bounds on the value of the radius $R(= 1/m)$ of the fifth dimension. The limits are obtained from precision electroweak experiments and require $m > 7$TeV. This is for models with one extra dimension, in which the gauge bosons propagate in the bulk but the fermions and Higgs are confined to four dimensions. More on this shortly. There are also bounds from astrophysics they are not model independent bounds. They are important for large dimensions in which only the graviton propagates, and they depend on number of large dimensions. The strongest bounds arise from supernova emissions, giving $1/m < 10^{-4}$ mm for the case of two large extra dimensions. Although researchers such as Einstein and Pauli had studied such theories for dozens of years this path has been abandoned. Extra dimensions resurfaced when string theory was formulated. It was found out that strings are much fussier than particles, (super) strings can propagate quantum mechanically only in a limited number of dimensions. In fact, not only was the number of allowed possible dimensions dramatically reduced, the allowed values of the number of space-time dimensions did not include the value 4. These dimensions are usually required to be 10 or 26 (The origin of this basic number is, at this stage, disappointingly technical.) In fact while the numbers 10 and 26 result from the theory they need not always be related directly to extra dimensions. When the extra dimensions emerged in string theory it was suggested in the spirit of KK that any extra dimensions are very small. As interest was diverted from string theory back to field theory the ideas of KK were revised allowing one to take into account the extra interactions discovered since the original work of KK. It turned out that if one which to trade all the known standard model gauge interactions, i.e. the all SU(3) $*$ SU(2) $*$ U(1) interactions, for a theory of gravity alone one was led to consider eleven dimensional gravity. Such theories raised interest also for other reasons. The return of string theory brought back with it an intense study of extra dimensions. In particular extra dimensions in the form of compact Calabi-Yau manifolds which allowed to have four dimensional effective theories with N $=$ 1 SUSY. The topology of the extra dimensions determined in some cases the number of zero modes on them, that is the spectrum of massless particles. The number of generations of particles for example could be correlated thus to the topology of those extra dimensions offering a solution to the origin of the repetitive structure of the elementary particles. A vast amount of research on the possible extra dimensions is ongoing. There have also been attempts to understand dynamically the origin of the difference between the four large extra dimensions and the rest. Some efforts were directed to explain how a spontaneous breaking of space time symmetries could lead to such asymmetry in the properties of the different dimensions. Solid state systems such as liquid crystals also exhibit such differences. Other efforts focused on determining why an expanding universe would expand asymmetrically after a while only in four directions. Another development occurred following the realization that string theory in particular allows what are called brane configurations. Branes are solitons, extended objects embedded stably in a space time of larger dimension. Vortices and magnetic monopoles were mentioned as such lower dimensional objects. There is a consistent possibility that for example a four dimensional universe can be embedded in higher dimensions. The known gauge interactions living only on the brane while gravity extends to the full space. If our universe has this structure many things can be explained and in particular this has given rise to the possibility the extra dimensions could be much larger than previously expected. Actually they could extend up to the submicron region. This could lead to measurable deviations from Newton's gravity at those distance scales. Astrophysics gives upper bounds on how large such extra dimensions may be but in any case this is a very significant relaxation of a bound, in fact I am not aware of any bound on such a fundamental quantity in physics that has been altered to such an extent by a theoretical idea. All this said one should stress the obvious, also the bounds have been relaxed the true value of the extra dimensions may still be very small, perhaps even Planckian. Some suggestions of larger extra dimensions could well be tested at the LHC.

The fact that unification may occur not far from the planck scale brings quantum gravity to the front row and with it a theory which attempts to be able to indeed tame quantum gravity that is string theory.

8.4 String Theory [4]

8.4.1 No NOH Principle

String Theory is at this time far from being a complete theoretical framework, not to mention a phenomenological theory. That said, the theory of extended objects has evolved significantly and has shaped and was shaped by aspects of modern mathematics. In my opinion is it appropriate to highlight the qualitative aspects of string theory and the ideas behind it. This choice is not made for lack of formulas or precision in string theory. An essential catalyst in the process of the formation of string theory was urgency, the urgency created out of the near despair to understand the amazing novel features of the hadronic interactions as they unfolded in the 1960's. String theory was revitalized in the latter part of the twentieth century by the urgency to understand together all of the four known basic interactions. Here I shall briefly review some of the motivation to study a theory of extended objects and discuss the challenges string theory faces, the successes it has had, and the magic spell it casts. We start by reviewing a hardware issue, the hard wiring of some scientific minds. Researchers using string theory are faithful followers of an ancient practice, that of ignoring the NOH principle. The NOH principle is very generic, it states that Nobody Owes Humanity neither a concise one page description of all the basic forces that govern nature nor a concise description of all the constituents of matter. Actually no reductionist type description is owed. Time and again physicists driven by their hard wiring processed the experimental data utilizing theoretical frameworks and were able to come up with explicit formulae; formulae that were able to express on less than one page the essence of a basic force. A key assumption used and vindicated was that the basic constituents of nature are point-like. The electromagnetic forces, the weak interactions and even the powerful color forces were rounded up one by one, straddled by the rules of quantum mechanics and then exhibited on a fraction of a page. Using these formulae, an accomplished student abandoned on an isolated island can predict with an amazing accuracy the outcome of some very important experiments involving the basic forces. The level of accuracy ranges from a few percent in the case of some aspects of the color forces to better than 10^{-10} for some features of the electromagnetic interactions.

8.4.2 Why Change a Winning Team: Extended Constituents Are Called Upon to Replace Point-like Ones

Point particles have done a tremendous job in shouldering the standard model of particle interactions. Why replace them by extended objects? In the section on SUSY we have outlined several of the reasons for that. The standard model does reflect the enormous progress made in the understanding of particle physics but it is not perfect, at least as seen by a critical theorist through his TPs. The model is afflicted with tens of parameters as well as what are termed naturality problems. The values of some physical quantities, such as the mass of the Higgs particle, are required to be much lighter than the theory would naturally suggest. Ignoring the NOH principle, scientists find themselves once again on the path of searching for a more concise description. One effort was directed towards unifying the three interactions, thus following the conceptual unification of the electromagnetic and weak interactions, described each by a gauge theory. That direction led to possible unifications at scales not so distant from where quantum gravity effects are definitely supposed to become important. In fact, gravity, the first of the basic forces that was expressed by a mathematical formula, is the remaining unbridled known force which is considered currently as basic. Here the quantum breaking methods based on the concept of a basic point like structure have failed. Sometimes what is considered failure is also a measure of intellectual restlessness, still it made sense to search for alternatives. Actually even with no apparent failure lurking upon basic physics, a study of a theory of fundamentally extended objects is called for. After all, why should the basic constituents be just point like? String theory is a natural extension of the idea that the basic constituents are point-like, it is an investigation into the possibility that basic constituents are ab initio extended objects. The simplest such extended objects being one-dimensional, that is strings. In a sense one may regard also point particles as extended objects dressed by their interactions. The direct consequences in this case are however totally different. It turned out that such a generalization did eventually reproduce many properties of point particle interactions and in addition enabled one to formulate what seems as a theory of gravity. This it should do, following the tradition of subjecting any new theory to the test of the correspondence principle. The theory is well defined up to several orders and perhaps to all orders in perturbation theory. The perturbations are small at the vicinity of a very small distance scale called the string scale, as small as that distance is, it is sometimes expected to be larger than the Planck length and thus is associated with lower energies than the Planck energy (it is related to the plank scale by the string interaction coupling). Interactions among strings are, to a large extent, softer and fuzzier than those among point particles. Recall that a large number of successes of particle physics consisted of explaining and predicting the physics of the various interactions at scales at which they were weakly coupled, otherwise, a large dose of symmetry was essential. This is perhaps the most significant achievement of string theory. It is also its major source of frustration, if the string scale is indeed of the order of 10^{-32} cm or even slightly larger, it is not known today how to verify, experimentally, any prediction resulting from perturbative string theory. It is very difficult to imagine, for example, how to set up an experiment measuring the differential cross section of graviton-graviton scattering at the string scale. We will return to this extremely important issue later on and continue to follow for a while the path of the theoreticians.

8.4.3   New Questions

The impact of scientific progress can be tested by the type of questions it makes us aware of, as well as by the answers it eventually provides for them. By studying string theory new questions do arise. Values taken for granted are subjected to query and downgraded to being parameters. String theory suggests questioning the values of some such physical quantities and in some cases offers scientific answers to the question. A prime example is that of the number of space-time dimensions. In a point like constituent theory there is a very large degree of theoretical freedom. As far as we know spin zero point particles, for example, could have propagated in any number of dimensions, their interactions would have been different depending on the dimension of space-time but nevertheless they would have been allowed. A theory of spin one half particles can turn out to be in consistent on certain type of manifolds but they do not restrict the dimensionality of the space in which they propagate. Strings are much fussier than particles, (super) strings can propagate quantum mechanically only in a limited number of dimensions. In fact, not only are the number of allowed possible dimensions dramatically reduced, the allowed values of the number of space-time dimensions do not include the value 4. These dimensions are usually required to be 10 or 26 depending on the amount of supersymmetry the string is endowed with. The origin of this basic number is, at this stage, disappointingly technical and one should note that though, strictly speaking, the number 26 does indeed appear in string theory, it is not always possible to associate this number with the number of dimensions. In fact the numbers appearing are 15 and 26 and they reflect from one point of view a conformal anomaly in the theory describing first quantized string theory. In first quantized field theory of particles can be expressed as a sum over one dimensional field theories, i.e. quantum mechanics, of a particle moving in space time and interacting along its space time trajectory. The action of the quantum mechanical system is geometrical and describes in its simplest form the length of the particle's trajectory. The result should not depend on the parameterization of trajectory and this should be general coordinate invariant. In a more fancy manner the action describes a particle coupled to gravity evolving in one space time dimension. The generalization to a description of the motion of a one dimensional object, a string, follows. As the string is a one dimensional object, its motion in space time spans a two dimensional manifold. The theory describing its motion is that of a particle, coupled to gravity, moving in two space-time dimensions. The two dimensions are called the world sheet and the theory describing them using first quantization is called the World Sheet theory. The usual counting of quantum degrees of freedom leads to the conclusion that gravity has minus one degrees of freedom in both one and two dimensions. In other words the system is over constrained and will require a higher degree of symmetry to be consistent. In addition the two dimensional gravitational system has an anomaly which restricts the algebraic structure of the conformal system describes by the string. A motion of a (super) string in (10) 26 flat dimensions removes the anomaly. There are many other ways to remove it and perhaps many more to be discovered. Sometimes the background on which the string can propagate has no obvious geometrical description, just an algebraic one. The space in which the strings moves is called the Target Space. The interactions of strings are described by them parting or joining, this is described (in a Euclidean formulation) by a compact two dimensional surface with non-trivial topology. These manifolds are called Riemman manifolds and are distinct by their topology. The freely propagating string is described by a world sheet 2 sphere, the first interactions occur when the world sheet is a torus. The theory is defined perturbatively by summing with a certain weight over all such two dimensional Riemann surfaces. The non-perturbative structure is non-trivial but less understood at this stage. But setting that aside, one would like to detect extra dimensions experimentally. As the idea of having extra dimensions was raised originally in field theory, upper bounds on the length of such dimensions were already available, more recently under the security net of string theory these bounds have been revised (they have been relaxed by more than 10 orders of magnitude and brought up to the sub micron regime), in fact I am not aware of any bound on a fundamental quantity in physics that has been altered to such an extent by a theoretical idea. This fact about strings illustrates the tension inherent in attempts to search for experimental verification of extremely basic but very weakly coupled phenomena. Sitting in our chairs we sometimes forget how frail gravity is. It is all the planet earth that gravitationally pulls us towards its center, yet all this planetary effort is easily counterbalanced by the electromagnetic forces applied by the few tiles beneath our chair. Although string theory sets constraints on the possible number of dimensions, that is not to say that point particles allow everything. We mentioned that spin one half particles can't be defined on all manifolds but the number of known restrictions is much smaller. The interactions could be of an infinite (sometimes classifiable) variety, the color group, which happens to be SU(3), could have been for all the theory cares, SU(641), the electromagnetic force could have been absent or there could have been 17 photons. Not everything is allowed, but an infinite amount of variations could have been realized instead of what one actually sees in nature today. The same goes for many (but not all) of the properties of the space-time arena in which all the interactions are occurring, in particular effective low energy theories (and one seems to be constrained to use only such theories for a long time to come) contain many more free parameters such as masses and some of the couplings. In a string theory one may have expected that all or most of these parameters are fixed in some manner. Actually is it more complicated, for some string backgrounds, some parameters are fixed, while other parameters are not. In addition there seemed to be several string theories leading to even more possibilities.

8.4.4   The One and Only?

As we have discussed the desire to find the one and only theory encompassing all fundamental physical phenomena seems hard wired in many scientific minds. Is string theory an example of such a theory? Before addressing this question let us reflect upon the limitations of the methods used. What would in fact satisfy the purest (or extremist) of reductionists? A one-symbol equation? Be her desires, what they may be, one should recall that one is using mathematics as our tether into the unknown. This tool, which has served science so well, has its own severe limitations, even ignoring the issue of using differential equations as tool to probe short distance, a generic problem in mathematics can't be solved, maybe the NOH principle will eventually have its day, one will run out of interesting questions which have answers, maybe that day is today! (Although some mathematicians suggest to use physicists as hound dogs that will sniff one's way towards interesting and solvable problems). Having this in mind the key question is posed in hope it does have an answer: find the unique theory describing the fundamental forces. There have been ups and downs for the proponents of the one and only string theory. The understanding of the situation has passed several evolutionary stages. From the start it was recognized that one may need to settle for more than only one string theory. It was thought that a distinction could been drawn between a theory of only closed strings and a theory containing both open and closed strings. A theory with only open strings was realized to be perturbatively inconsistent. Even those theories needed not only a fixed dimension but in addition an extra symmetry, supersymmetry, to keep them consistent. It was very quickly realized that actually there are an infinite varieties of backgrounds in which a string could move. For example, consider 26 dimensional spaces in which one dimension is actually a circle of radius R. It turned out that all possible values for the radius, R, of the circle are allowed. It was then realized that

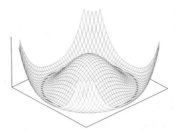

Fig. 8.7 The potential, familiar from the standard model, has a shape of a sombrero. In the figure, its minima lie on a circle, each point along the circle, can be a basis for a ground state. The physics around each is equivalent. In string theory and in supersymmetric systems flat potential arise

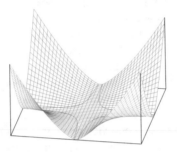

Fig. 8.8 The potential has two orthogonal flat directions. The landscape of the valleys is generically more elaborate. Each point along the flat directions can serve as a basis of a ground state. In many cases. The physics around each ground state is different

there is a large variety of 22 dimensional compact manifolds in the case of bosonic strings (6 dimensional compact dimensions in the case of the supersymmetry), which could accompany the four dimensional Minkowski space-time one is familiar with. Each such compact manifold was called a string compactification. Next it was suggested that actually all possible compactifications are nothing but different solutions/ground states of a single string theory. Each of the solutions differing from each other by the detailed values of the physical parameters it leads to. But the ground states have also many common features, such as generically the same low energy gauge group. To appreciate this consider the difference between the potential of the Higgs particles which may describe the mass generation of the carriers of the weak interactions and the effective potential leading to the ground states in string theory. For the electro-weak interaction case the potential has the form of a Mexican hat, the ground state may be described by any point in the valley (Fig. 8.7) but all choices are equivalent, they describe exactly the same physics.

In the case at hand the valley of minima is flat and extends all the way to infinity. Any point along the valley can be chosen as a ground state as they all have the same energy and yet each describes different physics. The manifold describing the infinity of these degenerate ground states is called the moduli space when they are continuously connected (Fig. 8.8).

These connected regions of valleys trace out very elaborate geography and describe different possible solutions of string theory. All known stable ones share in common the property of describing a supersymmetric theory. In such theories each bosonic particle has a fermionic particle companion. It then turned out that there actually are several different types of string theories, each having its own infinite set of degenerate ground states. What made the string theories different were details, most of which are rather technical, which seemed to affect the particle content and gauge symmetries of the low energy physics. There was a stage, in the eighties of the last century, at which some researchers imagined they were on the verge of being able to perform reliable perturbative calculations of a realistic theory containing all the aspects of the standard model. That turned out not to be the case. However, in the process, the geography of many interesting ground states was surveyed. Next, a large amount of circumstantial evidence started to accumulate hinting that all these theories are after all not that distinct and may actually be mapped into each other by a web of surprising dualities. This led once again to the conjecture that there is but one string theory, this time, the theory that was supposed to encompass all string theories and bind them, was given a special name, M theory. Even in this framework all string theories brought as dowries their infinite moduli space. M theory still had an infinite number of connected different degenerate supersymmetric ground states, they were elaborated and greatly enriched by what are called brane configurations. Branes are a special type of solitons that appear in string theory. They come in various forms and span different dimensions. What is common to several of them, is that open strings may end on them.

Thus what was thought to be a theory of only closed strings contains open strings excitations in each of its sectors that contain branes. Moreover, every allowed brane configuration leads to its own low energy physics. In fact there are suggestions that our four dimensional experience results from us living on one or several branes all embedded in a larger brane configuration placed in a higher dimensional space-time. How does that fit with the expectations to have the One and Only string theory? Well, in this framework there is one theory but an infinite amount of possible ground states. Are we owed only once such state? There are those who hope that the vacua degeneracy will be lifted by some yet to be discovered mechanism leaving the one unique ground state. Others are loosing their patience and claiming that the NOH principle takes over at this stage, the theory will retain its large number of ground states.

Moreover a crucial ingredient of string theory at present is super symmetry which was discussed obove, a beautiful symmetry not yet detected in nature, and even if detected, it seems at this stage to play only a role of a broken symmetry. It is not easy to obtain even remotely realistic models of string theory in which supersymmetry is broken (that is the symmetry between fermions and bosons is broken), a nearly vanishing cosmological constant and light particles whose mass values are amendable to a calculation within a reliable approximation. Candidates for such ground states have been suggested recently, these particular candidates consist of very many isolated solutions living off the shore of the supersymmetric connected valleys. Many of these solutions actually have different values for the vacuum energy and perhaps may thus decay from one to the other. In particular bubbles may form within one configuration, bubbles that contain in them a lower energy configuration. Some researchers would like to see the universe itself, or the bubbles that form in it, as eternal tourists who will end up visiting many of these metastable states. Some suspect that many of these metastable states are stable. Some would encourage brave sailors to take to sea again, as they did in the 1980s, and charter this unknown geography, some are trying to make this endeavor quantitative by counting vacua which have a given property, such as a given cosmological constant. This subject is still at its infancy, many hopes and opinions are currently expressed, it would be nice if this issue will turn out to fall within the realm of solvable problems.

The challenges to string theory on the experimental side have been to reproduce the standard model. This has not been done to this day. Many new ideas and insights have been added. Bounds have been shifted several times, one has felt being very near to obtaining the desired model, each new approach seemed to bring one nearer to that goal, but obtaining one completely worked-out example, which is the standard model, remains a challenge for string theorists. Another important issue is that of the value of the cosmological constant. It is usually stated that the small value of the cosmological constant is a major problem. These arguments are centered around some version of a low energy effective field theory. That theory could be valid below a few tens of electron volt thus describing only aspects of electromagnetism or it could be at a TeV scale describing both the electroweak interactions. The arguments goes on to say that the vacuum energy of such an effective theory should be, on dimensional arguments grounds, proportional to the scale of the cutoff, (the cutoff is set at that energy beyond which the ingredients of the physics start to change) that is evs or TeVs in the above examples. In either case, the value of this vacuum energy, which is the source of the cosmological constant, is larger by an astonishing number than the known bounds on the value of the cosmological constant. The argument is not correct as stated, it is not only the cutoff that contributes to the cosmological constant, all scales do actually contribute to the vacuum energy and if a cutoff should be invoked, it is the highest one, such as the Planck scale, that should be chosen. Moreover, this argument does not take into account the possibility that the fundamental theory has a certain symmetry which could be the guardian of vanishingly small value of the cosmological constant. Such a symmetry exists, it is called scale invariance, and it has shown to be a guardian in several settings. Scale invariance is associated with the absence of a scale in the theory. There are many classical systems which have this symmetry and there are also quantum systems which retain it. In a finite, scale invariant theory not only does the vacuum energy get contributions from all scales, they actually add up to give rise to a vanishing vacuum energy. This remains the case even when the symmetry is spontaneously broken. Being a consequence of a symmetry, the vacuum energy contribution to the cosmological constant vanishes also is the appropriate effective theory. What does string theory have to say about that? String theory could well be a finite theory, it is not scale invariant, as it contains a string scale, if that scale would be spontaneously generated perhaps one would be nearer to understanding the value of the cosmological constant. If that scale is indeed not elementary, one would be searching for the stuff the strings themselves are made off. Such searches originating from other motivations are underway, some go by the name of Matrix Model. Particle Physics as we know it is described by the standard model evolving in an expanding universe. The methods used to study string theory are best developed for supersymmetric strings, and for strings propagating on time independent background, nature is explicitly neither. What then are the successes of string theory?

8.4.5 Successes: Black Holes, Holography and All That ...

A major success described above was to obtain a theory of gravity well defined to several orders in perturbation theory, in addition, no obvious danger signals were detected as far as the higher order in perturbation theory are concerned, non perturbative effects are yet to be definitely understood. This issue involves the short distance structure of string theory. String theory is supposed to be a consistent completion of General Relativity (GR). General relativity suffers from several problems at low energies. String theory, which according to a correspondence principle, is supposed to reproduce GR in some long distance limit, will thus have inherited in this merger, all the debts and problems of GR. If it does solve them it will need to do it with a "twist" offering a different point of view. That perspective could have been adopted in GR but was not. This has occurred in several circumstances, one outstanding problem in GR is to deal with the singularities that classical gravity is known to have. These include black holes, big bangs and big crunches as well as other types of singularities. String theory can offer a new perspective on several of these issues. Let us start with black holes. The first mystery of black holes is that they seem to posses thermodynamical properties such as temperature and entropy. This is true for charged black holes as well as for uncharged Schwarzschild ones. In the presence and under the protection of a very large degree of supersymmetry, string theory tools enabled to provide a detailed microscopical accounting of the entropy of some black holes. More precisely it was shown in these special cases that the number of states of a black hole is identical to the number of states of essentially a gauge theory. It was possible to count the microscopic number in the gauge theory case and the resulting number of states was exactly that predicted for black holes! These are mostly higher dimensional very cold black holes. This has not yet been fully accomplished for uncharged black holes. Black holes have several more non-conventional properties. The above mentioned entropy of a mass M black hole is much smaller, for many types of black holes, than that of a non gravitating system of particles which have the same energy, M. In fact one may suspect that such black holes dominate by far the high energy spectrum of any gravitating system. The Schwarzschild radius of these black holes actually increases with their mass. In general one associates large energies with short distances, in the case of black holes large energies are related also to large distances. A meticulous separation of long and short distances is a corner stone of the renormalization group approach, this is not evident in the case of gravity. One possible implication of this behavior is that in gravitational systems the amount of information stored in a spatial volume is proportional to the boundary's surface area rather then to the bulk volume. String theory provides explicit and concrete realizations of this amazing behavior. A system of strings moving in particular ten dimensional space-times (of the anti de Sitter type) are equivalent to special highly symmetric supersymmetric non-Abelian gauge theories living in four dimensions and experiencing no gravitational forces. Moreover modifying the backgrounds on which the strings propagate by adding defects to them, one finds that strings propagating on certain backgrounds are equivalent to gauge theories in which color is confined. String theory methods are used to perform difficult calculations in hadronic physics. String theory, which was born out of hadronic physics, has returned to pay back its original debt! In addition, remarkable structures have been uncovered in supersymmetric gauge theories using string theory methods. Attempts have been made to confront strings with other singularities. Why should strings respond to singularities in any other way than particles do? A Talmudic story helps explain the manner in which extended objects modify problems associated with point particles. During a seminar, the following question came up: Assume a pigeon (a very valuable animal at the time) is found somewhere, to whom does it belong? The rule is, finders-keepers, as long as the pigeon is found further away, than some determined distance, from the entrance to an owned pigeon-hole. A student raised his hand and asked: "The pigeon is, after all, an extended object, what if one of its legs is nearer than the prescribed distance to the pigeon-hole but the other leg is further away?" The student was ejected from his place of study for raising the question. Strings definitely turn out to soften some harsh singularities which are time independent (called time like singularities). Strings are also found to punch through singularities which form at particular times such as a big crunch. The analysis of such systems is still at very early stages, still I would like to note a very interesting feature in common up the present to all attempts to study using string theory methods universes that are spatially closed. In these cases the compact universe is found to be immersed in one way or another in a non compact universe. The study of strings near these singularities could well be essential to appreciate their properties at short distances. A key ingredient driving some researchers is a new panoramic vista opening up to them. For some it is the precise calculations of experimentally measurable quantities, for others it is the Magic.

8.4.6 Magic

The universe as viewed with string probes is full of magic ambiguities and symmetries unheard of before. In fact most mathematical concepts used to describe the universe are veiled under symmetry. There are cases for which each of the following concepts becomes ambiguous: distances, the number of dimensions, topology, singularity structure, the property of being commutative or not being commutative. Let us somewhat elaborate on the magic.

- Distance—The simplest example is that of a universe in which one of the dimensions is extended along identified with a circle of radius R. It turns out that an experimentalist using strings as his probes will not be able to determine if the universe is indeed best described but one of it dimensions being a circle of radius R (in the appropriate string scale) or by being a circle of radius $1/R$. For point particles moving a circle of radius R, the energy spectrum has large gaps, when R is small, and very small gaps, when the value of R is large. For strings, which are extended objects, the spectrum consists of an additional part, that of the strings wrapping around the circle. A point particle can't wrap around the circle. This part of the spectrum is narrowly gapped for small values of R and widely gapped for large values of R, this is because the energy required to wrap a string, which has its tension, around the circle, is proportional to the length of the wrapped string. For an experimentalist, who can use point particles as probes, the distinction is clear, not so for one using string probes. This is but the tip of the iceberg of an infinite set of ambiguities. Some geometries, judged to be different by point probes, are thus identified by extended probes.

- Dimensions—One could at least expect that the value describing the number of spacetime dimensions to be unique. It turns out not to be the case. Ambiguities in calling the "correct" number of space-time dimensions come in several varieties. In string theory there are examples of a string moving in certain ten dimensional space-times which measures exactly the same physics as that measured by a certain point particle gauge theory in four dimensions. Each system is made out of totally different elementary constituents, one system is defined to exist in ten dimensions, the other in four and yet both reproduce the same observables, the same physics. Another example consists of a strongly coupled string theory in ten dimensions whose low energy physics is reproduced by an eleven dimensional supergravity theory. This is true for large systems. For small, string scale, systems, more magic is manifested. A string in some cases can't distinguish if it is moving on a three dimensional sphere or a one dimensional circle. So much for non-ambiguous dimensions.

- Topology—Objects that can be deformed to each other without using excessive violence (such as tearing them) are considered to have the same topology. The surface of a perfect sphere is the same, topologically, as that of the surface of a squashed sphere, but different, topologically, than that of a bagel/torus. Well, once again that is always true only as far as point particles probes are concerned. In string theory there are examples in which the string probes can describe the same set of possible observations as reflecting motions on objects with different topologies. In addition in string theory there are cases were one may smoothly connect two objects of different topology in defiance of the definition given above for different topologies. Topology can thus be ambiguous.

- Singularities—A crucial problem of General Relativity is the existence of singularities. Bohr has shown how such singularities are resolved in electrodynamics by quantum mechanics. In GR one is familiar with singularities which are stationary (time-like) such as the singularity of a charged black hole and those which form or disappear at a given instance instant such as the big bang and the big crunch (space-like). What happens in string theory to black holes big crunches or big bangs? A good question. Consider first the motion a particle on a circle of some radius and the motion of the string on a segment of some length. The circle is smooth, the segment has edges. For a point particle probe the first is smooth and the second is singular. In some cases, viewed by stringy probes, both are actually equivalent, i.e. both are smooth. The string resolves the singularity. Moreover there are black holes for which a string probe can't distinguish between the black hole's horizon and its singularity. There are many other cases in string theory where time-like singularities are resolved. The situation in the case of space-like singularities is more involved and is currently under study. Some singularities are in the eyes of the beholder.

- Commutativity—The laws of quantum mechanics can be enforced by declaring that coordinates do not commute with their conjugate momenta. It is however taken for granted that the different spatial coordinates do commute with each other and thus in particular can be measured simultaneously. The validity of this assumption is actually also subject to experimental verification. A geometry can be defined even when coordinates do not all commute with each other. This is called non-commutative geometry. There are cases when strings moving on a commutative manifold would report the same observations as strings moving on a non-commutative manifold. It seems many certainties can become ambiguities, when observed by string probes. Stated differently, string theory has an incredible amount of symmetry. This may indicate that the space-time one is familiar with, is in some sense suited only for a "low" energy description. At this stage each of these pieces of magic seems to originate from a different source. The hard-wired mind seeks a unified picture for all this tapestry. This review may reflect ambiguous feelings, challenges, success and magic each have their share. All in all, string theory has been an amazingly exciting field, vibrating with new results and ideas, as bits and pieces of the fabric of space time are uncovered and our idea of what they really are shifts.

8.4.7 Human Effort and Closing Remarks

One may estimate that more than 10,000 human years have been dedicated up to now to the study of string theory. This is a global effort. It tells an amazing story of scientific research. In the mid eighties a regime/leadership change occurred. The scientific leaders which had nursed the standard model, have been replaced by a younger generation of string theorists. Some of those leaders complained that the field had been kidnapped from them. Mathematical methods of a new type were both applied and invented, many pieces of a vast puzzle were uncovered and a large number of consistency checks were used to put parts of the puzzle together tentatively. As in any human endeavor alternative paths are suggested and critique is offered. I have incorporated in this note some of the serious challenges string theory faces. There are additional complaints, complaints that the new style leaves too many gaps in what is actually rigorously proven, complaints that a clean field theory description is needed and lacking. Research has been democratized and globalized to a quite unprecedented extent. This has mainly been achieved by making research results available simultaneously to most of the researchers by posting them on the web daily. There are some drawbacks inherent to that evolution, research attitudes have been largely standardized and the ranges of different points of view have been significantly focused/narrowed. The outside pressures to better quantify scientific production have found an easily accessible statistical database. In particular, that is leading to assigning a somewhat excessive weight for the citation counting in a variety of science policy decisions. The impact of that is yet to stabilize. All that said, one should notice that phenomenologists and experimentalists do turn to string theory and its spin offs in search of a stimulus for ideas on how to detect possible deviations from the standard model. The problems string theory faces are difficult, one approach is to accept this and plunge time and again into the dark regions of ignorance using string theory as an available guide, hoping that it will be more resourceful than its practitioners are. Another approach is to break away from the comfort of physics as we formulate it today and replace some of its working axioms by new ones, such as holography, the entropic principle or eventually perhaps something else. It seems that the question of the exact nature of the basic constituents of matter will remain with us for still quite awhile. My favorite picture is that it will turn out that asking if the basic constituents are point like, are one, two or three-dimensional branes is like asking whether matter is made of earth or air. A theory including the symmetries of gravity will have different phases, some best described by stringy excitations, some by point particles, some by the various branes, and perhaps the most symmetrical phase will be much simpler. Several of these will offer the conventional space time picture, other will offer a something new. String theory either as a source of inspiration, or as a very dynamic research effort attracting criticism leaves few people indifferent.

8.5   Hindsight from 2018

The years from 2007 to 2018 were yet another example of the well-known difficulty of predicting the future. Luckily what I wrote in 2007 contained, on this front, enough disclaimers to satisfy the most strict of lawyers.

During those years one could celebrate a gigantic engineering triumph. After succumbing to a very significant early setback the LHC outperformed its expectations. The experimental triumph was as impressive. The detectors performed well beyond what one could dream. They obtained results from picking out the Higgs particle out of haystack through rediscovering particles which were first detected at electron-positron machines to finding more hadrons in the heavier flavour sectors.

This success was not matched by validating any of the TPs (Theoretical Prejudices) which I had discussed in some detail. The discovery of the quite likely absence of new particles in the new energy vistas opened by the LHC had an impact. Of course one can still hope that positive new discoveries will be made and I for one do hope so; still it is more and more considered as a possibility that we will need to settle for a long time for accelerators that will validate the Standard Model and not uncover treasures beyond it. A school of thought is developing, since the activation of the LHC, that beyond the Standard Model lurks non other than the Standard Model itself. At least for a while.

This requires soul searching of the theoretical physics community. What could be the result of such an examination of the routes taken so far? Recall that NOH (Nobody Owes Humanity), it may be that using elegance and beauty as our guiding principles, and indeed SUSY and string theory are beautiful, has failed us. Perhaps our esthetics taste does not confirm with that of nature. Perhaps these TPs will manifest themselves at some higher energy and perhaps we are on a totally wrong track. Falsifying ideas is a most important component of the scientific method, but many would not agree that the TPs have been indeed falsified. New ideas should be encouraged and welcomed, there is nothing as good as a good new idea to revitalize our thinking. The results of such debates cannot be dictated. Each researcher will draw her/his own conclusions. Be that as it may, such an open debate is required. Hopefully there will be something to report on in the next edition of this book.

One aspect of "Beyond" physics that is considered more actively now involves testing the possibility that the Standard Model has scale or even conformal invariant features. Some of them I suggested long ago and pointed out in this volume, especially its relation to the puzzle of the value of the vacuum energy.

The fact that, depending on the exact value of the top quark mass, the Universe is living dangerously and may eventually self-destruct as it decays to some other place was also noted and investigated.

The dark matter elephant in the room remains very present.

The mathematical aspects of SUSY have been considerably sharpened during these years and have led to many new surprising exact results in more regions of strong coupling where no one ventured before. I was most attracted to a branch of research trying to attempt to use quantum information aspects as ingredients that build up semiclassical geometry and also quantum space time. These attempts are in their infancy but they have already covered quite some ground. Sometimes it has happened that a change of perspective and bringing in new tools was a way for a younger generation to take over the leadership of a field. The absence of experimental data to crown new leaders serves as a fertile ground for this mechanism.

Finally, back to experiment; for years black holes seemed as removed from experiment as the detection of stringy effects are 2017 has seen the hundred years search for the detection of gravitational waves reach a discovery moment. Humans observed particles collide at the LHC and black holes collide in the Universe—a great moment. Theorists are seriously considering the properties of those very black holes in terms of information processing. Seeds for a new TP have been sown.

Over this decade we have learned or should have learned the importance of patience, humility, and diversity of ideas. Not the type of lessons that theoretical physicists appreciate too much.

8.6 References for 8

We mention here only text books and reviews and only those which were used to prepare this entry. Most of the appropriate references for the works reviewed here can be found in them.

(1) For Super Symmetry:

Different narratives on the history of Super Symmetry by a large number of its pioneers has been collected in the book The Supersymmetric World, edited by G. Kane and M. Shifman. Published by World Scientific (2000). As most of them were still alive at the time this offers a unique perspective into how theoretical ideas develop.

- Supersymmetry and Supergravity by J. Wess and J. Bagger. Princeton Series in Physics (1992).
- The Quantum Theory of Fields, Volume 3: Supersymmetry by S. Weinberg. Cambridge University Press (2000).
- Lectures on Supersymmetric Gauge Theories and Electric—Magnetic Duality by K.A. In-triligator and N. Seiberg. Published also in Proceedings of Cargese Summer School lectures, Cargese.
- Supersymmetric Gauge Theories by D. Berman and E. Rabinovici. Published in: Unityfrom Duality: Gravity, Gauge Theory and Strings, Les Houches Session LXXVI. Editors: C. Bachas, A. Bilal, M. Douglas, N. Nekrasov and F. David. Publisher: Springer.
- Weak Scale Super Symmetry: From Superfields to Scattering Events by H. Baer and X.Tata. Cambridge University Press (2006).
- Lectures on Supersymmetry Breaking by K. Intriligator and N. Seiberg. Published in Class.Quant. Grav. (2007).

(2) For gauge theory:

This subject was barely touched upon here. A group theory reference is: Group Theory for Unified Model Building by R. Slansky. Published in Phys. Rept. **79** 1–128 (1981).

(3) For extra dimensions:

Kaluza-Klein Theory in Perspective. M.J. Duff (Newton Inst. Math. Sci., Cambridge). NI-94-015, CTP-TAMU-22-94, Oct. 1994. 38 pp. Talk given at The Oskar Klein Centenary Symposium, Stockholm, Sweden, 19–21 Sept. 1994. In *Stockholm 1994, The Oskar Klein Centenary* 22–35, hep-th/9410046.

(4) For strings:

For text books and many references see for example:

- M.B. Green, J.H. Schwarz, E. Witten: Superstring Theory. Cambridge Monographs On Mathematical Physics (1987).
- J. Polchinski: String Theory, Cambridge University Press (1998).
- E. Rabinovici: String Theory: Challenges, Successes and Magic. Published in Europhys. News (2004).

Chapter 9
Symmetry Violations and Quark Flavour Physics

Konrad Kleinknecht and Ulrich Uwer

9.1 Introduction

9.1.1 Matter–Antimatter Asymmetry in the Universe

One of the surprising facts in our present understanding of the development of the Universe is the complete absence of "primordial" antimatter from the Big Bang about 13.7 billion years ago. The detection of charged cosmic-ray particles by magnetic spectrometers borne by balloons, satellites, and the space shuttle has shown no evidence for such primordial (high-energy) antibaryons; nor has the search for gamma rays from antimatter–matter annihilation yielded any such observation. In the early phases of the expanding Universe, a hot (10^{32} K) and dense plasma of quarks, antiquarks, leptons, antileptons and photons coexisted in equilibrium. This plasma expanded and cooled down, and matter and antimatter could recombine and annihilate into photons. If all interactions were symmetric with respect to matter and antimatter, and if baryon and lepton numbers were conserved, then all particles would finally convert to photons, and the expansion of the Universe would shift the wavelength of these photons to the far infrared region.

This cosmic microwave background radiation was indeed observed by Penzias and Wilson in 1965 [1], and its wavelength distribution corresponds exactly to Planck black-body radiation at a temperature of 2.73 K (see Fig. 9.1). The density of this radiation is about 5×10^2 photons/cm^3.

K. Kleinknecht
Johannes Gutenberg-Universität, Mainz, Germany

Ludwig-Maximilians-Universität, München, Germany

U. Uwer (✉)
Heidelberg University, Heidelberg, Germany
e-mail: uwer@physi.uni-heidelberg.de

© The Author(s) 2020
H. Schopper (ed.), *Particle Physics Reference Library*,
https://doi.org/10.1007/978-3-030-38207-0_9

Fig. 9.1 Frequency
distribution of the cosmic
microwave background
variation, as measured by the
COBE satellite and
earth-based experiments

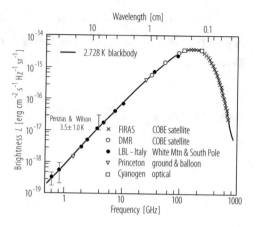

However, this radiation is not the only remnant of the Big Bang; there is also a small amount of baryonic matter left over, at a density of 6×10^{-8} nucleons/cm^3, about 10^{-10} of the photon density. This phenomenon can only be explained if the three conditions of Sakharov [2] are fulfilled:

- there must be an interaction violating CP invariance, where C is the particle–antiparticle transformation and P the space inversion operation;
- there must be an interaction violating the conservation of baryon number;
- there must be phases of the expansion without thermodynamic equilibrium.

The first condition was shown to be fulfilled when, in 1964, J. Christenson, J. Cronin, V. Fitch and R. Turlay discovered CP violation [3] in decays of neutral K mesons. The second criterion would imply that protons are not stable; searches for such a decay have been unsuccessful, showing that the lifetime of the proton is longer then 10^{31} years. The third condition can be met in cosmological models by inflationary fast expansion or by a first-order phase transition in the electroweak interaction of the Standard Model.

In the following, we shall concentrate on the observed CP violation, which could in principle lead to a small surplus of matter, the observed baryon asymmetry of 10^{-10} in the Universe.

9.2 Discrete Symmetries

Symmetries and conservation laws have long played an important role in physics. The simplest examples of macroscopic relevance are the laws of conservation of energy and momentum, which are due to the invariance of forces under translation in time and space, respectively. Both are continuous transformations. In the domain of quantum phenomena, there are also conservation laws corresponding to discrete transformations. One of these is reflection in space (the "parity operation") P

[4]. The second discrete transformation is particle–antiparticle conjugation C [5]. This transforms each particle into its antiparticle, whereby all additive quantum numbers change their sign. A third transformation of this kind is time reversal T, which reverses momenta and angular momenta [6]. This corresponds formally to an inversion of the direction of time. According to the CPT theorem of Lüders and Pauli [7–9], there is a connection between these three transformations such that, under rather weak assumptions, in a local field theory all processes are invariant under the combined operation $C \cdot P \cdot T$.

9.2.1 Discrete Symmetries in Classical Physics

9.2.1.1 Parity P

The parity operation consists in reversing the direction of the position vector $\vec{r} = (X, Y, Z)$ in Cartesian coordinates. This corresponds to reflection in a plane mirror, followed by a rotation by $180°$. Symmetry under parity operation is therefore also called mirror symmetry.

The parity operation reverses the direction of all polar vectors derived from the position vector; in particular, this is the case for the momentum $\vec{p} = m\vec{v} = m\mathrm{d}\vec{r}/\mathrm{d}t$ and the acceleration $\vec{a} = \mathrm{d}^2\vec{r}/\mathrm{d}t^2$. Therefore, the Newtonian force $\vec{F} = \mathrm{d}\vec{p}/\mathrm{d}t$ is also reversed under the parity operation.

This must be also the case for the Lorentz and Coulomb forces on a particle with charge q moving with velocity $\vec{v}$:

$$\vec{F} = q(\vec{E} + \vec{v} \times \vec{B}) . \tag{9.1}$$

Since the charge q is invariant under P, and the force $\vec{F}$ and the velocity $\vec{v}$ change sign, the electric field strength $\vec{E}$ must change sign and the magnetic field strength $\vec{B}$ must remain unchanged.

For the electric potential $\vec{A}$, we obtain from the relations

$$\vec{E} = -\mathrm{grad}\, V - \partial \vec{A}/\partial t, \tag{9.2}$$

$$\vec{B} = \mathrm{rot}\, \vec{A}, \tag{9.3}$$

the result that $\vec{A}$ changes sign and V remains invariant, since the spatial differential operator changes sign under the parity operation.

We therefore have four classes of quantities with different transformation behavior under P: axial vectors or pseudovectors such as $\vec{B}$ and the angular momentum $\vec{J} = \vec{r} \times \vec{p}$, and scalars such as V, which remain invariant under P; and polar vectors such as $\vec{r}$, $\vec{p}$, $\vec{F}$, $\vec{E}$ and $\vec{A}$, and pseudoscalars such as $\vec{E} \cdot \vec{B}$, which change sign under P.

9.2.1.2 Time Reversal $\mathcal{T}$

This operation consists in reversing the sign of the time axis t. Under this operation $t \rightarrow -t$, the velocity $\vec{v}$, the momentum $\vec{p}$ and the angular momentum $\vec{J}$ change sign, while the force $\vec{F}$ remains unchanged under $\mathcal{T}$. From the fact that the Coulomb and Lorentz forces are invariant, we derive the result that $\vec{E} \rightarrow \vec{E}$ and $\vec{B} \rightarrow \vec{B}$ under $\mathcal{T}$; for the potentials, $V \rightarrow V$ but $\vec{A} \rightarrow -\vec{A}$.

9.2.1.3 Dipole Moments

Elementary particles with spin may have electric or magnetic dipole moments. The spin $\vec{s}$ has the dimensions of an angular momentum, and therefore remains unchanged under parity and changes sign under time reversal.

The potential energy of an electric or magnetic dipole in an external field is proportional to the scalar product of the electric or magnetic moment with the strength of the external electromagnetic field. Since the moments must be parallel to the spin, the potential energy is given by

$$- d_e \vec{s} \cdot \vec{E} \quad \text{for the electric case} \tag{9.4}$$

and

$$- d_m \vec{s} \cdot \vec{B} \quad \text{for the magnetic case} \tag{9.5}$$

Here d_e and d_m are the electric and magnetic dipole moments, respectively. If we consider the transformation properties of $\vec{s}$, $\vec{E}$ and $\vec{B}$ under $\mathcal{P}$ and $\mathcal{T}$, it turns out that for both operations $d_m \rightarrow d_m$ and $d_e \rightarrow -d_e$. This means that observation of a nonvanishing electric dipole moment would violate any invariance under parity and time-reversal transformations.

In classical physics, all processes are invariant under parity and under time reversal. In the case of mirror symmetry, this means that a physical experiment will lead to the same result as a mirror-imaged experiment, since the equations of classical physics are left-right symmetric. In a similar way, the classical motion of one particle can be reversed, e.g. by playing a film backwards, and this inversion of the motion corresponds formally to time reversal. Again, the laws of motion are invariant under $\mathcal{T}$, and the reversed motion follows the same path backwards as forwards.

Of course, this is no longer the case if many particles move and interact with each other; in this case the second law of thermodynamics ensures that entropy is increasing, thus defining an arrow of time.

9.2.2 Discrete Symmetries in Quantum Systems

9.2.2.1 Particle–Antiparticle Conjugation

In relativistic quantum mechanics, the Dirac equation requires that for each solution describing a particle, there is a second solution with opposite charge, describing the antiparticle. The antiparticle of the electron, the positron, was found in 1933 [10], and the antiproton was found in 1955 [11]. The particle–antiparticle conjugation C transforms the field ϕ of the particle into a related field $\phi^\dagger$ which has opposite quantum numbers: the charge, lepton number, baryon number, strangeness, beauty, etc., for the antiparticle are opposite in sign to the values for the particle.

Invariance under the C transformation is always valid in the strong and electromagnetic interactions. This means, in particular, that the visible spectral lines from atoms and their antiatom partners are identical, and we cannot use these lines to identify antimatter in the Universe.

This would be especially important in the science-fiction scenario in which a man-made spacecraft sets out to meet a distant civilization, where it would be advisable to know whether the other planet was made of matter or antimatter. In this case another means of differentiation would have to be found.

9.2.2.2 Violation of Mirror Symmetry: Parity Violation in Weak Interactions

Lee and Yang [13] suggested that of the four interactions—strong, electromagnetic, weak and gravitational—the weak interaction might violate mirror symmetry when it was described by a combination of vector and axial-vector currents in the Lagrangian (V–A theory). The interference of these two currents could lead to pseudoscalar observables which would change sign under the parity operation. One such observable is the scalar product of an axial vector (such as the spin of a particle) with a polar vector (such as the momentum of another particle in the final state). If the expectation value of this pseudoscalar is measured to be nonvanishing, then parity is violated.

An experiment on the beta decay of cobalt-60 [14] measured exactly such an observable, the scalar product of the spin $5\hbar$ of the ^{60}Co nucleus and the direction of the electron from its beta decay into an excited state of ^{60}Ni with nuclear spin $4\hbar$. The ^{60}Co nuclei were polarized by embedding them in a cerium–magnesium crystal, where the magnetic moments were aligned by a weak external magnetic field of about 0.05 T. In the strong magnetic field inside this paramagnet, the ^{60}Co nuclei are polarized through hyperfine interactions if the temperature is low enough (0.01 K) to avoid thermal demagnetization. The polarization was measured through the asymmetry of γ rays from the cascade decay of the ^{60}Ni state. The measurement then required the detection of the electron direction relative to the polarization of the Co nuclei. The experimenters found that the electron was emitted preferentially in a direction opposite to the external magnetic field, and that this effect disappeared when the crystal was warmed and the nuclear polarization disappeared. Thus, at low temperature, a nonzero pseudoscalar is observed, demonstrating parity violation.

By comparing nuclear beta decays having an electron and an antineutrino in the final state with their counterpart with emission of a positron-neutrino pair, it was shown that the helicity $h = \vec{s} \cdot \vec{p}/|\vec{s} \cdot \vec{p}|$ of neutrinos is opposite to the one of antineutrinos [18].

Other experiments lead to similar results. The helicity $h = \vec{s} \cdot \vec{p}/|\vec{s} \cdot \vec{p}|$ of the neutrino emitted in the weak electron capture by ^{152}Eu was measured in an experiment by Goldhaber and Grodzins [19] to be negative; the neutrino is "left-handed", i.e. the spin is aligned antiparallel to the momentum. Similarly, measurements of the polarization of electrons from β^- decay showed a negative value, with a modulus increasing with the velocity of the electron, v/c. Positrons from β^+ decay were found to have a positive polarization that increases with v/c.

9.2.2.3   Violation of C Symmetry, and CP Invariance

In the realm of weak decays of particles, supporting evidence for the violation of mirror symmetry came from the observation that parity is violated in the decay $\pi^+ \to \mu^+ \nu_\mu$, and that the muon neutrino from this decay is left-handed [15, 16]. The P-conjugate process, i.e. $\pi^+ \to \nu_\mu \mu^+$, with a right-handed neutrino, does not occur. The same is true for the C-conjugate process, $\pi^- \to \mu^- \overline{\nu_\mu}$, with a left-handed antineutrino. However, if we combine the C and P operations, we arrive at a process $\pi^- \to \mu^- \overline{\nu_\mu}$ with a right-handed antineutrino, which proceeds at the same rate as the original π^+ decay, with a left-handed muon neutrino. Evidently, in weak interactions, P and C are violated, while it seemed at the time of those experiments that the process was invariant under the combined operation $C \cdot P$. This argument can be visualized as in Fig. 9.2. Here the P mirror and the C mirror act on a left-handed neutrino, both leading to unphysical states, a right-handed neutrino and a left-handed antineutrino. Only the combined CP mirror leads to a physical particle, the right-handed antineutrino. This argument was made by Landau [17], suggesting that the real symmetry was CP invariance.

9.2.2.4 CP Invariance and Neutral K Mesons

One consequence of this postulated CP invariance was predicted by Gell-Mann and Pais [12] for the neutral K mesons: there should be a long-lived partner to the known V^0 (K_1^0) particle of short lifetime $(10^{-10}$s$)$. According to this proposal, these two particles are mixtures of the two strangeness eigenstates $K^0 (\mathbb{S} = +1)$ and $\overline{K^0}(\mathbb{S} = -1)$ produced in strong interactions. Weak interactions do not conserve strangeness, and the physical particles should be eigenstates of CP if the weak

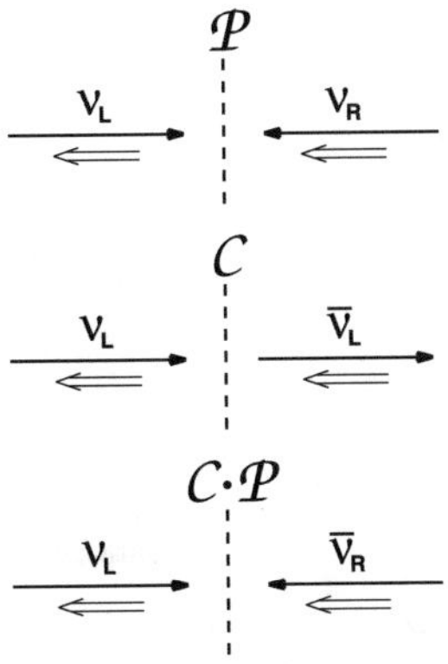

Fig. 9.2 The mirror image of a left-handed neutrino under $\mathcal{P}$, $\mathcal{C}$ and $\mathcal{CP}$ mirror operations

interactions are $\mathcal{CP}$-invariant. These eigenstates are described as follows (where we choose the phases such that $\overline{K^0} = \mathcal{CP}K^0$):

$$\mathcal{CP}K_1 = \mathcal{CP}\left[\frac{1}{\sqrt{2}}\left(K^0 + \overline{K^0}\right)\right] = \frac{1}{\sqrt{2}}\left(\overline{K^0} + K^0\right) = K_1 , \qquad (9.6)$$

$$\mathcal{CP}K_2 = \mathcal{CP}\left[\frac{1}{\sqrt{2}}\left(K^0 - \overline{K^0}\right)\right] = \frac{1}{\sqrt{2}}\left(\overline{K^0} - K^0\right) = -K_2 . \qquad (9.7)$$

Because $\mathcal{CP}\,|\pi^+\pi^-\rangle = |\pi^+\pi^-\rangle$ for π mesons in a state with angular momentum zero, i.e. the two-pion state has a positive $\mathcal{CP}$ eigenvalue, a decay into $\pi^+\pi^-$ is allowed for the K_1 but forbidden for the K_2; hence the longer lifetime of K_2, which was indeed confirmed when the K_2 was discovered [20, 21].

9.2.2.5 Discovery of $\mathcal{CP}$ Violation

In 1964, however, Christenson et al. [22] discovered that the long-lived neutral K meson also decays to two charged pions with a branching ratio of 2×10^{-3}.

The motivation of this experiment was twofold: the experimenters wanted to check on an effect found by Adair et al. [23] when the latter observed interactions of long-lived kaons (K_2) in hydrogen, and they wanted to test $\mathcal{CP}$ invariance by searching for the decay of K_2 into two pions. Adair et al. had found anomalous regeneration of short-lived kaons (K_1) above expectation. "Regeneration" is an effect due to the different strong interactions of the two components of a long-lived kaon, K^0 and $\overline{K^0}$. This leads to a creation of a coherent K_1 component when a K_2 beam traverses matter. The anomalous effect above expectation was still observed in the experiment of Christenson et al. when the (K_2) beam hit a hydrogen target.

Therefore, Christenson et al. emptied the target and looked for $K_2 \rightarrow \pi^+\pi^-$ decays from the vacuum. To their surprise, they found such decays, which meant that $\mathcal{CP}$ invariance was broken in this decay.

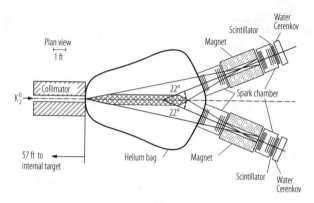

Fig. 9.3 The experimental setup used by Christenson et al. [22] for the discovery of $\mathcal{CP}$ violation

The magnetic spectrometer used by Christenson et al. is shown in Fig. 9.3. On each side of the spectrometer, one charged particle is detected through spark chambers in front of and behind the magnet. The two vector momenta $\vec{p}_i$ ($i = 1, 2$) of the two particles are measured. Assuming the mass of the particles to be the pion rest mass m_π their energies can be obtained from

$$E_i^2 = \vec{p}_i^2 + m_\pi^2 .\tag{9.8}$$

The invariant mass of the pair is

$$m_{\pi\pi} = \sqrt{\left[(E_1 + E_2)^2 - (\vec{p}_1 + \vec{p}_2)^2\right]} ,\tag{9.9}$$

and the kaon momentum is

$$\vec{p}_K = \vec{p}_1 + \vec{p}_2 .\tag{9.10}$$

From the reconstructed kaon momentum, the intersection of the kaon flight path with the target plane gives an indication of whether this was a two-body decay coming from the target or a three-body decay with an escaping neutrino.

In the latter case, the direction of $\vec{p}_K$ does not point back to the target source. The result of the experiment is shown in Fig. 9.4. A significant peak of $K \to \pi^+\pi^-$ decays coming from the target direction ($\cos\theta = 1$) is seen, while the background of three-body decays outside the peak can be extrapolated under the signal, and represents only $\sim$20% of the data in the signal region: there is a signal at the level of 2×10^{-3} of all decays, and $\mathcal{CP}$ is violated.

From then on the long-lived K meson state was called K_L because it was no longer identical to the $\mathcal{CP}$ eigenstate K_2. However, the physical long-lived state K_L was a superposition of a predominant K_2 amplitude and a small admixture of a K_1 amplitude, $K_L = (K_2 + \varepsilon K_1)/\sqrt{1 + |\varepsilon|^2}$ where the admixture parameter

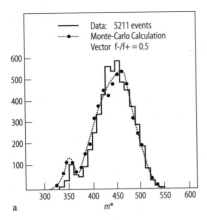
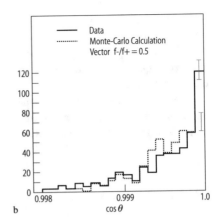

Fig. 9.4 (a) Experimental distribution of m^* compared with Monte Carlo calculation. The calculated distribution is normalized to the total number of observed events. (b) Angular distribution of those events in the range $490 < m^* < 510$ MeV. The calculated curve is normalized to the number of events in the total sample

ε is determined by experiment to satisfy $|\varepsilon| \sim 2 \times 10^{-3}$. Similarly, the short-lived state was called K_S, and $K_S = (K_1 + \varepsilon K_2)/\sqrt{1 + |\varepsilon|^2}$. The $\mathcal{CP}$ violation that manifested itself by the decay $K_L \to \pi^+\pi^-$ was confirmed subsequently in the decay $K_L \to \pi^0\pi^0$ [24, 25], and by a charge asymmetry in the decays $K_L \to \pi^\pm e^\mp \nu$ and $K_L \to \pi^\pm \mu^\mp \nu$ [26, 27].

9.2.3 Discrete Symmetries in Quantum Mechanics

The three discrete symmetries $\mathcal{P}$, $\mathcal{C}$ and $\mathcal{T}$ are described by the operators $\mathcal{P}$ for the parity transformation, $\mathcal{C}$ for particle–antiparticle conjugation and $\mathcal{T}$ for time reversal. Invariance of an interaction described by a Hamiltonian $\mathcal{H}$ under a symmetry operation means that $\mathcal{H}$ commutes with the relevant operator, e.g. $[\mathcal{H}, \mathcal{P}] = 0$. According to experimental evidence, the strong and electromagnetic interactions are $\mathcal{P}$- and $\mathcal{C}$-invariant. The corresponding operators are unitary, i.e. the Hermitian conjugate is equal to the inverse:

$$\mathcal{C}^\dagger = \mathcal{C}^{-1}, \tag{9.11}$$

$$\mathcal{P}^\dagger = \mathcal{P}^{-1}. \tag{9.12}$$

For two states $|\psi\rangle$ and $|\varphi\rangle$, such a unitary transformation does not change the product:

$$\langle \psi \mid \varphi \rangle = \langle \psi' \mid \varphi' \rangle, \tag{9.13}$$

where $|\psi'\rangle$ and $|\varphi'\rangle$ are the transformed states.

By defining the intrinsic parity of the proton as +1, a phase convention for the fields can be chosen, such that the parity operator $\mathcal{P}$ has only eigenvalues of +1 or −1, and all particles have intrinsic parities ±1 as quantum numbers. These conserved quantities, which correspond to discrete symmetries, are multiplicative quantum numbers.

The third discrete symmetry, time reversal, is a special case. The corresponding operator $\mathcal{T}$ is not unitary, but antiunitary. Here the bracket $\langle \psi | \varphi \rangle$ is not conserved by the $\mathcal{T}$ transformation, but rather

$$\langle \psi' | \varphi' \rangle = \langle \psi | \varphi \rangle^* . \tag{9.14}$$

Probability is still conserved, i.e.

$$|\langle \psi' | \varphi' \rangle| = |\langle \psi | \varphi \rangle| , \tag{9.15}$$

but the phases are not. The fact that $\mathcal{T}$ is antiunitary can be deduced from the Schrödinger equation for a free particle, where the time derivative is odd under $\mathcal{T}$ while the Laplace operator Δ is even. This can be reconciled with $\mathcal{T}$ invariance only if $\mathcal{T}$ makes the changes i $\rightarrow$ −i and $\psi \rightarrow \psi^*$.

$\mathcal{CPT}$, the product of all three discrete transformations, being a product of two unitary and one antiunitary operator, is also antiunitary. According to the $\mathcal{CPT}$ theorem of Lüders [7] and Pauli [8], and Jost [9], a field theory with Lorentz invariance, locality, and the commutation relations given by the spin-statistics theorem, is $\mathcal{CPT}$-invariant. At present there is no realistic field theory which violates $\mathcal{CPT}$ invariance.

As a consequence of this theorem, violation of one of the three discrete symmetries implies a violation of a complementary one. If $\mathcal{CP}$ is violated, then $\mathcal{T}$ is also violated.

The experimental consequences of $\mathcal{CPT}$ invariance are the equality of the masses, lifetimes and magnetic dipole moments of a particle and its antiparticle. These equalities have been tested with great precision, as shown in Table 9.1.

A very special case in this context is that of the masses of the neutral K mesons. The mass difference between the long-lived K_L and the short-lived K_S can be measured in interference experiments. This difference is due to second-order weak

Table 9.1 Comparison of masses m, lifetimes τ, and magnetic g-factors of particle and antiparticles

Particle	$|m - \overline{m}|/m$	$|\tau - \overline{\tau}|/\tau$	$|g - \overline{g}|/g$
e	$< 4 \times 10^{-8}$		$(-0.5 \pm 2.1) \times 10^{-12}$
μ		$(1 \pm 8) \times 10^{-5}$	$(-2.6 \pm 1.6) \times 10^{-8}$
π	$(2 \pm 5) \times 10^{-4}$	$(5.5 \pm 7.1) \times 10^{-4}$	
p	$< 7 \times 10^{-10}$		$(0.3 \pm 0.8) \times 10^{-6}$
K^0	$< 4.7 \times 10^{-19}$		

interactions and, therefore, is very small, about $\Delta m = (3.480 \pm 0.007) \times 10^{-6}$ eV, which means that $\Delta m / m_K < 10^{-14}$. From this, one can deduce very stringent limits on the mass difference between the K^0 and $\overline{K^0}$, of order 10^{-18}.

Thus, from experimental evidence, there is no doubt about the validity of $\mathcal{CPT}$ invariance.

9.3 Mixing and Decay of Neutral Flavoured Mesons

9.3.1 Particle–Antiparticle Mixing

Neutral mesons (represented by N^0 in this chapter) with a characteristic quantum number, such as the strangeness $\mathbb{S}$ for K^0 mesons, charm $\mathbb{C}$ for D^0 mesons and beauty $\mathbb{B}$ for B_s^0 and B_d^0 mesons, have the particular property that they can mix with their antiparticles, which carry an opposite-sign quantum number. Weak interactions do not conserve any of these quantum numbers ($\mathbb{S}$, $\mathbb{C}$, $\mathbb{B}$); consequently N^0 and $\overline{N^0}$ can mix by second-order weak transitions through intermediate states such as $2\pi, 3\pi, \pi\mu\nu, \pi e\nu$ (for K^0), or πK (for B^0). The states that obey an exponential decay law are linear superpositions of N^0 and $\overline{N^0}$,

$$\alpha \, |N^0\rangle + \beta \, |\overline{N^0}\rangle = \begin{pmatrix} \alpha \\ \beta \end{pmatrix} . \tag{9.16}$$

The time-dependent Schrödinger equation then becomes a matrix equation

$$i\frac{d}{dt}\begin{pmatrix} \alpha \\ \beta \end{pmatrix} = X\begin{pmatrix} \alpha \\ \beta \end{pmatrix} , \tag{9.17}$$

where $X_{ik} = M_{ik} - i\Gamma_{ik}/2$, and M_{ik} and Γ_{ik} are Hermitian matrices, called mass matrix and decay matrix, respectively. Both of the latter two matrices are Hermitian: $M = M^\dagger$ and $\Gamma = \Gamma^\dagger$. However, X is not Hermitian. The elements of the matrix X are

$$X_{11} = \langle N^0|\mathcal{H}|N^0\rangle , \quad X_{22} = \langle \overline{N^0}|\mathcal{H}|\overline{N^0}\rangle ,$$

$$X_{12} = \langle N^0|\mathcal{H}|\overline{N^0}\rangle , \quad X_{21} = \langle \overline{N^0}|\mathcal{H}|N^0\rangle \tag{9.18}$$

where $\mathcal{CPT}$ invariance requires the diagonal elements to be equal: $X_{11} = X_{22}$. The matrix has the form

$$X = \begin{pmatrix} m - \frac{i}{2}\Gamma & m_{12} - \frac{i}{2}\Gamma_{12} \\ m_{12}^* - \frac{i}{2}\Gamma_{12}^* & m - \frac{i}{2}\Gamma \end{pmatrix} . \tag{9.19}$$

The off-diagonal elements of the matrices are given by

$$\Gamma_{21} = 2\pi \sum \varrho_F \langle \overline{N^0}|\mathcal{H}_W|F\rangle \langle F|\mathcal{H}_W|N^0\rangle \,, \tag{9.20}$$

where the sum runs over all possible physical intermediate states F, which have a phase space density ϱ_F. Similarly,

$$M_{21} = \langle \overline{N^0}|\mathcal{H}_W|N^0\rangle + \sum_n \frac{\langle \overline{N^0}|\mathcal{H}_W|n\rangle \langle n|\mathcal{H}_W|N^0\rangle}{m_{N^0} - m_n} \,, \tag{9.21}$$

where the sum extends over all possible virtual intermediate states n.

The eigenvalue equations for X yield two eigenstates, which can be labeled by their mass: h for the higher mass, l for the lower mass. These eigenstates are the physical particles with a definite mass and an exponential lifetime distribution. The eigenvalues M_h and M_l of the matrix X are

$$M_h = m_h - \frac{i}{2}\Gamma_h \,,$$

$$M_l = m_l - \frac{i}{2}\Gamma_l \tag{9.22}$$

We denote the differences between the physical quantities by $\Delta\Gamma = \Gamma_h - \Gamma_l$ and $\Delta m = m_h - m_l > 0$ and denote the average values by

$$\Gamma = \frac{\Gamma_h + \Gamma_l}{2} \,, \tag{9.23}$$

$$m = \frac{m_h + m_l}{2} \,. \tag{9.24}$$

For the B^0—$\overline{B^0}$ system and the D^0—$\overline{D^0}$ system, the two decay widths Γ_h and Γ_l are expected to be nearly equal (because the numbers of final states for the decay are very similar). In these cases it is customary to introduce the dimensionless quantities

$$x = \frac{\Delta m}{\Gamma} \tag{9.25}$$

and

$$y = \frac{\Delta\Gamma}{2\Gamma} \,. \tag{9.26}$$

Here x is positive by definition, and y varies between -1 and $+1$. For heavy systems such as B^0, $|y|$ is expected to be much less than 1, while for the K^0 system, y is found experimentally to be close to -1, since here the decay width of the lighter state is 600 times larger then that of the heavier state. Therefore, for the K^0 system,

Table 9.2 Parameters of the four neutral oscillating meson pairs [28] (see Sects. 9.5.5.7 and 9.6.4)

	$K^0/\overline{K^0}$	$D^0/\overline{D^0}$	$B^0/\overline{B^0}$	$B_s^0/\overline{B_s^0}$
τ [ps]	89.59 ± 0.04	0.4101 ± 0.0015	1.519 ± 0.004	1.510 ± 0.007
	$51\,160 \pm 210$			
Γ [10^9s^{-1}]	5.61	2.44×10^3	(658 ± 2)	(662 ± 3)
$y = \Delta\Gamma/(2\Gamma)$	-0.9965	$(0.645 \pm 0.008) \times 10^{-2}$	$|y| \lesssim 0.01$	$-(0.062 \pm 0.005)$
Δm [10^9s^{-1}]	(5.286 ± 0.011)	<70	(507 ± 2)	$(17.76 \pm 0.02) \times 10^3$
Δm [10^{-6}eV]	(3.480 ± 0.007)	<5	(334 ± 3)	$(11.69 \pm 0.13) \times 10^3$
$x = \Delta m/\Gamma$	0.945 ± 0.002	<0.03	0.769 ± 0.004	26.81 ± 0.10

the lighter state is called K_S ("short-lived") and the heavier state K_L ("long-lived"). Table 9.2 gives a summary of various parameters of oscillating meson pairs.

9.3.2   Decays of Neutral Mesons

9.3.2.1   Time-Dependent Schrödinger Equation

From the time-dependent Schrödinger equation for mixed states given above, it follows that

$$\frac{\mathrm{d}}{\mathrm{d}t}\left(|\alpha|^2 + |\beta|^2\right) = -\left(\alpha^*\beta^*\right)\Gamma\begin{pmatrix}\alpha\\\beta\end{pmatrix}. \qquad (9.27)$$

Since both of the neutral mesons N^0 and $\overline{N^0}$ decay, the left-hand side of this equation is negative for any α or β. Therefore Γ is positive definite, in particular Γ_{11}, Γ_{22} and det Γ are positive.

The physical particles, which have a definite mass and lifetime, are mixtures of the eigenstates N^0 and $\overline{N^0}$ of the strong interaction, which carry definite values of their characteristic quantum numbers strangeness $\mathbb{S}$, charm $\mathbb{C}$, and beauty $\mathbb{B}$.

If the weak interaction through which these particles decay is invariant under a discrete symmetry, say $\mathcal{CP}$, then the physical particles are eigenstates of this symmetry because $\mathcal{H}_W$ commutes with $\mathcal{CP}$.

The effect of discrete symmetries on N^0 and $\overline{N^0}$ is the following: $\mathcal{CP}$ is unitary, and there is an arbitrary phase a:

$$\mathcal{CP}|N^0\rangle = \mathrm{e}^{\mathrm{i}a}|\overline{N^0}\rangle\,,$$

$$\mathcal{CP}|\overline{N^0}\rangle = \mathrm{e}^{-\mathrm{i}a}|N^0\rangle\,. \qquad (9.28)$$

$\mathcal{CPT}$, however, is antiunitary, and, with an arbitrary phase b,

$$\mathcal{CPT}|N^0\rangle = e^{ib}|\overline{N^0}\rangle \ ,$$

$$\mathcal{CPT}|\overline{N^0}\rangle = e^{ib}|N^0\rangle \ . \tag{9.29}$$

For $\mathcal{T}$, which is also antiunitary, we obtain

$$\mathcal{T}|N^0\rangle = e^{i(b-a)}|\overline{N^0}\rangle$$

$$\mathcal{T}|\overline{N^0}\rangle = e^{i(b+a)}|N^0\rangle \tag{9.30}$$

We choose the arbitrary phase a to be equal to 0 here, such that the eigenstates of $\mathcal{CP}$ are

$$|N_+\rangle = \frac{1}{\sqrt{2}}\left(|N^0\rangle + |\overline{N^0}\rangle\right) \ ,$$

$$|N_-\rangle = \frac{1}{\sqrt{2}}\left(|N^0\rangle - |\overline{N^0}\rangle\right) \ , \tag{9.31}$$

with the property that they have $\mathcal{CP}$ eigenvalues $+1$ and -1:

$$\sqrt{2}\mathcal{CP}|N_+\rangle = \mathcal{CP}|N^0\rangle + \mathcal{CP}|\overline{N^0}\rangle = |\overline{N^0}\rangle + |N^0\rangle = \sqrt{2}|N_+\rangle$$

$$\sqrt{2}\mathcal{CP}|N_-\rangle = \mathcal{CP}|N^0\rangle - \mathcal{CP}|\overline{N^0}\rangle = |\overline{N^0}\rangle - |N^0\rangle = -\sqrt{2}|N_-\rangle \tag{9.32}$$

Historically, in the K^0 system, $|N_+\rangle$ was designated by $|K_1\rangle$, and $|N_-\rangle$ by $|K_2\rangle$.

Discrete symmetries impose certain conditions on the elements of the mass and decay matrix. $\mathcal{CPT}$ invariance requires the masses and lifetimes of the particle and antiparticle to be equal, i.e. $X_{11} = X_{22}$, or $M_{11} = M_{22}$ and $\Gamma_{11} = \Gamma_{22}$, for the diagonal elements. $\mathcal{CP}$ invariance requires that $|X_{12}| = |X_{21}|$. In the following, we assume $\mathcal{CPT}$ invariance.

The eigenvalue equation for the matrix X yields

$$\Delta\mu = \Delta m - \frac{i}{2}\Delta\Gamma = 2\sqrt{X_{12}X_{21}} \ . \tag{9.33}$$

The corresponding eigenvectors of X are written

$$|N_h\rangle = p|N^0\rangle - q|\overline{N^0}\rangle \ ,$$

$$|N_l\rangle = p|N^0\rangle + q|\overline{N^0}\rangle \ , \tag{9.34}$$

or in the form of the corresponding relations

$$|N^0\rangle = \frac{1}{2p} (|N_h\rangle + |N_l\rangle) \, ,$$

$$\overline{|N^0\rangle} = \frac{-1}{2q} (|N_h\rangle - |N_l\rangle) \, . \qquad (9.35)$$

Unitarity requires $|p|^2 + |q^2| = 1$, and CP invariance would mean $p = q = 1/\sqrt{2}$. In the case of CP noninvariance, an asymmetry parameter can be defined by

$$\varepsilon = \frac{p - q}{p + q} \quad \text{or} \quad \frac{p}{q} = \frac{1 + \varepsilon}{1 - \varepsilon} \qquad (9.36)$$

and we obtain

$$\varepsilon = \frac{\frac{1}{2} \Im m \, \Gamma_{12} + i \Im m \, M_{12}}{\Delta m - \frac{i}{2} \Delta \Gamma} \, , \qquad (9.37)$$

where $\Im m \, M_{12} \gg \Im m \, \Gamma_{12}$ for the K meson system, and therefore

$$\arg \varepsilon \simeq \arctan \frac{2 \, \Delta m}{\Gamma_S} \qquad (9.38)$$

In this case the two physical states are not orthogonal, and we obtain

$$\langle N_l | N_h \rangle = \frac{2 \, \Re e \, \varepsilon}{1 + |\varepsilon|^2} \, . \qquad (9.39)$$

For the eigenstates of the time-dependent Schrödinger equation, the time evolution obeys an exponential decay law, as can be shown in the Wigner–Weisskopf approximation. Here, the time t is measured in the rest frame given by the common mass defined by the strong and electromagnetic interactions. The time evolution is given by

$$|N_h(t)\rangle = e^{-im_h t - \frac{1}{2} \Gamma_h t} \, |N_h(0)\rangle \, ,$$

$$|N_l(t)\rangle = e^{-im_l t - \frac{1}{2} \Gamma_l t} \, |N_l(0)\rangle \, . \qquad (9.40)$$

In this way, N_h decays as $\exp(-\Gamma_h t)$ and N_l as $\exp(-\Gamma_l t)$, while the phases of the two states evolve with different frequency, and this difference will show up in any interference effect between the two decaying mesons.

On the other hand, if initially a pure flavor state N^0 or $\overline{N^0}$ is produced, the decay law is not exponential but shows oscillations. If we define the complex quantities

$$
\begin{aligned}
\gamma_h &= im_h + \frac{\Gamma_h}{2}, \\
\gamma_l &= im_l + \frac{\Gamma_l}{2},
\end{aligned}
\tag{9.41}
$$

for the heavy (h) and light (l) meson states, respectively, then the amplitude for an initially pure state N^0 at time $t = 0$ is given by (9.34), and at a finite time t the two components evolve according to (9.40). At this time, the state is

$$
\psi_N = \frac{1}{2}\left(N^0\left(e^{-\gamma_h t} + e^{-\gamma_l t}\right) - \frac{q}{p}\overline{N^0}\left(e^{-\gamma_h t} - e^{-\gamma_l t}\right)\right).
\tag{9.42}
$$

The probability of finding an $\overline{N^0}$ after a time t, starting from an initially pure N^0 state is

$$
P(N^0 \to \overline{N^0}) = \frac{1}{4}\left|\frac{q}{p}\right|^2\left[e^{-\Gamma_h t} + e^{-\Gamma_l t} - 2e^{-\Gamma t}\cos(\Delta m\, t)\right].
\tag{9.43}
$$

If we express this in the unified variables $T = \Gamma t$, x and y, this reads

$$
P(N^0 \to \overline{N^0}) = \frac{1}{2}\left|\frac{q}{p}\right|^2 e^{-T}\left(\cosh yT - \cos xT\right).
\tag{9.44}
$$

Similarly, the probability of finding an N^0 in an initially pure N^0 state is

$$
P(N^0 \to N^0) = \frac{1}{2}e^{-T}\left(\cosh yT + \cos xT\right).
\tag{9.45}
$$

The difference between these two probabilities is then

$$
\begin{aligned}
P(N^0 &\to N^0) - P(N^0 \to \overline{N^0}) \\
&= \frac{1}{2}e^{-T}\left(\cosh yT + \cos xT\right) - \frac{1}{2}e^{-T}\left(\left|\frac{q}{p}\right|^2\cosh yT - \left|\frac{q}{p}\right|^2\cos xT\right) \\
&= \frac{1}{2}e^{-T}\left[\cosh yT\left(1 - \left|\frac{q}{p}\right|^2\right) + \cos xT\left(1 + \left|\frac{q}{p}\right|^2\right)\right].
\end{aligned}
\tag{9.46}
$$

However, the two states N_h and N_l are not orthogonal, but their overlap δ is

$$\delta = \langle N_h | N_l \rangle = |p|^2 - |q|^2 = \frac{1 - \left|\frac{q}{p}\right|^2}{1 + \left|\frac{q}{p}\right|^2} = \frac{2\,\Re e\,\varepsilon}{1 + |\varepsilon|^2} \,. \tag{9.47}$$

From this result, we obtain

$$P(\mathrm{N}^0 \to \mathrm{N}^0) - P(\mathrm{N}^0 \to \overline{\mathrm{N}^0}) = \frac{1}{2}\mathrm{e}^{-T}\left(1 + \left|\frac{q}{p}\right|^2\right)(\delta \cosh yT + \cos xT) \,. \tag{9.48}$$

Similarly,

$$P(\mathrm{N}^0 \to \mathrm{N}^0) + P(\mathrm{N}^0 \to \overline{\mathrm{N}^0}) = \frac{1}{2}\mathrm{e}^{-T}\left(1 + \left|\frac{q}{p}\right|^2\right)(\cosh yT + \delta \cos xT) \,, \tag{9.49}$$

and the flavor asymmetry at time T in an initially pure flavor state becomes

$$A(T) = \frac{P(\mathrm{N}^0 \to \mathrm{N}^0) - P(\mathrm{N}^0 \to \overline{\mathrm{N}^0})}{P(\mathrm{N}^0 \to \mathrm{N}^0) + P(\mathrm{N}^0 \to \overline{\mathrm{N}^0})} = \frac{\cos xT + \delta \cosh yT}{\cosh yT + \delta \cos xT} \,. \tag{9.50}$$

This function behaves very differently for the neutral K, D, and B meson systems.

9.3.2.2 Decay Asymmetries and $\mathcal{CP}$

We define the decay amplitudes of neutral mesons to a final state f as

$$A_f = \langle f | T | \mathrm{N}^0 \rangle \,,$$
$$\overline{A_f} = \langle f | T | \overline{\mathrm{N}^0} \rangle \,. \tag{9.51}$$

The decay amplitudes of the mass eigenstates are then

$$A_f^h = pA_f - q\overline{A_f} \,,$$
$$A_f^l = pA_f + q\overline{A_f} \,, \tag{9.52}$$

and we define the complex quantity

$$\lambda_f = \frac{q\overline{A_f}}{pA_f} \,. \tag{9.53}$$

The moduli for the decay of N^0 to f and the decay of $\overline{N^0}$ to its CP conjugate state $\overline{f}$ are equal if CP is conserved, and vice versa:

$$|A_f| = |\overline{A_{\overline{f}}}|,$$

$$|A_{\overline{f}}| = |\overline{A_f}|. \qquad (9.54)$$

(If f is a CP eigenstate, this is simplified to $|A_f| = |\overline{A_f}|$).
Now CP violation may occur in three different ways:

1. CP violation in the mixing, if $|q/p| \neq 1$, called "indirect CP violation".
2. CP violation in the decay amplitudes, when (9.54) is not valid, called "direct CP violation".
3. CP violation in the interference, when the phase of the expression

$$A_f \overline{A_f^*} A_{\overline{f}} \overline{A_{\overline{f}}^*} \, p^2/q^2 \qquad (9.55)$$

is not zero.

These three types of CP violation are characterized by the following details:

1. *CP violation in the mixing.* This type of CP violation will show up if the mass eigenstates of a neutral meson system are different from the CP eigenstates, i.e. if $|q/p| \neq 1$ (or $\varepsilon \neq 0$) and if there is a relative phase between M_{12} and Γ_{12}. For the neutral kaon system, this is evident from the existence of the decay $K_L \to \pi^+\pi^-$, where $|\varepsilon| \sim 2 \times 10^{-3}$, and from the charge asymmetry in semileptonic decays δ_L which is proportional to $2 \, \mathfrak{Re}\, \varepsilon$. For the neutral B system, this effect could be seen also in the charge asymmetry of semileptonic decays

$$a_{\rm SL} = \frac{\Gamma(\overline{B^0}(t) \to l^+\nu X) - \Gamma(B^0(t) \to l^-\bar{\nu}X)}{\Gamma(\overline{B^0}(t) \to l^+\nu X) + \Gamma(B^0(t) \to l^-\bar{\nu}X)}. \qquad (9.56)$$

This asymmetry is expected to be small in the Standard Model of order $\Delta\Gamma/\Delta m$ or $\mathcal{O}(10^{-3})$.

2. *CP violation in the decay amplitude.* This effect appears if the decay amplitude A_f of the neutral meson N^0 to a final state f is different from the amplitude $\overline{A_{\overline{f}}}$ of the antiparticle $\overline{N^0}$ to the charge-conjugate state $\overline{f}$, i.e. $|\overline{A_{\overline{f}}}/A_f| \neq 1$. In the neutral-kaon decay to two π mesons, this is realized by the interference of two decay amplitudes, one with $\Delta I = 1/2$ to an isospin $I = 0$ state, and another with $\Delta I = 3/2$ to an isospin $I = 2$ state. The amplitude of direct CP violation is denoted by ε' and proceeds through penguin diagram processes. The observed magnitude of this amplitude is $|\varepsilon'| \sim 4 \times 10^{-6}$. In the neutral B meson system, the required two decay amplitudes with different weak phases and different strong phases could be a penguin diagram and a tree diagram, e.g. for the decay to the final state $K^-\pi^+$ or $K^+\pi^-$. The $b \to s$ penguin diagram has a dominant

contribution from a top quark loop, with a weak coupling $V_{tb}^* V_{ts}$ and an isospin-1/2 (Kπ) state. The tree diagram for $b \to u + (\bar{u}s)$ has a coupling $V_{ub}^* V_{us}$ and leads to $I = 1/2$ or $3/2$ states. The observed quantity is the decay asymmetry:

$$a = \frac{N(\overline{B^0} \to K^-\pi^+) - N(B^0 \to K^+\pi^-)}{N(\overline{B^0} \to K^-\pi^+) + N(B^0 \to K^+\pi^-)}. \tag{9.57}$$

Asymmetries in the order of 10% and 025% have been observed for the B^0 and the B_s^0 meson respectively.

3. *CP violation in the interference.* Here the time dependence of the decay of an initially pure flavor state to a final state f is different for an initial particle or antiparticle. The final state can be a CP eigenstate such as $\pi^+\pi^-$ ($CP = +1$) or $J/\Psi K_S$ ($CP = -1$). In the kaon system the observed effect is $\Im m\,\varepsilon \sim 1.6 \times 10^{-3}$, while in the B^0 system it is a very large asymmetry of order $\mathcal{O}(1)$.

9.4 Models of CP Violation

After the discovery of CP violation in K decay, a host of theoretical models was proposed to allocate this phenomenon to known interactions. Assuming CPT invariance of all interactions, the observed CP-violating effects in K decay imply also T violation (the experimental data of Sect. 9.5 are even sufficient to prove T violation without CPT invariance). In general, with CPT invariance, there are four combinations of violations possible:

(a) T-conserving, C-violating and P-violating;
(b) T-violating, C-conserving and P-violating;
(c) T-violating, C-violating and P-conserving;
(d) T-violating, C-violating and P-violating.

Parity conservation in strong and electromagnetic interactions has been tested, for example by looking for a circular polarization in γ rays from nuclear transitions. The presence of a wrong-parity admixture in one of the nuclear states involved will cause a small amplitude for a γ transition with abnormal multipolarity that can interfere with the dominant amplitude and cause such a circular polarization. In the experiments of Lobashov et al. [46], polarizations of the order of 10^{-5} have been measured. These are consistent with being due to the two-nucleon force np $\to$ pn induced by the weak interaction (see [47] for a review).

From many experiments of a similar nature, one can infer that the strong and electromagnetic interactions are not of type (a), (b) or (d). Therefore, if the source of the CP-violating phenomena is located in the strong or the electromagnetic interaction, there must be a part of one of those interactions that belongs to class (c), i.e. C- and T-violating, but P-conserving.

The proposed models can be grouped into the following four categories:

1. Millistrong CP violation models [29–31] postulate the existence of C- and T-violating terms of order 10^{-3} in the strong interaction. The process $K_L \to \pi^+\pi^-$ is supposed to occur by the interference of two amplitudes: first, the K_L decays via the normal CP-conserving weak interaction, with $\Delta S = 1$, into an intermediate state X, and then this state decays into $\pi^+\pi^-$ by a T-violating strong interaction. The amplitude of the process is of order $G_F a$, where G_F is the Fermi coupling constant and a is the coupling of this CP-violating strong interaction. From the experimental value of $|\eta_{+-}|$, one can conclude that $a \approx 10^{-3}$.

2. Electromagnetic CP violation models [32–35] require large parts of the electromagnetic interaction of hadrons to be C- and T-violating, but P-conserving. A two-step process $K_L \to X \to 2\pi$ could then occur through the interference of a weak and an electromagnetic CP-violating amplitude. The product of G_F with the fine structure constant α is not too far from $G_F \times 10^{-3}$, as required by the magnitude of $|\eta_{+-}|$.

3. Milliweak models assume that a part, of the order of 10^{-3}, of the weak interaction is CP-violating and is responsible for the observed effects. The decay $K_L \to 2\pi$ is then a one-step process, hence the name "direct CP violation", and CP or T violations of the order of 10^{-3} should show up in other weak processes [36–44]. In these models, based on two doublets of quarks, CP violation is introduced in different ways. In one example [44], CP violation is due to the Higgs couplings, with flavor-changing neutral currents allowed; in another one [42], it is due to right-handed weak currents. A bold alternative was considered in 1973 by Kobayashi and Maskawa [48]: they saw that if there are three doublets of quarks, there is a possibility of CP violation in the 3×3 weak quark mixing. Today, with six quarks observed, this seems the most natural model as discussed below.

4. The superweak model [45] postulates a new $\Delta S = 2$ CP-violating interaction that has a coupling (coupling constant g) smaller than the second-order weak interaction. This interaction could induce a transition $K_L \to K_S$, with a subsequent decay $K_S \to 2\pi$. More precisely, this interaction would cause a first-order transition matrix element

$$\mathcal{M}_{SW} = \langle \overline{K} | \mathcal{H}_{SW} | K \rangle \sim g G_F . \tag{9.58}$$

The mass difference itself is related to the second-order weak matrix element

$$\mathcal{M}_{\overline{K}K} = \sum_n \frac{\langle \overline{K} | \mathcal{H}_w | n \rangle \langle n | \mathcal{H}_w | K \rangle}{E_K - E_n + i\varepsilon} . \tag{9.59}$$

where n is an intermediate state with energy E_n and $\mathcal{H}_w$ is the weak Hamiltonian. In order that the CP-violating amplitude for $K_L \to 2\pi$ relative to the CP-conserving amplitude should be of the observed magnitude, the ratio $\mathcal{M}_{SW}/\mathcal{M}_{\overline{K}K}$ must be of the order of 10^{-3}. Since $\mathcal{M}_{SW} \approx g G_F$ and

$\mathcal{M}_{\overline{K}K} \sim G_F^2 m_p^2$ where the proton mass m_p is used as a cutoff in the integration, this yields $g \sim G_F m_p^2 \approx 10^{-8}$.

This superweak interaction can be detected only in the K_L–K_S and B^0–$\overline{B^0}$ systems because these are the only known pairs of states with such a small difference in energy that they are sensitive to forces weaker than the second-order weak interaction. The clear prediction of this model is that there is no direct CP violation in the decay.

For models other than the superweak one, violations of CP or T should manifest themselves in other reactions of particles or nuclei. One observable is the electric dipole moment (EDM) of the neutron. Most milliweak models predict this EDM to be of order 10^{-23} ecm to 10^{-24} ecm, while the superweak model predicts 10^{-29} ecm. The present experimental upper limit is 0.63×10^{-25} ecm.

One of the milliweak models mentioned above under item 3 of the enumeration which is rather clear in its predictions should be noted: this is the idea of Kobayashi and Maskawa (KM) dating from 1973 [48]. At the time of the discovery of CP violation, only three quarks were known, and there was no possibility of explaining CP violation as a genuine phenomenon of weak interactions with left-handed charged currents and an absence of flavor-changing neutral currents. This situation remained unchanged with the introduction of a fourth quark because the 2×2 unitary weak quark mixing matrix has only one free parameter, the Cabibbo angle, and no nontrivial complex phase. However, as remarked by Kobayashi and Maskawa, the picture changes if six quarks are present. In this case the 3×3 unitary mixing matrix V_{ik} naturally contains a phase δ, in addition to three mixing angles (Sect. 9.7). It is then possible to construct CP-violating weak amplitudes from "box diagrams" of the form shown in Fig. 9.5.

In the K^0–$\overline{K^0}$ system, this amplitude is proportional to the product of the four weak coupling constants $G_F^2 V_{ts} V_{ts}^* V_{td} V_{td}^*$. If there is a nontrivial phase δ in the unitary mixing matrix, then the product is a complex number, with the imaginary part depending on the phase δ. This leads to time-reversal (T) violation and to CP violation. The CP-violating mixing parameter for the kaon system, ε, is given by

$$\varepsilon = \frac{G_F^2 \, f_K^2 \, m_K \, m_W^2}{G\sqrt{2}\,\pi^2 \Delta m} B_K \, \Im m(V_{td} \, V_{ts}^*) F(m_t^2, m_c^2) \, . \tag{9.60}$$

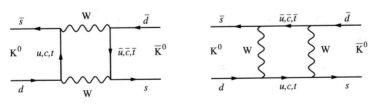

Fig. 9.5 Box diagram for K^0–$\overline{K^0}$ mixing connected with the CP-violating parameter ε

Here G_F is the Fermi constant, f_K the kaon decay constant, B_K the kaon bag factor (0.80 ± 0.15), and $F(M_t^2, m_c^2)$ the loop function due to interference of the top and charm graphs, given by

$$F(m_t^2, m_c^2) = \Re e \left(V_{cs}^* V_{cd} \right) \left[\eta_1 S_0 \left(m_c^2 \right) - \eta_3 S_0 \left(m_c^2, m_t^2 \right) \right] - \Re e \left(V_{ts}^* V_{td} \right) \eta_2 S_0 \left(m_t^2 \right),$$

$$(9.61)$$

where $S_0(m_t^2)$ is a kinematical factor, and η_1, η_2, and η_3 are QCD correction factors.

For the B_d^0–$\overline{B}_d^0$ mixing, a similar box graph applies, with the s quark replaced by a b quark. Here the amplitude is proportional to $G_F^2 V_{tb} V_{tb}^* V_{td} V_{td}^*$. Analogous diagrams can be calculated for B_s^0 $(b\overline{s})$ mixing and for D^0 $(c\overline{d})$ mixing.

All CP-violating amplitudes in the KM model are proportional to the following product of the three mixing angles and the phase δ (Sect. 9.7),

$$J = |V_{us} V_{ub} V_{cb} \sin \delta| \qquad (9.62)$$

A necessary consequence of this model of CP violation is the non-equality of the relative decay rates for $K_L \to \pi^+\pi^-$ and $K_L \to \pi^0\pi^0$. This "direct CP violation" is due to "penguin diagrams" of the form given in Fig. 9.6 for kaon decays. The amplitude for this direct CP violation is denoted by ε'. In kaon decays, it will show up in the interference of two decay amplitudes, with the final two-pion state having isospin 0 or 2 (A_0 and A_2 in (9.91)). With six quarks, the weak quark mixing through flavor change can carry a nontrivial phase δ in the mixing matrix, and therefore can induce a CP-violating difference between weak decay amplitudes, such that $|A_f| \neq |\overline{A}_f|$. This model gives an explicit origin of direct CP violation with a predictable size. In the kaon system, these asymmetries are very small because of the small value of $|J| \sim 3 \times 10^{-5}$, the suppression of $\Delta I = 3/2$ currents, and the partial cancellation of two penguin graphs, called Q_6 and Q_8, shown in Fig. 9.6. However, in the B^0 system, the asymmetries of the decay rates to CP eigenstates can be very large. Examples are the decays $B^0 \to J/\Psi K_S$ and $B^0 \to \pi^+\pi^-$.

The main models which could be tested experimentally were the KM model and the superweak model, and the decisive question was the existence or non-existence of direct CP violation. For the kaon system in the superweak model $\varepsilon' = 0$, and

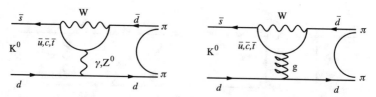

Fig. 9.6 Penguin diagrams for $K^0 \to 2\pi$ decay with direct CP violation (amplitude ε'). The graphs correspond to the Wilson operators Q_8 and Q_6 and give rise to amplitudes with opposite signs

the decay rates of K_L to $\pi^+\pi^-$ and to $\pi^0\pi^0$ are equal. The predicted value of ε' within the KM model can be estimated if one infers the magnitude of the mixing angles from other experiments and if the hadronic matrix elements for box graphs and penguin graphs are calculated. Typical values of $|\varepsilon'/\varepsilon|$ are in the range $+(0.05-2.0)\times 10^{-3}$ for three generations of quarks. A measurement of this quantity to this level of precision therefore becomes the *experimentum crucis* for our understanding of $\mathcal{CP}$ violation. If ε' is orthogonal to ε, then a measurement of the phases of η_{+-} and of η_{00} (see Eqs. (9.79) and (9.80), respectively) can help to detect a finite value of $\Im m(\varepsilon'/\varepsilon)$. If, however, the phase of ε' is close to that of ε, and since $|\varepsilon'/\varepsilon| \ll 1$ to a good approximation, we obtain

$$\frac{\varepsilon'}{\varepsilon} \simeq \Re e\left(\frac{\varepsilon'}{\varepsilon}\right) = \frac{1}{6}\left(1 - \frac{|\eta_{00}|^2}{|\eta_{+-}|^2}\right). \tag{9.63}$$

Various methods have been used to calculate the value of $\Re e(\varepsilon'/\varepsilon)$. Owing to the difficulties in calculating hadronic matrix elements in the penguin diagrams, which involve long-distance effects, the task turns out to be very difficult. In particular, the electroweak penguin diagram (corresponding to the Wilson operator Q_8) and the QCD penguin diagram (operatir Q_8) yield contributions of opposite sign and lead to a partial cancellation in the result ε'/ε. At the time of this review, great progress has been made by lattice calculations which now can be compared to the two groups of earlier analytic calculations.

The lattice calculation of the RBC-UKQCD-collaboration [49, 50] imply the following values of the penguin graph matrix elements [51, 52]:

$$B_6 = 0.57 \pm 0.19 \quad \text{and} \quad B_8 = 0.76 \pm 0.05. \tag{9.64}$$

From this the lattice method obtains:

$$\varepsilon'/\varepsilon = (1.9 \pm 4.5) \times 10^{-4} \tag{9.65}$$

On the side of analytic calculations, one method was based on the limit of large N, where N is the numbers of colours in QCD [53] ("dual QCD" or "DQCD"). At large N, QCD becomes a theory of three mesons. Here one gets $B_6 = B_8 = 1$ at the pion mass scale. Considering the meson evolution of these matrix elements to the mass scale of 1 GeV [54], Buras and Gerard obtain a suppression of both B_6 and B_8. At the scale of the charm quark mass, they find

$$B_6 \leq 0.6 \quad \text{and} \quad B_8 = 0.8 \pm 0.1, \tag{9.66}$$

in agreement with the lattice results. using the lattice values for B_8 and the relation $B_6 \leq B_8$, they find an upper bound;

$$\varepsilon'/\varepsilon \leq (6.0 \pm 2.4) \times 10^{-4} \tag{9.67}$$

However, an alternative analytic calculation [55] based on chiral perturbation theory which emphasizes final state interactions and neglects the meson evolution, obtains a much larger value

$$\varepsilon'/\varepsilon = (15 \pm 7) \times 10^{-4}. \tag{9.68}$$

Further improvement of the lattice calculations will hopefully clear-up this important discrepancy of the standard model prediction for ε'/ε.

9.5 The Neutral K Meson System

9.5.1 Mass Eigenstates and CP Eigenstates

The eigenstates of strangeness are K^0 ($\mathbb{S} = +1$) and $\overline{K^0}$ ($\mathbb{S} = -1$), the CP eigenstates are K_1 (with CP eigenvalue $+1$) and K_2 (CP eigenvalue -1), and the mass eigenstates are

$$|K_S\rangle = p\,|K^0\rangle + q\,|\overline{K^0}\rangle \quad \text{(Short-lived)}, \tag{9.69}$$

$$|K_L\rangle = p\,|K^0\rangle - q\,|\overline{K^0}\rangle \quad \text{(Long-lived)}, \tag{9.70}$$

where, from experiment, K_L is the heavier state (h). The lifetimes of the two eigenstates are very different. While the short-lived particle (K_S) has a mean lifetime of $(0.8959 \pm 0.0004) \times 10^{-10}$ s, the long-lived particle K_L has a lifetime of $(5.17 \pm 0.04) \times 10^{-8}$ s, i.e. 600 times larger. This is due to the fact that the dominant CP-conserving decays are $K_S \rightarrow 2\pi$ and $K_L \rightarrow 3\pi, \pi e\nu, \pi\mu\nu$, with a much smaller phase space for the three-body decays. Using the parameter $\varepsilon = (p - q)/(p + q)$, (9.69) and (9.70) can also be written

$$|K_S\rangle = \frac{1}{\sqrt{1 + |\varepsilon|^2}}(|K_1\rangle + \varepsilon|K_2\rangle), \tag{9.71}$$

$$|K_L\rangle = \frac{1}{\sqrt{1 + |\varepsilon|^2}}(|K_2\rangle + \varepsilon|K_1\rangle). \tag{9.72}$$

The long-lived state is therefore mainly a state with CP eigenvalue -1, with a small (2×10^{-3}) admixture of a CP $+1$ state K_1. The two mass eigenstates are not orthogonal if CP is violated, because $\langle K_S|K_L\rangle = 2\,\Re e\,\varepsilon$.

If the validity of CPT symmetry is not assumed, the expressions are generalized to

$$|K_S\rangle = \frac{1}{\sqrt{1 + |\varepsilon + \delta|^2}}(|K_1\rangle + (\varepsilon + \delta)|K_2\rangle), \tag{9.73}$$

$$|K_L\rangle = \frac{1}{\sqrt{1 + |\varepsilon - \delta|^2}} \left(|K_2\rangle + (\varepsilon - \delta)|K_1\rangle \right) . \tag{9.74}$$

with a $\mathcal{CPT}$ violating parameter δ.

9.5.2   Isospin Decomposition

In $K_{S,L} \rightarrow 2\pi$ decays, the angular momentum of the pions vanishes. The spatial part of the wave function is therefore symmetric, and since pions are bosons, the isospin wave function must be symmetric too. The two symmetric combinations of two $I = 1$ states have $I = 0$ and $I = 2$, and the four transition amplitudes that exist are

$$\langle 0|T|K_S\rangle , \quad \langle 2|T|K_S\rangle , \quad \langle 0|T|K_L\rangle , \quad \langle 2|T|K_L\rangle . \tag{9.75}$$

These can be reduced to three complex numbers by normalizing to the amplitude $\langle 0|T|K_S\rangle$:

$$\varepsilon_0 = \frac{\langle 0|T|K_L\rangle}{\langle 0|T|K_S\rangle} , \tag{9.76}$$

$$\varepsilon_2 = \frac{1}{\sqrt{2}} \frac{\langle 2|T|K_L\rangle}{\langle 0|T|K_S\rangle} , \tag{9.77}$$

$$\omega = \frac{\langle 2|T|K_S\rangle}{\langle 0|T|K_S\rangle} . \tag{9.78}$$

The experimentally observable quantities are

$$\eta_{+-} = \frac{\langle \pi^+\pi^-|T|K_L\rangle}{\langle \pi^+\pi^-|T|K_S\rangle} , \tag{9.79}$$

$$\eta_{00} = \frac{\langle \pi^0\pi^0|T|K_L\rangle}{\langle \pi^0\pi^0|T|K_S\rangle} , \tag{9.80}$$

$$\delta_L = \frac{\Gamma(K_L \rightarrow \pi^- l^+ \nu) - \Gamma(K_L \rightarrow \pi^+ l^- \bar{\nu})}{\Gamma(K_L \rightarrow \pi^- l^+ \nu) + \Gamma(K_L \rightarrow \pi^+ l^- \bar{\nu})} . \tag{9.81}$$

Relating the isospin states to the physical 2π states

$$\langle 0| = \frac{1}{\sqrt{3}} \langle \pi^-\pi^+| - \frac{1}{\sqrt{3}} \langle \pi^0\pi^0| + \frac{1}{\sqrt{3}} \langle \pi^+\pi^-| , \tag{9.82}$$

$$\langle 2| = \frac{1}{\sqrt{6}} \langle \pi^-\pi^+| + \sqrt{\frac{2}{3}} \langle \pi^0\pi^0| + \frac{1}{\sqrt{6}} \langle \pi^+\pi^-| . \tag{9.83}$$

we obtain

$$\eta_{+-} = \frac{\varepsilon_0 + \varepsilon_2}{1 + (1/\sqrt{2})\omega}, \tag{9.84}$$

$$\eta_{00} = \frac{\varepsilon_0 - 2\varepsilon_2}{1 - \sqrt{2}\omega}. \tag{9.85}$$

Because of the validity of the $\Delta I = 1/2$ rule for $\mathcal{CP}$-conserving weak nonleptonic decays, $\omega \ll 1$ and therefore can be neglected.

A suitable choice for the phase of the $K^0 \to 2\pi(I = 0)$ amplitude is obtained by choosing this amplitude to be real except for final-state interactions between two pions, leading to a phase shift δ_0:

$$\langle 0|T|K^0 \rangle = e^{i\delta_0} A_0 \quad \text{and } A_0 \text{ real.} \tag{9.86}$$

Similarly,

$$\langle 2|T|K^0 \rangle = e^{i\delta_2} A_2 . \tag{9.87}$$

With these choices, we obtain

$$\varepsilon_0 = \varepsilon , \tag{9.88}$$

$$\varepsilon_2 = \frac{i}{\sqrt{2}} e^{i(\delta_2 - \delta_0)} \frac{\Im m \, A_2}{A_0} = \varepsilon' . \tag{9.89}$$

Therefore, representing ε and ε' in the complex plane, we obtain the triangle relations

$$\eta_{+-} = \varepsilon + \varepsilon' , \qquad \eta_{00} = \varepsilon - 2\varepsilon' . \tag{9.90}$$

In this way, η_{+-}, η_{00} and $3\epsilon'$ form a triangle in the complex plane, the Wu–Yang triangle. The $\mathcal{CP}$-violating decay amplitude ε' is due to interference of $\Delta I = 1/2$ (A_0) and $\Delta I = 3/2$ (A_2) amplitudes:

$$\varepsilon' = \frac{i \Im m \, A_2}{2A_0} e^{i(\delta_2 - \delta_0)} . \tag{9.91}$$

Its phase is given by the $\pi\pi$ phase shifts in the $I = 0$ and $I = 2$ states, δ_0 and δ_2, respectively, assuming $\mathcal{CPT}$ invariance:

$$\arg(\varepsilon') = (\delta_2 - \delta_0) + \frac{\pi}{2} . \tag{9.92}$$

The $\pi\pi$ phase shifts have been measured precisely in pion-scattering experiments. The results obtained are $\delta_2 = (-7.2 \pm 1.3)°$ [151] and $\delta_0 = (39 \pm 5)°$ [152]. Using

dispersion relation calculations or chiral perturbation theory, these results can be
used to extract arg(ε'). The results are $(42.3 \pm 1.5)°$ [149] and $(46.0 \pm 3.6)°$ [150].

The decomposition of the observable decay amplitude into ε and ε' corresponds
to a separation of the CP-violating effects due to the mass and decay matrices
(represented by ε), which are seen also in the impurity of the K_L and K_S states,
from CP violation in the transition matrix element (represented by ε').

The phase of ε is given by (9.37) and (9.38):

$$\arg \varepsilon = \Phi_D + \arctan \left(\frac{2\Delta m}{\Gamma_S} \right) , \tag{9.93}$$

where $\Delta m = m_L - m_S$ and

$$\Phi_D = -\arctan \left(\frac{\Im m\ \Gamma_{12}}{2\,\Im m\ M_{12}} \right) . \tag{9.94}$$

If there is no strong CP violation in the channels $K \rightarrow 2\pi$ ($I = 2$), $K \rightarrow \pi l \nu$,
and $K \rightarrow 3\pi$, Φ_D is very small. This can be deduced from the Bell–Steinberger
unitarity relation [156]. If the final states of the K_L and K_S decays are designated by
$|F\rangle$, then

$$\Gamma_S = \sum_F |\langle F|T|K_S\rangle|^2 , \tag{9.95}$$

$$\Gamma_L = \sum_F |\langle F|T|K_L\rangle|^2 . \tag{9.96}$$

Unitarity leads to the relation

$$i\left(M_S - M_L^*\right) \langle K_S|K_L\rangle = \sum_F \langle F|T|K_L\rangle^* \langle F|T|K_S\rangle . \tag{9.97}$$

If CPT invariance is not assumed, the left side of this unitarity relation includes a
contribution of the CPT violating parameter δ. It then has the form [156]:

$$(1 + i \tan \Phi_{SW})[\Re e\ \varepsilon - i\Im m\ \delta] \tag{9.98}$$

The mass matrix elements are then (with CPT invariance assumed)

$$X_{11} = X_{22} = \frac{M_S + M_L}{2} \tag{9.99}$$

$$X_{12} = \frac{(M_S - M_L)(1 + \varepsilon)}{2(1 - \varepsilon)} \tag{9.100}$$

$$X_{21} = \frac{(M_S - M_L)(1 - \varepsilon)}{2(1 + \varepsilon)} . \tag{9.101}$$

Ignoring all final states but 2π or assuming $\Im m\, \Gamma_{12} = 0$, we obtain the following from the unitarity relation or from (9.37):

$$\arg \varepsilon = \arctan\left(\frac{2\Delta m}{\Gamma_{S}}\right) = \Phi_{sw} \qquad (9.102)$$

where Φ_{sw} designates the phase in the superweak model

When we add other final states of CP-violating decays, the phase is shifted by Φ_{D}, and an upper limit can be obtained from the unitarity relation:

$$\Phi_{D} \leq \frac{0.75}{\Gamma_{S}|\eta_{+-}|} \sum_{F} \sqrt{\Gamma_{F,CPV} \cdot \Gamma_{F,CPC}} \qquad (9.103)$$

where the sum runs over all states $F \neq 2\pi$ and the root is taken of the product of the CP-violating (CPV) and CP-conserving (CPC) decay rates.

Present limits on CP-violating processes in these decays show that contributions from semileptonic decays are negligible. Using the limits on $(\Delta Q = \Delta \mathbb{S})$-violating amplitudes, we obtain

$$|\Phi_{D}(K_{e3})| < 0.07° , \qquad |\Phi_{D}(K_{\mu 3})| < 0.05° . \qquad (9.104)$$

In the same way, the measurement of the CP-violating part of the $K_S \rightarrow \pi^+\pi^-\pi^0$ decay [153–155],

$$\frac{\Gamma(K_S \rightarrow \pi^+\pi^-\pi^0)}{\Gamma_S} = 3.5^{+1.1}_{-0.9} \times 10^{-7} , \qquad (9.105)$$

allows us to set the limit

$$|\Phi_{D}(\pi^+\pi^-\pi^0)| < 0.05° . \qquad (9.106)$$

Similarly, from the limit [158, 159] $\Gamma(K_S \rightarrow 3\pi^0)/\Gamma_S < 2.8 \times 10^{-8}$ it follows that

$$|\Phi_{D}(3\pi^0)| < 0.02° . \qquad (9.107)$$

New, more sensitive experiments will improve this limit. If we use the experimental values of Δm and Γ_S from Sect. 9.5.5, then $\arg \varepsilon = (43.4 \pm 0.1 \pm 0.17)°$, where the first error comes from the uncertainties of Δm and Γ_S and the second error from the uncertainty of Φ_{D}.

Another independent observable is the charge asymmetry

$$\delta_L = \frac{1 - |x|^2}{|1 - x|^2} 2 \Re e\, \varepsilon , \qquad (9.108)$$

where $x = g/f$ is the ratio of the $\Delta Q = -\Delta S$ to the $\Delta Q = \Delta S$ amplitude (Sects. 9.5.3.2 and 9.5.5.6).

9.5.3 Interference Between Decay Amplitudes of K_L and K_S

An arbitrary coherent mixture of K_L and K_S states will show interference phenomena when decaying into 2π and in other common decay channels. According to Sect. 9.3.2.1 the eigentime development of K_L is

$$|K_L\rangle \rightarrow |K_L\rangle\, e^{-iM_L\tau} , \tag{9.109}$$

where $M_L = m_L - (i/2)\,\Gamma_L$, and correspondingly for K_S. An arbitrary mixture

$$|\psi(0)\rangle = a_S|K_S\rangle + a_L|K_L\rangle \tag{9.110}$$

will develop into

$$|\psi(\tau)\rangle = a_S\, e^{-iM_S\tau}\, |K_S\rangle + a_L\, e^{-iM_L\tau}\, |K_L\rangle . \tag{9.111}$$

We call the ratio $a_S/a_L = V$.

9.5.3.1 2π Decay

The 2π decay amplitude is therefore

$$\begin{aligned}
\langle 2\pi|T|\psi(\tau)\rangle &= a_S\, e^{-iM_S\tau}\, \langle 2\pi|T|K_S\rangle + a_L\, e^{-iM_L\tau}\, \langle 2\pi|T|K_L\rangle \\
&= \langle 2\pi|T|K_S\rangle a_S\, e^{-iM_S\tau} + a_L\eta\, e^{-iM_L\tau} ,
\end{aligned} \tag{9.112}$$

where $\eta = \eta_{+-}$ for $\pi^+\pi^-$ decay and $\eta = \eta_{00}$ for $\pi^0\pi^0$ decay. The observed decay rate is proportional to

$$R(\tau) = |a_S|^2\, e^{-\Gamma_S\tau} + |a_L\,\eta|^2\, e^{-\Gamma_L\tau} + 2|a_S||a_L||\eta|\, e^{-(\Gamma_L+\Gamma_S)(\tau/2)} \cos(\Delta m\,\tau + \Phi) . \tag{9.113}$$

where $\Phi = \arg(a_S) - \arg(\eta a_L)$. We obtain for various initial conditions of the mixture:

1. For an initially pure K^0 state ($a_S = 1 = a_L$),

$$R_1(\tau) = e^{-\Gamma_S\tau} + |\eta|^2\, e^{-\Gamma_L\tau} + 2|\eta|\, e^{-(\Gamma_L+\Gamma_S)(\tau/2)} \cos(\Delta m\,\tau - \arg\eta) . \tag{9.114}$$

2. For an initially pure $\overline{K^0}$ state, the interference term changes sign.
3. For an incoherent mixture of K^0 (intensity N_K) and $\overline{K^0}$ (intensity $N_{\overline{K}}$), the interference term is multiplied by the "dilution factor"

$$\frac{N_K - N_{\overline{K}}}{N_K + N_{\overline{K}}}. \tag{9.115}$$

Measurement of the interference term under these conditions is called the *vacuum interference method*.
4. For the coherent mixture behind a regenerator, $a_S = \varrho$, $a_L = 1$, and we obtain

$$R_2(\tau) = |\varrho|^2\,e^{-\Gamma_S \tau} + |\eta|^2\,e^{-\Gamma_L \tau} + 2|\varrho||\eta|\,e^{-(\Gamma_L + \Gamma_S)(\tau/2)} \cos\left(\Delta m\,\tau + \Phi_\varrho - \arg\eta\right). \tag{9.116}$$

9.5.3.2 Semileptonic Decays

Interference phenomena and $\mathcal{CP}$ violation can also be observed in the decay of a coherent mixture of K^0 and $\overline{K^0}$ mesons into semileptonic final states. In particular the time-dependent charge asymmetry $\delta(\tau) = (N^+ - N^-)/(N^+ + N^-)$ shows an oscillatory behavior, where N^+ denotes decays into $\pi^+ e^- \nu$ final states and N^- into $\pi^- e^+ \nu$. Assuming $\mathcal{CPT}$ invariance, we obtain

$$\delta(\tau) = 2\frac{1 - |x|^2}{|1 - x|^2}\left[\Re e\,\varepsilon + |V|\,e^{-(1/2)(\Gamma_S - \Gamma_L)\tau} \cos\left(\Delta m\,\tau + \Phi_V\right)\right]. \tag{9.117}$$

where x is the ratio of amplitudes with $\Delta S = -\Delta Q$ and $\Delta S = \Delta Q$. x is consistent with zero, in agreement with the $\Delta S = \Delta Q$ rule.

For an initially pure K_L beam ($R = 0$), the asymmetry is independent of the decay time:

$$\delta_L = 2\,\Re e\,\varepsilon\,\frac{1 - |x|^2}{|1 - x|^2}. \tag{9.118}$$

For an initial incoherent mixture of K^0 (N_K) and $\overline{K^0}$ ($N_{\overline{K}}$) the quantity $|R|$ has to be replaced by $(N_K - N_{\overline{K}})/(N_K + N_{\overline{K}})$, i.e. by the same dilution factor as for 2π interference in a short-lived beam.

For the coherent mixture created by a regenerator, R is given by the regeneration amplitude ϱ, and Φ_R by the regeneration phase Φ_ϱ.

9.5.4 Detection of K^0 Decays

The main decay modes originating from K^0's in a neutral beam and their respective branching ratios are [163]

$$
\begin{aligned}
K_L &\to \pi^{\pm} e^{\mp} \nu && (40.55 \pm 0.11)\% && K_{e3}\,; \\
K_L &\to \pi^{\pm} \mu^{\mp} \nu && (27.07 \pm 0.07)\% && K_{\mu3}\,; \\
K_L &\to \pi^+ \pi^- \pi^0 && (12.54 \pm 0.05)\% && K_{\pi3}\,; \\
K_L &\to \pi^0 \pi^0 \pi^0 && (19.52 \pm 0.12)\% && K_{\pi3}\,; \\
K_S &\to \pi^+ \pi^- && (69.20 \pm 0.05)\% && K_{\pi2}\,; \\
K_S &\to \pi^0 \pi^0 && (30.69 \pm 0.05)\% && K_{\pi2}\,.
\end{aligned}
$$

The experimental problem is to detect the rare CP-violating decay modes $K_L \to \pi^+\pi^-$ and $K_L \to \pi^0\pi^0$, with branching ratios of order 10^{-3}, in this overwhelming background of other decays, and to measure their decay rate, and, by interference, their phase relation to CP-conserving decay amplitudes. In addition, the CP impurity in the K_L state can be obtained by measuring the charge asymmetry in the semileptonic decay modes.

9.5.4.1 Charged Decay Modes

The two charged decay products in $\pi^+\pi^-$ and semileptonic decays are usually recorded in a magnetic spectrometer consisting of a wide-aperture magnet and at least three layers of position-measuring detectors. The vector momenta $\vec{p}_i$ $(i = 1, 2)$ of the charged decay products are measured and the energies of the particles are obtained from the calculated vector momenta $\vec{p}_i$, assuming their rest mass to be m_π, as

$$
E_i = \sqrt{\vec{p}_i^{\,2} + m_\pi^2}\,. \tag{9.119}
$$

The invariant mass of the pair is

$$
m_{\pi\pi} = \sqrt{(E_1 + E_2)^2 - (\vec{p}_1 + \vec{p}_2)^2}\,, \tag{9.120}
$$

and the kaon momentum $\vec{p}_K = \vec{p}_1 + \vec{p}_2$. The lifetime of the kaon from the target to the decay vertex (z_V) in the kaon rest system is given by $\tau = (z_V - z_T)m_K/(cp_z)$, where m_K is the kaon mass, c the light velocity, and p_z the component of $\vec{p}_K$ along the beam line.

Two sets of information can be used to separate 2π and leptonic decays. First, the invariant mass $m_{\pi\pi}$ is required to be equal to m_K within the experimental resolution. Second, all experimenters use lepton identification.

The most frequently used methods for electron identification at intermediate energies, around 10 GeV, are Cerenkov counters, and identification through comparison of the energy deposition in electromagnetic and hadronic showers. At high energies, i.e. for electrons with energies between 10 GeV and 100 GeV, electron identification in calorimetric detectors works on the principle that for a particle of momentum p, the energy E deposited in a calorimeter by an electron (or photon) is much higher than for a hadron of the same momentum.

For the identification of muons one uses their penetration through several (~ 8) interaction lengths of material in order to distinguish them from pions interacting in this absorber.

Once the 2π decay mode has been identified, one has to know, in general, whether the K_S or K_L from which the decay products originate has undergone scattering on its way from its production to the decay point. In the case of a short-lived beam produced by protons interacting in a target near to the detector, this is done by calculating the distance of the intercept of the reprojected kaon momentum p_K in the target plane from the target center, p_t. Unscattered events cluster around $p_t = 0$. In the case of a long-lived beam, one uses the component of p_K transverse to the beam, p_t, or the angle θ between the kaon direction p_K and the beam direction in order to separate transmitted and coherently regenerated ($\theta = 0 = p_t$) kaons from events due to kaons that have undergone scattering, or diffractive, or inelastic, regeneration.

9.5.4.2 Neutral Decay Modes

The detection of the neutral decay mode $K_L \rightarrow \pi^0\pi^0 \rightarrow 4\gamma$ is complicated by the presence of the decay $K_L \rightarrow 3\pi^0 \rightarrow 6\gamma$ with a 21% branching ratio. This decay can simulate 4γ events for kinematic reasons, e.g. if two γ rays are missed by the detector. Very specific kinematic features of the $2\pi^0$ decay were therefore used in the early medium-energy experiments in order to obtain a clean $K_L \rightarrow 2\pi^0$ signal [167–170, 172].

For kaon energies between 40 GeV and 200 GeV, totally absorbing electromagnetic calorimeters are used. These calorimeters consist of scintillating crystals, Cerenkov lead glass counters, or liquid-noble-gas detectors with or without lead radiators. Their longitudinal thickness is around 25 radiation lengths, and their transverse segmentation corresponds to the transverse width of an electromagnetic shower, given by the Molière radius R_M of the material. In this way, the energies E_i and the transverse positions (x_i, y_i) of each of the four photon-induced showers are measured in the calorimeter. This is the only information available for reconstructing all variables describing the decay. In principle, the invariant mass of the four photons can be calculated using the relation

$$M^2 = E^2 - \vec{p}^2 = \left(\sum_{i=1}^{4} E_i\right)^2 - \left(\sum_{i=1}^{4} \vec{p}_i\right)^2 \tag{9.121}$$

$$= \sum_{i \neq j} (E_i E_j - E_i E_j \cos\theta_{ij}) \tag{9.122}$$

$$= 2 \sum_{i<j} E_i E_j \frac{\theta_{ij}^2}{2} . \tag{9.123}$$

The opening angle θ_{ij} between two photons can be obtained from the transverse distance r_{ij} between the impact points in the calorimeter,

$$r_{ij} = \sqrt{(x_i - x_j)^2 + (y_i - y_j)^2} , \tag{9.124}$$

and the distance z of the K meson decay point from the calorimeter. Using these variables, the invariant four-photon mass can be written as

$$M = \frac{1}{z} \sqrt{\sum_{i<j} E_i E_j r_{ij}^2} . \tag{9.125}$$

This relation can be used to calculate the distance of the decay point of the kaon from the calorimeter by using the kaon mass as a constraint:

$$z = \frac{1}{M_K} \sqrt{\sum_{i<j} E_i E_j r_{ij}^2} . \tag{9.126}$$

With this knowledge about the decay point, the invariant mass of any pair (i, j) of photons can then be calculated:

$$M_{ij} = \frac{1}{z} r_{ij} \sqrt{E_i E_j} . \tag{9.127}$$

Of the three possible combinations, the one where both masses are closest to the π^0 mass is chosen. A scatter plot of m_{12} versus m_{34} shows a signal at (m_{π^0}, m_{π^0}) if the four photons come from the decay $K_L \rightarrow \pi^0\pi^0$, while for events from the decay $K_L \rightarrow 3\pi^0$, with four detected photons, the invariant masses are spread over a large region around this point (Fig. 9.20). It is possible to extract the amount of background in the signal region by extrapolating the observed level of background events into this signal region, with the help of Monte Carlo simulations of the $K_L \rightarrow$

$3\pi^0$ background. In high-energy experiments, the background can then be reduced to a level below one percent.

9.5.4.3 Detectors Measuring Charged and Neutral Decay Modes Simultaneously

For the measurement of the parameter ε' of direct $\mathcal{CP}$ violation, the ratio of the decay rates of K_L into charged ($\pi^+\pi^-$) and neutral ($\pi^0\pi^0$) two-pion states has to be measured with great precision. For this purpose, experimentalists reduce systematic normalization uncertainties by measuring charged and neutral decays (from K_L and K_S mesons) simultaneously.

Four such experiments have been constructed for high-energy K meson beams, two of them at CERN (NA31 and NA48) and two at Fermilab (E731 and kTeV).

One experiment has been designed to detect K mesons of a few hundred MeV/c momentum arising from the annihilation of stopping antiprotons in a hydrogen target at the Low Energy Antiproton Ring (LEAR) at CERN (CPLEAR).

9.5.4.4 NA31

This detector (Fig. 9.7) [171], situated in a K_L or K_S beam from the CERN SPS, was based on calorimetry and was designed for good stability and high efficiency. The K_L and the K_S beam were produced by a 450 GeV proton beam with a production angle of 3.6 mrad.

A schematic illustration of the beam layout and the apparatus is shown in Fig. 9.7. The principal features can be summarized as follows:

- to adjust the steeply falling vertex distribution of the K_S decays to the almost flat vertex distribution of the K_L decays, the K_S target is located on a train that can be positioned at 41 stations in the decay volume;
- an anticounter with a 7 mm lead converter in the K_S beam is used to veto decays in the collimator, defines the upstream edge of the decay region, and provides for the relative calibration of the $2\pi^0$ and $\pi^+\pi^-$ energy scales to a precision better than $\pm 10^{-3}$;
- two wire chambers spaced 25 m apart, with ± 0.5 mm resolution in each projection, track charged pions;
- a liquid-argon/lead sandwich calorimeter with strip readout detects photons with ± 0.5 mm position resolution and an energy resolution of
  $\sigma_E/E = 10\%/E \oplus 7.5\%/\sqrt{E} \oplus 0.6\%$ (E in GeV);
- an iron/scintillator sandwich calorimeter measures the energy of charged pions with $\pm 65\%/\sqrt{E}$ (E in GeV) energy resolution.

The energies of the two pions and their opening angle are used to measure the invariant mass of the charged pair.

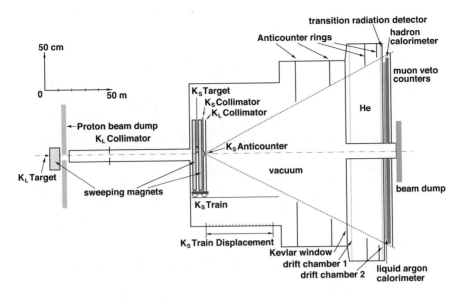

Fig. 9.7 Beam layout and detector in the NA31 experiment. The setup contains a movable target for the production of the K_S mesons, and a liquid-argon calorimeter; it does not contain a magnet

The NA31 experiment collected K_L and K_S decays in alternating time periods. In each case the charged and neutral decays were collected simultaneously.

9.5.4.5 NA48

This experiment (Fig. 9.8) was also built at CERN [201]. The detector was exposed to a simultaneous nearly collinear K_L/K_S beam, derived from a 450 GeV/c proton beam from the SPS.

The K_S beam was produced by using a fraction of the protons that did not interact with the K_L target.

Charged particles were measured by a magnetic spectrometer composed of four drift chambers with a dipole magnet between the second and the third chamber. The average efficiency per plane was 99.5%, with a radial uniformity better than $\pm 0.2\%$. The space point resolution was $\approx 95\,\mu$m. The momentum resolution was $\sigma_p/p = 0.48\% \oplus 0.009\% \times p$, where the momentum p is in GeV/c. The $\pi\pi$ invariant mass resolution is 2.5 MeV.

A liquid-krypton (LKr) calorimeter was used to reconstruct $K \to 2\pi^0$ decays. Cu–Be–Co ribbon electrodes of size 40 μm $\times$ 18 mm $\times$ 125 cm defined 13212 cells (each with a 2 cm $\times$ 2 cm cross section) in a structure of longitudinal projective towers pointing to the center of the decay region. The calorimeter was $\sim$27 radiation lengths long and fully contained electromagnetic showers with energies up to

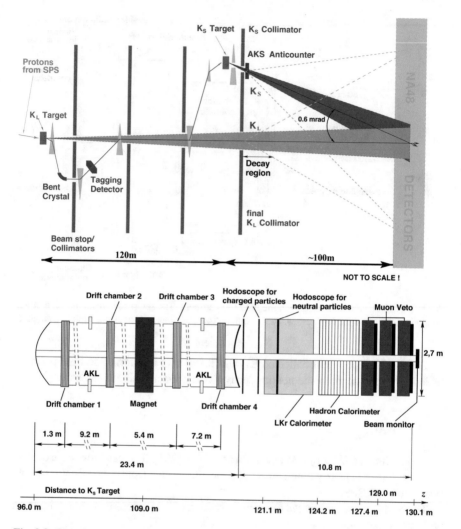

Fig. 9.8 Beam layout and detector in the NA48 experiment

100 GeV. The energy resolution of the calorimeter was $\sigma_E/E = (3.2 \pm 0.2)\%/E \oplus (9 \pm 1)\%/\sqrt{E} \oplus (0.42 \pm 0.05)\%$ where E is in GeV.

9.5.4.6 kTeV

The kTeV experiment at the 800 GeV/c Tevatron [180] uses a regeneration technique to produce the K_S beam (Fig. 9.9). The K_L "double beam" entered from the left, one half continuing as K_L, the other half producing a K_S beam by regeneration. The regenerator in the kTeV experiment was made of blocks of plastic scintillator.

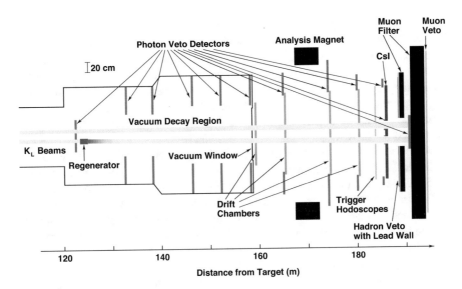

Fig. 9.9 kTeV beam and detector

These blocks were viewed by photomultiplier tubes to reject inelastically or quasi-elastically scattered kaons.

The evacuated decay volume extended over 40 m, or about 7 K_S mean lifetimes at 100 GeV/c, followed by the first drift chamber. The analysis magnet was located between the second and the third drift chamber. Each of the four drift chambers had two horizontal and two vertical planes of sense wires. The typical single-hit position resolution of the drift chambers was about 110 μm in either direction, which lead to a mean mass resolution of 2.2 MeV for the reconstructed kaon mass obtained from the $\pi^+\pi^-$ decay mode. The momentum resolution for a charged particle was $\sigma_p/p = 0.17\% \oplus 0.0071\% \times p$ (in GeV/c).

The four photons from the $2\pi^0$ were detected in an electromagnetic calorimeter made of pure cesium iodide. The calorimeter consisted of 3100 blocks arranged in a square array of 1.9 m side length. The blocks had two sizes: 2.5×2.5 cm^2 in the central region and 5×5 cm^2 in the outer region. All blocks are 50 cm, or ~27 radiation lengths, long. Two 15 cm square holes allowed the passage of the K_L and the K_S beam through the calorimeter. The calorimeter had an energy resolution of $\sigma_E/E = 2\%/\sqrt{E} \oplus 0.4\%$ (E in GeV). The average position resolution for electrons was about 1.2 mm for clusters in the smaller crystals and 2.4 mm for the larger crystals.

9.5.4.7 CPLEAR

In contrast to the detectors described in the previous sections, the CPLEAR detector measured decays from kaons produced in $\bar{p}p$ annihilations at rest obtained from the low energy $\bar{p}$ ring LEAR at CERN. The antiprotons were stopped in a 16 bar hydrogen gas target and formed a protonium before annihilation. In annihilation reactions of the type $\bar{p}p \rightarrow K^0 K^- \pi^+$ and $\bar{p}p \rightarrow \overline{K^0} K^+ \pi^-$, the charged kaon was identified through the time of flight and track curvature in a solenoidal magnetic field ("tagging"). This tag was used as a trigger for detecting the decay products of the neutral K meson associated with the K^+ or K^-. A unique property of this scheme is that the strangeness of the neutral K meson is known from the charge of the tagged K^+ or K^-.

As shown in Fig. 9.10, the experiment had a cylindrical, onion-type, setup. Six cylindrical drift chambers, starting at a radius of 25.46 cm and going out to a radius of 50.95 cm, provided the main tracking information for charged particles. The offline track-finding efficiency was better than 99% and the wire positions were determined with an accuracy of about 10 µm. The mean mass resolution achieved for the invariant kaon mass in the $\pi^+\pi^-$ final state was 13.6 MeV/c^2. By applying kinematically and/or geometrically constrained fits, the K^0 momentum resolution σ_{p_t}/p_t was improved from 5.5% to 0.25%.

The tracking detectors were followed by the particle identification detector (PID), used for charged-kaon identification and e/π separation. It was located at radii between 62.5 cm and 75.0 cm. and was composed of two layers of plastic scintillators with an 8 cm thick liquid threshold Cerenkov detector in between. The two charged tracks from the decays $\overline{K^0} \rightarrow \pi^+\pi^-$, $\pi^+ e^- \nu$, $\pi^+\mu^-\nu$ were reconstructed and the decay vertex was calculated. Using this vertex and the annihilation point in the hydrogen target, the proper time for decay of the kaon was obtained.

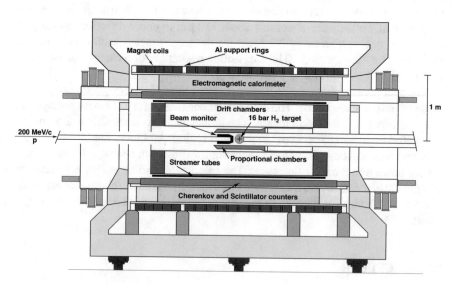

Fig. 9.10 CPLEAR detector

The electromagnetic calorimeter had the form of a barrel and was an assembly of 18 lead plates alternating with sampling chambers. It was located at radii between 75 cm and 100 cm. The calorimeter had a total thickness of ~6 radiation lengths, an energy resolution of $\sigma_E/E \approx 13\%/\sqrt{E}$ (E in GeV) and a position resolution of ~5 mm for the photon conversion points. The photon detection efficiency was $(90 \pm 1)\%$ for photon energies above 200 MeV. For photons with energies below 200 MeV the efficiency dropped significantly and was only about 60% for 100 MeV photons.

All subdetectors were embedded in a 3.6 m long, 2 m diameter solenoid magnet, which provided a 0.44 T uniform field.

9.5.5 Elucidation of $C\mathcal{P}$ Violation in K^0 Decays (I): Search for $\Im m(\varepsilon'/\varepsilon)$

9.5.5.1 The Significance of the Phase Φ_{+-}

The phase of η_{+-}, $\Phi_{+-} = \arg(\eta_{+-})$, was a possible clue that would help to disentangle the two components of $C\mathcal{P}$ violation, since

$$\eta_{+-} = \varepsilon + \varepsilon' . \tag{9.128}$$

If ε', the parameter of direct $C\mathcal{P}$ violation, was comparable in size to ε, and if its phase was orthogonal to ε, then the phase of η_{+-} would deviate in a detectable way from the phase of ε, which is mainly determined by the experimentally measurable values of Δm and $\Gamma_S = 1/\tau_S$ (see Sect. 9.5.2):

$$\arg(\varepsilon) = \arctan\left(\frac{2\Delta}{\Gamma_S}\right) + \Phi_D = \Phi_{SW} + \Phi_D \tag{9.129}$$

With the present values of Δm and Γ_S (Sect. 9.5.5.7), $\Phi_{SW} = (43.4 \pm 0.1)°$.

A significant deviation of the measured value of Φ_{+-} from Φ_{SW} would be evidence for a nonvanishing component $\Im m(\varepsilon'/\varepsilon)$ and against the superweak model of $C\mathcal{P}$ violation.

9.5.5.2 Measurements of the Phase Φ_{+-} in Interference Experiments Behind a Regenerator

The relative phase between the two amplitudes of the decays $K_L \to \pi^+\pi^-$ and $K_S \to \pi^+\pi^-$ has been measured by two distinct methods.

The first consists of measuring the interference of the $K_L \to \pi^+\pi^-$ amplitude with the coherently regenerated $K_S \to \pi^+\pi^-$ amplitude behind a slab of material (the regenerator). The experiments require (*a*) the measurement of the $\pi^+\pi^-$

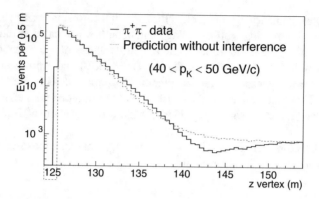

Fig. 9.11 z decay distribution of K $\rightarrow \pi^+\pi^-$ decays in the regenerator beam, for the restricted momentum range 40–50 GeV/c. The Monte Carlo prediction (*dashed line*) is without the interference term that is proportional to "$2|\varrho||\eta|$" (kTeV experiment [180])

intensity as a function of the K^0 eigentime behind the regenerator, which is given in Sect. 9.5.3.1, and (*b*) an independent determination of the regeneration phase.

The latest and most precise experiment of this type has been done by the kTeV collaboration [180] (see also [176]) in an experiment at Fermilab with an 800 GeV/c proton beam. The detector is described in Sect. 9.5.4.6; 5×10^9 events were recorded in 1996–1997, among those were about 9×10^6 $K^0 \rightarrow \pi^+\pi^-$ events. Their z decay distribution in the restricted kaon momentum interval from 40 to 50 GeV/c is shown in Fig. 9.11. The corresponding decay time distribution was fitted with the formula (9.116). The nuclear regeneration amplitude $F = i(f(0) - \overline{f}(0))/k$ was assumed to decrease with the kaon momentum $\vec{p}$ according to a power law $F(p) = F(70 \text{ GeV}/c) \times (p/70 \text{ GeV}/c)^\alpha$. This was motivated by a Regge model where the difference between the K and $\overline{\text{K}}$ scattering amplitudes would be described by one single ω meson exchange trajectory. In this model, the phase of the regeneration amplitude is given by $\Phi_F = -(\pi/2)(1+\alpha)$. In the fit, Φ_{+-}, Δm, Γ_S and α were free parameters, and Φ_F was assumed to be given by the Regge model. The results are

$$\Phi_{+-} = 44.12° \pm 0.72° \text{ (stat)} \pm 1.20° \text{ (syst)} = 44.12° \pm 1.40° , \quad (9.130)$$

$$\Delta m = (5288 \pm 23) \times 10^6 \text{ s}^{-1} , \quad (9.131)$$

$$\tau_S = (89.58 \pm 0.08 \text{ (stat)}) \times 10^{-12} \text{ s} , \quad (9.132)$$

$$\chi^2 = 223.6 \quad \text{for 197 degrees of freedom} . \quad (9.133)$$

The systematic error in Φ_{+-} includes a 0.25° uncertainty from the fact that the relation between the regeneration phase and the momentum dependence of the regeneration amplitude through a dispersion relation integral is incomplete. It has been argued [178] that this uncertainty is larger, more than one degree, because of the limited momentum range in which the regeneration amplitude was measured.

9.5.5.3 Measurements of Φ_{+-} in Vacuum Interference Experiments

The other method for measuring Φ_{+-} is the vacuum interference method mentioned above (Sect. 9.5.3.1), where one observes the $K \to \pi^+\pi^-$ distribution obtained from an initially pure strangeness state. The information on Φ_{+-} is contained in the interference term proportional to $\cos(\Delta m\, \tau - \Phi_{+-})$, and the time at which the two interfering amplitudes are equal is $\sim 12\tau_S$, so that the correlation of Φ_{+-} with Δm is rather strong.

Three experiments of this type have been done in the intermediate-energy domain [166, 182, 183]. An analysis of the latest and most precise of those has been performed by the CERN–Heidelberg group [166]. The apparatus was situated in a 75 mrad short neutral beam derived from 24 GeV/c protons. The time distribution of 6×10^6 $K_{S,L} \to \pi^+\pi^-$ decays is shown in Fig. 9.12: (*curve a*), together with the fitted time distribution, as given in Sect. 9.5.3.1. The result of this fit is

$$\Phi_{+-} = (49.4° \pm 1.0°) + 305° \frac{(\Delta m - 0.540 \times 10^{10}\ \text{s}^{-1})}{\Delta m}\,, \tag{9.134}$$

$$|\eta_{+-}| = (2.30 \pm 0.035) \times 10^{-3}\,, \tag{9.135}$$

$$\Gamma_S = (1.119 \pm 0.006) \times 10^{-10}\ \text{s}^{-1}\,, \tag{9.136}$$

$$\chi^2 = 421 \quad \text{for 444 degrees of freedom}\,. \tag{9.137}$$

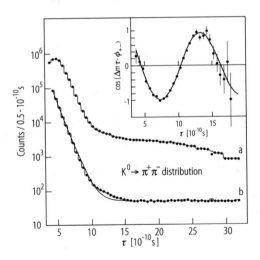

Fig. 9.12 Time distribution of $K \to \pi^+\pi^-$ events from a coherent mixture of K_L and K_S produced in pure strangeness states [166]. *Curve a*: events (*histogram*) and fitted distribution (*dots*). *Curve b*: events corrected for detection efficiency (*histogram*), and fitted distribution with interference term (*dots*) and without interference term (*curve*). *Inset*: interference term as extracted from data (*dots*) and fitted term (*line*). (CERN–Heidelberg experiment [166])

In the high-energy domain, this method has also been used by the NA31 collaboration [184]. In a 3.6 mrad neutral beam derived from 450 GeV protons, kaons of mean momentum around 100 GeV were allowed to decay. Two different target positions were chosen, at distances of 48 m and 33.6 m from the defining collimator of the neutral beam, which marked the upstream end of the decay volume. Kaons that decayed along 50 m in an evacuated tank were detected further downstream at about 120 m from the final collimator by the NA31 detector (Sect. 9.5.4.4, Fig. 9.7). The measured time distribution of $\pi^+\pi^-$ decays is similar to the one in Fig. 9.12. There are 2.24×10^6 and 0.57×10^6 $\pi^+\pi^-$ events in the data for the target in the near and far positions, respectively; the corresponding numbers of $\pi^0\pi^0$ events are 1.81×10^6 and 0.31×10^6. The phases were extracted from a fit to the time distribution of the ratio of the data in the near and far positions of the target.

The results are

$$\Phi_{+-} = (46.9° \pm 1.4°) + 310° \frac{(\Delta m - 0.5351 \times 10^{10}\,\text{s}^{-1})}{\Delta m}$$

$$+ 270° \frac{(\tau_S - 0.8922 \times 10^{-10}\,\text{s})}{\tau_S} \tag{9.138}$$

and

$$\Phi_{00} = (47.1° \pm 2.1°) + 310° \frac{(\Delta m - 0.5351 \times 10^{10}\text{s}^{-1})}{\Delta m}$$

$$+ 225° \frac{(\tau_S - 0.8922 \times 10^{-10}\,\text{s})}{\tau_S} . \tag{9.139}$$

The difference between the two phases comes out to be

$$\Phi_{00} - \Phi_{+-} = 0.2° \pm 2.6° \pm 1.2° . \tag{9.140}$$

9.5.5.4 Measurements of the Phase Difference $\Phi_{00} - \Phi_{+-}$

For small $|\varepsilon'/\varepsilon|$, this phase difference is related to ε'/ε by the equation

$$\Phi_{00} - \Phi_{+-} = -3 \Im m \left(\frac{\varepsilon'}{\varepsilon}\right) . \tag{9.141}$$

In this way, the component of ε' orthogonal to the direction of ε can be measured. In the absence of CPT violation and for small $|\varepsilon'/\varepsilon|$, both of the phases Φ_{00} and Φ_{+-} are close to the superweak phase Φ_{sw} (9.102).

The measurement of this phase difference by the NA31 experiment (Sect. 9.5.5.3) was improved by the simultaneous measurement of the time distributions of $\pi^+\pi^-$

and $\pi^0\pi^0$ decays behind a regenerator in the kTeV experiment [180] (Sect. 9.5.5.2). Here, the uncertainty arising from the phase of coherent regeneration (which is the determining uncertainty in the Φ_{+-} measurement) cancels in the comparison of the two decay modes. The authors of [180] conclude that

$$\Phi_{00} - \Phi_{+-} = (0.39 \pm 0.50)° . \qquad (9.142)$$

Together with an earlier measurement by E731/E773 [179] and the NA31 measurement (9.140), this gives

$$\Phi_{00} - \Phi_{+-} = (0.36 \pm 0.43)° . \qquad (9.143)$$

9.5.5.5 Measurement of Φ_{+-} from a Tagged Pure Strangeness State

The CPLEAR experiment (Sect. 9.5.4.7) offers the unique feature of tagging a pure neutral strangeness state K^0 or $\overline{K}^0$ produced in a $\overline{p}p$ annihilation at rest by identifying a charged kaon produced in the same reaction. Compared with the vacuum interference experiments (Sect. 9.5.5.3), this offers the advantage of a full-size interference term, whereas in the vacuum interference experiments, an incoherent mixture of (predominantly) K^0 and $\overline{K}^0$ forms the initial state, and the interference term is diluted.

The interference term is visualized by measuring the decay-rate asymmetry for decays into $\pi^+\pi^-$,

$$A_{+-}(\tau) = \frac{\overline{N}(\tau) - N(\tau)}{\overline{N}(\tau) + N(\tau)} \qquad (9.144)$$

$$= 2\,\Re e\,\varepsilon - 2\,|\eta_{+-}|e^{-(1/2)\Gamma_S\tau}\cos(\Delta m\,\tau - \Phi_{+-}) . \qquad (9.145)$$

In the corresponding experimental distribution, the background was subtracted and the events were appropriately weighted taking into account the tagging efficiencies for K^0 and $\overline{K}^0$. The result is shown in Fig. 9.13 [181]. The result of a fit to these data gives values for Φ_{+-} and $|\eta_{+-}|$, which are correlated with the values chosen for Δm and τ_S, respectively.

The correlation parameters are

$$\delta\Phi_{+-} = 0.30(\Delta m - 0.5301 \times 10^{10} \text{ s}^{-1}) \qquad (9.146)$$

and

$$\delta|\eta_{+-}| = 0.09(\tau_S - 0.8934 \times 10^{-10} \text{ s}) . \qquad (9.147)$$

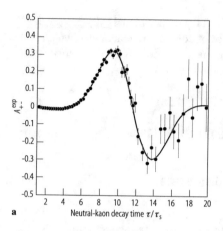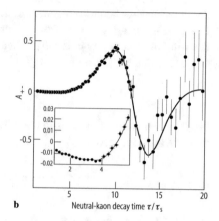

Fig. 9.13 (**a**) Measured decay-rate asymmetry, $A_{+-}^{\exp}(\tau)$; the data points include residual background. (**b**) Decay-rate asymmetry, $A_{+-}(\tau)$; the data points are background subtracted. In both cases the continuous curve is the result of the fit (CPLEAR experiment, [185])

For the values of Δm and τ_S chosen by the authors of [181], the results are

$$\Phi_{+-} = 43.19° \pm 0.53° \text{ (stat)} \pm 0.28° \text{ (syst)} \pm 0.42° \text{ } (\Delta m) , \quad (9.148)$$

$$|\eta_{+-}| = [2.264 \pm 0.023 \text{ (stat)} \pm 0.026 \text{ (syst)}] \times 10^{-3} . \quad (9.149)$$

A similar asymmetry is obtained for decays to $\pi^0\pi^0$, although with less statistical weight.

The following values of the CP parameters in the neutral mode have been extracted:

$$\Phi_{00} = 42.0° \pm 5.6° \text{ (stat)} \pm 1.9° \text{ (syst)} , \quad (9.150)$$

$$|\eta_{00}| = [2.47 \pm 0.31 \text{ (stat)} \pm 0.24 \text{ (syst)}] \times 10^{-3} . \quad (9.151)$$

9.5.5.6 Charge Asymmetry in Semileptonic Decays

This asymmetry δ_L is the third manifestation of CP violation (9.108). This asymmetry measures the CP impurity of the long-lived kaon state: $\delta_L = 2\Re e\,\varepsilon(1 - |x|^2)/(|1 - x|^2)$, where x is the $\Delta S = \Delta Q$ violation parameter. Considerable precision was achieved in the first ten years of experimentation after 1964: for the K_{e3} mode, two experiments, by the Princeton group [186] and by the CERN–Heidelberg group [187], have been reported; for the $K_{\mu3}$ mode, results have been obtained from Stanford [189] and one from the CERN–Heidelberg group [187], and a Brookhaven–Yale group [190] obtained a result for a mixture of both decay modes. Recently, two new results on this asymmetry have been reported by the kTeV and NA48 collaborations. The most significant features of these results are:

1. Event numbers of up to 298 million events in the K_{e3} mode and 15 million events in the $K_{\mu 3}$ mode.
2. An apparatus design such that the decay products (π and electron) traverse only minute amounts of matter (about 0.3–$0.4\,\mathrm{g\,cm^{-2}}$), thus diminishing corrections due to secondary interactions of these particles.
3. The precision of the $K_{\mu 3}$ asymmetry measurements is a factor of 4 below the value for K_{e3}, making a comparison between the two possible. Table 9.3 gives the results; the average $\delta_L = (3.316 \pm 0.053) \times 10^{-3}$.

The charge asymmetries for K_{e3} and $K_{\mu 3}$ decays are equal to within 8%: $\delta_L^e / \delta_L^\mu = 1.04 \pm 0.08$. Assuming the validity of the $\Delta Q = \Delta S$ rule, which is supported by the present experiments, we obtain $\Re e\, \varepsilon = (1.658 \pm 0.026) \times 10^{-3}$. If we use the most precise tests of the $\Delta Q = \Delta S$ rule by the CPLEAR experiment [185], there is a small correction

$$\frac{1 - |x|^2}{|1 - x|^2} = 0.996 \pm 0.012 . \tag{9.152}$$

9.5.5.7 Parameters of $\mathcal{CP}$ Violation in the $\mathbf{K^0}$ System: $\Im m(\varepsilon'/\varepsilon)$

$\mathbf{K_S}$ **Lifetime** We take the average of the older measurements combined with the two most recent measurements by the kTeV collaboration, $(0.8965 \pm 0.0007) \times 10^{-10}$ s [180], and by the NA48 collaboration, $(0.89598 \pm 0.0007) \times 10^{-10}$ s [197]. Our grand average is

$$\tau_S = (0.8959 \pm 0.0004) \times 10^{-10}\ \mathrm{s} . \tag{9.153}$$

Table 9.3 Charge asymmetry measurements in K_{l3} decays

Group and reference	Year	Decay mode	Result [$\times 10^3$]	δ [$\times 10^3$]
Columbia [164, 192]	1969	K_{e3}	2.46 ± 0.59	
Columbia–Harvard–CERN [193]	1970	K_{e3}	3.46 ± 0.33	K_{e3} average
San Diego–Berkeley [194]	1972	K_{e3}	3.6 ± 1.8	3.322 ± 0.055
Princeton [186]	1973	K_{e3}	3.18 ± 0.38	
CERN–Heidelberg [187]	1974	K_{e3}	3.41 ± 0.18	
kTeV [188]	2002	K_{e3}	$3.322 \pm 0.058 \pm 0.047$	
NA48 [157]	2003	K_{e3}	$3.317 \pm 0.070 \pm 0.072$	
Brookhaven–Yale [190]	1973	$K_{e3} + K_{\mu 3}$	3.33 ± 0.50	
SLAC–Berkeley [165, 195]	1969	$K_{\mu 3}$	5.8 ± 1.7	
Berkeley [196]	1972	$K_{\mu 3}$	6.0 ± 1.4	$K_{\mu 3}$ average
Stanford [189]	1972	$K_{\mu 3}$	2.78 ± 0.51	3.19 ± 0.24
CERN–Heidelberg [187]	1974	$K_{\mu 3}$	3.13 ± 0.29	

The $K_{\mu 3}$ and K_{e3} average value is $\delta_L = (3.316 \pm 0.053) \times 10^{-3}$

We use this value in the following sections.

Mass Difference Δm Combining the values from the CERN–Heidelberg experiments [173–175] with those from Fermilab E731 and E773 [176, 179], from CPLEAR [181, 198, 199] and from the most precise single measurement by kTeV [180], we obtain

$$\Delta m = (0.5286 \pm 0.0011) \times 10^{10} \text{ s}^{-1} \, . \tag{9.154}$$

Superweak Phase From the two parameters given above, we obtain the phase of ε in the superweak model,

$$\Phi_{\text{sw}} = \arctan\left(\frac{2\,\Delta m}{\Gamma_{\text{S}}}\right) = (43.4 \pm 0.1)° \, . \tag{9.155}$$

Moduli of the Amplitudes η_{+-} and η_{00} New measurements of η_{+-} from the ratio of $\pi^+\pi^-$ decays and semileptonic decays yield precise values of η_{+-}. We combine these with previous results and obtain [160–162].

$$|\eta_{+-}| = (2.230 \pm 0.006) \times 10^{-3} \, . \tag{9.156}$$

Absolute measurements of the amplitude η_{00} are much less precise. A recent average including the CPLEAR result is

$$|\eta_{00}| = (2.23 \pm 0.11) \times 10^{-3} \, . \tag{9.157}$$

Phase Φ_{+-} In all measurements, this phase is extracted from an interference term with a beat frequency Δm. Taking the results from NA31 [184], E731 [176], E773 [179], and kTeV [180], together with the pre-1975 data and the result from CPLEAR [181], we obtain the world average

$$\Phi_{+-} = (43.3 \pm 0.4)° \tag{9.158}$$

using the values for Δm and τ_{S} above. This result is in excellent agreement with the value of $\Phi_{\text{sw}} = (43.4 \pm 0.1)°$. Since the interference experiments were evaluated without assuming $\mathcal{CPT}$ invariance, this constitutes a stringent test of $\mathcal{CPT}$ invariance. The difference is $\Phi_{+-} - \Phi_{\text{sw}} = -(0.1 \pm 0.4)°$. At the same time, this result can again be used to constrain the component of ε' orthogonal to ε:

$$\Im m\left(\frac{\varepsilon'}{\varepsilon}\right) = -(1.7 \pm 7.0) \times 10^{-3} \, . \tag{9.159}$$

Therefore, at this level of 10^{-2} relative to the amplitude for $\mathcal{CP}$ violation by mixing, ε, there is no evidence for a direct $\mathcal{CP}$ violation amplitude ε' orthogonal to ε. Our interest now shifts to the component of ε' parallel to ε, i.e. $\Re e(\varepsilon'/\varepsilon)$.

9.5.6 Elucidation of CP Violation in K^0 Decays (II): Discovery of Direct CP Violation in $\Re e(\varepsilon'/\varepsilon)$

9.5.6.1 Significance of the Double Ratio R

The real part of ε'/ε is connected with the amplitude ratios η_{00} and η_{+-} of CP-violating K_L decays to CP-conserving K_S decays. A measurement of decay rates with the required precision of 10^{-3} is only possible by measuring the ratio of rates in the same beam in the same time interval. From the Wu–Yang triangle relations (9.90) $\eta_{00} = \varepsilon - 2\varepsilon'$ and $\eta_{+-} = \varepsilon + \varepsilon'$, we obtain

$$\Re e \left(\frac{\varepsilon'}{\varepsilon} \right) = \frac{1}{6} \left(1 - \left| \frac{\eta_{00}}{\eta_{+-}} \right|^2 \right) . \tag{9.160}$$

A measurement of the double ratio

$$R = \frac{|\eta_{00}|^2}{|\eta_{+-}|^2} = \frac{\Gamma(K_L \to 2\pi^0)/\Gamma(K_L \to \pi^+\pi^-)}{\Gamma(K_S \to 2\pi^0)/\Gamma(K_S \to \pi^+\pi^-)} \tag{9.161}$$

to a precision of about 0.3% is therefore required to distinguish between the two remaining models, the KM milliweak model and the superweak model. Since the KM model predicts values of $\Re e(\varepsilon'/\varepsilon)$ in the range between 0.2×10^{-3} and 3×10^{-3}, the precision required for detecting a signal of direct CP violation depends on the actual value. If the largest prediction is realized in nature, a precision of $\delta R = 5 \times 10^{-3}$ would be sufficient for a three-standard-deviation observation. If, however, the lowest value is realized, a precision of $\delta R = 0.4 \times 10^{-3}$ would be needed, corresponding to samples of several million events for each of the four decay modes.

9.5.6.2 The NA31 Experiment: First Evidence for Direct CP Violation

The first observation of direct CP violation was made by a collaboration of physicists at CERN in 1988 [191]. The experiment, called "North Area No. 31", or NA31, was based on the concurrent detection of $2\pi^0$ and $\pi^+\pi^-$ decays. Collinear beams of K_L and K_S were employed alternately. The beam layout and the apparatus were described in Sect. 9.5.4.4. The $K^0 \to 2\pi^0 \to 4\gamma$ decays were reconstructed and separated from the background primarily due to $K_L \to 3\pi^0 \to 6\gamma$ decays as described in Sect. 9.5.4.2. This background is uniformly distributed in a two-dimensional scatter plot of photon-pair masses, while the $2\pi^0$ signal peaks at a point S where both photon pairs have the π^0 mass, with a 2 MeV resolution. Signal and background events were counted in equal-area χ^2 contours around S. Figure 9.14 shows the χ^2 distribution of events in the K_S beam and in the K_L beam. The signal region was taken as $\chi^2 < 9$. Background in the K_L data was subtracted by linear

Fig. 9.14 Number of accepted 4γ events as a function of χ^2 for the $K_S \rightarrow \pi^0\pi^0$ (*left*) and $K_L \rightarrow \pi^0\pi^0$ (*right*) data, and a Monte Carlo calculation of the background originating from $K_L \rightarrow 3\pi^0$ decays (*dotted*). The signal region was taken as $\chi^2 < 9$ (NA31 experiment [191])

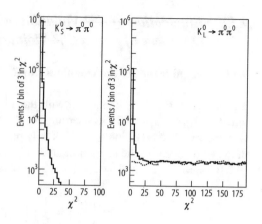

extrapolation into the signal region, and amounts to about 4%, while it is negligible in the K_S data.

The $K^0 \rightarrow \pi^+\pi^-$ decays were reconstructed from the four space points of the two pion tracks in two wire chambers. From these tracks, the position of the decay vertex along the beam was reconstructed with a precision of better than 1 m. The energies E_1 and E_2 of the two pions were obtained from the energies deposited in the liquid-argon electromagnetic calorimeter and the iron-scintillator hadronic calorimeter. The K^0 energy was then calculated using the kaon mass and the opening angle θ of the two tracks as constraints from the ratio E_1/E_2:

$$E_K = \sqrt{\frac{A}{\theta^2}\left(m_K^2 - A m_\pi^2\right)}, \qquad (9.162)$$

where

$$A = \frac{E_1}{E_2} + \frac{E_2}{E_1} + 2. \qquad (9.163)$$

Background from $K^0 \rightarrow \pi e \nu$ (K_{e3}) decay was reduced by comparing, for each track, the energy deposited in the front half of the electromagnetic calorimeter with the energy deposited in the hadron calorimeter.

After cuts on the invariant $\pi^+\pi^-$ mass and on the transverse location of the center of energy relative to the center of the neutral beam, a residual background of three-body decays was subtracted.

Figure 9.15 shows the transverse distance d_T between the decay plane, as reconstructed from the two tracks, and the K^0 production target, at the longitudinal position of the target. For K_S decays, this distributions peaks at $d_T = 0$ with a resolution given by the measurement error and multiple scattering. For K_L decays, in addition to this component of two-body decays from the target, there is a broader distribution mixed in due to three-body decays. The signal region was taken to be $d_T < 5$ cm, and the three-body background was extrapolated from a control region

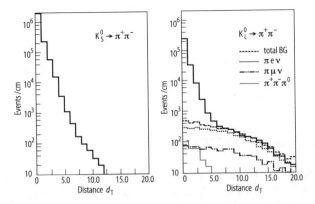

Fig. 9.15 Event distribution for charged decays as a function of the distance d_T (in cm) between the decay plane and the production target for K_S (*left*) and K_L (*right*) decays, and for various background components (*right*) (NA31 experiment [191])

$7 < d_T < 12$ cm. This background amounts to $(6.5 \pm 2.0) \times 10^{-3}$ of the signal, including systematic uncertainties.

The remaining event sample contained 109×10^3 of $K_L \to \pi^0\pi^0$, 295×10^3 of $K_L \to \pi^+\pi^-$, 932×10^3 of $K_S \to \pi^0\pi^0$, and 2300×10^3 of $K_S \to \pi^+\pi^-$. In order to equalize the acceptance for K_S decays (with an average decay length of 6 m) to that for the uniformly distributed K_L decays, the K_S data were taken with the K_S target displaced in 1.2 m steps over 48 m such that the distribution of K_S decays became effectively uniform in the fiducial region. This makes the double ratio essentially insensitive to acceptance corrections. The double ratio was evaluated in 10×32 bins in energy and vertex position. The weighted average, after all corrections, is

$$R = 0.980 \pm 0.004 \text{ (stat)} \pm 0.005 \text{ (syst)} . \tag{9.164}$$

The corresponding result for the direct-$\mathcal{CP}$-violation parameter is [191]

$$\Re e\left(\frac{\varepsilon'}{\varepsilon}\right) = (33.0 \pm 11.0) \times 10^{-4} . \tag{9.165}$$

This, with three-standard-deviation significance, shows that the $\mathcal{CP}$-odd K_2 decays to two pions, and was the first evidence of direct $\mathcal{CP}$ violation.

In further measurements with the NA31 detector, the event numbers recorded were considerably increased, by a factor of four, thus decreasing the statistical error. In addition, the background from K_{e3} decays was reduced by introducing a two-stage transition radiation detector as an additional identifier for electrons. With these improved data, the double ratio was measured with reduced uncertainty. Including the former result, the double ratio obtained is

$$R = 0.982 \pm 0.0039 , \tag{9.166}$$

leading to a value

$$\Re e \left(\frac{\varepsilon'}{\varepsilon} \right) = (23.0 \pm 6.5) \times 10^{-4} , \tag{9.167}$$

or 3.5 standard deviations from zero.

9.5.6.3 The Experiment E731 at Fermilab

In this experiment [177] charged and neutral decays were registered in separate runs with a slightly different detector. On the other hand, in E731 K_L and K_S decays were collected simultaneously using a split beam: in one half of the beam cross section, K_L mesons from the target were allowed to decay over a long decay region of 27 m (charged decays) or 42 m (neutral decays), while in the other half, the K_L beam hit a block of B_4C, whereby a beam of K_S mesons was regenerated.

The vertex distribution of the events from regenerated K_S was concentrated in a small region behind the regenerator, positioned at 123 m from the target, owing to the typical K_S decay length of 5 m. On the other hand, the vertex distribution of $K_L \rightarrow \pi^0 \pi^0$ decays extended from 110 to 152 m from the target. The detector acceptances for K_L and K_S decays therefore were very different, and since the decay volumes for $K_L \rightarrow \pi^+ \pi^-$ and $K_L \rightarrow \pi^0 \pi^0$ were different, this acceptance correction does not cancel in the double ratio.

The $\pi^+ \pi^-$ decays were selected by requiring the invariant $\pi\pi$ mass to be near the kaon mass and the transverse kaon momentum to satisfy $p_t^2 < 250 \text{ MeV}^2/c^2$. The background from incoherent kaon regeneration amounted to $(0.155 \pm 0.014)\%$. The extrapolated $3\pi^0$ background under the peak is 1.78% and 0.049% in the vacuum and regenerator beams, respectively. Neutral background from incoherent scattering in the regenerator was subtracted by evaluating the distribution of the transverse center of energy of each event around the center of each beam. After this background subtraction, the event numbers in the vacuum beams were 327×10^3 ($\pi^+ \pi^-$) and 410×10^3 ($\pi^0 \pi^0$). In the regenerator beams, there were 1.06×10^6 $\pi^+ \pi^-$ events and 0.800×10^6 $2\pi^0$ events. The regeneration amplitude was assumed to fall in accordance with a power of the kaon momentum, p^α. In the fit, the parameter α, the regeneration amplitude at 70 GeV/c momentum, and $\Re e(\varepsilon'/\varepsilon)$ were varied. The results were

$$\alpha = -0.6025 \pm 0.0065 \tag{9.168}$$

and

$$\Re e \left(\frac{\varepsilon'}{\varepsilon} \right) = (7.4 \pm 5.2 \text{ (stat)} \pm 2.9 \text{ (syst)}) \times 10^{-4} , \tag{9.169}$$

where the systematic uncertainty includes a part from acceptance calculations (1.19×10^{-4}) and from the energy calibration (1.6×10^{-4}). The authors of [177] deduce an upper limit $\Re e(\varepsilon'/\varepsilon) < 17 \times 10^{-4}$ (95% C.L.), which is at variance with the observation of the NA31 experiment.

9.5.6.4 The kTeV Experiment at Fermilab

The disagreement between the positive result of NA31 and the null result of E731 left an unsatisfactory state of affairs. For this reason, new experiments with a tenfold increase in data-taking capacity and reduced systematic uncertainty were designed, at both Fermilab and CERN.

The Fermilab experiment at the 800 GeV Tevatron, called kTeV, was described in Sect. 9.5.4.6 [180]. The main improvements compared with E731 were:

- all four decay modes were measured concurrently;
- the electromagnetic calorimeter was made of CsI, with much improved energy resolution;
- the regenerator at 123 m from the target was made of plastic scintillator, viewed by photomultipliers such that inelastic regeneration could be detected by the recoiling nucleus;
- the kaon momentum range from 40 to 160 GeV/c and the decay vertex region from 110 m to 158 m from the target were the same for all decay modes.

As in the E731 experiment, a double beam of K_L and regenerated K_S entered the decay volume. $K^0 \rightarrow \pi^+\pi^-$ decays were identified by their invariant mass. Semileptonic K_{e3} events were reduced by a factor of 1000 by requiring the ratio of the calorimetric energy E of a track to its momentum p to be less than 0.85. $K_{\mu3}$ events were rejected by registering the muon penetrating the 4 m iron wall. The invariant-$\pi\pi$-mass shows a rms mass resolution of 1.6 MeV, and events in the range 488–508 MeV were selected. Background from K_S produced in incoherent regeneration was suppressed mainly by vetoing events with a signal generated in the active regenerator indicating the recoil of a nucleus in the scattering process. Further reduction of this background was achieved by extrapolating the kaon direction back to the regenerator exit face and calculating the transverse momentum of the kaon relative to the line connecting this intercept with the target position. After a cut against the backgrounds from semileptonic decays and from collimator scattering, 11.1 million and 19.29 million $\pi^+\pi^-$ events remain in the vacuum beam and regenerator beam samples, respectively.

The selection of $2\pi^0$ events follows the lines described in Sect. 9.5.4.2. Events with an invariant $\pi^0\pi^0$ mass between 490 and 505 MeV were selected. Events in which a kaon scatters in the collimator or the regenerator were reduced by a cut in the ring number (RING), defined by the maximum deviation $\Delta x_{\rm coe}$ or $\Delta y_{\rm coe}$ (in cm) of the center of energy of all showers from the center of the corresponding beam spot at the CsI position, to which the event was assigned by use of the x-position of the center of energy:

$$RING = 4 \times Max(\Delta x_{coe}^2, \Delta y_{coe}^2) . \tag{9.170}$$

The signal was selected by the cut RING < 110 cm^2. The largest background comes from regenerator scattering in the regenerator beam, 1.13%, adding up to a total of 1.235% in that beam. Also, in the vacuum beam, the events scattered in the nearby regenerator make the largest contribution to the background, 0.25%, which is 0.48% in total. After all cuts and background subtraction, the remaining signal consists of 3.3 million and 5.55 million events in the vacuum and regenerator beams, respectively. The 3.3 million $K_L \rightarrow \pi^0\pi^0$ events are the limiting factor in the statistical uncertainty in the double ratio.

Since the vertex distributions of K_L decays (flat) and K_S decays (concentrated behind the regenerator) are very different, the raw double ratio has to be corrected by the double ratio of acceptances (Fig. 9.16). The quality of the Monte Carlo simulations for the acceptances was checked by reproducing the z vertex distributions of the vacuum beam data for different decay modes (Fig. 9.17). In general the agreement is good, except that the $\pi^+\pi^-$ data show a slope of $(-0.70 \pm 0.30) \times 10^{-4}$/m.

The result for $\Re e(\varepsilon'/\varepsilon)$ is [180]

$$\Re e \left(\frac{\varepsilon'}{\varepsilon}\right) = (20.71 \pm 1.48 \text{ (stat)} \pm 2.39 \text{ (syst)}) \times 10^{-4} = (20.7 \pm 2.8) \times 10^{-4} .$$

$$\tag{9.171}$$

The systematic uncertainty for the neutral decays is mainly due to background, CsI energy calibration and acceptance corrections; for the charged decays, it is mainly due to uncertainties in the acceptance and trigger efficiency.

The acceptance correction that has to be applied is about 5×10^{-2} for R, or $\sim 80 \times 10^{-4}$ for $\Re e(\varepsilon'/\varepsilon)$, four times larger than the signal.

9.5.6.5 The NA48 Experiment

When the NA31 observation of a nonvanishing $\Re e(\varepsilon'/\varepsilon)$ was not confirmed by the result from the E731 experiment, the CERN-based collaboration set out to construct a new, improved detector with the goal of achieving a precision measurement of $\Re e(\varepsilon'/\varepsilon)$ with a total uncertainty of 0.2×10^{-4}.

The new experiment was designed

- to measure all four decay modes concurrently by using two incident proton beams;
- to improve on neutral-background rejection by developing a liquid-krypton electromagnetic calorimeter with substantially better energy resolution;

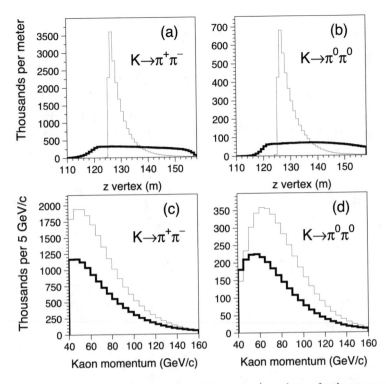

Fig. 9.16 (**a**) z vertex distribution for reconstructed $K \to \pi^+\pi^-$ decays for the vacuum beam (*thick*) and regenerator beam (*thin histogram*). (**b**) z vertex distribution for reconstructed $K \to \pi^0\pi^0$ decays. (**c**) Kaon momentum distribution for reconstructed $K \to \pi^+\pi^-$ decays. (**d**) Kaon momentum distribution for reconstructed $K \to \pi^0\pi^0$ decays. All $K \to \pi\pi$ analysis cuts have been applied and background has been subtracted (see [180])

– to improve on charged-background rejection by using a magnetic spectrometer.

The resulting beam and detector have been described in Sect. 9.5.4.5. Data were taken in 1997, 1998 and 1999 with 450 GeV protons. In the design of the NA48 detector, the cancellation of systematic uncertainties in the double ratio was exploited as much as possible [201]. Important properties of the experiment are

– two almost collinear beams, which lead to almost identical illumination of the detector, and
– the K_S lifetime weighting of the events defined as K_L events.

The K_L target was located 126 m upstream of the beginning of the decay region, and the K_S target 6 m upstream of the decay region. The beginning of the K_S decay region was defined by an anti-counter used to veto kaon decays occurring upstream of the counter and defining the global kaon energy scale.

The identification of K_S decays was done by a detector (tagger) consisting of an array of scintillators situated in the proton beam directed on to the K_S target. To

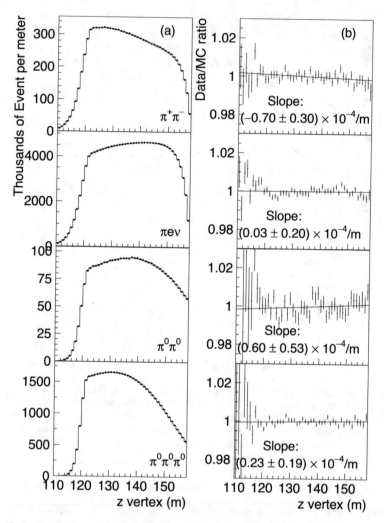

Fig. 9.17 (**a**) Comparison of the vacuum beam z distributions for data (*dots*) and MC calculations (*histograms*). The data-to-MC ratios (**b**) have been fitted to a line, and the z slopes are shown. The neutral distributions are for the combined 1996 + 1997 samples; the charged distributions are for 1997 only (kTeV experiment [180])

identify events coming from the K_S target a coincidence window of ± 2 ns between the proton signal in the tagger and the event time was chosen (see Fig. 9.18). Owing to inefficiencies in the tagger and in the proton reconstruction, a fraction α_{SL} of true K_S events are misidentified as K_L events. On the other hand, there is a constant background of protons in the tagger which have not led to a good K_S event. If those protons accidentally coincide with a true K_L event, this event is misidentified as a K_S decay. This fraction α_{LS} depends only on the proton rate in the tagger and the width of the coincidence window.

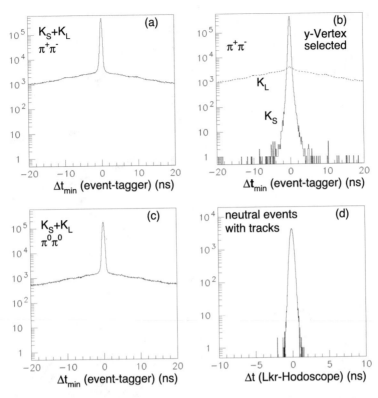

Fig. 9.18 (**a**), (**c**) Minimal difference between tagger time and event time ($\Delta t_{\min}$). (**b**) $\Delta t_{\min}$ for charged K_L and K_S events. (**d**) Comparison between charged and neutral event times. For this measurement, decays with tracks selected by the neutral trigger were used (γ conversion and Dalitz decays $K_S \to \pi^0 \pi^0_D \to \gamma\gamma\gamma e^+ e^-$) (NA48 experiment [201])

Both effects, α_{SL}^{+-} and α_{LS}^{+-}, can be measured (see Fig. 9.18b) in the charged mode, as K_S and K_L events can be distinguished by the vertical position of the decay vertex. The results are $\alpha_{SL}^{+-} = (1.63 \pm 0.03) \times 10^{-4}$ for the data from 1998/1999 and $(1.12 \pm 0.03) \times 10^{-4}$ for the data from 2001. For the accidental-tagging rate, the value measured was $\alpha_{LS}^{+-} = (10.649 \pm 0.008)\%$ for the 1998/1999 data sample and $(8.115 \pm 0.010)\%$ for the 2001 sample, owing to the lower instantaneous beam intensity. This means that about 11% or 8% of true K_L events are misidentified as K_S events; however, this quantity is precisely measured to the 10^{-4} level here. For the measurement of R the difference between the charged and the neutral decay modes, $\Delta\alpha_{LS} = \alpha_{LS}^{00} - \alpha_{LS}^{+-}$, is important. Proton rates in the sidebands of the tagging window were measured in both modes to determine $\Delta\alpha_{LS}$. The result is $\Delta\alpha_{LS} = (4.3 \pm 1.8) \times 10^{-4}$ for the 1998/1999 event sample and $(3.4 \pm 1.4) \times 10^{-4}$ for the 2001 event sample. Several methods have been used to measure $\Delta\alpha_{SL}$, leading to the conclusion that there is no measurable difference between the mistaggings measured by different methods within an uncertainty of $\pm 0.5 \times 10^{-4}$.

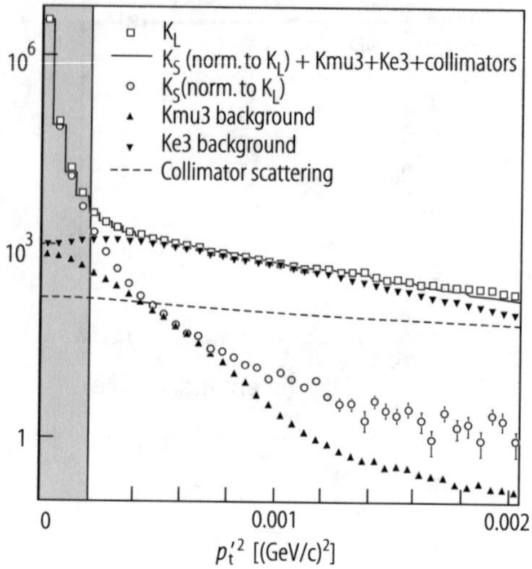

Fig. 9.19 Comparison of the $p_t'^2$ tail of the $K_L \rightarrow \pi^+\pi^-$ candidates with the sum of all known components (NA48 experiment [201])

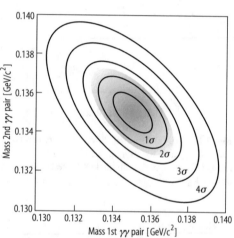

Fig. 9.20 Distribution of the $K_S \rightarrow \pi^0\pi^0$ candidates in the space of two reconstructed values of $m_{\gamma\gamma}$. The contours correspond to increments of one standard deviation (NA48 experiment [201])

Another important correction is the background subtraction. Decays of the types $K_L \rightarrow \pi e\nu$ and $K_L \rightarrow \pi\mu\nu$ can be misidentified as $K \rightarrow \pi^+\pi^-$ decays, as the ν is undetectable. These events were identified by their high transverse momentum p_t' and their reconstructed invariant mass. The remaining background can be measured by extrapolating the shape of the background in the $p_t'^2$ distribution into the signal region (Fig. 9.19). In this way, the charged background fraction leads to an overall correction to R of $(16.9 \pm 3.0) \times 10^{-4}$ for the 1998/1999 data and $(14.2 \pm 3.0) \times 10^{-4}$ for the 2001 sample.

The reconstruction of $\pi^0\pi^0$ decays followed the principles described in Sect. 9.5.4.2. The two $\gamma\gamma$ masses for the best pairing are anticorrelated because of the constraint of the kaon mass (Fig. 9.20). The ellipses in Fig. 9.20 designate contours with increments of one standard deviation. The background from $3\pi^0$

Table 9.4 Event numbers of the NA48 experiment after tagging correction and background subtraction [201, 202]

Event statistics [$\times 10^6$]					
	1998/1999	2001		1998/1999	2001
$K_S \to \pi^+\pi^-$	22.221	9.605	$K_L \to \pi^+\pi^-$	14.453	7.136
$K_S \to \pi^0\pi^0$	5.209	2.159	$K_L \to \pi^0\pi^0$	3.290	1.546

decays with two undetected photons is distributed with constant probability over each ellipse, as shown by Monte Carlo calculations. This leads to a correction to R of $(-5.9 \pm 2.0) \times 10^{-4}$ for the 1998/1999 sample and $(-5.6 \pm 2.0) \times 10^{-4}$ for the 2001 sample.

The numbers of signal events after these corrections are summarized in Table 9.4.

The efficiency of the triggers used to record neutral and charged events has been determined. In the neutral decay mode the efficiency was measured to be 0.99920 ± 0.00009, without any measurable difference between K_S and K_L decays. The $\pi^+\pi^-$ trigger efficiency was measured to be $(98.319 \pm 0.038)\%$ for K_L and $(98.353 \pm 0.022)\%$ for K_S decays. Here, a small difference between the trigger efficiencies for K_S and K_L decays was found. This leads to a correction to the double ratio of $(-4.5 \pm 4.7) \times 10^{-4}$ for the 1998/1999 sample and $(5.2 \pm 3.6) \times 10^{-4}$ for the 2001 sample.

Other systematic uncertainties include the limited knowledge of the energy scale, nonlinearities in the calorimeter, and small acceptance corrections.

Summing all corrections to and systematic uncertainties in R, the authors find the amount to $(35.9 \pm 12.6) \times 10^{-4}$ for the 1998/1999 data and $(35.0 \pm 11.0) \times 10^{-4}$ for the 2001 data.

The corresponding result for the direct-$\mathcal{CP}$-violation parameter is

$$\Re e \left(\frac{\varepsilon'}{\varepsilon} \right) = (15.3 \pm 2.6) \times 10^{-4} \tag{9.172}$$

for the data from 1997 [203] and 1998/1999 [201], and

$$\Re e \left(\frac{\varepsilon'}{\varepsilon} \right) = (13.7 \pm 3.1) \times 10^{-4} \tag{9.173}$$

for the data from 2001 [202].

A comparison of these two values is significant because they were obtained at different average beam intensities. The combined final result from the NA48 experiment is

$$\Re e \left(\frac{\varepsilon'}{\varepsilon} \right) = (14.7 \pm 1.4 \, (\text{stat}) \pm 0.9 \, (\text{syst}) \pm 1.5 \, (\text{MC})) \times 10^{-4} = (14.7 \pm 2.2) \times 10^{-4} . \tag{9.174}$$

Fig. 9.21 Time sequence of published measurements of the parameter $\Re e(\varepsilon'/\varepsilon)$ of direct CP violation. The experiments at CERN are marked by *filled circles* and the experiments at Fermilab are marked by *open squares*. The kTeV result from 2003 is a reanalysis of the data from the kTeV 1999 result (see [204] and also [180, 200–203])

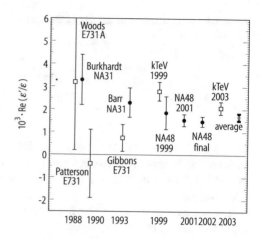

9.5.6.6 Conclusions About Direct CP Violation, $\Re e(\varepsilon'/\varepsilon)$ and the Wu–Yang Triangle

The two experiments kTeV and NA48 have definitively confirmed the original observation of the NA31 team that direct CP violation exists. The results of all published experiments on ϵ'/ϵ are shown in Fig. 9.21. Therefore, CP violation as observed in the K meson system is a part of the weak interaction due to weak quark mixing. Exotic, new interactions such as the superweak interaction are not needed. We therefore have a very precise experimental result for ε'/ε. The theoretical calculations of ε'/ε within the Standard Model, however, are still not very precise. This does not change the main conclusion of the experiments that ε' is different from zero and positive, i.e. direct CP violation exists.

If we take into account the four relevant experiments NA31, E731, NA48, and kTeV, the weighted average comes out to be

$$\Re e\left(\frac{\varepsilon'}{\varepsilon}\right) = (16.7 \pm 1.6) \times 10^{-4} . \tag{9.175}$$

The consistency of the result is not completely satisfactory, since $\chi^2/\text{ndf} = 6.3/3$. If the phase of ε' as defined in (9.92), arg $\varepsilon' = (42.3 \pm 1.5)°$ [149], is inferred, then a more precise value for the component of ε' transverse to ε can be derived.

We have done a complete fit to the Wu–Yang triangle (9.90), using as input

$$\Re e\left(\frac{\varepsilon'}{\varepsilon}\right) = (16.7 \pm 1.6) \times 10^{-4}, \quad \Re e\,\varepsilon = (1.658 \pm 0.0265) \times 10^{-3} ,$$

$$\Phi_{+-} = 43.3° \pm 0.4°, \quad \Phi_{00} - \Phi_{+-} = (0.36 \pm 0.43)° ,$$

$$|\eta_{+-}| = (2.230 \pm 0.006) \times 10^{-3}, \quad |\eta_{00}| = (2.225 \pm 0.007) \times 10^{-3} ,$$

$$\text{arg } \varepsilon = 43.4 \pm 0.1°, \quad \text{arg } \varepsilon' = 42.3° \pm 1.5° .$$

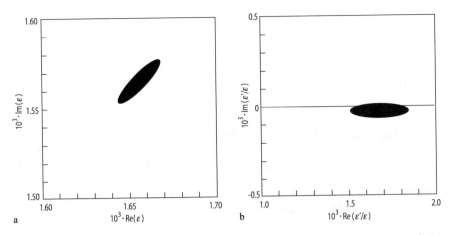

Fig. 9.22 Result of the fit to the Wu–Yang triangle relations between η_{+-}, η_{00}, ε and ε'. (**a**) Correlation of $\Re e\,\varepsilon$ and $\Im m\,\varepsilon$; (**b**) correlation of $\Re e(\varepsilon'/\varepsilon)$ and $\Im m(\varepsilon'/\varepsilon)$. The boundaries of the black areas correspond to one-standard-deviation uncertainties

The result of this fit is, with $\chi^2/\mathrm{ndf} = 3/4$

$$\Im m\left(\frac{\varepsilon'}{\varepsilon}\right) = (-3.2 \pm 4.4) \times 10^{-5} \tag{9.176}$$

and

$$\Im m\,\varepsilon = (1.530 \pm 0.005) \times 10^{-3} . \tag{9.177}$$

Also, the real part of ε is constrained by the fit:

$$\Re e\,\varepsilon = (1.619 \pm 0.005) \times 10^{-3}$$

The result of this fit is expressed in correlation plots for $\Re e\,\varepsilon$ and $\Im m\,\varepsilon$, and for $\Re e(\varepsilon'/\varepsilon)$ and $\Im m(\varepsilon'/\varepsilon)$, in Fig. 9.22.

The result for direct $\mathcal{CP}$ violation (9.175) can also be quoted as a decay asymmetry between the K^0 and $\overline{K^0}$ decay rates to a $\pi^+\pi^-$ final state. If the amplitudes are called $a = \mathrm{amp}(K^0 \to \pi^+\pi^-)$ and $\bar{a} = \mathrm{amp}(\overline{K^0} \to \pi^+\pi^-)$, then this asymmetry is

$$A = \frac{\Gamma(K^0 \to \pi^+\pi^-) - \Gamma(\overline{K^0} \to \pi^+\pi^-)}{\Gamma(K^0 \to \pi^+\pi^-) + \Gamma(\overline{K^0} \to \pi^+\pi^-)} = \frac{|a|^2 - |\bar{a}|^2}{|a|^2 + |\bar{a}|^2} . \tag{9.178}$$

Since $\bar{a}/a = 1 - 2\varepsilon'$, we obtain

$$\left| \frac{\bar{a}}{a} \right|^2 = 1 - 4\,\Re e\,\varepsilon' \tag{9.179}$$

and

$$A = 2\,\Re e\,\varepsilon' = 2\left(\Re e\,\varepsilon\,\Re e\left(\frac{\varepsilon'}{\varepsilon} \right) - \Im m\,\varepsilon\,\Im m\left(\frac{\varepsilon'}{\varepsilon} \right) \right) = (5.5 \pm 0.6) \times 10^{-6}\,. \tag{9.180}$$

This very small decay rate asymmetry can be compared with the large values of some similar observables in the B system.

9.6   The Neutral B Meson System

The KM model today often also referred as CKM^1 mechanism of the Standard Model predicts direct CP violation for the neutral K meson system and it was thus a strong support for the theory when a finite value of $\Re e(\varepsilon'/\varepsilon)$ was observed. The ultimate test of the quark-mixing paradigm however was the precise study of CP violation in the neutral B meson system for which the theory predicted large CP violation. The discovery of $B^0 - \overline{B^0}$ mixing in 1987 [58] made the B^0 system the prime candidate for observing CP violation in a physical system different from the neutral kaon. The observation of large time-dependent CP asymmetries in B^0 decays in 2001 by the B-factories [59, 60] allowed the determination of the complex phase δ of the CKM matrix and as a consequence provided a first sensitive test of the unitarity of the quark mixing matrix. With the observation of the B_s^0 meson mixing in 2006 [61] a second B meson system became available for precise mixing studies and studies of CP violating effects. With the start of the LHCb experiment [62] the focus of the B meson studies have changed. Precision measurements no longer aim to prove the CKM paradigm but look for small deviations from the theory predictions as possible smoking-guns for physics beyond the description of the Standard Model.

---

[1]N. Cabibbo introduced the concept of quark mixing for two quark generations [56]. It was extended to more generations by M. Kobayashi und T. Maskawa who also explained CP violation in case of three and more generations [57]. The corresponding 3×3 quark mixing matrix is today referred as KM or CKM matrix. In the remaining part of this chapter the term CKM matrix is used.

9.6.1 Phenomenology of Mixing in the Neutral B Meson System

The parameters of mixing in the two neutral B mesons—the B^0 with quark content $(\bar{b}d)$ and the B_s^0 with quark content $(\bar{b}s)$, in the following they are called both generically B^0—are very different from those observed in the neutral K meson system. In Sect. 9.3 the complex decay parameters γ_h and γ_l are defined for the heavy (h) and light (l) meson states (see Eq. (9.41)):

$$\gamma_h = im_h + \frac{\Gamma_h}{2} ,$$

$$\gamma_l = im_l + \frac{\Gamma_l}{2} .$$

With these parameters the time evolution of the neutral B states can be written as (see Eqs. (9.34) and (9.40))

$$|B_h(t)\rangle = (p\left|B^0\right\rangle - q\left|\overline{B^0}\right\rangle)e^{-\gamma_h t} ,$$

$$|B_l(t)\rangle = (p\left|B^0\right\rangle + q\left|\overline{B^0}\right\rangle)e^{-\gamma_l t} . \tag{9.181}$$

The two initial states at $t = 0$, with a definite quantum number $\mathbb{B}$, are

$$\psi_B(0) = B^0 \quad \text{and} \quad \psi_{\overline{B}}(0) = \overline{B^0} . \tag{9.182}$$

Their decay law is not any longer exponential and results to a finite probability P for a flavor change given by Eqs. (9.44) and (9.45). In the Standard Model, the CP violation in mixing of neutral B mesons is expected to be very small, such that $|q/p| = 1$ within $\mathcal{O}(10^{-4})$ and $\mathcal{O}(10^{-5})$ for the B^0 and the B_s^0 system respectively [63]. In this approximation the mixing probabilities for the two B^0 and $\overline{B^0}$ are equal, i.e.

$$P(B^0 \to \overline{B^0}) = P(\overline{B^0} \to B^0) , \tag{9.183}$$

and Eq. (9.44) can be written as

$$P(B^0 \to \overline{B^0}) = \frac{1}{2}e^{-T} (\cosh yT - \cos xT) = \frac{1}{2}e^{-\Gamma t} \left(\cosh \frac{\Delta\Gamma t}{2} - \cos(\Delta m\, t)\right) . \tag{9.184}$$

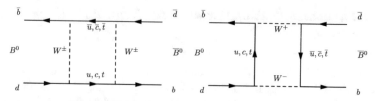

Fig. 9.23 Feynman diagrams of the B^0 mixing

Correspondingly, the probability for remaining in the original beauty state is,

$$P(B^0 \rightarrow B^0) = P(\overline{B^0} \rightarrow \overline{B^0}) = \frac{1}{2}e^{-\Gamma t}\left(\cosh\frac{\Delta\Gamma t}{2} + \cos(\Delta m\, t)\right). \qquad (9.185)$$

One thus obtains for the flavor asymmetry at time t of an initially pure flavor state (see Eq. (9.50)):

$$A(t) = \frac{P(B^0 \rightarrow B^0) - P(B^0 \rightarrow \overline{B^0})}{P(B^0 \rightarrow B^0) + P(B^0 \rightarrow \overline{B^0})} = \frac{\cos(\Delta m\, t)}{\cosh\frac{\Delta\Gamma t}{2}} \qquad (9.186)$$

The mixing parameter x is determined through short-range interactions given by box diagrams of the type shown in Fig. 9.23. For the B^0 mixing the diagrams with a virtual t quark dominate and one obtains [63, 65] for the mixing parameter x,

$$x \equiv \frac{\Delta m}{\Gamma} = \frac{G_F^2}{6\pi^2}\mathcal{B}_{B_q}f_{B_q}^2 m_{B_q}\tau_{B_q}|V_{tb}^*V_{tq}|^2 M_W^2\, F\left(\frac{m_t^2}{M_W^2}\right)\eta_{QCD} \qquad (9.187)$$

with index $q = d(s)$ for the B^0 (B_s^0) respectively. Here, m_{B_q} and τ_{B_q} are the mass and the lifetime of the neutral B meson, $\mathcal{B}_{B_q}$ is the *bag factor* parametrizing the probability that the d (s) and the $\overline{b}$ quarks will form a B^0 (B_s^0) hadron, f_{B_q} is the B meson decay constant, F is the calculated loop function, increasing with the top quark mass squared m_t^2, and $\eta_{QCD} \approx 0.8$ is a QCD correction. The parameters G_F and M_W are the Fermi coupling constant and the mass of the W boson and V_{ij} are the CKM matrix elements (see Sect. 9.7).

For the B^0 meson the lifetimes of the heavy and light state are approximately equal, $\Gamma_h = \Gamma_l = \Gamma$, and $y = \Delta\Gamma/2\Gamma \approx 0$. In this case the flavor asymmetry simplifies to $A(t) = \cos(\Delta mt)$. For the B_s^0 meson a significant lifetime difference between the heavy and light states arises due to different decay channels as result of the different CP eigenvalues of the two mass states. The lifetime or width difference can thus be measured using final states with defined CP values.[2] From recent theory calculations one expects for the B_s^0 meson $\Delta\Gamma/\Gamma \approx 13\%$ [64, 66]. It should further

---

[2]This is strictly true only if one ignores the very small CP violation in mixing.

be noted that applying the definition of Sect. 9.3 for $\Delta\Gamma$ leads to a negative width differences for the B_s^0 mesons. It is therefore common practice [67] to redefine $\Delta\Gamma$ to be positive,

$$\Delta\Gamma = \Gamma_l - \Gamma_h . \tag{9.188}$$

To test the approximation $|q/p| \approx 1$, i.e. the assumption of vanishing $\mathcal{CP}$ violation in the mixing,

$$a_{\text{mix}} = \frac{P(\overline{B^0} \to B^0) - P(B^0 \to \overline{B^0})}{P(\overline{B^0} \to B^0) + P(B^0 \to \overline{B^0})} = \frac{1 - |q/p|^4}{1 + |q/p|^4} \approx 0 ,$$

one can measure the semi-leptonic $\mathcal{CP}$ asymmetry for the two B^0 species (see Eq. (9.56)),

$$a_{\text{SL}} = \frac{\Gamma(\overline{B^0}(t) \to \ell^+ \nu X) - \Gamma(B^0(t) \to \ell^- \bar{\nu} X)}{\Gamma(\overline{B^0}(t) \to \ell^+ \nu X) + \Gamma(B^0(t) \to \ell^- \bar{\nu} X)} = a_{\text{mix}} , \tag{9.189}$$

where the detection of a B^0 decaying into a *wrong-sign muon* $B^0 \to \overline{B^0} \to \ell^- \bar{\nu} X$ indicates mixing.

The measurement of the time-dependent $\mathcal{CP}$ asymmetry between the decays of a B^0 and a $\overline{B^0}$ to a common final state f probes the $\mathcal{CP}$ violation in the interference between the decay with and without mixing. Following Eq. (9.51) one introduces the decay amplitudes

$$A_f = \langle f|T|B^0\rangle , \quad A_{\bar{f}} = \langle \bar{f}|T|B^0\rangle ,$$

$$\overline{A_f} = \langle f|T|\overline{B^0}\rangle , \quad \overline{A_{\bar{f}}} = \langle \bar{f}|T|\overline{B^0}\rangle . \tag{9.190}$$

The theoretical description simplifies if decays to $\mathcal{CP}$ eigenstates $B \to f_{CP}$ are used. These final states fulfill $\mathcal{CP}|f_{CP}\rangle = |\overline{f_{CP}}\rangle = \eta_{CP}|f_{CP}\rangle$ with $\eta_{CP} = \pm 1$ and $\overline{A_f} = \eta_{CP}\overline{A_{\bar{f}}}$.

Using Eq. (9.42), one obtains the decay rates of initially pure B^0 and $\overline{B^0}$ states,

$$\frac{dN(t)}{dt} = \frac{1}{4} \left| \left(e^{-\gamma_h t} + e^{-\gamma_l t}\right) A_f - \frac{q}{p} \left(e^{-\gamma_h t} - e^{-\gamma_l t}\right) \overline{A_f} \right|^2 ,$$

$$\frac{d\overline{N}(t)}{dt} = \frac{1}{4} \left| \left(e^{-\gamma_h t} + e^{-\gamma_l t}\right) \overline{A_f} - \frac{p}{q} \left(e^{-\gamma_h t} - e^{-\gamma_l t}\right) A_f \right|^2 . \tag{9.191}$$

The time dependent $\mathcal{CP}$ asymmetry is defined as

$$a_{CP}(t) = \frac{d\overline{N} - dN}{d\overline{N} + dN} . \tag{9.192}$$

Neglecting the very small deviation of $|q/p|$ from unity (i.e. $\mathcal{CP}$ violation in mixing) and using the ratio defined in Eq. (9.53),

$$\lambda_f = \frac{q\overline{A}_f}{pA_f},$$ (9.193)

one obtains for $a_{CP}(t)$ [68],

$$a_{CP}(t) = -\frac{A_{CP}^{dir}\cos(\Delta mt) + A_{CP}^{mix}\sin(\Delta mt)}{\cosh(\frac{\Delta\Gamma}{2}t) + A_{\Delta\Gamma}\sinh(\frac{\Delta\Gamma}{2}t)}$$ (9.194)

with

$$A_{CP}^{dir} = \frac{1 - |\lambda_f|^2}{1 + |\lambda_f|^2}, \quad A_{CP}^{mix} = -\frac{2\Im\lambda_f}{1 + |\lambda_f|^2}, \quad A_{\Delta\Gamma} = -\frac{2\Re\lambda_f}{1 + |\lambda_f|^2}.$$ (9.195)

The first term accounts for possible direct $\mathcal{CP}$ violation while the other two terms encode properties of the mixing. Experimentally, by measuring the time dependent $\mathcal{CP}$ violation $a_{CP}(t)$ one can determine the coefficients of $\cos(\Delta mt)$ and $\sin(\Delta mt)$ and thus determine $|\lambda_f|$ and $\Im\lambda_f$. In general, non-perturbative QCD effects prevent to relate these quantities to $\mathcal{CP}$ phases originating from the quark mixing. However, in case of so called *golden modes* which are dominated by a single decay amplitude and thus a single combination $\mathcal{V}_{CKM}$ of CKM elements such an association is possible. Since the strong interaction respects the $\mathcal{CP}$ symmetry, golden modes fulfill:

$$\frac{\overline{A}_f}{A_f} = \eta_{CP}\frac{\overline{A}_{\bar{f}}}{A_f} = \eta_{CP}\frac{\mathcal{V}_{CKM}^*}{\mathcal{V}_{CKM}}$$ (9.196)

From this equation one sees that $|\overline{A}_f| = |A_f|$ and that assuming $|q/p| = 1$, one obtains $|\lambda_f| = 1$. Thus golden modes satisfy

$$A_{CP}^{dir} = 0, \quad A_{CP}^{mix} = -\Im\lambda_f.$$ (9.197)

For golden decays of B^0 mesons for which the decay width difference $\Delta\Gamma$ is negligible, expression (9.194) further simplifies to

$$a_{CP}(t) = -\Im\lambda_f\sin(\Delta mt).$$ (9.198)

Thus, if λ_f carries a non-trivial weak phase ϕ ($\Im\lambda_f \neq 0$) the time dependent $\mathcal{CP}$ asymmetry will show a sinusoidal time behaviour with an amplitude given by $\sin\phi$.

9.6.2 *Production and Detection of B-Mesons*

The precise measurement of oscillation and $\mathcal{CP}$ violation in the neutral B meson system requires a high number of produced B mesons. A good reconstruction of the B meson decay vertex, necessary for an excellent decay time resolution, is needed to resolve the time-dependent effects. To perform the $\mathcal{CP}$ measurements the determination of the production flavor of the B meson, often referred as flavor-tagging, is necessary.

9.6.2.1 e^+e^- **B-Factories**

The first experiments which have systematically addressed these requirements are the BABAR experiment [69] at the e^+e^- collider PEP-II at Stanford and the Belle experiment [70] at the Japanese e^+e^- collider KEKB. Both e^+e^- machines were operating at a centre-of-mass energy of $\sqrt{s} = 10.58$ GeV, corresponding to the mass of the $\Upsilon(4S)$, an excited $b\bar{b}$ resonance which decays to $B^0\overline{B^0}$ ($\sim 50\%$) and B^+B^- ($\sim 50\%$). The cross section for the $\Upsilon(4S)$ production is 1.1 nb which is about a quarter of the total hadronic cross section.

BABAR was operated from 1999 to 2008 and collected a data set corresponding to an integrated luminosity of 550 fb^{-1}. The Belle experiment has also been started in 1999 and was taking data until the end of 2009, collecting about 1 ab^{-1} of data. Due to the large number of B mesons the two experiments have recorded, they are often referred as B-factories.[3] To continue this successful path, the KEKB collider as well as the Belle experiment have both undergone an upgrade. The new Belle-II experiment has started data-taking in 2018 and will be operated at a 40 times larger instantaneous luminosity [72].

The mass of the $\Upsilon(4S)$ lies only 11 MeV/c^2 above the sum of the two produced B mesons which would therefore be produced essentially at rest and would not fly. The vanishing decay length would prevent a measurement of the decay time. To overcome this problem both machines have been operated with slightly asymmetric beam energies resulting in a small boost of the $\Upsilon(4S)$ and thus of the produced B mesons. The boost factor $\beta\gamma$ was 0.56 (0.425) for the BABAR (Belle) experiment, resulting into typical decay lengths of about 250 μm for the two B mesons with lifetimes of 1.5 ps.

If the $\Upsilon(4S)$ with $J^P = 1^-$ decays into a pair of neutral B mesons, the two mesons are produced in a coherent $B^0\overline{B^0}$ state with negative parity,

$$B^0(\theta_1)\overline{B^0}(\theta_2) - B^0(\theta_2)\overline{B^0}(\theta_1) \, , \tag{9.199}$$

---

[3] A comprehensive description of the two e^+e^- machines, the two detectors, the operation and the physics results can be found in [71].

where $\theta_{1,2}$ are the B meson production angles relative to the e^+ direction. The flavors of the two mesons are thus fully correlated until the first meson decays at time t_1. If at that time this meson is a B^0 the flavor of the second meson is fixed to be a $\overline{B^0}$ and vice versa. The time evolution of the second B meson with decay time t_2 is then given by

$$\psi_2(t_2) = \overline{B^0} \left(e^{-\gamma_h(t_2-t_1)} + e^{-\gamma_l(t_2-t_1)} \right) - \frac{q}{p} B^0 \left(e^{-\gamma_h(t_2-t_1)} - e^{-\gamma_l(t_2-t_1)} \right) ,$$

(9.200)

i.e. the time evolution of the flavor of the second B meson is defined by the time difference $\Delta t = t_2 - t_1$ between the two decays. Depending on which of the two B mesons is studied, Δt can also be negative.

The decay topology of a $B^0\overline{B^0}$ pair produced at $t = 0$ from the decay of the $\Upsilon(4S)$ is illustrated in Fig. 9.24, where one of the neutral B mesons (signal B) is decaying at t_2 into the golden mode $J/\psi K_S$ used to measure the time-dependent $\mathcal{CP}$ violation, and the other B meson (tagging B) decays at t_1 into a flavor-specific final state. The charge of the electron indicates the flavor of the B meson at the time of the decay (B^0). At this time the signal B was thus a $\overline{B^0}$.

Due to the boost of the $\Upsilon(4S)$ the different decay times result into different z-positions of the two decay vertices along the beam direction. The difference Δz is about 250 μm and is related to the decay time difference Δt:

$$\Delta z = z_2 - z_1 = \beta\gamma c (t_2 - t_1) = \beta\gamma c \Delta t .$$

(9.201)

Both B-factories are very similar. Therefore only the Belle experiment at the KEKB collider in Tsukuba is discussed here. To produce a boost $\beta\gamma = 0.425$ of the

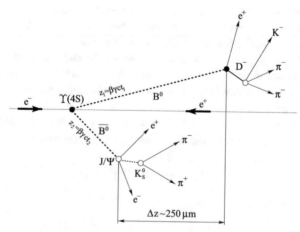

Fig. 9.24 Decay topology of $B^0\overline{B^0}$ pairs

$\Upsilon(4S)$ in the laboratory, the KEKB electron beam has an energy of 8 GeV while the positrons collide with an energy of 3.5 GeV. The highest instantaneous luminosity achieved at KEKB was $\mathcal{L} = 2 \times 10^{34} \mathrm{s}^{-1}\,\mathrm{cm}^{-2}$ [71].

The Belle detector (Fig. 9.25) followed the typical onion-shape design. After an exchange in 2003, a silicon vertex detector (SVD) made from four layers of double-sided silicon strip sensors was used as inner component. The innermost (outermost) SVD layer was located at a radial distance of only 20 (88) mm from the collision point. The impact parameter resolution for charged particle tracks obtained is [71]

$$\sigma_r = \left[21.9 \oplus \frac{35.5}{p} \right] \mu\mathrm{m} \quad [p \text{ in GeV/c}] \ ,$$

$$\sigma_z = \left[27.8 \oplus \frac{31.9}{p} \right] \mu\mathrm{m} \ .$$

The SVD was followed by the central drift chamber (CDC) which extended to a radial distances of 88 cm. The combined tracking system of SVD and CDC provided a good momentum resolution, especially for low momentum tracks, thanks to the minimization of material. For the transverse momentum a resolution of $\sigma_{p_T}/p_T = 0.0019 \cdot p_T \oplus 0.0030/\beta$ [p_T in GeV/c] was achieved [71]. Particle identification was provided by the time-of-flight (TOF) system and by the aerogel

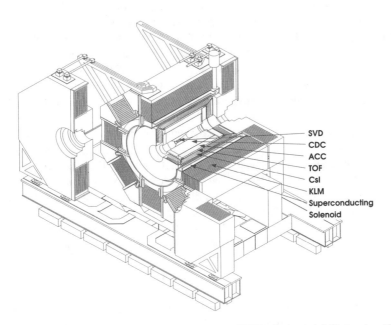

SVD
CDC
ACC
TOF
CsI
KLM
Superconducting
Solenoid

Fig. 9.25 Belle detector with the silicon vertex detector (SVD), the central drift chamber (CDC), the time-of-flight-system (TOF) and the aerogel Cherenkov counter (ACC), the CsI electromagnetic calorimeter and the K_L/μ detector (KLM). The detector is embedded in a magnetic field of 1.5 T produced by a solenoid with a length of 4.4 m and a diameter of 3.4 m [70]

Cherenkov counter (ACC). Photon and electron energies were measured in the electromagnetic calorimeter (ECL) consisting of thallium-doped CsI crystals. The energy resolution varied from 4% at 100 MeV to about 1.6% at 8 GeV [71]. All detector components were placed inside a super-conducting solenoid magnet of cylindrical shape which was providing a magnetic field of 1.5 T. The iron return yoke is instrumented with resistive plate chambers (KLM) and served to identify muons and K_L with a momentum above 600 MeV/c.

9.6.2.2 The LHCb Experiment at the Large Hadron Collider

The multi-purpose experiments CDF and D0 [73–76] operated at the $p\bar{p}$ collider Tevatron at Fermilab established that—despite of the harsh and high-multiplicity environment of the hadron collisions—precision measurements of B mesons competitive with those performed at e^+e^- B-factories are possible at hadron machines. Both experiments have pioneered studies of the B_s^0 system which was hardly studied at the e^+e^- machines, as kinematically the heavier B_s^0 system is only accessible through the decays of the $\Upsilon(5S)$. Only the Belle collaboration collected a small data sample of $\Upsilon(5S)$ decays.

At the Large Hadron Collider (LHC) at CERN, the LHCb experiment [77] and to a lesser extent also the ATLAS and CMS experiments [78, 79] followed the successful path of precision B meson studies explored at Tevatron. The cross section for $b\bar{b}$ pair production in the pp collision of the LHC is huge, about 500 µb for proton-proton center-of-mass energies of 13 TeV. As only small fractions of the proton energies are needed to produce the $b\bar{b}$ pair, the momentum factions x_1 and x_2 of the colliding partons are in general very different and the relatively light $b\bar{b}$ system is boosted in the laboratory frame into either the forward or the backward direction. The LHCb experiment has therefore been designed as a single-arm forward spectrometer. With a pseudo-rapidity coverage of $2 < \eta < 5$ about 35% of all produced $b\bar{b}$ pairs lie within the detector acceptance. The average boost of the b hadrons is large, $\beta\gamma \approx 25$, resulting into average flight distances of about 1 cm. Although the $b\bar{b}$ cross section at the LHC is large, the total inelastic cross section is about a factor 200 larger. The huge rate of non b events together with the fact that the decay products of the b hadrons are comparably soft makes the trigger to the primary challenge for any B meson experiment. Typical combined trigger and reconstruction efficiencies vary between 0.1 and 10% depending on the decay channel. Channels with muons in the final state are in general easier to trigger and reconstruct.

At a hadron collider the two produced b quarks hadronise separately and no quantum correlation between the two b hadrons exists. It is therefore harder to conclude from the flavor of the second b hadron (tagging B), which for the case that it is a neutral B meson oscillates independently, on the production flavor of the signal B. Effective tagging efficiencies therefore stay significantly below 10%.

The LHCb forward spectrometer is shown in Fig. 9.26 and includes a high-precision tracking system consisting of a silicon-strip vertex detector (Velo) sur-

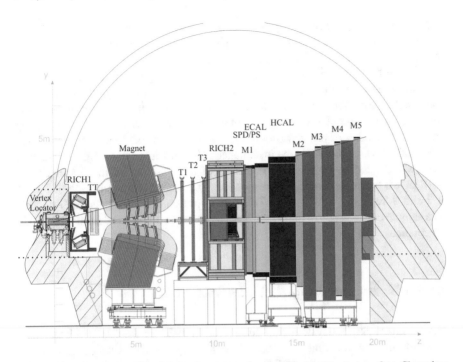

Fig. 9.26 LHCb detector consisting of a vertex detector (Vertex Locator), a first Cherenkov detector (RICH1), a large-area silicon-strip detector (TT), the main tracking stations (T1, T2, T3) with an inner part built from silicon-strip detectors and outer part build with straw drift tubes, a second Cherenkov detector (RICH2), a scintillating-pad detector (SPD) and a preshower detector (PS) in front of an electromagnetic (ECAL) and hadronic calorimeter (HCAL, and the muon system with 5 stations (M1 to M5). A dipole magnet (4 Tm) is placed between the large-area silicon-strip detector and the main tracking stations [77]

rounding the pp interaction region, a large-area silicon-strip detector (TT) located upstream of a dipole magnet with a bending power of about 4 Tm, and three stations of silicon-strip detectors (IT) and straw drift tubes (OT) placed downstream of the magnet. The tracking system provides a measurement of momentum of charged particles with a relative uncertainty that varies from 0.5% at low momentum to 1.0% at 200 GeV/c. The minimum distance of a track to the primary proton-proton vertex, the impact parameter (IP), is measured with a resolution of

$$\sigma_{IP} = (15 + 29/p_T)\mu m \quad [p_T \text{ in GeV/c}]$$

The average boost and the vertex resolution translates into a decay-time resolution of about 50 fs. Different types of charged hadrons are distinguished using information from two ring-imaging Cherenkov detectors (RICH). Photons, electrons and hadrons are identified by a calorimeter system consisting of scintillating pads (SPD) and a preshower detector (PS), an electromagnetic calorimeter (ECAL) and a hadronic calorimeter (HCAL). Muons are identified by a system composed of alternating layers of iron and multiwire proportional chambers. The online event selection is performed by a trigger system consisting of a hardware stage, based on information from the calorimeter and the muon systems, followed by a software stage, which applies full event reconstruction.

The displacement of the B decay vertex from the primary interaction vertex is an important identification requirement of B events. However, due to the large boost in forward direction, the association of b hadrons to their production vertex is challenging, in particular if the number of primary vertices is large. In order to keep the average number of primary proton-proton interactions per bunch crossing at the most optimal value (below 2) the instantaneous luminosity for the LHCb interaction point was leveled by displacing the centres of the colliding beams slightly. Until 2018 LHCb has recorded data corresponding to an integrated luminosity of 3 fb^{-1} at 7 and 8 TeV, and 6 fb^{-1} at 13 TeV proton-proton centre-of-mass energy. This data sample corresponds to more than 1×10^{12} $b\bar{b}$ events produced inside the LHCb acceptance.

9.6.3 Measurements of B Oscillations

Mixing in the $B^0 - \overline{B^0}$ system was discovered in 1987 by the ARGUS collaboration [80] and the first time-integrated determinations of the mixing parameter x_d have been performed by the ARGUS and the CLEO collaborations [81, 82]. Time-dependent measurements of the mixing frequency Δm_d became possible at the electron-positron collider LEP. The silicon vertex detectors and the large boost of the B mesons produced at the Z resonance allowed to observe the B^0 oscillations. With their high statistics and low background B samples the BABAR and Belle experiments improved the errors on the mixing frequency Δm_d significantly [83, 84]. Exploiting the excellent time resolution as well as the large statistics of recorded B mesons the LHCb experiment performed the so far most precise measurement of Δm_d [88].

After the first B^0 mixing measurement it was clear that mixing was an important effect also in the $B_s^0 - \overline{B_s^0}$ system and theory predicted a much faster oscillation of the B_s^0 meson. While limits on x_s existed from LEP and the two Tevatron experiments it took until 2006 that the CDF collaboration resolved the fast mixing and performed the first measurement of the mixing frequency [85]. The measurement was repeated with much smaller uncertainties by the LHCb collaboration in 2012 [86].

The oscillation frequency of neutral B mesons is measured using flavor-specific final states, i.e. final states like $B^0 \rightarrow D^{(*)-}\ell^+\nu_\ell$ or $B^0 \rightarrow D^-\pi^+$ where the charge of one of the final-state particles (e.g. the lepton or pion charge) indicate the flavor of the decaying signal B^0 (e.g. $\bar{b} \rightarrow \bar{c}\ell^+\nu_\ell$). In case of incoherent production of the two b hadrons, a flavor-specific decay of the second B (B_{tag}) is used to tag the flavor at the time $t = 0$ of the production. For the coherent production at the B factories the tagging B defines the flavor of the signal B at time Δt with respect to its decay—see Fig. 9.24 and Eq. (9.201). B-factory experiments have also used explicitly so called dilepton events where both B mesons decay semi-leptonically. Flavor-mixing of one of the B mesons is indicated by the presence of same-sign lepton events.

For LHCb, where the boost factor of the decaying B is not known and must be reconstructed from the final-state particles, semi-leptonic signal decays with an undetected neutrino present an additional complication compared to full hadronic decays. To account for the missing neutrinos in the calculation of the proper decay-time correction factors are used.

An important performance number for the mixing measurement and even more for $\mathcal{CP}$ violation measurements is the tagging power or effective tagging efficiency $Q = \epsilon_{tag} D_{tag}^2$ which corresponds to the efficiency to correctly tag the flavor of the signal B meson. The tagging power is the product of the tagging efficiency ϵ_{tag}, the probability that a specific algorithm delivers tagging information, and the square of the dilution factor $D_{tag} = 1 - 2\omega$, where ω is the mistag probability, i.e. the probability that a tagging decision is wrong. At a hadron collider, B mesons are produced independently of each other and mistag probabilities are large (about 40%). The B-factories profit from the coherent $B^0\overline{B^0}$ production and mistag probabilities of only a few percents are achieved for some tagging algorithms. Consequently, the tagging power at the B-factories is as high as 30% while for LHCb values only up to 6% are obtained.

If one considers the effect of the tagging dilution D_{tag} as well as the effect of a finite time resolution expressed by a dilution factor D_t one expects that the measured flavor asymmetry A_{meas} parameterizes as,

$$A_{meas}(t) = \frac{N_{unmixed}(t) - N_{mixed}(t)}{N_{unmixed}(t) + N_{mixed}(t)} = D_{tag} D_t \frac{\cos \Delta mt}{\cosh \Delta \Gamma /2t}, \qquad (9.202)$$

where t is the decay time of the signal B. For the measurement at BABAR and Belle with coherently produced B mesons the time difference $\Delta t = t_{sig} - t_{tag}$ between the decay times of the signal and the tagging B needs to be used instead. $N_{unmixed}$ and N_{mixed} are the observed numbers of unmixed ($B^0\overline{B^0}$) or mixed (B^0B^0 or $\overline{B^0}\,\overline{B^0}$) events for different decay times. The effect of the finite time resolution is treated in the fit by a convolution with a resolution function. For LHCb, the effect of the time resolution on the measurement of the slow B^0-$\overline{B^0}$ mixing is negligible ($D_t \approx 1$). For the fast oscillating B_s^0-$\overline{B_s^0}$ the measured decay-time resolution of 44 fs leads to a dilution factor of ≈ 0.73.

9.6.3.1 Measurement of the B^0-$\overline{B^0}$ Oscillation Frequency

BABAR and Belle have used large event samples of dilepton events where both B mesons decay leptonically with either a muon or an electron in the final-state to measure Δm_d. Figure 9.27 shows the measured decay time difference Δt ($t_2 - t_1$, see above) of opposite-sign and same-sign di-lepton events as measured by BABAR,

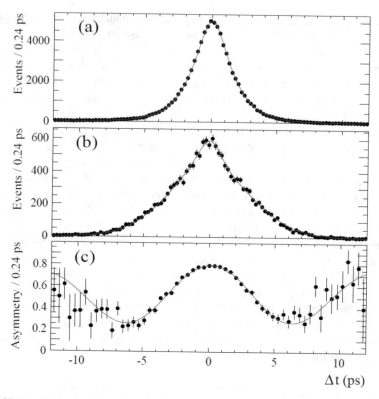

Fig. 9.27 Distributions of time difference Δt of the two neutral B mesons for (**a**) opposite-sign and (**b**) same-sign dilepton events; (**c**) asymmetry between opposite- and same-sign dilepton events. Points are data and the lines correspond to result of a fit. Figure taken from [83]

as well as the mixing asymmetry $A_{meas}(\Delta t)$. The fit to the data resulted in a value of $\Delta m_d = (0.493 \pm 0.012 \pm 0.009)$ ps^{-1} [83]

The most precise determination of Δm_d at the B-factories was performed by Belle analyzing simultaneously semi-leptonic decays $B^0 \rightarrow D^{*-}\ell\nu$ and a set of four different hadronic decays in a data sample of 140 fb^{-1} and resulting into the value $\Delta m_d = (0.511 \pm 0.005 \pm 0.006)$ ps^{-1} [84]. The various Δm_d measurements at the B-factories have been averaged [87] to

$$\Delta m_d = (0.509 \pm 0.003 \pm 0.003) \text{ ps}^{-1} .$$

LHCb has performed a mixing measurement using fully reconstructed hadronic final states as well as semi-leptonic decays. The most precise LHCb measurement was performed using semi-leptonic B^0 decays. A total of 1.923×10^6 $B^0 \rightarrow D^-\mu^+\nu_\mu X$

candidates and 0.829×10^6 $B^0 \rightarrow D^{*-}\mu^+\nu_\mu X$ candidates (charge conjugated decays included) are used [88] and yield

$$\Delta m_d = (0.5050 \pm 0.0021 \pm 0.0010) \text{ ps}^{-1} .$$

Dominated by the LHCb result, the current world average of the existing Δm_d measurements yields

$$\Delta m_d = (0.5064 \pm 0.0019] \text{ ps}^{-1} \text{ [130]}.$$

Using the B^0 lifetime average of 1.520 ± 0.004 ps [130] one obtains the mixing parameter,

$$x_d = \frac{\Delta m_d}{\Gamma_d} = 0.770 \pm 0.004 \text{ [130]},$$

which is of similar magnitude as the value of the neutral K system, $x_K = 0.945 \pm 0.002$ (see Sect. 9.5). As predicted, the decay width difference of the B^0 turns out to be very small. The world average [130] of y_d is,

$$y_d = \frac{\Delta\Gamma_d}{2\Gamma_d} = -0.001 \pm 0.005$$

compared to $y_K = -0.9965$ for the K meson.

9.6.3.2 Measurement of the $B_s^0 - \overline{B_s^0}$ Oscillation Frequency

Inserting the corresponding CKM elements in Eq. (9.187) one expects that the oscillation of the neutral B_s^0 meson is by a factor $|V_{ts}|^2 / |V_{td}|^2$ faster than the oscillation of the B^0. To resolve the fast oscillation pattern represented a challenge and the first measurement of Δm_s was achieved only in 2006 by the CDF experiment [85].

The by far most precise determination of Δm_s was performed by LHCb [86] using about 34,000 $B_s^0 \rightarrow D_s^-\pi^+$ decays (charge conjugated decays are included). Figure 9.28 shows the decay time distribution dependent number of observed B_s^0 meson decays which decay with the same (unmixed) or the different flavor (mixed) with respect to their production. One can nicely observe the rather fast oscillation pattern. A fit accounting for the tagging dilution and for the finite time resolution results into

$$\Delta m_s = (17.768 \pm 0.023 \pm 0.006) \text{ ps}^{-1} \text{[86]} .$$

This value agrees well with the theoretical prediction $\Delta m_s = (18.3 \pm 2.5) \text{ ps}^{-1}$ which however, exhibits large errors due to the uncertainties of the hadronic

Fig. 9.28 Decay time
distribution for B_s^0 candidates
tagged as mixed (different
flavour at decay and
production; red, continuous
line) or unmixed (same
flavour at decay and
production; blue, dotted line).
Figure adopted from [86]

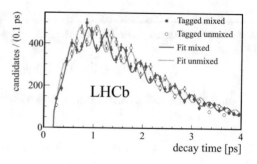

parameters $f_{B_s} \sqrt{B_{B_s}}$ [66]. With the mean B_s^0 life-time, $\tau = 1.509 \pm 0.004$ ps^{-1},
one obtains for the mixing parameter

$$x_s = 26.72 \pm 0.09 \, [130] \, .$$

9.6.4 CP Violation in Neutral B Meson Mixing

$\mathcal{CP}$ violation in mixing, $a_{mix} \neq 0$, is predicted to be $\mathcal{O}(10^{-4})$ for B^0 mesons and
$\mathcal{O}(10^{-5})$ for B_s^0 mesons [66]. As the $\mathcal{CP}$ violation in mixing is probed using semi-
leptonic B decays it is often also referred to as a_{SL} (see Eq. (9.56)). In 2010, the D0
collaboration reported an anomalous charge asymmetry in the inclusive production
rates of like-sign dimuon events [89] indicating a significant deviation from the
prediction. Their most recent study [90] shows a discrepancy to the theory prediction
of about three standard deviations .

B mesons containing a $\bar{b}$ quark undergo decays with a positively charged leptons
in the final-state ($\bar{b} \rightarrow \bar{c} \ell^+ \nu_\ell$), while $\bar{B}$ mesons decay into negatively charged
leptons ($b \rightarrow c \ell^- \bar{\nu}_\ell$). In case of pair production of B mesons with subsequent
semi-leptonic decay, a pair of negatively charged leptons indicates the mixing of
B^0 into a $\overline{B^0}$, while the observation of a positively charged lepton pair signals the
transformation of $\overline{B^0}$ into B^0. The dilepton asymmetry is measured with dimuon
events and is defined as

$$\mathcal{A}_{\ell\ell} = \frac{N(\mu^+\mu^+) - N(\mu^-\mu^-)}{N(\mu^+\mu^+) + N(\mu^-\mu^-)}, \tag{9.203}$$

where $N(\mu^+\mu^+)$ and $N(\mu^-\mu^-)$ are the number of events with two positively or
two negatively charged muons, respectively. At a hadron collider the detection of
two muons does not distinguish between initial B^0 or initial B_s^0 meson and the above
dimuon asymmetry as determined by the D0 experiment measures a combination of
the B^0 and B_s^0 mixing asymmetries a_{SL}^d and a_{SL}^s defined in Eq. (9.189).

Measurements performed by BABAR, Belle, D0 and LHCb determine separately the mixing asymmetries a_{SL}^d and a_{SL}^s for the two neutral B mesons by using partially reconstructed semi-leptonic decays or with fully reconstructed hadronic decays. Measuring the expected very tiny $\mathcal{CP}$ asymmetries requires a precise understanding of various experimental and instrumental effects, e.g. asymmetries arising from different detection efficiencies ($\mathcal{A}_D$) or different material interaction ($\mathcal{A}_I$) for the two charge-conjugated final states. The so far most precise determination of both mixing asymmetries has been performed by the LHCb experiment. At the LHC the initial proton-proton collision is not a particle-antiparticle symmetric state. The number of produced b and $\bar{b}$ hadrons of a given species is not necessarily the same and the production asymmetry $\mathcal{A}_P$ has also to be taken into account when the measurement of the mixing asymmetry is performed. The measured asymmetry of a neutral B meson, independent whether it is a B^0 or a $\overline{B^0}$, into a final state f or its conjugated state $\bar{f}$ can be expressed by the mixing asymmetry a_{SL}^q,

$$
\begin{aligned}
\mathcal{A}_{meas} &= \frac{N(\overline{B_q^0}/B_q^0 \to f)(t) - N(\overline{B_q^0}/B_q^0 \to \bar{f})(t)}{N(\overline{B_q^0}/B_q^0 \to f)(t) + N(\overline{B_q^0}/B_q^0 \to \bar{f})(t)} \\
&= \frac{a_{SL}^q}{2} - \left(\mathcal{A}_P + \frac{a_{SL}^q}{2} \right) \cdot \frac{\cos\left(\Delta m_q t\right)}{\cosh\left(\Delta\Gamma_q t/2\right)} + \mathcal{A}_D + \mathcal{A}_I .
\end{aligned}
\tag{9.204}
$$

By omitting the flavor tagging of the decaying neutral B the sensitivity to a_{SL} is a factor $1/2$ smaller with respect to a tagged measurement. At the same time one dramatically wins in statistical power as no tagging is required.

For the B^0 system the time dependent analysis together with the precise determination of $\mathcal{A}_D$ and $\mathcal{A}_I$ allows to eliminate the production asymmetry $\mathcal{A}_P$. LHCb has performed this complicated analysis [91] and obtained for the mixing asymmetry of the B^0 system the value

$$
a_{SL}^d = (-0.02 \pm 0.19 \pm 0.30) \ \%
$$

with a total precision of 3.6 per mill. In case of the fast oscillating B_s^0 system the measurement simplifies as the term involving the production asymmetry cancels in a time integrated measurement. Furthermore, the measurement was performed using $D_s^- \mu^+ \nu_\mu X$ final states with the D_s^- meson decaying to $\phi(\to K^+ K^-)\pi^-$ [92]. For the charge symmetric $K^+ K^-$ state the interaction asymmetry is negligible and one only needs to correct for the pion detection asymmetry. The final result obtained for a_{SL}^s is

$$
a_{SL}^s = (0.39 \pm 0.26 \pm 0.20) \ \% \ [92],
$$

with a total precision of 3 per mill.

Figure 9.29 shows the different a_{SL} measurements [91–97] for the two neutral B systems as well as the averages. The D0 measurement indicating large mixing

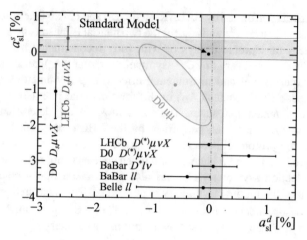

Fig. 9.29 Mixing asymmetries a^s_{SL} and a^d_{SL} as measured by the different experiments [91–97]. The D0 dimuon measurement (contour) [90] determines a combination of a^s_{SL} and a^d_{SL}. Green bands show the averages. The theory prediction is marked by the black dot. Figure adopted from [92]

asymmetries is not confirmed by the other experiments. The measured a_{SL} values can be used to determine the deviation of $|q/p|$ from unity,

$$a_{SL} = \frac{1 - |q/p|^4}{1 + |q/p|^4} \approx 2\,(1 - |q/p|)\,, \tag{9.205}$$

or equivalently

$$\left|\frac{q}{p}\right| \approx 1 - \frac{a_{SL}}{2}\,. \tag{9.206}$$

The most recent world averages as provided by the HFLAV group[4] and published in [130] are for the B^0 system,

$$a^d_{SL} = 0.0021 \pm 0.0017\,,$$
$$|q/p| = 1.0010 \pm 0.0008\,,$$

and correspondingly for the B^0_s system,

$$a^2_{SL} = 0.0006 \pm 0.0028\,,$$
$$|q/p| = 1.0003 \pm 0.0014\,.$$

---

[4]Heavy Flavor Averaging Group

The values so far show no evidence for CP violation in mixing and are consistent with the theory prediction of very small CP asymmetries [66].

9.6.5 CP Violation in the Interference of Mixing and Decay

The B-factories BABAR and Belle were built with the primary goal to discover CP violation in the interference of B^0 decays to the CP eigenstate $B^0 \rightarrow J/\psi K_S$ with and without mixing. Large CP violation had been predicted for this channel. Shortly after the start of data-taking the two collaborations presented first measurements in summer 2000. A clear evidence of the first CP violation outside the neutral K meson was established in 2001 [59, 60].

9.6.5.1 B^0 Meson

The expected time-dependent CP asymmetry is defined in Eq. (9.192) and parametrised by Eq. (9.194). Taking into account the negligible width difference, $\Delta\Gamma \approx 0$, for B^0 mesons the time-dependent CP asymmtery for the channel $B^0 \rightarrow J/\psi K_S$ can be written as

$$\mathcal{A}_{J/\psi K_S}(t) = A^{mix}_{J/\psi K_S} \sin(\Delta m_d t) - A^{dir}_{J/\psi K_S} \cos(\Delta m_d t), \tag{9.207}$$

where the coefficients $A^{mix}_{J/\psi K_S}$ and $A^{dir}_{J/\psi K_S}$ have been introduced in Eq. (9.195). The decay $B^0 \rightarrow J/\psi K_S$ is a *golden mode*, dominated by a single tree-level amplitude. CP violation in the decay amplitudes can therefore be neglected, i.e. $A^{dir}_{J/\psi K_S} \approx 0$. The coefficient $A^{mix}_{J/\psi K_S}$ is given by the CKM matrix elements involved in the short range box diagrams responsible for the B^0 mixing and by the CKM matrix elements appearing in the decay amplitudes. With $A^{mix}_{J/\psi K_S} = -\Im\lambda_{J/\psi K_S}$ (see Eq. (9.195)) and the definition of $\lambda_{J/\psi K_S}$ of Eq. (9.193),

$$\lambda_{J/\psi K_S} = \frac{q}{p} \frac{\overline{A}_{J/\psi K_S}}{A_{J/\psi K_S}},$$

one can calculate the time-dependent CP asymmetry according to Eq. (9.197). For the decay $B^0 \rightarrow J/\psi K_S$ one finds for the ratio q/p describing the mixing and for the ratio of the amplitudes the following CKM factors,[5]

$$\frac{q}{p} = \frac{V^*_{tb}V_{td}}{V_{tb}V^*_{td}} \quad \text{and} \quad \frac{\overline{A}_{J/\psi K_S}}{A_{J/\psi K_S}} = \eta_{J/\psi K_S} \left(\frac{V_{cb}V^*_{cs}}{V^*_{cb}V_{cs}}\right) \left(\frac{V_{cs}V^*_{cd}}{V^*_{cs}V_{cs}}\right), \tag{9.208}$$

---

[5]True only up to a phase factor which cancels later on.

The second CKM term in the expression for the amplitude ratio is a result of the K^0–$\overline{K^0}$ mixing necessary to produce the K_S. As the V_{cs} terms all cancel one obtains for the $A^{mix}_{J/\psi K_S}$

$$A^{mix}_{J/\psi K_S} = -\Im\left\{\eta_{J/\psi K_S}\left(\frac{V^*_{tb}V_{td}}{V_{tb}V^*_{td}}\right)\left(\frac{V_{cb}V^*_{cd}}{V^*_{cb}V_{cd}}\right)\right\} = -\eta_{J/\psi K_S}\sin(2\beta)\,, \qquad (9.209)$$

where the last equality uses the definition of the CKM angle β of Sect. 9.7. Finally one obtains for the time-dependent CP asymmetry,

$$\mathcal{A}_{J/\psi K^0_S}(t) = -\eta_{J/\psi K_S}\sin(2\beta)\sin(\Delta m_d t)\,. \qquad (9.210)$$

For the decay $B^0 \to J/\psi K_S$ the CP eigenvalue of the final state is $\eta_{J/\psi K_S} = -1$. Due to opposite CP eigenvalue of the K_L the CP eigenvalue of the decay $B^0 \to J/\psi K_L$ is $\eta_{J/\psi K_L} = +1$. Both statements are strictly true only if one ignores the very small CP violation in the $K^0\overline{K}^0$ mixing. The decay $B^0 \to J/\psi K_L$ is interesting as one expects exactly the same magnitude of the CP violation as for $B^0 \to J/\psi K_S$ but an opposite time behavior due to the opposite CP eigenvalue.

Experimentally the decay $B^0 \to J/\psi K^0_S$ is easy to access. The J/ψ meson decays into two leptons and can be easily triggered and selected. Additional requirements on the displacement of the B vertex from the primary vertex allow an almost background-free reconstruction even at a hadron collider. The measurement has been extended to include other ($c\bar{c}$) resonances for which the same CP behavior as for the golden mode is expected.

At the B-factories the time-dependent CP asymmetry is measured as function of the time difference $\Delta t = t_2 - t_1$. The most precise determination of the quantity $\sin(2\beta)$ has been performed by the Belle experiment using about 32,000 signal candidates with an average signal purity of 79% [98],

$$\sin(2\beta) = 0.667 \pm 0.023 \pm 0.012\,.$$

The time-dependent CP asymmetry for the CP-odd and CP-even final states as measured by Belle are shown in Fig. 9.30. The BABAR experiment has recorded less signal candidates (15,481 candidates with a purity of 76%) and obtained [99],

$$\sin(2\beta) = 0.687 \pm 0.028 \pm 0.012\,,$$

with a slightly larger statistical error than the Belle measurement.

The LHCb collaboration has also performed the measurement of $\sin(2\beta)$ by analysing the time dependent CP violation in $B^0 \to J/\psi K_S$ and $B^0 \to \psi(2S)K_S$ decays with 52,000 and 8000 signal candidates, respectively [100, 101]. As for the measurement of the CP violation in mixing the observed time dependent asymmetry has to be corrected for the production asymmetry between B^0 and $\overline{B^0}$ of about 1%.

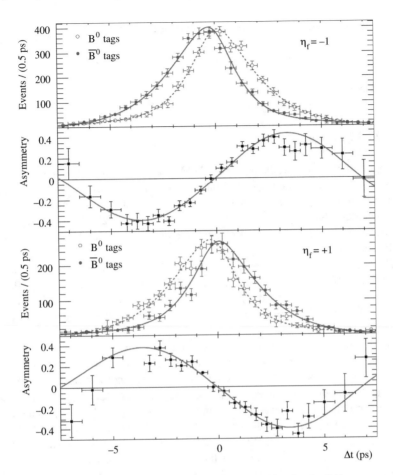

Fig. 9.30 Flavor-tagged Δt distributions and raw CP asymmetries for the Belle measurements of $\sin 2\beta$. The top plots show the $B^0 \rightarrow (c\bar{c})K_S$ ($\eta_f = -1$) measurement and the bottom plots show the measurement for $B^0 \rightarrow J/\psi K_L$ ($\eta_f = +1$) The distributions are background subtracted. Figure adapted from [98]

The resulting value $\sin (2\beta) = 0.760 \pm 0.034$ [101] is slightly larger than the ones measured at the B-factories. It is important to notice that despite the much larger number of signal candidates used for the LHCb measurement the total error is comparable with the ones of Belle and BABAR. The reason is the significantly larger mistag probability at a hadron collider which leads to a reduction of the statistical power of the events. Using all available measurements of $\sin (2\beta)$ an average value of

$$\sin (2\beta) = 0.691 \pm 0.017$$

is obtained [87] which corresponds to a value of the phase β of,

$$\beta = (21.9 \pm 0.7)° \quad \text{or} \quad \beta = (68.1 \pm 0.7)°,$$

where the ambiguity is not resolved.

9.6.5.2 B_s^0 Meson

Analogue to the measurement of the CKM phase β in the B^0 system, the measurement of a corresponding $\mathcal{CP}$ violating phase β_s, defined in Sect. 9.7, can be performed in the B_s^0 system, assuming that the B_s^0 oscillation can be resolved. The *golden* channel used is the decay $B_s^0 \rightarrow J/\psi\phi$. Due to its higher mass the B_s^0 system cannot be produced at the $\Upsilon(4S)$ resonance and β_s was therefore not accessible at the B-factories.

The two Tevatron experiments CDF and D0 have performed explorative studies of the decay $B_s^0 \rightarrow J/\psi\phi$ [102, 103] and also first measurements [104, 105] but the results have been limited by the number of recorded signal events. The first significant measurement of the phase[6] $\phi_s = -2\beta_s$ has been presented by the LHCb experiment in 2011 [106].

In the decay $B_s^0 \rightarrow J/\psi\phi$ the final-state is composed of two vector particles and angular momentum conservation allows for the relative angular momentum values $L = 0, 1$ or 2. The final-state $J/\psi\phi$ is thus not a pure $\mathcal{CP}$-state but a linear combination of $\mathcal{CP}$-even and $\mathcal{CP}$-odd eigenstates depending on the relative angular momentum L of the two vector mesons: $\eta_{CP} = \eta_{CP}(J/\psi)\eta_{CP}(\phi)(-1)^L = (-1)^L$. The measurement of the time dependent CP asymmetry requires a separate treatment of the CP-odd and CP-even states which can be achieved statistically by analyzing the angular distribution of the final-state particles $J/\psi \rightarrow \mu^+\mu^-$ and $\phi \rightarrow K^+K^-$. To measure the angular distribution of the final state particles it is common to use the helicity basis. The three decay angles necessary to describe the decay are denoted by $(\theta_K, \theta_\mu, \varphi_h)$ and are defined in Fig. 9.31. The polar angle θ_K (θ_μ) is the angle between the K^+ (μ^+) momentum and the direction opposite to the B_s^0 momentum in the K^+K^- ($\mu^+\mu^-$) centre-of-mass system. The azimuthal angle between the K^+K^- and $\mu^+\mu^-$ planes is φ_h.

The analysis of the $B_s^0 \rightarrow J/\psi\phi$ events is further complicated by a small fraction of non-resonant $B_s^0 \rightarrow J/\psi K^+K^-$ decays ($\mathcal{CP}$-odd) which interfere with the ϕ meson decaying to two charged kaons. These non-resonant events need to be considered. Effectively one therefore studies B_s^0 decays to $J/\psi K^+K^-$.

---

[6]In the Standard Model the phase ϕ_s determined by the measurement of the time-dependent $\mathcal{CP}$ asymmetry in the decay $B_s^0 \rightarrow J/\psi\phi$ equals in leading order to $-2\beta_s$ with β_s being defined in Sect. 9.7. However, in case of new physics effects the two phases could differ. In the context of this textbook these differences are ignored.

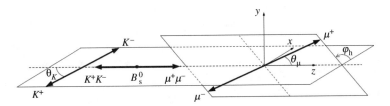

Fig. 9.31 Definition of helicity angles as discussed in the text. Figure taken from [107]

In the absence of $\mathcal{CP}$ violation in mixing the $\mathcal{CP}$-odd and $\mathcal{CP}$-even decays of the B_s^0 meson correspond to the heavy and light mass eigenstates. Thus separating statistically the polarization amplitudes in an angular analysis allows the measurement of the decay width Γ_H and Γ_L and thus the measurement of the different lifetimes of the two B_s^0 mass states.

With a data-sample of 96,000 $B_s^0 \rightarrow J/\psi K^+ K^-$ events LHCb has performed a simultaneous analysis of the measured decay time and the three decay angles to determine the time-dependent $\mathcal{CP}$ asymmetry and the related $\mathcal{CP}$ phase ϕ_s [108]. Figure 9.32 shows the decay time distribution as well as the distributions of the three decay angles. One clearly sees how the $\mathcal{CP}$-odd and $\mathcal{CP}$-even components can be statistically separated using the decay angles. The decay time distribution shows the different decay behavior of the two $\mathcal{CP}$ components with the $\mathcal{CP}$-even component decaying visibly faster. For the average decay width Γ_s and for the decay width difference $\Delta\Gamma_s$ one obtains [108],

$$\Gamma_s = (\Gamma_L + \Gamma_H)/2 = 0.6603 \pm 0.0027 \pm 0.0015 \, \text{ps}^{-1} \, ,$$

$$\Delta\Gamma_s = \Gamma_L - \Gamma_H = 0.0805 \pm 0.0091 \pm 0.0032 \, \text{ps}^{-1} \, .$$

The simultaneous determination of the time-dependent $\mathcal{CP}$ asymmetry results in the following value for the $\mathcal{CP}$ violating phase ϕ_s,

$$\phi_s = -0.058 \pm 0.049 \pm 0.006 \, .$$

The time-dependent $\mathcal{CP}$ violation in the B_s^0 system is thus very small and within the sensitivity of the measurement no $\mathcal{CP}$ violation has been seen. The phase ϕ_s can also be measured in the mode $B_s^0 \rightarrow J/\psi \pi^+ \pi^-$. The $J/\psi \pi^+ \pi^-$ final-state has been shown to be an almost pure (97.7%) CP-odd final state [109] and the CP violating phase can thus be extracted without the complication of an angular analysis. A much smaller signal sample with 27,000 events results into a comparable precision for ϕ_s, $\phi_s = -0.070 \pm 0.068 \pm 0.008$ [110]. The analysis of these events however, requires the knowledge $\Delta\Gamma_s$ as external input. Performing a combined analysis of the two decay channels [108] LHCb reports a value of ϕ_s of

$$\phi_s = -0.010 \pm 0.039 \, .$$

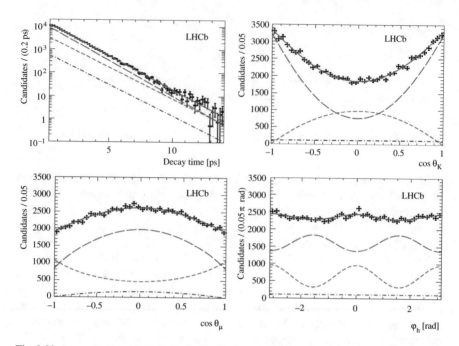

Fig. 9.32 Decay-time and helicity-angle distributions for $B_s^0 \to J/\psi K^+ K^-$ decays (data points). The solid blue line shows the total signal contribution, which is composed of $\mathcal{CP}$-even (long-dashed red), $\mathcal{CP}$-odd (short-dashed green) and S-wave (dotted-dashed purple) contributions. Figure taken from [108]

The ATLAS and CMS collaborations have also performed measurements of the phase ϕ_s and of the width difference $\Delta\Gamma_s$ [111, 112]. The overall experimental situation is summarized in Fig. 9.33 where also the theoretical expectation is shown. It should be noted that there is no ab-initio theory prediction for the phase ϕ_s. The prediction results from the measurement of $\sin(2\beta)$ and exploits the unitarity of the CKM matrix, see also Sect. 9.7. The average value for ϕ_s, $\phi_s = -0.021 \pm 0.031$ [130], agrees very well with the expectation $\phi_s = -0.0370 \pm 0.0006$ [113].

9.6.6 Direct CP Violation

B meson decays offer a broad decay phenomenology and are an ideal system to look also for large direct $\mathcal{CP}$ violation. Topologically very different tree-level and penguin decay amplitudes can result into the same final state and their interference can cause significant $\mathcal{CP}$ asymmetries. Considering the B meson decay $B \to f$

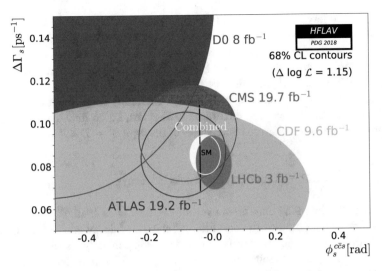

Fig. 9.33 68% confidence level contours in the ϕ_s-$\Delta\Gamma_s$ plane, showing the measurements from CDF, D0, ATLAS, CMS and LHCb together with their combination (white ellipse). The thin black line represents the Standard Model predictions. Figure taken from [130]

and its $\mathcal{CP}$ conjugates $\bar{B} \rightarrow \bar{f}$, direct $\mathcal{CP}$ violation or $\mathcal{CP}$ violation in the decay is measured as time-integrated asymmetry of the observed signal yields:

$$A_{CP} = \frac{\Gamma(B \rightarrow f) - \Gamma(\bar{B} \rightarrow \bar{f})}{\Gamma(B \rightarrow f) + \Gamma(\bar{B} \rightarrow \bar{f})}$$

For a given B decay $B \rightarrow f$, direct $\mathcal{CP}$ violation can only arise if at least two interfering decay amplitudes A_1 and A_2 exist and if these decay amplitudes carry different weak and strong phases, φ_i and δ_i. The total decay amplitude A_f is given as sum

$$A_f = |A_1| e^{\varphi_1 + \delta_1} + |A_2| e^{\varphi_2 + \delta_2} .$$

For the amplitudes of the charge conjugated process $\bar{A}_{\bar{f}}$ the weak phases change sign while the strong phases stay unchanged,

$$\bar{A}_{\bar{f}} = |A_1| e^{-\varphi_1 + \delta_1} + |A_2| e^{-\varphi_2 + \delta_2} .$$

The expected signal yields depend on the interference of the amplitudes and the observable $\mathcal{CP}$ asymmetry is determined by the difference of the interference terms. The $\mathcal{CP}$ asymmetry is a function of the weak and strong phases as well as a function of the ratio R of the decay amplitudes,

$$A_{CP} = \frac{2 \sin(\phi_1 - \phi_2) \sin(\delta_1 - \delta_2)}{R + R^{-1} + \cos(\phi_1 - \phi_2) \cos(\delta_1 - \delta_2)} \quad \text{with} \quad R \equiv \left| \frac{A_1}{A_2} \right| . \tag{9.211}$$

To observe large asymmetries the interfering amplitudes should have different weak and strong phases. While the weak phases, and in particular the weak phase difference, are determined by the CKM parameters, the theoretical determination of the strong phases is difficult. The B meson are subject to both, short and long-distance QCD effects which cannot be treated perturbatively.

In the following, two classes of B decays exhibiting large direct CP violation will be discussed: B decays to charmless final-states, i.e. to final-states without D mesons, and decays to charmed final-states of the type $B \to DK$.

9.6.6.1 Direct CP Violation in Charmless B Decays

As sketched in Fig. 9.34 charmless two-body B decays such as $B^0 \to K^+\pi^-$ can proceed via two topologically very different classes of decay amplitudes, tree and penguin processes, which carry different weak and in generally also different strong phases. Large direct CP violation is therefore expected for these decays.

First experimental evidence for direct CP violation in B meson decays was reported by the Belle collaboration in 2004 for the decay $B^0 \to \pi^+\pi^-$ [114]. As the neutral B meson decays here to a CP eigenstate the formalism of (9.194) and (9.207) has to be applied, i.e. the observable time dependent asymmetry can be parametrized by the coefficient $A_{\pi\pi}^{dir}$ describing the effect of direct CP asymmetry and the coefficient $A_{\pi\pi}^{mix}$ to describe the effect of indirect CP violation through interference of diagrams with and without mixing. Belle observed a deviation of the coefficient $A_{\pi\pi}^{dir}$ from zero, i.e. direct CP violation, at the level of 3.2σ. The result suggested large interference effects between the relevant tree and penguin diagrams and was confirmed by subsequent measurements. Today, the most recent value of this coefficient is $A_{\pi\pi}^{dir} = 0.32 \pm 0.04$ [87] and confirms the early hints for direct CP violation in $B^0 \to \pi^+\pi^-$.

The two B-factory experiments, BABAR and Belle, also studied $B^0 \to K^+\pi^-$ decays where large penguin contributions were expected. Both collaborations performed time-independent analyses and reported the observation of large direct CP violations for $B^0 \to K^+\pi^-$ decays (Babar: $A_{CP} = -0.133 \pm 0.030 \pm 0.009$ [115]; Belle: $A_{CP} = -0.101 \pm 0.025 \pm 0.005$ [116]). Both measurements represent the first doubtless observation of large direct CP violation in the B^0 system. It is

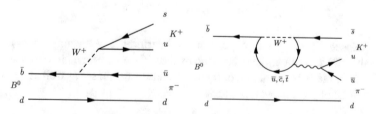

Fig. 9.34 Tree-level (left) and penguin (right) contribution to the charmless two-body decay $B^0 \to K^+\pi^-$

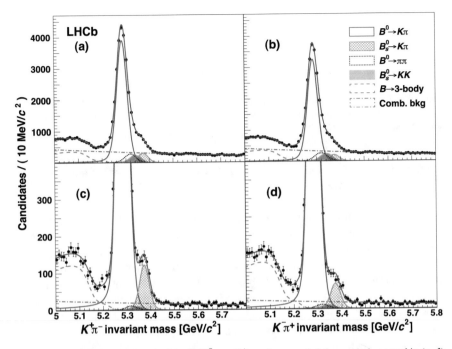

Fig. 9.35 Invariant mass spectra of (**a, b**) $B^0 \to K^+\pi^-$ the recorded decays and zoomed in (**c, d**) to show the $B_s^0 \to K^-\pi^+$ decays. Panels (**a**) and (**c**) represent the $K^+\pi^-$ invariant mass, whereas panels (**b**) and (**d**) represent the $K^-\pi^+$ invariant mass. The results of a mass fit describing the signals are overlaid. The main components contributing to the fit model are also shown. Figure taken from [118]

useful to remember that direct $\mathcal{CP}$ violation in the neutral kaon system is many orders smaller ($O(10^{-6})$). It is the effect of the large weak phases entering the B^0 decay amplitudes which produces the much larger direct $\mathcal{CP}$ violation.

Already with the very first data-set recorded in 2011 and corresponding to only 0.35 fb^{-1} of data, the LHCb experiment has repeated the measurement of A_{CP} for the channel $B^0 \to K^+\pi^-$ [117]. LHCb's excellent resolution of the reconstructed $K^\pm\pi^\mp$ invariant mass allows to distinguish between a decaying B^0 and B_s^0 and thus also probed the decay $B_s^0 \to K^-\pi^+$. The measurement of $A_{CP}(B_s^0 \to K^-\pi^+) = 0.27 \pm 0.08 \pm 0.02$ provided the first evidence (3.3σ) for direct $\mathcal{CP}$ violation in the B_s^0 system. The measurement was repeated with more data and the observed direct $\mathcal{CP}$ violation $A_{CP}(B_s^0 \to K^-\pi^+) = 0.27 \pm 0.04 \pm 0.01$ [118] confirmed the earlier result. Figure 9.35 shows the invariant mass distribution for the recorded $B_{(s)}^0 \to K\pi$ decays. Clear differences in the number of recorded decays for the B and the anti-B decays are observed.

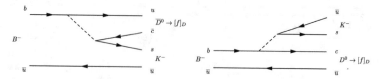

Fig. 9.36 Tree-level Feynman diagrams of the decays $B^- \to D^0(\overline{D}^0)K^-$. The left diagram implies a $b \to u\bar{c}s$ transition, and is strongly suppressed by the small value of $|V_{ub}|$. The right diagram proceeds via a transition $b \to c\bar{u}s$

9.6.6.2 Direct CP Violation in $B \to DK$ Decays and Measurement of CKM Phase γ

While direct CP violation in B decays is large and experimentally easy accessible, the theoretical interpretation is non-trivial. The calculation of the non-perturbative hadronic effects and the resulting strong phases is difficult. A way to use direct CP asymmetries to constrain the CKM parameters is therefore a simultaneous determination of the weak (CKM) phases and the hadronic nuisance parameters including the strong phases. This approach is followed to determine the CKM phase γ as defined in Sect. 9.7. Ignoring higher order terms, γ is in good approximation equal to the phase of the CKM matrix element V_{ub} ($V_{ub} = |V_{ub}|e^{-i\gamma}$).

The phase can be probed using $B^- \to D^0(\overline{D}^0)K^-$ decays with either a $b \to u\bar{c}s$ ($\overline{D}^0 K^-$) or $b \to c\bar{u}s$ ($D^0 K^-$) tree-level quark transition as depicted in Fig. 9.36. The two decays can be described by three parameters: r_B ($\approx O(0.1)$), the absolute value of the ratio of both amplitudes; δ_B the strong phase difference; and the CKM phase γ. In order to observe the CP violating interference between the two amplitudes the D^0 and the $\overline{D}^0$ mesons emerging in case of the two different amplitudes should decay into a common final state f_D, i.e. $D, \bar{D} \to f_D$. The size of the CP violating interference which provides the sensitivity to γ is proportional to r_B.

The possibility of observing direct CP violation in $B^- \to DK^-$ decays was first discussed in the 1980s [119, 120] using decays of the $D(\bar{D})$ to neutral kaons and pions. Since then, several methods have been proposed which can be grouped according to the choice of the final state.

- The Gronau-London-Wyler (GLW) method [121, 122] considers the decays of D mesons to CP eigenstates, such as the CP-even decays $D^0 \to K^+K^-$ and $D^0 \to \pi^+\pi^-$.
- The Atwood-Dunietz-Soni (ADS) approach [123, 124] extends this to include final states that are not CP eigenstates, for example $D^0 \to K^+\pi^-$ together with its doubly Cabibbo-suppressed counterpart $D^0 \to K^-\pi^+$. The interference between Cabibbo-allowed and doubly Cabibbo-suppressed decay modes in both the B and D decays gives rise to large charge asymmetries. However, the different D decays require additional parameters r_D and δ_D to describe the ratio of suppressed and favored D decay amplitudes as well as their phase difference.

- The Grossman-Ligeti-Soffer (GLS) method [125] is similar to the ADS method but uses singly Cabibbo-suppressed decays such as $D \to K_S K^+ \pi^-$ decays.
- The Giri-Grossman-Soffer-Zupan (GGSZ) method [126] uses self-conjugate multibody D meson decay modes like $K_S \pi^+ \pi^-$ or $K_S K^+ K^-$ and requires an analysis of the Dalitz plot to account for the varying D decay parameters. A model-dependent analysis assumes specific D decay amplitudes while a model-independent approach uses external input for the strong-phase difference δ_D and the D amplitude ratio r_D in bins of the Dalitz space.

Simultaneous fits to several observables, $\mathcal{CP}$ asymmetries or ratios of suppressed to favored modes, allow the determination of the decay parameters including the CKM phase γ. The main issue with all methods is the small overall branching fraction of the observable decays which range from 5×10^{-5} to 5×10^{-9} [127]. The precise determination of the CKM phase γ therefore requires a very large data-sample. The B factories have pioneered different methods to determine γ, however the achieved overall statistical precision on γ was limited. Combining the different methods and quoting a single result for γ, BABAR and Belle report the following values [127]:

$$\gamma = (69 \pm 17)^\circ \ (\text{BABAR}),$$

$$\gamma = (68 \pm 14)^\circ \ (\text{Belle}).$$

The breakthrough towards a precision determination of the phase γ came when the LHCb experiment was able to measure even very rare doubly Cabibbo-suppressed decays such as $B^- \to [\pi^- K^+]_D K^-$ and its charge-conjugated counter-parts with sufficiently high statistics to observe $\mathcal{CP}$ asymmetries. Figure 9.37 shows as an example the decay $B^- \to [\pi^- K^+]_D K^-$ as well as the decay $B^- \to [\pi^- K^+]_D \pi^-$ which also provides sensitivity to γ together with the corresponding charge conjugated decays [128]. The observed yields clearly signal direct $\mathcal{CP}$ violation for both channels. LHCb has analysed $B^+ \to DK^+$ decays with a multitude of different decay modes of the neutral D, $B^+ \to D^* K^+$ and $B^+ \to DK^{*+}$ decays and $B^0 \to DK^*$ decays. In addition time-dependent analyses of $B_s^0 \to D_s^\mp K^\pm$ and $B^0 \to D^\mp \pi^\pm$ are performed and measure combinations of the CKM phases γ and β_s or β, respectively. The combination of the different LHCb measurements [129] finally result into a value of

$$\gamma = (74.0^{+5.0}_{-5.8})^\circ \ (\text{LHCb}), \tag{9.212}$$

in perfect agreement with the early measurements by BABAR and Belle. For the average of all γ measurements one finds [87],

$$\gamma = (73.5^{+4.2}_{-5.1})^\circ. \tag{9.213}$$

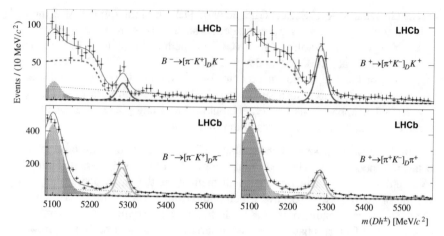

Fig. 9.37 Invariant mass distributions of selected $B^{\pm} \rightarrow \left[\pi^{\pm} K^{\mp}\right]_D h^{\pm}$ decays, separated by charge. The dashed pink line left of the signal peak shows partially reconstructed $B_s^0 \rightarrow \left[K^+\pi^-\right]_D K^-\pi^+$ decays, where the bachelor pion is missed. For the two channels, one clearly observes a yield difference, i.e. $\mathcal{CP}$ asymmetry, between the two charge-conjugated channels. Figure taken from [128]

9.7 Weak Quark Mixing and the CKM Matrix

The observation of direct $\mathcal{CP}$ violation in the neutral K meson and the confirmation of large $\mathcal{CP}$ violating effects in the neutral B meson system provided the experimental evidence of weak quark-mixing as the primary source of $\mathcal{CP}$ violation in the hadron sector. A multitude of precision measurements confirmed the prediction of the quark-mixing paradigm of the Standard Model [57] and led to the Nobel prize for Kobayshi and Maskawa. Despite this success of experimental and theoretical quark flavor physics we know today that other sources of $\mathcal{CP}$ violation must exists to explain the baryon asymmetry of our universe.

9.7.1 Quark-Mixing Matrix

Historically quark-mixing was introduced by Cabibbo [56] to explain the different coupling strength of hadronic currents in weak decays of neutrons and pions compared to strangeness changing processes, such as the decay of K mesons and Λ hyperons. Cabibbo postulated—expressed in today's notation—that the weak eigenstates of the then known quarks with charge $-1/3$ were not the flavor eigenstates d and s but a linear combination, rotated by an angle θ, the Cabibbo angle:

$$d_c = d \cos \theta + s \sin \theta .$$

(9.214)

The GIM mechanism [131], introduced to cancel the $K_L \rightarrow \mu^+\mu^-$ amplitude, required the orthogonal state

$$s_c = -d \sin\theta + s \cos\theta , \tag{9.215}$$

and an additional charge $+2/3$ quark, the charm quark, complementing the two quark doublets.

We know today that three generations of up-type quarks u^i ($i = 1, 2, 3$: u-, c-, t-quark) with electrical charge $+2/3$ and three generations of down-type quarks d^i ($i = 1, 2, 3$: d-, s-, b-quark) with electrical charge $-1/3$ exist. In the Standard Model, the masses and mixing of quarks arise from the Yukawa interaction with the Higgs condensate which couples the left-handed quark fields u_L and d_L to the right-handed quark fields u_R and d_R (for better readability the generation index is suppressed). After spontaneous breaking of the electroweak symmetry the Yukawa terms give rise to masses and mixing:

$$\mathcal{L}_{\text{Yukawa}} = -\frac{v}{\sqrt{2}} \left(\bar{d}_L \mathbf{Y_d} d_R + \bar{u}_L \mathbf{Y_u} u_R \right) + h.c.$$

The Yukawa matrices $\mathbf{Y_d}$ and $\mathbf{Y_u}$ are complex 3×3 matrices in generation or flavor space and do not need to be diagonal. Indeed, in the Standard Model they are not, and as a consequence the flavor states are not equal to the mass eigenstates of the quarks. The mass eigenstates $\tilde{u}_{L,R}$ and $\tilde{d}_{L,R}$ are obtained by unitary transformations of the above quark flavor states: $\tilde{u}_A = \mathbf{V_{A,u}} u_A$ and $\tilde{d}_A = \mathbf{V_{A,d}} d_A$ (with chirality index $A = L, R$ for left- and right-handed quark fields, and suppressed generation indices). The unitary matrices $\mathbf{V_{A,u}}$ and $\mathbf{V_{A,d}}$ diagonalize the Yukawa matrices and one obtains the diagonal quark mass matrices for the up- and down-type quarks,

$$\mathbf{M_u} = \text{diag}(m_u, m_c, m_t) = \frac{v}{\sqrt{2}} \mathbf{V_{L,u}} \mathbf{Y_u} \mathbf{V_{R,u}^\dagger} ,$$

$$\mathbf{M_d} = \text{diag}(m_d, m_s, m_b) = \frac{v}{\sqrt{2}} \mathbf{V_{L,d}} \mathbf{Y_d} \mathbf{V_{R,d}^\dagger} .$$

The quark masses will appear as usual Dirac mass terms in the above Yukawa part of the Lagrangian:

$$\mathcal{L}_{\text{Yukawa}} = -\bar{\tilde{d}}_L \mathbf{M_d} \tilde{d}_R - \bar{\tilde{u}}_L \mathbf{M_d} \tilde{u}_R + h.c.$$

If the up-type and and down-type Yuakawa matrices $\mathbf{Y_u}$ and $\mathbf{Y_d}$ cannot be diagonalized simultaneously by the same transformations there is a net effect of the change of the quark basis. The charged current terms of the Standard Model Lagrangian combining left-handed up and down-type quarks therefore get a flavor structure imprinted which is described by the Cabibbo Kobayshi Maskawa (CKM) quark-mixing matrix

$$\mathbf{V}_{\mathbf{CKM}} \equiv \mathbf{V}_{\mathbf{L},\mathbf{u}}\mathbf{V}_{\mathbf{L},\mathbf{d}}^{\dagger} \; .$$

The charge current terms expressed in the mass eigenstates have the form

$$\mathcal{L}_{CC} = -\frac{g}{\sqrt{2}} \left(\bar{\tilde{u}}_{L}\gamma^{\mu}W_{\mu}^{+}\mathbf{V}_{\mathbf{CKM}}\tilde{d}_{L} + \bar{\tilde{d}}_{L}\gamma^{\mu}W_{\mu}^{-}\mathbf{V}_{\mathbf{CKM}}^{\dagger}\tilde{u}_{L} \right) \; .$$

Here, the matrix element $(\mathbf{V}_{\mathbf{CKM}})_{ij}$ connects a left-handed up-type quark of the ith generation to a left-handed down-type quark of the jth generation. The matrix elements are therefore expressed using flavor indices:

$$\mathbf{V}_{\mathbf{CKM}} = \begin{pmatrix} V_{ud} & V_{us} & V_{ub} \\ V_{cd} & V_{cs} & V_{cb} \\ V_{td} & V_{ts} & V_{tb} \end{pmatrix}$$

For a non-diagonal CKM matrix the charged quark currents are thus inter-generation flavor changing currents. In analogy to the flavor eigenstates introduced by Cabibbo it is usual to absorb the CKM matrix by introducing for the down-type quarks the weak quark eigenstates $\tilde{d}' = \mathbf{V}_{\mathbf{CKM}}\tilde{d}_{L}$.

Since it is the product of unitary matrices the CKM matrix itself is unitary, i.e. $\mathbf{V}_{\mathbf{CKM}}\mathbf{V}_{\mathbf{CKM}}^{\dagger} = \mathbf{1}$, and its elements are in general complex. The number of parameters of a general unitary 3×3 matrix is nine, three rotation angles and six phases. By rephasing the quark mass eigenstates $\tilde{q} \rightarrow e^{i\alpha_{q}}\tilde{q}$ one can remove five phases, corresponding to the five independent phase differences between the quarks, and leaving one $\mathcal{CP}$ violating phase δ. The usual parametrization of the CKM matrix uses the three rotation angles $\theta_{12}, \theta_{23}, \theta_{13}$ and the phase δ:

$$\mathbf{V}_{\mathbf{CKM}} = \begin{pmatrix} 1 & 0 & 0 \\ 0 & c_{23} & s_{23} \\ 0 & -s_{23} & c_{23} \end{pmatrix} \begin{pmatrix} c_{13} & 0 & s_{13}e^{-i\delta} \\ 0 & 1 & 0 \\ -s_{12}e^{i\delta} & 0 & c_{13} \end{pmatrix} \begin{pmatrix} c_{12} & s_{12} & 0 \\ -s_{12} & c_{13} & 0 \\ 0 & 0 & 1 \end{pmatrix} \quad (9.216)$$

$$= \begin{pmatrix} c_{12}c_{13} & s_{12}c_{13} & s_{13}e^{-i\delta} \\ -s_{12}c_{23} - c_{12}s_{23}s_{13}e^{i\delta} & c_{12}c_{23} - s_{12}s_{23}s_{13}e^{i\delta} & s_{23}c_{13} \\ s_{12}s_{23} - c_{12}c_{23}s_{13}e^{i\delta} & -c_{12}s_{23} - s_{12}c_{23}s_{13}e^{i\delta} & c_{23}c_{13} \end{pmatrix} ,$$

where $s_{ij} = \sin\theta_{ij}$ and $c_{ij} = \cos\theta_{ij}$. in the Standard Model, the phase δ is responsible for all $\mathcal{CP}$ violating phenomena in quark-flavor changing processes.

The rotation angles are defined and labeled in a way which relates to the mixing of two specific generations. In the limit $\theta_{23} = \theta_{13} = 0$ the third generation decouples, and the situation reduces to the usual Cabibbo mixing of the first two generations, with θ_{12} identified as the Cabibbo angle. The angles θ_{12}, θ_{23}, θ_{13} can all be chosen to lie in the first quadrant, i.e. $s_{ij}, c_{ij} > 0$, by appropriate redefinition of the quark field phases.

From measurements it is known that $1 \gg s_{12} \gg s_{23} \gg s_{13}$. It is therefore common to use a parametrization of the CKM matrix that emphasizes this hierarchy. In the Wolfenstein parametrization one defines

$$\lambda = s_{12} = \frac{|V_{us}|}{\sqrt{|V_{ud}|^2 + |V_{us}|^2}}, \quad A\lambda^2 = s_{23} = \lambda \left| \frac{V_{cb}}{V_{us}} \right|, \quad A\lambda^3 (\rho + i\eta) = s_{13} e^{i\delta} = V_{ub}^*,$$

where λ is the sine of the Cabibbo angle ($\sin\theta \approx 0.22$) and the real numbers A, ρ and η are of order unity. With these parameters the CKM matrix can be expressed in powers of λ and takes the convenient form

$$V = \begin{pmatrix} 1 - \lambda^2/2 & \lambda & A\lambda^3(\rho - i\eta) \\ -\lambda & 1 - \lambda^2/2 & A\lambda^2 \\ A\lambda^3(1 - \rho - i\eta) & -A\lambda^2 & 1 \end{pmatrix} + \mathcal{O}(\lambda^4) . \tag{9.217}$$

As the definition of $(\rho + i\eta)$ depends on the phase convention, one often introduces the parameters $(\bar{\rho} + i\bar{\eta})$, defined by

$$A\lambda^3 (\rho + i\eta) = \frac{A\lambda^3 (\bar{\rho} + i\bar{\eta})\sqrt{1 - A^2\lambda^4}}{\sqrt{1 - \lambda^2} \left[1 - A^2\lambda^4(\bar{\rho} + i\bar{\eta}) \right]} . \tag{9.218}$$

This definition ensures that $(\bar{\rho} + i\bar{\eta}) = -(V_{ud} V_{ub}^*)(V_{cd} V_{cb}^*)$ is phase convention independent and the CKM matrix written in the parameters λ, A, $\bar{\rho}$ and $\bar{\eta}$ is unitary to all orders in λ.[7]

The elements of the CKM matrix are fundamental parameters of the Standard Model and need to be experimentally determined. The unitarity condition of the CKM matrix imposes a set of relations between the matrix elements:

$$\sum_i V_{ij} V_{ik}^* = \delta_{jk} \quad \text{with} \quad jk = \{dd, ds, db, ss, sb, bb\} \tag{9.219}$$

$$\sum_j V_{ij} V_{kj}^* = \delta_{ik} \quad \text{with} \quad ik = \{uu, uc, ut, cc, ct, tt\} \tag{9.220}$$

---

[7]To $\mathcal{O}(\lambda^2)$ one finds $\bar{\rho} = \rho(1 - \lambda^2/2)$.

Fig. 9.38 Sketch of the
Unitarity Triangle (UT).
Figure taken from [133]

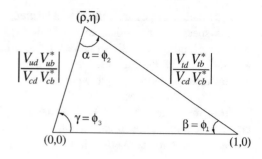

The six vanishing combinations ($j \neq k$, $i \neq k$) describe triangles in the complex
plane. The area of all triangles is given by half of the Jarlskog invariant J [132],

$$\Im\left[V_{ij}V_{kl}V_{il}^*V_{kj}^*\right] = J \sum_{m,n=1}^{3} \epsilon_{ikm}\epsilon_{jln}, \qquad (9.221)$$

where one representation of Eq. (9.221) reads for instance $J = \Im\left[V_{us}V_{cb}V_{ub}^*V_{cs}^*\right]$.
Expressed in the parameters of the standard CKM representation one finds $J = -c_{12}c_{13}^2c_{23}s_{12}s_{13}s_{23}\sin\delta$. A nonvanishing CKM phase and hence $\mathcal{CP}$ violation
requires $J \neq 0$.

While four of the six unitarity triangles are degenerated and rather slim, only two
triangles have approximately equal sides, of which one is usually referred as the
Unitarity Triangle (UT),

$$V_{ud}V_{ub}^* + V_{cd}V_{cb}^* + V_{td}V_{tb}^* = 0. \qquad (9.222)$$

Commonly one normalizes the triangle basis to unity by dividing each side by
$V_{cd}V_{cb}^*$ to obtain a triangle with vertices exactly at $(0,0)$, $(0,1)$ and the apex at
$(\bar{\rho}, \bar{\eta})$. The UT is sketched in Fig. 9.38. As can be seen from Fig. 9.38 the angles of
the UT are

$$\alpha = \arg\left(-\frac{V_{td}V_{tb}^*}{V_{ud}V_{ub}^*}\right) \qquad (9.223)$$

$$\beta = \arg\left(-\frac{V_{cd}V_{cb}^*}{V_{td}V_{tb}^*}\right) \qquad (9.224)$$

$$\gamma = \arg\left(-\frac{V_{ud}V_{ub}^*}{V_{cd}V_{cb}^*}\right) \qquad (9.225)$$

Often a different naming convention, also shown in Fig. 9.38, is used to label the three angles. The UT angles are constraint by the $\mathcal{CP}$ observables discussed in Sect. 9.6:

$$\beta = (21.9 \pm 0.7)^\circ \tag{9.226}$$

$$\gamma = (73.5^{+4.2}_{-5.1})^\circ \tag{9.227}$$

The measurement of the angle α has not been discussed in Sect. 9.6. It is measured using the observation of time-dependent $\mathcal{CP}$ violation in the decays $B^0 \rightarrow \pi\pi$, $\rho\rho$, and $\pi\rho$ by BABAR and Belle. The average of this measurements result into [133, 134],

$$\alpha = (84.5^{+5.9}_{-5.2})^\circ . \tag{9.228}$$

The sum of three angles, $\alpha + \beta + \gamma = (180 \pm 7)^\circ$, is consistent with the expectation and represents a first test of the unitarity of the CKM matrix. An additional constraint of the CKM phases and thus of the UT angles comes from the measurement of time-dependent $\mathcal{CP}$ violation in B_s^0 decays and the measurement of the phase β_s (with $\phi_s = -2\beta_s$, see Sect. 9.6),

$$\beta_s = \arg\left(-\frac{V_{ts}V_{tb}^*}{V_{cs}V_{cb}^*}\right) = (0.60 \pm 0.89)^\circ . \tag{9.229}$$

In addition to the measurements of $\mathcal{CP}$ violation in the B meson systems also the measurements of $\mathcal{CP}$ violation in the $K^0\bar{K}^0$ mixing, $|\epsilon_K| = (2.233 \pm 0.015) \times 10^{-3}$ (see Sect. 9.5) provides information about the CKM matrix. The measurement of $|\epsilon_K|$ can be translated into an approximate hyperbolic constraint on the apex $(\bar{\rho}, \bar{\eta})$ of the UT [135].

The sides of the unitarity triangle are accessible by measuring decay rates and mixing frequencies. In the following the experimental determination of the CKM elements as well as a test of the unitarity of the CKM matrix is discussed.

9.7.2 Determination of the CKM Matrix Elements

1. $|V_{ud}|$. Precise determinations of $|V_{ud}|$ are available from nuclear beta decays, from the decay of the free neutron and from semileptonic pion decays ($\pi^+ \rightarrow \pi^0 e^+ \nu$). The most precise value results from an analysis of superallowed $0^+ \rightarrow 0^+$ nuclear beta decays which are pure vector transitions. The measurements of the transition energies, the partial branching fractions, and the half-lives of the parent nuclei together with radiative and isospin-symmetry-breaking corrections allow the determination of the corrected $\mathcal{F}t$-value, from which, by using the muon life-time, $|V_{ud}|$ can be determined. The average of the fourteen most

precise determinations yield [136],

$$|V_{ud}| = 0.97417 \pm 0.00021 \,,$$

where the error is dominated by the theoretical uncertainty stemming from the nuclear structure and radiation correction.

The theoretical uncertainties in extracting a value of $|V_{ud}|$ from neutron decays are significantly smaller than those for the superallowed decays, however the value depends on the ratio between axial-vector and vector couplings ($g_A \equiv G_A/G_V$) and on the neutron lifetime. Using the most recent measurements [133] the following value for $|V_{ud}|$ is obtained,

$$|V_{ud}| = 0.9763 \pm 0.0016 \,,$$

with the error dominated by the g_A uncertainty.

An alternative approach is the measurement of the very small ($\mathcal{O}(10^{-8})$) branching ratio of the pion beta-decay $\pi^+ \rightarrow \pi^0 e^+ \nu$. The value normalized by a very precise theoretical prediction for $\pi^+ \rightarrow e^+ \nu$ and yields [137]

$$|V_{ud}| = 0.9749 \pm 0.0026 \,.$$

The error here stems mainly from the measurement of the rare process.

2. $|V_{us}|$. Earlier measurements of $|V_{us}|$ from kaon decays have used $K_L \rightarrow \pi e \nu$ to extract the product of $|V_{us}|$ and the form factor $|V_{us}| f_+(0)$ at $q^2 = 0$. The most recent data provide enough experimental constraints to also use decays to muons as well as decays of K_S and $K^{\pm}$. Averaging results from $K_L \rightarrow \pi e \nu$, $K_L \rightarrow \pi \mu \nu$, $K^{\pm} \rightarrow \pi^0 e^{\pm} \nu$, $K^{\pm} \rightarrow \pi^0 \mu^{\pm} \nu$ and $K_S \rightarrow \pi e \nu$ yields the value $|V_{us}| f_+(0) = 0.2165 \pm 0.0004$ [133]. Lattice QCD calculations of $f_+(0)$ have been carried out for different numbers of quark flavors. The form-factor average, $f_+(0) = 0.9704 \pm 0.0032$, of the (2+1)-flavor lattice calculations [138] is in good agreement with a classical calculation [139]. With this value one obtains,

$$|V_{us}| = 0.2231 \pm 0.0008 \,.$$

The lattice calculation of the ratio of kaon and pion decay constant, $f_K/f_\pi = 1.1933 \pm 0.0029$ [138], allows the determination of the ratio $|V_{us}/V_{ud}|$ from $K \rightarrow \mu \nu$ and $\pi \rightarrow \mu \nu$ decays. Using the precise measurement of the $K \rightarrow \mu \nu$ branching fraction by the KLOE collaboration [140] results in $|V_{us}| = 0.2253 \pm 0.0007$.

An alternative determination of $|V_{us}|$ uses hadronic τ decays to strange hadrons. The average of the measured inclusive and exclusive branching fractions yields $|V_{us}| = 0.2216 \pm 0.0015$ [134].

3. $|V_{cd}|$. First determinations of $|V_{cd}|$ came from neutrino scattering experiments. The difference of the ratio of double-muon to single-muon production for neutrino and antineutrino scattering is depending on the charm production cross

section and thus on $|V_{cd}|^2$, as well as on the semileptonic (muonic) branching ratio of the produced charm mesons, $\bar{\mathcal{B}}_\mu$. The method was first used by the CDHS group [141] but has also been applied by CCFR [142] and by CHARM II [143]. Averaging the results is complicated, also because $\bar{\mathcal{B}}_\mu$ is an effective quantity which depends on the specific neutrino beam characteristics. One finds $\bar{\mathcal{B}}_\mu |V_{cd}|^2 = (0.463 \pm 0.034) \times 10^{-2}$ [144] and using the average value of $\bar{\mathcal{B}}_\mu = 0.087 \pm 0.005$ one obtains [133]

$$|V_{cd}| = 0.230 \pm 0.011 .$$

Similar to $|V_{us}|$, $|V_{cd}|$ can also be extracted from semileptonic $(D \to \pi \ell \nu)$ and leptonic $(D^+ \to \mu^+ \nu)$ charm decays. Also here, QCD lattice calculations are used to determine the relevant form factors $f_+^{D\pi}(q^2 = 0)$ and f_D [138]. Using the average of the branching fraction measurements from BABAR, Belle, BES III and CLEO-c for $D \to \pi \ell \nu$ [134] results into [133]

$$|V_{cd}| = 0.2140 \pm 0.0029 \pm 0.0093,$$

where the first uncertainty is experimental and the second stems from the theoretical uncertainty of the form factor calculation. The measurements of the leptonic branching ratio $D^+ \to \mu^+ \nu$ by BES III and a CLEO-c results into [134]

$$|V_{cd}| = 0.2164 \pm 0.0050 \pm 0.0015 ,$$

where also here the first error is experimental while the second describes the uncertainty of the form factor calculation. For the average of the three different determinations of $|V_{cd}|$ the Particle Data Group [133] quotes a value of

$$|V_{cd}| = 0.218 \pm 0.004 .$$

4. $|V_{cs}|$. Measurements of semileptonic decays of D mesons to kaons $D \to K \ell \nu$ as well as the measurement of leptonic decays of D_s mesons, $D_s^+ \to \mu^+ \nu$, together with the corresponding form factors from lattice calculations allow the determination of $|V_{cs}|$. Branching fraction measurements have been performed by Belle, BABAR, CLEO-c and BES III, and are averaged in [134]. From the semileptonic measurements one obtains $|V_{cs}| = 0.967 \pm 0.025$ where the error is dominated by the theoretical uncertainty of the form factor calculations. The average of the leptonic measurements results into $|V_{cs}| = 1.006 \pm 0.019$ where the dominating uncertainty is experimental. For the average of both values the Particle Data Group [133] reports a values of

$$|V_{cs}| = 0.997 \pm 0.017.$$

5. $|V_{cb}|$. This matrix element is determined from semileptonic B decays to D or D^* mesons. Two experimentally and theoretically different approaches are used. The *inclusive* approach measures the inclusive semileptonic decay rate to any charmed final state together with the moments of the leptonic energy and the hadronic invariant mass spectra. An operator product expansion within the Heavy Quark Effective Theory (HQET) allows the calculation of decay rates and the energy and mass spectra in dependence of expansion parameters α_s and the inverse of the heavy quark mass. The simultaneous measurement of several distributions over-constrains the physical parameters and allows a determination of $|V_{cb}|$. An analysis and an averaging of existing measurements is performed in [145] and leads to $|V_{cb}| = (42.2 \pm 0.8) \times 10^3$ with uncertainties arising mainly from higher-order perturbative and non-perturbative corrections.

In the *exclusive* approach semileptonic B decays to exclusive channels containing D and D^* mesons are studied. In the infinite quark mass limit with m_b , $m_c \gg \Lambda_{QCD}$, heavy quark symmetry predicts that all form-factors are given by a single function which depends on the product of the four-velocities v of the B and v' of the $D^{(*)}$ (mesons) and are normalized at the point of maximum momentum transfer to the lepton system ($v \cdot v' = 1$). The matrix element $|V_{cb}|$ is obtained from an extrapolation to this point. The precise determination of normalization and the shape of the form factor function requires additional corrections calculated using HQET. Reference [145] quotes $|V_{cb}| = (41.9 \pm 2.0) \times 10^3$ as the average exclusive value of $|V_{cb}|$.

The inclusive and the exclusive determination of $|V_{cb}|$ are in agreement and [145] performs the average of both values:

$$|V_{cb}| = (42.2 \pm 0.8) \times 10^3.$$

6. $|V_{ub}|$. Similar to $|V_{cb}|$ also $|V_{ub}|$ is determined analysing inclusive and exclusive semileptonic B decays with a $b \rightarrow u\ell\bar{\nu}$ transition. The determination of $|V_{ub}|$ from inclusive decays however suffers experimental and theoretical difficulties. The total inclusive decay rate is hard to measure due to the large background from CKM-favored $b \rightarrow c\ell\bar{\nu}$ transitions. Therefore, strong kinematic cuts are introduced to suppress these background contributions. The restriction to tight kinematic regions however complicates the theoretical description. The calculation of partial decay rates in the various kinematic regions requires the introduction of non-perturbative distribution functions—the so called shape functions—to describe the effect of hadronic physics. At leading order there is only a single shape function which can be determined using inclusive $\bar{B} \rightarrow X_s\gamma$ decays. Subleading effects are considered using different theoretical models based on Heavy Quark Expansion (HQE). A recent summary of the $|V_{ub}|$ values extracted within different models from measurements by BABAR, BELLE and CLEO can be found in [145]. All calculations give similar values for $|V_{ub}|$ and similar error estimates. As average [145] quotes

$$|V_{ub}| = (4.49 \pm 0.15^{+0.16}_{-0.17}) \times 10^{-3}.$$

To consider different theoretical treatments the authors assign an additional quadratic error of $\pm 0.17 \times 10^{-3}$.

The determination of $|V_{ub}|$ from exclusive decays, such as $B \to \pi \ell \nu$ decays, suffers experimentally from very small signal yields and requires the theoretical determination of the corresponding form factors. Lattice form factor calculations are available for the high q^2 regions. So called light-cone QCD sum rules (LCSR) are applicable for the low q^2 region. A simultaneous fit to experimental $B \to \pi \ell \nu$ data and lattice results as function of q^2 together with additional constraints from LCSR results into

$$|V_{ub}| = (3.67 \pm 0.09 \pm 0.12) \times 10^{-3} \text{ [134]}.$$

The inclusive and exclusive determinations of $|V_{ub}|$ are largely independent and the large discrepancy between both methods remains a puzzle. Inflating the errors to account for this discrepancy [145] quotes an average of

$$|V_{ub}| = (3.94 \pm 0.36) \times 10^{-3}$$

The LHCb experiment has used the ratio of the two baryonic decays $\Lambda_b \to p\mu^-\bar{\nu}$ and $\Lambda_b \to \Lambda_c\mu^-\bar{\nu}$ to extract the ratio $|V_{ub}/V_{cb}| = 0.083 \pm 0.006$ [146]. The q^2-dependent form factor ratio had to be taken into account to consider the different kinematical ranges of the two decays. Using the above average for $|V_{cb}|$ one obtains

$$|V_{ub}| = (3.50 \pm 0.26) \times 10^{-3}.$$

7. $|V_{tb}|$. An experimental determination of $|V_{tb}|$ without assuming unitarity is possible using the production of single-top quarks. Single-top quark production cross-sections have been measured by the Tevatron experiments CDF and D0, and at LHC by ATLAS and CMS. Using these measurements [133] quotes the value

$$|V_{tb}| = 1.019 \pm 0.025.$$

8. $|V_{td}|$ and $|V_{ts}|$. The two CKM elements $|V_{td}|$ and $|V_{ts}|$ are expected to be very tiny and tree-level decays of the top-quark to a d-quark or s-quark will be very difficult to measure. However, both elements are accessible through the measurements of the mixing frequency of B^0 and B_s^0 mesons (see Sect. 9.6). Using the most recent lattice QCD results for the hadronic factors, $f_{B_d}\sqrt{\hat{B}_{B_d}} = (219 \pm 14)$ MeV and $f_{B_s}\sqrt{\hat{B}_{B_s}} = (270 \pm 16)$ MeV [138], together with the world averages for the mixing frequencies Δm_d and Δm_s results into [133]

$$|V_{td}| = (8.1 \pm 0.5) \times 10^{-3} \quad \text{and}$$

$$|V_{ts}| = (39.4 \pm 2.3) \times 10^{-3},$$

where the uncertainties are dominated by the lattice uncertainties of the hadronic factors. Several uncertainties are reduced when calculating the ratio

$$\xi = \left(f_{B_s} \sqrt{\hat{B}_{B_s}} \right) / \left(f_{B_d} \sqrt{\hat{B}_{B_d}} \right) = 1.239 \pm 0.046 .$$

The ratio $|V_{td}|/|V_{ts}|$ is therefore stronger constraint [133],

$$|V_{td}|/|V_{ts}| = 0.210 \pm 0.001 \pm 0.008 .$$

Using the independent measurements of the CKM elements the unitarity of the CKM matrix can be checked. One obtains for the first two rows [133],

$$|V_{ud}|^2 + |V_{us}|^2 + |V_{ub}|^2 = 0.9994 \pm 0.0005 ,$$
$$|V_{cd}|^2 + |V_{cs}|^2 + |V_{cb}|^2 = 1.043 \pm 0.034 ,$$

which agrees well with the unitarity assumption. In addition, the direct measurement of $|V_{tb}|$ leaves little room for mixing of the top into unknown other states.

9.7.3 Global Analysis and Test of the Unitarity of the CKM Matrix

The available information on the magnitude and the phases of the CKM elements can be analysed by a global fit to all CKM parameters. The most precise information is obtained imposing Standard Model constraints such as the unitarity of the CKM matrix and the existence of exactly three quark generations. Input parameters of the global analysis are, beside the experimental measurements, also theoretically determined hadronic parameters with sometimes large errors.

Different approaches exist to combine the experimental data and to treat the experimental and theoretical errors. The CKM-Fitter group [135, 147] is using a frequentist's framework based on a χ^2 analysis. In these fits the frequentist's treatment is also applied to the theoretical errors. The UTfit group [148] uses a baysian approach for all errors. The two different statistical approaches lead to very similar results and here only the results of [135] are presented.

Figure 9.39 shows the experimental constraints in the $(\bar{\rho}, \bar{\eta})$ plane. Indicated is the unitarity triangle with its angles α, β and γ. The different measurements clearly limit the apex of the triangle to a small (dashed) region. For the Wolfenstein parameters introduced in Eq. (9.217) the global fit [133, 135] gives

$$\lambda = 0.22453 \pm 0.00044 , \qquad A = 0.836 \pm 0.015 , \qquad (9.230)$$
$$\bar{\rho} = 0.122^{+0.018}_{-0.017} , \qquad\qquad \bar{\eta} = 0.355^{+0.012}_{-0.011} . \qquad (9.231)$$

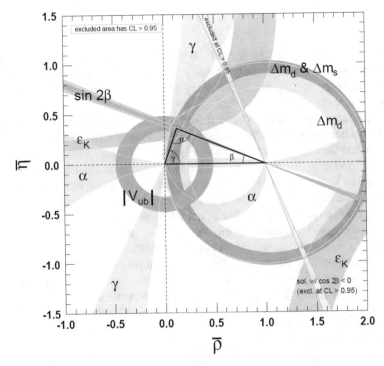

Fig. 9.39 Experimental constraints of the unitarity triangle. Shaded areas correspond to 95% C.L. Figure taken from [133]

Alternatively one can give the fit results of the magnitudes of all nine elements of the CKM matrix:

$$
\mathbf{V_{CKM}} = \begin{pmatrix} 0.97446 \pm 0.00010 & 0.22452 \pm 0.00044 & 0.00365 \pm 0.00012 \\ 0.22438 \pm 0.00044 & 0.97359^{+0.00010}_{-0.00010} & 0.04214 \pm 0.00076 \\ 0.00896^{+0.00024}_{-0.00023} & 0.04133 \pm 0.00074 & 0.999105 \pm 0.000032 \end{pmatrix}
$$
(9.232)

Figure 9.39 represents an impressive confirmation of the CKM paradigm which describes the flavor transition and the $\mathcal{CP}$ violation in the quark sector. The global fit currently does not point to deviations from the Standard Model picture. However, the effect of *New Physics* might be small. Over-constraining measurements of $\mathcal{CP}$ asymmetries, mixing and flavor changing decays will put further bounds on possible *New Physics* contributions.

9.8   Conclusion

Nearly 40 years after the discovery of CP violation, the nature of the phenomenon has been clarified experimentally. In the K meson system, CP violation has been discovered in the mixing (ε), in the decay (ε'/ε, direct CP violation), and in the interference between mixing and decay ($\Im m \, \varepsilon$); in the B meson system, CP violation in the interference between mixing and decay ($\sin 2\beta$) as well as CP violation in the decay have been observed. While finishing this article, the LHCb collaboration has also reported the first observation of CP violation in the decay of neutral D mesons to K^+K^- and $\pi^+\pi^-$ ($\mathcal{O}(1.5 \times 10^{-3})$) [205]. All observations are consistent[8] with the model of Kobayashi and Maskawa where the 3×3 mixing matrix of six quarks has one non-trivial complex phase $\delta = (71.0 \pm 0.3)°$. This leads to complex weak coupling constants of quarks, and to $\mathcal{T}$ and CP violation.

In cosmology, CP violation together with a large departure from thermal equilibrium via a first-order electroweak phase transition [206] and a baryon number violation by instanton processes [207] could explain the observed baryon asymmetry. The source would be the asymmetric interactions of quarks and antiquarks with the Higgs field [208]. The size of the observed CP violation in the quark sector, expressed by the Jarlskog determinant, however is insufficient to explain the matter asymmetry by several orders of magnitude [209]. Moreover, the low mass of the Higgs boson disfavours baryogenesis during the electroweak phase transition [210]. A way out of this dilemma is the hypothesis that CP violation in the neutrino sector could cause a lepton asymmetry. $(B+L)$-violating processes before the electroweak phase transition could then convert the lepton asymmetry into a baryon asymmetry. Leptogenesis would proceed through the production of heavy Majorana neutrinos in the early Universe [211]. While the experimental proof of leptogenesis is difficult, the observation of CP violation in the neutrino sector and the exploration of the nature of the neutrino (Dirac or Majorana fermion) could establish strong hints.

In quark flavor physics the focus has changed after the successful establishment of the Kobayashi-Maskawa paradigm. Today, the precise measurement of flavor changing processes and CP asymmetries are used as tools to test the Standard Model predictions. Tiny deviations between observations and theory predictions could hint to additional quantum corrections modifying the size and the phase of flavor changing amplitudes. The origin of these additional corrections could be new heavy particles at mass scales much higher than the energies accessible by today's collider experiments. Experimental results on neutral meson mixing already constrain the ratio between mass scale and the coupling strength of new particles to values above 10^2 to 10^4 TeV, where the strongest bounds come from the neutral K meson system [212]. The next generation of precision quark-flavor experiments will

---

[8]For the D mesons the theoretical predictions still have large uncertainties. For K mesons recent theoretical results on the value of the CP violating amplitude $\Re\left(\epsilon'/\epsilon\right)$ both from dual QCD [213] and from lattice QCD [214] deviate from the experimental value reported in Sect. 9.5, thus further clarification is needed.

thus open a window to *New Physics* complementary to the direct searches at highest energies.

References

1. A.A. Penzias, R. Wilson: Astrophys. J. **142**, 419 (1965).
2. A.D. Sakharov: JETP Lett. **5**, 24 (1967).
3. J.H. Christenson, J.W. Cronin, V.L. Fitch, R. Turlay: Phys. Rev. Lett. **13**, 138 (1964).
4. E. Wigner: Z. Phys. **43**, 624 (1927).
5. G.C. Wick: Ann. Rev. Nucl. Sci. **8**, 1 (1958).
6. E. Wigner: Nachr. Akad. Wiss. Göttingen **31**, 546 (1932).
7. G. Lüders: Kgl. Danske Videnskab. Selskab, Matfys. Medd. **28(5)**, 1 (1954).
8. W. Pauli, in *Niels Bohr and the Development of Physics* ed. W. Pauli, Pergamon, Oxford 2nd edn. (1955), p. 30.
9. R. Jost: Helv. Phys. Acta **30**, 209 (1957).
10. C.D. Anderson: Science **76**, 238 (1932); Phys. Rev. **43**, 491 (1933).
11. O. Chamberlain: E. Segre, C. Wiegand, T. Ypsilantis, Phys. Rev. **100**, 947 (1955).
12. M. Gell-Mann, A. Pais: Phys. Rev. **97**, 1387 (1955).
13. T.D. Lee, C.N. Yang: Phys. Rev. **104**, 254 (1956).
14. C.S. Wu, et al.: Phys. Rev. **105**, 1413 (1957).
15. R.L. Garwin, L.M. Lederman, M. Weinrich: Phys. Rev. **105**, 1415 (1957).
16. J.I. Friedman, V.L. Telegdi: Phys. Rev. **105**, 1681 (1957).
17. L.D. Landau: Nucl. Phys. **3**, 127 (1957).
18. H. Schopper: Phil. Mag. **2**, 710 (1957).
19. M. Goldhaber, L. Grodzins, A.W. Sunyar: Phys. Rev. **109**, 1015 (1958).
20. M. Bardon, et al.: Ann. Phys. NY **5**, 156 (1958).
21. D. Neagu, et al.: Phys. Rev. Lett. **6**, 552 (1961).
22. J.H. Christenson, J.W. Cronin, V.L. Fitch, R. Turlay: Phys. Rev. Lett. **13**, 138 (1964).
23. R. Adair, et al.: Phys. Rev. **132**, 2285 (1963).
24. J.M. Gaillard, et al.: Phys. Rev. Lett. **18**, 20 (1967).
25. J.W. Cronin, et al.: Phys. Rev. Lett. **18**, 25 (1967).
26. S.Bennett, et al.: Phys. Rev. Lett. **19**, 993 (1967).
27. D. Dorfan, et al.: Phys. Rev. Lett. **19**, 987 (1967).
28. Particle Data Group: M. Tanabashi et al., Phys. Rev. D **98**, 030001 (2018).
29. J. Prentki, M. Veltman: Phys. Lett. **15**, 88 (1965).
30. L.B. Okun: Sov. J. Nucl. Phys. **1**, 670 (1965).
31. T.D. Lee, L. Wolfenstein: Phys. Rev. B **138**, 1490 (1965).
32. J. Bernstein, G. Feinberg, T.D. Lee: Phys. Rev. B **139**, 1650 (1965).
33. S. Barshay: Phys. Lett. **17**, 78 (1965).
34. F. Salzman, G. Salzman: Phys. Lett. **15**, 91 (1965).
35. B. Arbuzov, A.T. Filipov, Phys: Lett. **20**, 537; Phys. Lett. **21**, 771 (1965).
36. L. Wolfenstein: Nuovo Cimento **42**, 17 (1966).
37. W. Alles: Phys. Lett. **14**, 348 (1965).
38. S.L. Glashow: Phys. Rev. Lett. **14**, 35 (1964).
39. F. Zachariasen, G. Zweig: Phys. Lett. **14**, 794 (1965).
40. S.N. Lotsoff: Phys. Lett. **14**, 344 (1965).
41. R.G. Sachs: Phys. Rev. Lett. **13**, 286 (1964).
42. R.N. Mohapatra: Phys. Rev. D **6**, 2023 (1972); R. Mohapatra, J. Pati, Phys. Rev. D **11**, 566 (1975).
43. A. Pais: Phys. Rev. D **8**, 625 (1973).

44. T.D. Lee: Phys. Rev. D **8**, 1226 (1973); P. Sikivie, Phys. Lett. **65B**, 141 (1976).
45. L. Wolfenstein: Phys. Rev. Lett. **13**, 562 (1964).
46. V.M. Lobashov, et al.: Phys. Lett. B **25**, 104 (1967).
47. E.M. Henley: Ann. Rev. Nucl. Sci. **19**, 367 (1969).
48. M. Kobayashi, T. Maskawa: Prog. Theor. Phys. **49**, 652 (1973).
49. Z. Bai et al., Phys. Rev. Lett. 115, 21212001 (2015).
50. T. Blum et al., Phys. Rev. D 91, 7074505 (2015).
51. A. J. Buras et al., JHEP 11, 033 (2015).
52. A. J. Buras et al., JHEP 11, 202 (2015).
53. W. A. Bardeen, A. J. Buras and J.-M. Gerard, Phys. Lett. B 180, 133, (1986).
54. A. J. Buras and J.-M. Gerard, JEP 12, 008 (2015).
55. H. Gisbert and A. Pich, Reports on Progress in Physics, Vol. 81, 7 (2018).
56. N. Cabibbo, Phys. Rev. Lett. **10**, 531 (1963).
57. M. Kobayashi, T. Maskawa: CP-Violation in the Renormalizable Theory of Weak Interaction. In: Progress of Theoretical Physics. Vol. 49, Issue 2, 652 (1973).
58. H. Albrecht et al., ARGUS Collaboration, Phys. Lett. B 192 (1987) 245.
59. BABAR Collaboration, B. Aubert et al., Phys. Rev. Lett. **87**, 091801 (2001).
60. Belle Collaboration, K. Abe et al., Phys. Rev. Lett **87**, 091802 (2001).
61. CDF Collaboration, T. Aaltonen et al., Phys. Rev. Lett. **97**, 242003 (2006).
62. LHCb Collaboration, A. Alves Jr. et al., J. of Inst **3** S08005 (2008).
63. A. Lenz and U. Nierste, Journal of Higher Energy Physics **06**, 72 (2007).
64. A. Lenz and U. Nierste, arXiv:1102.4274 (2011).
65. A. J. Buras, W. Slominski, H. Steger Nucl. Phys. B **245**, 369 (1984).
66. M. Artuso, G. Borissov and A. Lenz, Rev. Mod. Phys. **88** 045002 (2016).
67. K. Anikeev et al., *B Physics at the Tevatron*, arxiv:hep/0201071.
68. U. Nierste, Z. Ligeti and A. S. Kronfeld in *B Physics at the Tevatron*, arxiv:hep/0201071.
69. BaBaR Collaboration, B. Aubert et al., Nucl. Instrum. Meth. A479 (2002) 1–11;
70. Belle Collaboration, A. Abashian et al., Nucl. Instrum. Meth. A479, 117–232 (2002).
71. A. J. Bevan et al., Eur. Phys. J. C74 3026 (2014).
72. Belle-II Collaboration, I. Adachi et al., Nucl. Instrum. Meth. A907 46–59 (2018).
73. CDF Collaboration, F. Abe et al., Nucl. Instrum. Meth. A271 387–403 (1988).
74. CDF Collaboration, A. Sill for the collaboration, Nucl. Instrum. Meth. A447 (2000) 1–8.
75. D0 Collaboration, S. Abachi et al., Nucl. Instrum. Meth. A338 185–253 (1994);
76. D0 Collaboration, V.M. Abazov et al., Nucl. Instrum. Meth. A565 463–537 (2006).
77. LHCb Collaboration, A. Augusto Alves Jr. et al., Journal of Inst. 3 S08005 (2008).
78. ATLAS Collaboration, G. Aad et al., Journal of Inst. 3 S08003 (2008).
79. CMS Collaboration, CMS Collaboration, S. Chatrchyan et al., Journal of Inst. 3 S08004 (2008).
80. H. Albrecht, et al.: Phys. Lett. B192 245 (1987).
81. ARGUS Collaboration, H. Albrecht, Phys.Lett. B324 249 (1994).
82. CLEO Collaboration, John E. Bartelt et al., Phys.Rev.Lett. 71 1680 (1993).
83. BABAR Collaboration, B. Aubert et al., Phys. Rev.Lett. 88, 221803 (2002).
84. Belle Collaboration: K. Abe et al., Phys. Rev. D71, 072003 (2005).
85. CDF Collaboration, A. Abulencia et al., Phys.Rev.Lett. 97 242003 (2006).
86. LHCb collaboration: R. Aaij et al., New J. Phys. 15 053021 (2013).
87. Heavy Flavor Averaging Group (HFLAV), Y. Amhis et al. Eur. Phys. J. C77 895 (2017) and updates at www.slac.stanford.edu/xorg/hflav/.
88. LHCb collaboration: R. Aaij et al., Eur. Phys. J. C 76 412 (2016).
89. D0 collaboration, V. M. Abazov et al., Phys. Rev. Lett. 105 081801 (2010).
90. D0 collaboration, V. M. Abazov et al., Phys. Rev. D89 012002 (2014).
91. LHCb collaboration: R. Aaij et al., Phys. Rev. Lett. 114 041601 (2015).
92. LHCb collaboration: R. Aaij et al., Phys. Rev. Lett. 117 061803 (2016).
93. Belle collaboration, E. Nakano et al., Phys. Rev. D73 112002 (2006).
94. BaBar collaboration, J. P. Lees et al., Phys. Rev. Lett. 111 101802 (2013).

95. BaBar collaboration, J. P. Lees et al., Phys. Rev. Lett. 114 081801 (2015).
96. D0 collaboration, V. M. Abazov et al., Phys. Rev. D86 072009 (2012).
97. D0 collaboration, V. M. Abazov et al., Phys. Rev. Lett. 110 011801 (2013)
98. Belle Collaboration, I. Adachi et al., Phys. Rev. Lett. 108, 171802 (2012).
99. BABAR Collaboration, B. Aubert et al., Phys. Rev. D79, 072009 (2009).
100. LHCb Collaboration, R. Aaij, Phys. Rev. Lett. 115, 031601 (2015).
101. LHCb Collaboration, R. Aaij, Journal of High Energy Physics 11 170 (2017).
102. CDF collaboration, T. Aaltonen et al., Phys. Rev. Lett. 100 161802 (2008).
103. D0 collaboration, V. Abazov et al., Phys. Rev. Lett. 101 241801 (2008).
104. D0 Collaboration, V. M. Abazov et al., Phys. Rev. D 85, 032006 (2012).
105. CDF Collaboration, T. Aaltonen et al., Phys. Rev. D 85, 072002 (2012).
106. LHCb Collaboration, R. Aaij et al., Phys. Rev. Lett. 108, 101803 (2012).
107. LHCb Collaboration, R. Aaij et al., Phys. Rev. D 87, 112010 (2013). .
108. LHCb Collaboration, R. Aaij et al., Phys. Rev. Lett. 114, 041801 (2015).
109. LHCb Collaboration, R. Aaij et al., Phys. Rev. D86 052006 (2012).
110. LHCb Collaboration, R. Aaij et al., Phys. Lett. B 736 186 (2014).
111. ATLAS Collaboration, G. Aad et al., JHEP 08, 147 (2016).
112. CMS Collaboration, V. Khachatryan et al., Phys. Lett. B 757, 97 (2016).
113. J. Charles et al. (CKMfitter Group), Phys. Rev. D91, 073007 (2015); 2016 updated results at http://ckmfitter.in2p3.fr/.
114. Belle Collaboration, K. Abe et al., Phys. Rev. Lett. 93, 021601 (2004).
115. BaBar Collaboration, B. Aubert et al., Phys. Rev. Lett. 93, 131801 (2004).
116. Belle Collaboration, Y. Chao, et al., Phys. Rev. Lett. 93, 191802 (2004).
117. LHCb Collaboration, R. Aaij et al., Phys. Rev. Lett. 108, 201601 (2012).
118. LHCb collaboration: R. Aaij, Phys. Rev. Lett. 110, 221601 (2013).
119. A. B. Carter and A. I. Sanda, Phys. Rev. D23, 1567 (1981).
120. I. I. Y. Bigi and A. I. Sanda, Phys. Lett. B211, 213 (1988).
121. M. Gronau and D. Wyler, Phys. Lett. B265 (1991) 172.
122. M. Gronau and D. London, Phys. Lett. B253 (1991) 483.
123. D. Atwood, I. Dunietz, and A. Soni, Phys. Rev. Lett. 78, 3257 (1997).
124. D. Atwood, I. Dunietz, and A. Soni, Phys. Rev. D63, 036005 (2001).
125. Y. Grossman, Z. Ligeti, and A. Soffer, Phys. Rev. D67, 071301 (2003).
126. A. Giri, Y. Grossman, A. Soffer, and J. Zupan, Phys. Rev. D68, 054018 (2003).
127. F. Martinez-Vidal, K. Trabelsi, I. Bigi in Chap. 17.8 of Eur. Phys. J. C74 3026 (2014).
128. LHCb collaboration, R. Aaij et al., Phys. Lett. B 760, 117 (2016).
129. LHCb collaboration, R. Aaij et al., CERN-LHCb-CONF-2018-002.
130. Particle Data Group: M. Tanabashi et al., Phys. Rev. D 98, 030001 (2018).
131. S.L. Glashow, J. Iliopoulos, L. Maiani (1970) Physical Review D 2, 1285 (1970).
132. C. Jarlskog, Phys. Rev. Lett. 55, 1039 (1985).
133. Particle Data Group: M. Tanabashi et al., Phys. Rev. D 98, 030001 (2018).
134. Heavy Flavor Averaging Group, Y. Amhis et al., , Eur. Phys. J. C77, 895 (2017) and updates at www.slac.stanford.edu/xorg/hflav/.
135. CKMfitter Group, J. Charles et al., Eur. Phys. J. C41, 1 (2005) and updates at http://ckmfitter. in2p3.fr/.
136. J.C. Hardy, I.S. Towner, Phys. Rev. C 91, 025501 (2015).
137. D. Pocanic et al., Phys. Rev. Lett. 93, 181803 (2004).
138. FLAG Working Group, S. Aoki et al., Eur.Phys.J. C77, 122 (2017).
139. H. Leutwyler and M. Roos, Z. Phys. C25, 91 (1984).
140. KLOE Collaboration,F. Ambrosino et al., Phys. Lett. B632, 76 (2006).
141. H. Abramowicz et al., Z. Phys. C15, 19 (1982).
142. S.A. Rabinowitz, et al.: Phys. Rev. Lett. 70, 134 (1993).
143. CHARM II Collaboration, P. Vilain et al., Eur. Phys. J C11, 19 (1999).
144. F.J. Gilman et al., Phys. Lett. B592, 793 (2004).
145. R. Kowalewski and T. Mannel in [133].

146. LHCb Collaboration, R. Aaij et al., Nature Physics 10, 1038 (2015).
147. A. Höcker et al., Eur. Phys. J. C21, 225 (2001).
148. UTfit Collaboration, M. Bona et al., JHEP 507, 28 (2005), and updates at http://www.utfit.org/.
149. G. Colangelo: J. Gasser, H. Leutwyler, Nucl. Phys. B **603**, 125 (2001).
150. W. Ochs: π N Newsletter **3**, 25 (1991); J.L. Basdevant, C.D. Frogatt, J.L. Petersen, Phys. Lett. B **41**, 178 (1972).
151. M.F. Losty, et al.: Nucl. Phys. B **69**, 185 (1974); W. Hoogland, et al.: Nucl. Phys. B **126**, 109 (1977).
152. G. Grayer, et al.: Nucl. Phys. B **75**, 189 (1974).
153. J.R. Batley, et al.: Phys. Lett. B **630**, 31 (2005).
154. R. Adler, et al.: Phys. Lett. B **407**, 193 (1997).
155. Y. Zou, et al.: Phys. Lett. B **369**, 362 (1996).
156. J.S. Bell, J. Steinberger, in Proc. Oxford Int. Conf. on Elem. Part. 1965, p. 195.
157. R. Wanke, presented at 38th Rencontres de Moriond, Electroweak Interactions and Unified Theories (2003).
158. F. Ambrosino, et al.: Phys. Lett. B **619**, 61 (2005).
159. D. Babusci, et al., Phys. Lett. B **723**, 54 (2013).
160. T. Alexopolous, et al.: Phys. Rev. D **70**, 092006 (2004).
161. F. Ambrosino, et al.: Phys. Lett. B **638**, 140 (2006).
162. A. Lai, et al.: Phys. Lett. B **645**, 26 (2007).
163. Particle Data Group: M. Tanabashi et al., Phys. Rev. D 98, 030001 (2018).
164. S. Bennett, et al.: Phys. Rev. Lett. **19**, 993 (1967).
165. D. Dorfan, et al.: Phys. Rev. Lett. **19**, 987 (1967).
166. C. Geweniger, et al.: Phys. Lett. B **48**, 487 (1974).
167. M. Holder, et al.: Phys. Lett. B **40**, 141 (1972).
168. M. Banner, et al.: Phys. Rev. Lett. **28**, 1597 (1972).
169. B. Wolff, et al.: Phys. Lett. B **36**, 517 (1971).
170. J.C. Chollett, et al.: Phys. Lett. B **31**, 658 (1970).
171. H. Burkhardt, et al.: Nucl. Instrum. Methods Phys. A **268**, 116 (1988).
172. G. Barbiellini, et al.: Phys. Lett. B **43**, 529 (1973).
173. M. Cullen, et al.: Phys. Lett. B **32**, 523 (1970).
174. C. Geweniger, et al.: Phys. Lett. B **52**, 108 (1974).
175. S. Gjesdal, et al.: Phys. Lett. B **52**, 113 (1974).
176. L.K. Gibbons, et al.: Phys. Rev. Lett. **70**, 1199 (1993).
177. L.K. Gibbons, et al.: Phys. Rev. Lett. **70**, 1203 (1993).
178. K. Kleinknecht, S. Luitz: Phys. Lett. B **336**, 581 (1994); K. Kleinknecht, Phys. Rev. Lett. **75**, 4784 (1995).
179. B. Schwingenheuer, et al.: Phys. Rev. Lett. **74**, 4376 (1995).
180. A. Alavi-Harati, et al.: Phys. Rev. D **67**, 012005 (2003).
181. A. Apostolakis et al.: Phys. Lett. B **458**, 545 (1999).
182. A. Böhm, et al.: Nucl. Phys. B **9**, 605 (1965).
183. D.A. Jensen, et al.: Phys. Rev. Lett. **23**, 615 (1965).
184. R. Carosi, et al.: Phys. Lett. B **237**, 303 (1990).
185. A. Angelopoulos, et al.: Phys. Rep. **374**, 165 (2003).
186. V.L. Fitch, et al.: Phys. Rev. Lett. **31**, 1524 (1973).
187. C. Geweniger, et al.: Phys. Lett. B **48**, 483 (1974).
188. A. Alavi-Harati, et al.: Phys. Rev. Lett. **88**, 181601 (2002).
189. R. Piccioni, et al.: Phys. Rev. Lett. **29**, 1412 (1972).
190. H.H. Williams, et al.: Phys. Rev. Lett. **31**, 1521 (1973).
191. H. Burkhardt, et al.: Phys. Lett. B **206**, 169 (1988).
192. H. Saal: PhD thesis. Columbia University, New York (1969).
193. J. Marx, et al.: Phys. Lett. B **32**, 219 (1970).
194. V.A. Ashford, et al.: Phys. Rev. Lett. **31**, 47 (1972).

195. M.A. Paciotti: PhD thesis, UCRL-19446, University of California Radiation Laboratory, Berkeley (1969).
196. R.L. McCarthy, et al.: Phys. Rev. D **7**, 687 (1973).
197. A. Lai, et al.: Phys. Lett. B **537**, 28 (2002).
198. A. Angelopoulos, et al.: Phys. Lett. B **503**, 49 (2001).
199. A. Apostolakis et al.: Phys. Lett. B **444**, 38 (1998).
200. A. Alavi-Harati, et al.: Phys. Rev. Lett. **83**, 22 (1999).
201. A. Lai, et al.: Eur. Phys. J. C **22**, 231 (2001).
202. J.R. Batley, et al.: Phys. Lett. B **544**, 97 (2002).
203. V. Fanti, et al.: Phys. Lett. B **465**, 335 (1999).
204. M. Woods, et al.: Phys. Rev. Lett. **60**, 1695 (1988); H. Burkhardt, et al.: Phys. Lett. B **206**, 169 (1988); J.R. Patterson, et al.: Phys. Rev. Lett. **64**, 1491 (1990); L.K. Gibbons, et al.: Phys. Rev. Lett. **70**, 1203 (1993); G.D. Barr, et al.: Phys. Lett. B **317**, 233 (1993).
205. LHCb collaboration, R. Aaij et al., Phys. Rev. Lett. 122, 211803 (2019).
206. V.A. Rubakov, M.E. Shaposhnikov: Phys. Uspekhi **39**, 461 (1996).
207. G. t'Hooft: Phys. Rev. Lett. **37**, 8 (1976); Phys. Rev. D **14**, 3422 (1976)
208. A. Riotto, M. Trodden: Ann. Rev. Nucl. Part. Sci. **49**, 35 (1999).
209. M.B. Gavela, P.Hernandez, J. Orloff and O.Pene, Mod. Phys. Lett. A9, 795, (1994).
210. A. I. Bochkarev and M. E. Shaposhnikov, Mod. Phys. Lett. A 2, 417 (1987); K.Kajantie, M. Lainea, K. Rummukainend and M. Shaposhnikov, Nucl. Phys. B466 189–258 (1996).
211. M. Fukugita, T. Yanagida: Phys. Lett. B **174**, 45 (1986); M. Flanz, et al.: Phys. Lett. B **345**, 248 (1995); W. Buchmüller: Ann. Phys. (Leipzig) **10**, 95 (2001).
212. G. Isidori, Y. Nir, and G. Perez, Ann. Rev. Nucl. and Part. Sci. 60, 355 (2010).
213. A.J. Buras and J.-M. Gerard, J. of High Energy Physics 12, 008 (2015).
214. T. Blum et al., Phys. Rev. D 91, 7074502 (2015).

Chapter 10
The Future of Particle Physics: The LHC and Beyond

Ken Peach

10.1 2017 Update

I have been asked to submit a revised version of this chapter, published almost a decade ago. However, I think that it is better to leave the historical record as it was—this was an article written in its time and for its time. If I was writing this article today, I would call it "The Future of Particle Physics—Beyond the LHC", in recognition of the fact that, when originally written, the LHC was still under construction and now it has been operating for several years. The other key event which informed the original article was the recently-developed European Strategy for Particle Physics, adopted by the CERN Council in July 2006; the strategy was updated in 2013 and formally adopted in May of that year [1]; as I write, the process of updating the strategy is under way.

The intervening decade has seen significant progress, most spectacularly in the observation of the Higgs scalar [3] at 125.10 ± 0.14 GeV/c^2 [2] which completes the Standard Model of the particles and their interactions (excluding gravity) which dominate the local region of the universe. This means that all of the free parameters of the Standard Model are now known with reasonable precision, which can be chosen as (see Table 10.1):

1. The quark masses (m_u, m_d), (m_c, m_s), (m_t, m_b);
2. the charged lepton masses m_e, m_μ, m_τ;
3. the Z^0 mass M_Z;
4. the Higgs Mass M_H;
5. the electromagnetic coupling constant α;
6. the strong coupling constant $\alpha_s(M_Z)$;

K. Peach (✉)
John Adams Institute for Accelerator Science, Oxford, UK
e-mail: ken.peach@adams-institute.ac.uk

© The Author(s) 2020
H. Schopper (ed.), *Particle Physics Reference Library*,
https://doi.org/10.1007/978-3-030-38207-0_10

Table 10.1 Parameters of the Standard Model [2]

m_u $2.14^{+0.49}_{-0.26}$ MeV/c²	m_c 1.27 ± 0.02 GeV/c²	m_t 172.9 ± 0.4 GeV/c²	
m_d $4.67^{+0.48}_{-0.17}$ MeV/c²	m_s 93^{+11}_{-5} MeV/c²	m_b $4.18^{+0.03}_{-0.02}$ GeV/c²	
m_e $0.5109989461(31)$ MeV/c²	m_μ $105.6583745(24)$ MeV/c²	m_τ $1776.86(0.12)$ MeV/c²	
M_Z 91.1876 ± 0.0021 GeV/c²	M_H 125.10 ± 0.14 GeV/c²		
α $7.2973525664(17)$ $\times 10^{-3}$	G_F $1.1663787(6) \times 10^{-5}$ GeV⁻²	$\alpha_s(M_Z)$ $0.1179(10)$	
λ 0.22453 ± 0.00044	A 0.836 ± 0.015	ρ 0.122 ± 0.0175	η 0.355 ± 0.0115
Derived quantities	$\sin^2\theta_W$ 0.23129 ± 0.00005	$M_W = M_z \cos\theta_w$ 80.385 ± 0.015 GeV/c²	$\sin^2 2\theta_w = \frac{\alpha\pi}{8G_F M_Z^2}$

7. the weak coupling constant G_F; and
8. the Cabibbo-Kobayashi-Maskawa (CKM) parameters in the Wolfenstein parametrization λ, A, ρ and η.

In addition, two other useful quantities are given—M_W and $\sin^2\theta_W$—which can be written in terms of the other parameters, as shown in the last line of Table 10.1.

This is a total of 18 parameters which can only (so far) be determined by experiment. However, once determined, they describe the observed interactions with remarkable precision; for example, the muon anomalous magnetic moment (averaged between $\mu+$ and $\mu-$) is $(11,659,209\ 1 \pm 54_{stat} \pm 33_{syst}) \times 10^{-11}$, to be compared with the theoretical expectation of $(11,659,180\ 3 \pm 1_{EW} \pm 49_{Had}) \times 10^{-11}$, where the first error in the theoretical calculation comes from electroweak corrections and the second comes from a combination of low-order (42) and higher order (26) hadronic corrections [4]. The difference between these is $(288 \pm 80) \times 10^{-11}$ (3.5 standard deviations) and it is, of course, hotly debated whether this is statistically significant and, if it is, what that significance might be. There is a new experiment at Fermilab [5] which aims to reduce the total error in the muon anomalous magnetic moment to 14×10^{-11}, which, even without improvements to the theoretical calculations, would (if the central value of the discrepancy remained unchanged) increase the significance to more than 5 standard deviations.

Within the Standard Model, because the neutrinos have only one helicity state, they must be strictly massless. However, the phenomenon of neutrino oscillations, now well established, requires the neutrinos to have mass—essentially, the oscillation is a beating in the propagation through time of the different mass eigenstates that form the flavour eigenstates. The past decade has seen enormous progress in

Table 10.2   Neutrino Oscillation Parameters [6, 7][a]

	Normal hierarchy	Inverted hierarchy
Δm_{12}^2 $7.37^{+0.20}_{-0.14} \times 10^{-5}$ eV²/c⁴	$\|\Delta m_{23}^2\|$ $2.50 \pm 0.04 \times 10^{-3}$ eV²/c⁴	$2.46 \pm 0.04 \times 10^{-3}$ eV²/c⁴
$\sin^2\theta_{12}$ $0.297^{+0.019}_{-0.016}$	$\sin^2\theta_{23}$ $0.439^{+0.060}_{-0.019}$	$0.569^{+0.023}_{-0.060}$
Sign (Δm_{23}^2) ?	$\sin^2\theta_{13}$ $0.0214^{+0.0011}_{-0.0010}$	$0.0218^{+0.0010}_{-0.0011}$
Σm_ν < 0.183 eV (95% CL)	δ/π $1.35^{+0.21}_{-0.14}$	$1.32^{+0..32}_{-0.22}$

[a]The 3-sigma limits have been converted into 1-sigma asymmetric errors

the determination of the neutrino oscillation parameters (see Table 10.2). Although the mathematics of the flavour mixing is similar in the quark and neutrino sectors, the phenomenology is very different. In the quark sector, the flavour oscillations—described by the CKM matrix—take place in either mixing or decay of bound states. In the neutrino sector—described by the Pontecorvo-Maki-Nakagawa-Sakata (PMNS) matrix—the flavour oscillations take place in vacuum, although there is an additional term for neutrino transmission in matter to take account of the different interaction cross-sections of the electron neutrinos, and the muon and tau neutrinos (and, of course, their antineutrinos). The main uncertainty is whether the third neutrino mass is heavier than the other two (normal hierarchy) or lighter (inverted hierarchy), with an emerging preference [8] for normal hierarchy. Most impressive is the evidence that the CP-violating phase angle in the PMNS matrix (δ) is non-zero –0/2π is excluded at the 3-standard deviation level, and π at 2.5 (1.5) standard deviations for the normal (inverted) hierarchy. The best limits on the absolute neutrino mass scale come from cosmology, giving an upper limit on Σm_ν of 0.183 eV at the 95% confidence level [7]. There are essentially two ideas for extending the Standard Model to include finite neutrino masses. The first notes that, since the neutrino has no electric charge, it is possible to construct a mass-like term from the left-handed neutrino and right-handed-antineutrino fields, which changes its nature to that of a Majorana particle [9]. (If the neutrinos are indeed Majorana particles, there are two additional phase angles (α and β) in the PMNS matrix which are, however, extremely difficult to measure.) The second postulates the existence of a heavy right-handed neutrino which, in combination with a very high mass scale (like the GUT scale), produces through the "see-saw" mechanism one light left-handed neutrino (<< 1 eV/c²) and a heavy right-handed neutrino (>> 10^{12} GeV/c²)—as the left-handed neutrino becomes lighter the right-handed neutrino becomes heavier through the "see-saw". One consequence of these is that there are interactions that allow a change of lepton number by 2 units, thus enabling neutrinoless double-β decay to occur. (For a recent review, see [10].)

The other area where the Standard Model obviously fails is in the Dark Sector—Dark Matter and Dark Energy. The cosmological and astronomical evidence for these is overwhelming (see reference [11] for a recent review) with a recipe for the current state of the Universe containing about 70% of Dark Energy, 25% Dark Matter and 5% "normal" matter, the only part addressed by the Standard Model. From a particle physics perspective, Dark Matter is more tractable. First, there is the challenge of directly detecting it using large volume, sensitive detectors deep underground. The past decade has seen the limits steadily improved (see reference [12] for a recent review), while the small number of claimed signals have still not been independently confirmed. There is an ongoing programme of experiments at the major underground laboratories which should improve these limits by several order of magnitude over the next few years. Second, there are straightforward extensions of the Standard Model which provide candidates for Dark Matter. Unfortunately, there is no reliable estimate for the mass scale at which these new particles might appear. Such evidence might come from the Dark Matter searches themselves. The other place where a mass scale might be forthcoming is in the analysis of the deficiencies of the Standard Model. For example, Supersymmetry was proposed to address some hierarchy problems within the model, and also to assist in Grand Unification, both of which pointed to a TeV energy scale for appearance of new particles. However, so far these have failed to materialize at the LHC; unfortunately, the parameter space for Supersymmetry is extensive and so, while this might be disappointing it may not be conclusive. Dark Energy, on the other hand, is still mysterious and it is unclear where it fits in to the global picture; its understanding might require the reconciliation of quantum mechanics with general relativity.

The "open questions within the Standard Model" identified in the original article have been largely resolved, although there are still details to be addressed, for example the transition from perturbative to non-perturbative QCD. However, the "open questions beyond the Standard Model" remain, refined somewhat; for example, the "pattern of fermion masses" might nowadays be cast as "understanding the Yukawa couplings of the fundamental fermions to the Higgs". The improvements in the knowledge of the neutrino oscillation parameters do not explain how the mixing comes about, nor how the neutrinos derive their (tiny) masses. Again, there has been much progress in the measurement of the various cosmological parameters, including those which either determine or are determined by particle physics, but the "open questions in particle physics and cosmology" are still far from resolved. For example, the arguments for Grand Unification of the strong and electroweak forces remain intact (see Fig. 10.1) even though there is, as yet, no direct evidence for the existence of Supersymmetric particles. A second area which points to new physics is to note that, of the 18 parameters that define the Standard Model, 11 are directly related to mass (the quarks, the charged leptons, the Z and the Higgs), which becomes 12 if the Fermi constant is replaced (as it can be) by the mass of the W. Given this, there are three mass hierarchies to be explained. First, within a generation, what determines the relative masses of the charged lepton, the down-type quark (charge $-1/3$) and the up-type quark (charge $+2/3$) (and why is the up-quark lighter than the down quark, while the charm and top quarks are heavier than the strange and bottom quarks)? Second, what determines the large (several

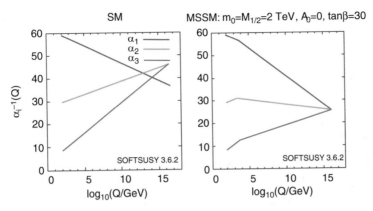

Fig. 10.1 The evolution of the fundamental forces as a function of energy in the standard model (left) and a particular set of parameters in the Minimal Supersymmetric Standard Model (right). (Adapted from [13])

orders of magnitude) mass differences between the generations of the same type of fundamental fermion. Third, what determines the masses of the heavy field bosons (Higgs and the W and Z)? (The other gauge bosons—the photon and the gluon—remain massless, protected by an unbroken gauge symmetry.) The origin of the mixing matrix and CP-violation in the quark sector is still unresolved, as are the masses and oscillation parameters in the neutrino sector.

The detection of Gravitational Waves from the merger of two black holes [14] and the subsequent detection of gravitational waves from a binary neutron star inspiral [15] open the door to a new era of observational astronomy of matter under extreme conditions, and may also provide insight into the long-standing issue of the incompatibility of general relativity with quantum mechanics. Resolving this is likely to require understanding of physics at the Planck scale.

Given this, the main conclusions of the "Way Forward" are largely unchanged. Clues to the physics that must exist beyond the Standard Model could come directly from the observation of new states of matter (e.g. Supersymmetry), either at the LHC or at a new energy frontier machine, or indirectly from precision measurements of Standard Model parameters, for example at a Linear Collider built to produce large numbers of Higgs particles under clean conditions. Alternatively, ultra-high precision measurements of low energy parameters (g-2, $\mu \rightarrow$ e conversion, neutrinoless double β decay...) as well as new astrophysical observations (Dark Matter, gravitational waves) could give strong pointers toward the new physics. The tremendous progress in the neutrino sector from, particularly, T2K and Nova probably means that the arguments in favour of a neutrino factory have weakened somewhat, but the arguments for the other facilities discussed (including the luminosity upgrade for the LHC) remain and have strengthened.

A recent publication by the T2K collaboration (*Nature* 580, 339–344 (2020)) reports that "The 3σ confidence interval for δ_{CP}, which is cyclic and repeats every

2φ, is $[-3.41, -0.03]$ for the so-called normal mass ordering and $[-2.54, -0.32]$ for the inverted mass ordering".

References

1. The European Strategy for Particle Physics, http://cds.cern.ch/record/1567258/files/esc-e-106.pdf?subformat=pdfa
2. Patrignani, C *et al.* (Particle Data Group), Chin. Phys. C **40**, 100001 (2016) and 2017 update; M. Tanabashi *et al.* (Particle Data Group), Phys. Rev. D **98**, 030001 (2018) and 2019 update
3. Aad, G *et al.* (ATLAS Collaboration) Phys. Lett. B **716** 1-29 (2012); Chatrchyan, S *et al.* (CMS Collaboration) Phys. Lett. B 716 30-61 (2012)
4. Hoecker, A and Marchiano, WJ, in J. Beringer *et al.* (Particle Data Group), Phys. Rev. D **86**, 010001 (2012), updated in 2013
5. The Muon g-2 collaboration, co-spokespersons Roberts, BL and Hertzog, DW, http://muon-g-2.fnal.gov/
6. Nakamura, K, Petcov, ST in Patrignani, C *et al.* (Particle Data Group), Chin. Phys. C 40, 100001 (2016)
7. Giusarma, E *et al.*, Phys. Rev. D **94**, 083522 (2016); a recent analysis by Couchot F *at al.*, to be published in *Astronomy and Astrophysics*, reports an upper limit of 0.17 eV.
8. Adamson, P *et al*, Phys. Rev. Lett. 118, 231801 (2017); Abe,K arXiv:1707.01048 [hep-ex] (2017)
9. Majorana, E, Il Nuovo Cimento 14 171-184 (1937)
10. Dell'Oro, S, Marcocci, S, Viel, M and Vissani, F, Advances in High Energy Physics 2016 2162659 (2016)
11. Arun, K, Gudennavar, SB and Sivaram, C, Advances in Space Research 60 166-186 (2017)
12. Liu, J, Chen, X and Ji, X, Nature Physics 13 212-216 (2017)
13. Hebecker, A and Hisano, J in in Patrignani, C *et al.* (Particle Data Group), Chin. Phys. C 40, 100001 (2016)
14. Abbott BP *et al.* (LIGO Scientific Collaboration and VIRGO Collaboration), Phys. Rev. Lett. **116** 061102 (2016)
15. Abbott BP *et al.* (LIGO Scientific Collaboration and VIRGO Collaboration), Phys. Rev. Lett. **119** 161101 (2017)

Correction to: Particle Physics Reference Library

Herwig Schopper

Correction to:
H. Schopper (ed.),
Particle Physics Reference Library,
https://doi.org/10.1007/978-3-030-38207-0

The online version of this book was inadvertently published with a figure in chapter 1 though the print version didn't hold any figure. In addition, the online version of Chapter 2 was published with incorrect section heading "2.1 Introduction to Chaps. , 3 and 4". The figure in chapter 1 has now been deleted and the heading in chapter 2 has been updated as: "2.1 Introduction to Chaps. 2, 3 and 4".

The updated versions of these chapters can be found at
https://doi.org/10.1007/978-3-030-38207-0_1
https://doi.org/10.1007/978-3-030-38207-0_2

Printed in the United States
By Bookmasters